POSTHARVEST PLANT PATHOLOGY

About the Author

Dr. N.G. Ravichandra, Professor of Plant Pathology and Scheme Head, All India Co-ordinate Research Project (Nematodes), Department of Plant Pathology, University of Agricultural Sciences, GKVK, Bengaluru, Karnataka, India, is involved in teaching, research and extension activities. He has been a major advisor to the Post-Graduate students majoring in Plant Pathology for M.Sc., (Agri.) and Ph.D. Degree Programmes.

Dr. N.G. Ravichandra has authored seven text books in Plant Pathology & Nematology, has 251 scientific research papers to his credit, published in National and International journals, contributed chapters to eleven text books, edited five books and also authored ten books in Kannada language.

Dr. N.G. Ravichandra has identified and reported a new species of root-knot nematode (*Meloidogyne enterolobii*) on Guava and a new host, Lac plant (*Flemingia semialalta* L.) for root-knot nematode (*Meloidogyne incognita*) from Karnataka. These are the first records in India.

Dr. N.G. Ravichandra has been bestowed with the "Eminent Scientist Award" for the year 2018 for his research accomplishments and contributions. He has been recognized as the International Editor for the prestigious International Journal 'Research Journal of Plant Pathology' published from California. He is also rewarded as the National Editor for the Eminent Journal 'Indian Phytopathology'. He has developed eight technologies for the management of root-knot nematodes infecting tomato, brinjal and rice, which have been included in the Package of Practices of University of Agricultural Sciences, Bengaluru, Karnataka, India.

Dr. N.G. Ravichandra has participated and presented original research papers on various aspects of plant diseases in National and International conferences. He has operated research projects funded by the Department of Bio-Technology & Indian Council of Agricultural Research, New Delhi, India as Principal Investigator and Co-Principal Investigator. He has undergone advanced training programs sponsored by the Department of Biotechnology & Indian Council of Agricultural Research, New Delhi, India. He is an active life member of professional societies including Nematological Society of India, Indian Phytopathological Society, Society of Mycology & Plant Pathology, National Environmental Science Academy and Institute of Agricultural Technology.

POSTHARVEST PLANT PATHOLOGY

N.G. Ravichandra

Professor of Plant Pathology & Scheme, Head, AICRP (Nematodes)
Department of Plant Pathology
College of Agriculture
University of Agricultural Sciences, GKVK
Bengaluru-560 065, Karnataka, India

NEW INDIA PUBLISHING AGENCY
New Delhi-110 034

NEW INDIA PUBLISHING AGENCY
101, Vikas Surya Plaza, CU Block, LSC Market
Pitam Pura, New Delhi – 110 034, India
Email: info@nipabooks.com
Web: www.nipabooks.com

For customer assistance, please contact
Phone: + 91-11-27 34 17 17 Fax: + 91-11- 27 34 16 16
E-Mail: feedbacks@nipabooks.com

ISBN: 978-93-90175-12-3

Composed and Designed by NIPA.

Digitally printed all across the globe.

Dedicated to
The Indian Farmer

Preface

Horticultural products for human consumption are an important part of the market economy. Any waste due to pathogen attack and spoilage in the field but mostly during the postharvest phase, results in significant economic losses. Postharvest diseases are very significant under the present scenario of postharvest management, which is insufficient to meet the requirements in several countries. Such losses are more pronounced as the losses occur closer to the time of produce sale.

Microbial decay is one of the main factors that determine losses compromising the quality of the fresh produce. Optimal distribution of fresh horticultural products entails prolonging their freshness and nutritional quality as long as possible after harvest. Losses caused by postharvest diseases are greater than generally realized because the value of fresh fruits and vegetables increases several-fold while passing from the field to the consumer.

Owing to its significance, 'Postharvest Plant Pathology', as a subject is being studied either in Undergraduate or Postgraduate degree programmes at Agricultural and Horticultural Universities, for which a specific syllabus has been prescribed by the Indian Council of Agricultural Research (ICAR), New Delhi, India. At present, there is scarcity of a book for the use by the students that covers the entire syllabus prescribed by the ICAR, New Delhi for 'Postharvest Plant Pathology' course.

This book explicates the fundamental aspects of postharvest diseases of crops and is conveniently divided into ten chapters, providing the latest information on the concept & types of postharvest diseases, economically significant postharvest pathogens & diseases of major crops, factors governing postharvest diseases, storage conditions, food safety issues, quiescence in post harvest pathogens, detailed & recent information on major mycotoxins, various approaches of postharvest disease management, integrated management strategies, biochemical & molecular aspects of postharvest diseases, apart from which, an exclusive chapter for discussing the postharvest nematode diseases and their management is also furnished.

Impressive diagrams at appropriate places, convincing tables and suitable graphs / illustrations have been furnished. A bibliography providing the list of references cited has also been included.

The information presented in this book, reflecting an extensive literature search, will be useful for teachers, researchers, students in several departments including Plant Pathology, Microbiology, Food Technology, Postharvest Technology, Environmental Sciences, officials of the State Departments of Horticulture & Agriculture, personnel of Commercial Production Centers, Plant Quarantine, Certification agencies and Policy planners who are concerned with the production and supply of horticultural and agricultural produce of high quality and acceptable for human consumption.

The author appreciates receiving suggestions, remarks and constructive criticism that would improve the quality of the book.

Bangaluru, May 2020 **N.G. Ravichandra**

Acknowledgement

My respectful thanks to,

Dr. S. Rajendra Prasad, Hon'ble Vice Chancellor, University of Agricultural Sciences, GKVK, Bangalore, Karnataka, India.

Dr. Y.G. Shadakshari, Director of Research, University of Agricultural Sciences, GKVK, Bengaluru

Dr. Umarao, Project Coordinator, AICRP (Nematodes), Indian Agricultural Research Institute, New Delhi.

My special thanks to Dr. T. Narendrappa, Professor & University Head Dr.A.Nagaraja, Professor & Head, Dr. H.A.Prameela, Dr. N. Nagaraju Dr.H.K.Ramappa, and Mr. B.R. Nagendra, Department of Plant Pathology UAS, GKVK, Bengaluru.

I would like to express my deep sense of gratitude to my beloved parents Sri. N. Gurushankar and Smt. Parvathamma. I am grateful to my in-laws for their affection and encouragement.

I wish to record my appreciation and affection to my wife, Ms. Deepthi and daughter, Ms. Anusha for their care, patience, keen interest and affection throughout the preparation of the manuscript.

My heartfelt thanks are due to my brother-in-law Mr. B.Sudesh Kumar and my beloved sister Ms. N.G.Tara, for their sustained moral support, affection and special care. I fondly acknowledge their keen interest, continued encouragement and blessings being showered on me.

My appreciation to the publisher, New India Publishing Agency, New Delhi, India, who deserves commendation for their professional contribution.

Contents

List of Colour Plates

1

Concept and Types of Postharvest Diseases

Food waste and loss is an enormous issue worldwide. The FAO states that 'Hunger is still one of the most urgent development challenges, yet the world is producing more than enough food', but a vast amount of what is produced gets lost on its way from the field to the consumer. According to the FAO, about 30% of cereals, 20% of dairy products, 35% of fish and seafood, 45% of fruits and vegetables, 20% of meat, 20% of oilseed and pulses and 45% of roots and tubers are lost or wasted. Postharvest food loss is defined as measurable qualitative and quantitative food loss along the supply chain, starting at the time of harvest till its consumption or other end uses. Food losses can either be the result of a direct quantitative loss, for example, during the process of harvest or transport or arise indirectly due to losses in crop quality such as undesired sprouting or water loss. Postharvest food loss consists of many factors, which have not changed much in the last 40 years and so postharvest food loss in general has not changed much. Figure 1 illustrates the main reasons for food loss along the supply chain and is based on 'The food pipeline' published by Bourne in 1977. Food loss is highest during storage due to pathogens (insects, bacteria and moulds), environmental conditions (e.g. rain, humidity, heat and frost), sprouting and quality loss (rancidity, water loss and saccharification) or animals (rodents and birds). Recently published data for Switzerland have shown that about 53%–55% of the initial fresh potato production and 41%–46% of the initial processing potato production are lost mostly due to pathogen infection, saccharification, water loss of tubers and early sprouting during storage (Buchholz *et al*, 2018).

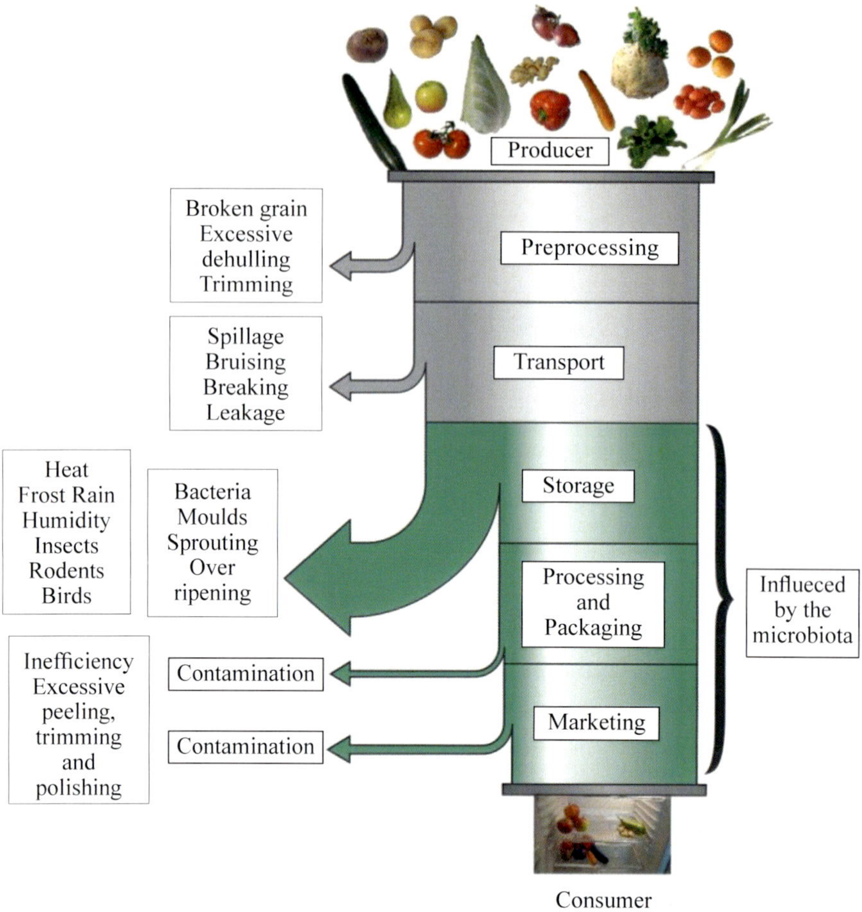

Fig. 1: Along the food pipeline that takes the food from the field to the table, there are many opportunities for food to be lost the result that much of the food that has been produced, never reaches the consumer for whom it was intended. The size of the arrows reflects the relative food loss at each postharvest handling step.

Diseases caused by microbial pathogens - fungi, bacteria, and viruses - account for substantial losses of grains, fruits, and vegetables at both pre- and postharvest stages of crop production. It has been emphasized that "the responsibilities of the plant pathologists do not end with the harvest of satisfactory yields of plant products and that harvesting marks the termination of one phase of plant protection and the beginning of another". This statement clearly indicates that the second phase of plant protection - of seeds, fruits, vegetables, and other economic plant parts, from the time of harvest until they reach the consumer - is equally important. Postharvest pathology, earlier termed "market pathology", deals with the science of, and practices for, the protection of harvested produce during harvesting, packing, transporting, processing, storing, and distribution.

Several studies have shown that losses occurring between harvesting and ultimate utilization of agricultural/horticultural products by the consumer are substantial, though there is no precise estimate. The harvested produce may be classified as: durables (consisting of cereal grains, oilseeds, and grain legumes) and perishables (comprising succulent storage organs such as fleshy fruits, vegetables, rhizomes, and tubers). The harvested commodities are essentially dormant plant organs with physiological functions quite different from tissues of the mother plant. As the ripening/senescence process sets in, the susceptibility of the stored products to decay caused by microbial pathogens increases progressively. Furthermore, storage is an abnormal state for living plants or plant organs and it requires crowding of high volumes of plant produce in intimate contact in the limited available space. These unnatural conditions are likely to predispose them to different kinds of diseases which may be due to physiologic or pathogenic causes. The discussion in this book is confined to diseases caused by microbial pathogens.

It is now well recognized that the most economically feasible and expedient means to increase the world food supply is to reduce losses in food crops that occur after they are harvested as the area for cultivation and resources are limited. In India, the diverse Agro climatic conditions permit cultivation of about fifty kinds of fruit crops. The total production of fruits in India is approximately 21 million tons. This can easily be doubled if adequate attention is given to control diseases, both in the field and in storage, which account for over 25-30 per cent loss. There is an urgent need for undertaking detailed studies on the various methods of controlling the diseases and to minimize the loss of fruits in the field, in transit as well as in storage. Postharvest losses are much more painful and costlier than pre harvest losses both in terms of money. Postharvest losses involve parasitic diseases caused by fungi, bacteria, nematodes and viruses, environmental factors, physiological and mechanical factors.

1.1 Postharvest food loss due to diseases : Postharvest food loss is defined as measurable qualitative and quantitative food loss along the supply chain, starting at the time of harvest till its consumption or other end uses. Food losses can either be the result of a direct quantitative loss, for example, during the process of harvest or transport or arise indirectly due to losses in crop quality such as undesired sprouting or water loss (Aulakh and Regmi, 2013). Postharvest food loss consists of many factors, which have not changed much in the last 40 years and so postharvest food loss in general has not changed much.

The reduction of losses in perishable food crops because of postharvest diseases has become a major objective of efforts by international organizations to increase the food supplies of the world . Accurate data are not available on the magnitude of economic losses resulting from diseases affecting susceptible fruit, vegetable and root crops during storage and in transit to market. However, there is general agreement that heaviest postharvest losses occur in developing countries where storage facilities are inadequate and refrigerated storages and vehicles are lacking. Worldwide losses in the postharvest period are probably in the range of 10-30% on most crops. In some perishable crops such as tomato, losses as high as 30-50% are not unusual (Yadav *et al*, 2013). Based on relatively conservative estimates, the amount of food lost annually during the postharvest period could feed between 200 and 300 million people for one year.

The postharvest diseases that cause spoilage of both durable and perishable commodities are widespread. Greater losses occur in developing countries due to nonavailability of proper storage and transportation facilities and improper handling methods, resulting in greater levels of injuries or wounds during harvesting and transit. Durable commodities are generally stored in a dry state with moisture level below 12%, whereas perishable products have higher moisture levels (about 50% or more) at the time of storage. The harvested produce might have been infected by pathogens prior to harvest under field conditions or they may get infected during transit and storage. It is estimated that, in the tropics, about 25% of all perishable food crops harvested are lost between harvest and consumption. Losses in durable commodities, such as cereals, oilseeds, and pulses, may be about 10% on a worldwide basis. The losses may be both quantitative and qualitative. Under conditions favoring pathogens, loss caused by postharvest diseases may be greater than the economic gains achieved by improvements in primary production. Studies on postharvest diseases are primarily directed at preventing economic loss from spoilage of harvested commodities during transit and storage and at eliminating the adverse effects of mycotoxins produced by fungal pathogens contaminating both durables and perishables. The mycotoxins are known to be carcinogenic, causing several serious ailments in humans and animals (Narayanasamy, 2005). There is an imperative need to gather information on the microbial pathogens involved in various postharvest diseases, conditions favoring disease development, and methods of developing effective systems of disease management.

Although losses in developing countries are often at unacceptable levels, the evidence is strong that the adverse economic impact of postharvest disease is much greater in developed countries than has been recognized. The mechanization of most phases of harvesting, storing, grading, washing, and moving perishable food crops has increased the prospects for bruising and enhanced opportunities for infection by fungal , bacterial, viral and nematode pathogens. The need to increase emphasis on research on control of postharvest diseases is based on the fact that it will be difficult to increase world food supplies by traditional approaches designed to increase crop yields and expand the areas devoted to crop production. Most of the land best suited for crop production is already under cultivation, and large acreages of highly productive land are being lost every year because of urbanization. In addition the losses resulting from water and wind erosion of good agricultural land are at a staggering level. There is a bright prospect that crops with improved disease resistance and higher yields will be available in the near future; even with these added inputs plus increased use of fungicides, bactericides, nematicides and insecticides, it will be difficult to reach the desired objective of doubling total world food production in the next two to three decades. Such an increase is needed if we are to meet the pressure of continuing population growth. It should be recognized that the least cost intensive effort to increase food supplies would value of the harvested crop, relatively small percentage losses in shipments may result in a total loss in a total loss in profit since profit margins are so small in this industry.

In view of the fact that most pathologists will agree that estimates of postharvest losses of 10-30 % in perishable food crops are not unreasonable, the question may be raised as to why so little effort is directed in management of postharvest diseases. Few departments of plant pathology have courses concerned with postharvest diseases, most basic texts devote little or no space to this topic area, and only a relatively small number of plant pathologists have been involved in postharvest disease research. Unlike other disease problems, little or no attention has been given by plant breeders to the possibility that breeding for postharvest resistance to diseases should be an integral part of their breeding programs. In part this gap reflects a lack of recognition of the potential payoff of such breeding efforts and it also can be attributed to a lack of the availability of standardized inoculation tests that would facilitate screening for disease resistance to diseases causing most of the losses after harvest. Inasmuch as the prospects are good that many of the fungicides now on the market for management of postharvest diseases may be restricted or banned, it is essential that increased attention needs to be given to alternative approaches to disease control. major avenues of research that warrants increased attention-increasing resistance to postharvest pathogens and the potential use of biocontrol agents as substitutes for fungicides. It is obvious that present losses justify a substantial increase in the investment of intellectual and financial resources to gain an improved understanding of the causal factors and to improve the current management methods.

The diseases which develop on harvested parts of the plants like seeds, fruits are known as postharvest diseases. Pathogen attack may take place during harvesting and subsequent handling, storage, marketing, and after consumer purchase. The plant parts may get infected in the field, but expression of symptoms may take place later, at any stage before final consumption. The plant products may get infected by microorganisms and cause rotting or decaying, partially or totally. The quantity of plant products becomes reduced due to the above infection. The seeds or grains of cereals, pulses and other crops may get damaged by accumulation of toxic substance, the mycotoxin produced by the infected microorganism. The fleshy fruits, vegetables like tomato, banana, citrus, strawberries, rhizome of zinger, bulb of onion, tuber of potato etc., may get damaged. This results in reduction of quantity, quality or both of the affected parts or products as a whole. The amount or extent of damage depends mainly on the pathogen(s) involved, on the condition of the products and the condition of storage. The postharvest diseases that cause spoilage of both durable and perishable commodities are widespread. Losses inflicted throughout the supply chain due to pathogen-induced diseases are the major component of food wastage and may occur at any time from preharvest to consumption. The capability of a microorganism to initiate postharvest diseases, as well as its final outcome, depend on a number of factors that can conveniently be associated with, microorganism, the host and/or the environment. The integrative strategies for control of postharvest diseases include effectively inhibiting pathogens growth, enhancing resistance of hosts and improving environmental conditions resulting favourable to the host and unfavourable to the pathogen growth (Nabi *et al*, 2017).

The possibility of integrating the different effective strategies to achieve higher level of control of postharvest pathogens and to minimize or replace the use of synthetic fungicides has to be explored in certain host–pathogen systems. Postharvest losses of perishables in developing countries have been estimated to be in the range of 5-50 % or more of the harvest. Many scientists estimate that 20 to 30% losses in fruits and vegetables are due to postharvest losses of perishables in developing countries have been estimated to be in the range of 5-50 % or more of the harvest. Many scientists estimate that 20 to 30% losses in fruits and vegetables are due to postharvest diseases. In India nearly 20-50 % of perishable are lost due to postharvest diseases (Table 1) (Yadav *et al*, 2013).

Table 1: Postharvest disease loss in major crops

Crop	Loss	Crop	Loss
Banana	20-80 %	Sweet potato	35-95 %
Papaya	40-100 %	Onion	16-35 %
Grapes	27 %	Tomato	5-50%
Apple	14 %	Cabbage	37%
Potato	5-40%	Cauliflower	49%

Durable and perishable commodities are prone to widespread postharvest diseases that cause huge spoilage to the produce. Greater losses in developing countries, is due to non-availability of proper storage and transportation facilities and improper handling methods, resulting in greater levels of injuries or wounds during harvesting and transit. It is estimated that about 20– 25% of the harvested fruits and vegetables are decayed by pathogens during postharvest handling even in developed countries. In developing countries, postharvest losses are often more severe due to inadequate storage and transportation facilities.

In a report by the United Nations Food and Agricultural Organization, it was estimated that one-third of the food produced worldwide for human consumption is lost after harvest. The harvested produce might have been infected by pathogens prior to harvest under field conditions or they may get infected during transit and storage. Fruits and vegetables are considered as the perishable crops than cereal, pulses and oil seed crops. Most of them contain very high moisture content (about 70-95 % water), usually have large size (5g-5kg), exhibit higher respiration rate, and usually have soft texture, which favour the growth and development of several diseases by the microorganisms between harvest and consumption. Postharvest pathology, earlier termed "market pathology", deals with the science of, and practices for, the protection of harvested produce during harvesting, packing, transporting, processing, storing, and distribution. Diseases caused by microbial pathogens – fungi, bacteria, and viruses account for substantial losses of grains, fruits, and vegetables at both pre- and postharvest stages of crop production. "The responsibilities of the plant pathologists do not end with the harvest of satisfactory yields of plant products and that harvesting marks the termination of one phase of plant protection and the beginning of another". This statement clearly indicates that the second phase of plant protection ,of seeds, fruits, vegetables and other economic plant parts, from the time of harvest until they reach the consumer, is equally important. Postharvest losses for fresh commodities like fruits and vegetables are about 5% in developed countries and may average 30-50% in underdeveloped countries. In developing countries losses are more, due to non-availability of proper handling methods, transportation and storage facilities, which results in greater levels of injuries or wounds during harvesting and transit.

Marketable life of fruits is largely affected by the prevailing temperature, relative humidity, and the composition of the atmosphere during and after harvest, and type and degree of infection by the microorganisms. They deteriorate during storage through loss of moisture, decay caused by pathogens, rodents, loss of stored energy, loss of nutrients and vitamins, physical losses through pests and disease attack, loss in quality from physiological disorders, fibre development etc. Economic gains achieved by improvements in primary production can be less as losses caused by the postharvest diseases are more under conditions favourable for postharvest pathogens to act. Studies on postharvest diseases are primarily directed at preventing economic loss from spoilage of harvested commodities during transit and storage. So there is an imperative need to collect information on the microbial pathogens involved in various postharvest diseases of temperate fruits, favourable conditions for disease development, and effective methods for disease management.

The role of the plant microbiota in plant establishment, growth and health is well studied, but the dynamics of postharvest crop microbiota and its role in postharvest crop quality are largely unexplored, although food loss is an enormous issue worldwide. The microbiota might be especially important during crop storage by either preventing or favouring rots, or quality loss due to, for example, sprouting, saccharification, water loss or spoilage. We need more research on plant–microbe interactions in postharvest crops to be in future able to provide microbial solutions for plant production along the whole food chain from field to fork. Food waste and loss is an enormous issue worldwide. The FAO states that 'Hunger is still one of the most urgent development challenges, yet the world is producing more than enough food.', but a vast amount of what is produced gets lost on its way from the field to the consumer (Pierre Pétriacq *et al*, 2018). According to the FAO, about 30% of cereals, 20% of dairy products, 35% of fish and seafood, 45% of fruits and vegetables, 20% of meat, 20% of oilseed and pulses and 45% of roots and tubers are lost or wasted. This is one-third of all the food produced for human consumption. However, instead of decreasing postharvest losses to ensure food supply, the efforts in recent years focused at increasing food production. This is also reflected in plant microbiome research, where the focus is clearly on the exploitation of plant–microbe interactions for improved plant productivity.

1.2 Categories of postharvest diseases: The postharvest diseases can also be grouped into three categories, *viz.*, seed infection (infection through floral parts /fruits/ natural opening and injuries, may be fungal, bacterial, viral and nematodes ; seed deterioration (effect of seed germination/quality and discoloration of seeds and abnormalities, toxin contamination) (fungal, bacterial and viral) and infection through fruits and vegetables (spore germination, entry into the host tissue, colonization of host tissue, replication of viral particles, reproduction in nematodes, symptom expression may be fungal, bacterial, viral and nematode infection).

Pathogen induced decay is certainly the most obvious reason for postharvest crop loss caused by microbial activity (bacteria and moulds). Plant microbiomes consist of complex communities of potentially mutualistic, commensal and pathogenic microbes colonizing the same niches in plants, including, for example, grains. Moreover, a clear distinction between plant beneficial or harmful microorganisms is not always possible. The outcome of plant–microbe interactions depends on biotic and abiotic environmental factors but also the manifold interactions between members of the plant microbiota, including pathogens, strongly influence the phenotype of plant–microbe interactions (Pierre Pétriacq *et al*, 2018). Often disease outbreak in plants correlates with shifts in the microbiome composition, resulting in a microbial dysbiosis and a response of specific microbes, which can act as antagonists or synergists towards plant pathogens. Only a few studies have investigated the dynamics of the microbiota during postharvest disease development. Several changes were described in the bacterial community composition of potato tubers in response to infection with the soft rot pathogen Pectobacterium atrosepticum. It was observed that pathogenicity is triggered by the pathogen, but the endophyte community strongly contributes to the development of the disease.

Shifts in the fungal and oomycete community composition associated with storage soft rot development were reported in different sugar beet hybrids from different environments stored at different temperatures. Interestingly, while the genotypes showed differences in the susceptibility to soft rot, the shifts in the microbial community of sugar beet were genotype independent. The authors suggested an unspecific resistance mechanism slowing down the spread of pathogens in more resistant genotypes but not preventing infection. However, the study focused on fungi and oomycetes and the dynamics of bacteria community during storage soft rot development of sugar beet and its putative role in disease expression remains elusive. In summary, pathogen infestation and spoilage may often not be caused by a single organism but is likely to result from the interplay of individual members of the microbial community in crops. On the other hand, stored plant organs such as seeds may contain bacteria with antagonistic activity against plant pathogens with the potential to protect plants not only in the field but also postharvest.

Furthermore, it was suggested that the diversity of a microbial community in plants determines pathogen establishment (Berg *et al.*, 2017). The seed microbiome of oilseed rape, for example, is cultivar-specific and cultivars hosting a higher indigenous microbial diversity showed better resistance towards colonization by pathogenic microorganisms.

Losses caused by postharvest diseases are greater than generally realized because the value of fresh fruits and vegetables increases several-fold while passing from the field to the consumer. Postharvest losses are estimated to range from 10 to 30% per year despite the use of modern storage facilities and techniques (Harvey, 1978). Postharvest diseases affect a wide variety of crops particularly in developing countries which lack sophisticated postharvest storage facilities. Infection by fungi and bacteria may occur during the growing season, at harvest time, during handling, storage, transport and marketing, or even after purchase by the consumer (Dennis, 1983). The reduction of losses in perishable food crops because of postharvest diseases has become a major objective of international organizations. The reality is that there is a portending food crisis that will require the concerted efforts of all who are involved in food production to double their efforts. In fact, to adequately feed the world's expected 10 billion people within the next 40 to 50 years, food production efficiency and distribution will need to be improved immensely. Specific causes of postharvest losses of fruits and vegetables may be classed as parasitic, nonparasitic, or physical (Cappellini and Ceponis, 1984).

Fungi are more commonly found attacking fruit and bacteria are more common as postharvest pathogens of vegetables. Weather affects many factors related to plant diseases, from the amount of inoculum that overwinters successfully to the amount of pesticide residue that remains on the crop at harvest (Conway, 1984). Abundant inoculum and favorable conditions for infection during the season often result in heavy infection by the time the produce is harvested. For example, conidia of the fungus that causes bull's-eye rot are rain dispersed from cankers and infected bark to fruit especially if rainfall is prolonged near harvest time, causing rotten fruit in cold storage several months later (Spotts, 1990).

1.3 Pathogens causing postharvest diseases: Several pathogens such as fungi, bacteria, viruses and nematodes are responsible for causing postharvest diseases. However, it is well known that the major postharvest losses are caused by fungi such as *Alternaria, Aspergillus, Botrytis, Colletotrichum, Diplodia, Monilinia, Penicillium, Phomopsis, Rhizopus, Mucor* and *Sclerotinia* and bacteria such as *Erwinia* and *Pseudomonas*. Most of these organisms are weak pathogens in that they can only invade the damaged produce.

Fruits and vegetables in their fresh forms contain high percentage of water. They are living and hence, carry out their physiological function of respiration thereby absorbing and releasing gasses and other materials from and to their environment. These activities lead to their deterioration in storage and transit which is more rapid under conditions of high temperature and humidity. One factor that imparts negatively on the economic value of fruits is that they have a short shelf-life. This is as a result of many factors, prominent among which is the activity of Pathogens. They are susceptible to pathogenic attack due to their low pH, high moisture content and nutrient composition. These make them rot and unfit for consumption due to the production of mycotoxins. Their presence in these food produce also constitute health risks. Fruits and vegetables are exposed to contamination by microbes through contact with soil, dust and water and by handling at harvest or during postharvest processing. This makes them harbor a wide range of microorganisms including plant and human pathogens. Differences in microbial profiles of diverse fruits and vegetables could be due to varying factors including resident micro-flora in the soil, application of non-resident micro-flora through animal manures, sewage or irrigation water, transportation and handling by individual retailers (Onuoha *et al.*, 2017).

Many of the fungi which cause postharvest diseases belong to the phylum Ascomycota and the associated Fungi Imperfecti. In the case of the Ascomycota, the asexual stage of fungus (the anamorph) is usually encountered more frequently in postharvest diseases than the sexual stage of the fungus (the teleomorph). Important genera of anamorphic postharvest pathogens include *Penicillium*, *Aspergillus*, *Geotrichum*, *Botrytis*, *Fusarium*, *Alternaria, Colletotrichum, Dothiorella, Lasiodiplodia* and *Phomopsis*. Some of these fungi also form ascomycete sexual stages. In the phylum Oomycota, the genera *Phytophthora* and *Pythium* are important postharvest pathogens, causing a number of diseases such as brown rot in citrus (*Phytophthora citrophthora* and *P. parasitica*) and cottony leak of cucurbits (*Pythium* spp.). *Rhizopus* and *Mucor* are important genera of postharvest pathogens in the phylum Zygornycota. *R. stolonifer* is a common wound pathogen of a very wide range of fruit and vegetables, causing a rapidly spreading watery soft rot (Coates *et al*, 1995). Genera within the phylum Basidiomycota are generally not important causal agents of postharvest disease, although fungi such as *Sclerotium rolfsii* and *Rhizoctonia solani*, which have Basidiomycete sexual stages, can cause significant postharvest losses of vegetable crops such as tomato and potato (Lindy Coates and Greg Johnson, 1989). While diseases caused by these pathogens are primarily field diseases, the development of symptoms often accelerates after harvest. The major causal agents of bacterial soft rots are various species of *Erwinia*, *Pseudomonas*, *Bacillus*, *Lactobacillus* and *Xanthomonas*. Bacterial soft rots are very important postharvest diseases of many vegetables, although they are generally of less importance in most fruit. This is because most fruits have a low pH which is inhibitory to the majority of bacterial plant pathogens.

1.4 Infection process : Microorganisms either infect the produce while still immature on the plant (pre-harvest infection) or during the harvesting and subsequent handling and marketing operations (postharvest infection). Postharvest infection process is greatly aided by mechanical injuries to the peel of the produce. The infection may occur by direct penetration of the cuticle or entry through stomata, lenticels, wounds or abscission scar tissue. Furthermore, the physiological condition of the produce, the temperature and the formation of the periderm significantly affect the two types of infection i.e. pre-harvest infection and postharvest infection. Sources of infection include at the field or orchard, greenhouses, harvesting tools, buckets, carts, boxes packing house facilities, storage facilities, markets etc.,

1.4.1 Pre-harvest infection : Infection often takes place before harvest. Several fungi infect products during flowering while other pathogens stick to be product skin. Pathogens that are relatively dormant during plant growth can develop and cause symptoms after harvest.

Pre-harvest infection of fruit and vegetables may occur through several avenues, such as direct penetration of the peel, infection through natural openings on the produce and infection through damaged portion. Several types of pathogenic fungi are able to initiate the infection process on the surface of floral parts, and developing fruits. The infection is then arrested, which remains quiescent until after harvest, when the resistance of the host decreases and conditions become favorable for the growth of the pathogen *i.e.*, when the fruit begins to ripen or its tissues become senescent. Such latent infections are important in the postharvest wastage of many temperate fruits, such as apple, pear, apricot etc. Weak parasitic fungi and bacteria may also gain access to immature fruits through natural opening such as stomata, lenticels and growth cracks (Anonymous, 2015). Again, this infection may not develop until the host becomes less resistant to the invading organism, such as when the fruits ripen. It appears that sound fruits and vegetables can suppress the growth of these organisms for a considerable time. For example, spores of *Phlyctaena vagabunda* penetrate apple lenticels before harvest, which cause fruit rotting around the lenticels during the storage. Infection of some fruiting vegetables (eg. internal rots in capsicums) actually occurs during flowering. Fungi such as *Alternaria* and *Botrytis* are able to infect flower petals. However, they only start to develop once the vegetable has matured.

Many plant pathogens are present in the soil or on plant debris. Vegetables that are grown in contact with the soil are likely to be carrying spores or mycelium that cause infection, even though these are not apparent at harvest. In the example shown below, pakchoy grown hydroponically on raised benches appeared similar at harvest to the same product grown in the ground (Fig.2). However, the hydroponic product had consistently longer shelf life, largely because of postharvest development of leaf spots and rots on the field-grown product. The pathogens responsible for these diseases are present in the soil but transfer to the leaves by water splash during rain or irrigation.

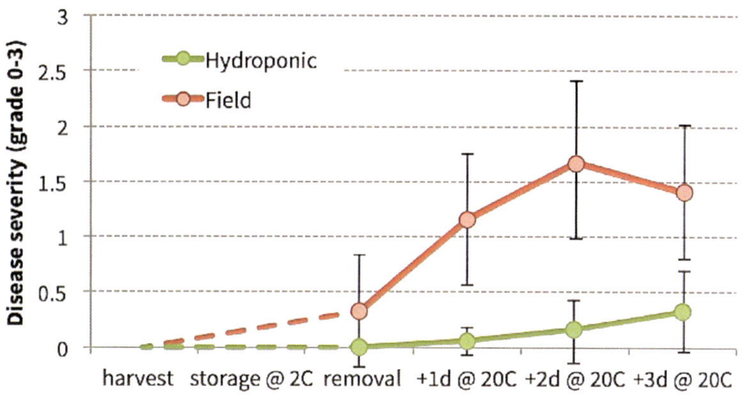

Fig. 2: Postharvest development of leaf diseases in pak choy grown in the soil or hydroponically on run-to-waste benches. Bunches were evaluated at harvest, on removal after one week at 2°C, then daily during simulated retail at 20°C. Differences are likely due to the amount of inoculum present at harvest.

Fungal spores are also carried on the wind. Nearby crops infected with fungal pathogens such as white blister (*Albugo candida*), or blue mould (*Penicillium* spp.) release huge numbers of spores, which can land on the intact vegetables. These spores are unable to germinate while the product remains dry. However, condensation during cooling or storage can allow germination, growth and infection.

The citrus fruit is attacked by a number of pathogens from bloom to harvesting stage and subsequently by postharvest pathogens that affect the production of the crop and considerably deteriorate the fruit quality. The pre-harvest pathogens like *Colletotrichum gloeosporioides, C. acutatum, Botryodiplodia theobromae, Alternaria citri, Phomopsis citri etc.* attack the fruit from fruit set till harvest and cause considerable damage to its production and quality. The incipient infection of pre-harvest pathogens subsequently also manifest in the form of postharvest diseases besides the attack of other postharvest wound pathogens viz. *Penicillium digitatum, P. italicum, Geotrichum candidum etc*. during postharvest handling, transport, storage and marketing (Naqvi, 2014).

The timing of *B. cinerea* infection of pear stems and calyxes was determined in the orchard during the growing season, to investigate the development of gray mold in storage, and to determine whether preharvest levels of *B. cinerea* in pear stems and calyxes can be used as predictors of gray mold levels observed in storage. Very low levels of *B. cinerea* were isolated from stem tissue prior to harvest. In a single year repeat experiment, stems sampled at harvest had higher levels of infection than those sampled earlier in the season. Little or no stem end gray mold was detected in fruit after 3 months in air-storage; however, incidence increased between 6 and 8 months. Calyx end gray mold was detected at low levels in fruit stored for up to 8 months. The mean incidence of stem end gray mold was 3.6 and 2.0%, and incidence of calyx end gray mold was 1.2 and 0.2%, in 1996 and 1997, respectively. Calyxes were susceptible to infection soon after full bloom; however, inoculation of calyxes in April or May did not result in higher levels of calyx end gray mold in storage (Lennox and Spotts, 2004). Therefore, preharvest level of calyx infection is a poor predictor of calyx end gray mold in storage. In addition, application of benomyl in the orchard reduced the level of *B. cinerea* in blossoms but had no effect on levels of calyx end gray mold of fruit in storage. Packing and shipping fruit within 3 to 6 months of harvest may mitigate economic losses due to gray mold.

Field and greenhouse studies were used to determine the nature of preharvest infection of corn by *Aspergillus flavus*. Inoculation of external silks that were yellow-brown resulted in more extensive colonization of the silks and a greater number of infected kernels than inoculation of brown silks (Stephen F. Marsh, 1984). In silk-inoculated and uninoculated ears, the pattern of tissue colonization was very similar: growth generally proceeded from the ear tip towards the base, colonizing the silks first, then the glumes and (by the late milk stage) the kernel surfaces but rarely penetrating the cob pith. Silk senescence and the subsequent growth of *A. flavus* down the silks was rapid in environment chambers at 30/34 C; the fungus reached the base of some ears in 4 days. Under field conditions, equivalent progress took between 4 and 13 days. The mycelium of *A. flavus* spread quickly from the silks onto the kernel surfaces, forming a clustered distribution. The percentage of kernels colonized within an ear half was correlated ($r \geq 0.67$ for six of seven harvest dates) with the extent of contamination of the associated silks by *A. flavus* but showed no relationship to the location or extent of visible insect damage. Internal infection of kernels did not appear until early dent stage.

1.4.2 Effect of bruising in infection: Fruits have high potential to be mechanically damaged during their developmental stages and/or before harvest. Generally, the chances that fruit can be damaged while still on the tree are quite substantial, and this can happen from a variety of sources. Several ways in which fruit can be mechanically damaged whilst on tree include forceful contact of fruit with other fruit or parts of the tree such as branches during growth which may cause abrasion, puncture and bruising predation by slugs, insects, birds and mammals can also puncture the skin and consume a portion of the tissue, and) effect of weather, such as wind and hail that can aggravate damage caused by contact of fruit with other parts of the tree, causing mechanical injury such as bruising, cleavage, slip and buckling (Zaharan Husseina *et al*, 2017). An average preharvest loss of up to 30.4% is found in litchi fruit during sorting at harvest, which mainly comprised losses due to sunburn, cracking, bruising anthracnose, and fruit borer infestation among others.

The physical evidence of bruising onto a produce is usually indicated by discolouration of injured tissue which marks the damaged spot . Mechanical impact or compression (due to loading) onto a biological produce provokes mechanical stress that induces cell wall and membrane rupture and hence bruising. Bruise susceptibility (BS) is the measure of produce response to external loading. Hence, the extent of dynamic or static loading onto a produce is considered the most important bruise factor, usually expressed in terms of loading or absorbed energy. The former comprises all impacts likely to occur during harvesting and handling operations such as fruit dropping into the picking buckets or during sorting, or a vibration, mainly occurring during transportation. On the other hand, static or compression loading can occur during harvesting, transportation or storage when poorly designed bins are overfilled and stacked such that the produce in the lower bins support the weight, which possibly causes damage.

The formation of external bruising is associated mainly with the breakage of cell structures and the failure of membranes. Damage of cells and fruit tissue initiates the contact between polyphenol oxidase (PPO) and peroxidase (POD) cytoplasmic oxidizing enzymes and phenolic contents originally stored in the vacuole. In the presence of oxygen, the enzymatic oxidation in the damaged cells transforms phenolic substances into quinones, which polymerize to form dark/ brown pigments on damaged part of the fruit. The formation of brown pigment on the surface of the damaged region provides the external sign of an impact or compression bruising. The difference in the concentration of phenolic contents and the activity of oxidizing enzymes between the fruit exocarp, mesocarp and endocarp tissues makes the browning inhomogeneous. For instance, in fruits such as litchi, the phenolic content and oxidizing enzymes (PPO and POD) activity are higher in external tissues, and hence the browning predominantly occurs on the external surface of the fruit's bruise damaged region. On the contrary, other fruits such pear, tomato and longan, the phenolic content and PPO and POD activity are lower in the external tissues and therefore browning occurs internally.

Bruise damage of fresh fruit is not limited to visual aspects, but may also accelerate other biological processes such as microbial spoilage. Fruit bruising aggravates the risk microbial contamination, hence providing potential causes for fruit quality losses and lower shelf-life. Postharvest rots and decay are more prevalent in bruised or otherwise mechanically damaged fruits and vegetables than in undamaged produce (Zaharan Husseina *et al*, 2017). Decay pathogens such as bacteria and fungi easily enter through dead or wounded tissues before contaminating the rest of the fruit or vegetables, resulting in significant losses during storage and long-distance transportation. The problem of postharvest loss due to decay and subsequent economic losses could be reduced by applying proper measures to reduced bruise damages both at prior to harvest, at harvest and during subsequent postharvest handling.

Banana fruits are highly susceptible to mechanical injury owing to their tender texture and high moisture content. If they are exposed to undesirable environmental conditions during handling, the tissue will soften and easily bruise, causing rapid microbial deterioration. In the study area, quality and safety assurance problems such as lack of temperature management, uniformity of quality within containers, sanitation problems in the market, transportation related problems, careless handling during loading and unloading were identified as the main factors, which favored fungal pathogen development and associated banana fruit losses (Chala G.Kuyu and Yetenayet B.Tola, 2018). Disease intensity had similar trends with percentage of fruit damage with more damage, and infections were recorded in retailers' shop. The percentages of fruit damage were as high as 56.2%, and the associated disease incidence and severity were 54.2% and 34%, respectively, in retailers' shop. *Colletotrichum* spp. and *Fusarium* spp were identified as the most important fungal pathogens causing fruit loss in Jimma town of Bishishe market. Mechanical injury due to mishandling along supply chains and sanitary problem in the market could be the possible causes for observed fungal pathogens. In order to reduce mechanical injury and associated microbial deterioration, a close integration of all stakeholders along the value chain of banana fruits becomes necessary.

It wa observed that losses of banana fruits due to spoilage and physical injury were common problems for all fruit vendors. Practices such as lack of sanitation temperature management, and improper packaging and transportation problems were identified among the common causes for observed losses. A total of 48 fruits were purposively selected from open market, wholesales and retailers and fruit damage, disease incidence, disease severity and identification of diseases causing fungal pathogen were conducted under laboratory. The highest fruit damage (56.2%) was recorded in sample taken from retailers' shop, and the associated disease incidence and severity were 54.2% and 34%, respectively. Morphological identification of pure culture revealed that anthracnose caused by *Colletotrichum musae* and crown rot caused by *Fusarium* spp. are the most important disease causing agents and result in large percentage of fruit loss in Jimma town of bishishe market.

The possible relationships between bruising, irrigation, nitrogen and calcium fertilization, and stone fruit decay incidence were studied. The fruit skin (cuticle and epidermis) provides barrier protection against infection by pathogens. Most fungi are unable to penetrate healthy fruit skin and must enter through wounds. Fruit resistance to infection resulting from the biochemical action of tissues is not the result of a single compound or a single mechanism. Multiple systems with complex interactions may be involved in decay resistance. Immature fruit appear to be resistant to pathogens due in part to fungitoxic compound(s) already present at the infection point. Even if immature fruit are penetrated by fungi, in many cases the fungus is unable to colonize the tissue. Commonly, a high degree of resistance is maintained until the fruit approaches maturity. This loss of resistance occurs at the same time that fruit firmness is decreasing (Carlos H. Crisosto, 1999). This means that softer, over-mature fruit not only are more susceptible to mechanical injury and infection but also lack the biochemical resistance of immature fruit. Cultural practices may condition fruit to different susceptibilities to disease, however there is little information on the relationship between preharvest factors and bruising on decay incidence. Also, postharvest handling will affect decay development.

1.4.3 Disease associated with excessive preharvest fruit drop: Huanglongbing (HLB), presumably caused by the bacterium "*Candidatus Liberibacter asiaticus*," is a devastating citrus disease associated with excessive preharvest fruit drop. *Lasiodiplodia theobromae* (*diplodia*) is the causal organism of citrus stem end rot (SER) (Wei Zhao *et al*., 2015). The pathogen infects citrus fruit under the calyx abscission zone (AZ-C) and is associated with cell wall hydrolytic enzymes similar to plant enzymes involved in abscission. By means of DNA sequencing, *diplodia* was found in "*Ca. Liberibacter asiaticus*"-positive juice from HLB-symptomatic fruit (S) but not in "*Ca. Liberibacter asiaticus*"-negative juice. Therefore, the incidence of *diplodia* in fruit tissues, the impact on HLB-related postharvest decay, and the implications for HLB-related preharvest fruit drop were investigated in Hamlin and Valencia oranges. Quantitative PCR results (qPCR) revealed a significantly ($P < 0.001$) greater incidence of *Diplodia* in the AZ-C of HLB-symptomatic (S; "*Ca. Liberibacter asiaticus*" threshold cycle [C_T] of <30) than in the AZ-C of in asymptomatic (AS; "Ca. Liberibacter asiaticus" C_T of ≥30) fruit. In agreement with the qPCR results, 2 weeks after exposure to ethylene, the incidences of SER in S fruit were 66.7% (Hamlin) and 58.7% (Valencia), whereas for AS fruit the decay rates were 6.7% (Hamlin) and 5.3% (Valencia). *Diplodia* colonization of S fruit AZ-C was observed by scanning electron microscopy and confirmed by PCR test and morphology of conidia in isolates from the AZ-C after surface sterilization. *Diplodia* C_T values were negatively correlated with ethylene production ($R = -0.838$ for Hamlin; $R = -0.858$ for Valencia) in S fruit, and positively correlated with fruit detachment force ($R = 0.855$ for Hamlin; $R = 0.850$ for Valencia), suggesting that *Diplodia* colonization in AZ-C may exacerbate HLB-associated preharvest fruit drop.

In citrus fruits, invasion of the fruit tissue by *Diplodia* can cause stem end rot (SER), which is typically a postharvest disease. The occurrence of SER is greatly enhanced by exposure to ethylene, which is commonly used to enhance the color development of the fruit (Sdiri *et al*, 2012). Ethylene is a gaseous plant hormone that promotes chlorophyll degradation, ripening of climacteric fruits, senescence, and fruit abscission. Symptoms of *Diplodia* SER on citrus fruit include rind softening around the button, followed by a brown discoloration of affected areas. Typically, decay develops at the stem end followed by the stylar end before involving the entire fruit, since the fungus progresses rapidly through the spongy central axis of the fruit, reaching the stylar end much sooner by this route than through the rind. In citrus fruits, *Diplodia* infects the calyx and floral disk (button) during development. The infection then normally remains quiescent, while the fruit is on the tree, and the fungus does not usually invade the fruit until after harvest. Like many other fruit-rotting fungi, *Diplodia* has been reported to produce and secrete cell wall-digesting enzymes, including polygalacturonase, cellulase, and β-glucosidase, similar to citrus enzymes involved in abscission. In addition, it has been reported that the increased activity of the cell wall-digesting enzymes was associated with increased incidence of SER.

The citrus greening disease, huanglongbing (HLB), is a devastating condition that has been reported in 40 countries across the globe. Citrus trees infected with *Candidatus Liberibacter asiaticus* (CLas) develop HLB symptoms such as reduced plant height, yellow shoots, blotchy mottled leaves, canopy thinning, and premature abscission of fruits. Citrus fruits severely affected by HLB can be small, lopsided, and poorly coloured, with bitter or sour tasting juice (Chen *et al,* 2016). As HLB severity increases, preharvest fruit drop becomes increasingly problematic, contributing to declining crop yields. HLB is a very serious issue for the citrus industry, as it can kill or debilitate citrus trees quickly, and no effective treatment has been identified.. It was suggested that fruits from severely HLB-symptomatic sweet orange trees are more likely to have problems with preharvest fruit drop and postharvest pressure damage and breakdown, but may have less puncture damage in harvesting, transportation, packing, and juice processing,

Fruit drop is a natural, physiological phenomenon that occurs in all citrus varieties. However, pre-harvest fruit drop often causes distress and concern as it can result in significant yield reduction as well as loss of tree resources in a non-harvested fruit (Vashisth, 2018). For citrus trees grown in Florida, increased drop of mature fruit prior to harvest is one of the many symptoms of Huanglongbing (HLB) disease, caused by Candidatus Liberibacter asiaticus (CLas). This HLB-associated pre-harvest fruit drop, in addition to tree decline, results in a concomitant reduction in fruit production. For example, for the late-season 'Valencia' sweet orange, the rate of pre-harvest fruit drop was 14% for the 2009-2010 production season, whereas the rate rose to 30% during the 2016-2017 season. To date, how HLB escalates pre-harvest drop in citrus remains unclea

1.4.4 Postharvest infection: Many fungi that cause considerable wastage of produce are unable to penetrate the intact peel of produce, but readily invade via any injury point in the peel. The damage is microscopic, but is sufficient for pathogens present on the crop to grow on it. For infection of postharvest produce, different parts of the plants are infected i.e., floral infection, stem-end infection and quiescent infection. The floral infection occurs through various parts (sepals, petals, stigma etc) in many temperate fruits. The cut stem is a frequent point of entry for microorganisms and stem-end rots are important forms of postharvest spoilage of many fruit. For example, postharvest infection by *Sclerotina* and *Colletotrichum* is very common in many fruits through direct penetration of the peel. Quiescent Infection, the time between initial infection and appearance of disease symptoms is known as the latent or quiescent period. The term refers to "quiescent" or dormant parasitic relationship, which after some time, changes to an active one. A fungus may become quiescent at initiation of germination, germ tube elongation, appressorium formation, penetration or subsequent colonization. The failure to germinate or to develop beyond any subsequent stage is due to adverse physiological conditions temporarily imposed by the host, either directly on the pathogen or indirectly by modification in its pathogenic capability.

A previously undescribed spotting of tomato fruit was observed in the packinghouse and storage facility of a commercial organic grower in the Fraser Valley of British Columbia during 2009–2011. Symptoms initially appeared as small black spots on the tomato fruit, barely visible after harvest (Syama Chatterton *et al*, 2012). The black spots expanded to form larger grey-black lesions with evidence of mycelial growth in the centre after 5–10 days of storage at 12 °C or after tomatoes were shipped to retail outlets. Isolations from symptomatic skin and pericarp tissues from early and expanded lesions primarily yielded species of *Penicillium*. Healthy tomato fruit inoculated with conidia of *Penicillium* isolates developed black spots which progressed to form larger grey lesions, typical of symptoms observed in the commercial greenhouse. Several *Penicillium* species were identified using morphological characteristics and DNA sequencing of the ITS region and partial β-tubulin gene. Isolates identified as *P. olsonii* produced the characteristic spotting of fruit followed by expanding lesions, while other isolates identified as *P. solitum* and *P. polonicum* caused a rapid soft rot and decay. Ripe fruit inoculated with *P. olsonii* had a significantly higher disease severity index compared with those that were inoculated at the breaker or flush stages. Inoculations on the tomato shoulder resulted in significantly higher disease severity than inoculations at the middle or base of the fruit. Swabs of tomato fruit surfaces and calyx tissues taken while fruit were still on the vine followed by plating onto PDA revealed that high populations (>40 CFU) of resident *P. olsonii* were present on the fruit. Similar populations were detected when calyx tissues were plated directly onto PDA. The mode of entry of *P. olsonii* is likely through naturally occurring cracks or russeting of the fruit cuticle from resident populations. Wounding of the skin did not significantly enhance disease severity compared with unwounded treatments. Disease severity increased significantly when tomatoes were incubated at 12 °C compared with 22 °C following inoculation. This appears to be the first report of a fruit spotting and postharvest decay of tomato fruit caused by *P. olsonii*.

The activity of microorganisms could also affect sprouting, ripening and quality losses such as moisture loss or saccharification during crop storage. Premature sprouting of tubers or onions is a problem during storage, especially in industrialized countries, where consumers want to buy plant produce that does not show signs of sprouting, all over the year. In potato, for example, sprouting is a complex physiological process involving usage of storage reserve metabolites such as carbohydrates and proteins (Aksenova *et al*., 2013). Recently, it was shown that improved seed germination and seedling growth of *A. bifolim* in response to inoculation with endophytic bacteria were coupled with increased storage reserve mobilization and the degradation of proteins, lipids and sucrose. Phytohormones play a key role in the regulation of sprouting – cytokinins and indole-3-acetic acid signalling induce the onset of sprouts, and gibberellin stimulates sprout growth. Many plant associated bacteria are able to produce plant hormone like metabolites, and the role of auxins, cytokinins and gibberellins produced by bacteria in plant growth regulation is well documented. An inhibitory effect of six bacterial strains on potato sprouting was also observed. The bacteria, which were originally studied for their antagonistic activity towards potato rot fungi, significantly reduced sprouting in postharvest tubers of the cultivar Russet Burbank during storage. Those bacteria exhibiting sprout inhibitory effects were found to produce the plant hormone indole-3-acetic acid. It appears likely that the microbiome composition in storage crops influences dormancy duration.

Sprouting of crops could be prevented easily by storing at low temperatures (4°C). This, however, causes enormous energy consumption and favours quality loss because of 'cold-induced sweetening' (different from senescent sweetening), a phenomenon which is typically observed in stored starchy crops such as potato or cereals and was already documented in 1882. The sweeting is based on saccharification of starch, which means that starch is converted to the reducing sugars glucose and fructose. This process is favoured at lower storage temperatures. The actual problems arise during processing of starchy crops, such as frying, roasting or baking, when the reducing sugars react with free amino acids (asparagine) to acrylamide , a potentially carcinogenic substance. Proposed strategies to limit the accumulation of reducing sugars in tubers include preventing sucrose accumulation and preventing sucrose conversion into reducing sugars. Many bacteria and fungi possess enzymes such as α-amylases, which hydrolyze starch molecules into polymers composed of glucose units and so the activity of plant associated microorganisms could play a role in the accumulation of reducing sugars in plants. Whether modulating the microbiome with the aim of suppressing amylase activity could contribute to solving this problem remains elusive but is worth investigating (Pierre Pétriacq *et al*, 2018).

Moisture loss and consequently weight loss through transpiration is one of the main reasons for optical quality penalties with vegetables during storage. Drought stress during plant growth results in transpiration reduction and this continues postharvest. It is well known that inoculation with certain plant beneficial bacteria can minimize symptoms of drought stress in plants and one of the effects is that the water content in plant tissue remains higher during drought stress periods than in non-inoculated control plants. The application of microbial inoculants that reduce drought stress on the field could therefore also have positive effects on water content of crops during storage.

Microorganisms could also play a role in the ripening process of crop plants. About 20% of postharvest loss of fruits and vegetables is attributed to prolonged climacteric ripening that leads to senescence, apoptosis, lesions, spotting, bruising, infection and spoilage (Perry and Williams, 2014). During ripening, climacteric plants release various volatile compounds, including ethylene, a hydrocarbon, which can be metabolized by many microorganisms, thereby affecting plant development. However, fungi and bacteria, which are living inside fruits and vegetables, can also produce ethylene themselves. Besides direct modulation of ethylene levels in plants, microorganisms can also indirectly affect ethylene in plants. Plants produce ethylene from 1-aminocyclopropane-1-carboxylate (ACC) via ACC oxidase activity. Many bacteria in soil and plants possess ACC deaminase activity, which cleaves ACC to 2-oxobutyrate and ammonia and can contribute to ethylene balancing in plants. An alternative role of ethylene in microbial-driven modulation of ripening was introduced by a recent study exploring the potential of a *Rhodococcus rhodochrous* strain to delay ripening in different species of climacteric fruits. In this case, the proposed mechanism is that ethylene induced nitrile hydratase and/or nitrilase activity in the bacterium, which resulted in maintenance of fruit firmness and reduced spotting and spoilage (Perry, 2011).

More recently, the same authors demonstrated that application of ethylene gas to soil induced changes in the soil microbial activity, resulting in the release of a nitrile compound, which negatively affected climacteric ripening of organically grown bananas and peaches. This is a good example for the impact of the soil microbiota on crop quality and subsequent storage properties. The soil, in which crop plants have been grown, could generally affect storage stability of crops, that is, via the nutrient and water content but also via the soil microbiota. The soil is the main reservoir for the recruitment of microorganisms colonizing plants. Bacteria and fungi, attracted by root exudates migrate to the rhizosphere, further into the rhizoplane and some of them may also invade and colonized inner plant tissue. The soil microbiota is therefore also the main driver in the composition of the crop-associated microbiota. However, the impact of the soil microbial community composition on storage stability of crops is still a black box.

Available literature search has clearly revealed that the potential of the plant microbiome to influence postharvest losses is largely unexplored. There might be issues causing postharvest losses, which are independent from the microbiome, for example, a mouse might eat an apple no matter how the microbiome looks like. However, exploring the role of the plant microbiota in storage stability of crops could in future ensure food supply and food quality by employing microbial-based solutions from the field to the fork

Matured fruits and vegetables often become more susceptible to injury, therefore, more vulnerable to the attack of those microorganisms which require an injury or damaged tissue for penetration. Natural disease resistance, which acts in developmental stages of the plants, weakens as a result of separation of fruits and vegetables from parent plant. In addition picked fruits and vegetables are rich in moisture and nutrients; consequently, their prolonged storage leads to the senescence of tissues that suits the development of pathogens. Fungi and bacteria responsible for storage decay often originate in field or orchard. Many fungi viz. *Alternaria*, *Fusarium*, *Penicillium*, *Aspergillus*, *Rhizopus*, *Mucor*, *Botrytis* are among the most important cause of decaying which affect harvested fruits and vegetables. Postharvest pathogens perpetuate on crop debris in the field and under suitable conditions produce abundant spores.

These fungal spores are easily carried by winds, rain, or dispersed by insects to flowers and young fruits at various stages of development and form a potential source of infection. Soil, irrigation water and infected plant debris forms an important source of infection. Soil-residing fungi and bacteria can attack the bulb, tuber, root and other vegetal parts through contact with the soil, lifting of soil particles by winds, rains or by arriving in storage with attached soil residues. Host infection can occur on pre harvest, harvest or any of the postharvest handling stages. Harvesting instruments, containers, hands of the labours, packing and storage rooms are all prolific sources of fungal and bacterial spores. Fruits and vegetables carry the microorganisms into storage, only a few species can attack naturally while other species inhabits on the surface in large quantities will not penetrate or cause any decay. The development of disease during storage depends primarily on the existence of the appropriate microorganisms along with particular host. However, fungal spores or bacterial cells that have reached the suitable host should be capable of penetration or infection into the host tissues (Lindy Coates and Greg Johnson, 1989). The right environmental temperature, available moisture and presence of nutrients transferred from the host into water are the most important congenial factors for spore germination which is a preliminary stage for fungal penetration into the host.

Postharvest disease may occur at any time during postharvest handling, from harvest to consumption. Losses caused by postharvest diseases are greater than generally realized because the value of fresh fruits increases several fold while passing from the field to the consumer. Postharvest losses are estimated to range from 10 to 30% per year despite the use of modern storage facilities and techniques. When estimating postharvest disease losses, it is important to consider reductions in fruit quantity and quality, as some diseases may not render produce unsalable yet still reduce product value. Postharvest diseases affect a wide variety of crops particularly in developing countries which lack sophisticated postharvest storage facilities. Infection by fungi and bacteria may occur during the growing season, at harvest time, during handling, storage, transport and marketing, or even after purchase by the consumer (Bijendra Kumar Singh *et al*, 2017). It is also important to take into account costs such as harvesting, packaging and transport when determining the value of produce lost as a result of postharvest wastage. Aside from direct economic considerations, diseased produce poses a potential health risk. A number of fungal genera such as *Penicillium*, *Alternaria* and *Fusarium* are known to produce mycotoxins under certain conditions. Generally speaking, the greatest risk of mycotoxin contamination occurs when diseased produce is used in the production of processed food or animal feed.

Losses due to postharvest disease are affected by a great number of factors including commodity type, cultivar susceptibility to postharvest disease postharvest environment (temperature, relative humidity, atmosphere composition, etc.), produce maturity and ripeness stage, treatments used for disease control, produce handling methods, postharvest hygiene etc. Virtually all postharvest diseases are caused by fungi and bacteria. In general, however viruses are not an important cause of postharvest disease. Nematodes do cause postharvest disease in several crops.

The so-called 'quiescent' or 'latent' infections are those where the pathogen initiates infection of the host at some point in time (usually before harvest) but then enters a period of inactivity or dormancy until the physiological status of the host tissue changes in such a way that infection can proceed. Examples of postharvest diseases arising from quiescent infections include anthracnose of various tropical fruit caused by *Colletotrichum* spp. and grey mould of strawberry caused by *B. cinerea.* The other major groups of postharvest diseases are those which arise from infections initiated during and after harvest. Often these infections occur through surface wounds created by mechanical or insect injury. Wounds need not be large for infection to take place and in many cases may be microscopic in size. Common postharvest diseases resulting from wound infections include blue and green mould (caused by *Penicillium* spp.) and transit rot (caused by *Rhizopus stolonifer*). Bacteria such as *Erwinia carotovora* are also common wound invaders. Many pathogens, such as the banana crown rot fungi, also gain entry through the injury created by severing the crop from the plant.

The development of postharvest disease is intimately associated with the physiological status of the host tissue. To create the right environment for minimising postharvest losses due to disease it is important to understand the physiological changes that occur after produce is harvested. All plant organs undergo the physiological processes of growth, development and senescence. Growth and development generally only occur while the organ is attached to the plant (with the exception of seed germination and sprouting of storage organs), but senescence will occur regardless of whether the organ is attached or not (albeit at different rates). When an organ such as a fruit is harvested from a plant, it continues to respire and transpire depleting both food reserves and water. Such changes ultimately lead to senescence. Treatments which slow respiration and water loss, such as cool storage, therefore help to delay senescence. Different commodity types vary greatly in the processes which precede senescence, making it difficult to draw general conclusions. Postharvest decay organisms may penetrate through wounds, and is most common. Only minor wounds are required (micro-wounds). Wounds commonly occur before harvest by various means including insect injuries, implement injuries, wind damage, etc. or more frequently during and after harvest during handling, transport, packaging. Penetration may also be through intact fruit /vegetable, through the surface of mature fruit. Also quiescent infections that are established early during fruit growth but remain inactive until the fruit matures. It may also be the colonization of flower parts and invasion of maturing fruit.

1.5 Latent infections in the pre- and postharvest environment: For the epidemiologist, latency is the period from infection until the infected tissue becomes infectious, "spore-to-spore", but it generally has come to mean a period of pathogen quiescence, sometimes even extended to cover the time when microorganisms lie quiescent on the host surface before infection occurs. In managing diseases of horticultural crops, pathogen quiescence has important implications in timing prophylaxis, the reduction of stresses that trigger the transition of quiescence to aggression, and managing the harvested crop to prolong quiescence to the point at which yield is no longer affected. The inception of pathogen quiescence and its maintenance on or within the host implies a dynamic equilibrium between the host, the pathogen, and their environment. Physiological and physical changes in the host, its environment, or both trigger changes in that equilibrium to permit the pathogen to resume aggression (William R. Jarvis, 1994). Microorganisms lying on the host surface constitute a potential latent infection; as can be the case with *Botrytis* cinerea conidia that are prevented from germinating by antagonistic bacteria on leaf surfaces. Quiescence can run the spectrum from the ungerminated spore; through symptomless, internal infections; to visible but nonexpanding lesions, such as ghost spot of tomato; to expanding lesions on banana, caused by *Colletotrichum musae* and on avocado caused by *C. gloeosporioides*. With the exception of tomato ghost spot, most quiescent infections eventually become aggressive rots in the field and, if undetected, in stored and marketed produce. Quiescent infections are common in a wide variety of fruit, vegetable, and floral crops. They are caused by bacteria and fungi. Soft rot bacteria are common in apparently healthy tomatoes, cucumbers. Also in potatoes, quiescent *Clavibacter michiganense* subsp. *sepedonicum* is a serious impediment to export trade. The world's three most important small-berry crops, grapes, strawberries and raspberries, all have quiescent infections of *Botrytis* cinerea established in their floral parts; these infections usually become aggressive in ripe fruit. However, aggression may be significantly advanced in wet weather, when, for example, normally resistant green strawberries succumb to gray mold. *Botrytis cinerea* also can establish a quiescent infection in apple) calyx, which becomes eye rot and in fruit wounds.

Quiescent petal flecking in many floral crops is also frequently caused by *B. cinerea* and other host-specific *Botrytis* spp. in the monocotyledonous Corolliferae and the Ranunculaceae. In the *Sclerotiniaceae*, *Monilinia fructicola* Honey is quiescent in stone fruits and *Sclerotinia sclerotiorum* is quiescent in many vegetables, including carrots. *Botrytis aclada* is quiescent in onions and is responsible for neck rot in storage. Outside the *Sclerotinia*ceae, *Colletotrichum* spp. are common quiescent pathogens of tropical fruit.

1.5.1 Mechanisms of quiescence: In most unripe fruit and stem and root vegetables, the mechanisms limiting pathogen aggression are associated with either preformed antimicrobial substances (phytoncides) or with phytoalexins, enzymes, or physically resistant structures. A wide variety of postulates have been, or can be, advanced to explain quiescence (William R. Jarvis , 1994). Most of these postulate that aggression begins when certain physiological and physical changes occur, particularly in the host cell wall. The changes in dynamic equilibrium between climacteric metabolism and the strains induced by the stress of infection involve ethylene production. Other stresses are imposed by such factors as wounding, an unduly wet flowering season in strawberries, harvesting in potatoes, and inappropriate storage conditions.

Setting aside the epiphytic populations of potential pathogens on plant surfaces, which may be controlled by pesticides and biological agents, the infection process, as defined by breaching the host epidermis, may occur several days or weeks before symptoms appear. This means that it is essential to know what environmental conditions permit infection to occur. In strawberries, white flower buds and fully open flowers are the most susceptible to infection, and the calyx and corolla are more susceptible than the receptacle. Strawberries and raspberries present a difficult problem in timing prophylaxis because of the cymose inflorescences long development sequence (William R. Jarvis

1994). This situation is further complicated by the finding that first-formed berries in the strawberry inflorescence are more susceptible to gray mold than successively later berries on the same inflorescence. That microorganisms must be prevented from breaching the host epidermis means that the choice of pesticide is restricted to those that act on contact with the pathogen on the host surface; systemic, eradicant pesticides are less successful since they usually act imperfectly after infection.

Phonological and epidemiological studies permit formulating forecasting methods. The complex phenology of strawberries and raspberries makes forecasting difficult, but it has been done for grapes, which have a well-defined bloom period, and for which the microclimates conducive to *B. cinerea* infection are known. Precise control over greenhouse microclimates permits designing effective disease prevention strategies for protected crops.

Prophylaxis: Once the climatic and phonological conditions that permit initial infection to occur are established, pesticide application may be timed accurately usually well in advance of harvesting, so that undue residues are avoided. Several studies have shown that fruit crops are better protected by applying pesticides during bloom rather than later, and that vegetable crops store better when infections are controlled in the field. Disease prevention measures can also be devised for crops in the field. Thus, green beans escape gray mold and white mold infection if the plant architecture and row orientation permits floral parts to dry rapidly. Tomatoes may escape gray mold if the correct soil N and Ca levels are maintained.

Prolonging quiescence : Even if tissues are infected quiescently, it still maybe possible to obtain blemish-free floral crops or rot-free fruit and vegetables by manipulating the crop so that aggression is not triggered. How this may be done, however, is not widely understood. Water content of tissues seems to be important; for example, green strawberries are very susceptible to *B. cinerea* in wet weather, an effect simulated by supplying water directly into the fruit pedicel. The development of quiescent neck rot infections (*Botrytis aclada*) in onions depends on adequate moisture in the senescing green leaf tissue. If onions are dried adequately in windrows or cured artificially before storage, neck rot ceases to be a problem. Similarly, root crops with green leaves, like carrots, or senescent outer leaves, like cabbage, should be carefully trimmed and dried before storage. Correct mineral nutrition prolongs the quiescent phase. The effect of soil N in prolonging the quiescent phase of *B. cinerea* in tomato stem tissue seems to be one of delaying tissue senescence. Biological control can be applied rationally to prolong quiescence. The temperature of insolated fruit is also important; in picked fruit left in the sun, the temperature maybe 8 to 10 ^{0} C higher than ambient air temperature. Experience has shown that most fruits are best picked in the early morning; they have a longer shelf life if field heat is removed before shipping. Quiescent infections are common in a wide variety of horticultural crops, but while it is well established that bloom period prophylaxis gives protection to ripening fruit crops, the manner in which aggression is triggered is poorly understood. Devising disease prevention measures by delaying the onset of pathogen aggression depends on better knowledge of host–parasite interactions during ripening. In particular, environmental stresses that trigger aggression have to be identified and controlled in the field and in storage.

1.5.2 Major types of penetration of postharvest pathogens: Seven main types of penetration of postharvest pathogens include the following

i. **Infield penetration:** Fungi and bacteria responsible for in-storage decay often originate in the field or orchard. Late blight of potatoes caused by *Phytophthora infestans* is an example of decay originating in tuber infection in the field. The infection is caused by the zoospores found in soil or that fall onto the tubers from infected foliage during harvest. Following germination, the zoospores penetrate into the tubers through eyes, lenticels, growth cracks, wounds, or via the point of attachment to the plant. Tubers that were infected a few days prior to harvest or during the harvest itself are brought to storage carrying the disease in its early developmental stages, with no visible symptoms of decay. In a warm and damp climate, the blight fungus can also attack the tomato fruit at its various ripening stages in the field. The attack usually takes place at the edges of the fruit stalk or directly through the skin. In course of epidemics the entire crop may rot in the field.

   However, when the disease is less severe, fruits with no visible symptoms or with slight blemishes might be picked and the fungus will continue to develop during storage. The brown rot, which develops in citrus fruits during storage, originates in preliminary infections initiated in the orchards by *Phytophthora citrophtliora* and other *Phytophthora* species. During the rainy season, the fungal zoospores descend on the lower fruits of the tree and penetrate them directly.The fungus can develop in the orchard, but zoospore infection occurs a few days prior to picking and the external symptoms are not visible, the fruits will later rot and might comprise a serious problem during storage or transportation.

   The principle concern is that the harvest ends fungus development within the host while it is still on the parent plant and causing the decay after harvest or during storage. In many vegetables such as carrot, cucumber, lettuce, celery, cabbage, cauliflower and others, the soft watery rot caused by *Sclerotinia sclerotiorum* is common. This fungus, common in the soil might attack the vegetable when still in the field or during harvest and continue to develop in its tissues throughout storage or marketing.

ii. **Latent infections:** Pathogen penetration takes place in the field and will get to storeroom within the host tissue without eliciting any symptoms of decay known as early or quiescent stages of infection. The fungi that penetrate into the host in fields may cause latent or quiescent infections. These pathogens arrive at fruits or vegetables that are still on parent plant. However, during one of the phases, between their reaching the host and development of progressive disease, the growth is arrested until after the harvest, when physiological and biochemical changes occurring within the host will enable their renewed growth. Such arrested infections are referred to as "latent infections" in which the pathogen growth is temporarily inhibited. The latent state of the pathogen, whether involving spores that have landed on the host surface have commenced germination, or primary hyphal development within the host tissues, is linked to a dynamic balance among the host, the pathogen and the environment. The physiological and biochemical changes occurring in the host tissues after and during storage might affect the pathogen and host interrelationship lead to the activation of the latent pathogen. The brown rot, caused by *Monilinia fructicola* in stone fruits and the decay caused by Nectria galligena in apples (Swinburne, 1983) originate in the infections of young fruit in orchard.

The Nectria fungus forms a preliminary mycelium within the young tissues which is the dormant stage of disease, while the interrupted stage in the development of the brown rot caused by *Monilinia* occur during spore penetration through the stomata or even at the pre-germination stage. Another example of quiescent infection is anthracnose caused by *Colletotrichum gloeosporioides* in many tropical and subtropical fruits such as mango guava, papaya and various citrus fruits. The fungal conidia are found in large quantities on the surface of fruits during their development on the tree. In presence of free water upon the fruit, the spores germinate and form an appressorium at the tip of the germ tube. The appressoria function as resting spores on the fruit, since they resist environmental conditions much better than the conidia and remain vital for long periods while embedded in the natural wax of the fruit or bound to its surface.

iii. **Contact infections:** Fruits or vegetables which are free of a pathogen invasion by any means of penetration might be infected through contact with infected produce during storage. Contact infection is a significant factor in spreading white watery rot (*Sclerotinia* spp.) and bacterial soft rot (*Erwinia* spp.) in lettuce, carrot, cabbage, celery or squash during storage.

The development of *Botrytis* infection in stored strawberries, which turns into "mummies" covered with a gray layer of spore-bearing mycelium, causes contact infection and makes fruits susceptible to the disease. Similarly, one strawberry or tomato or a single grape infected by *Rhizopus* can spread the infection within the container. In fact, contact infection by *Botrytis* or *Rhizopus* is typical of many fruits and vegetables and account for major losses caused by these pathogens during long-term storage. In citrus fruits, contact infection of green and blue mould caused by *Penicillium digitatum* and *P. italicum* is very common; it often occurs during shipment and under certain conditions disqualifies the entire shipment

iv. **Penetration through natural openings:** Some pathogenic fungi and bacteria that cannot normally penetrate the directly, without the presence of wound in its surface, can penetrate through natural openings such as stomata and lenticels. The penetration of germ tubes of *Colletotrichum gloeosporioides* spores into young papaya fruits and the penetration of *Monilinia fructicola* spore germ tubes into young stone fruits can take place through stomata while the fruit is still in the orchard. Penetration through lenticels has also been described for *Alternaria* alternata spores in mango and *Gloeosporium perennans* in apples. The lenticels in potato tubers are prone to develop bacterial soft rot during storage. Most of lenticels are infested with the cells of *Erwinia cartovora* during harvest. The bacteria remain inactive in the lenticels until the favourable conditions such as mechanical pressure, existence of free water or low oxygen pressure within the tuber enhancing the tuber sensitivity to decay.

v. **Penetration through wounds:** In contrast to pathogens that attack the fruit and vegetable in the field, most of the storage pathogens are incapable of penetrating directly through cuticle or epidermis of the host, but require a wound or an injury to facilitate their penetration. Therefore, the fungi and bacteria that develop during storage are often called wound pathogens. Growth cracks present on harvested fruits and vegetables are natural avenues of infections. Careless separation of the fruit or vegetable from the parent plant might result in an injury liable to be attacked by the pathogen. The extent of injury caused by mechanical harvesting is far greater than that caused by a manual operation.The harvesting is accompanied by mechanical injuries enables the weak pathogens to penetrate. It has turned out that the depth of the injury, combined with the humid conditions during storage determine the fate and extent of the infection.

Each scratch, incision or other mechanical injuries inflicted on the fruit or vegetable during handling, harvesting, transporting, sorting, packing and storing might present adequate penetration points for the storage pathogens. A possible penetration point is the stem end separation area where damage often occurs during fruit picking and in this regard, the separation area is no different from any other injury.It regularly happens that simultaneously with the injury, a large amount of fungal spores and bacterial cells arrive at the injured area, some of which use the injury site to penetrate and infect the host. Penetration through wounds is a characteristic of *Penicillium digitatum* and *Penicillium italicum* pathogen conidia in citrus fruits. Wound infection is also characteristic of *Rhizopus stolonifer* that causes watery soft rot in many species of fruits and vegetables; of *Alternaria alternata* that causes a dark, rather dry decay in a large number of fruits and vegetables stored for a prolonged period; of *Geotriclium candidum* which causes the sour rot in citrus, melons and tomato fruits; and of various *Aspergillus*, *Cladosporium* and *Trichothecium* species, and other storage fungi of various fruits and vegetables.

vi. **Penetration owing to result of senescence:** Tissue senescence during continuing storage reduces disease resistance. Commonly at the end of storage period, the sensitivity of melons to blue green mould caused by various species of *Penicillium* and pink mould caused by *Trichothecium roseum* is increased. A senescencing onion that has commenced sprouting often harbours base decay, caused by various species of *Fusarium*. Generally, the rate of decay during storage increases with the duration of storage as tissue senescence progresses. Increasing the tissue sensitivity to diseases during storage also contributes to contact infection of a healthy product by an infected one covered with spore-bearing mycelium.

vii. **Penetration after physical damage:** Injuries caused by low temperature, heat, oxygen shortage or any other environmental stress increases the fruit or vegetable sensitivity to storage fungi. The physiological damage can be externally expressed through tissue browning and splitting, thus forming site vulnerable to invasion of wound pathogens. Extreme environmental conditions enhance sensitivity to an attack without any visible external signs of damage. Development of sun scald lesions in apples leads to *Alternaria alternata* and *Stemphylium hotryosum* attack while *Alternaria* may also be associated with other physiological disorders on apples such as bitter pit or soft scald.

1.5.3 Modes of infection by postharvest pathogens: Infection of fruit and vegetables by postharvest pathogens can occur before, during or after harvest. Infections which occur before harvest and then remain quiescent until some point during ripening are particularly common amongst tropical fruit crops. Anthracnose, which is the most serious postharvest disease of a wide range of tropical and sub-tropical fruit such as mango, banana, papaya and avocado, is an example of a disease arising from quiescent infections established prior to harvest. Anthracnose is also an important postharvest disease of a number of vegetables and temperate fruit (Droby, 2011). Various species of *Colletotrichum* can cause anthracnose. Some species are host-specific, whereas others can attack a wide range of fruit and vegetables. The infection process begins when conidia germinate on the surface of host tissue to produce a germ tube and an appressorium. Although it is known that there is a quiescent phase in the life cycle of the fungus, it is not entirely clear whether the ungerminated or the germinated appressorium represents the quiescent stage, as different studies have reported conflicting results. It may be that the fungus behaves differently on different hosts, or perhaps some researchers have been unable to detect appressorial germination due to the limitations of the techniques used. In avocado for example, early studies reported that ungerminated appressoria were the quiescent phase of *C. gloeosporioides*. Studies conducted two decades later however showed that appressoria germinated to produce infection hyphae prior to the onset of quiescence. In any case, the fungus ceases growth soon after appressorium formation and remains in a quiescent state until fruit ripening occurs. During ripening, the fungus resumes activity and colonizes the fruit tissue, leading to the development of typical anthracnose symptoms.

Natural antifungal compounds present in fruit tissue may be involved in regulating the quiescence of *Colletotrichum* infections. In avocados, antifungal compounds are present in the peel of unripe fruit at concentrations inhibitory to *Colletotrichum* gloeosporioides, the avocado anthracnose pathogen. During ripening, levels of these compounds decline to sub-fungitoxic levels coincidentally with the development of anthracnose symptoms on fruit. Treatments which stimulate the production of compounds also delay symptom development, suggesting that these compounds have a role in regulating quiescence. Grey mould of strawberry caused by *Botrytis cinerea* is another important postharvest disease sometimes arising from quiescent infections established before harvest. Conidia of *B. cinerea* on the surface of necrotic flower parts germinate in the presence of moisture. The fungus colonises the necrotic tissue and then remains quiescent in the base of the floral receptacle. Several months later when fruit are harvested, infections develop as a stem end rot in ripe fruit.

Many postharvest diseases develop from the stem end of fruit. The mode of infection involved in this group of diseases can however vary considerably. In the example of *B. cinerea*, lesions occurring at the stem end of fruit arise from quiescent floral infections. Stem end rots of citrus caused by *Lasiodiplodia* theobromae and *Phomopsis citri* result from quiescent infections in the stem button of fruit. These infections can be initiated at any stage of fruit development, and remain quiescent until the button begins to separate from the fruit during abscission. In other stem end rot diseases, infection occurs during and after harvest through the wound created by severing the fruit from the plant. Recent studies have brought attention to another important mode of infection involved in the development of some stem end rot diseases. Endophytic infection, whereby the fungus symptomlessly and systemically colonises the stem, inflorescence and fruit pedicel tissue, is important in a number of stem end rot diseases of tropical fruit. Mango stem end rot caused by *Dothiorella domintcana* is one example of a postharvest disease arising from endophytic colonisation of fruit pedicel tissue. In this case, the fungus colonises the pedicel and stem end tissue of unripe fruit, where it remains quiescent until fruit ripening commences.

A considerable number of postharvest diseases develop from infections which occur shortly before harvest. Such infections may not be visible at the time of harvest, yet at the same time are not necessarily inactive. Symptoms may develop more rapidly after harvest, particularly if storage conditions favour pathogen development (Droby, 2011). Examples of postharvest diseases which can arise from late season infections include brown rot of peach (caused by *Monilinia fructicola*), grey mould of grape (caused by *Botrytis* cinerea) yeasty rot of tomato (caused by *Geotrichum candidum*) and sclerotium rot of various vegetables (caused by *Sclerotium rolfsii*).

Many common postharvest pathogens are unable to directly penetrate the host cuticle. Such pathogens therefore infect through surface injuries or natural openings such as stomata and lenticels. Injuries can vary in size from microscopic to clearly visible and may arise in a number of ways. Mechanical injuries such as cuts, abrasions, pressure damage and impact damage commonly occur during harvesting and handling. Insect injuries may occur before harvest yet remain undetected at the time of grading, providing ideal infection sites for many postharvest pathogens. Infection of lychee fruit by yeasts commonly occurs through insect injuries which are difficult to see at harvest. Some chemical treatments used after harvest, such as fumigants used in insect disinfestations and disinfectants such as chlorine, may also injure produce if applied incorrectly. Various types of physiological injury such as chilling and heat injury can predispose produce to infection by postharvest pathogens. Tropical fruit in particular are very sensitive to low temperatures, many developing symptoms of chilling injury below 13°C (depending on storage duration). Symptoms of chilling injury include abnormal ripening, peel and flesh discolouration, water-soaking and pitting. Chilling injury can increase the susceptibility of produce to postharvest disease considerably. For example, the incidence of *Alternaria* rot in papaya, apple and various vegetable crops is increased by exposure to excessive cold. High temperatures can also increase susceptibility of harvested produce to disease. For example hot water dipping of mangoes for excessive times and/or temperatures can result in increased levels of stem end rot, caused by *Dothiorella* spp.

The natural resistance of fruit and vegetables to disease declines with storage duration and ripeness. Weak pathogens which normally require a wound in order to infect can become a problem in produce that has been stored for long periods of time. Treatments which help to maintain the natural 'vitality' of fruit and vegetables aid in delaying the onset of disease in stored produce.

i. **Surface damage:** Most postharvest pathogens of tropical and subtropical fruits are incapable of infecting the fruit directly through the intact fruit surface. These pathogens are largely dependent on the presence of surface injuries for penetration, and hence are called 'wound pathogens'. The nature of surface wounds may vary from cracks in the cuticle and underlying tissues created during fruit growth, maturation and ripening, to mechanical damage inflicted during harvesting and subsequent handling of the fruit. For example, pathogens such as *Penicillium digitatum* and *P. italicum*, the cause of green and blue mold of citrus, respectively, are totally dependent on surface wounds (Macarisin *et al*, 2007). For this reason, the method of harvesting (hand vs. mechanical) can significantly impact postharvest decay development. To minimize mechanical damage of fruit skin, proper management decisions that include the selection of an optimum time to harvest in relation to produce maturity, and the appropriate climatic conditions for harvest, must be considered. Black rot, caused by *Thielaviopsis paradoxa* is the most common and best-known postharvest pathogens of pineapple fruit. The pathogen gains entrance through natural growth cracks and wounds on the fruit surface generated during picking or packing. After penetration through the cut stem-end the pathogen grows rapidly upwards and into the flesh. Pathogens can also invade fruit tissue though physiological damage of the tissue caused by cold, heat, and CO_2. In most cases the physiological damage is manifested by localized browning and death of fruit peel tissue forming vulnerable sites for entry of wound pathogens.

ii. **Infection through natural openings:** Without the presence of surface injuries, some pathogens can penetrate fruit through natural openings such as stomata and lenticels. For example, germinating spores of *C. gloeosporioides* can penetrate young papaya fruits in the orchard through stomata at various developmental stages. Penetration through lenticels was described also for *Alternaria alternate* spores in mango and persimmon fruit.

iii. **Insect mediated infection:** The activity of various types of invertebrates can play a critical role in contaminating produce with inoculum. The most substantial evidence in this regard has been shown in relation to *B. cinerea* infection of kiwifruit through flowers facilitated by thrips (*Thrips obscuratus*) and honey bees (*Apis* sp). High incidence of *Botrytis* infections were also reported on kiwifruit when garden snail damage was present; slime secreted by the snail stimulated the germination of the conidia. Honey bees and other types of insects visiting flowers potentially have the capability to disseminate *Botrytis.*

iv. **Infection through flower parts:** Pathogen infection of tropical fruits through flower parts has rarely been studied. The sporadic nature of floral infections by pathogens combined with extended latent periods prior to decay development has made this a difficult topic to study. The colonization of persistent sepals and fruit receptacles in California kiwifruit by *B. cinerea* starts as early as 30 days after fruit set, and this colonization correlates significantly with postharvest decay after three to five months of cold storage. The authors found that harvest infections along with infections through wounds could explain only 10% of the postharvest decay. In contrast, colonization of sepals of kiwifruit in New Zealand plays a minor role (if any) in postharvest grey mold decay. Instead the fresh stem scar becomes infected during harvest. Pineapple is another example where floral infection followed by an extended latent period can occur. The flower of pineapple is the portal for several major pathogens, and the period of latency ranges from four to six months.

v. **Attack mechanisms :** Fungal pathogens can be divided into three classes: necrotrophic, biotrophic, or hemibiotrophic fungi. Necrotrophs are opportunistic pathogens that invade wounds to access dead tissue. They grow intercellularly producing lytic enzymes and toxins to decompose surrounding tissue and utilize the dead tissue as a source of nutrients (Wolpert *et al*, 2002).. In contrast, biotrophs, which are not known to attack tropical and subtropical fruits, do not kill their host immediately. They are in fact dependent on living tissue to complete their development. Generally these fungi do not produce large quantities of extracellular enzymes or toxins during their parasitism on the host. Hemibiotrophic fungi are regarded as a subgroup of nectrotrophic fungi that require living host tissue during part of their life cycle and then later switch to a necrotrophic stage.

Colletorichum is a hemibiotroph on many tropical and subtropical fruits. *Colletorichum* spores adhere to and germinate on the plant surface, where they produce appresoria that give rise to germ tubes. The tip of the germ tube, developing from the appressorium, penetrates through the cuticle with an infection peg. After penetration and colonizing one or more host cells, the biotrophic hyphae subsequently gives rise to secondary necrotrophic hyphae. Both *Botrytis* and *Monilinia* can either invade host tissue through wounds or breach the fruit cuticle by extending an infection peg from an appressorium. These initial infections can then remain quiescent or latent for long periods. Depending on the physiological status of the fruit, these hyphae may continue the infective process or remain quiescent. The quiescence of the latent infections established by biotrophic pathogens can be attributed to several factors, viz., lack of nutrients in the fruit needed for pathogen development after infection, the presence of preformed or inducible antifungal compounds in resistant unripe fruits and an unsuitable environment for the activation of fungal pathogenicity factors.

Activation of quiescent biotrophic infections seems to involve a set of coordinated events including physiological and biochemical changes that occur in the host during ripening and senescence that lead to decreased host resistance and increased susceptibility. Concomitant to the developmental loss of resistance that accompanies fruit ripening, there also appears to be an activation of processes by the pathogen that compromise host defenses, directly or indirectly, by the detoxification of antifungal agents produced by the host (Prusky and Lichter, 2007). The physiological changes that accompany fruit ripening and host senescence, for example, changes in host pH, sugar content, cell wall components, and the oxidation of wounded tissue trigger responses by the infecting fungus. The acidification of the tissue by organic acids (oxalic and gluconic) or its alkalinization by ammonia, and the possible interference with the production of reactive oxygen species (ROS) by the host in response to fungal infection, as well as fungal ROS production, may contribute to rapid necrosis of the tissue. Further amplification of the decay can result from activation of fungal genes coding for cell wall degrading enzymes. Regardless of the mechanisms that are involved in the onset and termination of quiescence, however, quiescent infections appear to be a case of coevolution, as they can be seen as advantageous to both pathogens and hosts in a natural ecosystem. In relation to a pathogen, quiescent infection of climacteric fruits by *C. gloeosporioides* represents an example of pathogen adaptation to host. The involvement of ethylene in the termination of quiescence strongly suggests that *Colletotrichum* spp. must have coevolved to develop a mechanism to use the host's ripening hormone as a signal to reactivate the infection process. This mechanism prevents contact of the pathogen with host tissues that have high levels of antifungal compounds.

Citrus fruit volatiles released after wounding of the fruit rind play an important role in host recognition by *P. digitatum* and *P. italicum*. When exposed to volatiles from citrus fruit peel tissue, the percentage of germinated spores of *P. digitatum* and *P. italicum* markedly increased compared to the unexposed controls. In contrast, *B. cinerea* and *P. expansum*, non-pathogens of citrus fruit, were either not affected or inhibited by the citrus peel volatiles. The citrus example is illustrative of a situation where a pathogen has become uniquely adapted to the physiology of its host. *P. digitatum* in its early stages of infection suppresses a defense-related hydrogen peroxide (H_2O_2) burst in host tissue, thereby compromising fruit defenses. In contrast, *P. expansum*, a non-pathogen of citrus, triggers production of a significant amount of H_2O_2 in citrus fruit exocarp. The results of this study demonstrated that H_2O_2 is a critical element in citrus fruit defense. Whether it acts directly on the fungus or as a signaling molecule needs to be elucidated. An example of a pathogen that attacks a wide range of tropical and subtropical fruits is *Alternaria*. *A. alternata* is known to cause black spot disease on a variety of fruits after harvest and causes significant economic loss. On mango for example, the pathogen penetrates fruit tissue through lenticels during fruit growth. In persimmon fruit it penetrates either through surface wounds or directly through the cuticle. In both cases the pathogen remains quiescent until fruit ripen, after which symptoms become visible (Prusky and Lechter, 2007). It has been suggested that the transition from the quiescent to the active necrotrophic stages is modulated by changes in ambient pH.

Black rot, caused by *Alternaria citri*, can affect all species of citrus, where it develops through the central columella of the fruit. Penetration of the pathogen takes place either at the stylar or stem end of the fruit around the calyx (button) of immature fruit. The pathogen remains quiescent until after harvest, after which the decay process is initiated and is evidenced during storage as an internal black rot spreading from the fruit core. External symptoms are not often apparent. The ability of the pathogen to produce an endopolygalcturonase appears essential for *Alternaria* isolates to cause black rot. None of the black rot strains isolated from citrus to date have been found to produce host specific toxins (HST). In general, affected citrus fruit are more brightly colored than normal fruit due to the ethylene generated in response to infection (Timmer *et al* 2003). Another mechanism of pathogen attack is via the fruit's stem end. Stem-end rots are usually caused by a complex of pathogens that varies with the fruit type. These rots usually originate from quiescent infections at the stem end of the fruit. In citrus for example, stem-end rots caused by *Botrydiplodia theobromae* and *Phomopsis citrii* develop from quiescent infections located in the stem end button. Infections occur in the orchard under wet conditions that allow the release of fungal spores from pycnidia. These spores infect flowers in the initial stages of fruit development and establish primary infections at the stem end of the developing fruit. Infections remain quiescent until fruit is harvested and then become active. Visible decay becomes evident, however, only after senescence of the fruit's button. Stem-end rot pathogens of mango include

Dothiorella dominicana, *Dothiorella* mangiferae, *Lasiodiplodia* theobromae, *Phomopsis mangiferae*, *Cytosphaera mangiferae*, *Pestalotiopsis* sp. as well as other fungi. These fungal species were found to occur endophytically in the stem tissue of mango trees prior to inflorescence emergence.

vi. **Fruit defense mechanisms:** The development of postharvest disease is a process that involves contamination of a fruit surface with a pathogen's propagules, followed by attachment, germination, penetration and establishment of the pathogen. This is followed by pathogen development in the tissue and the production of decay symptoms. To be capable of producing a disease, the pathogen needs to overcome constitutive and induced host-defense mechanisms, and exert pathogenicity factors (cutinases, pectolytic enzymes, cellulytic enzymes, toxins, etc.) that facilitate its development through the use of the released nutrients. Plants have evolved a number of strategies to resist fungal infection. These strategies include the following.

a. **Preformed antifungal compounds:** The ability to provide direct evidence for the role of preformed compounds in resistance of tropical and subtropical fruit to postharvest diseases is problematic due to the difficulty in determining the concentration of inhibitory compounds that may actually come into contact with the invading pathogen. In several systems studied, changes in concentrations of inhibitory compounds did not coincide with changes in susceptibility. Dopamine, for example, was isolated from the peel of unripe banana in concentrations that inhibited *C. musae in vitro* and was therefore presumed to be a possible preformed antifungal compound involved in fruit resistance. However, changes in the concentration of dopamine were not correlated with changes in decay development. In mango, a mixture of antifungal compounds consisting of 5-12-cis-heptadecenyl resorcinol and 5-pentadecenyl resorcinol was found at a fungitoxic concentration in the peel of unripe mango fruit that were resistant to *A. alternata* and markedly decreased in ripe and susceptible fruit (Droby *et al*, 1987). The resistance of unripe avocado fruit to the activation of latent infections of *C. gloeosporioides* is correlated with the presence of high concentrations of preformed antifungal compounds. The major antifungal compound has been shown to be 1-acetoxy-2-hydroxy-4-oxo-heneicosa-12,15-diene. The amount of this compound decreases tenfold, to sub-fungitoxic concentrations, during fruit ripening. The decrease correlates with the activation of quiescent infections. Similarly, it has been suggested that in citrus the resistance of young mature green lemons is related to the presence of a preformed monoterpene aldehyde, citral, that decreases in older yellow fruit, enabling decay to develop rapidly.

b. **Induced antifungal compounds:** Inducible defense responses include the early formation of physical barriers to pathogen development and/ or the activation of biochemical defenses. Physical barriers may include strengthening of the plant cell wall via phenolic deposition and lignification to prevent the pathogen progressing beyond the first few penetrated cells. Biochemical defenses include the activation of pathogenesis related (PR) proteins, some of which are enzymes (chitinase and β-1-3- glucanase) known to hydrolyze components of fungal cell walls, and the *de novo* synthesis by the plant of antimicrobial secondary metabolites, known as phytoalexins. In green banana, resistance to *C. musae* has been associated with a growing necrotic reaction within the peel. Five antifungal compounds not present in healthy tissue have been isolated from these necrotic tissues. In some cases, the induction of antifungal compounds during fungal penetration can lead to the confinement of the pathogen by an antimicrobial barrier. 'Wound gum' accumulated in injured citrus exocarp , predominantly consisted of the antifungal compound, 3-[4-hydroxy- 3-(methyl-2-butenyl) phenyl]-2-(E)-propenal. This compound increases in concentration as a function of time resulting in the complete quiescence of the invading pathogen.

Levels of antimicrobial secondary metabolites may also be induced by various physical treatments. For example, grapefurits exposed to UV irradiation and subsequently inoculated with *P. digitatum*, exhibited increased resistance to the development of green mold decay. The increased resistance following UV irradiation was associated with the induction and accumulation of chitinase proteins in flavedo tissues. UV irradiation induced the accumulation of a 25 kD chitinase protein but had almost no effect on the levels of β-1,3-endoglucanase proteins. The induction of the 25 kD chitinase protein by UV irradiation was detected after 48 hours, and increased further after 72 and 96 hours (Droby *et al*, 1993). Resistance of citrus fruit to development of green mold has also been attributed to increased levels of phytoalexins produced in response to fungal infection.

Heat may also inhibit pathogen spread by inducing defense mechanisms in the outer cell layers of the epicar. Hot water rinsing and brushing (HWRB) induced the accumulation of 21, 22 and 25 kDa proteins that cross-reacted with citrus and tobacco chitinase antibodies and 38, 42 and 43 kDa proteins that cross-reacted with citrus and tobacco β-1,3- glucanase antibodies. The increase in the accumulation of glucanase and chitinase proteins may be part of a complex of fruit disease resistance mechanisms induced by the HWRB treatment (Pavoncello, *et al*, 2001). Exposure of harvested mangoes to CO_2 at a flow rate of 100 mL of CO_2 min^{-1} for 24 hours significantly reduced anthracnose development. However, the optimum effective dosage of

CO_2 varied according to cultivar, 20% CO_2 for 'Keitt', 60% CO_2 for 'Tommy Atkins'. A concomitant increase of 5-(7,12-heptadecadienyl) resorcinol was also detected in 'Keitt' fruits in response to the 20% CO_2 treatment. Dipping fruit in hot water (55°C) for five minutes also resulted in an increase in 5-(7,12-heptadecadienyl) resorcinol.

Crops are attacked by a wide range of microorganisms in the postharvest phase. Actual disease only occurs when the attacking pathogen starts to actively grow in the host. Diseases are loosely classified according to their signs and symptoms. Signs are visible growths of the causal agents, and symptoms the discernible responses produced by the host. Postharvest diseases are caused primarily by microscopic bacteria and fungi, with fungi the most important causal agent in fruit crops. Fungi are further subdivided into classes and are described as lower fungi, characterized by the production of sporangia which give rise to numerous sporangiospores, or higher fungi, described as ascomycetes, deuteromycetes and basidiomycetes (Bijendra Kumar Singh *et al*, 2017).. Ascomycetes are exemplified by fruiting bodies that release sexual spores when mature. Deuteromycetes, a form of ascomycetes, only release asexual spores. They are more common than the sexual ascomycetes stage in postharvest crops. Deuteromycetes are further subdivided into hyphomycetes and coelomycetes based on spore and structural characteristics. The agonomycetes contain important soil pathogens that form survival structures known as sclerotia that allow them to survive in the absence of the host.

Precise identification of the pathogen causing postharvest disease is central to the selection of an appropriate disease control strategy. Many of the fungi which cause postharvest disease belong to the phylum Ascomycota and the associated fungi Anamorphici. In the case of the Ascomycota, the asexual stage of fungus (anamorph) is usually encountered more frequently in postharvest diseases than the sexual stage of the fungus (teleomorph). Important genera of anamorphic postharvest pathogens include *Penicillium*, *Aspergillus*, Geotrichum, *Botrytis*, *Fusarium*, *Alternaria*, *Colletotrichum*, *Dothiorella*, Lasio*diplodia* and *Phomopsis*. In the phylum Oomycota, the genera *Phytophthora* and *Pythium* are important postharvest pathogens, causing a number of diseases such as brown rot in citrus (*Phytophthora citrophthora* and *P. parasitica*). *Rhizopus* and *Mucor* are important genera of postharvest pathogens in the phylum Zygornycota. *R. stolonifer* is a common wound pathogen of a very wide range of fruit, causing a rapidly spreading watery soft rot. Genera within the phylum Basidiomycota are generally not important causal agents of postharvest disease, although fungi such as *Sclerotium rolfsii* and *Rhizoctonia solani*. The major causal agents of bacterial soft rots are various species of *Erwinia*,

Pseudomonas, Bacillus, Lactobacillus and *Xanthomonas*. Bacterial soft rots are very important postharvest diseases generally of less importance in most fruit.

1.5.4 Host physiological status: The development of postharvest disease is intimately associated with the physiological status of the host tissue. To create the right environment for minimizing postharvest losses due to disease it is important to understand the physiological changes that occur after produce is harvested. All plant organs undergo the physiological processes of growth, development and senescence. Growth and development generally only occur while the organ is attached to the plant (with the exception of seed germination and sprouting of storage organs), but senescence will occur regardless of whether the organ is attached or not. When an organ such as a fruit is harvested from a plant, it continues to respire and transpire depleting both food reserves and water. Maturity is a term often used in reference to fruit and is frequently confused with the term ripeness. Fruit are often classified into two groups on the basis of how they ripen. Climacteric fruit exhibit a pronounced increase in respiration and ethylene production coincidentally with ripening. Climacteric fruit can be harvested in an unripe state and providing they are sufficiently mature, will ripen to an acceptable quality. Non-climacteric fruit do not exhibit a rapid increase in respiration during the ripening process. The eating quality of non-climacteric fruit does not improve after harvest, although they may undergo some changes in colour development and softening. For this reason they should not be harvested until they are ready to eat.

Infection mode: Infection of fruit by postharvest pathogens can occur before, during or after harvest. Infections which occur before harvest and then remain quiescent until some point during ripening are particularly common amongst tropical fruit crops. Anthracnose, which is the most serious postharvest disease of a wide range of tropical and sub-tropical fruit such as mango, banana, papaya and avocado, is an example of a disease arising from quiescent infections established prior to harvest. Various species of *Colletotrichum* can cause anthracnose. In avocado for example, early studies reported that ungerminated appressoria were the quiescent phase of *C. gloeosporioides*. Studies conducted two decades later however showed that appressoria germinated to produce infection hyphae prior to the onset of quiescence. In any case, the fungus ceases growth soon after appressorium formation and remains in a quiescent state until fruit ripening occurs. In avocados, antifungal dienes are present in the peel of unripe fruit at concentrations inhibitory to *Colletotrichum gloeosporioides*, the avocado anthracnose pathogen. Grey mould of strawberry caused by *Botrytis cinerea* is another important postharvest disease sometimes arising from quiescent infections established before harvest. Conidia of *B.*

cinerea on the surface of necrotic flower parts germinate in the presence of moisture. The fungus colonises the necrotic tissue and then remains quiescent in the base of the floral receptacle. Many postharvest diseases develop from the stem end of fruit. The mode of infection involved in this group of diseases can however vary considerably. In the example of *B. cinerea*, lesions occurring at the stem end of fruit arise from quiescent floral infections. Stem end rots of citrus caused by Lasio*diplodia theobromae* and *Phomopsis citri* result from quiescent infections in the stem button of fruit. In other stem end rot diseases, infection occurs during and after harvest through the wound created by severing the fruit from the plant (e.g. banana crown rot). Endophytic infection, where by the fungus symptomlessly and systemically colonises the stem, inflorescence and fruit pedicel tissue, is important in a number of stem end rot diseases of tropical fruit. Mango stem end rot caused by *Dothiorella dominicana* is one example of a postharvest disease arising from endophytic colonisation of fruit pedicel tissue. In this case, the fungus colonises the pedicel and stem end tissue of unripe fruit, where it remains quiescent until fruit ripening commences. Postharvest diseases which can arise from late season infections include brown rot of peach (*Monilinia fructicola*) and grey mould of grape (*Botrytis cinerea*). Mechanical injuries such as cuts, abrasions, pressure damage and impact damage commonly occur during harvesting and handling. Some chemical treatments used after harvest, such as fumigants used in insect disinfestations and disinfectants such as chlorine, may also injure produce if applied incorrectly. Tropical fruit in particular are very sensitive to low temperatures, many developing symptoms of chilling injury below 13^{0}C (depending on storage duration). For instance, the incidence of *Alternaria* rot in papaya and apple is increased by exposure to excessive cold ; hot water dipping of mangoes for excessive times or temperatures can result in increased levels of stem end rot (*Dothioretla* spp.).

Postharvest decay of fruits, vegetables, and grains by pathogens causes significant economic losses. Infected produce presents a potential health risk since some decay fungi produce mycotoxins that are hazardous to human health. Infections are the result of the interplay between host resistance and pathogen virulence. Both of these processes, however, are significantly impacted by environmental factors, such as temperature, UV, oxidative stress and water activity. The impact of various physical postharvest treatments (e.g., heat and UV) on the viability and virulence of postharvest pathogens has been reviewed by Liu *et al*,(2018). Oxidative injury, protein impairment, and cell wall degradation have all been proposed as the mechanisms by which these abiotic stresses reduce fungal viability and pathogenicity. The response of decay fungi to pH and the ability of pathogens to modulate the pH of the host environment also affect pathogenicity. The effects of the manipulation of the postharvest environment by ethylene, natural edible coatings and controlled atmosphere storage on fungal viability are of importance.

Activities on vegetables: After harvest, different organisms have been isolated from carrot diseases ranging from *Pythium* sp , *Thielaviopsis basicola*, *Fusiarium* sp. and *Alternaria* sp . Ewekeye *et al.* (2013) reported that *Mucor* sp, *Rhizopus* sp and *Aspergillus* niger, *A. fumigatus* and *Fusiarium* sp collected from different market places and isolated from diseased carrots induced similar disease symptoms when inoculated on healthy carrots. The variations on the isolates obtained from different researchers may be connected to the fact that different varieties of carrots may be used as well all different experimental procedures. It was observed that *Hemithosporuim* sp had 57% disease frequencies in *Telfaria* spp while *Curvularia* spp had 38% disease. The reason for this could be because fungi are quite ubiquitous. Its high prevalence could have resulted from the fact that the seeds from which the vegetables were grown were not treated before sowing into the nursery. Fungi and bacterial diseases are spread by microscopic spores which are distributed in the air and soil and via decaying plant material. Infection after harvest can occur at any time. It is usually the result of harvesting or handling injuries.

Fungi, especially moulds are important pathogens of fruits and vegetable particularly under tropical and subtropical conditions. The importance of storage rate includes reduction in the value. Other important consequence often over looked, is mycotoxin contamination of the affected materials (Adebayo and Diyaulo, 2003). Black mould disease caused by *Aspergillus* niger var *Tieghem* (An) is a limiting factor in onion production worldwide, *Aspergillus niger* also primarily reported to survive between onion crops as a soil saphrophyte (on decaying matter) in or on onion bulbs or on cull onions in field or storage and being ubiquitous in occurrence it attacks/infects bulbs of onions in field/storage whenever they find injured tissues by producing various enzymes or toxins. Fungal contamination of onion bulbs especially black moulds constitute a menace in the production and storage of onion particularly in the tropics. Apart from toxins productions, presence of moulds in onion bulbs eventually leads to disease development, deterioration and reduction in market value.

Activities on other food crops: Many fungal diseases have been affecting the production of Irish potatoes but the most challenging are the postharvest diseases. These losses were as a result of physical, physiological or pathogenical factors or a combination of the three factors especially when there are mechanical injuries due to physical factors. Postharvest food loses occur during harvesting and handling due to grain shattering, spillage during transport and biodeterioration at all steps in the postharvest chain. The principal agents of biodeterioration could be moulds, insects, rodents and birds. Also, losses of roots and tubers can be physiological (caused by the effect of environmental conditions), pathological (caused by the attack of pathogens such as fungi, bacteria, insects) (Obayelu, 2014) Exposition to extreme temperatures (high, low) during pre and postharvest and rough handling does not only reduce the value of the crop through damage created in appearance, it also leads to invasions of pathogens causing decay of the attacked crop in the storage. Most organisms that cause postharvest disease are weak pathogens. Only a few fungi (such as *Colletotrichum*) can directly invade through healthy skin. Most fungi enter vegetables through wounds in the skin, or infect products already damaged by unsuitable storage conditions.

Preharvest infection occurs before harvest. Some fungi infect product during flowering, while other pathogens may stick to the product skin. Microbes that are relatively dormant during plant growth can develop and cause symptoms after harvest. Pre-harvest infection can occur by direct penetration of the skin, entry through natural openings (stomata and lenticels), transmission through the parent plant, and through damaged areas. Plant defence responses are able to hold such infections in check while the plant is actively growing. However, once removed from sources of water and nutrient, the disease can take hold. Infection of some fruiting vegetables (Ex:. internal rots in capsicums) actually occurs during flowering. Fungi such as *Alternaria* and *Botrytis* are able to infect flower petals. However, they only start to develop once the vegetable has matured. Postharvest infection commonly occurs through injuries or other breaks in the skin, and is increased if products are wet.

Fungal spores are present in the air, on equipment, on containers and on the hands of harvest workers and packers. However, for those spores to infect the vegetables there must be suitable conditions for spore germination and growth as well as openings that allow them to penetrate the plant tissue. Moisture is essential for spores to germinate. While most can germinate in pure water, if nutrients are present then germination is increased and growth is more vigorous. For example, studies have found that spores of *Rhizopus* spp. and *Botrytis cinerea* need both water and nutrients to allow spores to germinate and penetrate a host. Bacteria also need water plus nutrients for cells to divide and multiply.Wounding allows cell contents to leakout. Broken cells provide both water and nutrients, in the form of sugars, acids and other substrates. Wounds therefore provide ideal environments for spores to germinate and bacteria to grow. Moreover, while waxy epidermal (skin) cells represent a significant barrier to infection, the cells underneath are less protected. Once established in a wound site, pathogens can directly attack these cells and spread through the plant tissue.

References

Adebajo, L.O. and Diyaolu, S.A., 2003. Mycology and spoilage of retail cashew nuts. *African J. Biotech.,* **2**:369-373

Aksenova, N,P., Lidiya I Sergeeva & Konstantinova, T. N. and Georgy A Romanov,2013. Regulation of potato tuber dormancy and sprouting. *Russian J. Pl. Physiol.*, **60** : 301-312.

Anonymous, 2015. Postharvest management of vegetables. Horticulture Innovation Australia, Alexandria NSW 1435

Aulakh, J. and Regmi, A., 2013. Postharvest food losses estimation development of consistent methodology. *Selected Poster Prepared for Presentation at the Agricultural & Applied Economics Association's 2013 AAEA & CAES Joint Annual Meeting,* Washington DC.

Berg G., Köberl M., Rybakova D., Müller H., Grosch R., and Smalla K., 2017. Plant microbial diversity is suggested as the key to future biocontrol and health trends. *FEMS Microbiol Ecol.*, **93**: fix050

Bijendra Kumar Singh, Kulveer Singh Yadav and Akhilendra Verma. 2017. Impact of Postharvest Diseases and Their Management in Fruit Crops: An Overview. *J.Bio.Innov.*, **6** : 749-760

Buchholz, F., Kostić, T., Sessitsch, A. and Mitter, B., 2018. The potential of plant microbiota in reducing postharvest food loss. *Microb Biotechnol.*, **11**:971–975.

Cappellini, R. A. and Ceponis, M. J., 1984. Postharvest losses in fresh fruits and vegetables. In: Postharvest Pathology of Fruits and Vegetables: Postharvest Losses in Perishable Crops. (Eds. Moline, H. E.). UC, Berkeley: Agriculture Experiment Station. pp. 24–30

Chala G.Kuyu and Yetenayet B.Tola, 2018. Assessment of banana fruit handling practices and associated fungal pathogens in Jimma town market, southwest Ethiopia. *Food Sci. Nutr.*, **6:** 609-616

Chen, H., McCollum, G., Baldwin, E. and Bai, J.., 2016. Impacts of huanglongbing symptom severity on fruit detachment force and mechanical properties of sweet oranges (*Citrus sinensis*). *HortScience*, **51:** 356-361.

Coates, L., Cooke, A., Persley, D., Beattie, B., Wade, N. and Ridgeway, R. (1995). Postharvest diseases of horticultural produce, Volume 2: Tropical Fruit. DPI, Queensland.

Droby S., 2011. Postharvest pathology of tropical and subtropical fruit and strategies for decay control. In: Postharvest biology and technology of tropical and subtropical fruits. Fundamental Issues (Ed). Woodhead Publishing Series in Food Science, Technology and Nutrition. 194-223

Droby, S., Chalutz, E., Horev, B., Cohen, L., Gaba, V., Wilson, C.L., Wisniewski, M.E., 1993. Factors affecting UV-induced resistance in grapefruit against the green mold decay caused by *Penicillium* digitatum. *Plant Pathol.* **42** : 418–424

Droby, S., Prusky, D., Jacoby, B. and Goldman, A. (1987). Induction of antifungal resorcinols in unripe mango fruits and its relation to latent infection by *Alternaria* alternata. *Physiol. Mol. Plant .Pathol.* **30**: 285- 292.

Ewekeye, T.S., Oke, O.A., Quadri AI, Isikalu, A.O., Umenwaniri, M.O. and Durosinmi, M.L., 2013. Studies on Postharvest Deterioration of Some Fruits and Vegetables in Selected Markets in Lagos State, Nigeria. *American J. Res. Commn.*, **1**:209-23

Liu,J., Sui,Y., Wisniewski,M., Xie,Z., Liu,Y., You,Y., Zhang,X., SAun,Z., Li,W., Li,Y. Wang, Q., 2018. The impact of the postharvest environment on the viability and virulence of decay fungi. *Crit.Rev.Food Sci.Nutr.*, **58**: 1681-1687

Macarisin, D., Cohen, L., Eick, A., Rafael, G., Belausov, E., Wisniewski, M.E., Droby, S., 2007. *Penicillium* digitatum suppresses production of hydrogen peroxide in host tissue during infection of citrus fruit. *Phytopathology*, **97**, 1491–1500.

Nabi, S.U., Raja, W.H., Kumawat, K.L., Mir, J.I., Sharma, O.C., Singh, D.B. and Sheikh, M.A., 2017. Postharvest Diseases of Temperate Fruits and their Management Strategies-A Review, *Int. J. Pure App. Biosci.* **5**: 885-898

Narayanasamy, P., 2005. Postharvest Pathogens and Disease Management. John Wiley & Sons. ISBN: 0-471-74303-8

Obayelu, A.E., 2014. Post -harvest losses and food waste: The key contributing factors to African food insecurity and environmental challenges. *African J. Food, Agri. Nutr. & Devt.*,**14**:1 -8.

Onuoha , Chioma Progress and Opara Emma Umunna, 2017. Phytopathogenic Activities on Postharvest Vegetable Fruit Crops a Case Study in Nigeria. *Sch J Agric Vet Sci* , **4:**186-190

Pavoncello, David & Lurie, Susan & Droby, Samir & Porat, Ron. (2001). A hot water treatment induces resistance to *Penicillium digitatum* and promotes the accumulation of heat shock and pathogensis-related proteins in grapefruit flavedo. *Physiologia Plantarum.* **111:** 17 - 22. 10.1034/j.1399-3054.2001.1110103.x.

Perry, G. and Williams, D., 2014. Ethylene induces soil microbes to delay fruit ripening. *Peer J. PrePrints.,* **2**: e506v501.

Perry, G.D., 2011. Enhancing the expression of enzymes used to degrade hydrocarbons and cyanohydrins in *Rhodococcus* sp. DAP 96253 by using inducers such as cobalt, urea, and propylene gas; also enhances the ability of the bacteria to delay the ripening of several fruit species. Dissertation. Georgia State University.

Pierre Pétriacq, Ana López and Estrella Luna, 2018. Fruit Decay to Diseases: Can Induced Resistance and Priming Help? *Plants*, **10**: 3390

Prusky Dov and Lichter Amnon.,2007. Activation of quiescent infections by postharvest pathogens during transition from the biotrophic to the necrotrophic stage. *FEMS Microbio. Letters*. 268. 1-8. 10.1111/j.1574-6968.2006.00603.x.

Spotts, R. A., 1990. Bull's eye rot. Page 56 *in: Compendium of Apple and Pear Diseases*. A. L. Jones and H. S. Aldwinckle (Eds). American Phytopathological Society, St. Paul, MN

Timmer, L.W. and Peever, Tobin & Solel, Z & Akimitsu, K. (2003). *Alternaria* diseases of citrus - Novel pathosystems. *Phytopathologia Mediterranea,* **42** : 99-112..

Vashisth, T.,2018. Fruit Drop in Citrus. UF IFAS Extension, Univ. of Florida, *EDIS*,

Wolpert, T.J., Dunkle, L.D. and Ciuffetti, L.M., 2002. Host□selective toxins and avirulence determinants: what›s in a name? *Annu. Rev. Phytopathol.* **40**: 251– 285

Zaharan Husseina , Olaniyi Amos Fawoleb , Umezuruike Linus Opara, 2017. Preharvest factors influencing bruise damage of fresh fruits – a review. *Scientia Horticulturae*, **229**:45-58

2

Postharvest Diseases

Fruits, nuts and vegetables play a significant role in human nutrition, especially as sources of vitamins, minerals, dietary fibre, and antioxidants. Increased consumption of a variety of fruits and vegetables on a daily basis is highly recommended because of associated health benefits, which include reduced risk of some forms of cancer, heart disease, stroke, and other chronic diseases. Both quantitative and qualitative losses occur in horticultural commodities between harvest and consumption. Qualitative losses, such as loss in edibility, nutritional quality, caloric value, and consumer acceptability of fresh produce, are much more difficult to assess than are quantitative losses. Quality standards, consumer preferences and purchasing power vary greatly across countries and cultures and these differences influence marketability and the magnitude of postharvest losses. Postharvest losses vary greatly across commodity types, with production areas and the season of production. Losses of fresh fruits and vegetables in developed countries are estimated to range from 2 percent for potatoes to 23 percent for strawberries, with an overall average of 12 percent losses between production and consumption sites. In contrast, the range of produce losses in developing countries varies widely. Losses at the retail, food-service, and consumer levels are estimated at approximately 20 percent in developed countries and about 10 percent in developing countries (Adel A. Kader and Rosa S. Rolle, 2004). Overall, about one third of horticultural crops produced are never consumed by humans. Reduction of postharvest losses can increase food availability to the growing world population, decrease the area needed for production, and conserve natural resources. Strategies for loss prevention include use of genotypes that have longer postharvest-life, use of integrated crop management systems and good agricultural practices that result in good keeping quality of the commodity and use of proper postharvest handling practices in order to maintain the quality and safety of fresh produce.

Although minimizing postharvest losses of already produced food is more sustainable than increasing production to compensate for these losses, less than 5 percent of the funding of agricultural research and extension programs worldwide is devoted to activities related to maintenance of produce quality and safety during postharvest handling. This situation must be changed if success is to be achieved in reducing postharvest losses of horticultural perishables. A wide variety of pathogens *viz*., fungi, bacteria, viruses and nematodes cause postharvest disease in various crops. Some of these infect produce before harvest and then remain quiescent until conditions are more favorable for disease development after harvest. Other pathogens infect produce during and after harvest through surface injuries. Matured fruits and vegetables often become more susceptible to injury, therefore, more vulnerable to the attack of those microorganisms which require an injury or damaged tissue for penetration. Natural disease resistance, which acts in developmental stages of the plants, weakens as a result of separation of fruits and vegetables from parent plant. In addition picked fruits and vegetables are rich in moisture and nutrients; consequently, their prolonged storage leads to the senescence of tissues that suits the development of pathogens.

Fungi and bacteria responsible for storage decay often originate in field or orchard. Many fungi *viz. Alternaria*, *Fusarium*, *Penicillium*, *Aspergillus*, *Rhizopus*, *Mucor*, *Botrytis* are among the most important cause of decaying which affect harvested fruits and vegetables. Postharvest pathogens perpetuate on crop debris in the field and under suitable conditions produce abundant spores. These fungal spores are easily carried by winds, rain, or dispersed by insects to flowers and young fruits at various stages of development and form a potential source of infection. Soil, irrigation water and infected plant debris forms an important source of infection. Soil-residing fungi and bacteria can attack the bulb, tuber, root and other vegetal parts through contact with the soil, lifting of soil particles by winds, rains or by arriving in storage with attached soil residues. Some soil microorganisms such as species of the fungi *Botrytis*, *Sclerotinia* and *Fusarium* or the bacterium *Erwinia carotovora*, are among the main decay agents in stored vegetables and fruits. Host infection can occur on pre harvest, harvest or any of the Postharvest handling stages.

Harvesting instruments, containers, hands of the labours, packing and storage rooms are all prolific sources of fungal and bacterial spores. Fruits and vegetables carry the microorganisms into storage, only a few species can attack naturally while other species inhabits on the surface in large quantities will not penetrate or cause any decay. The development of disease during storage depends primarily on the existence of the appropriate microorganisms along with particular host. However, fungal spores or bacterial cells that have reached the suitable host should be capable of penetration or infection into the host tissues. The right environmental temperature, available moisture and presence of nutrients transferred from the host into water are the most important congenial factors for spore germination which is a preliminary stage for fungal penetration into the host.

2.1 Seed borne diseases : The seed primordium or the maturing seed may be infected either directly from the infected plant through the flower or fruit stalk and the seed stalk or directly from the seed surface, or infection from outside may be introduced through stigma or ovary wall or pericarp, and the Dower or fruit stalk, and later through the seed coat. A pathogen may penetrate several of these parts of the seed and in turn infect them. The infestation / contamination of the seed may occur during harvesting, threshing and processing. The pathogen may, thus, be carried with the seeds in three ways *viz*., admixture-pathogens are independent of seeds but accompany them. Ergot sclerotia are mixed with healthy seeds during threshing *viz*., external, where the pathogen may be present on seed surface as spores, oospores and chlamydospores (as in case of karnal bunt of wheat, covered smut of barley, downy mildew of pear millet *etc*.) and internal, where pathogens establish within the seed with definite relationship with seed parts.

The seed borne pathogens may result in loss in germination, discoloration and shriveling, development of plant diseases, distribution of pathogen to new areas, introduction of new strains or physiologic races of the pathogen along with new germplasm from other countries, toxin production in infected seed *etc*. Visualizing the importance of seed borne diseases the Central Seed Certification Board has prescribed certification standards for foundation and certified seed for several diseases. Seed borne microorganisms may be either saprophytic, pathogenic or opportunistic. They may be fungi, bacteria, viruses and nematodes

Classes of seed borne microorganisms : Infected seed (the primary inoculum source. If seed infection is controlled, the disease is controlled), important pathogen, but infected seed (the minor source of inoculum), seed borne microorganisms never demonstrated to cause disease, pathogens that infect seed in fields or in storage & reduce seed quality (Table 2).

Attributes of seed transmitted organisms : Ability to gain access to seed , ability to survive commercial processes (harvest, cleaning, treatment, storage), ability to establish on emerging seedlings (primary or secondary transmission).

Biological factors that affect these attributes: Infection method(s), timing of infection of seed crop, nature of pathogen(biotroph, necrotroph), survival structures or states (nematodes, cryptobiosis, cysts, fungi and Chromistans (sclerotia, microsclerotia, oospores, chlamydospores)

Ability to gain access to seed : Active process of infection within seed, passive access (present on seed surface , contaminant in seed lot), affects ability to clean or treat seed.

Seed borne pathogens: Implications for risk analysis, epidemiology of the pathogen (Potential for seed transmission, conditions for disease development & spread, ease of eradication), In-field control measures (Resistant cultivars fungicides & forecasting, crop rotation, plant spacing, row orientation, irrigation) Threshold(s) for seed borne inoculums (environmental conditions, resistant vs. susceptible cultivars), Alternative sources of inoculum (infested residues, soil borne inoculum, infected adjacent or overwintering crops or weeds.

Potential for seed transmission: Environmental factors (temperature, moisture, light, pH, soil microflora, degree of internal infection of seed Genotype (host resistance, maternal vs. paternal pathogen isolate/strain) Inoculum potential (amount of inoculum & conduciveness of environment transmission rate threshold for seed borne inoculum, complex, lack of adequate field research).

Challenges presented by seed borne pathogens: To determine definitively if a seed borne pathogen is seed borne & seed transmitted, to develop thresholds for seed borne inoculum applicable to a range of environments, regulatory implications for seed transmitted vs. seed borne status, management of diversity of seed borne pathogens under diverse production environments.

Table 2: Some examples of seed borne diseases

S.No.	Crop	Disease	Pathogen
1.	Rice	Blast	*Magnaporthe oryzae*
		Bacterial blight,	*Xanthomonas oryzae pv. oryzae*
		False smut	*Ustilaginoidea virens*
		Sheath rot	*Sarocladium oryzae and Sacroladium attenuatum*
		Brown spot	*Helminthosporium oryzae*
		Stalk burn	*Alternaria padwickii*
2.	Wheat	Loose smut	*Ustilago tritici.*
		Glume Blotch	*Stagonospora (or Parastagonospora) nodorum*
		Bacterial streak / Black chaff	*Xanthomonas campestris pathovar undulosa*
		Fusarium head blight / scab	*Fusarium graminearum*
		Wheat gall nematode	*Anguina tritici*
		Karnal Bunt	*Tilletia indica*
		Flag smut	*Urocystis agropyri*
3.	Maize	Head smut	*Sphacelotheca reiliana*
		Stalk rot / ear rot	*Colletotrichum graminicola*
		Diplodia ear rot	*Stenocarpella maydis (Diplodia maydis).*
4.	Bajra	Downy mildew	*Sclerospora graminicola*
		Ergot	*Claviceps purpurea*
		Smut	*Tolyposporium penicillariae*

5.	Sorghum	Grain smut	*Sphacelotheca sorghi*
		Loose smut	*Sphacelotheca cruenta*
		Long smut	*Tolyposporium ehrenbergii*
		Head smut	*Sphacelotheca reiliana*
6.	Cotton	Anthracnose	*Colletotrichum gossypii*
		Wilt	*Fusarium oxysporum f.*sp. *vasinfectum*
		Alternaria blight	*Alternaria macrospora*
7.	Sugar cane	Red rot	*Colletotrichum falcatum*
		Wilt	*Fusarium sacchari*
		Sett rot	*Ceratocystis paradoxa*
		Smut	*Sporisorium scitamineum.*
		Mosaic	*Sugarcane Mosaic Virus (SCMV)*
		Grassy stunt virus	*Phytoplasma (Candidatus phytoplasma)*
8.	Ground nut	Charcoal rot	*Macrophomina phaseolina*
		Crown rot	*Aspergillus niger and A. pulverulentum*
		Yellow mold	*Aspergillus flavus*
		Virus diseases	Groundnut bud necrosis virus (GBNV), Groundnut rosette assistor virus (GRAV), Groundnut rosette virus (GRV), Peanut clump virus (PCV), Peanut stripe virus (PStV), Peanut mottle virus (PeMoV), *etc.*,
9.	Sunflower	*Alternaria* blight	*Alternaria helianthi*
		Downy mildew	*Plasmopara halstedii.*
		Charcoal rot	*Macrophomina phaseolina*
10.	Soybean	Anthracnose	*Colletotrichum truncatum*
		Pod and stem blight	*Diaporthe phaseolorumsojae*
		Purple seed stain	*Cercospora kikuchii,*
11.	Chickpea	Ascochyta blight	*Phoma rabiei (formerly Ascochyta rabiei)*
		Wilt	*Fusarium oxysporum f.* sp. *ciceris*
		Grey mold	*Botrytis cinerea.*
		Alternaria blight	*Alternaria alternata*
12.	Barley	Yellow dwarf	Barley yellow dwarf virus
13.	Tobacco	Mosaic virus	Tobacco mosaic virus (TMV)
		Ring spot virus	Tobacco ring spot virus (TRSV)

In addition, yellow leaf curl, blistering, streaks/striates, yellows, vein yellowing, ring spots, *etc*hing, necrosis , colour breaking/variegation *etc*. are also seed borne in several crops

Seed contamination: It may occur through seed contamination or infestation (Pathogen itself or parts of it stick or mix with seeds during harvesting, extraction, threshing, selection, packing), accompanying contamination (Physical mixing of the seed with pathogen's propagation organs *viz*., spores, sclerotium, nematode's galls, contaminated plant parts or soil particles containing pathogens).

Location of pathogen in seed : Infection of the embryo Under the seed coat In the endosperm or cotyledon On the surface of seed

Pathogens may infect seed through flowers, fruits or funiculus, the stigma, the wall of the ovary or immature seed covers, wounds & natural openings or systemically.

2.2 Classification of postharvest diseases: Cristensen and Kaufmann (1965) divided the pathogens into field pathogens and storage pathogens.

i. **Field pathogens:** The field pathogens are those, which cause infection during development of plants or their products before harvest.

ii. **Storage pathogens:** The pathogen which cause infection during storage are the storage pathogen.

Symptoms from infection caused by the 'field pathogens' may be very inconspicuous to be noted at the time of harvest. In fleshy and/or juicy fruits and vegetables, infection by field pathogen continues to develop even after harvest. They may become infected during storage by the same field pathogen(s) or by other pathogen(s). In seeds and grains, the disease caused by field pathogens ceases to develop further soon after harvest. But, they may be infected further by the other pathogens during storage.

Stackmann and Harrar (1957) divided the pathogenic storage diseases into two groups, *viz*., diseases of dry, bulk materials, such as seeds and grains and diseases of fleshy storage organs, such as vegetables (tubers, rhizomes, bulbs *etc*.) and fruits. Observations of many investigators indicate that the real cause of the spoilage of vegetables and fleshy fruits in transit and also in storage are due to high moisture, high temperature, and injuries caused during marketing. Due to high moisture content and nutrient in harvested vegetables and fruits, they are vulnerable to attack by the pathogenic organisms. Injuries of fruits and vegetables may be caused during harvesting, packing and transposition they help the pathogen to enter the host and cause damage. But the seeds and grains can be stored for long time due to low moisture content (about 12-14%), where most of the pathogens cannot grow favourably.

2.2.1 . Diseases of stored seeds and food grains: Field fungi like *Alternaria, Fusarium*, *Cladosporium*, Verticillium, *Helminthosporium, Colletotrichum etc*., attack seeds and grains on growing crops, but are unable to grow in storage due to low relative humidity i.e., below 90%. During storage or transit the seeds and grains are damaged by the different species of *Aspergillus* and *Penicillium*, which can grow well at a relative humidity ranges from 70-90%. The commonly available *Aspergillus* species are A. repens, A. ruber, *A. flavus*,

A. candidus, *etc.Aspergillus* and a number of other storage fungi invade the embryo of the seeds and grains and they discolour the embryo or seeds as a whole, thereby the germination percentage reduces markedly. In some cases, spoilage of stored grains and seeds results in drastic increase of temperature up to 70°C or more, which encourage the growth of different thermophilic and thermotolerant fungi such as *Aspergillus* fumigatus, Absidia spp., *Mucor* pusillus, *etc*.

In addition to storage fungi, other microorganisms may grow in/or on seeds and accelerate the deterioration process. During breeding period of insects, the moisture content and temperature of seeds increase, thereby rapid growth of the pathogen takes place producing enormous amount of spores. During storage, the fungi produce mycotoxins that cause great damage to both domestic animals and human beings. The important fungi in this respect are *Aspergillus* and *Penicillium*, which produce aflatoxin and other toxins.

Economic losses due to postharvest decays are higher than is often recognized. Because of the added costs of harvesting, packaging, transporting, and marketing, the value of fresh fruits and vegetables increases several-fold from the field to the consumer. These losses are accompanied by the nonproductive use of resources, such as land, fertilizer," pesticides, water, labor, and energy, during production. Pesticides are needed to produce and protect inexpensive fresh produce on a large scale. However, public concerns over food safety have become a major concern. Preharvest fungicide applications may have a beneficial carryover effect, i.e., in reducing postharvest decays. Postharvest fungicide applications are made routinely on many fresh market fruits and vegetables. Consequently, fresh produce may contain fungicide residues within established food and drug administration tolerances. In many cases, the cost of reducing postharvest diseases may be greater than the value of the fruits and vegetables at harvest (Benny D. Bruton, 1994). Farming operations that use integrated pest management, low-input agriculture, organic farming, and others will undoubtedly present us with new challenges in postharvest research. It becomes even more compelling that we as plant pathologists, horticulturists, and plant physiologists develop a cooperative team approach to undertake these challenges.

There are a number of fungi which can invade and cause damage to grains and seeds. In general terms we can divide these fungi into two groups, *viz*., field fungi and storage fungi (Laura Sweets, 2020).

i. **Field fungi:** They invade the seeds before harvest while the crop is still in the field. Field fungi may affect the appearance and quality of seed or grain. Usually damage caused by field fungi occurs before harvest, can be detected by routine inspection and does not continue to increase in storage if grain is stored at the proper moisture content and temperature. Most field fungi are more prevalent when rainfall is above normal during grain fill and harvest. Invasion by field fungi may be more severe is the crop has been damaged by insects, birds or hail. With corn, ears well covered by husks and maturing in a downwards position usually have less rot than ears with open husks or ears maturing in an upright position.They include species of *Alternaria, Cladosporium, Aspergillus, Penicillium, Diplodia, Fusarium* and *Gibberella.* Field fungi may also occur on soybean seed in the field with species of *Phomopsis* and *Cercospora* being the most common problems.

ii. **Storage fungi (Storage molds):** These are fungi which invade grains or seeds during storage. Storage fungi are usually not present to any serious extent before harvest. Small quantities of spores of storage fungi may be present on grain going into storage or may be present on spilled grain present in harvest, handling and storage equipment or structures. Under improper storage conditions this small amount of inoculum can increase rapidly leading to significant problems. The development of storage fungi in stored grain is influenced by the moisture content of the stored grain, the temperature of the stored grain, the condition of the grain going into storage, the length of time the is grain stored and the amount of insect and mite activity in the grain. The most common storage fungi are species of *Aspergillus* and *Penicillium.* These fungi are widely distributed and almost always present.

2.3 Storage fungi and major toxins: The distinction between field and storage fungi is based not on taxonomic classification, but rather on life habits. Storage fungi have adapted to grow well under dry conditions. Debris, such as dirt, chaff, and green tissues (pods and beans) are a reservoir for storage fungi and moisture. Although a few of the fungi from the field can grow under storage conditions, most will invade the grain after harvest. The fungi enter the grain through breaks in the seed coat caused by mechanical damage, insects, or preharvest diseases. Grain stored for only a few weeks at any combination of moisture content and temperature that permits even moderate invasion by storage fungi will be at high risk if kept in continued storage (Charles Woloshuk and Ernesto Moreno Martinez, 2012). Research has shown that grain moisture level will greatly influence the fungal species that can attack and grow on the grain. Under the driest conditions, species such as *Aspergillus* restrictus and A. glaucus can grow. As the grain moisture rises to 16 to 18% in corn and 15 to 17% in soybeans, *Penicillium* species and even the aflatoxin-producer *A. flavus* can grow. All of these fungi can cause germ damage, mustiness, caking, and attract fungal feeding insects. Determination of what fungal species are on the stored grain requires laboratory examination and culturing, which may take several days to determine. Once storage fungi become established in the grain, they continue to develop at moisture and temperature levels below those required for the initial invasion of sound grain. Preventing them from infecting the grain is essential (Anonymous, 2002).

Conditions under which storage fungi are likely to damage stored grains:

The following major factors determine when stored grains will be damaged by storage fungi

1. **Moisture content :** A moisture content below 13.5 percent in starchy cereal seeds such as wheat, barley, rice, corn and sorghum and below 12.5 percent in soybean prevents invasion by storage fungi regardless of how long the grains are stored. As the moisture content rises above these levels, invasion by storage fungi increases with temperature and time. It is also important to be aware that there is variation in moisture content through a grain mass. Storage fungi will grow where moisture is suitable and not according to the average moisture content of the grain mass. These moisture content limits for safe storage imply that nowhere in the bulk of grain is the moisture content higher than that specified.

2. **Temperature :** In the range of temperature between 40 to 50 degrees F, storage fungi grow very slowly. At 80 to 90 degrees F, they grow much more rapidly.

3. **Cracked and broken kernels and foreign material :** Broken or cracked kernels are more likely to be contaminated with storage fungi going into storage and more likely to be invaded once they are in storage than sound kernels. Foreign material may restrict air movement through the grain mass leading to temperature and moisture problems which may favor storage mold development.

4. **The extent to which grain in already invaded by storage fungi when it arrives at a given storage site :** Grain invaded by storage fungi, even if not detected in ordinary inspection, is partly deteriorated and is a much poorer storage risk than grain free of storage fungi and otherwise sound. Grain moderately invaded by storage fungi develops damage at a lower moisture content, at a lower temperature and in a shorter time than does grain free or almost free of storage fungi.

5. **Length of time the grain is to be stored :** Grain that is to be stored for only a few weeks before it is processed can be stored safely with a higher moisture content and more extensive invasion by storage fungi and can be

kept at a higher temperature than grain that is to be stored for months or years.

6. **Amount of insect and mite activity in grain:** Insects and mites may carry fungal spores on their bodies thus introducing storage fungi into the grain mass. Insect and mite activity in a grain mass tends to lead to an increase in both temperature and moisture content of the grain surrounding the insect infestation. In these 'hot spots' conditions may be favorable for mold growth.

High moisture corn can be protected from storage molds with propionic acid or other organic acids sold under various trade names. It is important to follow label directions on rate and application methods. Grain treated with propionic acid can only be used for animal feed and it is not permitted in commercial grain channels.

At harvest, grain contains populations of field microbes. Once the grain is placed into a storage facility, a succession of new microbial species begins to grow. Without intervention, microbial respiration will increase temperature and moisture, providing optimum growth conditions for even more diverse fungal species. The type of grain, moisture, and temperature will influence the specific fungi that are associated with grain. The fungi on the maturing seed in the field typically require free water (liquid) for spore germination and high moisture to grow and flourish. Some of these fungi are pathogenic to the plant, while other fungi exist serendipitously. When grain is dried after harvest, most of the fungi on or in the grain before harvest will not be able to grow under the low-moisture conditions within the storage environment. Over time these fungi will slowly die off. In an ideal grain ecosystem, spoilage can be prevented if grain mass moisture is maintained at levels below that at which microbes grow. In practice, temperature and air movement continually change within the grain mass. This causes moisture to migrate, resulting in areas with moisture conditions that allow storage fungi to grow and slowly impact grain quality.

In the temperature range of 40° to 50°F (5° to 10°C), storage molds grow very slowly; while at 80° to 110°F (27° to 44°C), growth is very rapid. Grain that is to be stored for only a few weeks before processing may contain a higher moisture content, have more extensive invasion by storage molds, and be kept at a higher temperature without serious problems than can grain stored for longer periods. However, grain stored for only a few weeks at any combination of moisture content and temperature that permits even moderate invasion by storage fungi (usually not detectable by grain inspectors, but easily seen with the aid of a microscope) will be a high risk if kept in continued storage.

Grain moderately invaded by storage fungi or molds develops damage at lower combinations of moisture content and temperature and in a shorter time than grain free or almost free of storage fungi. Once storage molds become established, they continue to develop at moisture and temperature levels below those required for the initial invasion of sound grain.

The damage done by fungi growing in stored grain is the end product of storage conditions. The responsibility for it lies with those who store the grain. Storage rots and germ damage reduce the feeding value of grain and lower the market grade. Occasionally, certain storage molds may produce toxins that seriously affect poultry and livestock. Wheat with discolored and damaged embryos is discounted by millers because the germs are brittle and crumble easily. Such embryos show up in the flour as unsightly, black specks. Flour milled from wheat containing more than 20 percent "sick" kernels may yield "offflavor" bread that is also smaller in loaf volume.

Cause: More than 25 different species of fungi cause germ damage and spoilage and are known to invade stored, small-grain seed. With the exception of *Fusarium* (*Gibberella*)–fungi that cause head blight or scab of wheat, barley, oats, and rye as well as ear and kernel rots of corn–storage molds do not normally enter undamaged seed until the seed matures. If the seed is sound at harvest, *Fusarium* will not attack it in storage, although *Fusarium* may invade corn of high-moisture content stored on the cob in cribs.

The fungi responsible for most of the damage in stored grain are species of *Aspergillus* and *Penicillium*. *Aspergillus* glaucus and *A. restrictus* colonize grain with over 13.5 percent moisture. At 15 percent moisture and above, *A. candidus*, *A. ochraceus* and *A. flavus* begin to develop. Species of *Penicillium* begin to grow once the grain moisture exceeds 16 percent. These fungi are common to and destructive in stored grain when the moisture content is 14 to 17 percent or more, and the temperature is above 50°F (10°C). At least one species of *Aspergillus* can grow slowly on and in wheat, oats, and barley that has a moisture content of only 13 percent. Small-grain seed stored at a relative humidity (RH) of 70 percent will have an equilibrium moisture content of 13.5 percent. At 85 percent RH, the moisture increases to 18 percent.

Within a moisture range of 13.5 to 18 percent, the majority of storage molds grow most rapidly at 85° to 91°F (30° to 33°C)–although *Aspergillus flavus* develops quickly on and in moist grain at 113°F (45°C); and *A. candidus*, at 131°F (54°C). Most of these fungi can grow between 41° and 104°F (5° and 40°C). A few species of *Penicillium* can grow slowly at temperatures lower than 32°F (0°C). Storage molds work like a "bucket brigade" at a fire. Each fungus is active within rather narrow limits. When those limits are reached, another fungi or other fungi begin to colonize the grain quickly, resulting in a succession of organisms colonizing the grain. All storage molds give off heat and moisture–utilized, in turn, by their successors. This accelerates the rotting process. Increased temperature and moisture content (within limits) leads to more rapid rotting. Insects and mites are often present in spoiled grain. They take advantage of and contribute to the heat and moisture given off by the molds. Insects and mites also bring in and promote the development of storage molds. Determining the number and kinds of fungi in a given lot of grain often indicates the moisture content and temperature at which the grain has been stored; and sometimes, the approximate length of the storage time.

Symptoms : Storage fungi produce a sharp decrease in germination; dark germs (also called "sick" or germ-"damaged"); toxins that may be a health hazard for man and animals; musty or sour odors; heating; "caking"; and bin burning. These are the end results, caused by storage molds invading the grain. However, a complete invasion and killing of the embryo commonly occurs within the kernel before growth or symptoms are visible from the outside. Germ discoloration can be detected by removing the pericarp (germ covering) and examining the embryo. Germs that are lightly discolored throughout (or discolored only on the tip) are likely to be moldy and may later turn dark brown to black. The mold growth may be tan, white, black, blue, bluish green, or pinkish red.

They can be recognized as white, grey, black or green discolouration on the grain surface. Other signs of the presence of mould are dustiness of grain; caking of grain; feed refusal by animals for no apparent reason; a bad musty (mouldy) smell; darkening of feed grain.

Management

1. Be certain that harvesting equipment is operating properly and set correctly. Cylinder speed and settings should be checked to insure the lowest possible damage to harvested grain.

2. Wherever possible, store grain (free of dirt, debris, chaff, and broken kernels) in a clean, tight bin at a moisture content below 13 percent and a temperature below 50°F (10°C). If the grain is sound and dry when stored, it can be kept for years under these conditions without damage.

3. Aerate high-moisture grain as soon as possible to provide a temperature of 35° to 50°F (2° to 10°C) throughout the bulk. At such temperatures, insects and mites are dormant; and most storage fungi grow very slowly if at all.

4. After the grain has been stored, take probe samples at weekly to monthly intervals from different portions. Use calibrated moisture meters, "official"

drying ovens, or the distillation methods of determining moisture. Avoid taking several samples from a bin and averaging them. The highest moisture content, not the average, determines storability. Examine each sample for moisture content, damage, and the number and kinds of fungi. No germ or heat damage will develop in grain when the moisture content is below 13 percent.

5. Measure and keep a record of temperatures in different parts of the grain. Even a slight rise in temperature means that some spoilage is occurring. Only vigorously feeding insects or rapidly growing storage fungi can cause stored grain to heat up. If a "hot spot" is found, determine its size and location, as well as the condition of the grain in and near it. Do this by sampling and examining the grain. Temperature detection cables or other devices to measure the temperature at various places within bulk grain are an aid to–but not a substitute for–the intelligent handling of stored grain. Relatively dry grain is a good insulator. A rise in temperature of even a few degrees (as indicated by a thermo-couple) may mean that the grain in the region of highest temperature is already spoiled.

6. When "hot spots" or a crust of moldy grain are found, take the following corrective measures:

a. The rotted and moldy grain should be removed and dried, and either fed or sold. Moldy grain should be fed with extreme caution to all classes of poultry and livestock, but if mixed with sound grain, it can be fed with less risk to livestock being finished for market. Moldy grain is considered unsafe for lactating cows and all breeding and potential breeding animals.

b. The moisture content of the remaining grain should be checked.

c. The remaining grain should be turned and thoroughly mixed to redistribute moisture and allow heat to escape.

7. Aeration fans should be installed to move small quantities of air through the grain. Doing this will maintain a uniform temperature and help prevent "wet" spots. It is cheaper and more effective to maintain the temperature and moisture throughout a bin with aeration than to transfer the grain from bin to bin. Such transfers also increase the amount of cracked and broken kernels.

Little can be done to prevent or reduce the invasion of crops in the field by field fungi. However, the following recommendations should help prevent storage fungi problems or minimize damage from storage fungi in stored grains. In general,following strategies may be followed.

1. Harvest as soon as the moisture content allows for minimum grain damage.
2. Adjust the harvesting equipment for minimum kernel or seed damage and maximum cleaning.
3. Clean all grain harvesting and handling equipment thoroughly before beginning to harvest. Clean bins or storage facilities thoroughly to remove dirt, dust and other foreign material, crop debris, chaff and grain debris.
4. Clean grain going into storage to remove light weight and broken kernels or seeds as well as foreign material and fines.
5. Moisture content is by far the most important factor affecting the growth of fungi in stored grain. After harvest grain should be dried to safe moisture contents as quickly as possible.
6. Aerate grain to safe and equalized temperatures through the grain mass.
7. Protect grain from insect and mite damage.
8. Check stored grain on a regular basis and aerate as needed to maintain low moisture and proper temperature.

Infections leading to postharvest diseases: Numerous factors influence postharvest quality and shelf life. The preharvest environment and production methods have a large impact on postharvest diseases. Produce grown under less than optimum conditions are more likely to be predisposed to postharvest diseases than those grown under optimum conditions. Proper storage temperature and humidity are probably the most important postharvest disease control measures available. Refrigeration and high humidity retard senescence and the growth of many postharvest pathogens. Some fungal pathogens are capable of penetrating intact surfaces of fruits and vegetables, although most enter at a wound site. Bacteria gain entry only through wounds or natural openings, such as stomata and lenticels. Fungi are typically involved in postharvest decays of fruit, whereas vegetables may be attacked by either fungi or bacteria. Infections leading to postharvest decay can be separated into three categories; incipient infections, wound infections, and latent or quiescent infections. Incipient infections occur at any time before or after harvest. Incipient infections may remain active, causing decay, although they are often arrested as a result of exogenous influences, i.e., fungicide application and refrigeration. By definition, incipient means just beginning to exist or to come to notice. Many incipient infection; on produce may pass through the culling process at any time during harvesting, packaging, or retail display.

The safety and integrity of the food supply are of paramount importance and are among the drivers of safe grain storage. Bacterial contamination of stored grains may be a significant consideration for users where grain will not undergo further processing that includes a microbiocidal step, but problems caused by moulds and mycotoxins are more usually associated with grain storage (Ailsa D. Hocking , 2003). With an increased knowledge of the origins of mycotoxins, we now understand that at least some mycotoxins in some commodities are formed either before harvest, or immediately after harvest. Mycotoxins can be formed during storage, but only if there is sufficient moisture. Fungal growth usually develops as a complex succession of species, beginning with the most drought-tolerant (xerophilic) species. The metabolic activity of these pioneer species raises the moisture content of the grain, which may allow growth of mycotoxigenic species and, ultimately, the formation of mycotoxins.

Some of the fungi associated with grain in the field (often referred to as 'field fungi') can form mycotoxins, either immediately before, or just after harvest. *Alternaria*, *Fusarium*, *Aspergillus* and *Penicillium* can all act as preor postharvest pathogens of grain, and may form mycotoxins. *Alternaria* and *Fusarium* do not compete strongly at reduced water activities (aw) so are unlikely to form mycotoxins once the grain is dry, or during storage. Conversely, *Aspergillus* and *Penicillium* are more often considered as 'storage fungi'. They are known to form mycotoxins in stored grains, and are usually not regarded as fungi that can produce mycotoxins before harvest.

The bacterial species that occur commonly on grain are generally non-pathogenic, though contamination with bacterial pathogens such as *Salmonella*, *Escherichia coli* and *Bacillus cereus* can occur. *Salmonella* and *E. coli* are enteric bacteria, and their presence on grain is usually an indication that it birds or rodents have contaminated it. This may occur during harvesting, but more often is a result of poor hygiene in road or rail trucks during transportation, or poor pest control during storage. Levels of contamination with enteric pathogens are usually very low.

Most important toxins : The genera of mycotoxigenic fungi are mainly represented by *Aspergillus*, *Penicillium* and *Fusarium*, but *Trichoderma*, *Trichothecium* and *Alternaria* are also important as food contaminants or pathogens for plants, among others (Ashiq,2015). Mycotoxins, particularly aflatoxins (AFTs) and ochratoxin A (OTA) pose a significant threat to human health. Aflatoxins are potent carcinogens and, in association with hepatitis B virus, are responsible for many thousands of human deaths per annum, mostly in non-industrialized tropical countries. AFTs are secondary metabolites, difuranocoumarins, produced by *Aspergillus* flavus and A. parasiticus, commonly found in food and feeds and have been associated with various diseases such as aflatoxicosis, in livestock, domestic animals and humans throughout the world.

General categories include Yellow Rice Toxins (Produced in grains of rice, barley *etc.*, by species of *Penicillium*), Tremorgenic Toxins (Produced on prepared food during storage in refrigeration or in other places and also on food produced from infected grains and/or seeds) and Penicillic Acid (It is a carcinogenic substance produced by the different species of both *Aspergillus* and *Penicillium* in molded cereal grains).

Diseases of vegetables and fruits: Different members of Ascomycotina and Deuteromycotina cause the major postharvest diseases of fruits and vegetables. These are *Alternaria*, *Botrytis*, *Fusarium*, *Penicillium*, *Sclerotinia etc.*:

i. ***Alternaria:*** Different species of *Alternaria* cause rot of many fresh fruits and vegetables, e.g., black rot of orange, tuber rot of potato, rot of sweet potato, purple blotch of onion, *Alternaria* rot of onion, *Alternaria* rot of cabbage, *etc.*

ii. ***Botrytis:*** It causes "grey mold rots" of fruits like pear, apple, citrus *etc.*, and vegetables like onion, tomato *etc.* Every year it causes great economic loss.

iii. ***Fusarium:*** It causes different diseases, commonly called "pink or yellow molds". Different species of *Fusarium* cause damage to tubers, bulbs, storage roots *etc.* and frequently on cucurbits, tomato *etc.* It also cause brown rot of fruits like lemon, orange *etc.*

iv. ***Penicillium:*** Species of *Penicillium* are commonly called "blue or green molds", these cause rots of different fruits like onion, sweet potato *etc.* They also cause spots on different fruits. Under storage, the spotted fruits bear tufts of spores. Though most of the *Penicillium* species prefer relatively high temperature for their growth in storage, they still remain active near freezing temperature at a slow rate. A few species produce ethylene which increases respiration of fruits, thereby it reduces the storage life of the fruits. It also produces patulin, a mycotoxins, which directly contaminates the sauces and fruit juices prepared from infected partly rotten fruits.

v. ***Sclerotinia:*** It infects different fruits and vegetables. Most common diseases are cottony rot of lemon, watery soft rot of bean pods, cucurbits *etc.* Storage diseases like bacterial soft rot of vegetables such as onion, carrot, potato *etc.* are mainly carried out by different species of *Erwinia*, such as *E. carotovora* *E. chrysanthemi* etc.

2.4 Major pathogens of postharvest fruits and vegetables: The important fungal and bacterial pathogens of harvested fruits and vegetables that cause infection during or after the harvest with disease and the primary hosts are listed in the Table 3.

Table 3: Some common postharvest diseases and pathogens of fruits and vegetables

Sl. No.	Disease	Pathogen
I. **1.**	**Temperate fruits** **Pome fruits**	
	Blue mould	*Penicillium* spp.
	Grey mould	*Botrytis cinerea*
	Bitter mould	*Colletotrichum gloeosporioides*
	Alternaria rot	*Alternaria* spp.
	Mucor rot	*Mucor pyriformis*
2.	**Stone fruits**	
	Brown rot	*Monilinia fructicola (syn. Sclerotinia fructicola)*
	Rhizopus rot	*Rhizopus* spp.
	Grey mould	*Botrytis cinerea*
	Blue mould	*Penicillium* spp.
	Alternaria rot	*Alternaria alternata*
3.	**Grapes**	
	Grey mould	*Botrytis cinerea*
	Blue mould	*Penicillium* spp.
	Rhizopus rot	*Rhizopus* spp.
4.	**Berries**	
	Grey mould	*Botrytis cinerea*
	Rhizopus rot	*Rhizopus* spp.
	Cladosporium rot	*Cladosporium* spp.
	Blue mould.	*Penicillium* spp
II.	**Subtropical fruits**	
1.	**Citrus Fruit**	
	Blue mould	*Penicillium italicum*
	Green mould	*Penicillium digitatum*
	Black centre rot	*Alternaria citri*
	Stem end rot	*Phomopsis citri*
	Brown rot	*Phytophthora citrophthora and/or P. parasitica*
2.	**Avocado**	
	Anthracnose	*Colletotrichum gloeosporioides*
	Stem end rot	*Dothiorella spp.* *Lasiodiplodia theobromae* *Stilbella cinnabarina* *Phomopsis perseae*

III.	**Tropical fruits**	
1.	**Banana**	
	Anthracnose	*Colletotrichum musae*
	Crown rot	*Various fungi including Fursanum* spp.,*Verticillium spp., Acremonium* sp. *and Colletotrichum musae*
	Black end	*Various fungi including Colletotrichum musae, Nigrospora sphaerica and Fusarium* spp.
	Ceratocystis fruit rot	*Thielaviopsis paradoxa*
	Burrowing nematode	*Radopholus similis*
2.	**Mango**	
	Anthracnose	*Colletotrichum gloeosporioides, C. acutatum*
	Stem end rot	*Dothiorella spp.* *Lasiodiplodia theobromae* *Phomopsis mangiferae* *Pestalotiopsis mangiferae*
	Rhizopus rot	*Rhizopus stolonifer*
	Black mould	*Aspergillus niger*
	Alternaria rot	*Alternaria alternata*
	Grey mould	*Botrytis cinerea*
	Blue mould	*Penicillium expansum*
	Mucor rot	*Mucor circinelloides*
3.	**Papaya**	
	Anthracnose	*Colletotrichum spp.*
	Black rot	*Phoma caricae-papagae*
	Phomopsis rot	*Phomopsis caricae-papagae*
	Rhizopus rot	*Rhizopus stolonifer*
	Phytophthora fruit rot	*Phytophthora palmivora*
4.	**Pineapple**	
	Water blister	*Thielaviopsis paradoxa*
	Fruitlet core rot	*Penicillium funiculosum* *Fusarium moniliforme var.subglutinans*
	Yeasty rot	*Saccharomyces* spp.
	Bacterial brown rot	*Erwinia ananas*
IV.	**Vegetables**	
1.	**Cucurbits**	
	Bacterial soft rots	*Various Erwinia* spp., *Bacillus polymyxa,* *Pseudomonas syringae* *Xanthomonas campestris*
	Grey mould	*Botrytis cinerea*
	Fusarium rot	*Fusarium* spp.
	Alternaria rot	*Alternaria* spp
	Charcoal rot	*Macrophomina phaseolina*
	Cottony leak	*Pythium* spp.
	Rhizopus rot	*Rhizopus* spp.

2.	**Tomato, Brinjal and Capsicum**	
	Bacterial soft rots	*Various Erwinia* spp., *Bacillus polymyxa, Pseudomonas* spp. *and Xanthomonas campestris*
	Grey mould	*Botrytis cinerea*
	Fusarium rot	*Fusarium* spp.
	Alternaria rot	*Alternaria alternata*
	Cladosporium rot	*Cladosporium* spp.
	Rhizopus rot	*Rhizopus* spp.
	Watery soft rot.	*Sclerotinia* spp
	Cottony leak	*Pythium* spp.
	Sclerotium rot	*Sclerotium rolfsii*
3.	**Vegetable legumes**	
	Grey mould	*Botrytis cinerea, B.fabae*
	White mould and Watery soft rot	*Sclerotinia* spp
	Cottony leak	*Pythium spp.*
	Sclerotium rot	*Sclerotium rolfsii*
4.	**Brassicas**	
	Bacterial soft rot	*Various Erwinia* spp. *Bacillus* spp. *Pseudomonas* spp. *and Xanthomonas campestris*
	Grey mould	*Botrytis cinerea*
	Alternaria rot	*Alternaria* spp.
	Watery soft rot	*Sclerotinia* spp.
	Phytophthora rot	*Phytophthora porri*
5.	**Leafy Vegetables**	
	Bacterial soft rot	*Various Erwinia* spp. *Pseudomonas* spp. *and Xanthomonas campestris*
	Grey mould	*Botrytis cinerea*
	Watery soft rot	*Sclerotinia* spp.
6.	**Onion**	
	Bacterial soft rot	*Various Erwinia* spp. *Lactobacillus* spp. *and Pseudomonas* spp.
	Black mould rot	*Aspergillus niger*
	Fusarium basal rot	*Fusarium oxysporum f.* sp.*cepae*
	Smudge	*Colletotrichum circinans*
	Stem & bulb nematode	*Ditylenchus dipsaci*

7.	**Carrot**	
	Bacterial soft rot	*Various Erwinia spp., Pseudomonas* spp.
	Rhizopus rot	*Rhizopus* spp.
	Grey mould	*Botrytis cinerea*
	Watery soft rot	*Sclerotinia* spp.
	Sclerotium rot	*Sclerotium rolfsii*
	Chalara rot	*Chalara thielavioides*
	Thielaviopsis rot	*Thielaviopsis basicola*
8.	**Potato**	
	Bacterial soft rot	*Erwinia spp.*
	Dry rot	*Fusarium spp.*
	Gangrene	*Phoma exigua var. exigua and var. foveata*
	Black scurf	*Rhizoctonia solani*
	Silver scurf	*Helminthosporium solani*
	Skin spot	*Polyscytalum pustulans*
	Root-knot nematode	*Meloidogyne incognita*

2.4.1 *Penicillium digitatum* (Green mold, Blue mold): Postharvest diseases caused by the necrotrophic ascomycetes *Penicillium* spp., on several crops are destructive.

Penicillium is a very large and ubiquitous genus which currently contains 354 accepted species. The name *Penicillium* comes from the resemblance of the spore producing structures (conidiophores) of the fungus to a paintbrush (penicillus is the Latin word for paintbrush) (Fig. 3).

Domain: Eukaryota
Kingdom: Fungi
Phylum: Ascomycota
Subphylum: Pezizomycotina
Class: Eurotiomycetes
Order: Eurotiales
Family: Trichocomaceae
Genus: *Penicillium*

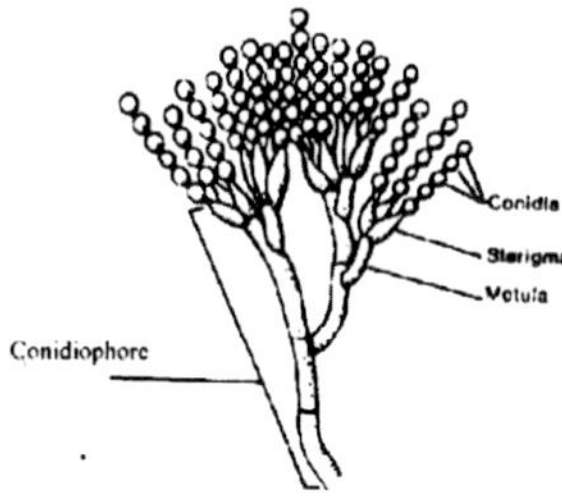

Fig. 3 : *Penicillium* digitatum

Many species are common contaminants on various substrates and are known as potential mycotoxin producers. Colonies are usually fast growing, in shades of green, sometimes white, mostly consisting of a dense felt of conidiophores. Microscopically, chains of single-celled conidia are produced in basipetal succession from a specialised conidiogenous cell called a phialide. The term basocatenate is often used to describe such chains of conidia where the youngest conidium is at the basal or proximal end of the chain. In *Penicillium*, phialides may be produced singly, in groups or from branched metulae, giving a brush-like appearance. The penicillus may contain both branches and metulae (penultimate branches which bear a whorl of phialides). All cells between the metulae and the stipes of the conidiophores are referred to as branches. The branching pattern may be either simple (non-branched or monoverticillate), one-stage branched (biverticillate-symmetrical), two-stage branched (biverticillate-asymmetrical) or three- to more-staged branched (Sivakumar, *et al*, 2014). Conidiophores are hyaline, smooth or rough-walled. Phialides are usually flask-shaped, consisting of a cylindrical basal part and a distinct neck, or lanceolate (with a narrow basal part tapering to a somewhat pointed apex). Conidia are in long dry chains, divergent or in columns, are globose, ellipsoidal, cylindrical or fusiform, hyaline or greenish, smooth or rough-walled. *Sclerotia* are produced by some species.

P. digitatum and *P. italicum* affect all citrus species and cultivars and are strictly originated from infections in peel wounds. Successful infection and disease development depend on intrinsic characteristics of the pathogen and the fruit host and on their interactions, modulated by the environmental conditions. Recent research on molecular and biochemical aspects of host–pathogen interaction has considerably increased the knowledge on the mechanisms of natural and induced fruit resistance to disease (Lluís Palou, 2014).

Symptoms : In infected fruit, very profuse sporulation can be seen , fruit is completely covered by white mycelium followed by green and bluish spores of *Penicillium* digitatum and *Penicillium* italicum respectively. The typical terpenous odor spreads in the surrounding area where these fungi infect the fruit. It is quite possible that these fungi produce ethylene in sufficient quantities, resulting in the rapid senescence of adjacent fruits.

Blue mold is more harmful because it spreads in the box and healthy fruits are directly attacked, regardless of injury. Blue mold is a nesting-type pathogen, meaning that it produces enzymes that soften the adjacent fruit and thus allow fungus to enter. Green mold does not spread by nesting, thus, if a single fruit is affected it remains as such without contaminating adjacent fruit. However, spores lead to soiling of fruits and thus require repacking with a box change.

Green mold (*P. digitatum*) is quite common in India and grows rather slowly at lower temperatures. At higher temperatures (25–30°C) it grows very rapidly. Green mold infects fruit through wounds. Orchard and packing house sanitation is required to restrict sporulation of *Penicillium* on fruits in orchards and packinghouses to minimize decay losses.

P. funiculosum infects the flower of pineapple fruits, before the characteristic yellow fruit is formed. When *P. funiculosum* infects the closed pineapple flowers, early symptoms include necrosis of the anthers, which are the male parts of the flower, and pistil, the female part, and cork formation and sporulation within the ovary of the flower (Rohrbach and Phillips,1990). This destruction of reproductive tissue prevents propagation of healthy fruit and ruins the growing crops. Later symptoms of the disease include a darkening of the septa between the locules. This discoloration can spread throughout the fruit. Extensive corking is what results in leathery pocket.In a healthy plant, fruitlets join together to produce one large fruit. Interfruitlet corking occurs when these fruitlets develop incorrectly and grow together unevenly.

2.4.2 *Botrytis cinerea* (Grey mold rot): '*Botrytis*' is derived from the ancient Greek botrys (βότρυς) meaning 'grapes', combined with the New Latin suffix -itis for disease.

Botrytis cinerea is a significant necrotrophic plant pathogen causing devastating diseases on more than 500 plant species, especially on fresh fruits and vegetables, resulting in the economic losses ranging from $10 billion to $100 billion worldwide. This fungal pathogen invades nearly all parts of plants including stems, leaves, flowers, fruits, and seeds at both pre-harvest and postharvest stages. Due to its wide host range and the huge economic losses that it causes, extensive investigations have been carried out to effectively control this plant pathogen. It is beneficial for exploring the pathogenic mechanisms of *B. cinerea* to provide fundamental basis for control strategies (Li Hua *et al*, 2018). In recent years, tremendous progress has been made in understanding these pathogenic genes and regulatory pathways, as well as the control strategies of *B. cinerea*. The asexual spores of *B. cinerea* are conidia, which are easily to be dispersed by wind or water, and the sexual spores of *B. cinerea* are sclerotia, which are essential for survival under adverse environment. *Botrytis cinerea* is characterized by abundant hyaline conida (asexual spores) borne on grey, branching tree-like conidiophores (Fig. 4).

Kingdom: Fungi
Phylum: Ascomycota
Subphylum: Pezizomycotina
Class: Leotiomycetes,
Order: Helotiales,
Family: *Sclerotiniaceae*,
Genus: *Botrytis*

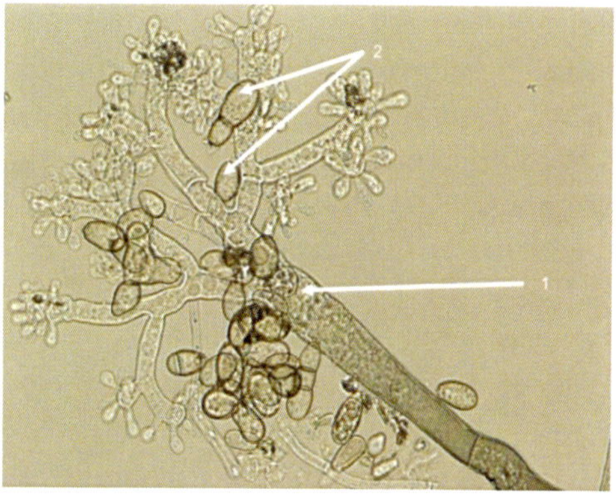

Fig. 4: The conidiophore (1) terminating in a botryose cluster producing multiple macroconidia (2)

The fungus belongs to the *Sclerotinia*ceae within the Ascomycetes. The anamorph stage *Botrytis cinerea* has a worldwide distribution. However, it is not clear how widespread and frequent the teleomorph *B. fuckeliana* is. It is concluded that sexual reproduction might be an important source of genetic variation. The fungus grows well on artificial media. The anatomy and morphology of the mycelium of *B. fuckeliana* are typical of the Ascomycota. Septa are present and are perforated by a simple pore. The mycelium is composed of brownish olive hyphae. Anastomoses between hyphae are often noted. Conidiophores are frequently 2 mm or more long, mostly 16-30 µm wide with often a swollen basal cell. The conidia are ellipsoidal or ovoid, generally with a slightly protuberant hilum. They are dry and hydrophobic, colorless to pale brown, smooth, 6-18 x 4-11 µm (mostly 8-14 x 6-9 µm). In mass the conidia are gray-brown.

Botrytis cinerea is an aggressive plant pathogen, which causes gray mould rot in fresh horticultural crops, resulting in heavy economical losses in the world. Due to the genetic variety of *B. cinerea*, resistant strains are frequently found. It is important to reveal molecular basis of pathogenicity and regulatory mechanisms of *B. cinerea*. Current understanding indicates that ROS and extracellular proteins are related to the regulation of growth, development, and virulence. The genome sequence of *B. cinerea* has been released and genetic methods such as construction of deletion mutants have been successfully used to investigate its pathogenic mechanisms. Since a few of target genes involved in pathogenicity of *B. cinerea* have been found, it is possible to develop and improve control technologies. However, more research effectors are still needed to understanding of comprehensive information about molecular pathogenic mechanisms and regulatory network to develop more precise and effective strategies for prevention and control of *B. cinerea*

Reactive oxygen species (ROS) are a collective of highly reactive molecules including superoxide anion (O2−), hydroxyl radical (OH), and certain non-radical oxidizing agents such as hydrogen peroxide (H2O2) and ozone (O3) that can be converted into radicals. ROS have an ambivalent role since they damage DNA, causing lipid peroxidation and protein oxidation, but also function as diffusible second messengers . In the early stage of infection, plant hosts usually trigger oxidative burst which generate large amounts of ROS transiently to counteract the invasive pathogen (Tian *et al*., 2013). However, as a necrotrophic fungus, *B. cinerea* can exploit the oxidative burst and even contribute to it by producing its own ROS. It was shown that aquaporin8 (BcAQP8) in *B. cinerea* could play a crucial role in the transmembrane transportation of ROS, and the distribution of mitochondria, which is the main source of ROS (An *et al*., 2016). Besides to regulate ROS transportation, BcAQP8 also affects the expression of BcNoxR, the regulatory subunit of NOX. Deletion of bcaqp8 results in decrease in growth, sporulation, and pathogenicity of *B. cinerea.*

As a necrotrophic fungus, the appressorium of *B. cinerea* was not strong enough to breach the plant cell wall. Therefore, it is necessary for *B. cinerea* to secrete a series of cell wall–degrading enzymes (CWDEs) to degrade the structural polysaccharides of the host cell wall. In *B. cinerea*, 1155 genes are predicted to encode enzymes to degrade, modify, or create glycosidic bonds. Among them, 275 have signal peptide sequence indicating their function in extracellular matrix (Fillinger and Elad, 2016).

Symptoms and signs: Gray mold rot, also called gray mold blight or *botrytis* blight, disease of plants growing in humid areas that is caused the genus *Botrytis*, usually *B. cinerea*. Most vegetables, fruits, flowers, and woody plants are susceptible. The disease primarily affects flowers and buds, though infections on fruits, leaves and stem can occur. Gray mold rot is characterized by tan to brown soft spots or blotches that become covered with a dusty mold and can cause seedlings, young shoots, and leaves to wither and collapse and buds, flowers, and fruits to become flecked and rotten. Symptoms vary across plant organs and tissues. *B. cinerea* is a soft rot that will have a collapsed and water soaked appearance on soft fruit and leaves. Brown lesions may develop slowly on undeveloped fruit Twigs infected with gray mold will die back. Blossoms will cause fruit drop and injury, such as ridging on developing and mature fruit Symptoms are visible at wound sites where the fungus begins to rot the plant. Gray masses with a velvety appearance are conidia on the plant tissues are a sign of plant pathogen.

2.4.3. *Colletotrichum gloeosporioides* (Bitter rot) : *Colletotrichum* is a genus of fungi that are symbionts to plants as endophytes (living within the plant) or phytopathogens.

Colletotrichum gloeosporioides (Anthracnose) is one of the most common *Colletotrichum* fungal plant pathogens. It causes bitter rot in variety of crops worldwide, particularly perennials in the tropical regions (Waller, 1992). Some of the important host plants include citrus, yam, papaya, avocado, coffee, eggplant, sweet pepper, and tomato. It produces substantial amount of pre- and postharvest loss in these crops worldwide. It acts as a secondary invader.

There are more than 600 synonyms of *C. gloeosporioides* showed many morphological and physiological variations. The acervuli were highly variable in size, shape and exude pink masses of conidia when mature under moist conditions (Sattar and Malik, 1939). It often result in fruit drop and fruit rot (infonet-biovision.com). Anthracnose is caused by fungi that produce conidia within black fungal fruiting bodies called acervuli. f injured tissue, but can also survive as a saprophyte.

Kingdom: Fungi
Phylum : Ascomycota
Class : Sordariomycetes
Order: Glomerellales
Family: Glomerellaceae
Genus: *Colletotrichum*

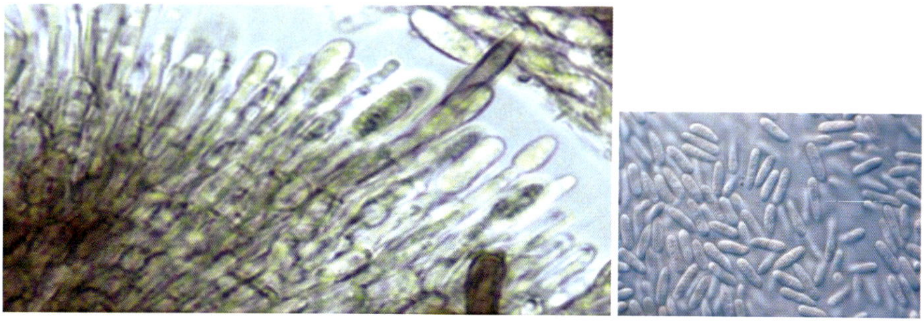

Fig. 5 : *Colletotrichum gloeosporioides* - Conidiophores and Conidia

Other species are also responsible for most anthracnose disease. The fungus produces hyaline, one-celled, ovoid to oblong, slightly curved or dumbbell shaped conidia, 10-15 µm in length and 5-7 µm in width (Fig.5) Masses of conidia appear pink or salmon colored. The waxy acervuli, that are produced in infected tissue, are subepidermal, typically with setae, and simple, short, erect conidiophores.

Environmental conditions favoring the pathogen are high temperatures, 28°C being optimal, and high humidity. Spores must have free water to germinate; germination is negligible below 97% relative humidity. Spores are only released from acervuli when there is an abundance of moisture. Splashing from rain is a common means of spread. Severity of disease is related to weather and the fungus is relatively inactive in dry weather. Sunlight, low humidity and temperature extremes (below 18°C or greater than 25°C) rapidly inactivate spores.

Symptoms and signs: *Colletotrichum gloeosporioides* produces a wide variety of symptoms depending on host species and tissue infected. In the fruits, black or brown lesions (e.g., on pods, scab, and pitting) are common. Infection on the inflorescence produces blight, necrosis, and lesions with flecks and streaks. Infected leaves produce abnormal colors and patterns with dark, necrotic, angular, or irregular areas. Dieback and discoloration with gummosis and resinosis occurs on infected stems. Sometimes, cankers are also seen in the infected stem. Fungal sporulation forms acervuli, a pinkish pinhead-sized structure, when humidity is high. The acervuli form in a concentric pattern around the necrotic tissue. The fruiting bodies can be seen as black flecks in the infected tissue. In some cucurbit host, more pale lesions with less necrosis are seen. Generally, necrosis spreads and later turns into blight in yam and tea. In stylosanthes and cassava, lesions are elliptical in shape and appear dark and sunken.

The fungus primarily colonizes damaged and senescent tissue and often grows as saprophyte on dead tissues. Short distance spore dispersal occurs by air currents, rain splash, overhead irrigation, and heavy dew. Spore deposition on young leaves and fruits is followed by colonization. Sexual spores, although produced in comparatively less number, are airborne and thus contribute to long distance spread. Upon germination, spores can form a dormant resting structure until injury or damage occurs in the host tissue. Degreening also breaks this dormancy. In the fruits that are harvested early, ethylene stimulates the fungal colonization. Insects, infested seed crops residues, and latent weed host also help further the persistence of the pathogen and spread of the disease. High temperature and high humidity are favorable for the growth of the pathogen. Germination of the spores requires high relative humidity. Only when the moisture content is highest, are acervuli able to release spores. The fungus is not active during dry weather.

2.4.4. *Alternaria alternata* (Black Rot, Black Spot): *Alternaria alternata* causes black spot in many fruits and vegetables around the world.

It is a latent fungus that develops during the cold storage of fruits, becoming visible during the marketing period thereby causing large postharvest losses. *Alternaria alternata* (Fr.) Keissl. is a ubiquitous necrotrophic fungus. There are at least seven pathogenic variants of this fungus, each producing unique host-selective toxins (HSTs) and causing disease on specific host plants Major pathotypes are A. alternate on apple, citrus, Japanese pear, strawberry, tobacco, and tomato, which cause *Alternaria* blotch of apple, brown spot of citrus (two different HSTs), black spot of Japanese pear, *Alternaria* black spot of strawberry, brown spot of tobacco, and stem canker of tomato, respectively (Gat *et al*, 2012) .

Conidiophores are pale brown to olive brown, 25–60 x 3–3.5 µm, Straight or flexuous, individual conidiophores arise directly from substrate forming bushy heads consisting of 4–8 large catenate conidia chains, secondary conidiophores are generally short and 1-celled. Conidia are pale brown to light brown, obclavate to obpyriform orellipsoid, short conical beak at the tip, or beakless, surface smooth to verruculose, 20–63 x 9–18 µm in size, (on PCA) mature conidia typically 10–30 x 5–12 µm, several vertical and −8 transverse septa, (on PCA) 3–7 transepta, 1–5 longisepta, produced in an often branched, long chain more than 5 conidia., (on PCA) individual chains of 5–15 conidia, complex of branching chains may contain up to 50–60 conidia. (Figs.6 & 7).

Kingdom: Fungi
Phylum : Ascomycota
Class : Dothideomycetes
Order: Pleosporales
Family: Pleiospraceae
Genus: *Alternata*

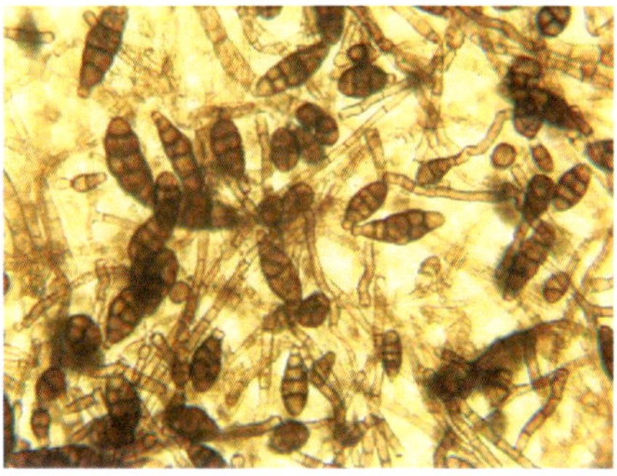

Fig. 6: *Alternaria alternata* spores

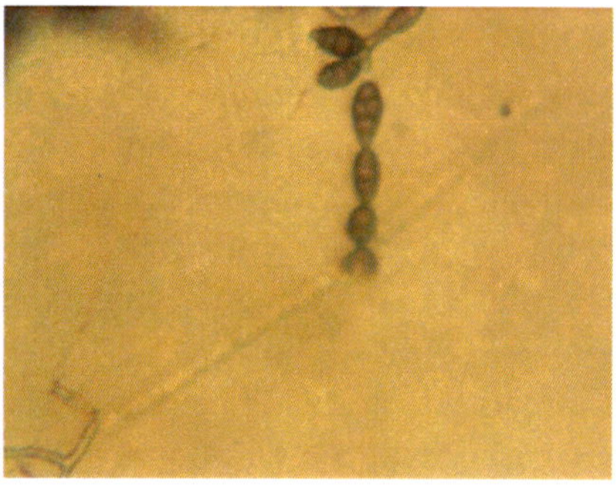

Fig. 7: Spores in a chain

Alternaria species is a hardy fungus and can live in extreme conditions. *A. alternata* can pass the winter in the soil, seed, infected crop debris or perennial host tissue, such as bark, nodes, and scaly leaves, as mycelia and/or conidia. Some strains can produce survival structures to resist the unfavorable conditions. While some *Alternaria* species need stimuli to initiate conidiophore formation and sporulation, *A. alternata* can sporulate easily without triggering. The spores of *A. alternata* can be propelled into the air by a shift from wetness to dryness, a rapid increase in humidity or exposure to red light. The germination of *Alternaria* species is improved in the presence of free moisture. At high relative humidity, germination may be facilitated by transient condensation of water vapor caused by fluctuating temperature. For example, Stevenson and Pennypacker (1988) observed that germination of *A. solani* spores in 96% relative humidity was associated with microscopic condensation of water. In another study, spores of *A. alternata* in containers inside temperature-controlled baths (20 ± 0.01°C) germinated only when wetted occasionally at 100% relative humidity, but never at 98% relative humidity (Rotem, 1998).

Dehpour *et al.* (2007) published that conidia of *A. alternata* germinate quickly if moisture is present and begin to produce toxin even before they penetrate the tissue. *Alternaria* brown spot infection is most associated with environmental conditions (Canihos *et al.*, 1999) and leaf or fruit maturity of tangerines (Whiteside, 1976). In addition, it has been reported that, in these fruits, the fungus sporulates most prolifically on necrotic lesions of mature or senescent leaves when the substrate is lightly moistened or at high relative humidity. Conidia are also released after rainfall and/or sudden changes in humidity and dispersed by wind (Timmer *et al.*, 2000).

The minimum wetting period needed for the establishment of various *Alternaria* species in host tissue ranges from 3 to 72 h. The requirements for a long wetting period may result from the need to weaken invaded tissue by means of toxic or enzymatic agents, and it is possible that this requirement is more common in infections by weakly pathogenic species. However, a long wetting period may also be required for infection of organs with high resistance by virulent species. Sawamura (1972) reported that the toxin-producing *A. alternata* f. sp mali infects apple leaves after a wetting of 3 h but needs a wetting of 48 h for infecting apple fruits. In another study, the effect of eight wetting periods (from 2 to 48 h) and nine temperatures (from 4 to 36°C) on the infection of 'Delicious' apple seedlings by *A. alternata* f. sp. mali was tested. The authors observed that at all temperatures infection increased with the duration of wetness and were most intensive between 12 and 28°C, 23.5°C being the optimum. At this temperature, 5.1 and 12.7 h of wetness were required for light and severe infection, respectively (Filajdič and Sutton, 1991).

In addition, the effect of temperature on the germination of *Alternaria* species has been reported. In almost all *Alternaria* species, the optimum temperature for germination is around 25°C, and the maximum temperature is around 35°C, although different values are given in different reports. The optimum temperature for germination of *A. alternata* in tomato is around 29°C, and the maximum temperature is 40°C while, in apple, the optimum temperature is 28°C. In 'Minneola' tangelo leaves, the optimum temperature for infection by *A. alternata* is 27°C. As temperature declines, longer wetting periods are needed for infection to occur. In citrus, the optimum conditions for *Alternaria* infections are temperatures from 23 to 27°C and 8–12 h of continuous leaf wetness in the form of dew or rainfall. From these studies, it is clear that *A. alternata* is able to germinate at a wide range of temperatures .

Under controlled conditions, infection occurs over a wide range of temperatures, from less than 10°C up to 35°C or above. Various researchers have sometimes arrived at different conclusions. Thus, the optimum temperature for infection of tobacco by *A. alternata* is 20°C, 25°C and 30°C in different studies. In some cases, the optimum temperature for infection is similar to the optimum for germination, which may reflect the general requirement of the fungus. In other cases, these temperatures can be different, which may possibly reflect a specific reaction of the host, succumbing to infection under conditions not necessarily optimal for spore germination or penetration. For instance, the optimum temperature for infection of tomato by *A. alternata* is 25°C while, in apple, it is 28–31°C.

Symptoms and signs: Symptoms include black spots on leaves and fruits, ranging from a single lesion to lesions that cover more than 50% of the fruit surface, as well as chlorosis and the abscission of some leaves. Lesions begin to appear on all cultivars at the beginning of summer and are round on fruits and round to irregular on leaves. They are surrounded by a green-yellow halo. On fruits like pomegranate, the damage is restricted to the peel surface while the edible tissue remains unaffected . This is in contrast to black rot of pomegranate, in which the fruit rot is restricted to the internal area whereas the peel and leaves remains unaffected. Thus, *A. alternata* that causes black spot of pomegranate is different from the one that causes internal rot of pomegranate and other crops' *Alternaria* spp.-incited diseases.

A. alternata can be found as an epiphytic saprophyte on all parts of the plant without causing any damage. Isolation of *A. alternata* from plants does not indicate whether pathogenic isolates of this species are present . Currently, the only method available for identification of pathogenic isolates is by pathogenicity tests, which can be time and labor consuming. Due to the increasing commercial importance of pomegranate and the desire to reduce the use of fungicides, a fast and simple method for early detection of *A. alternata* associated with infectious black spot is needed. Arbitrarily primed polymerase chain reaction (apPCR) is a simple and reproducible assay based on amplification of random genomic DNA segments with single primers of arbitrary nucleotide sequence. These primers detect polymorphisms in the absence of specific nucleotide sequence information, and the polymorphisms function as genetic marker). The goal of this study was to generate a PCR-based molecular marker capable of detecting genomic DNA of virulent isolates of *A. alternata* causing black spot disease of pomegranate.

2.4.5. *Mucor pyriformis* (Mucor Rot) : *Mucor pyriformis* causes a soft rot of several fruits known as *Mucor* rot . *Mucor* genus consists of an approximately 40 species of moulds commonly found in soil, digestive systems, plant surfaces, rotten vegetable matter and iron oxide residue in the biosorption process. Colonies are very fast growing, cottony to fluffy, white to yellow, becoming dark-grey, with the development of sporangia. Sporangiophores are erect, simple or branched, forming large (60-300 μm in diameter), terminal, globose to spherical, multispored sporangia, without apophyses and with well-developed subtending columellae. A conspicuous collarette (remnants of the sporangial wall) is usually visible at the base of the columella after sporangiospore dispersal. Sporangiospores are hyaline, grey or brownish, globose to ellipsoidal, and smooth-walled or finely ornamented (Fig. 8). Chlamydospores and zygospores may also be present *Mucor* spores or sporangiospores can be simple or branched and form apical, globular sporangia that are supported and elevated by a column-shaped columella. *Mucor* species can be differentiated from moulds of the genera Absidia , *Rhizomucor* and *Rhizopus* by the shape and insertion of the columella, and the lack of stolons and rhizoids. Some *Mucor* species produce chalmydospores. They form mold with irregular non-septate hyphae branching at wide angles (>90°).

Kingdom: Fungi
Phylum : Zygomycota
Class : Zygomeycetes
Order: *Mucor*ales
Family: *Mucor*aceae
Genus: *Mucor*

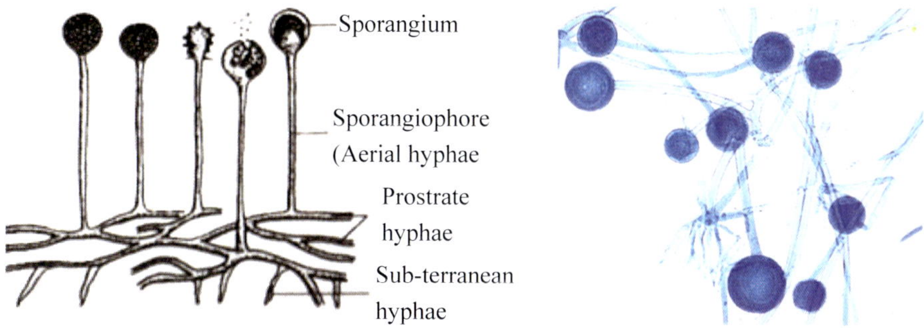

Fig. 8 : *Mucor pyriformis*

Mucor mucedo uses asexual reproduction. When erect, hyphal sporangiophores are formed. The tip of the sporangiophore swells to form a globose sporangium that contains uninucleate, haploid sporangiospores. An extension of the sporangiophore called the columella protrudes into the sporangium. The sporangium walls are easily ruptured to release the spores, which germinate readily to form a new mycelium on appropriate substrates. During sexual reproduction, compatible strains form short, specialized hyphae called gametangia. At the point where two complementary gametangia fuse, a thick-walled, spherical zygosporangium develops. The zygosporangium typically contains a single zygospore. Nuclear karyogamy and meiosis (sexual recombination) occur within the zygospores, which are thought to be long-lived and resistant to adverse conditions. They may germinate to form hyphae or a sporangium. *Mucor* includes both homothallic (self-compatible) and heterothallic species.

Symptoms and signs: *Mucor* spp. invade the fruit through the slightest wound. The fungus secretes an enzyme that rapidly liquifies the entire fruit. Under conditions of high humidity, the berry becomes covered with a coat of tough, wiry mycelium and black sporulation at the tips of long spore bearing structures. *Mucor* and *Rhizopus* fruit rots closely resemble each other and may be difficult to differentiate in the field. *Mucor* rot develops from infections through puncture wounds or cracks on the rind of the fruit In the early stage of symptom development, lesions were quasicircular or irregular, light to dark brown, and soft and watery and sporangiophores of *Mucor* spp. protruded from the wounds or cracks. At the advanced stage of disease development, infected fruit generally were completely covered with masses of white or white to gray shiny, erect sporangiophores, on the top of which black or pale brown sporangia were evident, and decayed tissues were very watery (Saito *et al*, 2016).

2.4.6. *Monilinia fructicola* (Brown Rot): Worldwide, brown rot caused by *Monilinia laxa* (Aderhold & Ruhland) Honey, Monilinia fructigena (Aderhold & Ruhland) Honey and *Monilinia fructicola* (Winter) Honey is responsible for severe losses in important commercial crops such as peach and nectarine (*Prunus persica* (L.) Batsch), apricot (*P. armeniaca* L.), plum (*P. domestica* L.), sweet (*P. avium* L.) and sour cherry (*P. cerasus* L.), almond (*P. amygdalus* Batsch), apple (*Malus pumila* Mill.) and pear (*Pyrus communis* L.) determining an important economic impact. *Monilinia fructigena* is prevalent on pome fruits and only a low percentage of isolates (less than 10%) derived from stone fruits (Chalkley, 2019). Direct yield losses result from blossom and twig blight, and from fruit rot before harvest and during the postharvest phase. Losses depend on weather conditions and are especially severe if high humidity, warm temperatures and abundant rainfall prevail prior to harvest. Indirect losses are due to the cost of fungicide applications during the bloom and the pre- and postharvest periods, however, the appearance of disease after fruit purchase, at consumer level, can negatively affect consumer satisfaction and propensity to consumption.

Conidia (asexual spores) are produced on tufts of conidiophores called sporodochia. The conidia are hyaline (colorless), lemon-shaped, and produced in a moniloid manner (resembling a string of beads with constricted ends). Under ideal conditions, conidia germinate within 3 to 5 hours. Extensive mycelial growth can occur within 24 hours. Brown rot fungi are ascomycetes. They produce ascospores (sexual spores) in tubular sacs termed asci that are produced on the upper surface of a cup- or disc-shaped structure, known as an apothecium.(Fig. 9). Apothecia can be 5-20 mm (up to nearly an inch) in diameter and are borne on mummified fruit that have fallen to the ground. Although commonly described in textbooks, apothecia are rarely observed in most areas.

Kingdom: Fungi
Phylum : Ascomycota
Class : Leotiomycetes
Order: Helotiales
Family: *Sclerotinia*ceae
Genus: *Monilinia*

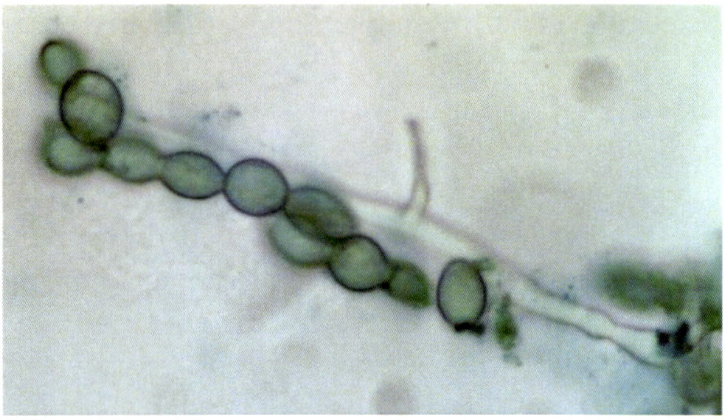

Fig. 9 : *Monilinia fructicola*

Symptoms and signs: Brown rot causes blossom blight, twig blight; twig canker and fruit rot.(3) Brown rot is caused by a fungus that produces spores, and can be a major problem during particularly wet seasons. Prolonged wet weather during bloom may result in extensive blossom infection. The length of wet periods required for blossom infection depends upon the temperature. Humid wet conditions are when the fruit trees are most at risk from infection. Young green fruit can be infected just before autumn, but the infection often remains inactive until near maturity of the fruit. Brown rot can spread after harvest. Mature fruit can decay in only 2 days under warm conditions. In blossom blight, infected blossoms wilt, shrivel and become covered with greyish mould. Petals may appear light brown or water-soaked. Blighted blossoms do not produce fruit. Dead blossoms may stick to spurs and twigs until harvest, providing a source of spores for the fruit rot phase. Twig Blight and Canker: On peaches and apricots the infection may spread to twigs, causing brownish, oval cankers that may girdle and kill twig.

2.4.7. *Rhizopus* spp. (Bread mold/Soft Rots): The genus *Rhizopus* contains several species. The most common ones are *Rhizopus arrhizus*, *Rhizopus azygosporus*, *Rhizopus microsporus*, *Rhizopus schipperae*, and *Rhizopus stolonifer*. Some morphological features, such as the length of rhizoids and sporangiophores, the diameter of sporangia, the shape of columellae, and the size, shape and surface texture of sporangiospores aid in differentiation of *Rhizopus* species from eachother. *Rhizopus* stolonifer is commonly known as black bread mold. Like other members of the genus *Rhizopus*, *R. stolonifer* grows rapidly, mostly in indoor environments (Baggio *et al*, 2016). *Rhizopus* species grow as filamentous, branching hyphae that generally lack cross-walls. They reproduce by forming asexual and sexual spores. In asexual reproduction, sporangiospores are produced inside a spherical structure, the sporangium.

This fungus was first discovered by the German Christian Gottfried Ehrenberg in 1818 as *Rhizopus nigricans*. The name was changed in 1902 to *Rhizopus stolonifer* by the French mycologist J. P. Vuillemin (Onions, *et al*, 1981). Like other species of *Rhizopus*, *R. stolonifer* grows rapidly and spreads by means of the stolons. The stolons provide an aerial structure for the growth of the mycelium and the occupation of large areas. They can climb vertically as well as horizontally. Sporangiophores of *R. stolonifer* can be up to 2.5 mm long and about 20 μm in diameter. The spores are shaped differently depending on the available nutrients (Fig. 10). They can be ovate, polygonal or angular. The optimal temperature for growth varies between 25-30°C. The thermal death point, which is defined as the lowest temperature that can kill all cells in ten minutes, is 60^0 C *R. stolonifer* can grow in acidic environments with a pH of as low as 2.2. The pH range can vary from 2.2 to 9.6. Uv irradiation can delay spore germination.

Kingdom: Fungi
Phylum : Zygomycota
Class : *Mucor*omycotina
Order: *Mucor*ales
Family: *Mucor*aceae
Genus: *Rhizopus*

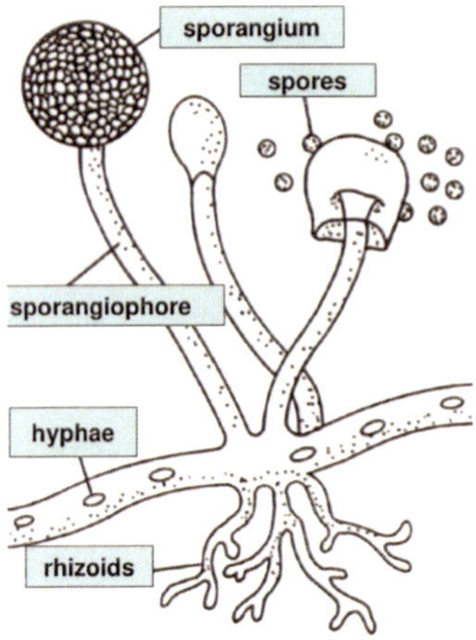

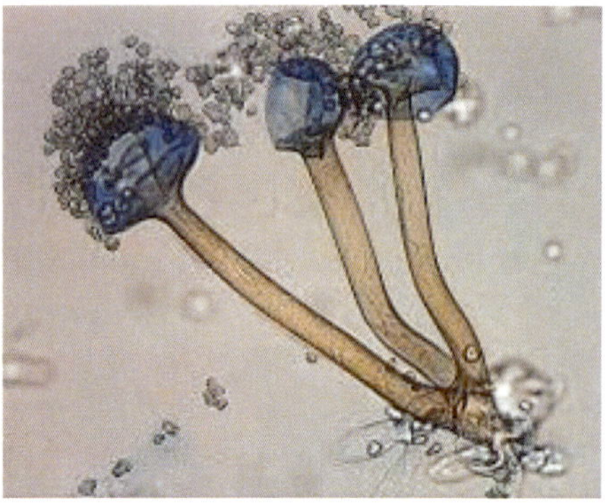

Fig. 10 : *Rhizopus stolonifer*

Rhizopus differs from the molds previously described, because it is nonseptate and produces sporangiospores rather than conidia. It is a very-fast-growing, spreading type of mold which has white mycelia and black sporangia. *Rhizopus* species form rhizoids at the base of the sporangiophores, and columella in the sporangium. Young sporangia are white before turning black with age. Probably the most common of the *Rhizopus* species is *R. stolonifer*, the so-called bread mold. Besides bread, *Rhizopus* causes spoilage of strawberries, other berries, fruits and vegetables.

Symptoms and signs: A water-soaked lesion starts to develop a few hours after the root has been wounded or bruised during harvest or handling. The lesion becomes covered by a web-like outgrowth with scattered pinhead-like structures which are the mycelia and sporangia of the fungus. In 4 to 5 days, the entire inner tissues of the root rot become soft, slimy and watery as they liquefy. However, the peridium remains almost intact except for small cracks and wrinkled appearance. Black brush- like domes, which are the sporangiospores, sporangia, and spores, emerge from these cracks, giving a whisker-like appearance. Rotting starts at either end where the fleshy root has been separated from the rest of the stem and the root system during harvest but any bruise or a slight scratch can initiate rotting. Soft rot caused by *Rhizopus* has some similarities with bacterial soft rot, as both produce the watery, flesh decaying rot and both attract fruit flies. However, bacterial rot produces an unpleasant smell while *Rhizopus* rot emits a pleasant fermented odour that attracts fruit flies. Also, bacterial rot does not produce whiskers unlike *Rhizopus* soft rot. Soft rot caused by *R. stolonifer* is a postharvest disease that can be economically important without proper harvest and postharvest practices. It can destroy storage roots that are bruised or wounded during harvest, left overnight in the field after harvest, and are not properly cured before storage.

2.4.8. *Cladosporium* spp., (Bread mold): *Cladosporium* is a genus of fungi including some of the most common indoor and outdoor molds.. Species produce olive-green to brown or black colonies, and have dark-pigmented conidia that are formed in simple or branching chains. Many species of *Cladosporium* are commonly found on living and dead plant material. Some species are plant pathogens, others parasitize other fungi. *Cladosporium* spores are wind-dispersed and they are often extremely abundant in outdoor air. Indoors *Cladosporium* species may grow on surfaces when moisture is present. *Cladosporium fulvum*, cause of tomato leaf mould, has been an important genetic model, in that the genetics of host resistance are understood. In the 1960s, it was estimated that the genus *Cladosporium* contained around 500 plant pathogenic and saprotrophic species, but this number has since been increased to over 772 species. The genus *Cladosporium* is very closely related to black yeasts in the order Dothideales. *Cladosporium* species are often highly osmotolerant.

Cladosporium species have been reported to cause infections of the skin and toenails as well as sinuses and lungs. The airborne spores of *Cladosporium* species are significant allergens and in large amounts they can severely affect asthmatics and people with respiratory diseases. *Cladosporium* species produce no major mycotoxins of concern, but do produce volatile organic compounds (VOCs) associated with odours. *Cladosporium* appears in clusters of black, yellow, or green spots. The spots spread quickly when not cleaned away. Colonies are slow growing, mostly olivaceous-brown to blackish-brown but also sometimes grey, buff or brown, suede-like to floccose, often becoming powdery due to the production of abundant conidia. The reverse is olivaceous-black. Vegetative hyphae, conidiophores and conidia are equally pigmented. Conidiophores are more or less distinct from the vegetative hyphae, being erect, straight or flexuose, unbranched or branched only in the apical region, with geniculate sympodial elongation in some species. Conidia are produced in branched acropetal chains, being smooth, verrucose or echinulate, one to four-celled, and have a distinct dark hilum. The term blastocatenate is often used to describe chains of conidia where the youngest conidium is at the apical or distal end of the chain (Fig.11). The conidia closest to the conidiophore and where the chains branch are usually "shield-shaped."

Kingdom: Fungi
Phylum : Ascomycota
Class : Dothidiomycetes
Order: Capnodiales
Family: Davidiellaceae
Genus: *Cladosporium*

Fig. 11 : *Cladosporium*

Symptoms: *Cladosporium* cladosporioides and C. herbarum cause *Cladosporium* rot of red wine grapevines. The incidence of infection is much higher when the harvest of the grapes is delayed. Over 50% of grape clusters can be affected at harvest, which greatly reduced the yield and affects the wine quality. This delay is required in order for the phenolic compounds in the grapes to ripen and contribute to the aroma and flavour development in wine of optimum quality. Symptoms of *Cladosporium* rot are typically observed on mature grapes and are characterized by dehydration, a small area of decay that is firm, and a layer of olive-green mould. Although leaf removal reduces the incidence of infection by many species of fungi, it leads to an increase in *C. cladosporioides* populations on grape clusters and an increase in rotten grapes at harvest (Wang, Xiaoning *et al*, 2013).

This species has also been involved in the rotting of strawberry blossoms Infection of strawberry blossoms by *C. cladosporioides* has been associated with simultaneous infections by *Xanthomonas fragariae* (in California) and more recently *C. tenuissimum* (in Korea). *C. cladosporioides* infects the anthers, sepals, petals and pistils of the strawberry blossom and is typically observed on older flowers with dehisced anthers and signs of senescence. From 1997-2000, there was a higher proportion of misshapen fruits due to *C. cladosporioides* infection, and their culling affected the strawberry industry in California. Infection leads to necrosis of the entire flower, or parts of it, as well as to the production of small and misshapen fruits and green-grey sporulation on the stigma. A higher occurrence of infection is observed in strawberry plants cultivated outdoors than cultivated in a greenhouse.

2.4.9. *Phomopsis* spp., (Stem end rot / Stem canker/ Spots): A number of small fruit crops are affected by fungi belonging to the genus *Phomopsis*. Diseases caused by *Phomopsis* species on economically important crops include *Phomopsis* cane spot of grapevines by *P. viticola*. *Phomopsis* leaf spot by *P . viticola* , Stem canker of sunflower caused by *P. helianthi*, Twig canker on *Prunus persica* (peach) caused by *P. amygdali, P. citri* causing stem-end rot, *P. vaccinii* causes twig blight and cane canker on blueberries as well as twig dieback and fruit rot on cranberries; *P. obscurans* causes leaf blight and fruit rot in strawberries, P.vexans causes fruit rot in brinjal (Lim *et al*, 2010). Yet other *Phomopsis* species cause tree fruit diseases, demonstrating that this fungal genus is versatile and well adapted to fruit crops. In particular, *Phomopsis* species like cool, rainy weather,

Phomopsis sp. is typically found sporulating abundantly on dried petioles that remain attached to the tree. During rainy periods, conidia are discharged and deposited on fruit surfaces. The fungus lacks cutinases and, therefore, the ability to penetrate through intact cuticle. It requires wounds such as the peduncle broken during harvesting, fruit fly oviposition puncture wounds, and abrasions on the fruit to infect fruit. The disease is rarely seen in the field on green fruits and is more usually seen on fruits that have fully ripened. The fast growing mycelium is initially white and gradually turns to a light gray color on potato-dextrose agar (PDA). Pycnidia are formed in PDA under continuous fluorescent light in about three weeks and sporulated about one to two weeks later. Conidia are of two types, hyaline, fusiform, unicellular, 6.4-8.0 x 2.7-3.1 µm and are 13.7-20.0 x 1-1.8 µm. These conidial dimensions differ from P. caricae-papayae, which has been reported from India causing a fruit and stem rot of papaya. (Figs.12 & 13) *Phomopsis* caricae-papayae also differs from wet fruit rot in the symptoms it causes on papaya fruit (lesions dark brown to black, depressed, and cracked at advanced stages) and the lack of formation of the conidia by pycnidia produced on host tissue. The *Phomopsis* sp. that causes wet fruit rot readily produces conidia on infected fruit and petioles. No perfect stage is known to be associated with either fungi.

Kingdom: Fungi
Phylum : Ascomycota
Class : Sordariomycetess
Order: Diaporthales
Family: Valsaceae
Genus: *Phomopsis*

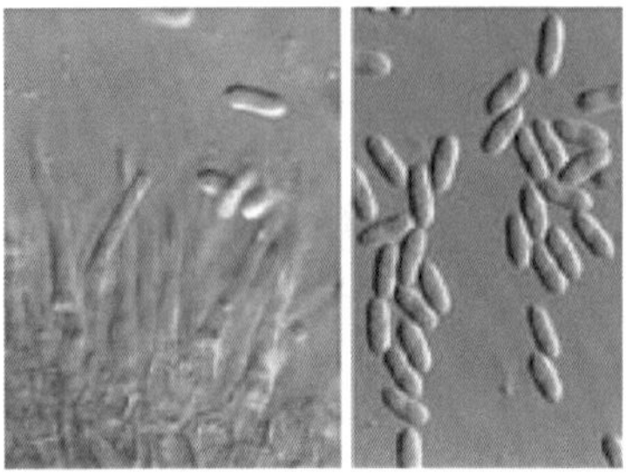

Fig.12 : Conidiophores and conidia

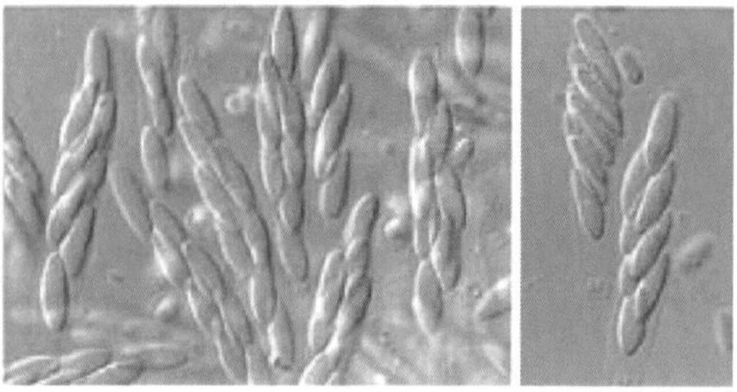

Fig.13 : Asci with ascospores

Symptoms: Wet fruit rot in its early stages resembles *Rhizopus* watery soft rot. Its incidence is sporadic but damage can be extensive when it does occur. Wet fruit rot occurs most frequently as a stem-end rot but any part of the fruit can be affected. It is not known to infect other parts of the papaya plant except senescing petioles. The infected area is slightly depressed, soft and translucent. The fungus is rapid growing causing lesions to expand very quickly with the infected area extending to the seed cavity. The cuticle over the infected area remains intact and develops a delicate, wrinkled pattern that is parallel to the leading edge of the lesion. Black, scattered pycnidia usually form in the infected area after about seven days. The infected tissue is soft, mushy and wet but does not usually leak liquids that is characteristic of *Rhizopus* watery soft rot. Under conditions of high humidity the infected area can become covered by white to gray mycelium. In brief, stems girdled slightly above the soil line, plant topples and dies; stem lesions - dark brown, becoming grey in the centre as pycnidia develops; leaf - irregular brown spots ; fruits soft, watery & decays, finally black, mummified as pycnidia develop abundantly over the surface.

2.4.10. *Phytophthora* (Rots); *Phytophthora* is a genus of plant-damaging oomycetes (water molds), whose member species are capable of causing enormous economic losses on crops worldwide. *Phytophthora* species may reproduce sexually or asexually. The cell wall of *Phytophthora* is made up of cellulose. Many species of *Phytophthora* are soil- and water-borne plant pathogens capable of causing enormous economic losses on crops worldwide, as well as environmental damage in natural ecosystems. *Phytophthora* species cause various diseases on a wide range of hosts, but discussion here will focus on root and crown rots of woody plants. Members within the Oomycetes, such as *Phytophthora*, resemble fungi morphologically with the production of filamentous hyphae and both sexual and asexual spores. Oomycetes, however, have cell walls made of cellulose and β-glucans, are diploid in their vegetative state, and at some point in their life cycle, produce heterokont zoospores with two unequal flagella (Lamour, 2013).

In many species, sexual structures have never been observed, or have only been observed in laboratory matings. In homothallic species, sexual structures occur in single culture. Heterothallic species have mating strains, designated as A1 and A2. When mated, antheridia introduce gametes into oogonia, either by the oogonium passing through the antheridium (amphigyny) or by the antheridium attaching to the proximal (lower) half of the oogonium (paragyny), and the union producing oospores (Anonymous, 2018). Like animals, but not like most true fungi, meiosis is gametic, and somatic nuclei are diploid. Asexual (mitotic) spore types are chlamydospores, and sporangia which produce zoospores. Chlamydospores are usually spherical and pigmented, and may have a thickened cell wall to aid in their role as a survival structure (Fig. 14). Sporangia may be retained by the subtending hyphae (noncaducous) or be shed readily by wind or water tension (caducous) acting as dispersal structures. Also, sporangia may release zoospores, which have two unlike flagella which they use to swim towards a host plant.

Kingdom: Fungi
Phylum : Heterokontophyta
Class : Oomycota
Order: Peronosporales
Family: Peronosporaceae
Genus: *Phytophthora*

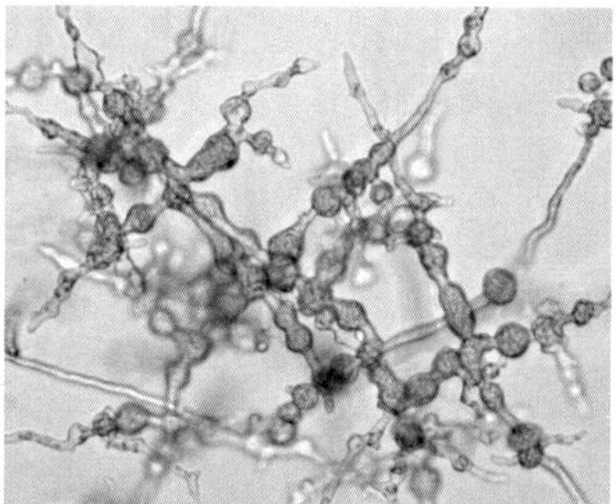
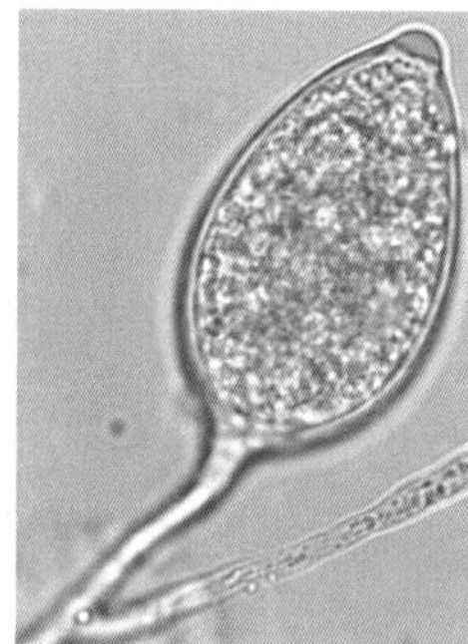

Fig. 14 : Hyphal swellings with chlamydospore Sporangium

P. fragariae survives in the soil in the form of oospores. These oospores can survive for up to 4 years; there are some reports of spores remaining viable 13 – 15 years. The oospores germinate to form usually one sporangium. Sporangia germinate in the presence of water to release motile zoospores. The presence of water is important because the zoospores have no way to reach a host without swimming. The zoospores swim through the water in the soil to find roots where they encyst and form germination tubes to infect the plant. This usually occurs in the late autumn or early spring. Growth of the disease occurs mostly in the stem and root. Hyphae protrude out of the root and produce new, secondary sporangia. The secondary sporangia are usually produced in winter months and the process takes a few days. The sporangia are produced more quickly in cold temperatures. The new sporangia move through water in the soil to infect new plants.

Symptoms and signs: *Phytophthora* symptoms are often generalized, causing weakness and slow collapse in affected plants. Early warning signs are few; most plants appear drought stressed regardless of adequate watering. Some plants wilt and die from water stress during the first warm weather after infection, but others linger for several years before succumbing. Leaves may appear dull or fade to yellow, red or purple long before fall sets in. The leaves of plants affected by *Phytophthora* rot appear drought stressed. Trees or plants often wilt and die rapidly with the first warm weather of the season. Leaves may turn dull green, yellow, or in some cases red or purplish. Often, only plants in the most poorly drained area of the field or garden are affected. *Phytophthora* infections typically kill young trees, because their root systems and crown areas are small compared to those of mature trees.

Symptoms may develop first on one branch or stem then spread to the rest of a tree or plant. Trees may decline over a period of years before finally dying or they may be killed in a single season. Slow decline occurs when the roots are attacked; rapid decline occurs when the crown or basal stem is attacked and girdled, the damage completely encircling the stem in a single season. Symptoms on roots and crowns may vary somewhat depending on the species of *Phytophthora* involved, the plant being attacked, the resistance of the plant variety, and soil moisture and temperature (Anonymous, 2019.). In general, trees affected by *Phytophthora* develop darkened areas in the bark around the crown and upper roots. Gum or dark sap may ooze from the margins of the diseased trunk area. If bark tissue is carefully cut away, reddish brown streaks or zones can be seen in the inner bark and outer layer of wood. No mycelium (slender filaments of a fungus body) is visible in between the bark and wood in trees affected by *Phytophthora*, distinguishing this disease from Armillaria root rot, which is caused by a true fungus. When tomatoes and eggplants are affected by *Phytophthora* root rot, roots of all sizes develop water-soaked spots that dry out and turn a chocolate brown as the disease becomes advanced. Early infections, caused by *Phytophthora* and other pathogens that cause damping-off diseases, kill seedlings.

2.4.11. *Aspergillus niger* (Black mould): *Aspergillus* genus consisting of a few hundred mold species found in various climates worldwide. Aspergillum is an asexual spore-forming structure common to all *Aspergillus* species; around one-third of species are also known to have a sexual stage. *Aspergillus* is defined as a group of conidial fungi that is, fungi in an asexual state. Some of them, however, are known to have a teleomorph (sexual state) in the Ascomycota, so with DNA evidence forthcoming, members of the genus *Aspergillus* can tentatively be considered members of the Ascomycota.(Swilaiman, *et al*, 2013). Given that a majority reproduce asexually, they are often described as conidial fungi. However, studies are yet to determine how a good number of others reproduce while some have been shown to reproduce sexually Conidial fungi are fungi that reproduce asexually through the production of spores from the fungi hyphae. The spore produced is referred to as conidium (Fig 15).

Members of the genus possess the ability to grow where a high osmotic pressure exists (high concentration of sugar, salt, *etc*.). *Aspergillus* species are highly aerobic and are found in almost all oxygen-rich environments, where they commonly grow as molds on the surface of a substrate, as a result of the high oxygen tension. Commonly, fungi grow on carbon-rich substrates like monosaccharides like glucose and polysaccharides such as amylase. *Aspergillus* species are common contaminants of starchy foods (such as bread and potatoes), and grow in or on many plants and trees. In addition to growth on carbon sources, many species of *Aspergillus* demonstrate oligotrophy where they are capable of growing in nutrient-depleted environments, or environments with a complete lack of key nutrients. A.niger is a prime example of this; it can be found growing on damp walls, as a major component of mildew. The most common pathogenic species are *A. fumigatus* and *A. flavus*, which produces aflatoxin which is both a toxin and a carcinogen, and which can contaminate foods such as nuts. The most common species causing allergic disease are *A. fumigatus* and *A. clavatus*. Other species are important as agricultural pathogens. *Aspergillus* spp. cause disease on many grain crops, especially maize, and some variants synthesize mycotoxins, including aflatoxin.

Aspergillus niger is not only a species of plant pathogen, but a group in the genus *Aspergillus* that is made up of 15 varieties, all with black conidia. The host range includes 37 genera of fruits and vegetables such as tomatoes, peanuts, grapes, onions, and mangoes. The fungus is commonly found in the soil, living as a saprophyte on decaying vegetation and leaves, compost piles, and stored grain. It is also found in indoor environments and as a contaminant of food. As the fungus is opportunistic, humans, birds, and animals can also be hosts, although this is extremely rare.

Kingdom: Fungi
Phylum : Ascomycota
Class : Eurotiomycetes
Order: Eurotiales
Family: Trichocomaceae
Genus: *Aspergillus*

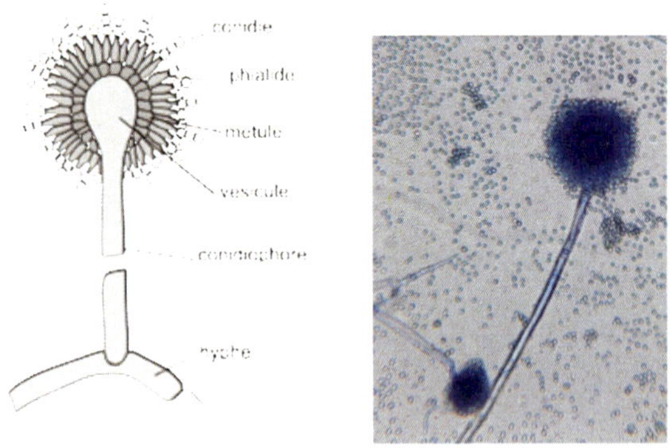

Fig. 15 ***Aspergillus***

Aspergillus niger has a plethora of strains. The size of different strains of *Aspergillus niger*, as individual organisms, can range from anywhere between 900-1,600μm in length, with the rough, spherical conidia measuring 3-5 μm.Some strains of *Aspergillus* niger are known to secrete ochratoxins, mycotoxins which can give rise to nephrotoxicity and renal tumours in a variety of animal species and are potentially hazardous to human health through their consumption. They are released by *Aspergillus* and *Penicillium* fungi species. Ochratoxin A, for example, is a problem organism during the bulk storage of grains if the conditions are in any way wet or damp. More recently, the strain ATCC 16404 strain of *Aspergillus niger* has had to be reclassified as *Aspergillus* brasiliensis (Hinton-Sheley Phoebe, 2018).

Symptoms and signs: *Aspergillus niger* causes several diseases including black mold (or black rot) of onions and garlic, crown rot of peanuts, and vine canker of table grapes. In black rot of onions and garlic, *A. niger* develops between the outer (dead, flaky) skin and the first fleshy scales of the bulb which become water soaked. If conditions are dry, the area dries and shrivels, revealing black spore masses between outer scales. Secondary invaders are common and transform the bulb into a soft, watery mass . Crown rot of peanuts is characterized by root curling and deformation of the upper parts of the plant. In vine canker of table grapes, preliminary symptoms include pinhead drops of reddish sap at the infection site. As the disease progresses, the trunk of the vine becomes girdled, spongy, and slightly larger. The vine displays fall colors prematurely and the inside of the canker contains powdery black spores

(Ghoshdastider *et al*, 2015). The cankers average about 8 cm and are commonly calloused. Other diseases attributed to *A. niger* include: tuber rot of yams, stem rot of Dracaena, black mold rot of cherry, kernel rot of maize, fruit rot of grapes, fruit rot of banana, rot of tomatoes, boll rot of cotton, root stalk rot of Sansevieria, and mango rotting. Postharvest decay results in discoloration, quality deterioration, and reduction in the commercial value of various crops.

2.4.12. *Galactomyces citri-aurantii* (Formerly *Geotrichum candidum*): *Geotrichum* is a common fungus found worldwide. It grows on fruits, milk products, cereals, soil, air and plants. *Geotrichum* is also considered a normal part of human flora. *Geotrichum* colonies are white, flat and have a powdery or waxy texture. *Geotrichum* has a rapid growth rate, and its dry spores disperse easily through the air, making them easy to inhale. It is common in soil and has been isolated from soil collected around the world, in all continents. The genus *Geotrichum* includes several species. The most common one is *Geotrichum* candidum. *G. clavatum* and *G. fici* are among other *Geotrichum* species. *G. fici* has an intense smell resembling that of pineapple.

G. candidum is also the causative agent of the human disease geotrichosis apart from the plant disease sour rot which infects citrus fruits, tomato, carrot and other vegetables. It can affect harvested fruit of durians such as Durio graveolens. *G. candidum* colonies are thin, spreading, soft, creamy and white in the anamorph state. The fungus *G. candidum* is characterized by hyphae that appear creeping, mostly submerged and septate. The hyphae colour appears to be hyaline or lightly pigmented. When the hyphae becomes airborne it changes shape from arthroconidia to cylindrical or barrel-shaped or ellipsoidal (Fig.16). Chlamydospores are subglobose, solitary, borne on undifferentiated hyphae. Blastoconidia sometimes develop on hyphae laterally. Conidia appear arthrosporous, terminal or intercalary, aerial on an agar surface. The conidia size ranges from 4.8–12.5 µm x 2.4–2.5 µm. In teleomorph stage, The fungus produces globose asci that contain a single, thick walled, uninucleated, globose to oval ascospore measuring 6–7 µm by 7–10 µm. The ascospores have a smooth inner wall and a furrowed outer wall. The septa are perforated by microspores, arranged in a ring structure. The colonies appear to be growing faster in the sexual stage than the asexual stage. Colonies grow at a rate of 5–7 mm daily at 24 °C (75 °F).(Domsch, *et al*, 1980)

Kingdom: Fungi
Phylum : Ascomycota
Class : Eurotiomycetes
Order: Eurotiales
Family: Trichocomaceae
Genus: *Aspergillus*

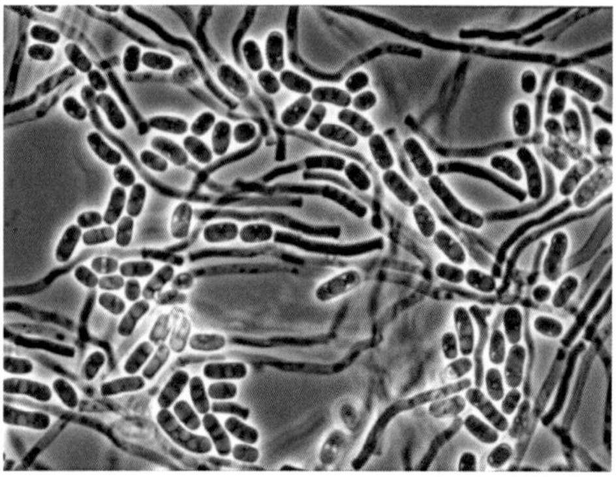

Fig. 16 : Unicellular arthroconidia

Symptoms and signs: As *Geotrichum* sour rot is very reliant on favorable environmental conditions, the symptoms produced are highly varied, and thought to be dependent on the duration of favorable conditions. The most obvious symptom of an active infection is a wet, soft rot of the storage root combined with a distinct sour odor. During active infections, white mycelial tufts may also become present on the root surface. When favorable conditions are removed, affected areas on the storage root will exhibit a 1-3 mm collapse of the outer root tissue, which will become firm over time. These collapsed lesions will have highly irregular borders and can greatly vary in size and shape, but rot and decay of further tissue will cease. *Geotrichum* sour rot lesions can be associated to chilling injury wounds and the tips of shrink wrapped roots. Most of the fruit of the Nagpur mandarin is spoiled by this rot in the March–April harvest season. The disease is not as common on crops harvested in November–December when temperatures are low. Temperatures lower than 10°C are known to suppress fungal development. This disease develops very rapidly at ambient temperatures of 28–30°C. The fruit becomes a soft, stinking, semi-solid mass as the fungi secrete very active extracellular enzymes that rapidly degrade the tissue. This disease spreads rapidly inside the package, and corrugated cartons may also lose strength as they get wet.

2.4.13. ***Lasiodiplodia (Botryodiplodia) theobromae:*** *Botryodiplodiatheobromae* (Syn: *Lasiodiplidia theobromae*) and its asexual state, *Botryosphaeria rhodina* are fungal pathogens of great economic importance. *B. theobromae* is an opportunistic plant pathogen that causes different types of plant diseases with worldwide distribution within tropical and subtropical regions *L. theobromae* is an economically important fungal pathogen with worldwide distribution in tropical and subtropical regions. It has over 500 different kinds of host plants and can cause many diseases such as tree dieback, root rot, fruit rots and leaf spot. This fungus is a key pathogen that causes postharvest decay on many important horticultural fruits such as citrus, avocados, mangos, papayas, bananas and guava in tropical and subtropical regions in the world (Peter Twumasl *et al*, 2014). This postharvest decay can lead to significant economic losses, and the fungus can also cause preharvest diseases on some horticultural crops, such as mango tree dieback disease.

Botryodiplodia is very active in warmer seasons and its symptoms are characteristic. The spores are lodged at the stem end because of wind or rains when fruit is very small and growing in the field. After picking, the fungus enters the fruit through scars at the stem-end. These stem-end rots are a serious problem in humid areas. In India, particularly the central and southern parts, postharvest losses of citrus are mainly due to this pathogen. *L. theobromae* is a facultative parasite that completes its life cycle on dead twigs in groves by growing and sporulating in the deadwood of citrus trees. Water transmits fungal spores from deadwood to the surfaces of immature fruit where the fungus colonizes the dead tissue of the button. The fungal colonization remains latent (quiescent) and does not cause any fruit decay before harvest.

Kingdom: Fungi
Phylum : Ascomycota
Class : Dothideomycetes
Order: Botryosphaeriales
Family: Botryosphaeriaceae
Genus: *Lasiodiplodia*

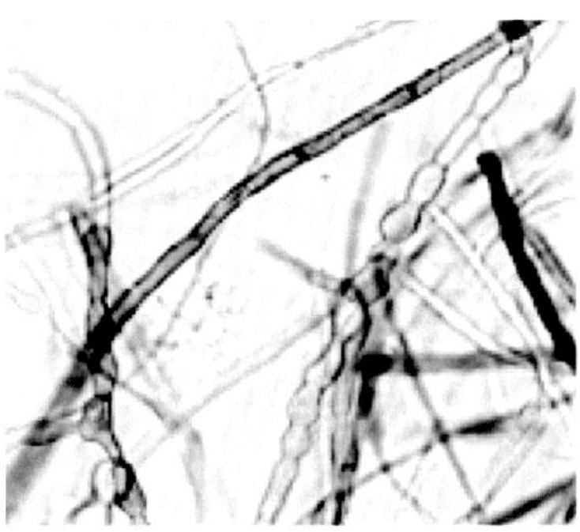

Fig. 17: Hyphae

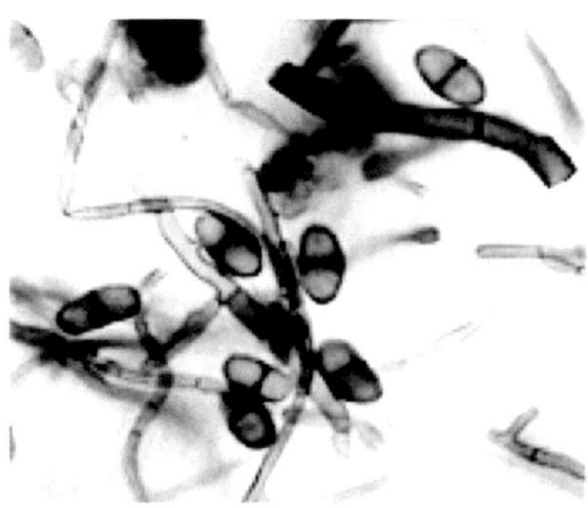

Two celled conidia

L. theobromae on potato dextrose agar medium grows rapidly at 20–30°C, with a white to light or dark gray aerial mycelia on PDA and later produces abundant black pigmentation, which can be clearly viewed from the reserve side of the PDA culture. In the case of PDA as a fungal cultural medium, mycelial growth, pigmentation and pycnidium productions are largely affected by the dextrose concentrations in the medium. The higher the dextrose concentration, the more pigment production and the less pycnidium formation. Conidiophores are hyaline, simple, sometimes septate, rarely branched cylindrical, and arising from the inner layers of cells lining the pycnidial cavity. Conidia (10–8 × 17–43 μm) are formed in the pycnidia (Fig. 17). Young conidial cells are hyaline and granular and have no septum, and mature conidia are striated and contain one septum (Brown and Eckert, 2000). Once the two-celled spores are formed, they readily survive from one season to the next on citrus trees and can cause latent infections on citrus fruit during the winter and spring in groves

Symptoms and sings: Infections develop after harvest, especially under conditions of high temperatures and relative humidity. The optimal temperature for fungal mycelial growth is about 30°C. The fungus usually infects citrus from the button (calyx and disk) at the stem-end of the fruit. It proceeds through the core more quickly than the rind, leading to the development of soft brown to black decay symptoms at both ends of the fruit. The pathogen usually develops unevenly in the rind, forming finger-like projections of black to brown discolorations at the lesion margin between the segments. The fungus may also infect fruit through injuries, giving a soft brown to black lesion that enlarges very rapidly. Decay develops rapidly during and after excessive fruit ethylene degrading treatment and can be observed in fruit at the packing house. It is often observed on arrival at market or shortly thereafter. Normally, this fungus does not sporulate on infected fruit, and the decay does not spread from infected fruit to healthy fruit in packed containers. Dieback in some crops is characterized as a 'dead arm' and a loss of spur positions. More symptoms include stunted shoots in the spring, delay or lack of growth in the spur positions of the bud burst, bleached canes and necrotic buds. Bud necrosis, bud failure, and the dieback of arms are all a result of the necrosis of the hosts vascular system (Rodríguez-Gálvez and Maldonado A. Alves, 2014).

2.4.14. *Fusarium* rot: *Fusarium* species is one of the common pathogens of postharvest disease to cause rot on tomato and other perishable vegetable fruits. *Fusarium* is a large genus of filamentous fungi, part of a group often referred to as hyphomycetes, widely distributed in soil and associated with plants. *Fusarium* species can appear as saprophytes or pathogens on plants, animals as well as humans. They are well associated with a wide range of plants in their natural habitats such as tomato, banana, asparagus, barley, mango, pineapple, carnation, coffee, corn, grasses, legumes, oats, pine, rice, sorghum, sugarcane and wheat. Harvested tomatoes are susceptible towards infections caused by *Fusarium* species due to its succulent epicarp which enable the fungal hyphae to penetrate deeply into the fruit . As a result, the yield of this economically important farm product is affected, hence lowering the production rate. *Fusarium* rot on tomato fruits are often caused by *Fusarium* species. Tomato contaminated with *Fusarium* species are lethal for human and animals consumption since several species of them produce mycotoxins (Abu Bakar *et al*, 2013).

The genus *Fusarium* currently contains over 70 species. The most common of these are *F.solani, F. oxysporum* and *F. chlamydosporum* (e Hoog, *et al*, 2000). *Fusarium* is a soil borne fungus and can persist for long periods in the soil. Transmission may occur via infested soil on tools or equipment, infected debris, infected seed, or run-off water. The pathogen enters the plant through stem plate or wounded tissue. The disease develops from the base of the bulb and progresses towards the tips of the cloves. Infection may occur at any time in the field, or in storage. The disease is favored by higher temperatures, 68° – 86° F (20-30 °C), and high humidity. Late season rains may favor the pathogen (Susan B. Jepson, 2008). Fruit rots caused by various *Fusarium* spp. are some of the most common pre- and postharvest diseases of cucurbit fruits. *F. solani* colonies grow rapidly, 4.5 cm in four days, aerial mycelium white to cream, becoming bluish-brown when sporodochia are present. Macroconidia are formed after 4-7 days from short multiple branched conidiophores which may form sporodochia. They are three to five-septate (usually three-septate), fusiform, cylindrical, often moderately curved, with an indistinct pedicellate foot cell and a short blunt apical cell, 28-42 x 4-6 µm. Microconidia are usually abundant, cylindrical to oval, one to two-celled and formed from long lateral phialides, 8-16 x 2-4.5 µm. Chlamydospores are hyaline, globose, smooth to rough-walled, borne singly or in pairs on short lateral hyphal branches or intercalary, 6-10 µm (Fig. 18).

Kingdom: Fungi
Phylum : Ascomycota
Class : Sordariomycetes
Order: Hypocreales
Family: Nectriaceae
Genus: *Fusarium*

Fig. 18 : Microconidia Macroconidia Chlamydospores

Symptoms and signs : Affected plants became stunted, the older leaves droop and curve downward and the plants frequently wilt and die; vascular tissue of a diseased plant is dark brown in color ; fruit infection occasionally occurs and can be detected by the vascular tissue discoloration within the fruit. Plants may or may not show symptoms in the field or at harvest, but bulbs / fruits may subsequently rot in storage. In the field affected plants may show reduced emergence, yellowing and/or browning (necrosis) of leaves beginning at tips. The discoloration will move toward the base of the leaf, which will eventually wither and die. Other symptoms include reduced bulb size, bulb decay, and brown, poorly developed root systems. This is a result of decay of the stem plates and storage leaves. The stem plate and dry outer scales may crack open. There may be reddish or reddish purple discoloration on stems Plants may or may not show symptoms in the field or at harvest, but bulbs may subsequently rot in storage. In the field affected plants may show reduced emergence, yellowing and/or browning (necrosis) of leaves beginning at tips. The discoloration will move toward the base of the leaf, which will eventually wither and die. Other symptoms include reduced bulb size, bulb decay, and brown, poorly developed root systems. This is a result of decay of the stem plates and storage leaves. The stem plate and dry outer scales may crack open. There may be reddish or reddish purple discoloration on stems and bulbs. In storage, bulbs show spongy, sunken yellow-brown rotting lesions. In the early stages, infected bulbs are softened, brown and watery when cut open. There may be a white, light pink or reddish fungal growth (mycelium) covering the cloves, or in the rot cavities. Deep cracks form in the cloves, followed by break down of the tissue, which will eventually dry down to a portion of its original size, the cloves becoming crinkled and small.

2.4.15. *Phoma* rot : *Phoma* species are phytopathogens that are widely distributed in the environment, most commonly found in aquatic systems and soil. *Phoma* spp. contaminates common food sources. *Phoma* spp. can change from opportunistic to pathogenic organisms once in contact with the appropriate host. *Phoma* spp. have been known to contaminate seeds, nuts, soybeans, potatoes, bananas, sorghum, maize, kiwi berries, lemons, tomatoes, eggplants, pomegranates *etc*.,

Phoma is a large genus whose species are difficult to distinguish (Williams and Williams Neaves, 2006). Distinguishing features of the *Phoma* spp. include the ability to form asexual fruiting bodies lined with conidiophores, which are spore-producing hyphae. The colonies tend to be powdery or velvety in texture and spread, while the pigmentation can vary from greenish gray to brown. *Phoma* spp. in vitro exhibit morphological features such as chlamydospores, conidia, and pycnidial conidiomata (fruiting structures that act as a means of dispersing conidia), which are unique to the genus, helping distinguish them from other dematiaceous fungi (Ashely Bennett, *et al*, 2018) (Fig.19). The isolation and growth of *Phoma* spp. best occurs at a pH close to 5.5. The growth media type aids in producing the best growth characteristics in *Phoma* spp. identification; oatmeal agar supports abundant pycnidia production and malt extract agar stimulates pigment production and crystal formation. Mycelium is septate, branching & hyaline, becoming dark with age, pycnidia are black, subcutaneous and later erumpent, dark subglobose, pycnospores are hyaline, unicellular and subglobous.. *Phoma* is arbitrarily limited to those species in which the spores are less than 15 mm as the larger spored forms have been placed in the genus *Macrophoma*.

Phoma glomerata is the cause of blight of vine flowers and grapes, leaf and fruit spot of apple and damping-off of conifers and is also a secondary invader causing a root rot of tomato, potato tubers and citrus. This fungus is also known to be associated with a number of mycotic diseases in man. *P. prunicola* is the cause of leaf spot of apple and pear, *P. medicaginis* var. *pinodella* is the cause of fruit-rot and collar rot of peas, *P. destructiva* is the cause of fruit-rot and leaf spot of tomato, *P. betae is* the cause of black leg and damping-off of beet, *P. lingam* (*Leptosphaera maculans*) is the cause of black leg of crucifers.

Kingdom: Fungi
Phylum : Ascomycota
Class : Dothideomycetes
Order: Pleosporales
Family: Didymellaceae
Genus: *Phoma*

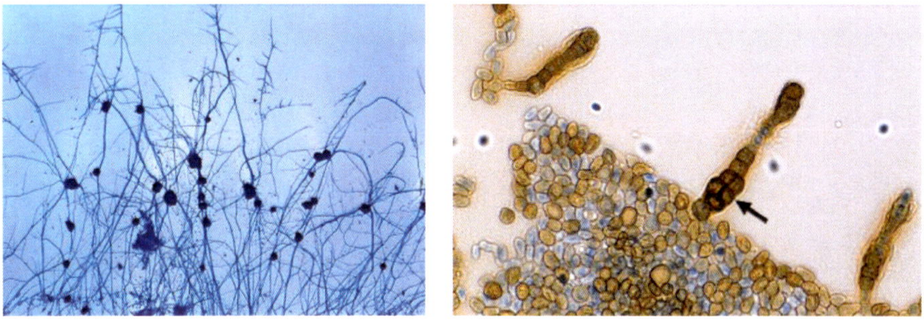

Fig. 19: *Phoma glomerata*- Hyphae with pycnidia Chlamydospores with septae

Symptoms and signs : The disease can affect all above-ground parts of the plant. On leaves, it causes numerous, small dark brown to black spots, which may develop concentric rings as they enlarge. Older leaves can be infected first, however, all leaves are susceptible and defoliation can result when disease is severe. The leaf spot looks very similar to that caused by early blight except that the *Phoma* lesions contain numerous minute black fungal fruiting bodies (pycnidia). Dark brown lesions with concentric rings form on the stems, and both the green and ripe fruit can be infected. Fruit lesions usually develop at the calyx end as small sunken lesions that later develop into sunken, black, leathery lesions with numerous pycnidia in the center. In tomato, major symptoms due to *Phoma* destructive include ditinguished from other rots by the black color of this spot ; small, black, pimple-like eruptions ; specks are the pycnidia or fruiting bodies of the fungus ; moderate temperature and high humidity.

Phomopsis (*Phoma*) blight of eggplant, rot may appear in fruit, in transit after harvest, whose symptoms include minute sunken greyish spots with a brownish halo, which later enlarge and coalesce, producing concentric rings of yellow and brown zones on fruits (Haggag, 2014). These spots increase in size and form large rotten areas on which conidiomata often develop concentrically, covering most of the rotten fruit surface. Pycnidia on fruit are larger than those on stems and leaves. If the infection enters the fruits through the calyx, the whole fruit may become mummified due to dry rot. In grapes, *P. glomerata* can cause scarce growth and chlorotic leaves in some vineyards, in the lst or 2nd year after grafting. The affected grapevines are alone or grouped and scattered here and there in the vineyard. The colour alteration generally starts with a noticeable yellowing of main leaf veins, that tend to spread gradually all over the leaf blade. Progressive wilting and necrosis phenomena till the death of the entire plant follow the initial symptoms. At the graft union heavy necroses may be seen on the affected vines.

2.4.16. *Rhizoctonia solani (Rots):* *Rhizoctonia* is a genus of anamorphic fungi. Species do not produce spores, but are composed ofhyphae and sclerotia (hyphal propagules). *Rhizoctonia* species are saprophytic, but are also facultative plant pathogens, causing commercially important crop diseases. They are also endomycorrhizal associates of orchids. The genus name was formerly used to accommodate many superficially similar, but unrelated fungi. One of the most harmful seed and soil borne pathogens, *Macrophomina phaseolina* is a fungus that infects nearly 500 plant species in more than 100 families. It develops during its life cycle in anamorphous stage; black pycnidia are formed under epidermis. The sterile mycelium stage (*R. bataticola*) is also present in the cycle of its development. Morphological structures of the pathogen include conidia (pycnospores) on very short threadlike conidiophores, septate vegetative mycelium, and black (charcoal) fine sclerotia, 50 mkm to 1 mm long.

***Rhizoctonia* solani (teleomorph:** Thanatephorus cucumeris) is a plant pathogenic fungus with a wide host range and worldwide distribution. It was discovered more than 100 years ago. *R. solani* frequently exists as thread-like growth on plants or in culture, and is considered a soil-borne pathogen. *R. solani* is best known to cause various plant diseases such as collar rot, root rot, damping off, and wire stem. *R. solani* attacks its hosts when they are in their early stages of development, such as seeds and seedlings, which are typically found in the soil.

R. solani does not produce spores, hence is identified only from mycelial characteristics or DNA analysis. Its hyphal cells are multinucleated (Fig. 20). It produces white to deep brown mycelium when grown on an artificial medium. The hyphae are 4–15 μm wide and tend to branch at right angles. A septum near each hyphal branch and a slight constriction at the branch are diagnostic. *R. solani* is subdivided into anastomosis groups based on hyphal fusion between compatible strains. The teleomorph of *R. solani* is Thanatephorus cucumeris. It forms club-shaped basidia with four apical sterigmata on which oval, hyaline basidiospores are borne (Cubeta, *et al*, 1997).

Kingdom: Fungi
Phylum : Basidiomycota
Class : Agaricomycetes
Order: Cantharellales
Family: Ceratobasidiaceae
Genus: *Rhizoctonia*

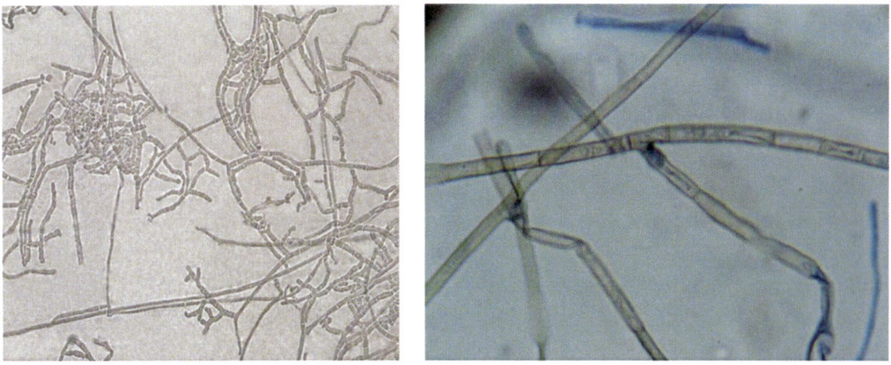

Fig. 20 : *Rhizoctonia* solani

Symptoms and signs: *Rhizoctonia solani* causes a variety of symptoms including damping off, stem lesions, stem rot, root rot, crown rot, and aerial web blighting. Infection causes wilting, stunting and possibly plant death. Some vegetatively propagated plants are susceptible to rot at the base of the cutting. This is a soil borne pathogen, it persists in soil as mycelium and sclerotia (small, brown, long-term survival structures). The disease is spread by the movement of contaminated soil, plant material, tools and equipment. Plants' roots are not typically visible as they are below the surface of the soil, so the symptoms of root rot are often only apparent when the disease is advanced. Roots of plants affected by root rot may turn from firm and white to black/brown and soft. Affected roots may also fall off the plant when touched.

M. phaseolina affects the fibrovascular system of the roots and basal internodes of its host, impeding the transport of water and nutrients to the upper parts of the plant. As a result, progressive wilting, premature dying, loss of vigor, and reduced yield are characteristic symptoms of *M. phaseolina* infection. The fungus also causes many diseases like damping off, seedling blight, collar rot stem rot, charcoal rot, basal stem rot, and root rot. Although brown lesions may form on the hypocotyls or emerging seedlings, many symptoms occur during or after flowering, including grey discoloration of the stem and taproots, shredding of plant tissue in the stem and top of the taproot, and hollowing of the stem (Subramanian Sundravadana *et al*, 2011). Small black dots may form beneath the epidermis of the lower stem and in the taproot, giving the stems and roots a charcoal-sprinkled appearance.[When the epidermis is removed small and black microsclerotia (a sign of the disease) may be so numerous that they give a greyish-black tint to the plant tissue. In addition, reddish-brown discoloration and black streaks can form in the pith and vascular tissues of the root and stem.

2.4.17. *Diplodia* stem-end rot: *Diplodia* stem-end rot incited by *Botryodiplodia theobromae* and *Diplodia* natalensis (synonym :*Lasiodiplodia* theobromae) is an economically important postharvest fruit decay that occurs on different types of fruits including citrus and mango (Jiuxu Zhang, 2014). The rot not only spoils 4 to 6 per cent fruits every year in the Indian Markets. But also affects the mango export. The fear of India losing the mango export may become quite real. The extent of damage in Indian consignments landed at the U.K. airports had been as much as nearly 50 per cent due to this disease alone. Besides India, where its occurrence was recorded as late as in 1964, the disease is known to occur in Burma, Ceylon, the Philippines, Mauritius and the U.S.A. In the initial stage, the epicarp darkens around the base of the pedicel. In the next few hours, the affected area enlarges to form a circular, black patch which under humid atmosphere extends rapidly and turns the whole fruit completely black within two to three days.

Kingdom: Fungi
Phylum : Ascomycota
Class : Dothideomycetes
Order: Botryosphaerales
Family: Botryosphaeriaceae
Genus: *Diplodia*

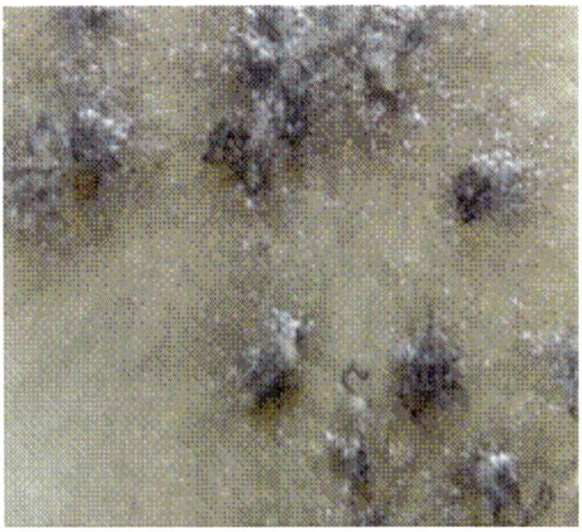

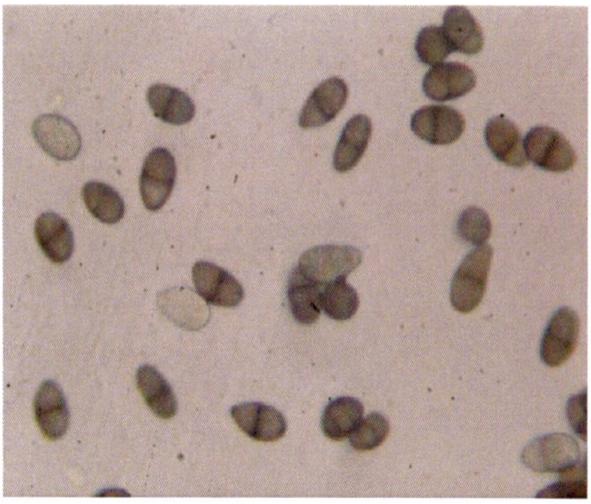

Fig. 21 : Mycelia and pycnidia **Conidia**

The main characteristic that distinguishes the genus *Lasiodiplodia* from other closely related genera is the presence of pycnidia, paraphyses and longitudinal striations in mature conidia. Around 20 species have been described based on the morphology of conidia and paraphyses. The most recent descriptions of these species, other than the morphology, are based on the sequencing of the intergenic spacer regions of the rDNA (ITS) and elongation factor 1 alpha (EF-1) (Netto *et al*, 2014). The morphology of its ascocarp is a dark brown to black, aggregate, with a thick dark brown wall and hyaline in internal layers of 250-400 μm in diameter. The asci is bitunicate, stipitate, with 8 spores, of 90-120 μm longitude. The ascospores are biseriated, hyaline, aseptated of 30-35 x 11-14 μm. The conidiomata is stromal, simple or aggregate, immersed in the host and once mature, it emerges a dark brown color, unilocular, thick or thin brown wall, with setoso frequency, up to 5 mm wide, central ostiolo,unique, papillado. Paraphyses hyaline, cylindrical, partitioned, occasionally branched with the ends rounded up to 55μm long and 3-4 μm wide (Phillips *et al*, 2013). The conidiophore are hyaline, simple, sometimes septated, rarely branched, cylindrical. The conidiogenous cells are hyaline, of a thick wall, smooth, cylindrical to sub-obpiriforms, holoblastic, with one or two rings (Fig. 21). The conidia are subovoids to ellipsoidal, with apex amply rounded, that narrow to truncate at the base, wider mid-upper third, of thick walls, with a granular content, at first hyaline and aseptated, turning a dark brown once mature, with 1 septum, showing deposits of melanin on the interior surface of the wall longitudinally disposed given a striated appearance with measurements 21.5-31-5 x 13-17 μm and one portion of 1.9 length / wide.

Symptoms and signs: The stem-end rot of mango in transit and storage is a serious problem which has received relatively limited attention in this country. The disease manifests only in the ripe mangoes. In the initial stage, the epicarp darkens around the base fo the pedicel. Later, the affected area enlarges to form a circular, brownish-black patch which under humid atmosphere extends rapidly and turns the whole fruit black within 2–3 days. The pulp of the diseased fruits becomes brown and somewhat softer. The pedicel, if present becomes dry. Sporulation does not occur unless the fruit is left over under humid condition for over a month. Under dry condition, the blackened fruit get mummified. The stem-end rot is quite distinct from the lateral rots caused by *Colletotrichum gloeosporioides*, *Aspergillus niger*, *Macrophoma mangifereae* and certain other fungi.

The pulp of the diseased fruit becomes brown and some-what softer. The pathogen incites a soft rot with the aid of pectinolytic and cellulolytic enzymes. The infected fruits lose ascorbic acid rapidly and they have low content of non-reducing sugars in the pulp. Epidemiological investigations in India have revealed that the dead twigs and bark of mango trees harbour the pathogen abundantly, and with the onset of rains, the atmosphere of the mango orchard gets contaminated with spores of the fungus. Cultures can infect readily when inoculated in cavities made on the pedicels of the fruits. The avenues of natural infection of uninjured mango fruits are the exposed surface of the attached pedicel or the stem-end scar when the pedicel is removed (Srivastava, 1972). Postharvest *Diplodia* stem-end rot develops from latent infections of the fungus established in necrotic tissues of the fruit button (calyx and disk) on citrus trees in groves. After harvest, *L. theobromae* reactivates, grows and infects the fruit through the natural openings formed between the fruit and button during the button separation upon abscission. Fungal latent infection levels on immature fruit in groves are favored by many factors such as high amounts of deadwood on trees, high temperature, high and frequent rainfall, and less fungicide applied in the grove. Postharvest *L. theobromae* infection and stem-end rot development can be enhanced by ethylene degreening treatment and excess delay and holding of the fruit before packing and treatment with appropriate postharvest fungicides.

2.4.18. *Ceratocystis paradoxa* (Black rot): *Ceratocystis paradoxa* (synonym- *Thielaviopsis paradoxa*) is a plant pathogen of the phylum Ascomycota. It is a soil-borne fungus that causes black root rot. It is characterized as the teleomorph or sexual reproduction stage of infection. This stage contains ascocarps, or sacs/fruiting bodies, which contain the sexually produced inoculating ascospores. These are the structures which are used primarily to survive long periods of time or overwinter to prepare for the next growing season of its host. Unfortunately, the sexual stage is not often seen in the natural field but instead the anamorph, or asexual stage is more commonly seen. This asexual stage name is *Thielaviopsis paradoxa* and is the common cause of Black rot or stem-end rot of its hosts. *T. paradoxa* forms thick walled resting spores, chlamydospores, and produces infective asexual spores or conidia. *T. paradoxa* produces two types of conidia; the first type is hyaline, cylindrical and small microspores of 8.5–12.7 × 2.5–4.7 μm in size (Fig. 22). The second type is noticeably bigger, brown colored, oval in shape (macrospores) and mainly responsible for the black color in the final stage of development of disease (Hewajulige and Wijesundera, 2014).

One of the most well-known diseases caused by *C. paradoxa* is black rot or stem-end rot of pineapple, but it can also infect tropical fruit plants such as banana and coconuts as well as sugarcane. The pathogen infects the fruits through wounds or other openings after harvest has already happened and the fruit is fresh. This is because the time to processing takes too long.

Kingdom: Fungi
Phylum : Ascomycota
Class : Sordariomycetess
Order: Microascales
Family: Ceraticystidiaceae
Genus: *Ceratocystis*

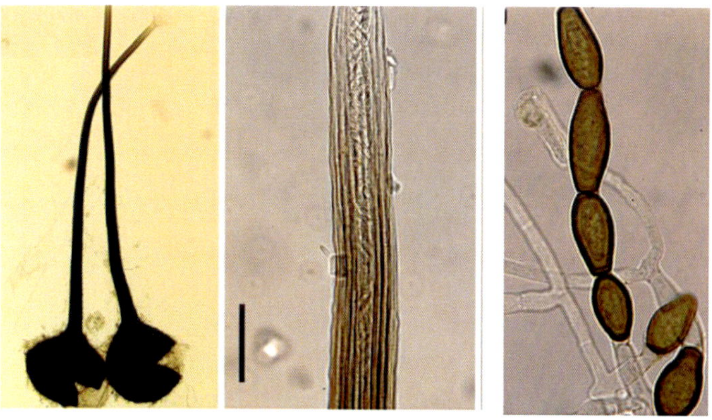

Fig. 22 : Perithecia **Hypha** **Conidia**

Symptoms and signs: Obvious black lesions on the fruit, the main infection part of the plant. If the pathogen infects the plant while fruits are still on it, they will prematurely drop. Other symptoms include discoloration of leaves as well as the seeds. The lesions on the fruit evolve to become soft rot spots that produce a heinous odor. The fruit can even get to the point of breakdown. The most important postharvest disease in pineapple is black rot, which can be found in all production areas around the world and the disease incidence could be as high as 70%. *Thielaviopsis paradoxa* infects through a wound or cracked surface. The flesh of infected fruit becomes soft and watery and later turns dark due to the growth of the fungal mycelium and its chlamydospore. Postharvest control can be achieved by careful harvest and handling to prevent wounding or bruising and treating the fruit with suitable fungicides within six to 12 hours after harvest. Other pineapple diseases are caused by fungi or bacteria already present on the fruit at preharvest. Infection usually occurs through wound or cracks on the surface or damage caused by insects. The incidence of these diseases after harvest is rather limited as compared to black rot and mostly cannot be controlled after harvest.

2.4.19. *Curvularia* rots: It is a serious postharvest disease in India. Injuries permit infection.*Curvularia* is a hypomycete, facultative mold fungus commonly found in soil. *Curvularia* defined by the type species *C. lunata* (Fruit rot of tomato) appears as shiny velvety-black, fluffy growth on the colony surface. *C. lunata* is distinguished by septate, dematiaceous hyphae producing brown, geniculate conidiophores. The poroconidia are curved slightly to distinctly, transversely septate, with an expanded third cell from the pore end of the conidium. The teleomorphic state of the type species *C. lunata* is *Cochliobolus lunatus*.

C. tuberculata is another major pathogen. Conidiophores in *C. tuberculata* arise singly or in groups, terminal or lateral on hyphae, stromata, and ascomata, simple or branched, straight or flexuous, smooth, pale to mid-brown, septate, up to 300 μm long, 2-7 μm thick. Conidia are straight, ovoid, obclavate or ellipsoidal, 3-5 (sometimes 8, but mostly 3) septate, intermediate cells brown to dark brown, end cells subhyaline to pale or dark brown, mature conidia tuberculate, 23-52 x 13-20 μm (Fig.23). Young conidia are smooth and subhyaline. First septum in the conidium is usually median, second septum often delimiting the basal cell but variations in septal formation may occur. Germination is both by bipolar and lateral germ tubes (Navi, *et al*, 1999). The species is heterothallic and the teleomorph is obtained by pairing monoconidial compatible isolates . Ascomata are black, globose, often borne on a columnar basal stroma or a flattened crust, 500-720 μm high, 400-490 μm wide, with a conical truncate beak up to 300 μm high, 115-140 μm wide at the base, often hairy in the globose part with simple, brown, septate hyphae. Conidiophores arise from the globose part of the ascoma but are not formed abundantly. Pseudoparaphyses are hyaline, filiform, and branched above. Asci are cylindrical, short-stalked, with 2-8 spored, vestigial bitunicate, 170-340 x 13.5 µm. *Curvularia* eragrostidis causes postharvest rot of pineapple in some regions (Ferreira *et al*, 2014).

Kingdom: Fungi
Phylum : Ascomycota
Class : Dothideomycetes
Order: Pleosporales
Family: Pleosporaceae
Genus: *Curvularia*

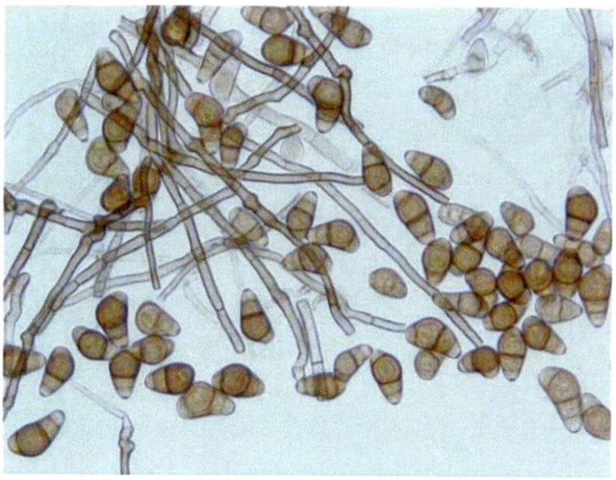

Fig. 23 : *C. lunata*-Conidiophores and conidia

Symptoms and signs : *Cochliobolus lunatus* is best known as the causative agent of seedling blight and seed germination failure in monocotyledon crops such as sugarcane, rice, millet, maize *etc*., *C. lunatus* also causes leaf spot on a wide variety of angiosperm hosts, where each lesion contains a sporulating mass of fungi at its center. The Clk1 gene plays an important role in fungal growth during the infection process, specifically conidiation, which is vital to the process of foliar infection. It also causes rots in crops like tomato, strawberry and bell pepper. The symptoms includ 1-to 2-cm water-soaked lesions on the tomato fruit which progress into light-brown to black lesions that enlarge gradually (3 to 4 cm). In advanced stages, the lesions become sunken, leathery, dark, and covered with fungal fructifications (Iftikhar *et al*, 2016). In bell pepper the disease symptoms appear near the apical portion of bell pepper fruit and rot with rapid discoloration and ultimate progression of internal decay.

2.4.20. *Sclerotium* rot (*Sclerotium rolfsii*): *Sclerotium rolfsii*, an omnivorous, soil borne fungal pathogen, causes disease onan extensive host range, at least 500 species in 100 families are susceptible. *Sclerotium rolfsii* produces abundant white mycelium on infected plants and in culture (Fig.24). Advancing mycelium and colonies often grow in a distinctive fan-shaped pattern and the coarse hyphal strands may have a somewhat ropy appearance. In culture, mycelium appears smooth at first, but some cultures may develop aerial mycelia that cover all or part of the culture after a few days. The fungus produces at least two types of hypae, large diameter (5 to 9 µm) main branch hyphae and smaller diameter (2 to 4 µm) branch hyphae. Cells are hyaline with thin cell walls and sparse cross walls. Main branch hyphae may have clamp connections on each side of the septum. In agar plate culture, sclerotia are not formed until the mycelium covers the plate. *In vitro* or *in vivo*, sclerotia begin as small tufts of white mycelium that form spherical sclerotia 0.5 to 1.5 mm in diameter. Sclerotia darken as they mature, becoming tan to dark brown in color. Young sclerotia often exude droplets of clear to pale yellowish fluids. Mature sclerotia are hard, slightly pitted, and have a distinct rind. Although most sclerotia are spherical, some are slightly flattened or coalesce with others to form an irregular sclerotium. *S. rolfsii* does not form asexual fruiting structures or spores.

The teliomorph of *S. rolfsii* (*Athelia rolfsii*) is rarely observed on hosts or in culture. *A. rolfsii* produces basidia on an exposed hymenium and basidia produce four haploid basidiospores. The appressed hymenium develops in small, thin, irregular patches. The clavate basidia are 4 to 6 µm x 7 to 14 µm; basidiospores are hyline, 1.0 to 5 µm x 5 to 12 µm (Rivard *et al*, 2010).

Kingdom: Fungi
Phylum : Basidiomycota
Class : Agaricomycetes
Order: Atheliales
Family: Atheliaceae
Genus: *Athelia*

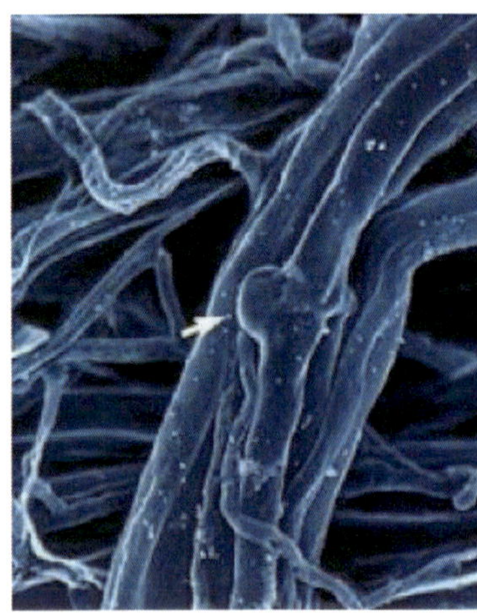

Fig. 24 : Mycelia and sclerotial bodies **Clamp connections**

Symptoms and signs: *Sclerotium rolfsii* causes disease on a wide variety of plants, including field, vegetable, fruit, and ornamental crops. Disease caused by *S. rolfsii* can be very destructive to numerous vegetable and fruit crops, especially tomato, pepper, melon, and watermelon. Signs of infection include the development of coarse white strands of mycelium growing in a fan-shaped pattern on lower stems, leaf litter, and soil. Even under dry conditions, at least a trace of white mycelium should be evident on the surface of the stem or crown. In some cases, mycelium will be found only underground. After 7 to 14 days, tan-to-brown, mustard-seed-sized (0.5 to 1.5 mm) sclerotia form on the mycelial mat. Although symptoms vary with the host affected, infection usually is restricted to plant parts in contact with the soil. Early symptoms consist of water-soaked lesions on crown and lower stem tissue. The disease usually is recognized by the yellowing and wilting of foliage, followed by a complete collapse of the plant. On tomato and pepper, dark water-soaked lesions on the lower stem at or near the soil surface are present and rapidly develop to completely girdle the stem. Infection of melon and watermelon is normally restricted to fruit lying in contact with the soil and runners where soil has been deposited over the vine. In watermelon, one or more runners may be affected and begin turning yellow; with cantaloupe, the melons are usually affected first. Fruit and other fleshy organs near the soil surface may become infected with *S. rolfsii*. Soft, water-soaked, sunken, slightly yellowish lesions develop (Mullen, 2001). These lesions quickly spread throughout most or all of the fruit, which will eventually become soft and collapse within 3 to 4 days of infection. The skin of the fruit often crack open and fine white mycelium and developing sclerotia spreads over the surface and quickly fills lesion cavities.

It is a major pathogen of peanut wherever the crop is grown. Southern blight is an important disease.. Strands of coarse white mycelium may also be visible on the surface of pods and thick rhizomorph-like strands are sometimes seen on below-ground portions of the plant. Occasionally, mycelium takes on a tufted appearance or forms sheaths around the main stem . In dry weather, strands or fans of mycelium are less apparent; instead, white to pale buff mycelium may be appressed to lower stems, looking like small patches of paint . Later, tan to brown round sclerotia 0.5 to 1.5 mm in diameter may be present on the mycelium, lesions, or soil, or on litter near infected plants. The signs of *S. rolfsii* are diagnostic of southern stem rot, but damage can occur even when signs are not evident (Keinath and DuBose, 2017). Early symptoms of southern stem rot include wilting or yellowing of individual branches or the main stem. Stem lesions usually are formed near the crown of the plant, usually starting at points of contact with the soil or previously infected stems. Lesions are the color of a brown paper bag and may have a canker-like appearance. As the disease progresses, lesions may extend into the crown of the plant, killing it. Colonized stems and crowns are dark brown, brittle, and shredded. Brown shredded pegs may rot without producing a pod, or pegs may separate from pods that have already formed, causing them to shed. Distinct lesions may be formed on pods during early stages of infection, but later pods are completely consumed with a dry brown rot. Rotted pods are thin and brittle and seeds may be stained, rotted, or absent. Pod shedding and rot is a major cause of yield loss. In potato, once harvested and stored, the infected tubers show spectacular symptoms of numerous sclerotia bursting out of rotten tubers.

2.4.21. *Choanephora cucurbitarum* (Wet Rot): *Choanephora cucurbitarum* is a plant pathogenic fungus causing fruit rots, flower rot and leaf blights on a variety of plants including okra,squash, pumpkin, pepper, pea and bean. This fungus is known to attack several other crops which include cereals such as millet, rice and sorghum. The fungus also causes pod blight known as wet rot, blossom blight and whisker rot . This disease is also common on squash and southern pea but occurs on the floral parts of many types of plants. It causes blossom blight, die back, wet rot and soft rot of stems or side shoots of chilli plants (Maeda *et al.*, 2010). The fungus is more successful under humid conditions and thrives best at a temperature of 25°C and relative humidity of about 100%. A temperature of about 31°C stimulates the production of large sporangia but unfavorable for conidia formation. The colonies of the *C. cucurbitarum* grown on potato dextrose agar are white at the early stage of growth with abundant aerial mycelium, but at later growth stages turn pale yellow. Sporangia and sporangiola are readily observed on the tip of upright sporangiophores, which are numerous on aerial mycelium (Fig. 25). Sporangia spherical in shape, 25-124 μm in diameter, white at first but turning dark brown at maturity. Sporangiola, containing only one spore, indehiscent, ellipsoid to broadly ellipsoid, 11-13 x 13-20 μm in size subtended by a short cylindrical pedicel. Sporangiospores, ellipsoid to broadly ellipsoid in shape, brown to pale brown, 8-12 x 20-24 μm in size and are characterized by longitudinal striatures on the wall surface and several fine hyaline appendages at both poles.

Kingdom: Fungi
Phylum : Zygomycota
Class : Zygomycetes
Order: *Mucor*ales
Family: Choanephoraceae
Genus: *Choanophora*

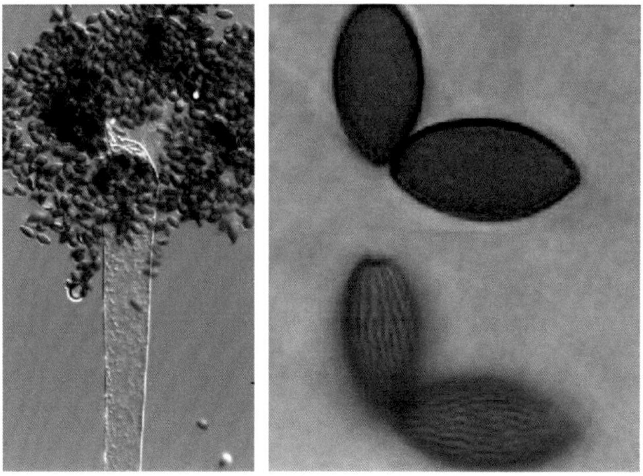

Fig. 25 : Sporangiophore bearing sporangiola Sporangiola

Symptoms and signs : Symptoms are usually found on the blossom end of the fruit. The initial symptom is a soft, wet rot. As the disease progresses, growth of the pathogen is easily found. A gray-to-black "whisker appearance will cover the rotted area of the fruit. *C. cucurbitarum* mostly attacks tissues that have been damaged by insects or mechanical means, or crops that are poorly adapted to a hot humid climate. The fungal strands are whitish-gray in color with tiny black spores on the tips that resemble pincushions. The fungus overseasons in the soil on plant debris and in the soil as dormant spores. When conditions are favorable, spores are released and dispersed to blossoms by wind, splashing water, bees, cucumber beetles or other insects (Melanie L. Lewis Ivey, 2015). Symptoms vary considerably on different crops, so it is important to be aware of the specific symptoms produced on the crop or crop produce in hand. On flower stalk rot of sugarbeet, dark brown and soft-rotted lesions appear on young flower stalks, and the stalks and flowers become covered with luxuriant metallic, clustered sporangiophores. Garden peas (*Pisum sativum*) attacked by *C. cucurbitarum* develop a dark-greenish rot of leaves, stems and pods with silk-like threads of sporangiophores growing densely on the lesions. Pole beans (*Phaseolus vulgaris*) show irregular, water-soaked leaf lesions, which later enlarged and became dry, resulting in severe blight. Peppers are attacked annually. The disease originates on declining flowers and spreads to the stems and leaves producing a wet rot and eventual dieback. Periwinkle in landscape plantings develops dark-grey spots on the flowers. These spots often coalesced, leading to blight. Crotalaria spectabilis develop 'twig blight'. On cowpea, potato and soyabean, symptoms begin as water-soaked lesions at the leaf margins and tips. These lesions became dry and turned olive-green to light brown. Numerous spiny, long sporangiophores developed during dry weather. This could result in total necrosis of entire plants.

2.4.22. *Phytophthora parasitica* (Rots): The oomycete *Phytophthora* parasitica is a very broad host range pathogen that causes destructive diseases of a wide variety of crop plants, nursery and ornamental plants and forest ecosystems. This pathogen can cause root rot, crown rot, fruit rot, leaf infection, and stem infection. Root rot symptoms are observed on tobacco, tomato, pineapple, watermelon, *etc*.,. Fruit rots occur ontomato, papaya, and brinjal. Onion shows a leaf and stem infection. In tobacco Black Shank affects the roots and basal stem area, but all parts of the plant can become infected. Damping-off symptoms can be observed in young seedlings. The first above ground symptom that will be observed is the wilting of plants, which leads to stunting. Roots will be blackened and decayed. In final stages of the disease the stem begins to turn black, hence the name 'Black Shank'. As this happens, tobacco leaves turn brown and become not marketable. Another symptom is disk-like appearance of the pith, although this is not a definitive symptom as it may also be the result of lightning strikes. On onion it causes the disease known as *Phytophthora* neck and bulb rot. Different stages of onion may be affected. Initially, tips of newly infected plants start to yellow and dry followed by softening of the "neck" of the plants that eventually fall over. Infected leaves may show grey lesions. Roots may become necrotic in late disease.

Phytophthora parasitica (syn. *Phytophthora* nicotianae) is a typical root pathogen. The hyphae are hyaline and aseptate, sometimes with hyphal swellings. *P. parasitica* produces asexual sporangia, zoospores and chlamydospores. The sporangia are difficult to be released from the hyphae, which is different from the released, airborne sporangia of *P. infestans*. Zoospores are produced by the sporangia and are wall-less cells with two flagella, which enable them to swim (Meng *et al*, 2014). Chlamydospores are thick-walled, multinucleated asexual spores, usually produced at the tips or in the middle of hyphae (Fig. 26). *P. parasitica* is predominantly het-erothallic, requiring A1 and A2 mating types for the production of oospores.

Kingdom: Fungi
Phylum : Heterokontophyta
Class : Oomycota
Order: Peronosporales
Family: Peronosporaceae
Genus: *Phytophthora*

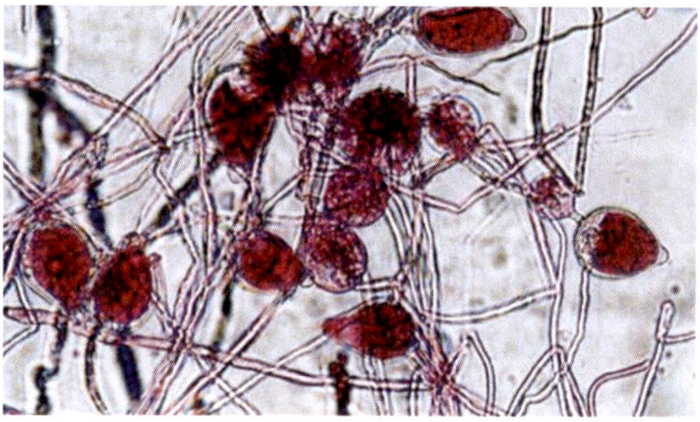

Fig. 26 : Mycelia with sporangia

Symptoms and signs: In most cucurbits, symptoms begin as brown, water-soaked lesions that quickly develop into soft, rotted areas on the part of the fruit that is in direct contact with the soil (Keinath *et al*, 2017). In watermelons, symptoms may begin at the blossom end of the fruit. During humid conditions, white, cottony pathogen growth may be produced and cover the rotted area of the fruit . The mycelial growth produced by the cottony leak pathogens may be confused with that produced by species of *Phytophthora*. However, the mycelium produced on rotting fruits infected by *Phytophthora* spp. is felt-like rather than cottony. In cucumbers, a dark green to brown blister may develop before the previously mentioned symptoms. In some crops, root rots are not initially noticeable. Small feeder roots are infected and larger roots become diseased with time as the fungus feeds on and destroys the root system. Infected plants may be stunted, lack vigor, or appear to be suffering from nutritional deficiency. On smaller plants, movement of the fungus from the roots to the collar causes rots at the soil line. These plants, wilt rapidly since water movement up the plant is inhibited. On larger plants, collar rots may progress into a lower branch and sections of the plant are gradually killed. Infected branches commonly have no leaves and only dead twigs remain. In tomato, the skin of the fruit stays firm at first. Under humid conditions white cottony growth of the oomycete develops on the surface of the fruit. Later, the fruit develops soft rots as bacteria and fungi invade. Water moulds survive in the soil as thick-walled resting spores called "chlamydospores". When conditions are right, the chlamydospores germinate and produce spores called "sporangia". Another spore is produced inside the sporangia called "zoospores", and these are capable of swimming short distances in the water between soil particles. If they reach the root or stem, they infect and the disease begins. The water mould spreads in rain splash, surface water, in water within the soil, and in soil on machinery and shoes over long distances.

2.4.23. *Pythium aphanidermatum* (Soft rots): It has a wide host range, and can have an economic impact on the cultivation several crops. The pathogen is an oomycete, related to algae. It is an aggressive species of *Pythium*, causing damping- off, root and stem rots, and blights of grasses and fruit. It is of economic concern on most annuals, cucurbits, and grasses. It is considered one of the water molds because it survives and grows best in wet soils. Warm temperatures favor the pathogen, making it an issue in most greenhouses. Soft rot is the most destructive disease in crops like ginger which results in total loss of affected clumps.

Pythium aphanidermatum overwinters in the soil as oospaore, hyphae and / or sporangia. Oospores can produce a germ tube and infect the plant directly, or, if the environment is favorable (that is an adequate amount of water is present), the oospore may produce sporangia, which in turn produce motile, biflagallete zoospores that swim to the host plant, encyst, and germinate. This infection can occur on seeds, which can rot, or produce a weak seedling. If it infects the roots of a seedling, the mycelium will grow throughout the plant tissue, releasing digestive enzymes which break down plant cell walls allowing the pathogen to absorb the nutrients, effectively killing the plant over time. *P. aphanidermatum* is a polycyclic disease. A polycyclic plant pathogen has several life cycles during a season, meaning it can reinfect the host plant, or travel to another plant. After infection, several things can happen that spreads the infection, *viz*., more asexual structures form, including sporangiophores and sporangia, which release more zoospores which can reinfect the host plant, or move to other plants ; there can also be sexual reproduction when two different mating hyphal types meet, creating an oogonium (Fig.27) and an antheridium. This results in genetic recombination and exists as an oospore- the original overwintering stage of the pathogen. Colonies grew with cottony aerial mycelium. Main hyphae were 5.7 to 9.6 μm wide. Globose sporangia consisting of terminal complexes of swollen hyphal branches were 11.4 to 18.3 μm wide. The average diameter of zoospores was 9.2 μm. The oogonia were globose and smooth, with a diameter of 21 to 33 μm (Li *et al*, 2014).

Phylum : Heterokontophyta
Class : Oomycota
Order: Peronosporales
Family: Peronosporaceae
Genus: *Phytophthora*

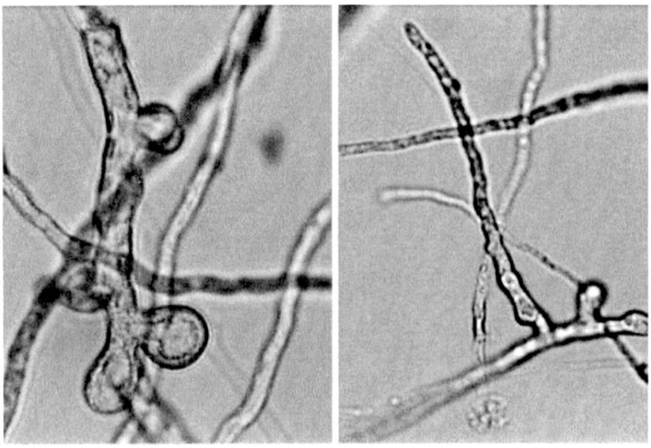

Fig. 27: Toruloid zoosporangia

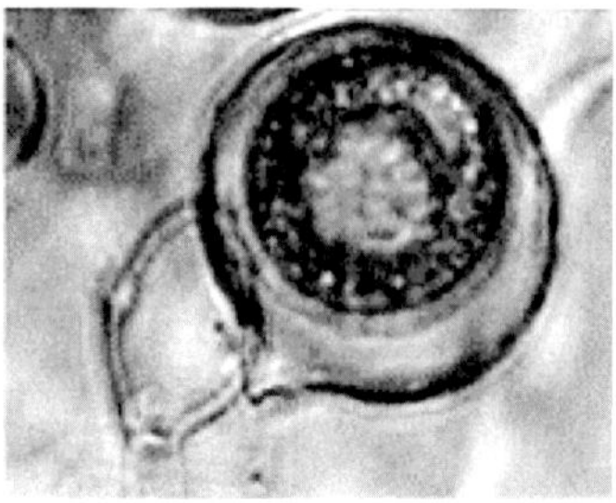

Aplerotic oospore

Symptoms and signs: The common name is water mould. The water mould lives in the soil on roots of ginger and other crops, the remains of dead plants, on or in "seed" (rhizomes) used for planting, and weeds. When conditions are not right for growth, the water mould produces thick-walled resting spores called "oospores" and these can stay alive in the soil or in the rhizomes for a long time, waiting for a susceptible host to stimulate their germination. The rot that occurs on ginger is typically a wet weather disease, influence by heavy rains after planting. Infection occurs when the water mould in the soil, or inside rots in the rhizome (the "seed" or planting piece), produces spores . Swimming spores are formed inside larger spores and, when released, find their way to the fine roots, the young buds (or "eyes") on the rhizome, or the junction of stem and rhizome. Damage to the rhizomes , roots and stems causes plants to yellow, and stems to collapse . At this stage, stems are easily pulled from the rhizomes (Anonymous, 2017). Spread of the water mould occurs when the spores swim short distance in the water between soil particles, or are carried for longer distances in rainwater through the soil or over the surface. Neighbouring plants are infected, and patches of yellowing plants develop in the field. Spread occurs over long distances in *Pythium*-infected rhizomes used for planting.

2.4.24. *Sclerotinia sclerotiorum* (White mold): It is a plant pathogenic fungus and can cause a disease called 'white mold' if conditions are conducive. *S. sclerotiorum* can also be known as cottony rot, watery soft rot, stem rot, drop, crown rot and blossom blight. White mold affects a wide range of hosts. It is known to infect 408 plant species. Its diverse host range and ability to infect plants at any stage of growth makes white mold a very serious disease. The fungus can survive on infected tissues, in the soil, and on living plants. It affects young seedlings, mature plants, and fruit in the field or in storage. White mold can spread quickly in the field from plant to plant. It can also spread in a storage facility throughout the harvested crop. Some crops it affects commonly are soybeans, green beans, sunflowers, canola, and peanuts (Bennett, *et al*, 2014).

A key characteristic of this pathogen is its ability to produce black resting structures known as sclerotia and white fuzzy growths of mycelium on the plant it infects. These sclerotia give rise to a fruiting body in the spring that produces spores in a sac which is why fungi in this class are called sac fungi (Ascomycetes). This pathogen can occur on many continents and has a wide host range of plants. When *S. sclerotiorum* is onset in the field by favorable environmental conditions, losses can be great and control measures should be considered. Common hosts of white mold are herbaceous, succulent plants, particularly flowers and vegetables. It can also affect woody ornamentals occasionally, usually on juvenile tissue. Sclerotia formed by *S. sclerotiorum* serve as resting structures that secure fungal survival in soil for prolonged periods (Mohammad Akram Husain and Choudhary, 2018). The pathogen produces hyaline aerial mycelium, branches well developed and cottony, consisting of closely septate hyphae which are both inter and intra cellular. The hyphae are 2.0 to 11.5 µm in width and contain dense granular protoplasm. The sclerotia are round to irregular in shape in culture and measure 1.5 to 7 mm in width and 2 to 15 mm in length. Cup-shaped apothecia develop on germination of sclerotia. Apothecia are brown in colour, round , funnel shape or globate type (Fig. 28). The length of apothecia measure from 5 to 21 mm, whereas diameter ranges from 0 to 7 mm with number ranging from 1 to 9 per Sclerotium.

Kingdom: Fungi
Phylum : Ascomycota
Class : Leotiomycetes
Order: Helotiales
Family: Sclerotiniaceae
Genus: *Sclerotinia*

Fig. 28 : Germinating ascospores Apothecium Mycelia and sclerotial bodies

Symptoms and signs: White mold can affect their hosts at any stage of growth, including seedlings, mature plants, and harvested products. It can usually be found on tissues with high water content and in close proximity to the soil. One of the first symptoms noticed is an obvious area of white, fluffy mycelial growth. Usually this is preceded by pale to dark brown lesions on the stem at the soil line. The mycelium then cover this necrotic area. Once the xylem is affected, other symptoms occur higher up in the plant. These can include chlorosis, wilting, leaf drop, and death quickly follows. On fruits, the initial dark lesions occur on the tissue that comes in contact with the soil (Bennett, *et al*, 2014). Next, white fungal mycelium covers the fruit and it decays. This can occur when the fruit is in the field or when in storage.

During late summer/ early fall the fungus will produce a survival structure called a sclerotium either on or inside the tissues of a host plant. The following spring the dormant sclerotia will germinate to produce fruiting bodies called apothecia, which are small, thin stalks ending with a cup-like structure about 5-15mm in diameter. The cup of the apothecium is lined with asci, in which the ascospores are contained. When the ascospores are released from the asci, they are carried by the wind until they land on a suitable host. The ascospores will then germinate on the host and begin to invade the host's tissues via mycelium, causing infection. *S. sclerotiorum* is capable of invading nearly all tissue types including stems, foliage, flowers, fruits, and roots. Eventually white, fluffy mycelium will begin to grow on the surface of the infected tissues. At the end of the growing season, *S. sclerotiorum* will once again produce sclerotia. The sclerotia will then remain on the surface of the ground or in the soil, on either living or dead plant parts until the next season. The lifecycle will then continue respectively.

2.5 Symptoms of Postharvest Diseases of Fruits

I. Pome fruits

1. Apple

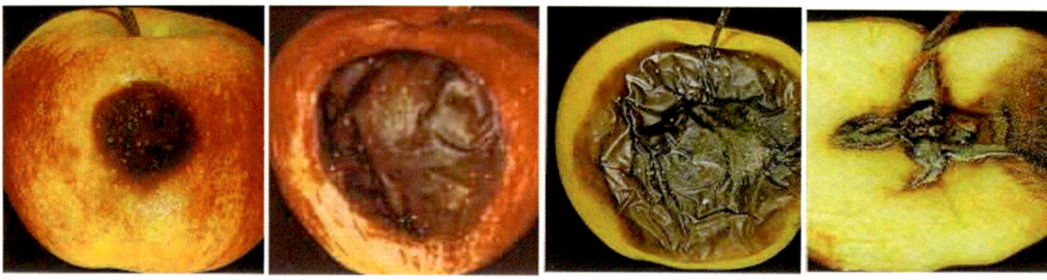

Rounded brown spot **Enlarged brown spot** **Rotten tissues** **Internal rotten tissue**

Alternaria rot (*Alternaria alternata*)

Fruit rind become watery **Fruit rot** **Affected fruits**

Blue mould (*Penicillium italicum*)

Internal watery tissue **External watery spots**

Green mold rot (*Penicillium* italicum)

Black circular lesions on skin

Enlarge spots

Ascospores

Apple scab (*Venturia inaequalis*)

Brown spot

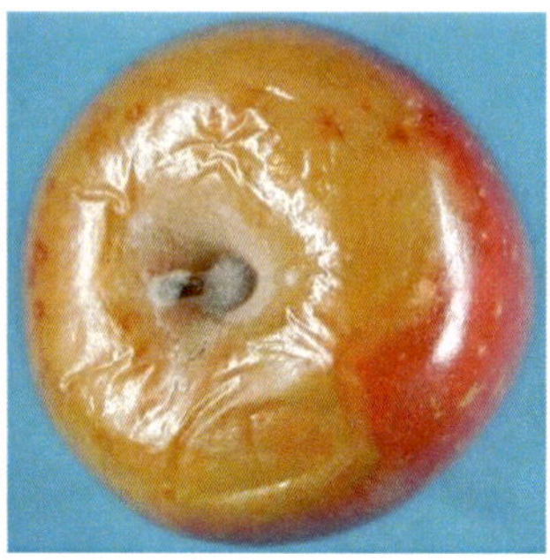

Watery mass

Grey mould

Grey mold (*Botrytis cinerea*)

Black concentric rings

Decayed brown spots

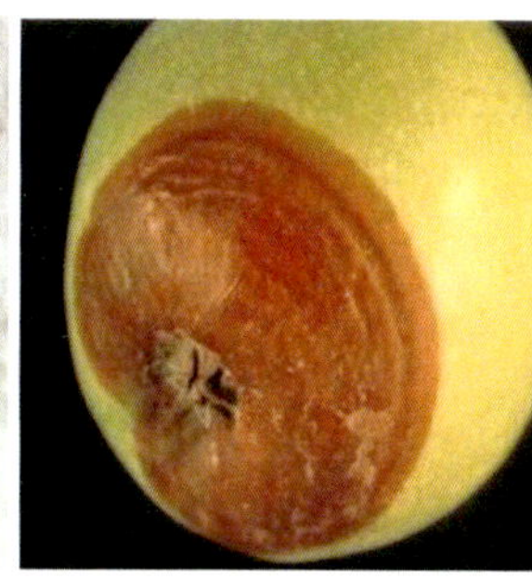

Reddish border

Black rot (*Botryosphaeria obtuse*)

Bitter rot (*Glomerella cingulata*)

White rot

Dark brown spot

White rot (*Botryosphaeria dothideaI*)

Circular and medium brown fruit

Decayed fruits

Brown rot (*Monilinia fructicola*)

Blister spot (***P. syringae*** pv. papulans)

Rot (***Mucor pyriformis***)

2. Pear

Cladosporium herbarum

Mucor piriformis

Coprinopsis psychromorbida

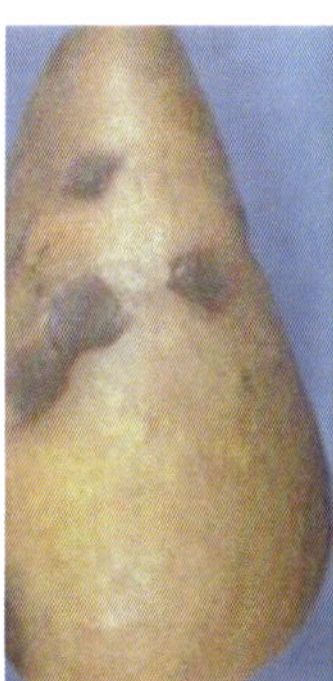

Cadophora malorum

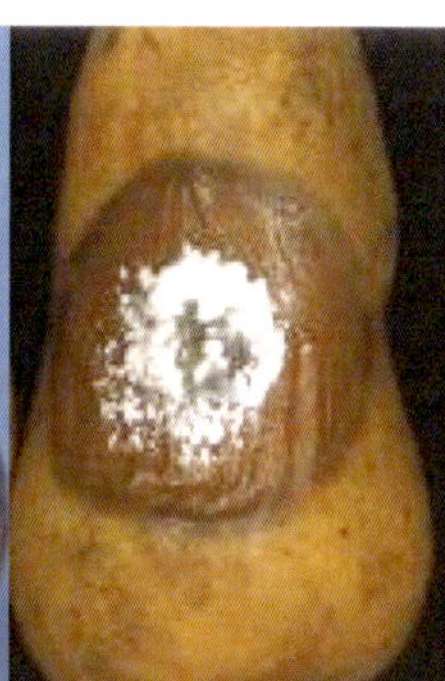

***Penicillium* spp**

Phytophthora cactorum

Alternaria alternata

II. Banana

Black bunch **Immature fruits** **Fruit turns black and shrivel**

Anthracnose (*Colletotrichum musae*)

Initial Symptom **Darkening of peduncle** **Fungus**

Crown rot (*Fusarium roseum, Lasidiplodia theobromae*, *Deightonialla torulosa*)

Blackening at the tip **Cigar end appearance** **Powdery grey conidia**

Cigar-end rot (*Verticillium theobromae*)

III. Citrus

Early stage

Grey lesions

Dark lesions

Fruit anthracnose

Anthracnose: *Colletotrichium gloeosporioides*

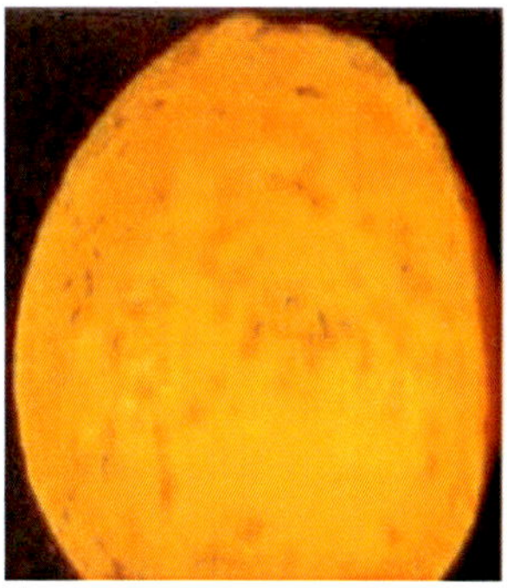

Early stage

Dark brown collapsed lesions

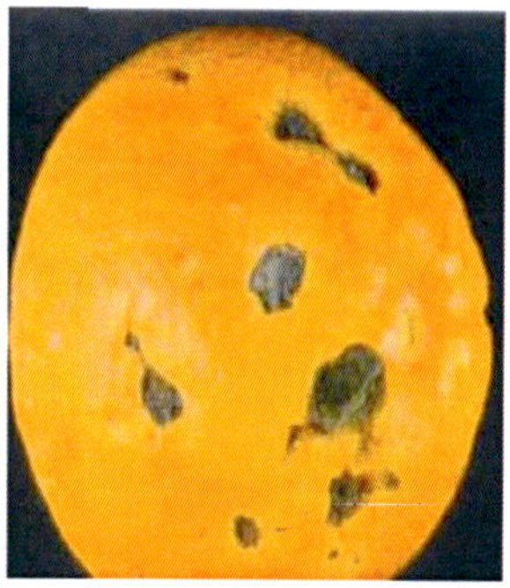

Black spots

Septoria spot (*Septoria depressa*)

Blue mold (*Penicillium italicum*) & Green mold (*P. digitatum*)

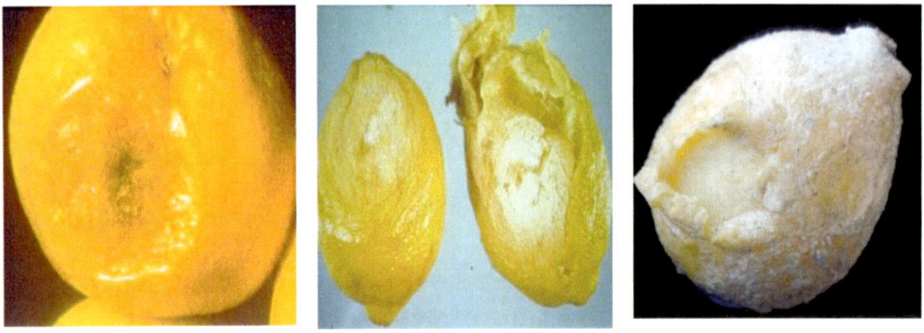

Soft fruit part **Watery decay** **Decayed fruit**

Sour Rot (*Galactomyces citri-aurantii* formally, *Geotrichum candidum*)

IV. Grapes

Anthracnose/ Bird's eye spot (*Elsinoe ampelina*)

***Botrytis* bunch rot or Gray mold (*Botrytis cinerea*)**

Blue mold rot (*Penicillium expansum*)

***Rhizopus* rot (*Rhizopus nigricans*)**

V. Guava

Initial symptom

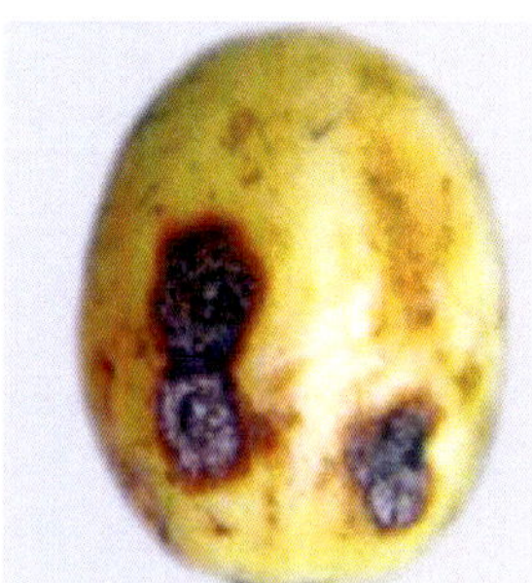

Circular black spots

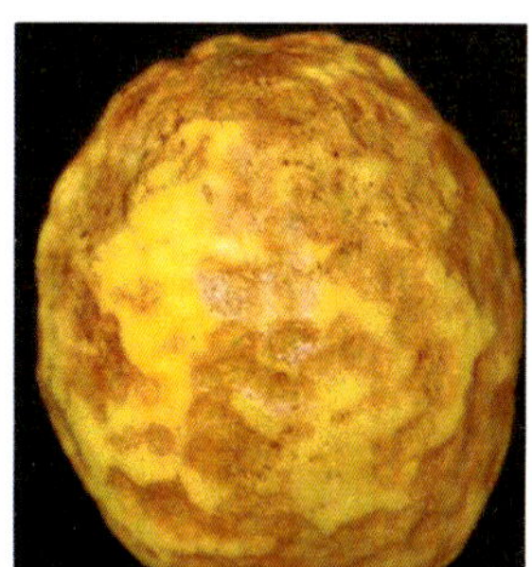

Corky fruit

Anthracnose (*Gloeosporium psidii*)

***Phomopsis* fruit / husk rot (*Phomopsis* psidii)** **_Botryodiplodia_ rot (*B. theobromae*)**

Brown color necrosis **Necrosis on epidermis** **Fruit canker**

Fruit Canker (*Glomerella psidii & Pestalotia psidii*)

VI. Mango

Sunken black spots **Severely affected fruits** **Infected fruits**

Anthracnose: *Colletotrichum gloeosporioides*

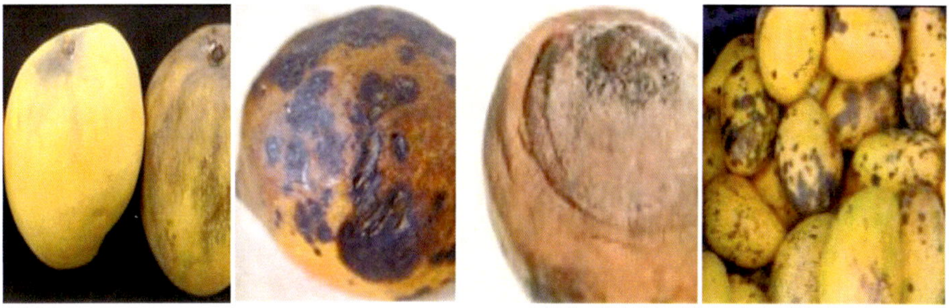

Darken pericarp **Circular black patch** **Severe infection** **Infected fruits**

Stem end rot (*Lasiodiplodia theobromae*)

***Diplodia* stem end rot (*Diplodia* natalensis)**

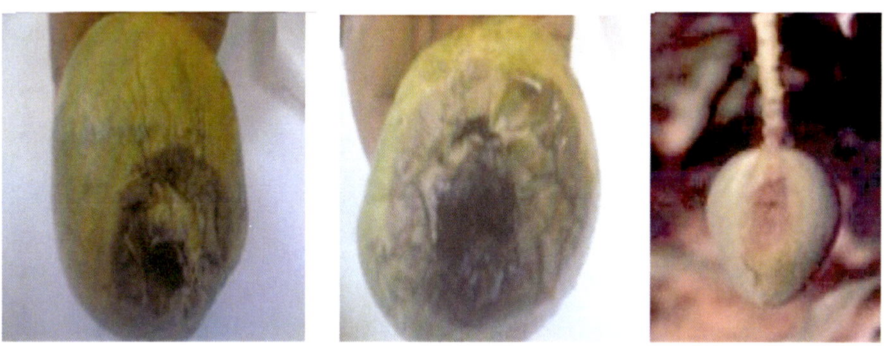

Yellowing of surface **Blackish fungal growth** **Infected fruit**

Black mold rot (*Aspergillus niger*)

Black soft rot (*Phomopsis mangiferae*)

Brown spot (*Pestalotia mangiferae*)

VII Papaya

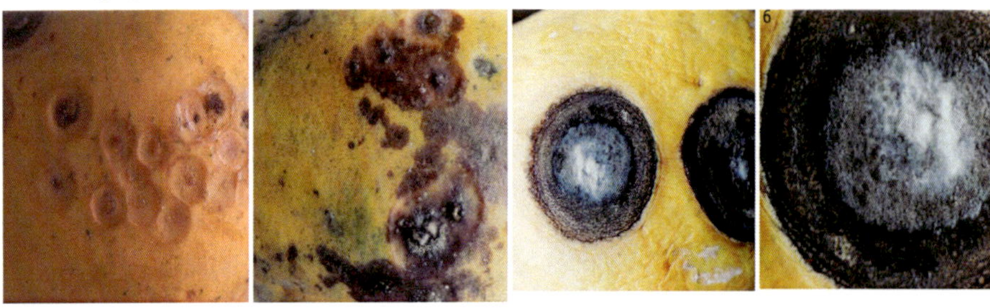

Circular infection ***Brown discolouration*** ***Anthracnose*** ***Mycelial growth***

Anthracnose (*Gloeosporium papaya*)

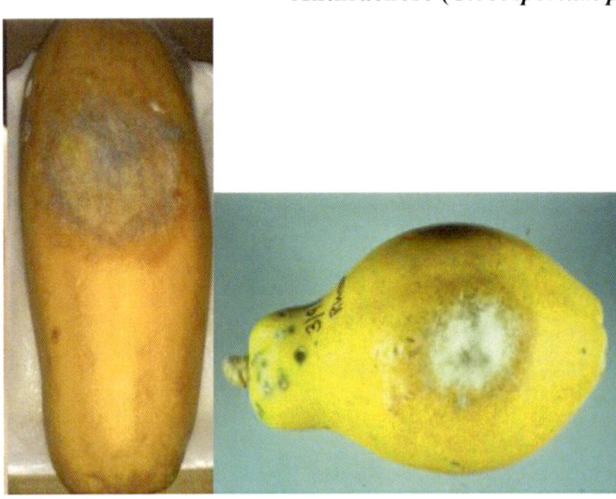

Fruit rots : *Rhizopus* stolonifer

Alternaria alternata

Rot **Water- soaked spots**

Phomopsis **fruit rot (*Phomopsis caricae-papayae*)**

Stem end rot **Water- soaked margin** **Black pycnidia on fruit**

Stem end rot (*Botryodiplodia theobromae*)

VIII. Strawberry

Grey mold (*Botrytis cinerea*)

Fruit decay (*Penicillium* spp.)

Anthracnose (*Colletotrichum acutatum*)

Soft rot (*Rhizopus stolonifer*)

IX. Avocado

Anthracnose (*Colletotrichum gloeosporioides*)

Stem end rot (*Dothiorella* spp.)

X. Pineapple

Water blister of pine apple (*Thielaviopsis paradoxa*)

Fruit let core rot / Green eye of pine apple (*Penicillium funiculosum*)

Yeasty rot of pine apple (*Saccharomyces* spp.)

Bacterial brown rot (*Erwinia ananas*)

XI. Passion fruit

Anthracnose (*C. gloeosporioides*)

Soft rot (*Fusarium oxysporum*)

XII. Amla (*Phyllanthus acidus*)

Soft rot (*Phomopsis* phyllanthi)

Blue mould (*Penicillium citrinum*)

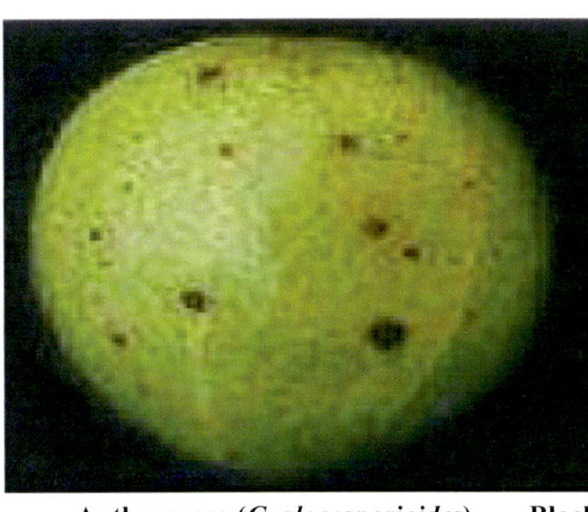

Anthracnose (*C. gloeosporioides*)

Black soft rot (*Syncephalostrum racemosum*)

XIII. Other fruits

Pomegranate **Kiwi**

Grey mold (*B. cinerea*)

Soft rot in Sapota (*Pestalotiopsis mangiferae*)

Black cottony rot in jack (*Rhizopus nigricuns*)

2.6 Symptoms of Postharvest Diseases of Vegetable and Other Crops

I. Tomato

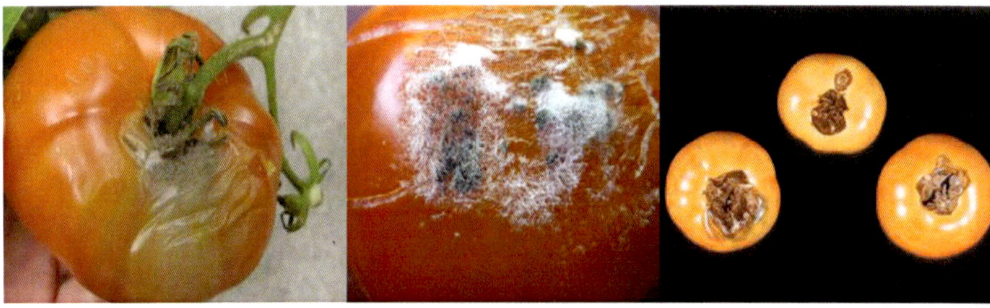

Grey Mold (*Botrytis cinerea*) Rot (*Rhizopus stolonifer*) Early blight (*Alternaria solani*)

Anthracnose (*Colletotrichum coccoides*, *C. gloeosporoides*, *C. dematium*)

Bacterial soft rot (*Erwinia carotovora* pv *carotovora*)

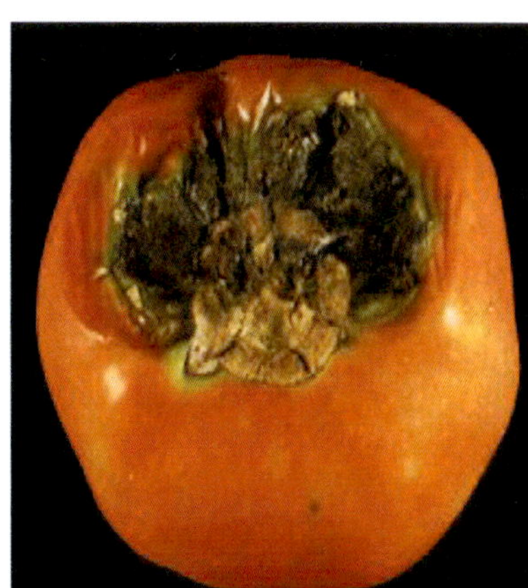

***Phoma Rot* (*P. destructive*)**

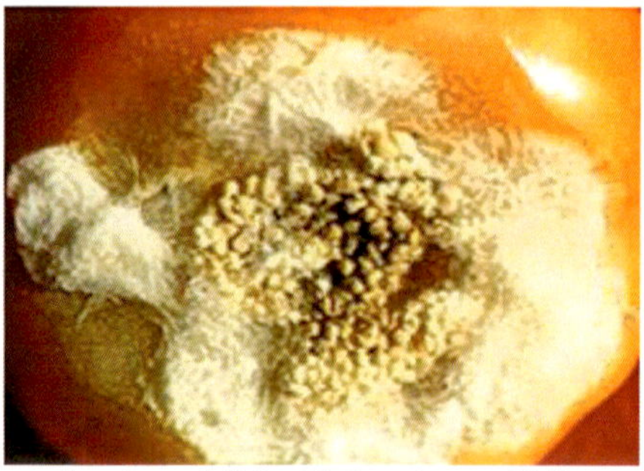

Southern blight (*Sclerotium rolfsii*)

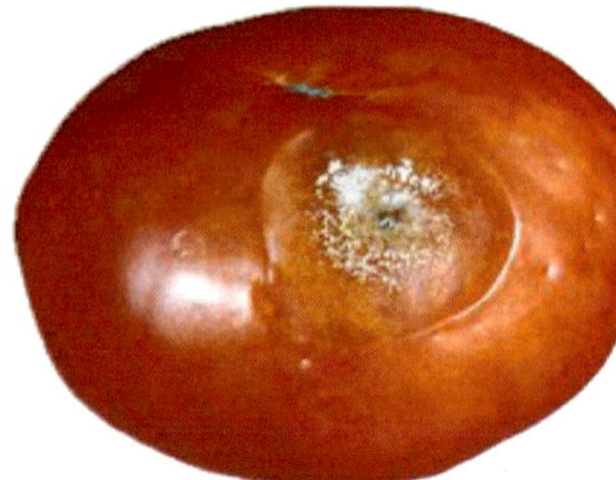

Fusarium* rot *(F. oxysporum* f. sp. *lycopersici)

II. Chilli

Alternaria rot (A. alternata)

Anthracnose (*Colletotrichum capsici*)

Bacterial soft rot (*Erwinia carotovora* sub sp. *carotovora*)

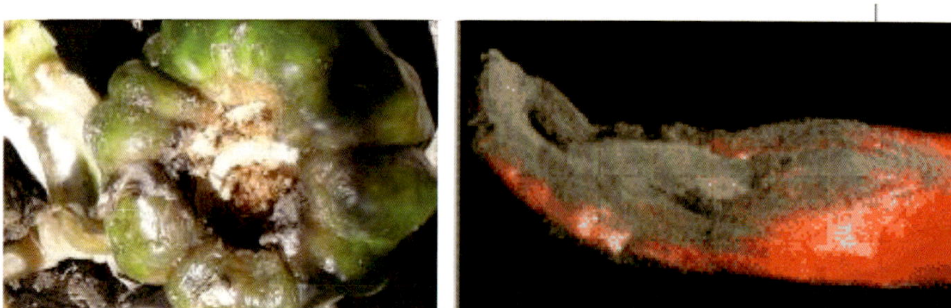

Grey mold (*Botrytis cinerea*)

Fruit rot (*Phytophthora infestans*)

III. Bhendi

Brownish pods (Internal tissue with mycelia)

***Rhizoctonia* rot (*Rhizoctonia* solani)**

Inner syndrome of pods (Necrotic tissues)

***Alternaria* rot (*Alternaria alternata*)**

IV. Brinjal

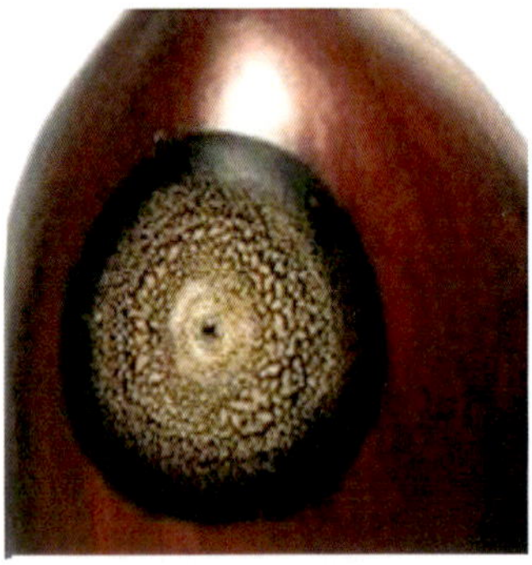

Anthracnose **Decayed fruit Fruit rot (*Phomopsis vexans*)**

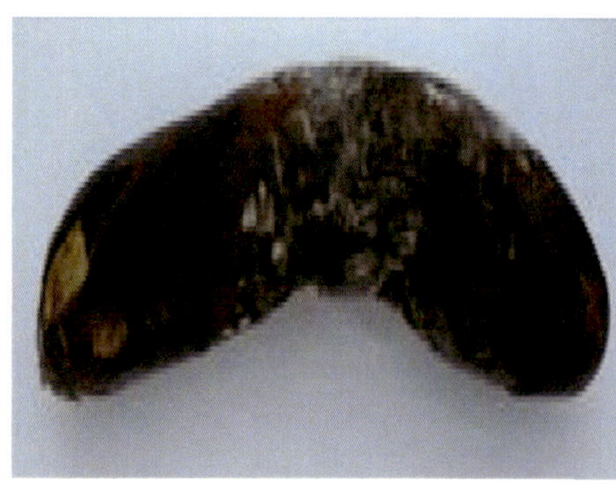

Fruit rot (*Phytophthora nicotianae* var *nicotianae*)

V. Carrot

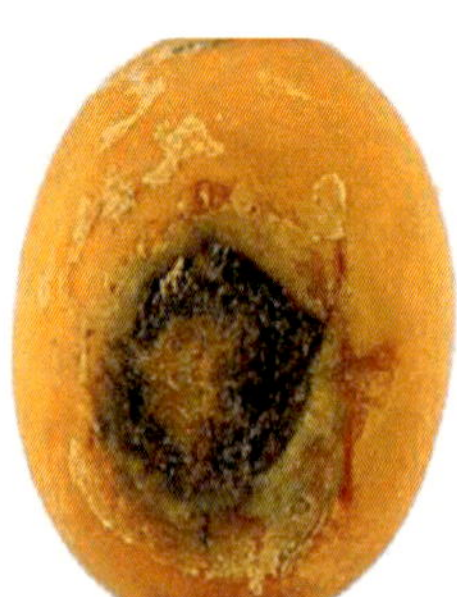

Sour rot (*Geotrichum candidum*) **Sunken lesions** **Crown rot (*Rhizoctonia solani*)**

Cottony soft rot / White mould (*Sclerotinia* sclerotiarum)

Dark brown rings

White mycelia on carrot

***Phytophthora* root rot / Rubbery brown rot (*Phytophthora megasperma*)**

Root dieback (*Pythium debaryanum*, *P. aphanidermatum*)

Crater rot (*Rhizoctonia carotae*)

Bacterial soft rot (*E. carotovora* sub sp. *carotovora*)

VI. Cucurbits

Anthracnose (*Colletotrichum lagenarium*)

Gummy stem blight & black spot (*Didymella bryoniae*)

Choanephora wet rot (*C. cucurbitarum*)

Fruit rot (*Pythium aphanidermatum*)

Belly rot (*Rhizoctonia* solani)

***Diplodia* fruit rot (*Diplodia natalensis*)**

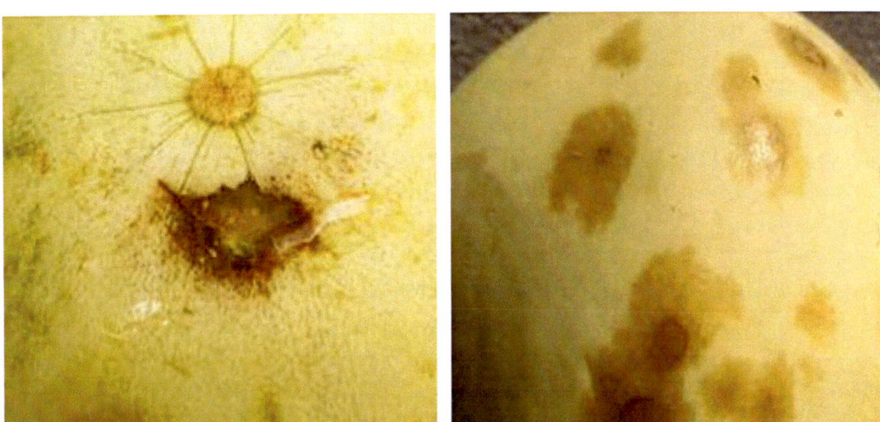

Bacterial soft rot (*Erwinia carotovora*)

Cucumber

Squash

Pumpkin

Bottle gourd rot

***Phytophthora* Fruit Rot (*Phytophthora capsici*)**

VII. Garlic

White rot (*S.cepivorum* , *S rolfsii*)

Black mould (*A. niger, A. alliaceous*)

Basal rot
(*Fusarium oxysporum* f. sp. *garlic*)

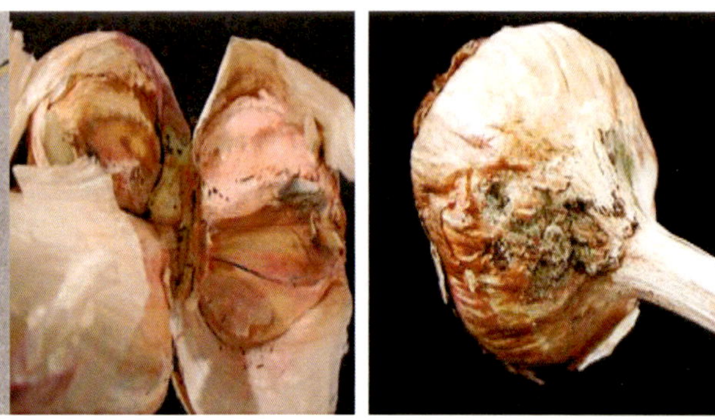

Blue mold rot (*Penicillium* spp.)

Pink rot (*Pyrenochaeta terrestis*)

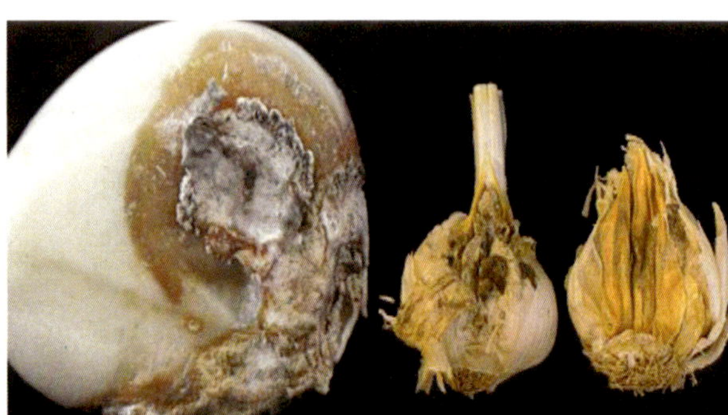

Neck rot (*Botrytis allii*)

VIII. Onion

Neck Rot (*Botrytis allii*)

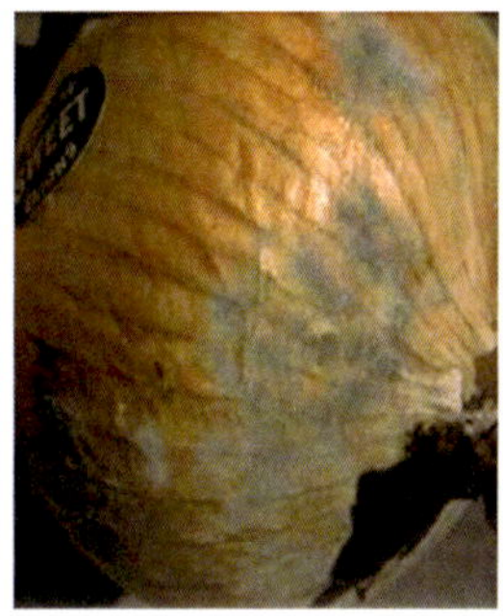

Blue mold (*Penicillium* spp.)

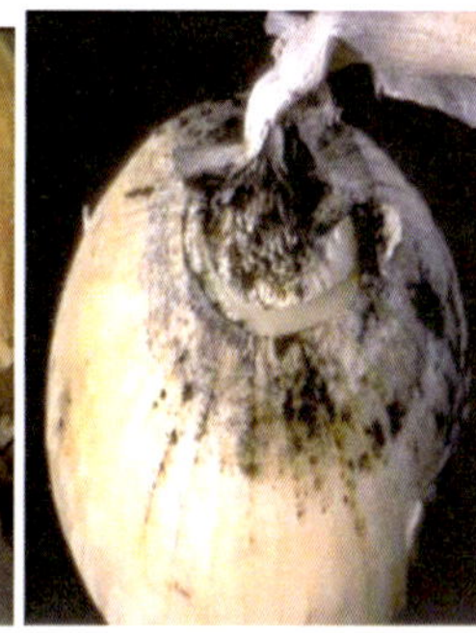

Black mold (*A. niger*)

White rot / Dry rot (*Sclerotium cepivorum*)

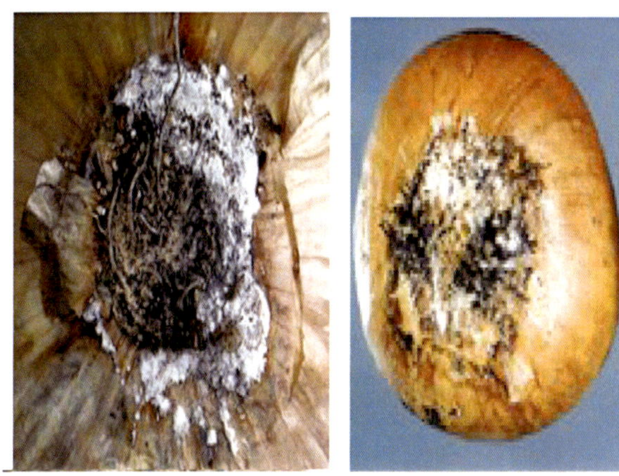

Bulb rot/ basal rot (*F.oxysporum* f.sp. *cepae*)

Black stalk rot (*Stemphylium botryosum*)

Purple blotch/scald (*Alternaria porri*)

IX. Potato

Black leg & soft rot (*Erwinia* spp.)

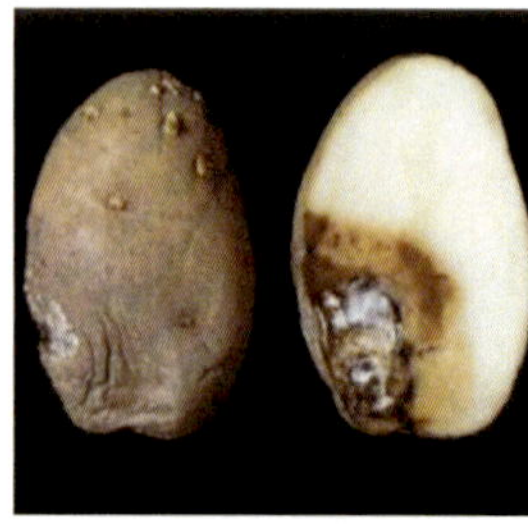

Dry rot (*F.solani* var. *coeruleum*)

Brown rot (*R. solanaceraum*)

Late blight (*Phytophthora* infestans)

Scab (*Streptomyces scabies*)

Potato wart (*Synchytrium endobioticum*)

Silver scurf (*Spondyocladium atrovirens*)

***Sclerotium* rot (*S.rolfsii*)**

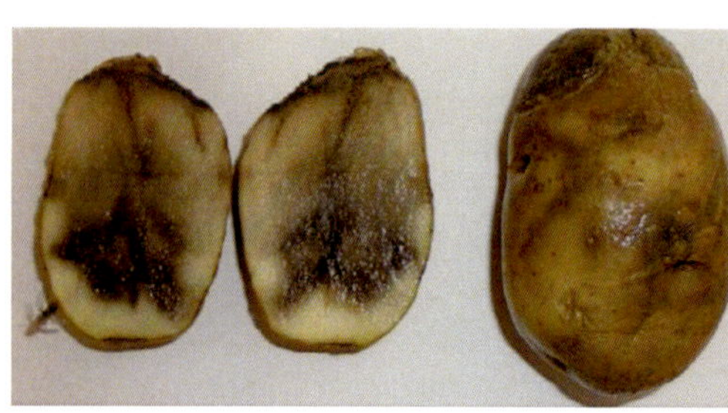

Charcoal rot (*Macrophomina phaseolina*)

X. Beans

Grey mold of beans (*Botrytis cinerea*)

White mold (*S. sclerotiorum*) Cottony leak (*Pythium* spp.) *Sclerotium* rot (*S. rolfsii*)

XI. Crucifers

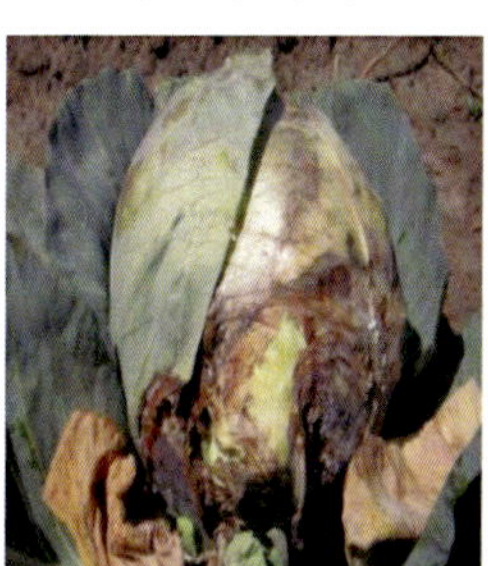

Bacterial soft rot **White mould (*Sclerotinia* spp.)** **Watery soft rot**

Grey mold

***Alternaria* rot**

Bacterial soft rot

XII. Maize

Head smut

Stalk /Ear rot

***Diplodia* ear rot**

XII. Sorghum

Grain smut

Loose smut

Long smut

Head smut

XIV. Wheat

Loose smut

Karnal bunt

Glume Blotch

Bacterial streak /Black chaff

***Fusarium* head blight /scab**

Flag smut

XV. Rice

Blast

Bacterial blight

False Smut

Sheath rot

Brown spot

XVI. Bajra

Downy mildew

Ergot

Smut

XVII. Cotton

Anthracnose

Alternaria blight

XIII. Sugar cane

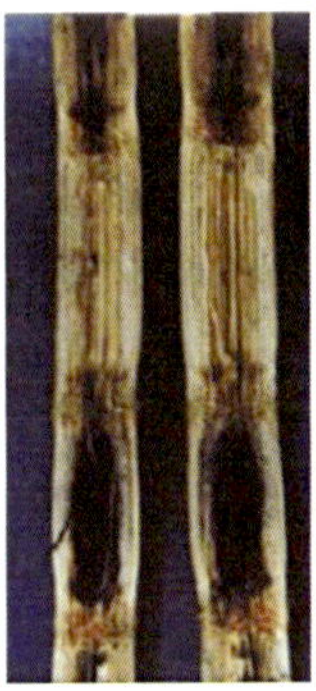

Pineapple disease

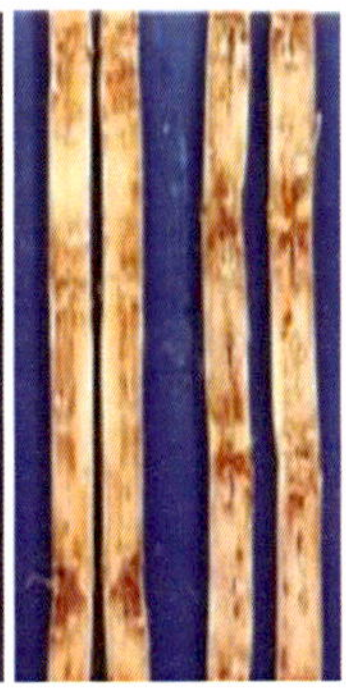

Red rot

Whip smut

Grassy stunt

XIX. Ground nut

Molds (*Aspergillus* spp.,)

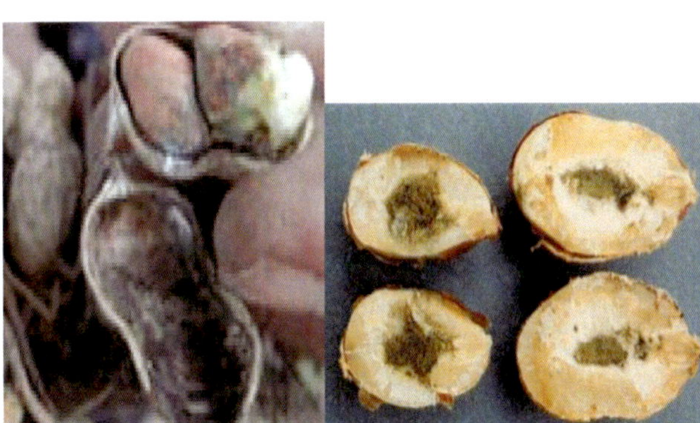

Molds (*Aspergillus* spp.,)

Charcoal rot

XX. Sunflower

Alternaria blight

Downy mildew

XXI. Pea

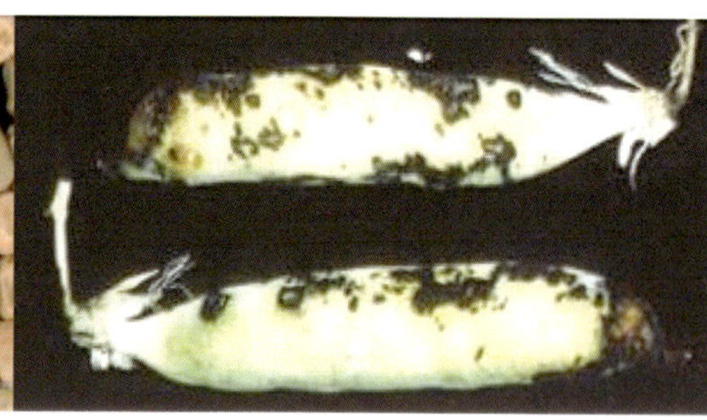

Aschochyta blight

XXII. Soybean

Anthracnose Pod & stem blight

Purple seed stain

XXIII. Chick pea

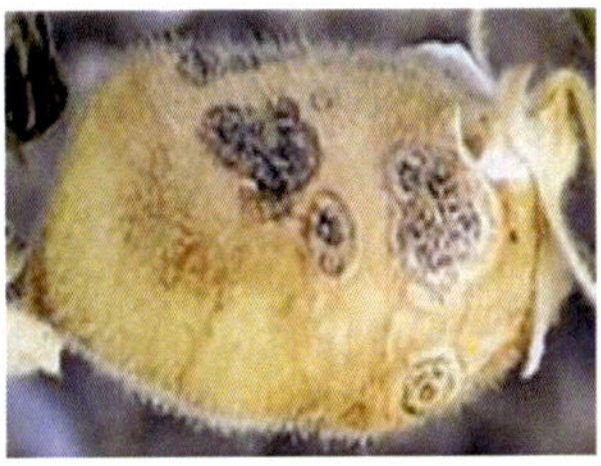

Aschochyta blight **Wilt of seedlings** **Grey mold**

References

Abu Bakar, A.I., Nur Ain Izzati, M.Z. and Umi Kalsom, Y., 2013. Diversity of Fusarium Species Associated with Postharvest Fruit Rot Disease of Tomato. *Sains Malaysiana,* **42:** 911–920

Adel A. Kader and Rosa S. Rolle, 2004. The role of postharvest management in assuring the quality and safety of horticultural produce. FAO Agricultural Services Bulletin 152, Food And Agriculture Organization of The United Nations, Rome.

Ailsa D. Hocking , 2003. Microbiological facts and fictions in grain storage. *In: Stored grain in Australia.* (E.J. Wright, M.C. Webb and E. Highley, (Eds.,). Proc. of the Australian Postharvest Technical Conference, Canberra, 25–27 June 2003. CSIRO Stored Grain Research Laboratory, Canberra.

Aktaruzzaman, M., Afroz, T., Lee, Y.G., 2 018. Postharvest anthracnose of papaya caused by *Colletotrichum truncatum* in Korea. *Eur. J .Pl. Pathol.,* **150**: 259.

An, B., Li, B. Q., Li, H., Zhang, Z. Q., Qin, G. Z., Tian, S. P. 2016. Aquaporin8 regulates cellular development and reactive oxygen species production, a critical component of virulence in *Botrytis cinerea*. *New Phytologist*, **209**:1668–1680.

Anonymous, 2017. Pacific Pests and Pathogens - Fact Sheets-Ginger soft rot. Mini Fact Sheet Index, Autralian Cnetre for International Agricultural Research.

Anonymous, 2018. "*Phytophthora*" *(HTML). NCBI taxonomy. Bethesda, MD: National Center for Biotechnology Information.*

Anonymous, 2019. Phytophthora Root and Crown Rot. Agriculture and Natural Resources, University of California, UCIPM,

Ashiq, S.,2015. Natural occurrence of mycotoxins in food and feed: Pakistan perspective. *Compr. Rev. Food Sci.& Food Safety,* **14** :159–175.

Ashely Bennett, Michelle M.Ponder and Julia Garcia-Diaz., 2018. *Phoma* Infections: Classification, Potential Food Sources, and Their Clinical Impact. *Microorganisms*, 6: 58

Baggio, J., Gonçalves, F., Lourenço, S., Tanaka, F., Pascholati, S., & Amorim, L..,2016. Direct penetration of *Rhizopus stolonifer* into stone fruits causing Rhizopus rot. *Plant Pathology*, **65:** 633-642.

Bennett, J. Michael; Rhetoric, Emeritus; Hicks, Dale R.; Naeve, Seth L.; Bennett, Nancy Bush (2014). The Minnesota Soybean Field Book (PDF). St Paul, MN: University of Minnesota Extension. p. 81. Retrieved 21 February 2016.

Benny D. Bruton, 1994. Mechanical Injury and Latent Infections Leading to Postharvest Decay. *Hort. Sci*., **29**: 747

Chalkley, D., 2019. Systematic Mycology and Microbiology Laboratory, ARS, USDA. . Invasive Fungi. Asian/European brown rot of rosaceae - *Monilinia fructigena*, USDA Agricultural Research Service

Charles Woloshuk and Ernesto Moreno Martinez, 2012. Molds and Mycotoxins in Stored Products. *Tech.Bull.,* K-State Research and Extension, Kansas State University

Cubeta, M. A., and R. Vilgalys, 2011. "Population Biology of the Rhizoctonia Solani Complex." Population Genetics of Soilborne Fungal Plant Pathogens 87.4 (1997): 480–84. The American Phytopathological Society. Web. 4 Oct. 2011.], The American Phytopathological Society.

Domsch, K.H., W. Gams, W.and Andersen, T.H., 1980. *Compendium of soil fungi* (2nd ed.). London, UK: Academic Press

e Hoog, G. S., J. Guarro, J. Gene, and M. J. Figueras. 2000. Atlas of Clinical Fungi, 2nd ed, vol. 1. Centraalbureau voor Schimmelcultures, Utrecht, The Netherlands.

Ferreira, A. P. S., Pinho, D. B., Machado, A. R. and Pereira O. L., 2014. First Report of *Curvularia eragrostidis* Causing Postharvest Rot on Pineapple in Brazil. *American Phytopathol.*, **98:** 1,277.2 - 1,277.2

Fillinger, S., Elad, Y.,2016. *Botrytis*-the Fungus, the Pathogen and its management in agricultural systems. Pub. Springer, New York

Gat, T., Liarzi, O., Skovorodnikova, Y., and Ezra, D., 2012. Characterization of *Alternaria alternata* causing black spot disease of pomegranate in Israel using a molecular marker. *Plant Dis.* **96**:1513-1518

Ghoshdastider U., Wu, R., Trzaskowski, B., Mlynarczyk, K., Miszta, P., Gurusaran, M., Viswanathan, S., Renugopalakrishnan, V. and Filipek, S.,2015. Nano-Encapsulation of Glucose Oxidase Dimer by Graphene. *RSC Advances*. **5** : 13570–13578.

Haggag, W.M., 2014. First report of Phomopsis blight of eggplant in Egypt. *Int. J. Agri. Innov. Res.*, **2**: 835

Hassan, A. and Siriphanich,J., 2011. Postharvest biology and technology of tropical and subtropical fruits. Mangosteen to White Sapote. Woodhead Publishing Series in Food Science, *Techn. & Nutrin.*, 536 p.

Hewajulige, I.G.N and Wijesundera,R.L.C., 2014. *Thielaviopsis paradoxa, Thielaviopsis basicola* (Black Rot, Black Root Rot),. P 287-308. *In: Postharvest Decay.* Academic Press,

Hinton-Sheley Phoebe, 2018. What is Aspergillus niger?. News-Medical. Retrieved on December 24, 2019 from https://www.news-medical.net/life-sciences/What-is-Aspergillus-niger.aspx

Humaira Rizwana, Iffat Siddiqui and Najat Bukhary, 2012. A postharvest disease of *Mangifera indica* fruit caused by *Pestalotiopsis mangiferae*, in Saudi Arabia. *African J. Microbio. Res.*, **6** : 5723-5724

Iftikhar, Sehrish & Shahid, Ahmad & Nawaz, Kiran & Ali, Shinawar. 2016. First Report of *Curvularia lunata* Causing Fruit Rot of Tomato (*Lycopersicum esculentum*) in Pakistan. *Plant Disease*. 100. 10.1094/PDIS-09-15-1066-PDN.

Jiuxu Zhang, 2014. Postharvest decay-Control strategies. *Lasiodiplodia theobromae* in Citrus Fruit (Diplodia Stem-End Rot). Academic Press, 309-335

Kamle, M., Kumar, P., Gupta, K. V., Tiwari, K. A., Misra, K. A., Pandey, K. Brijesh. 2013. Identification and phylogenetic correlation among *Colletotrichum gloeosporioides*pathogen of anthracnose for mango. *Biocatalysis & Agril. Biotech.*, **2** : 285-287

Keinath, A., Wintermantel, W. M. and Zitter, T. A., (eds). 2017. Compendium of Cucurbit Diseases and Pests. *American Phytopathol. Soc., Press*, St. Paul, MN. 220 p,

Keinath, A.P. and DuBose, V.B., 2017. *Management of southern blight on tomato with SDHI fungicides". Crop Protn.,* **101***:* 29–34

Lamour, K. H, ed. 2013. *Phytophthora*: A Global Perspective. CAB International, Oxfordshire, U.K.

Laura Sweets, 2020. Stored Grain Fungi. AgEBB, Univ. Missouri Extension, CAFNR

Li Hua, Chen Yong, Zhang Zhanquan, Li Boqiang, Qin Guozheng and Tian Shiping, 2018. Pathogenic mechanisms and control strategies of *Botrytis cinerea* causing postharvest decay in fruits and vegetables. *Food Quality and Safety*, **3**: 111–119

Li,Y., Mao,L.G., Yan,D.D., Liu,X.M., Ma,T.T., Shen,J., Liu, P.F., Li, Z., Wang,Q.X., Quyang,C.B., Guo,M.X. and Cao, A.C., 2014. First Report in China of Soft Rot of Ginger Caused by *Pythium aphanidermatum*. *Plant Dis.*, **98**: 1011

Lim, C., Kim, J, Choi, J.N., Ponnusamy, K., Jeon, Y., Kim, S.U., Kim, J.G. and Lee, C., 2010. Identification, fermentation, and bioactivity against Xanthomonas oryzae of antimicrobial metabolites isolated from Phomopsis longicolla S1B4. *J. Microbiol.Biotech.*, **20** : *494–500.*

Lluís Palou, 2014. *Penicillium digitatum* (Green mold, Blue mold). *Postharvest decay.* 10.1016/B978-0-12-411552-1.00002-8.

Maeda, K., Karahashi, Y., Ohsato, S and Yoneyama, K., 2010. Appearance of a new leaf rots disease on common ice plant. *J. Gen. Pl. Pathol.*, **54**: 207-212.

Melanie L. Lewis Ivey, 2015. Choanephora Flower and Fruit Rot. Louisiana State University Agricultural Center, Louisiana Agricultural Experiment Station Louisiana Cooperative Extension Service LSU College of Agriculture.

Meng, Y., Zhang, Q., Ding, W. and Shan, W., 2014. *Phytophthora parasitica*: a model oomycete plant pathogen. *Mycology,* **5**:43–51.

Mohammad Akram Husain and C. S. Choudhary, 2018. Morphological, Cultural and Physiological studies on Slerotinia sclerotiorum causing Stem rot of Oilseed brassica. *Int.J.Curr.Microbiol.App.Sci* (2018) Special Issue- **7**: 1044-1052

Mullen, J. 2001. Southern blight, Southern stem blight, White mold. The Plant Health Instructor. DOI: 10.1094/PHI-I-2001-0104-01.

Navi, S.S., Bandyopadhyay, R., Hall, A.J. and Bramel-Cox, P.J., 1999. A pictorial guide for the identification of mold fungi on sorghum grain. Information Bulletin no. 59. International Crops Research Institute for the Semi-Arid Tropics , Patancheru 502 324, Andhra Pradesh, India, 118 pp

Netto MSB, Assuncao IP, Lima GSA, Marques MW, Lima WG, Monteiro JHA, de Queiroz BV, Michereff SJ, Phillips AJLand Camara MPS. 2014. Species of *Lasiodiplodia* associated with papaya stem-end rot in Brazil. http://dx.doi.org/10.1007/s13225-014-0279-4 (Consulta, febrero 2014).

Onions, A.H.S., Allsopp, D., Eggins, H.O.W. (1981). Smith's introduction to industrial mycology (7th ed.). London, UK: Arnold. ISBN 0-7131-2811-9.

Peter Twumasl, Godfried Ohene-Mensah and Emmanuel Moses, 2014. The rot fungus *Botryodiplodia theobromae* strains cross infect cocoa, mango, banana and yam with significant tissue damage and economic losses. *African J. Agril. Res.*, **9**: 613-619

Phillips AJL, Alves A, Abdollahzadeh J, Slippers B, Wingfield MJ, Groenewald JZand Crous PW 2013. The *Botryosphaeriaceae*: genera and species known from culture. Studies in *Mycology* **76**: 51-167

Priyadharsini, Perumalsamy, Muthukumar and Thangavelu, 2017. The root endophytic fungus *Curvularia geniculata* from *Parthenium hysterophorus* roots improves plant growth through phosphate solubilization and phytohormone production. *Fungal Ecol.,* **27**: 69–77

Rivard, C. L., O'Connell, S., Peet, M. M., and Louws, F. J. 2010. Grafting tomato with inter-specific rootstock to manage diseases caused by *Sclerotium rolfsii* and southern root-knot nematode. *Plant Dis.* **94:**1015-1021.

Rodríguez-Gálvez, E. and MaldonadoA. E. Alves, 2014. "Identification and pathogenicity of Lasiodiplodia theobromae causing dieback of table grapes in Peru". *European Journal of Plant Pathology*. **141** : 477–489.

Rohrbach, K. G., and Phillips. D. J., 1990. Postharvest Diseases of Pineapple. *Acta Horticulturae*, **269**: 503–508

Saito, S., Michailides, T. J., and Xiao, C. L. 2016. Mucor rot—An emerging postharvest disease of mandarin fruit caused by *Mucor piriformis* and other *Mucor* spp. in California. *Plant Dis.,* **100:**1054-1063

Sattar, A. and Malik, S.A. 1989. Some studies on anthracnose of mango caused by *Glomerella* cingulata (Stonem.) Spauld. Sch. (*Colletotrichum gloeosporioides*). *Indian J.l Agri. Sci.,* **1**:511-521.

Sivakumar, Dharini, Bautista-Baños and Silvia, 2014. A review on the use of essential oils for postharvest decay control and maintenance of fruit quality during storage. *Crop Protn.,* **64**: 27–37.

Srivastava, D.N. (1972). Epidemiology and Prevention of Diplodia Stem-End Rot of Ripe Mango Fruits. *Acta Hortic*. **24**: 235-236.

Subramanian Sundravadana, Subramaniam Thirumurugan and Devadason Alice, 2011. Exploration of molecular variability in *Rhizoctonia bataticola*, the incitant of root rot disease of pulse crops. *J. Pl. Protn. Res*., **51**: 184-189

Susan B. Jepson, 2008. OSU Plant Clinic, 1089 Cordley Hall, Oregon State University, Corvallis, OR 97330-2903

Swilaiman, S.S., O'Gorman, C.M., Balajee, S.A. and Dyer, P.S., 2013. Discovery of a sexual cycle in *Aspergillus lentulus*, a close relative of *A. fumigates*. *Eukaryotic Cell.* **12:** 962–9.

Tian, S. P., Qin, G. Z., Li, B. Q., 2013. Reactive oxygen species involved in regulating fruit senescence and fungal pathogenicity. *Plant Mol. Biol.,* **82**: 593–602.

Waller, J. M., 1992. *Colletotrichum* diseases of perennial and other cash crops. In: *Colletotrichum: biology, pathology and control*. J.A.Bailey and M.J. Jeger (Edrs). CABI, Wallingford.

Wang, Xiaoning, Radwan, Mohamed, M., Taráwneh, Amer, H., Gao, Jiangtao, Wedge, David, E., Rosa, Luiz H., Cutler, Horace, G.., Cutler and Stephen, J., 2013. Antifungal activity against plant pathogens of metabolites from the endophytic fungus *Cladosporium cladosporoides. J.Agril. Food Chem.,* **61:** 4551–4555

Williams,A.P. and Williams Neaves, 2006. Food Spoilage Microorganisms. Woodhead Publishing Series in Food Science, Technology and Nutrition, 488-503 pp.

3

Factors Governing Postharvest Diseases

Postharvest loss of fruits and vegetables is defined as "that weight of wholesome edible product (exclusive of moisture content) that is normally consumed by human and that has been separated from the medium and sites of its immediate growth and production by deliberate human action with the intention of using it for human feeding but which for any reasons fails to be consumed by human." Not only quantity and quality but even the appearance of fruits and vegetables are affected, and their market value is reduced. Most 'skin deep' injuries such as 'fly speck' in apples do not affect the edible part. Some infections are not harmful as they occur on inedible parts and can be trimmed off before consumption ego *Colletotrichum* spots on the outer leaves of cabbage]. Fresh fruits and vegetables are perishable and highly prone to these losses because they are composed of living tissues. These tissues must be kept alive and health throughout the process of marketing. These are composed of thousands of living cells which require care and maintenance (Alao, 2000). The pathological rots are the most serious which is followed by mechanical injury. Pathological rots in combination with mechanical damage cause serious damage to the perishables. Environmental factors such as temperature, relative humidity and oxygen balance most especially during storage are also greatly responsible for damage also environmental conditions such as temperature and humidity are responsible for rendering fruits and vegetables susceptible for pathological attacks (Yahaya and Mardiyya, 2019). Fungi and bacteria are the major pathogens, which cause postharvest diseases in crops leading to about 30 % loss. Normally diseases will be more at improper temperature and relative humidity. Apart from causing disease, pathogens are also involved in ethylene production, which hastens the deterioration of the product.

Pathogenic filamentous fungi are the most important causal agents of postharvest decay of fresh fruits. Depending on the fruit species, cultivar and a wide range of pre- and postharvest factors and conditions, the incidence of fungal decay can cause considerable economic losses to growers and traders especially if the produce is intended for export markets. Wholesale buyers often reject fruit loads if decay is found in export shipments, and furthermore, they may charge the producer for the transport and handling costs. Every fresh fruit commodity is prone to decay caused by pathogenic fungi. Depending on the origin and characteristics of the infection, fungi that cause postharvest diseases can be classified into two general groups, *viz*., those that infect the fruit in the field and remain latent until their development after harvest, and those that infect the fruit through rind microwounds or injuries inflicted during harvest, transportation, postharvest handling, and commercialization. Typically, fungal species that cause latent infections can also cause wound infections near or after harvest in the same fruit commodity under certain conditions, but the opposite is not true, and many economically important wound pathogens can only infect fruit if its peel is broken, therefore they are known as strict wound pathogens (Lluis Palou, 2018). This is the case of *Penicillium* spp., the causal agents of blue or green molds on many relevant fruit commodities.

For example, blue mold, caused by the species *Penicillium* expansum, can lead to significant postharvest losses of apple, pear, stone fruits, kiwifruit, many berries, pomegranate, persimmon, and other subtropical and tropical fruits. The two postharvest diseases that account for the largest decay losses of citrus fruits worldwide are green and blue molds, caused by *Penicillium digitatum* and *Penicillium italicum*, respectively. It is clear from these denominations that the common names of postharvest diseases are based on the symptoms they produce. Other important wound pathogens are *Geotrichum* spp., which cause sour rot on different horticultural products, and *Rhizopus* spp., which are dangerous postharvest pathogens that cause soft rot or *Rhizopus* rot on a wide range of fruit hosts. The most important species are *Geotrichum candidum*, which attacks stone fruits such as peach, nectarine, plum, and sweet cherry, but also tomato and other fruit-like vegetables; and *Geotrichum citri-aurantii*, the cause of sour rot of citrus fruits. *Rhizopus stolonifer* is the most important species causing fruit soft rot.

Among latent pathogens, *Botrytis cinerea* , which causes gray mold, is the main causal agent of postharvest decay of grape, strawberry, blueberry and other small berries, kiwifruit, pomegranate, fig, etc., but also attacks many other fruits and vegetables. *Colletotrichum* spp. causes anthracnose on many subtropical and tropical fruits, such as citrus fruit, avocado, persimmon, banana, papaya, mango, and pineapple. The most important species are *C. gloeosporioides*, *C. acutatum* and *C. horii*. *Colletotrichum* spp., which causes black spot, and *Lasiodiplodia* theobromae and other Botryosphaeriaceae pathogens, which cause stem-end rot, are also economically relevant for the worldwide industry of subtropical and tropical fruit crops. The species *Colletotrichum alternata* is also an important postharvest pathogen of pome fruits, persimmon, loquat, and pomegranate, among many other fruits. Species of *Monilinia*, especially *M. fructicola*, *M. laxa* and *M. fructigena* cause field and postharvest brown rot of stone fruits and are generally considered among the most substantial limiting factors of the yield of these fruit crops (Lluis Palou, 2018). *Monilinia* spp. can also cause postharvest disease on pome fruits.

Postharvest diseases vary each year and a wide range of preharvest factors influences their development. These include the weather (rainfall, temperature, humidity, etc.), production area, cultivar, cultural practices (pesticide application, fertilization, irrigation, planting density, pruning, mulching, fruit bagging, etc.) and planting material. These factors may have a direct influence on the development of the disease by reducing inoculum sources or by discouraging infection. Alternatively, they may affect the physiology of the produce in a way that impacts on disease development after harvest. Examples are given below that demonstrate how these preharvest factors lead to disease in specific crop.

In general, storage environment influences the pathogen activity. Low temperature storage cannot prevent disease, but can limit the rate at which disease develops. The optimum temperature for spore germination of most fungal pathogens is 20–25°C. Although germination is possible at a much wider range, the further the temperature is from this optimum the slower germination will be. Cold temperatures also delay deterioration of the product, reducing susceptibility to decay. Fungi differ in their ability to grow at low temperatures, viz., *Aspergillus niger* is unable to grow below 11°C, *Colletotrichum* spp. cannot grow below 9°C and *Rhizopus stolonifer* cannot grow below 2°C. However, *Botrytis cinerea* can continue to grow down to –2°C and *Colletotrichum alternata* still develops at –3°C, making these diseases hard to control in storage. High relative humidity and free moisture on produce both increase opportunities for disease development. Dry conditions mean that fungal spores are unable to germinate and grow, and bacteria cannot multiply. Even if fungal germination does occur in a small amount of moisture, low humidity can inhibit further growth. Conversely, wet conditions and high humidity allow plant pathogens to spread and multiply.

It is to be noted that most organisms that cause postharvest disease are weak pathogens. Only a few fungi such as *Colletotrichum* can directly invade through healthy skin. Most fungi enter vegetables through wounds in the skin, or infect products already damaged by unsuitable storage conditions.

3.1 Preharvest and postharvest infections: Infection in this case occurs prior to the harvest. Some fungi infect products during flowering, while others may stick to the product skin. Pathogens that are relatively dormant during plant growth can develop and cause symptoms after harvest. Pre-harvest infection can occur by direct penetration of the skin, entry through natural openings (stomata and lenticels), transmission through the parent plant, and through damaged areas. Plant defence responses are able to hold such infections in check while the plant is actively growing. However, once removed from sources of water and nutrient, the disease can take hold. Infection of some fruiting vegetables (eg. internal rots in capsicums) actually occurs during flowering. Fungi such as *Colletotrichum* and *Botrytis* are able to infect flower petals. However, they only start to develop once the vegetable has matured. Many plant pathogens are present in the soil or on plant debris. Vegetables that are grown in contact with the soil are likely to be carrying spores or mycelium that cause infection, even though these are not apparent at harvest. Fungal spores are also carried on the wind. Nearby crops infected with fungal pathogens such as white blister (*Albugo candida*) or blue mould (*Penicillium* spp.) release huge numbers of spores, which can land on the intact vegetables. These spores are unable to germinate while the product remains dry. However, condensation during cooling or storage can allow germination, growth and infection.

Postharvest infection : This infection occurs through injuries or other breaks in the skin and is increases if products are in wet condition. Fungal spores are present in the air, on equipment, on containers and on the hands of harvest workers and packers. However, for those spores to infect the vegetables there must be suitable conditions for spore germination and growth as well as openings that allow them to penetrate the plant tissue. Moisture is essential for spores to germinate. While most can germinate in pure water, if nutrients are present then germination is increased and growth is more vigorous. For example, studies have found that spores of *Rhizopus* spp. and *Botrytis cinerea* need both water and nutrients to allow spores to germinate and penetrate a host. Bacteria also need water plus nutrients for cells to divide and multiply. Wounding allows cell contents to leak out. Broken cells provide both water and nutrients, in the form of sugars, acids and other substrates. Wounds therefore provide ideal environments for spores to germinate and bacteria to grow. Moreover, while waxy epidermal (skin) cells represent a significant barrier to infection, the cells underneath are less protected. Once established in a wound site, pathogens can directly attack these cells and spread through the plant tissue.

3.2 Factors affecting postharvest diseases: Knowledge of the time and mechanism of infection is essential for the development of an effective programme for the control of postharvest diseases. There are many factors that affect the postharvest losses due to pathogens.

Environmental factors affect postharvest diseases to a greater extent. In general, the factors that play a crucial role in the maintenance of quality and incidence of postharvest diseases are temperature and relative humidity (RH). The microbial attack to different crops becomes very slow at low temperature. But some microbes like the species of *Botrytis*, *Sclerotinia*, *Cladosporium* can grow slowly even near freezing temperature. Chilling injury has been reported in some tropical fruits when stored at 10-15°C. Chilling injuries are generally developed at temperatures below 12°C in grape fruits and which reduce host resistance against pathogen. The role of R.H. in the postharvest environment is equally important as temperature (Yadav *et al*, 2013). If the RH is high, the decay is closely relative but if RH is near saturation and temperature is near 0°C, it will lower decay losses. The surrounding environment of produce plays an important role in development of infection by the pathogens and subsequent postharvest wastage of the produce. The various factors which favor the postharvest diseases are discussed as under. It is common knowledge that most diseases appear and develop best during wet, warm and humid days. The postharvest environmental factors that most seriously affect the initiation and symptoms development of infectious fruits are temperature, moisture and air composition (particularly O_2 and CO_2 concentrations).

i. **Biochemical changes :** Fruits and vegetables are considered as the best sources of energy due to the presence of sugars, amino acids, organic acids, vitamins and other nutrients. During pathogenesis many of the fruit rot pathogens bring about a change in nutrients, cause depletion of nutrients and bring about the total loss of nutrients. The infection of fruit and vegetables both in storage and market by diversified pathogens or by storage fungi brings about many changes in the host fruit by modifying their biochemical composition. Pectolytic and cellulolytic enzymes produced by the pathogens break down the pectic and cellulosic substances of host cell wall. Ripening is a physiological process that involves numerous biochemical changes including change in sugar composition and change in pH of the fruit and vegetable tissues. These changes attract a variety of pathogens causing decay, thereby leading to losses up to 25-50% in harvested crops. The infection by the postharvest pathogens may remain quiescent during fruit or vegetable growth due to unfavourable growth conditions for pathogen's development to pathogenesis. Owing ripening and senescence the pathogen transform to necrotrophs, activation of the pathogen is primarily due to change in pH and cause typical decay symptoms. Some pathogens have developed a mechanism for altering the pH of host environment (tissue), thereby increasing virulence. The postharvest pathogens cause acidification by producing organic acids and alkalization by ammonification of host tissues (Kumar Ritesh *et al*, 2018). Management of postharvest tissue-decaying pathogens is usually performed by use of fungicides but due to raised environmental concerns and residual effects of fungicides on health alternative management strategies are required. This approach should be eco-friendly; the use of plant extract and biological agents are in limited use. Thus, it should be emphasized that management of postharvest food commodity (particularly those of direct consumption like fruit) pathogens can be achieved by altering the physiological environment and use of alternative management strategies to reduce the residual effect of inorganic fungicides and other chemicals.

ii.**Temperature:** Storage temperature is so critical for controlling postharvest diseases that all other control methods are sometimes described as "supplements" to refrigeration. The high temperature and high humidity favour the development of postharvest decay. However, the speed of either infection or resistant action is significantly related to storage temperature, being the most important environmental factor. With the advent of higher temperatures, pathogens become active as and, when other conditions are favourable, they can quickly infect hosts and cause serious disease. Obviously, temperature management is so critical to postharvest disease control that all remaining control methods can be described as supplements to refrigeration. Temperature is the most important environmental factor that influences the deterioration of harvested commodities. Most perishable horticultural commodities have an optimal shelf-life at temperatures of approximately 0 °C. The rate of deterioration of perishables however increases two to three-fold with every 10 °C increase in temperature. Temperature has a significant effect on how other internal and external factors influence the commodity, and dramatically affects spore germination and the growth of pathogens.

Temperatures either above or below the optimal range for fresh produce can cause rapid deterioration due to the following disorders: Freezing Perishable commodities are generally high in water content, and possess large, highly vacuolate cells. The freezing point of their tissues is relatively high (ranging from -3°C to -0.5°C), and disruption caused by freezing generally results in immediate collapse of their tissues and a total loss of cellular integrity. Freezing occurs in cold storage systems either due to inadequate refrigerator design, or to thermostat failure. Freezing can also occur upon exposure to inclement weather conditions as occurs when produce is allowed to remain for even short periods of time on unprotected transportation docks during winter. Some commodities (chiefly those native to the tropics and subtropics) respond unfavorably to storage at low temperatures which are well above their freezing points, but below a critical temperature termed their chilling threshold temperature or lowest safe temperature. Chilling injury is manifested in a variety of symptoms including surface and internal discoloration, pitting, water soaking, failure to ripen, uneven ripening, development of off flavors and heightened susceptibility to pathogen attack. High temperature conditions are also injurious to perishable crops. Transpiration is vital to maintaining optimal growth temperatures in growing plants. Organs removed from the plant, however, lack the protective effects of transpiration, and direct sources of heat, such as sunlight, can rapidly elevate the temperature of tissues to above the thermal death point of their cells, leading to localized bleaching, necrosis (sunburn or sunscald) or general collapse.

Temperature management is critical for the reduction of postharvest losses in banana. Storage of banana at 18 °C showed significant reduction in disease severity and grade of all postharvest rots *viz.* crown rot, anthracnose and cigar end rot on both 5th and 7th day of storage. The disease severity (%) was found highest at 35 °C.(Azeem, *et al*, 2016).

The effect of extended periods at low temperatures on spore germination, survival and appressorial formation in *Colletotrichum acutatum* and *C. gloeosporioides*, and on pre-formed appressoria of *C. gloeosporioides*, was studied on water agar on glass slides on a temperature gradient plate (Everett, 2003). *C. acutatum* was more tolerant of low temperatures than was *C. gloeosporioides*. The addition of peptone to simulate endogenous nutrients on the avocado fruit surface affected spore germination and survival only slightly, and the effect was not statistically significant. These trends *in vitro* were also apparent in isolations from fruit rots. Infections by *C. acutatum* were not reduced by storage at 5.5°C before ripening. Incidence of infections caused by *C. gloeosporioides* was significantly fewer in fruit cool stored at 5.5°C for 30 days prior to ripening at 20°C than in fruit stored at 20°C. In the two lines of fruit studied, coolstorage reduced body rots in one line but not the other. The rots on the line of fruit not affected by cool stored at were mainly caused by *C. acutatum*. It was observed that in cool stored fruit, the reduction in incidence of rots caused by *C. gloeosporioides*, but not of rots caused by the closely related fungus *C. acutatum*, was due to the greater sensitivity of spores of *C. gloeosporioides* to cold temperatures. Survival of pre-formed appressoria of *C. gloeosporioides* was not affected by temperatures used for cool storage of avocados.

Maximum growth of *Helminthosporium fulvum* was obtained at 25°C and 30°C temperatures. The fungus was also cultured on 100, 92.5, 85, 74 and 32.5% relative humidity regimes. The fungus showed maximum growth at 92.5 and 100% relative humidity. The growth of the fungus was observed to increase with increase in relative humidity and vice versa (Ibrahim *et al*, 2011).. There were significant differences ($P \leq 0.05$) in the growth of the fungus at different temperature and relative humidity regimes. It was concluded that low temperature and low relative humidity do not support the growth of *H. fulvum*. Therefore, it is highly recommended that fruits and vegetables should be stored at low temperature and low relative humidity regimes to avoid infections due to H. fulvum and this may contribute significantly in controlling the heavy losses our peasant farmers incured due to this fungus.

Conidial germination of *Colletotrichum gloeosporioides* and *Dothiorella aromatica* was affected by temperature. The optimal temperature for germination was between 25 and 30°C. No germination occurred at 5 and at 40°C respectively. *C. gloeosporioides* produced appressoria *in vitro* while *D. aromatica* produced long germ tubes without appressoria. Appresorium formation was more sensitive to temperature than was germination. Appressoria were produced at temperatures ranging from 10 to 30°C. Temperature requirements for mycelial growth of *C. gloeosporioides* and of *D. aromatica* were similar to those for conidial germination (Denner *et al*, 1986). The information is useful for the prediction of infection periods.

Temperature has significant influence on all components of infection process i.e. conidial germination, appressoria formation, germ tube growth in *Magnaporthe oryzae*. All the components of infection process were higher at optimal temperature (27°C) as compared to both suboptimal (22°C) and supraoptimal (32°C) temperatures for *M. oryzae*. Appressoria formation and spore germination were faster in optimal temperature and too slower at supra-optimal temperature, which is an indirect indication that *M. oryzae* is not capable to cause infection at higher temperature. Interestingly, change in temperature from sub optimal to optimal temperature had not affected rates of conidial germination and rate of appressoria formation. Similar kind of trend was observed for germ tube growth and after appressoria formation re-growth was observed, which is an indication of diauxic growth in *M. oryzae* (Rajput *et al*, 2017).

The grass seed gall nematode, *Anguina agrostis*, feeds and reproduces within the developing ovaries of bentgrass seeds and overwinters in seed galls as anhydrobiotic juveniles. These dormant juveniles can survive within the seed gall for many years. In this dehydrated state, they are more tolerant to extreme environmental conditions than are their hydrated counterparts. Nematodes in seed galls were exposed to various high temperatures (80 to 160°C) for time intervals of 5 to 30 min. Survival decreased as time and temperature increased. Remarkably, these nematodes survived exposure to 155°C for 5 min, higher than that recorded for any other metazoan. In contrast, seed galls that had been stored at room temperature and humidity for 5 yr also survived exposure to extreme temperatures; however, their survival rates were not as high as those for freshly collected galls (Eisenback, 2013). Juveniles within the seed gall were coiled and grouped together conforming to the shape of the seed gall.

The gross morphology of the cuticle of the juveniles was very smooth and relatively undistorted by the shrinkage from the loss water from their body tissues. Wherever the nematodes were cut with a razor blade, a small amount of their contents oozed out of the opening and coalesced with that of other nearby specimens and appeared gel-like.

Detached *Vitis vinifera* 'Cabernet Sauvignon' berries were inoculated either singularly or in combination with spore suspensions of *Botrytis cinerea*, *Colletotrichum acutatum* and *Greeneria uvicola* and the degree of disease expression examined at either 20 or 27 °C (Steel *et al*, 2011). Berries were more susceptible to *B. cinerea* at 20 °C and to *G. uvicola* at 27 °C but were highly susceptible to *C. acutatum* at either temperature. In experiments involving inoculation of berries with mixtures of fungal organisms, *B. cinerea* infection was diminished at 27 °C by either *C. acutatum* or *G. uvicola* but only by *C. acutatum* at 20 °C. *G. uvicola* infection was diminished by *C. acutatum* at both temperatures investigated. *B. cinerea* reduced the level of infection of both *C. acutatum* and *G. uvicola* at 20 °C. The findings have implications for seasonal bunch rot management of grapes in relation to predicted changes in global temperature.

Gawai and Mangnalikar (2018) observed that temperature 35 °C encouraged better growth of *Colletotrichum alternata* as compared to 30 °C and 40°C. The result also revealed that pH 6.5 encouraged better growth of *A. alternata* to the pH 7.5 and 5.5. It is concluded from the result that temperature 35 °C and pH 6.5 is optimum temperature and pH for the growth *A. alternata*.

Effect of different pH levels, temperature, light intensity and media were tested against the growth of *A. alternata* under *in vitro* conditions. The results of experiment indicated that the growth of *A. alternata* was maximum in pH range of 6.00-6.50 and temperature range of 25-30 °C. The exposure of the fungus to alternate cycles of 12 hour light and 12 hour darkness resulted in the maximum mycelial growth of *A. alternata* compared to continuous light and dark. Among the different media tested, host leaf extract medium supported significantly the maximum growth of all the fifteen isolates of *A. alternata* followed by potato dextrose agar (Manjunath Hubballi *et al*, 2010) . Incubation at 15, 20, or 25 °C caused differences in the size and aggressiveness of conidia from *Monilinia fructicola* cultured on peach or nectarine fruit. The volume and aggressiveness of the conidia grown in incubators were greatest at 15 °C and least at 25 °C. Temperature had less effect on the size of spores produced on nectarines than on peaches. Spores produced on potato-dextrose agar were smaller and less aggressive than those produced on fresh peaches (Phillips, 1984). Spores collected from fruit rotting in orchards near Fresno, CA, became smaller as the average 3-day maximum temperature increased.

Rhizopus stolonifer and *Gilbertella persicaria* are two economically important pathogens that are commonly associated in simultaneous infections of many stone and pome fruits. The ability of *R. stolonifer* and *G. pesicaria* to assimilate diverse carbon (C) and nitrogen (N) sources and the effect of pH and the interaction temperature/ water potential (Ψ) on their conidial germination and growth were investigated by Amiri *et al*, (2011). The fungi were grown on a liquid medium containing D-glucose, sucrose, fructose, maltose, or raffinose as C sources and urea, NH_4NO_3, $NaNO_2$, glycine, or glutamine as N sources. The biomass production of *R. stolonifer* was the greatest when sucrose and fructose were used as C sources, respectively, whereas biomass production of *G. persicaria* was greater on fructose and glucose than on other sources. Glutamine and sodium nitrite ($NaNO_2$) used as N sources yielded the highest biomass for *R. stolonifer* and *G. persicaria*, respectively. Sporangiospore germination and mycelial growth of *R. stolonifer* and *G. persicaria* were optimal between pH 3 and 10 and were totally inhibited at pH ≤2.5. The two fungi were affected by the interaction of temperature and Ψ. Both *R. stolonifer* and *G. persicaria* were sensitive to low temperatures (<10°C) and were completely inhibited at Ψ levels of -17 and -18.3, respectively. The nutrients utilization, pH, and Ψx temperature profiles for sporangiospore germination and mycelial growth were to some extent similar between the two species corroborating that *R. stolonifer* and *G. persicaria* occupy similar ecological niches. Identifying ecophysiological determinants of these commercially important pathogens will contribute to our understanding of their physiology and epidemiology and will aid in developing appropriate models for disease prediction.

Citrus scab caused by Elsinoe fawcettii was most severe at temperatures from 23.5 to 27°C and much less severe at 17, 20, 30, or 32°C (Agostini, *et al*, 2003). A leaf wetness duration of 4 h was sufficient for some infection, but 12 h of leaf wetness were needed for maximum infection with scab. Melanose was equally severe at 24 and 28°C, moderate at 20°C, and low at 32°C. Melanose infection was minimal with 4 h of leaf wetness, moderate with 8 to 16 h, and reached maximum levels at 24 h or more of leaf wetness. In field studies, grapefruit seedlings with new shoots were placed beneath trees weekly, and disease severity was evaluated in relation to environmental factors (Farrar, 2000).. Melanose severity increased sharply with an increase in total weekly rainfall, leaf wetness durations of greater than 80 h per week, and average temperatures above 22°C. The greatest degree of soft rot was caused by Erwinia chrysanthemi at 30 and 35°C. E. *carotovora* subsp. *carotovora* isolates were relatively less virulent than *E. chrysanthemi* at 30°C and none of the E. *carotovora* subsp. *carotovora* isolates reduced carrot disk weight at 35°C .

elevated temperature significantly increases the susceptibility of *Arabidopsis* to *Pseudomonas syringae* pv. *tomato* (Pst) DC3000 independently of the phyB/PIF thermosensing pathway (Huot *et al*, 2017). Instead, elevated temperature promotes translocation of bacterial effector proteins into plant cells and causes a loss of ICS1-mediated salicylic acid (SA) biosynthesis. Global transcriptome analysis reveals a major temperature-sensitive node of SA signalling, impacting ~60% of benzothiadiazole (BTH)-regulated genes, including ICS1 and the canonical SA marker gene, PR1. Remarkably, BTH can effectively protect *Arabidopsis* against Pst DC3000 infection at elevated temperature despite the lack of ICS1 and PR1 expression. *Xanthomonas arboricola* pv. *pruni* is the causal agent of bacterial spot disease of stone fruits and almond. The estimated optimal temperature for infection by *X. arboricola* pv. pruni was 28.9°C. Wetness periods longer than 10 h at temperatures close to 20°C, or 5 h at temperatures between 25 and 35°C were necessary to cause high disease severity. The predictive capacity of the model was evaluated using an additional set of data obtained from new wetness duration-temperature combinations. In 92% of the events the observed severity agreed with the predicted level of infection risk (Gerard Morales, *et al*, 2018). The risk chart derived from the reduced form of Duthie's model can be used to estimate the potential risk for infection of Prunus by *X. arboricola* pv. *pruni* based on observed or forecasted temperature and wetness duration.

Ralstonia solanacearum causes bacterial wilt on a wide range of plant hosts. Most strains of *R. solanacearum* are nonpathogenic below 20°C; however, Race 3 Biovar 2 (R3B2) strains are classified as quarantine pathogens because of their ability to infect crops, cause disease and survive in temperate climates. Ana M. Bocsanczy *et al*, (2012) identified race 1 biovar 1 Phylotype IIB Sequevar 4 strains present in Florida which were able to infect and produce wilt symptoms on potato and tomato at 18°C. Moreover they infected tomato plants at rates similar to strains belonging to R3B2. Strains naturally nonpathogenic at 18°C were able to multiply, move in planta, and caused partial wilt when inoculated directly into the stem, suggesting that low temperature affects virulence of strains differently at early stages of infection. Bacterial growth *in vitro* was delayed at low temperatures, however it was not attenuated. Twitching motility observed on growing colonies was attenuated in nonpathogenic strains at 18°C, while not affected in the cool virulent ones. Using pilQ as a marker to evaluate the relative expression of the twitching activity of R. solanacearum strains, it was confirmed that cool virulent strains maintained a similar level of pilQ expression at both temperatures, while in nonpathogenic strains pilQ was down regulated at 18°C.

Germination of conidia of eight botryosphaeriaceous fungi infecting grapevines was evaluated after 2, 4, 6, 12, and 24 h incubation under eight different temperatures (Úrbez-Torres, *et al*, 2010). The effect of temperature on conidial germination was also evaluated in different stages (hyaline versus pigmented conidia) of the species *Lasiodiplodia* theobromae. Conidial germination of Botryosphaeriaceae species infecting grapevines was significantly affected by temperature. Overall, conidial germination increased significantly with longer incubation times, especially from 2 to 12 h. In most cases, germination of conidia was not significantly different between 12 and 24 h incubation. Conidia of botryosphaeriaceous species did not germinate (with the exception of Botryosphaeria dothidea and Neofusicoccum parvum) at 5°C, and only *B. dothidea*, *Diplodia seriata*, and *L. theobromae* showed high levels of conidial germination at 40°C. Optimum conidial germination temperatures (defined as the temperature in which germination reached at least 50% in the shortest incubation time) were 25°C for *B. dothidea* and *Dothiorella iberica*, 25 to 30°C for Spencermartinsia viticola, 30°C for *Diplodia* corticola, *D. mutila*, *D. seriata, N. parvum*, and hyaline *L. theobromae*, and 40°C for pigmented *L. theobromae* conidia. Successful conidial germination of species of Botryosphaeriaceae infecting grapevines was always observed between 10 and 35°C with the exception of *Dothiorella iberica* and pigmented *L. theobromae* conidia, neither of which germinated at 35 and 10°C, respectively. Results of this study showed conidia of botryosphaeriaceous species infecting grapevines to be capable of germination under a broad range of temperatures including those considered to be extreme, which may explain the success of these species as pathogens.

To determine the optimum conditions for development of soft rot under controlled conditions, pots of carrots inoculated with *E. chrysanthemi* were saturated with water and incubated at 20, 25, 30, or 35°C (Farrar, *et al*, 2000). Plants were harvested and examined for disease 12, 24, 36, 48, 72, and 96 h after inoculation. Negligible disease occurred after 12 h. Disease severity and incidence increased with increasing temperature and duration of saturation from 24 to 96 h. In a second experiment, carrot disks were inoculated with three isolates each of *E. chrysanthemi* and E. *carotovora* subsp. *carotovora* and incubated at 15, 20, 25, 30, and 35°C. After 48 h, the disks were washed to remove rotted tissue and reweighed. Neither bacterium reduced carrot disk weight at 15°C. In general, moderate weight reduction occurred at 20 and 25°C. The greatest degree of soft rot was caused by *E. chrysanthemi* at 30 and 35°C. E. *carotovora* subsp. *carotovora* isolates were relatively less virulent than *E. chrysanthemi* at 30°C and none of the E. *carotovora* subsp. *carotovora* isolates reduced carrot disk weight at 35°C.

Multiplication of *Xanthomonas campestris* pv. *undulosa* in the leaf tissues of susceptible and field-olerant wheat genotypes was determined at the secdling stage, under different incubation conditions of temperature and air humidity (Etienne Duveiller and Henri M.M. Maraithe, 2008). Dry air conditions did not hamper the multiplication of the population trend. Symptoms were not observed at 15°C. The occurrence of lesions coincided with populations of bacteria that had reached an estimated threshold of 108 colony forming units/leaf. Differences in populations between genotypes Alondra and Pavon were only noticeable at 25°C and dry air conditions, without correlation to differences in symptom expression. Considering these two genotypes, resistance to bacterial leaf streak did not seem to operate at the seedling stage when plants were inoculated by infiltration. In both genotypes, lesions did not extend significantly. An active mechanism stopping bacterial growth which is apparently temperature-dependent or a saturation of the, target sites was possibly involved, as shown in the different experiments by the fit-curves calculated for Alondra and Pavon. This study provided information on the role of climatic factors in relation to epidemics. Temperature appeared to have a major effect on the disease and results confirmed that the risk of bacterial leaf streak prevails in warmer non-traditional wheat growing areas. Several types of resistance are likely to exist in the *X. e. pv. undulosa* /wheat system.

Humidity: Relative humidity is a very important environmental factor for harvested fruit in storage. The germination of some fungi and direct penetration to fruit are aided by saturated atmospheres or moisture on the fruit surface. In general, high humidity can promote diseases development if temperatures are favourable. Fresh fruits need high humidity levels (>95% relative humidity) in storage environment in order to minimize the loss of moisture and preventing them from shrivel and losing tissue turgidity. Mannaa and Kim (2018) evaluated the effect of different temperatures (10, 20, 30, and 40 °C) and relative humidities (RHs; 12, 44, 76, and 98%) on populations of predominant grain fungi (*Aspergillus* candidus, *Aspergillus* flavus, *Aspergillus* fumigatus, *Penicillium* fellutanum, and *Penicillium islandicum*) and the biocontrol activity of *Pseudomonas protegens* AS15 against aflatoxigenic *A. flavus* KCCM 60330 in stored rice. Populations of all the tested fungi in inoculated rice grains were significantly enhanced by both increased temperature and RH. Multiple linear regression analysis revealed that one unit increase of temperature resulted in greater effects than that of RH on fungal populations. When rice grains were treated with P. protegens AS15 prior to inoculation with *A. flavus* KCCM 60330, fungal populations and aflatoxin production in the inoculated grains were significantly reduced compared with the grains untreated with strain AS15 regardless of temperature and RH (except 12% RH for fungal population). In addition, bacterial populations in grains were significantly enhanced with increasing temperature and RH, regardless of bacterial treatment. Higher bacterial populations were detected in biocontrol strain-treated grains than in untreated control grains.

Longevity of conidia of *Penicillium digitatum* (cause of citrus green mold) and arthroconidia of *Geotrichum* citri-aurantii (cause of sour rot of citrus) was determined under controlled temperature and relative humidity (RH) or ambient summer conditions in central California (Smilanick and Mansour, 2007). Longevity at low RH was longer than at high RH. Hours to kill 99% of the conidia (LT99) of nine *P. digitatum* isolates were determined at 50°C and 75 or 95% RH. At 75 and 95% RH, the LT99 was 24.9 and 4.9 h, respectively. The LT99 was 30 and 42 days, respectively, for conidia of *P. digitatum* under ambient conditions at two outdoor locations. The LT99 of arthroconidia of G. citriaurantii from colonies cultured on potato dextrose agar was briefer than that of conidia of *P. digitatum*. At 45°C and 75 or 95% RH, the LT99 was about 4 and 2 h, respectively, whereas at 50°C, none was viable after 1 h at either humidity. Sanitation is an important practice for managing these diseases. Since there was little or no survival of either fungus after 1 day at 50°C and 75% RH or higher, it was concluded that commercial sanitation could employ a similar regime. To study the effect of weather parameters i.e., temperature and relative humidity on the development of blue mould rot (*Penicillium islandicum* Sopp.) on Indian gooseberry, *in vitro* experiments were carried out by Anil Kumar Saini *et al*, (2018). An interaction of three temperatures i.e. 10, 20 and 30ºC with relative humidity of 40, 60, 80 and 100 % was carried out at five and ten days after inoculation (DAI). Temperature of 30 ºC along with cent per cent relative humidity resulted in maximum disease incidence of 87 and 100% in both 5 and 10 DAI, respectively. Minimum disease intensity (15 and 22%) was recorded at 10 ºC temperature and 40 per cent relative humidity at 5 and 10 days after inoculation. However, with further increase in temperature and relative humidity, the per cent disease intensity was also increased and it also increased during the period of time (5 and 10 days after inoculation). it was elucidated that low temperature (10 ºC) with low relative humidity (40%) was ideal for the postharvest storage of Indian gooseberry.

The influence of temperature and relative humidity (RH) on germinability and viability of *Mucor* piriformis spores were determined (Regina B.Bernd, 1987). Spores did not survive when stored at 35 °C and their survival rate decreased rapidly at 30 °C; however, spores remained viable for more than 1 year at 0 °C. RH also significantly affected spore viability. Spores held at 26 °C and 100% RH no longer germinated after 35 days, while those held at 75 or 90% RH germinated for 65 days. At 20 °C, RH had little effect on spore germinability.

The effect of temperature and RH on percentage spore germination also varied. At all temperatures studied, spore viability decreased more rapidly with time at 100% RH than at 75 or 90% RH. The least favorable, temperature-humidity combination, 30 °C and 100% RH, decreased spore germination from 100% to less than 1% in 14 days.

The effects of temperature and relative humidity on growth rate of *Actinomucor elegans* and *Rhizopus oligosporus* were optimum at 25°C at RH 95-97°and 35°C at RH 95-97°respectively. Yields of protease (108 U/g pehtze), lipase (172 U/g) and glutaminase (176 U/g) by *A. elegans* were maximum after 48 h at 25°C and RH 95-97°and for α-amylase (279 U/g pehtze) and α-galactosidase (227 U/g) at 30°C and RH 95-97fter 48 and 60 h of incubation (Han Bei-Zhong *et al*, 2003). Highest protease (104 U/g pehtze), and lipase (187 U/g) activities of R. oligosporus were observed after 48 h at 35°C and RH 95-97°while maximum α-amylase (288 U/g pehtze) and glutaminase (187 U/g) activities were obtained after 36 h at 35°C and RH 95-97ÐMaximum α-galactosidase activity (226 U/g) by *R. oligosporus* was found after 36 h at 30°C and RH 95-97.

A study was conducted to define how environmental factors (viz. temperature, soil type, moisture and nutrition) as well as variety, influence the extent of damping-off and root disease as well as subterranean clover productivity under challenge by *Pythium irregulare*. Relationships were statistically modeled using linear and generalized linear models and boosted regression trees. Modeling found complex relationships between explanatory variables and the extent of *Pythium* damping-off and root rot. Linear modeling identified high-level (4 or 5-way) significant interactions for each dependent variable (dry shoot and root weight, emergence, tap and lateral root disease index) (You *et al*, 2017). Furthermore, all explanatory variables (temperature, soil, moisture, nutrition, variety) were found significant as part of some interaction within these models. A significant five-way interaction between all explanatory variables was found for both dry shoot and root dry weights, and a four way interaction between temperature, soil, moisture, and nutrition was found for both tap and lateral root disease index. A second approach to modeling using boosted regression trees provided support for and helped clarify the complex nature of the relationships found in linear models. All explanatory variables showed at least 5% relative influence on each of the five dependent variables. All models indicated differences due to soil type, with the sand-based soil having either higher weights, greater emergence, or lower disease indices; while lowest weights and less emergence, as well as higher disease indices, were found for loam soil and low temperature. There was more severe tap and lateral root rot disease in higher moisture situations.

A comparison of rates of germination and appressorium formation by an isolate of *Colletotrichum gloeosporioides* on mango leaves, fruit surfaces and cellophane membranes showed that behaviour was broadly similar on all three substrates (Estrada *et al*, 2001). Frequency of appressorium formation was slightly higher on cellophane membranes, and both hyaline and melanized appressoria were formed. Only melanized appressoria were formed on mango surfaces. *In vitro* experiments on membranes showed comparative differences in physiological behaviour with temperature for two Philippine isolates of *C. gloeosporioides*. The most stimulatory temperature for production of appressoria differed in isolates I-2 and I-4 (25 and 20°C, respectively). At 30°C more appressoria became melanized than at lower temperatures, but the frequency of formation of penetration pegs was highest at 25°C. Conidia of *C. gloeosporioides* germinated on cellophane membranes at relative humidities as low as 95%, but the percentage of conidia germinating and forming appressoria increased as the RH approached 100%. Approximately 18% of conidia of *C. gloeosporioides* I-2 held at 62 and 86% RH for 4 weeks retained viability, and some were capable of forming appressoria when placed at 100% RH. These results have implications for epidemiological models for disease control.

Experiments were conducted to investigate the effects of temperature and relative humidity (RH) on the *in vitro* germination and viability of conidia of the apple brown rot fungus (*Monilinia* fructigena), and on colonization and sporulation on detached fruits by *M. fructigena* (Xiangming Xn, *et al*, 2001). Conidia only germinated under near-saturation humidity (greater than or equal to 97% RH) and the rate of germination initially increased with temperature to a maximum at approximate to 23-25 degrees C and then decreased. Conidia germinated rapidly - more than 70% of viable conidia had germinated within 2 h at 20 and 25 degrees C. The rate of colonization on detached fruits increased log-linearly with increasing temperature. Sporulation on detached fruits was not observed at 5 or 25 °C; sporulation appeared to be unaffected by either temperature (10-20 °C) or RH (45-98%) once infection was established. Detached conidia remained viable for a long period of time, up to 20 days, the longest assessment time in this study, depending on storage temperature (10 or 20 °C) and RH (45 or 85%). Temperature appeared to be more important than RH in affecting conidial viability. Low temperature and high RH resulted in reduced loss of conidial viability. Storage at 10 degrees C and 85% RH for up to 20 days appeared not to affect conidial viability.

Experiments were carried out to determine the effects of inoculum concentration, relative humidity (RH), and temperature at these two infection sites. Infection of permanent flower parts increased as a function of inoculum concentration and both length of exposure to high RH (approximately 100% for 0–36 h) and specified continuous RH (56–100%). A low level of infection was still evident under continuous 56% RH (Eden, *et al*, 2003).. Interruption of periods of high RH with breaks of low RH did not reduce infection. Infection of stem wounds was less dependent on inoculum concentration or RH. Factorial combinations of inoculum concentration, RH, and temperature produced significant interactions. Higher temperature increased infection of flowers but reduced infection of stem wounds. The main implications for control in commercial crops are as follows. Lowering the aerial spore concentration by maintaining the disease at a low level will reduce flower infection. Lowering RH will reduce but not eliminate flower infection but will have only a small effect on stem infection. Raising the temperature (from 15 to 25°C) will reduce stem infection, and whilst flower infection increases, this is counteracted by increased flower production and a decrease in the proportion of infections reaching the peduncle and stem.

Among the external factors which influence the growth of *Colletotrichum gloeosporioides*, temperature and relative humidity plays an important role. Each pathogen has got its own cardinal temperature and relative humidity, also understanding about these factors will help to standardize the management strategies (Udayakumar, 2018). Among the six temperature levels tested, 25°C was found to be more conducive for the mycelial growth (88.33 mm), conidial germination (69.00%) of *C. gloeosporioides*, and maximum disease development (97.00 %) which was followed by room temperature (28+2°C). The maximum disease incidence of 97.00 per cent in the fruits was recorded at 25° C followed by room temperature (28±2° C) (83.33 %) after 10 days of inoculation. Among the seven relative humidity levels tested, the maximum mycelial growth (87.00 mm), conidial germination (72.00%) and maximum disease incidence (97.00 %) was recorded at 100 per cent relative humidity, which was followed by 92.9% relative humidity level (83.00 mm; 61.67% mycelial growth and conidia germination, respectively). The least mycelial growth (7.33 mm) and conidial germination (0.0%) was recorded at 36.8% relative humidity level.

Fusarium oxysporum f. sp. *erythroxyli* was investigated as a mycoherbicide for the narcotic plant coca. Sporulation of the fungus in seven formulations containing different organic substrates and movement of its propagules through soil were studied by Gracia-Garza and Fravel (1998). The formulations were a granular wheat flour/kaolin (pesta); an extruded wheat and rice flour (C-6); and five alginate pellet products containing corn cobs, soybean hull fiber, canola meal, rice flour, or rice flour plus canola oil. Formulations were incubated at 25 degrees C for 6 weeks in desiccators with various salt solutions to provide nine relative humidities (RH), ranging from 100% (pure deionized water) to 0% (anhydrous ($CaSO_4$). Hyphae of *F. oxysporum* f. sp. *erythroxyli* grew out of alginate pellets with canola meal, rice, and rice plus canola oil as early as 24 h at 100% constant RH. Alginate pellets of rice plus canola oil and granular C-6 and pesta formulations consistently produced more microconidia, macroconidia, and CFU than the other four formulations at all RH tested. The C-6 formulation produced more propagules than the other formulations at low RH (<53%). Canola meal pellets produced more spores than three other formulations when exposed to fluctuating RH (100 to 75%). The effect of percolating water on spore movement through soil was studied using a plant-pathogenic isolate of *F. oxysporum* f. sp. *niveum*. To determine the effect of water percolation on propagule movement, formulations were placed on soil columns and artificial rain was applied. In general, 10-fold fewer CFU were recovered at a 8- to 10-cm depth compared with a 0- to 2-cm depth.

Common bacterial blight (CBB), caused by *Xanthomonas axonopodis* pv. *phaseoli* Smith (Xap) and *Xanthomonas axonopodis* pv. *phaseoli* var. *fuscan* Burkholder (Xapf) is the most serious biotic constraint of common bean (*Phaseolus vulgaris* L.) production (Hailu *et al*, 2017). Variables temperature and moisture are dominant climate factors that affect common bean growth as well as the development of CBB epidemics. Two sets of experiments were conducted to assess the effect of temperature and moisture on the resistance level of common bean in 2014 and 2015. In the first experiment, two common bean varieties (Gofta and Mexican 142) were inoculated with two bacterial strains (Xap and Xapf) and a control were incubated at four temperature levels (28°C, 30°C, 32°C and 34°C) in growth chambers. In the second experiment, three-soil moisture levels (100%, 75% and 50%) were employed to that of experiment one. The treatment combinations were arranged in factorial completely randomized design (CRD) in the growth chambers for both series of experiments. The disease rating was significantly affected by common bean varieties and temperature levels at 17days after inoculation. Higher disease rating was recorded on the variety Mexican 142 than on Gofta. The highest (1.75) mean disease rating was recorded at 28°C and the lowest (1.44) at 34 °C. The mean disease ratings differed significantly among the moisture levels. The highest disease rating was recorded from 75% moisture content, while the lowest (1.80) disease rating was obtained from 50 % moisture content. These findings indicated that climate change effects above optimum level woud not be favorable for the pathogen development in the arid and semi arid agro ecologies unless new bacterial strains adapted to the drought tolerant common beans in the area.

A relative humidity of 95% and an average temperature of 26-27°C were optimum for infection and substantially favoured spore release of *Pyricularia oryzae*. The first symptoms of infection were detected on the leaves when the crop was at the mid-tillering stage, the number of lesions increasing as the plant developed. After the onset of heading the number of leaf lesions decreased while the number of node and panicle base lesions increased. A temperature increase of 1°C in early August (mid-tillering stage) led to an increase in the mean intensity of disease. The presence of spores in the air from July 15th predicted the appearance of leaf lesions a few days later. The peak spore concentration in August could be used to forecast panicle blast. Assessing the airborne concentration of *P. oryzae* could help in understanding the population dynamics of this pathogen (Castejón-Muñoz, 2008.)

A mathematical model for *Sclerotinia sclerotiorum* infection and disease development on lettuce is presented here for the first time, based on quantifying the effects of temperature, relative humidity (RH) and ascospore density in multiple controlled environment experiments. It was observed that disease can develop on lettuce plants inoculated with dry ascospores in the absence of apparent leaf wetness (required for spore germination) (Clarkson *et al*, 2014). To explain this, the model conceptualises an infection court area containing microsites (in leaf axils and close to the stem base) where conditions are conducive to infection, the size of which is modified by ambient RH. The model indicated that minimum, maximum and optimum temperatures for ascospore germination were 0.0, 29.9 and 21.7°C respectively and that maximum rates of disease development occurred at spore densities >87 spores cm^{-2}. Disease development was much more rapid at 80–100% RH at 20°C, compared to 50–70% RH and resulted in a greater proportion of lettuce plants infected. Disease development was also more rapid at 15–27°C compared to 5–10°C (85% RH). The model was validated by a further series of independent controlled environment experiments where both RH and temperature were varied and generally simulated the pattern of disease development well.

iii. Maturity: Fruits are usually harvested before they are completely ripe, in order to secure sufficient time for long distance transportation and marketing. As fruit ripen they become susceptible to a variety of fungi, whose attacks they were capable to resist, during their development on the tree. Much of decay that develops in storage is derived from spores that deposit on the surface during the growing season, but which are incapable of causing rotting until harvest. Ripe fruits are more susceptible to invasion by specific pathogenic micro-organisms as they are high in moisture and nutrients and no longer protected by the intrinsic factors which conferred resistance during their development phase. Many fruits become easily injured as they reach full maturing, and therefore, are more vulnerable to wound pathogens. Some fungi, such as *Monilinia* spp., *Botrytis cinerea*, *Rhizopus* spp. or *Penicillium* spp., are most likely to invade after the fruit is completely ripe or has become senescent. The maturity stage of fruits at harvest is an important determinant of many quality traits. For example, tomato, being a climacteric fruit, can be harvested at different stages during maturity, like mature green, half ripen, or red ripen stage. Each stage at harvest has its own postharvest attribute that the fruit will exhibit. The shelf life of all tomato cultivars under investigation is longest when harvested at green mature stage (Isaac Kojo Arah *et al*, 2015). Although shelf life has been the most important aspect in loss reduction biotechnology of fruit and vegetables, other aspects may be of interest rather than shelf life.

Fruit nutritional values and appearance may be affected when harvested green. For instance, sugar transport to fruits in a vine-ripened tomato appears to increase during the latter part of maturity and, therefore, when fruits are harvested immature or in a green state sugar import to fruits will be cut off making postharvest degradation of starch, the main source of carbohydrates, which is both undesirable and inadequate. Meanwhile, harvesting later also promotes higher sugar accumulation in riper fruits which are susceptible to mechanical injuries with a shorter shelf life. The pH of tomatoes is an important parameter in the tomato processing industry. Tomatoes are processed as high-acid foods and therefore the higher the acidity the better for processing. Cultivars with high pH therefore may not be suitable for processing. A pH of 4.4 has been suggested to be the maximum and the optimum of a target of 4.25. The acidity of tomatoes is highest at the pink stage of maturity with a rapid decrease as the fruit ripens. Tomato fruits can be harvested at mature green to give producers enough time for long distance marketing but for local marketing harvesting at the fully ripe stage is preferred to maximise nutritional value.

iv. Wounds and bruises : Fungal spore often infect fruit by invading wounds, including cuts punctures, bruises and abrasions, which easily occur during harvest and handling. Various types of injuries can be sustained before, and after the harvest of produce. Injury can be caused by weather, insects, birds, rodents and farm implements. Injuries to the fruits usually occur when produce is dropped on to a hard surface, before, during or after packing, but injury is not usually apparent immediately. Later, bruising may also take place, but it is seen only externally (as in apple) or it may be evident only on peeling (as in potatoes). Compression bruising may result from the overstocking of bulk produce in store houses or from the overfilling of the packaging, for example grapes. The vibration damage can occur in under-filled packs, especially during long distance road transportation. The damaged produce is attacked by various microorganisms, resulting in a progressive decay, which may affect the entire produce. Fresh wounds can support nutrients and humidity for spore germination and colonization by fungi. For example, conidia of *Monilinia fructicola* need moisture and nutrients for spore germination and growth if deposited in fresh wound of stone fruit46. Fungal growth and lesion development follow when temperature conditions are favourable. Healed wounds may no longer be highly prone to fungal invasion. Many studies of responses to wounding in a wide assortment of tissues including fruits, leaves, with common traits among plants have been carried out.

The terms controlled atmosphere (CA) and modified atmosphere (MA) refer to atmospheres in which the gas composition surrounding the commodity is different from air (i.e., 78.08% N2, 20.95% O_2, 0.93% argon and 0.03% CO_2). In practice, CA and MA usually involve reducing oxygen (O_2) levels below 5% and/or elevating carbon dioxide (CO_2) levels above 3%. Controlled atmosphere differs from MA only in the degree of precision in controlling the partial pressures of O_2 and CO_2; the control is more exact in CA than in MA. Other terms used to refer to certain CA storage conditions include reduced (or low) O_2 storage, elevated (or high) CO_2 storage, and nitrogen (N_2) storage. The last term is meaningless, since N_2 is an inert gas with no physiological effects.

Harvesting a quality crop is an important topic on everybody's mind for the upcoming season. Blackspot bruise and shatter bruise, or open wounds such as of nicks, cuts, abrasions, are quality issues to focus on and minimize at harvest. The level of shatter bruise can have a huge impact on the potential for disease development and higher weight loss in storage. The pathogens that cause *Fusarium* dry rot and *Pythium* leak need a wound to infect the potato tuber, whereas pathogens that cause pink rot, late blight, early blight and others may not need a wound, but infection is dramatically increased with wounding. Regardless of the type of wound, economic losses can occur due to disease invasion and shrinkage. Bruises can be reducedby considering the following factors.

a. **Soil condition:** Anything that makes soil separation from tubers at harvest more difficult has the potential to increase bruising. Clods are hard to deal with at harvest time because just about every device used to remove them (shakers, roller tables, etc.) has the potential to injure tubers. Soil moisture conditions at harvest are also important, as very wet or dry conditions make soil separation a challenge.

b. **Tuber condition:** Tuber temperature, hydration level (crispness) and maturity interact to influence bruise susceptibility. In general, the lower thc pulp temperature, the more susceptible tubers are to both shatter and black spot bruising. Pulp temperatures below 45°F make it very difficult to handle potatoes without causing significant shatter bruise damage. The more hydrated or crisp the potatoes are, the more susceptible they are to shatter bruise. In contrast, dehydrated potatoes are susceptible to black spot bruise. Likewise, as plants mature, the tubers become more susceptible to black spot bruise. Tubers that sit in dry soil under dead or dying vines can become very susceptible to black spot, so it is best to vine kill and harvest before vines reach that stage. Knowledge of the relative bruise susceptibility of the potato varieties you are growing allows you to use tuber hydration level to your advantage. For example, Bannock Russet is known to be relatively susceptible to shatter bruise, but resistant to black spot. Therefore, it is recommended that Bannock fields be harvested under slightly drier conditions to reduce shatter bruise. See the table at the end of this article for information on the bruise susceptibility of some of the most commonly grown potato varieties.

c. **Harvester operation :** The more susceptible the tubers are to damage, the more important it is to operate the harvester under optimum conditions. Achieving optimum harvester operation requires reducing drops, cushioning impact areas, and setting chain speeds in relation to ground speed so that harvester chains are filled to capacity. If the volume of material (tubers and soil) passing over each chain can be matched to the capacity of that chain, then damage due to excess tuber movement, rollback, and drops can all be reduced.

d. **Bruise after the harvester :** Getting potatoes into the truck with minimum damage is only the beginning of a bruise reduction program. It has been reported that about 30% of all bruise damage found on tubers in storage occurs after the harvester. Bruise reduction during the trucking and piling operations involves minimizing drops, matching conveyor speed to tuber volume, and piling in a progressive, stepwise fashion to minimize tubers rolling down the face of the pile. Systematically observe each area of the handling equipment to minimize drops.

e. **Bruise on the way to market:** Surveys show that potatoes can experience extensive damage during the packing, shipping and receiving operations. Fortunately, the level of damage caused during these handling steps can be significantly reduced by making minor modifications such as lowering drops, filling equipment to capacity and cushioning impact points. Most potato handling operations are performed by equipment, but there is almost always a human factor involved in managing that equipment in a way that minimizes bruising. Taking time to educate employees about bruise prevention should be a standard part of your harvest preparation. Remind employees to be constantly watchful of where a drop may be too high and make adjustments on the fly to remedy it. Monitoring bruise incidence throughout your harvest system is a useful way to determine whether bruise susceptibility is higher for a particular lot or field. Taking tuber samples at different points throughout the harvest and handling process (eg. hand dug, harvester, conveyors, any drop points or turns, etc.) and holding them at warm temperatures will allow you to assess the tubers for black spot damage. The higher the temperature, the quicker the pigment formation will occur and the sooner the results will be available. There are "hot boxes" commercially available for purchase, or rooms can be constructed for this same purpose. Shatter bruise will be apparent regardless of warming the tubers. The use of an instrumented sphere is another way to determine where unacceptable impacts are occurring in harvest, handling or packaging systems.

v. **Preharvest factors in the disease development:** A wide range of preharvest factors influences the development of postharvest diseases. These include the weather (rainfall, temperature, etc.), production locality, choice of cultivar, cultural practices (pesticide application, fertilization, irrigation, planting density, pruning, mulching, fruit bagging, etc.) and planting material (Fig. 29). These factors may have a direct influence on the development of disease by reducing inoculum sources or by discouraging infection. Alternatively, they may affect the physiology of the produce in a way that impacts on disease development after harvest. For example, the application of certain nutrients may improve the 'strength' of the fruit skin so that it is less susceptible to injury after harvest and therefore less prone to invasion by wound pathogens.

Maintaining the Cold Chain for Perishables

Harvest
- Protect the product form the sun
- Transport quickly to the packaging house

Cooling
- Minimize delays before cooling
- Cool the product thoroughly as soon as possible

Temporary Storage
- Store the product at optimum temperature
- Practice first in first out rotation
- Ship to market as soon as possible

Temporary to Market
- Use refrigerated loading area
- Cool truck before loading
- Load pallets towards the center of the truck
- Put insulating plastic strips inside door of reefer if truck makes multiple stops
- Avoid delays during transport
- Monitor product temperature during transport

Handling at destination
- Use a refrigerated unloading area
- Measure product temperature
- Move product quickly to the proper storage area
- Transport to retail markets or foodservice operations in refrigerated trucks
- Display at proper temperature range

Handling at home or foodservice outlet
- Storage product at proper temperature
- Use the product as soon as possible

Fig. 29: Preharvest and postharvest factors affecting disease development

Susceptibility of fruit to pathogen attack after harvest is very much influenced by fruit quality before and at harvest. Therefore, the relationship between production conditions and postharvest decay development needs to be well understood in order to have effective control strategies. Many of the production and postharvest factors affecting quality, as well as disease development after harvest, however, are still poorly understood in tropical and subtropical fruits.

a. Cultivar and rootstock genotype: Cultivar and rootstock have an important role in determining different physiological and quality characteristics of fruit that, in turn, also influence susceptibility to the development of postharvest disease. New cultivars and rootstocks with improved quality are developed through breeding programs that unfortunately only rarely take into consideration susceptibility to postharvest pathogens in the early stages of the breeding process (Droby *et al*, 2011). Rootstock studies conducted on 'Hass' avocado showed that rootstock had a significant impact on postharvest anthracnose susceptibility. The severity and incidence of anthracnose was significantly lower on 'Hass' grafted to 'Velvick' Guatemalan seedling rootstock compared to scions grafted on 'Duke 6' Mexican seedling rootstock. Anthracnose susceptibility in the two different rootstock-scion combinations was related to significant differences in the concentration of antifungal dienes in leaves, as well as the levels of mineral nutrients in both leaves and fruits. In another study, cultivar was also reported to affect susceptibility of avocado and banana to *Colletotrichum gleoesporiodes* and *C. musae*, respectively. In citrus rootstock significantly affected the development of postharvest decay on Navel orange, 'Ray Ruby' grapefruit, 'Oroblanco' and 'Valencia' orange.

The potential quality of fruit is dependent on the cultivar type. Different cultivars are characterised by different quality parameters making some more desirable to the producers and consumers than others. The choice of an adequate-yielding tomato cultivar with desired fruit qualities and longer shelf life is therefore a vital decision a producer must take. Failure to select an appropriate cultivar may lead to lower yield, low quality fruits, or less market acceptability. Fruits of different cultivars differ in size, colour, texture, and flavour as well as storage potential. Tomato cultivar Roma VF has higher sugar content whilst maintaining lower weight loss as compared to cultivar Marglobe. Cultivar selection is therefore critical to the postharvest storage life and eating qualities of tomatoes.

Grafting of vegetable seedlings is a unique horticultural technology, practised for more than 50 years, to overcome problems associated with intensive cultivation on limited arable land. Grafting vegetables can protect against soil-borne diseases and nematodes, against abiotic stresses such as high/low temperatures, salinity, drought or excessive soil-water content, and against elevated soil concentrations of heavy metals and organic pollutants. In addition, the grafted plant takes up water and nutrients from the soil more efficiently and retains its vitality for longer periods during the growing season (Elazar Fallik and Zoran Ilic, 2014). However, rootstock/scion combinations may affect and alter the final size, yield, and quality of fruits from grafted plants, both immediately postharvest and during prolonged storage. These alterations may be attributed in part to differing production environments and methods, the type of rootstock/scion combinations used, and harvest date.

b. Weather: Weather has a marked effect on the development of plant diseases, from the amount of inoculum that overwinters to the amount of pesticide residue that remains on the crop at harvest. Environmental conditions during growth, such as low temperature, wind, rain or hail can have a direct effect on a crop and not only determine the yield but also fruit quality after harvest. Wind and rain may be responsible for the dispersal of spores of many pathogens that survive in soil and dead plant parts in an orchid. Abundant inoculum and favorable conditions for infection during the growing season often result in heavy infection when the produce is harvested. For example, conidia of the fungus that causes brown rot on citrus fruit are dispersed from the soil surface and from rotten fruits after heavy rains especially near harvest time, causing infections that later develop during storage or shipment. Infection of many tropical and subtropical fruits by anthracnose in the orchard is greatly dependent on temperature, inoculum dispersal and relative humidity. Moisture conditions in the field at harvest are very important in rot development. Hence, fruit should not be harvested after prolonged rains. Moreover, relatively high temperatures in the field during harvest can render tropical fruit more susceptible to chilling injury and contribute to the development of disease. The combination of high temperature and relative humidity favors the growth of postharvest pathogens which in turn can significantly increase disease development when the fruit reaches a retail market (Droby *et al*, 2011). The effect of weather factors like temperature and moisture on postharvest diseases have been discussed in this book.

c. Cultural practices: Cultural practices such as tree pruning, handling of debris, and removal of decayed fruits may have a marked influence on the survival and quantity of pathogen inoculum in an orchard. Therefore, sanitation programs to reduce the inoculum in the orchard are an extremely important aspect of disease management. Picking bags contaminated with conidia can become a major source of inoculum, and research has shown that nylon mesh bags may reduce *Botrytis* build-up compared to enclosed bags. Korsten (2006) indicated that delaying harvesting of litchi fruit to as close to actual packing time as possible and developing harvesting techniques that caused less fruit injury could greatly increase fruit quality and extend shelf life. In grapes, early season horticultural practices have a substantial effect on bunch and berry structure that can affect fruit quality and influence its storage potential. 'Berry thinning' performed manually or by chemical induction of flower abscission with gibberellins prevents excessive fruit set. Failure in thinning may result in grape bunches that have a 'corn'-like appearance with tightly compressed berries, creating a favorable microclimate for infection and development of decay either before or after harvest. An open canopy in kiwifruit allows for plenty of air movement in the orchard which is an unfavorable condition for *Botrytis* conidial germination and establishment. Good orchard hygiene such as the removal of sources of *Botrytis* conidia including dead leaves, petals etc. can reduce the risk of *Botrytis* infection of the fruit and subsequent postharvest decay (Michailides and Elmer , 2006) In relation to postharvest decay, mechanical injuries such as scrapes, bruises or abrasions inflicted during harvesting and packing result in increased susceptibility to postharvest decay.

Bull's eye rot of pome fruits caused by Neofabraea perennans is characterized by infection occurring in the orchard throughout the growing season whereas rot lesions develop during long-term storage after harvest. Bull's eye rot was observed on pear fruit exposed to natural infection for any of six to nine sequential 1-to-2-week exposure periods during two growing seasons (Henriquez *et al*, 2008). Highest infection levels were associated with exposure closest to harvest. Over-tree irrigation and late harvest resulted in higher bull's eye rot incidence than under-tree irrigation and early or midseason harvest. Fruit were inoculated prior to harvest with Neofabraea perennans to determine the effect of environmental factors on the development of bull's eye rot. The effect of temperature was inconsistent; disease was greatest at 10°C in one year of study but greatest at 30°C in the second year. Bull's eye rot developed independently of wetness durations longer than 0.5 h.

d. Physiological and nutritional condition: The development of postharvest disease is intimately associated with the physiological status of the host tissue. To create the right environment for minimising postharvest losses due to disease it is important to understand the physiological changes that occur after produce is harvested. All plant organs undergo the physiological processes of growth, development and senescence. Growth and development generally only occur while the organ is attached to the plant (with the exception of seed germination and sprouting of storage organs), but senescence will occur regardless of whether the organ is attached or not (albeit at different rates). When an organ such as a fruit is harvested from a plant, it continues to respire and transpire depleting both food reserves and water. Such changes ultimately lead to senescence. Treatments which slow respiration and water loss, such as cool storage, therefore help to delay senescence. Different commodity types vary greatly in the processes which precede senescence, making it difficult to draw general conclusions. For this reason, the following discussion will be limited to fruit. 'Maturity' is a term often used in reference to fruit and is frequently confused with the term 'ripeness'. Put simply, physiological maturity is attained when the process of natural growth and development is complete. In practice, fruit are considered to be mature when they have reached a stage where, after harvesting and postharvest handling, they will ripen to an acceptable quality. In contrast, a ripe fruit is one which is ready to eat. The process of ripening basically signals the end of development and the beginning of senescence. It can involve a number of changes in the fruit, such as conversion of starch to sugars, increase in pH, increase in fruit softness, development of aromas, reduction in chlorophyll content and corresponding increase in carotenoid levels (yellow and orange pigments). Fruit are often classified into two groups on the basis of how they ripen. Climacteric fruit exhibit a pronounced increase in respiration and ethylene production coincidentally with ripening. Climacteric fruit can be harvested in an unripe state and providing they are sufficiently mature, will ripen to an acceptable quality. Non-climacteric fruit do not exhibit a rapid increase in respiration during the ripening process. The eatlng quality of non-climacteric fruit does not improve after harvest, although they may undergo some changes in colour development and softening. For this reason they should not be harvested until they are ready to eat.

The physiological condition of harvested fruit, as well as fruit maturity and ripening process, markedly affect its susceptibility to pathogen attack and disease development during storage and transit. It is a common principle that quality at harvest cannot be improved but merely maintained for a limited period of time. Harvesting fruits at the optimal maturity stage can therefore result in optimal quality and maximum shelf life. In most cases, advanced fruit ripening and the onset of senescence in various fruit renders them more susceptible to infection by pathogens (Kader, 2004). Either an excess or a shortage of water during the growing season can result in harvested fruit being more susceptible to postharvest diseases. Endophytic *Botryoshaeria* spp. is normally more prominent in trees exposed to continual water shortage and salt stress.

On the other hand, fruit can be made less prone to decay by management of crop nutrition. In this regard, calcium has been more closely associated with disease resistance than any other mineral cation associated with the cell wall. In general, produce containing adequate levels of calcium does not develop physiological disorders and can be stored longer with lower decay incidence (Hernández-Muñoz *et al*, 2006). Conversely, high nitrogen content in fruit predisposes them to decay. The lower the content of minerals such as calcium and magnesium in the avocado fruit, the higher is the incidence of anthracnose. The mineral content of immature and mature mango fruit (cv. Nam Dork Mai) grown under different soil conditions also had a direct relationship to postharvest fruit quality and rot development. Fruit grown in soil with higher Ca/N and (Ca + Mg)/K ratios remained firm, resistant to decay and had longer storage life.

The physiological component refers to the level of fruit susceptibility to crown rot, which reflects the physiological state of the fruit and is dependent on agrotechnical and pedoclimatic factors during plant growth. Only a few studies have been conducted on factors influencing the level of banana susceptibility to crown rot (Ludivine Lassois *et al*, 2010). In Guadeloupian conditions, it was shown that marked variations in fruit susceptibility could occur at the same production site over a period of 10 successive weeks. Variations in crown rot susceptibility among different Guadeloupian production areas during the same period have also been observed. Fruit age as expressed in accumulated degree day (dd) also seems to affect crown rot development. An intrabunch banana crown rot susceptibility gradient has been documented. Clusters growing on the first hands of a bunch, which are, on average,70 dd more advanced than the hands initiated last , are more susceptible than clusters developing on the last hands . Moreover, a linear relationship between fruit age (in dd) and the susceptibility of the fruit to crown rot was reported, with the oldest fruit being most susceptible to this disease . Some agricultural practices have a bearing on crown rot development. Modifying the source-sink ratio by trimming leaves and hands (where leaves are considered as sources and fruits as sinks) may induce variations in fruit susceptibility.

Nutrients could affect the disease tolerance or resistance of plants to pathogens. However, there are contradictory reports about the effect of nutrients on plant diseases and many factors that influence this response are not well understood. Dordas Christos (2008) summarized the most recent information regarding the effect of nutrients, such as N, K, P, Mn, Zn, B, Cl and Si, on disease resistance and tolerance and their use in sustainable agriculture. There is a difference in the response of obligate parasites to N supply, as when there is a high N level there is an increase in severity of the infection. In contrast, in facultative parasites at high N supply there is a decrease in the severity of the infection. K decreases the susceptibility of host plants up to the optimal level for growth and beyond this point there is no further increase in resistance. In contrast to K, the role of P in resistance is variable and seemingly inconsistent. Among the micronutrients, Mn can control a number of diseases as Mn has an important role in lignin biosynthesis, phenol biosynthesis, photosynthesis and several other functions. Zn was found to have a number of different effects as in some cases it decreased, in others increased, and in others had no effect on plant susceptibility to disease. B was found to reduce the severity of many diseases because of the function that B has on cell wall structure, plant membranes and plant metabolism. Cl application can enhance host plants' resistance to disease. Si has been shown to control a number of diseases and it is believed that Si creates a physical barrier which can restrict fungal hyphae penetration, or it may induce accumulation of antifungal compounds. Integrative plant nutrition is an essential component in sustainable agriculture, because in most cases it is more cost-effective and also environmentally friendly to control plant disease with the adequate amount of nutrients and with no pesticides. Nutrients can reduce disease to an acceptable level or at least to a level at which further control by other cultural practices or conventional organic biocides are more successful and less expensive.

e. Fungicide /chemical sprays : Fungicides /chemicals are used extensively to manage postharvest disease in fruits and vegetables. The timing of application and type of fungicide used depend primarily on the target pathogen, the timing of the infection, and the commodity. For postharvest pathogens that infect produce before harvest and generally remain quiescent until after harvest, field application of fungicides is often necessary. This may involve the repeated application of protectant fungicides during the growing season, and/or the strategic application of systemic fungicides. For example, in the control of mango anthracnose (*Colletotrichum gloeosporioides*) in Australia, trees are sprayed regularly with a protectant fungicide such as mancozeb during flowering and fruit development. If rain occurs during flowering, however, a systemic fungicide is also applied to inactivate infections already established and to guard against new infections. Protective sprays applied on a 7 to 14 day schedule have been extensively used to control anthracnose on mangoes in many crops.

Management of postharvest diseases of fresh fruits has relied for many years on the continuous use of conventional chemical fungicides. However, nonpolluting alternatives are increasingly needed because of human health and environmental issues related to the generation of chemical residues. Low-toxicity chemicals classified as food preservatives or as generally recognized as safe (GRAS) compounds have known and very low toxicological effects on mammals and minimal impact on the environment. Among them, inorganic or organic salts such as carbonates, sorbates, benzoates, silicates, etc., show significant advantages for potential commercial use, such as their availability, low cost, and general high solubility in water. Typically, these substances are first evaluated *in vitro* against target pathogens that cause important postharvest diseases (Lluis Palou, 2018). Selected salts and concentrations are then assayed as aqueous solutions in in vivo tests with target fresh fruit. Laboratory and small-scale experiments are conducted with fruit artificially inoculated with pathogens, whereas naturally infected fruit are used for large-scale, semicommercial, or commercial trials. Another approach that is increasingly gaining importance is evaluating GRAS salts as antifungal ingredients of novel synthetic edible coatings. These coatings could replace the fungicide-amended commercial waxes applied to many fruit commodities and could be used for organic or "zero-residue" fresh fruit production systems.

f. Fertilizer application: Consumers in the past few decades have become increasingly concerned about the quality of the food products they are consuming. Researchers have therefore investigated the impact of plant nutrition on the quality of fruits produced. The quality traits required or purpose for which the crop is grown will help in selecting not only the type of fertiliser but the quantity used during production. For instance, an adequate supply of potassium fertilizer in tomato production improves fruit colour and reduces the incidence of yellow shoulder, whilst enhancing the titratable acidity of the fruit. Yellow shoulder is a physiological disorder of tomatoes that is characterized by discoloured regions that border the stem scar (Isaac Kojo Arah *et al*, 2015). Insufficient supply of potassium in soilless tomato production can also result in ripening disorders. Unlike potassium, an increase in nitrogen supply to greenhouse-grown tomatoes, beyond a certain threshold level, may reduce fruit quality by decreasing the sugar content of the fruits. High nitrogen supply of about 250 kg/ha can impair some important quality traits of fruits, such as total soluble solids, glucose, fructose, and pH.

A supply of reduced forms of nitrogen, such as ammonium, can result in improved fruit flavours. However, the variation of phosphorus supply in soils for growing tomato crops does not significantly influence quality traits such as the total soluble solids, pH, acidity of the tomato juice, or the fruit colour characteristics. For trace elements use, the quality of tomato fruit is affected predominantly by the amount of boron used, although other micronutrients may affect fruit quality only when the plants show severe deficiency symptoms. Lower amounts of boron supply reduce fruit firmness which is of major concern during storage. Calcium application in tomato production has recently been studied and found to have a positive effect on the prevention of some diseases whilst slowing the reduction in fruit firmness during ripening. Spraying tomato leaves with different combinations of calcium salts is effective in controlling powdery mildew on the crop. Also an enhanced supply of calcium significantly increases the resistance to bacterial wilt caused by *Ralstonia solanacearum* in tomato. Calcium can be applied to the root zone in the soil or sprayed on the leaves. The type of application is determined by the growth stage of the crop and type of disease prevalent. Postharvest calcium application can also have a positive storage effect in calcium deficient harvested fruits.

Adequate mineral nutrition is central to crop production. However, it can also exert considerable Influence on disease development. Fertilizer application can increase or decrease development of diseases caused by different pathogens, and the mechanisms responsible are complex, including effects of nutrients on plant growth, plant resistance mechanisms and direct effects on the pathogen. The effects of mineral nutrition on plant disease and the mechanisms responsible for those effects have been dealt with comprehensively elsewhere (Parthasarathy, 2015). The influence of nitrogen, phosphorus, potassium, calcium and silicon on diseases is briefed hereunder.

Nitrogen : Nitrogen fertilizer applied above the recommended rates can result in increased disease incidence and lesion area. This has been demonstrated for biotrophic fungal pathogens such as powdery mildews and rusts and necrotrophic pathogens such as *Magnaporthe grisea*, the cause of rice blast. It is commonly thought that application of nitrogen fertilizer can increase disease severity via effects on crop canopy development. Thus, large canopies with high shoot densities may be more conducive to spore transfer and pathogen infection than sparse canopies. For example, nitrogen has been shown to increase the severity of *Fusarium* head blight in wheat, and it has been suggested that this might be the result of a nitrogen-induced increase in canopy size, leading to an altered microclimate. In contrast, work on yellow rust on winter wheat suggested that the impact of nitrogen on disease was the result of effects of nitrogenous substances in wheat leaves on pathogen growth, rather than effects on canopy growth and microclimate. However, nitrogen fertilization is not always associated with increased disease.

Several studies have reported no effect of nitrogen on disease severity, also the effect of nitrogen depended on the type of pathogen. Thus, nitrogen increased susceptibility of tomato to the powdery mildew pathogen *Oidium lycopersicum* and the bacterial pathogen *Pseudomonas syringae* pv. tomato, while it had no effect on susceptibility to the vascular wilt pathogen *Fusarium oxysporum* f. sp. *lycopersici* (Parthasarathy, 2015).. In contrast, tomato plants were more susceptible to *Botrytis cinerea* when grown under low nitrogen conditions. It is clear therefore that generalizing about the effects of nitrogen on plant disease is unwise and practically, although manipulation or assessment of crop nitrogen status might be used as part of disease control strategies, the approach adopted will depend on the crop and the pathogens from which it is most at risk.

Effects of preharvest nitrogen fertilization on the development of these postharvest diseases was studied by (Bally *et al*, 2009). Fruit from trees grown under high nitrogen (350 g $tree^{-1}$) applied in three applications during fruit growth had significantly higher anthracnose severity (20.6%) than fruit from trees grown without nitrogen (0 g $tree^{-1}$) (9.9%). When nitrogen was applied at rates between 0 and 417 g $tree^{-1}$ during fruit development, fruit from trees with more than 125 g N $tree^{-1}$ had significantly higher anthracnose severity (>8.2%) than trees receiving less nitrogen (2.8%). Fruit from trees with nitrogen applied at flowering and mid fruit development, had significantly higher anthracnose severity than trees with nitrogen applied 2 weeks prior to harvest, or no nitrogen. There were no treatment effects on the severity of stem-end rots in any of the experiments. These data confirmed the detrimental effects of high preharvest nitrogen on postharvest fruit anthracnose and suggested that nitrogen should not be applied during flowering or fruit development unless applied just prior to harvest, if nitrogen stimulated anthracnose is to be reduced.

In tomato plants grown with different regimes of N fertilization (0.5 to 20 mM), overall disease severity was lower for all isolates on plants with higher nitrogen inputs, but the effect of host N on *Botrytis* development depends on the strain and the concentration of inoculum leading to primary infection (Manzoor Ali Abro, 2013). One explanation for this could be that the fungal development depend both on trophic and defence processes, and that the availability of nutrients for the pathogen ultimately depends of its ability to explore new sources of unexplored plant tissues. Upon N scarcity, low substrate availability would impair growth of moderately aggressive isolates, those against which plant defences are more efficient, while resources available to aggressive strains would not be limiting, whatever the nutrition status of the host. More research is needed in order to find how N fertilization increase or decrese disease tolerance or resistance to different host-pathogen systems, what the changes are in plant metabolism and how this can be used to control plant disease. In addition, further research is required to use different isolates of the pathogen on different patho systems. The nutritional status of the host plant had a significant impact on the abundance and pathogenicity of secondary inoculum produced by plant pathogenic *B. cinerea*. This information could lead to exciting new studies to elucidate cellular mechanisms involved and suggests that fertilization can influence the development of epidemics on a field scale. Conversely, it opens the possibility of manipulating crop fertilization not only to reduce plant susceptibility but also to limit the impact of secondary inoculum and minimize the need for pesticide use.

The level of N fertilization provided to the tomato plant can influence not only the susceptibility of tomato stems to *B. cinerea* but also the efficacy of biocontrol. High N fertilization enhanced the efficacy of both biocontrol agents tested against *B. cinerea.* It suggested that the manipulation of N fertilization could provide a new useful tool for integrated protection of greenhouse tomato. To devise specific recommendations for growers, further work is needed to validate the present results in conditions of commercial production and to evaluate possible side effects on other diseases or pests in the tomato greenhouse. Thus, the novel results presented here on the effect of N fertilization on the efficacy of two biocontrol agents against *B. cinerea* should stimulate further work to better understand which mechanisms are involved at the cellular and molecular level.

At cyto-histological levels high N fertilization greatly influences *B. cinerea* mycelial development in tomato tissues. It is also enhanced the efficacy of fungal biological control agents. Thus the manipulation of N fertilization makes it possible not only to reduce the disease severity but also improve the efficacy of biological control against *B. cinerea* in green house tomato. To our knowledge, this is the first in planta observation of the interaction between N fertilization and biological control agents with *B. cinerea* (Manzoor Ali Abro, 2013). The effect of N fertilization on susceptibility appeared statistically mediated by the metabolism of sugars. As sugar accumulation is specific in each plant species, and for a given species, in its different organs, it might explain why the variation of plant susceptibility to necrotrophs in response to N fertilization is not unequivocal. Genotypes with contrasted sugar accumulation should be useful to test a general relationship between specific sugars and resistance to necrotrophs in lettuce and other species. Clearly more work is needed in this area and it is, of course, likely those different fungi will respond in different ways to the application of N fertilization. It is important that future studies combine measurements of fungal growth and symptom production with plant biomass and in plant nutrient levels. Further studies are thus required to understand the mechanisms involved at the level of gene expression. Trophic interactions related to primary metabolism are unlikely to be the only mechanisms implicated. Other compounds related to plant defense mechanisms were not examined in our study but would merit further attention as they could influence fungi indirectly through their effect on plant tissue colonization by the pathogen.

Most of the studies have clearly shown that optimum N fertilization of the host plant has decreased the susceptibility of tomato plants to *B. cinerea,* their sporulation and aggressiveness and also high N fertilization enhanced the efficacy of biocontrol agents against this pathogen for the protection of pruning wounds of tomato (Manzoor Ali Abro, 2013). This is probably because N nutrient is involved in the tolerance or resistance mechanism of the host plant. In sustainable agriculture balanced nutrition is an essential component of any integrative crop protection programme because in most cases it is more cost-effective and also environmentally friendly to control plant disease with the adequate amount of nutrients with no pesticides. Also, balanced nutrient fertilization can reduce disease to an acceptable level, or at least to a level at which further control by other cultural practices or conventional biocides are more successful and less expensive.

Phosphorus: In an analysis of some studies of the effects of fertilizer on more than 400 diseases and pests, found that, in general, phosphorus fertilization tended to improve plant health, with reductions in disease recorded in 65% of cases. Nevertheless, phosphorus fertilization increased disease and pest problems in 28% of the cases examined. As with nitrogen, the effects of phosphorus on plant disease may be the result of direct effects on the pathogen, host plant metabolism, leading to effects on pathogen food supply, and effects on plant defenses. Indeed, foliar application of phosphate salts has been shown to induce resistance to pathogens in a range of crop plants, including cucumber, broad bean, grapevine, maize and rice. Clearly, an adequate phosphorus supply is important for crop growth and in turn, may well help to reduce disease (Parthasarathy, 2015). However, the regime of phosphorus fertilizer used will depend on a range of factors, including the crop and the pathogens likely to be important. suggested that foliar-applied phosphate might be used as part of an integrated disease control programme, although grower adoption of such an approach will depend on the existence of other effective disease control measures and the economics of disease control in the particular crop.

Potassium : In an analysis of 181 papers reporting effects of potassium on plant disease, found that 120 (66%) reported reductions in disease, while 49 (27%) reported an increase in disease. Although this suggests that in many cases, potassium is associated with disease reductions, point out that inadequate consideration has been given to the effects of associated anions, nutrient balance and nutrient status to allow the definitive role of potassium to be determined. For example, it has been suggested that in some cases, the effects of potassium, applied as potassium chloride fertilizer, might be due to the chloride ion rather than potassium. Further, chloride fertilization has been shown to suppress disease in cereal crops (Parthasarathy,2015). Foliar-applied potassium chloride has been shown to control *Blumeria graminis* and *Septoria tritici* on wheat in field studies, probably due to osmotic effects on the fungal pathogens, disrupting pathogen development and subsequent infection. Application of potassium to deficient soils usually increases plant resistance to diseases. This might be partly related to the effect of potassium in increasing epidermal cell wall thickness or disease escape as a result of vigorous crop growth, although the mechanisms by which potassium affects plant disease are not well understood.

The effect of potassium sorbate (K-sorb), a low-toxicity chemical, to control *Monilinia* spp. was investigated (Gregori Roberto *et al*, 2008). Preliminary *in vitro* studies found the MIC of K-sorb for conidial germination and mycelial growth was, respectively, 260 and 1,250 mg/ liter. Immersion of naturally infected peach and nectarine fruit in a solution (15 g/liter) of K-sorb for 120 s reduced brown rot by over 80% in four of five trials. Although treated fruits showed a significant reduction in firmness with respect to the control, they did not reach the overripe stage and retained acceptable quality parameters. In an attempt to elucidate the mechanism of action for K-sorb, the inhibition of enzymatic activity by K-sorb was also tested. In a radial diffusion assay, the addition of K-sorb to agarose reduced polygalacturonase (PG) activity across the concentrations considered. The greatest reduction (54.3%, with respect to the control) was obtained at a sorbate concentration of 15 g/liter. PG kinetic activity of *Monilinia laxa* observed by a spectrophotometric assay peaked after 40 min in all samples tested. PG activity was significantly higher in the control than in the samples with increased K-sorb concentrations. In conclusion, based on these findings, K-sorb can be recommended as a low-toxicity antifungal compound against *Monilinia* spp. in peaches and nectarines with its mode of action probably depending in part on the inhibition of PG activity in M. laxa.

Since 2010, potassium silicate (PSi, K_2SiO_3) is included in the list of synthetic substances allowed for use in organic crop production in the USA (Moscoso-Ramírez and Palou, 2016). Thus, the curative activity of postharvest treatments with PSi against one of the most economically important citrus postharvest diseases, green mold caused by the fungus *Penicillium digitatum*, was evaluated in laboratory trials. Oranges (*Citrus sinensis* L. 'Lane Late') were artificially inoculated in rind wounds with the pathogen and 24 h later immersed in aqueous solutions of 90 mM PSi at 20 or 50°C for 60 or 150 s. This PSi concentration was selected as the most effective in previous in vivo primary screening tests. Treated fruit were incubated for 7 days at 20°C, and then disease incidence (% of infected wounds) and severity (lesion diameter) were recorded. Dips at 20°C for 60 s were selected as the most effective and practical treatment conditions. These treatments were then applied to 'Valencia' oranges inoculated 24 h before and subsequently stored at 5°C and 90% RH for up to 6 weeks. Control fruit were dipped in water. At the end of the cold storage, PSi dips significantly reduced the incidence of green mold by 45% with respect to control fruit. Lesion diameter at the end of the cold storage was 200 and 98 mm on control and PSi-treated oranges, respectively. In conclusion, postharvest PSi treatment showed potential as a new reduced-risk chemical treatment for cost-effective control of citrus green mold. It can be of use in integrated disease management programs to replace or reduce the use of conventional postharvest fungicides.

Calcium: Calcium is essential macronutrient involved in numerous biochemical and physiological processes in the plants. It acts as intracellular messenger and is responsible for maintaining the plant cell wall integrity. These properties instigate significant impact on the postharvest fruit quality, ripening, senescence, decay, and physiological disorders. Pre-and postharvest application of calcium has reportedly reduced incidence of different disorders, maintained quality and extended postharvest life in horticulture produce. In the present Chapter an attempt is made to provide an overview about different aspects of calcium application, mechanism of action and its impacts on the postharvest life of the fruits and vegetables (Tokala, Vijay Yadav and Mahajan, 2018).

There are many reports that application of calcium to soils foliage and fruit reduces the incidence and severity of a range of diseases of crops, including cereals, vegetable crops, legumes, fruit trees, as well as postharvest diseases of tubers and fruits. For example, calcium has been shown to inhibit anthracnose caused by *Colletotrichum gloeosporioides* or *C. acutatum* in apples and to decrease postharvest disease development on strawberry, while treatment of tomato with calcium carbonate reduced *Fusarium* crown rot disease. In contrast, could find no effect of calcium on anthracnose on strawberry. Because calcium increases resistance of plant cell membranes and cell walls to microbial enzymes, increasing calcium concentrations in storage organs could lead to enhanced resistance to pathogens. However, the form in which the calcium is applied can influence the mechanism by which calcium affects disease. For example, the addition of lime can affect disease by altering pH, while calcium salts (e.g. propionate) can be directly inhibitory to pathogens. Making general recommendations for the use of calcium in plant disease control would be unwise due to the range of crops and pathogens affected by calcium application. Instead, the appropriate amount and form of calcium to be applied needs to be determined for individual crop–pathogen interactions (Parthasarathy, 2015). The dwindling availability of fungicides, together with increasing public concern for the environment means that the use of calcium to control plant disease, especially postharvest, is attracting increased attention.

Hot water immersion (HW) and calcium chloride (Ca) treatments have been used to control papaya postharvest diseases, however, the effect of the combination HW-Ca on the pathogen growth and the development of the disease in infected papaya fruit has been scarcely studied (Ayón-Reyna *et al*, 2017). The aim of this study was to evaluate the effect of the HW-Ca treatment on the *in vitro* growth of *C. gloesporioides* conidia and the quality of infected papaya. *In vitro*, the HW-Ca treated conidia showed reduced mycelial growth and germination. *In vivo*, the HW-Ca treatment of infected papaya delayed for 5 days the onset of the anthracnose symptoms and improved the papaya postharvest quality. The combined treatment HW-Ca was better than any of the individual treatments to inhibit the *in vitro* development of *C. gloeosporioides* and to reduce the negative effects of papaya anthracnose.

Silicon: Although silicon (Si) is not recognized as an essential element for general higher plants, it has beneficial effects on the growth and production of a wide range of plant species. Si is known to effectively mitigate various environmental stresses and enhance plant resistance against both fungal and bacterial pathogens. In this review, the effects of Si on plant–pathogen interactions are analyzed, mainly on physical, biochemical, and molecular aspects. In most cases, the Si-induced biochemical/molecular resistance during plant–pathogen interactions were dominated as joint resistance, involving activating defense-related enzymes activates, stimulating antimicrobial compound production, regulating the complex network of signal pathways, and activating of the expression of defense-related genes. The most previous studies described an independent process, however, the whole plant resistances were rarely considered, especially the interaction of different process in higher plants. Si can act as a modulator influencing plant defense responses and interacting with key components of plant stress signaling systems leading to induced resistance (Wang *et al*, 2017). Priming of plant defense responses alterations in phytohormone homeostasis, and networking by defense signaling components are all potential mechanisms involved in Si-triggered resistance responses.

Although the effects of silicon in reducing disease severity have been known since 1940, it was not until the 1980s that more detailed work was carried out in this area. In this work, cucumbers grown in nutrient solutions supplemented with silicon were found to have significantly less powdery mildew infection than plants not receiving silicon supplementation. Indeed, silicon has been shown to suppress both foliar and soil-borne pathogens in cucurbits and to reduce susceptibility of rice to various pathogens. Wheat grown in soil amended with silicon showed reduced infection by several pathogens, including *B. graminis* f. sp. *tritici*, *S. tritici* and *Oculimacula yallundae*. (Parthasarathy, 2015). It has been suggested that the effects of silicon in providing disease control are due to the creation of a mechanical barrier to penetration. However, this has been disputed by studies which could find no evidence for the creation of a physical barrier following silicon treatment in wheat inoculated with powdery mildew and bitter gourd and tomato inoculated with *Pythium aphanidermatum*. Rather, several studies have suggested that silicon activates defences in plants. For example, in wheat inoculated with *B. graminis* f. sp. *tritici*, epidermal cells of silicon-treated plants were shown to react to attempted infection with specific defences, including papilla formation and callose production. In the rice M. grisea pathosystem, silicon-mediated resistance was found to be associated with accumulation of antimicrobial compounds at infection sites, including diterpenoid phytoalexins. In fact, phytoalexin accumulation occurs in silicon-mediated resistance in both dicots and monocots and since phytoalexins are highly specific to plant species, it has been suggested that silicon might be acting on mechanisms shared by all plant species, for example, those resulting in activation of plant stress genes.

Song, *et al*, (2016) reported for the first time that bacterial blight of rice can be alleviated by silicon (Si) added. In both inoculated and uninoculated plants, shoot dry weight was significantly higher in the +Si plants than in the −Si plants. A soil-cultured trial showed that disease severity was 24.3% lower in the Si-amended plants than in the non-Si-amended plants. Plants that were switched from −Si to +Si nutrient solution and simultaneously inoculated with Xoo also exhibited the same high resistance to bacterial blight as the plants that were treated continuously with Si, with control efficiencies of 52.8 and 62.9%, respectively. Moreover, total concentrations of soluble phenolics and lignin in rice leaves were significantly higher in the +Si plants than in the −Si plants. Polyphenoloxidase (PPO) and phenylalanine ammonia-lyase (PAL) activities in rice leaves were observed to be higher in the +Si plants than in the −Si plants. The expression levels of Os03g0109600, Prla, Rcht2 and Lox2osPil, were also higher in +Si plants than in −Si plants post-inoculation during the experimental time. Addition of Si resulted in increased Pal transcription, and inhibited CatA and Os03g0126000 expression in the earlier and later stages of bacterial inoculation, respectively.

Silicon (Si) has proven to enhance disease resistance in a wide range of tropical fruits and vegetables. It has been used in controlling various diseases: mildews, rots, moulds, wilt, blight, anthracnose and leaf spots. However, the highest number of records was available on reducing diseases of powdery mildews on fruits and vegetables belonging to the family Cucurbitaceae. Silicon mediated defense responses in plant pathosystems are mainly attributed to the physical resistance, which involves reduced penetrability and/or increased hardness and abrasiveness of plant tissues because of silica deposition. The main mechanism is the chemical resistance, which involves chemical defenses to pathogen attack through the enhanced production of defensive enzymes and the production of antifungal compounds such as phenolic metabolism products, phytoalexins and pathogenesis related proteins. Silicon has been applied as soluble silicates to the substrate or to soilless media at pre-harvest level on reducing disease susceptibility (Weerahewa Darshani and Somapala Kosal, 2016). Postharvest dips of fresh produce in silicate solutions and use of Si combined with a biocontrol agent are other aspects of silicon application in controlling diseases in fruits and vegetables.

g. Pruning: Controlling the number of flowers, fruits, or fruit trusses in tomatoes is an effective way of reducing the competition between fruits. Pruning therefore ensures nutrients are channelled to fewer fruits sinks which can lead to increased fruit size whilst increasing sugar content of fruits in some cases. Pruning clusters to three fruits increased total marketable yield and fruit weight and reduced cull yield of all cultivars under investigation. Meanwhile, the effect of pruning on other quality traits of the fruit produced depends on many factors including the sink developmental stage, fruit to leaf ratio, truss position, and genetic background (Isaac Kojo Arah *et al*, 2015). Pruning can result in increased fruit size in most cultivars, under the right growing conditions, and thus can be used to improve the marketability of high total soluble solids (TSS) fruits which in most cases tend to be smaller in size. Larger fruits which are within a certain size range are usually considered to be of better value by most consumers.

h. Irrigation: Proper irrigation scheduling production is therefore crucial to the crop development. However, with water being a scarce resource in most production areas, growers in recent years have therefore had to develop a more efficient water management scheme that maintains crop yield but has a moderate and controlled level of moisture stress on their crops. In a study it was revealed that deficit irrigation reduced fruit water accumulation and fresh fruit yield but increased fruit total soluble solids] levels. They also discovered that irrigating with saline water had no significant effect on total fruit yield but moisture content of fruits was slightly reduced. Ismail *et al.* Early morning irrigation every three days resulted in higher yields than daily irrigation (Isaac Kojo Arah *et al*, 2015). The use of trace elements or the practice of soilless tomato production can be made possible during irrigation where the fertilisers (trace elements) are added to the irrigation water in a form of solution and administered. These trace elements are selected depending on the specific postharvest quality traits needed in the fruits.

i. pH: In the case of postharvest pathogenic fungi, ambient pH is important because it contributes to the ability of the pathogen to successfully colonize the targeted host tissue. Pathogens may enhance their virulence by locally modulating the host's ambient pH either up or down. *Penicillium* spp.

colonization acidified citrus and apple tissues and was enhanced by low pH. The increase in ambient pH was shown to increase gene expression of hydrolases in *Colletotrichum gloeosporioides* and of glucanases in *Alternaria alternate* (Smilanick *et al*, 2005). pH is a major environmental factor that affects the interaction between *B. cinerea* and its hosts. Fruits usually present a pH ranging from 3.32 to 4.39, whereas leaves, stems, and roots exhibit a higher pH ranging from 5.81 to 6.3 (Li Hua, et. al., 2018). To explore the secretome of *B. cinerea* on different host tissues, we chose pH 4 and 6 to mimic the pH values of fruits and other tissues, cultured *B. cinerea* at different pH conditions and found that distinct differences exist in the secretome of *B. cinerea*. At pH 4, most of the identified proteins were proteolysis, whereas the major proteins detected at pH 6 were CWDEs. The proteases are usually utilized by the fungi to degrade the structural plant cell or antifungal proteins secreted by the plant host, and CWDEs are essential for fungi to decompose plant cell wall to achieve full virulence. Moreover, we found that the production of those extracellular proteins was regulated at the transcriptional level, suggesting that *B. cinerea* has the ability to fine-tune its secretome according to the predominant pH conditions to achieve successful infection, and that it might possess complicate regulatory mechanism to perceive and response to ambient pH at the transcriptional level (Figure 30).

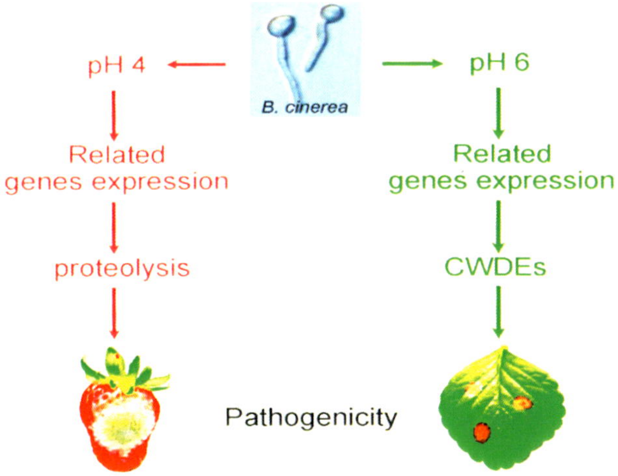

Fig. 30 : *Botrytis cinerea* utilize different extracellular enzymes to infect hosts according to the ambient pH conditions.

The most well characterized regulatory mechanism is pal signalling pathway. In *Aspergillus* nidulans, several genes are involved in the pal signalling pathway, namely, PacC, PalA, PalB, PalC, PalF, PalH, and PalI. Under alkaline conditions, the pH signal is sensed and transmitted from the plasma membrane to the endosomal membrane by PalH, PalI, and PalF. When the endosomal membrane complex including PalA, PalB, and PalC receives the signal, the complex will proteolyse PacC to a protease-accessible conformation. PacC will be further processed to its active form in a pH-independent manner. Under acidic conditions, most of PacC exist in a protease-inaccessible conformation and only trace amounts of PacC are in protease-accessible conformation. PacC is a zinc finger transcription factor which will activate the alkaline-expressed genes and repress acid-expressed genes (Penalva *et al*., 2008).

PacC has also been reported in many other plant pathogens, such as *Sclerotinia sclerotiorum* (Rollins and Dickman, 2001), *Colletotrichum alternata, Fusarium oxysporum* (Caracuel *et al*., 2003), *P. digitatum*, and *Penicillium* expansum. In *B. cinerea*, the gene expression of bcpacc was significantly higher at pH 6 than that at pH 4, indicating its role in ambient pH response of *B. cinerea*. BcACP1 is a G1-family endopeptidase which only functions in acid environment. Ambient pH regulates bcacp1 at both the transcriptional and post-translational levels; however, the pH regulation of BcACP1 was independent of the canonical PacC binding site (Rolland *et al*., 2009), suggesting other binding site of BcPacC or even other pH signalling pathway might exist in *B. cinerea*. Besides adjusting an arsenal of pathogenicity factors to fit for the wide range of environment conditions, most fungi hold the potential to actively altering the local pH environment to fulfil its infection. Fungi usually achieve environmental pH changes through secreting acids or alkali. The organic acids secreted by fungi include oxalic acid, gluconic acid, citric acids, butyric acid, malate and succinate, whereas the alkali usually refers to ammonia. *Botrytis cinerea* has been reported to acidify the plant tissue through secreting large amounts of oxalic acid. Acidic pH environment can in turn regulate the activity or expression of putative pathogenicity factors, such as endopolygacturonase, laccase, and protease . An oxalate-deficient mutant of *B. cinerea* grew normally *in vitro*, but failed to produce disease symptoms. These results demonstrated that *B. cinerea* can modulate environmental pH conditions via utilizing OA, thus contributing to its colonization of the host tissues.

In vitro, spores of *Penicillium digitatum* germinated without inhibition between pH 4 and 7, but were inhibited at higher pH. Estimated concentrations of imazalil (IMZ) in potato-dextrose broth–Tris that caused 50% reduction in the germination of spores (ED50) of an IMZ-sensitive isolate M6R at pH 4, 5, 6, and 7 were 0.16, 0.11, 0.015, and 0.006 μg/ml, respectively (Smilanick *et al.*, 2005). ED50 IMZ concentrations of an IMZ-resistant isolate D201 at pH 4, 5, 6, and 7 were 5.9, 1.4, 0.26, and 0.07 μg/ml, respectively. The natural pH within 2-mm-deep wounds on lemon was 5.6 to 5.1 and decreased with fruit age. IMZ effectiveness to control green mold and its residues increased with pH. The pH in wounds on lemon fruit 24 h after immersion in 1, 2, or 3% $NaHCO_3$ increased from pH 5.3 to 6.0, 6.3, and 6.7, respectively. $NaHCO_3$ dramatically improved IMZ performance. Green mold incidence among lemon fruit inoculated with M6R and treated 24 h later with IMZ at 10 μg/ml, 1% $NaHCO_3$, or their combination was 92, 55, and 22%, respectively. Green mold among lemon fruit inoculated with D201 and treated 24 h later with water, IMZ at 500 μg/ml, 3% $NaHCO_3$, or their combination was 96.3, 63.0, 44.4, and 6.5%, respectively. $NaHCO_3$ did not influence IMZ fruit residue levels.

j. Light: Light is a vital environmental factor acting as energy source, signal, and stress to various living organisms. Light responsive organisms sense light to schedule their development and adjust their adaption (Li Hua, *et al.*, 2018). *Botrytis cinerea* is a light responsive strain that actively sense light conditions to fine-tune its development and pathogenicity. Light is an essential developmental signal for *B. cinerea* as it triggers exclusively formation of conidia, whereas constant darkness initiates the solely formation sclerotia. Conidia and sclerotia are two types of survival structures of *B. cinerea*. Conidia are asexual spores produced under favourable environmental conditions and contribute to rapid growth and reproduction, whereas sclerotia are sexual spores that allow for the fungi to survive adverse environmental conditions. The balance between asexual and sexual development is tightly regulated to ensure better survival and drive adaptive responses. Extensive studies about light regulation on the development of *B. cinerea* have been conducted, especially those light responsive transcription factors. BcMADS1 is MADS-box family transcription factors that play crucial roles in regulation of various cellular functions. It was observed that deletion mutant of bcmads1 in *B. cinerea* resulted in reduced growth rate compared with the wild type B05.10. An always conidia phenotype which lost the ability to produce sclerotia was observed, and the conidia formed by *B. cinerea* in dark was even higher than that produced in light (Brandhoff *et al.*,2017). The complementation strain of bcmads1 restored the sclerotia formation ability of Δbcmads1, indicating the function of BcMADS1 in light perception and in the environment fitness of this pathogen. Other six light-responsive transcription factors that are significantly induced when exposed to white light were termed BcLTF1-6.

BcLTF1 is a GATA transcription factor participating in the regulation of light-dependent vegetative growth and differentiation. Deletion mutants of bcltfl (Δbcltfl) lost the ability to grow under light conditions on minimal medium, and the radial growth rates have a negative correction with the exposure time to light. However, Δbcltfl forms more conidia upon exposed to yellow light and more sclerotia under red light than wild type (Schumacher *et al*., 2014). BcLTF2 belongs to the C_2H_2 transcription factor family, and the overexpression of BcLTF2 is capable of switching the conidiation program on and suppressing sclerotia generation. BcLTF3 is another C_2H_2 transcription factor with dual functions and represses conidiophore development but is essential for the maturation of conidia. Deletion mutant of BcLTF3 produced abundant conidiophores but failed to produce mature conidia. BcLTF4-6 are Zn2Cys6 transcription factors and exhibit no obvious roles in vegetative growth or development. The expression of BcLTF2 is tightly regulated by BcLTF1, BcLTF3, BcREG1, and other proteins.. BcREG1 is a transcriptional factor previously known to be involved in the conidiogenesis and secondary metabolism in *B. cinerea* . In recent study, BcREG1 was re-identified as a light-responsive transcriptional regulator which influences conidia formation by repress BcLTF2-induced conidiation in the light, while active a conidiation program independent of BcLTF2 in the dark, suggesting that light is a determinant factor of generating conidia or sclerotia.

Some of the light-responsive transcriptional factors or regulators also function in the regulation of pathogenicity. They found that Δbcmads1 showed reduced disease symptoms on apple fruit when compared with the wild type. Based on proteomic analysis, they identified the potential downstream targets of BcMADS1, and proved that proteins changed abundance in Δbcmads1. Among them, two proteins (BcSEC14 and BcSEC31) were related to secretion and involved in the pathogenicity of *B. cinerea* . BcLTF1 was also demonstrated to play an important role in virulence, ROS homoeostasis, and secondary metabolism. Deletion of bcltfl resulted in reduced virulence on bean leaves, increased ROS accumulation, and overexpression of secondary metabolism related genes. The light-responsive transcriptional regulator BcREG1 is required for pathogenicity, and the deletion mutant of bcreg1 is hampered in causing necrotic lesions though capable to penetrate plant tissue. Δbcreg1 mutant displayed reduced ability to produce phytotoxins such as botryane, sesquiterpenes, and polyketides, which might facilitate pathogenesis. It is worth noting that the pathogenesis related light responsive transcriptional factors or regulators also function in protein secretion or second metabolism. In filamentous fungi, velvet proteins are among the most important components coordinating light signal and secondary metabolism. Velvet proteins have been identified as a heterotrimeric velvet complex VelB/VeA/LaeA in A. nidulans (Schumacher *et al.*, 2015). VeA physically interacts with VelB and bridges VelB to LaeA, a nuclear master regulator of secondary metabolism. In *B. cinerea*, BcVEL1/BcVEL2/BcLAE1 are, respectively, homologue to VelB/VeA/LaeA. BcVEL1 interacts with BcVEL2 and BcLAE1 to form the velvet complex. Deletion mutants of BcVEL1, BcVEL2, or BcLAE1 present similar phenotypes such as reduced virulence and altered secondary metabolism.

The effect of ultraviolet-C light (u.v.-C) at low doses on postharvest decay of strawberries caused by *Botrytis cinerea* and other pathogens was investigated by Franco Nigro *et al*, (2000). Pheny-lalanine ammonia-lyase (PAL) activity and ethylene production, as influenced by ultraviolet-C irradiation, were also determined. Strawberries (cv. 'Pajaro') from plants that had been treated with chemicals against grey mould were irradiated with u.v.-C doses ranging from 0.25 to 4.00 kJ m -2 and inoculated with *B. cinerea* at different times (0, 12, 24 and 48 hours) after irradiation. To assess the effect of u.v.-C light on the naturally occurring postharvest decay, organically grown straw-berries were also used. After treatment the strawberries were stored at 20±1°C or at 3°C. u.v.-C doses at 0.50 and 1.00 kJ m -2 significantly reduced *botrytis* storage rot arising from both artificial inoculations and natural infections in comparison with the unirradiated control. The doses shown to reduce *botrytis* rot produced an increase in PAL activity 12 h after irradiation; this result indicates the activation of metabolic a pathway related to the biosynthesis of phenolic compounds, which are usually characterized by antifungal activity. In addition, u.v.-C irradiation caused an increase in ethylene production proportional to the doses applied, reaching the highest value 6 h after treatment. The overall results from these investigations indicate that treatment with low u.v.-C doses produces a reduction in postharvest decay of strawberries related to induced resistance mechanisms. Moreover, a germicidal effect of reducing external contaminating pathogens cannot be excluded.

Application of technologies using light as an abiotic factor aiming to promote a regulatory and signaling role in the developmental and metabolic processes of plants is encouraged. Ultraviolet light (UV-C) is a continuous radiation that is widely reported in the induction of resistance mechanisms and control of postharvest diseases, thus extending the shelf-life of fruit and vegetables (Francisco Oiram Filho *et al*, 2019). However, studies involving pulsed light (PL) as radiation from the perspective of inducing resistance mechanisms are still discreet. PL is a new, nonthermal technology, where a lamp containing an inert gas such as xenon emits high frequency radiation pulses between wavelengths of 180 to 1100 nm. This technology is used to sanitize fruit and vegetable surfaces by acting on microorganism cells and breaking and altering their DNA sequences, thus inhibiting the pathogen reproductive capacity . Moreover, the use of PL as a physical method can stimulate the production of phytochemicals in plant tissues to minimize the possible deleterious effects caused by radiation. Metabolomics has been emphasized within the "omics" sciences, and efficiently evaluates a large part of the metabolites of an organism both quantitatively and qualitatively, at a given time and in a specific situation. Metabolomics can identify a change in the concentration of compounds involved in primary metabolism or production of secondary metabolites. These 81 metabolic alterations may arise from cellular lesions, metabolic adjustments to restore cellular homeostasis, or synthesis and/or accumulation of metabolites resulting from cellular pathways.

Pulsed light, as a postharvest technology, is an alternative to traditional fungicides, and can be used on a wide variety of fruit and vegetables for sanitization or pathogen control. In addition to these applications, other effects also are detected in vegetal cells, including changes in metabolism and production of secondary metabolites, which directly affect disease control response mechanisms. the possible applications of pulsed ultraviolet light in controlling postharvest rot, mainly caused by *Fusarium pallidoroseum* in yellow melon 'Goldex', in natura and its implications in the disease control as a function of metabolomic expression to effect fungicidal or fungistatic was taken up (Francisco Oiram Filho *et al*, 2019). The dose of pulsed light (PL) that inhibited *F. pallidoroseum* growth in melons (Cucumis melo var. Spanish) was 9 KJ m^{-2}. Ultra performance liquid chromatography (UPLC) coupled to a quadrupole time-of-flight (QTOF) mass analyzer identified 12 compounds based on the MS/MS fragmentation patterns. Chemometric analysis by Principal Components Analysis (PCA) and Orthogonal Partial Least Squared Discriminant Analysis (OPLS-DA and S-plot) were used to evaluate the changes in fruit metabolism. PL technology provided protection against postharvest disease in melons, directly inhibiting the growth of *F. pallidoroseum* through upregulation of specific fruit biomarkers such as pipecolic acid , saponarin and orientin , which acted as major markers for the defense system against pathogens. PL can thus be proposed as a postharvest technology to avoid chemical fungicides and may be applied to reduce the decay of melon quality during its export and storage.

Considering the high economic worldwide loss due to the postharvest decay of fruit and vegetables, and the frequent development of pathogen isolates that are resistant to synthetic fungicides, there is a need to develop more eco-friendly alternatives for controlling postharvest diseases. Irradiation (Gamma and UV-C) is one such physical method which is promising, but when used separately is not as effective as fungicides (Deepshikha *et al*, 2017). Therefore, to improve its effectiveness and persistence, irradiation method can be combined with other complementary control measures such as sodium carbonate, heat treatment, chemicals, cold storage and bio-control agents as a part of integrated disease management strategies. Low doses of UV radiation, particularly UV-C have proven to be effective in delaying ripening and senescence, diminishing decay and even in increasing the content of beneficial compounds owing to its germicidal properties. It mainly inactivates microorganisms, especially pathogens by creating damage in the genetic material (DNA) of the cell. These methods can be efficiently used to delay the ripening of fruits, to inhibit germination, improve nutritional quality, minimize insect infestation and deactivate viruses. Short shelf life of mushroom can be extended by inhibiting cap opening and browning, stalk elongation, reducing the level of microbial contamination and finally by increasing the concentration of vitamin D2 significantly, without causing any adverse effect on its taste.

Ensuring food safety and freshness of fresh produce is always challenging because of a variety of microbial contamination in the field by soil and water, wild or domesticated animals, as well as during harvest and postharvest operations by contaminated workers, tools and packaging materials (Arosha Loku Umagiliyage and Ruplal Choudhary,2018). A thermal kill process in fresh produce is not practical due to loss of freshness and quality. A nonthermal process to decontaminate and improve food safety is urgently required to improved food safety in fresh produce Industry. Although the Food Safety Modernization act recommends general agricultural practices and farm food safety rules, a kill step for human pathogens and spoilage organisms in food produce will help the fresh produce industry preventing recalls. Ultraviolet light is a promising sanitation technology to enhance food safety in produce industry. The article reviews the principles and applications of germicidal ultraviolet in fresh produce in general and berries in particular. A wide range of research on fresh produce shows promising use of ultraviolet light in berries with significant inhibition of pathogenic and spoilage bacteria and fungi on fresh produce. Innovative design of ultraviolet light delivery equipment that enables exposure of whole surface of produce will promote rapid adoption of this low cost sanitation technology readily available for the produce industry.

k. Inoculum load : Organic matter including soil, plant parts and remnants of rotten fruit in a packinghouse processing area can serve as substrates for decay-causing pathogens. Dump tanks, if not frequently cleaned and disinfected, become sources of contamination due to the accumulation of high quantities of spores. For these reasons, it is important to have and routinely apply a rigorous sanitation program in packinghouse sheds, cold storage rooms, and on the packing line itself (Kader and Rolle, 2004).. Inoculum sources may also come from trucks and other transport vehicles used to move the fruit from the orchard to the packinghouse. Additional sources of inoculum are container or cool trucks, port terminals, distribution centers, repack facilities, markets and other retail locations .

Several different sanitizers are usually used to disinfect surfaces, packing lines and dip tanks. These include chlorine (sodium and calcium hypochlorite), chlorine dioxide, peracetic acid and different quaternary ammonium-based products. Caution, however, should be exercised when using some of these sanitizers as they may be corrosive to metal or may be phytotoxic and cause injury to the fruit if used improperly in dip tanks.

Alfalfa, maize, sorghum and sugarbeet plants were inoculated with zoospores of *Phytophthora* and *Pythium* species in order to assess the effects of inoculum density, plant age and temperature on disease severity. Seedlings were grown axenically in test tubes and inoculated with zoospore suspensions (Raftoyannis *et al*,. 2002). Disease severity was assessed by measuring the root growth and discoloration of treated and control seedlings. The incremental root length of all plants decreased and root discoloration increased as inoculum concentration of the pathogen increased. Changes were more intensive among low levels of zoospore concentrations and no significant differences in disease severity were found for inoculum densities higher than 104 zoospores ml-1. Disease severity was negatively related to plant age. Disease development on sugarbeet seedlings infected with *Pythium* and *Phytophthora* species was affected by temperature, but the pattern of response was determined by the pathogen's temperature preferences. The incremental root length decreased as temperature increased up to 25°C. The effect of *Pythium* dissimile and *Phytophthora cactorum* on root length was significantly lower at 35°C than at 25°C, whereas *Pythium aphanidermatum* and *Phytophthora nicotianae* caused significant damage to roots even at 35°C.

The effect of inoculum load of postharvest pathogen (*Penicillium islandicum*) on disease initiation, threshold point and population build up of blue mould rot in Indian gooseberry was studied (Saini *et al*, 2019). Inoculations were made on the fruits of Indian gooseberry at five concentrations (10, 50, 100, 200 and 500 spores/ml.). The disease initiation commenced after 54 hrs. of inoculation with inoculum load of 500 spores/ml followed by 200 spores/ml where the initiation of disease was observed after 72 hrs. At lower spore load (10 spores/ml), there was no disease initiation even after 108 hrs of inoculation. Minimum number of spores required to initiate infection after 102 hrs was 50 spores/ml which can be considered as the threshold point. Maximum population build up (2 x 106 and 2 x 108) of the pathogen occurred at 500 spores/ml after 5 and 10 days of inoculation, respectively.

vi. Postharvest factors

a. Temperature and relative humidity: Penetration of the host by a pathogen does not automatically result in the development of decay, unless favorable conditions exist. These conditions include appropriate temperature and relative humidity. Spore germination and mycelial growth largely depend on temperature, and the optimal and minimal temperature may vary according to fungal species. In fact, most of the decay that occurs in the marketplace, particularly that of tropical fruits, results from overexposure to damaging temperatures. Temperatures either below or above the optimal range (e.g. freezing, chilling, high temperature) for a specific commodity can cause accelerated decay development. Temperature management is the most effective tool to minimize decay development and extend shelf life. It begins with rapid removal of field heat followed by refrigerated transport and storage. In this regard, maintaining the cold chain during all the postharvest processing from packing to the marketplace, is crucial to preserving quality(Kader and Rolle, 2004). High relative humidity (RH), required to protect fresh produce from dehydration and weight loss, also carries the potential to stimulate pathogen development during storage. Prospects for decay development are enhanced by water films or water condensation on a fruit surface. These conditions may arise when fruit is packaged with sealed plastic films having no permeability to water vapor or result from temperature fluctuations during shipment which can lead to condensation of water. Many fruits are more susceptible to pathogen attack when their tissue is in a turgid state as a result of being kept under high RH. Under these conditions fungal spores use the moisture to germinate and penetrate the tissue.

Chocolate spot (*Botrytis* fabae) is a devastating disease of faba bean and reduces its production and productivity. Three controlled condition experiments were conducted to assess the effect of temperature on growth and sporulation of *B. fabae*, and faba bean resistance reaction against the pathogen using a single *B. fabae* isolate and Degaga and Bulga-70 faba bean varieties. For cultural experiment, a circular block of actively growing *B. fabae* mycelia was placed on faba bean dextrose agar medium and arranged in a completely randomized design (CRD) with four replications. For resistance reaction evaluation, fresh culture of isolate suspension was prepared (2 × 105 spores ml-1) and inoculated on to three weeks-old faba bean seedling detached leaves and the whole plant. Inoculated leaflets and seedlings were factorial arranged in a CRD with four replications. Both sets were incubated at 20, 22, 24 and 26°C. The maximum (84.00 mm) radial growth on 5 days after inoculation (DAI); average conidial size (24.86 × 16.32 µm), sporulation (2.48 × 103 conidia ml-1) on 12DAI and growth rate (1.058 mm day-1) were recorded at 22°C.

The least values of these parameters and nil sporulation were obtained from 26°C. The highest average lesion size (17.67 mm in Degaga and 22.83 mm in Bulga-70), AUDPC for lesion sizes (30.92 mm in Degaga and 42.08 mm in Bulga-70) and severity (2.13 score) values were recorded at 22°C on 5DAI in detached leaf test. Infection and disease development was reduced at 26°C. The trend was similar in the whole plant test. Such parameters were linearly increased with temperature to maximum and declined progressively in both reaction evaluation tests (Terefe *et al*, 2015). The two evaluation experiments indicated that the optimum temperature for *B. fabae* growth, sporulation, infection and disease development was at 22°C.

b. Postharvest factors affecting structural deterioration : The major postharvest problem with storage of fruits and vegetables is the excessive softening. Ripening of many fruits is mainly orchestrated by biosynthesis of ethylene that triggers a serial biochemical and physiological process inducing the softening in texture. After harvesting, the fruit still remains alive and performs all functions of a living tissue. The climacteric burst of ethylene which makes the fruit palatable also triggers senescence and subsequent ripening in the fruits. The goal of any postharvest handling practice or treatment is to manage the concentration and timing of ethylene synthesis so that the fruit reaches the consumer at optimal eating quality The most important factors affecting structural deterioration of fruits and vegetables can be summarized as follows.

Processing : At low extraction percentages (up to 33 %), pectic polysaccharides and hemicellulosic xyloglucans were the main type of polymers affected, suggesting the modifi cation of the cell wall matrix, although without breakage of the walls (Negi and Handa, 2008). At higher extraction rates (up to 64 %), a major disruption of the cell wall occurred as indicated by the losses of all major types of cell wall polysaccharides, including cellulose. At higher extraction rates, fatty acid chains are able to exit the cells either through unbroken walls, or the modifi cation of the pectin-hemicellulose network might have increased the porosity of the wall. Due to high pressure, a progressive breakage of the cell walls was observed, which allows the free transfer of the fatty acid chains from inside the cells.

Heat: Postharvest heat treatments lead to an alteration of gene expression, and fruit ripening can sometimes be either delayed or disrupted. Cell wall–degrading enzymes and ethylene production are frequently the most disrupted, and their appearance is delayed following heating. Fruit sensitivity to heat treatments is modifi ed by preharvest weather conditions, cultivar, rate of heating, and subsequent storage conditions (Negi and Handa, 2008)..

Prestorage heat treatment appears to be a promising method of postharvest control of decay. Heat treatments against pathogens may be applied to fresh harvested commodities by hot water dips, by vapor heat, by hot dry air, or by a very short hot water rinse and brushing. Prestorage heat treatment could delay the ripening of "Gala" and "Golden Delicious" apples and maintain storage quality. Heating "Golden Delicious" apples for 4 days at 38 °C reduced decay and maintained fruit fi rmness during months of storage at 0°C. Cooking resulted in an increase in the water-soluble pectins and a decrease in the pectins associated with cellulose. The total cell wall polysaccharide and galactose content of the squash cultivars remained unchanged for up to 2 months of storage and decreased later.

Physiological disorders: Water soaking developed during the late stages of fruit ripening. The major changes were observed in a protein implicated in calcium signaling processes. While the amount of total calmodulin, the ubiquitous calcium-binding protein, was not modifi ed, a particular calmodulin-binding protein (CaM-BP) was absent in water-soaked but not in sound mature tissues. This CaM-BP may be a marker or a determinant of this physiological disorder. Gel breakdown in inner mesocarp tissue of plums was associated with high viscosities of water-soluble pectin with low levels of extractable juice (Negi and Handa, 2008). In outer mesocarp tissue where extractable juice levels were higher, over- ripeness developed. Cell walls of inner tissue were thicker and had a better developed middle lamella than outer tissue. Inner mesocarp tissue was composed of larger cells than outer tissue.

Chilling and freezing injury : Insoluble pectin levels declined during ripening and cold storage of plum fruit with a concomitant increase in soluble pectin levels. Neither harvest maturity nor storage time had a signifi cant effect on the concentration of calcium pectate, and this pectic fraction did not appear to infl uence development of gel breakdown (GB). Water-soluble pectin and availability of cell fl uids indicated a high gel potential in plums. Signifi cant levels of GB developed only in plums harvested at postoptimum maturity. In GB fruit, higher sugar levels and loss ofcell membrane integrity probably enhanced formation of pectin sugar gels as cell fl uids bind with pectins in cell walls. The initial response to low temperature is considered to involve physical factors such as membrane alteration and protein/enzyme diffusion, but physiological changes that lead to losses of structural integrity and overall fruit quality also occur. During softening, dissolution of the ordered arrangement of cell wall and middle lamella polysaccharides occur. As the fruit ripens, a substantial portion of its cell wall pectins are converted to a water-soluble form affecting the texture (Lurie *et al.*, 2003). The major changes involved in softening and chilling injury in peaches are the catabolism of cell walls and the development of an intercellular matrix containing pectins. Gel- like structure formation in the cell wall due to the deesterification of pectins without depolymerization leads to the development of woolliness in peach.

The combination of chitosan coating, calcium chloride and intermittent warming partially inhibited PG activity, slowed down the increase in soluble pectinefi c substances (Ruoyi *et al.* 2005). Addition of calcium chloride andintermittent warming could keep the intactness of cell wall and reduce fruit sensitivity to injury in peach. Endo-PG, PE, and endoglucanase (EGase) activities of delayed- storage nectarines fruit were same as the control fruit at the beginning of storage, although exo-PG was higher. Endo-PG activity was lower in control than delayed- storage fruit at the end of storage, while PE activity was higher, and exo-PG and EGase activities were similar. Prevention of chilling injury by delayed storage (DS) appears to be due to the ability of the fruit to continue progressive and slow cell wall degradation in storage, which allows normal ripening to proceed when the fruits were rewarmed.

Modified atmosphere: Fruit softening is associated with the disassembly of primary cell wall and middle lamella structure. The changes in cell wall structure and composition result from the composite action of hydrolytic enzymes produced by fruits, which include PG, PE, PL, βGAL, and cellulases. High oxygen atmosphere retards the decrease in firmness in grapes, sweet cherries Decrease in firmness of grapes under different oxygen storage was accompanied by a dramatic decrease in hemicellulose and moderate decrease in cellulose and total pectins, which indicates that the softening in grapes is due to increased depolymerization and degradation of cell wall polysaccharides. At higher oxygen storage, grapes maintained firmness, which coincided with higher retention of cell wall polysaccharides (Negi and Handa, 2008). The lower levels of water soluble pectins in high oxygen atmosphere were correlated with delayed softening, and the activities of PG, PE, β -GAL increased to lower extent than air storage, which indicates that higher oxygen might have inhibited relative enzyme activities, reducing the degradation and depolymerization of pectin substances. Cellulase activity in oxygen compare to air. In controlled atmosphere (CA) stored apples, ripening related softening was inhibited after an initial loss of firmness. However, softening resumed after transfer of apples to normal atmosphere storage at 8 °C.

Pathogen attack : In apple and tomato fruits, *Penicillium expansum* infection cause reduction in the molecular mass of hemicelluloses, particularly in the xyloglucan. Xyloglucan endotransglucosylase/hydrolase (XTH) activity decreased drastically during the infection process in both fruits. XTH reduction during the infection might be related with the fungus attack mechanism (Miedes and Lorences, 2004). Decrease in activity and the consequent lower xyloglucan endotransglucosylation, together with the increase in endoglucanases, would permit fungal access to the cellulosexyloglucan network, increase the efficiency of cellulose hydrolysis, and thus facilitate the progress of the fungal infection. Hemicellulose degradation is important in the breakdown of plant cell walls, causing cell wall loosening, increasing the porosity of the wall, and allowing the colonization of plant tissue.

Irradiation : The biological effect of gamma rays is based on the interaction with atoms or molecules in the cell, particularly water, to produce free radicals, which can damage different important compounds of plant cell. The UV-B/C photons have enough energy to destroy chemical bounds, causing a photochemical reaction. Gamma rays accelerate the softening of fruits, causing the breakdown of middle lamella in cell wall. They also influence the plastid development and function, such as starch sugar interconversion. The penetration of UVB light into the cell is limited, while gamma rays penetrate through the cells. For this reason, UV-B light has a strong effect on surface or near-to-surface area in plant cells. Plant pigments, such as carotenoids and flavonoids, save plant cells against UV-B and gamma irradiation. UV lighthas been used as a postharvest treatment to enhance shelf life of various fruits and vegetables. The beneficial doses of UV-C are reported to induce the accumulation of phytoalexins and activate genes, encoding pathogenesis-related proteins. Prestorage exposure of peaches with UV-C irradiation significantly reduced chilling injury.

A higher accumulation of spermidine and spermine was found in peaches after UV exposure, and it is postulated that these higher levels of polyamines apparently are a response to the UV-C irradiation and might be beneficial in increasing the resistance of fruit tissue to deterioration and chilling injury (Negi and Handa, 2008). Major postharvest problem with storage of fruits and vegetables is the excessive softening. Ripening of many fruits is mainly orchestrated by biosynthesis of ethylene that triggers a serial biochemical and physiological process inducing the softening in texture. The most important factors affecting structural deterioration of fruits and vegetables are processing, heating, physiological disorders, chilling and freezing injury, pathogen attack, modified atmosphere and irradiation.

Combination gases : The combination of different gases in a storage environment is very important in extending the storage life of tomato fruits. The optimal atmosphere needed to inhibit senescence in mature green and ripe fruit of tomatoes is 3–5% (v/v) of oxygen but for carbon dioxide it is 1–3% (v/v) and 1–5% (v/v) in mature green and ripe fruit, respectively], whilst 94–96%

(v/v) of nitrogen gas is required. A very low supply of oxygen can have a detrimental effect on fruits by causing anaerobic respiration. Carbon monoxide (CO) has been investigated as a gas for treating fruits and has been found to speed up ripening. It is therefore necessary to balance the carbon monoxide with low oxygen to delay senescence in the fruits. Carbon monoxide slows down postharvest pathogenic infestations whilst improving some quality traits of tomatoes. For instance, tomatoes stored in 5–10% (v/v) carbon monoxide with 4% (v/v) oxygen were found to have superior total soluble solids (TSS) and titratable acid (TA) profiles as compared to control samples stored in air (Isaac Kojo Arah *et al*, 2015). The use of carbon monoxide in the food industry is however complicated because of its health hazards to humans and must be used with great caution.

Postharvest calcium chloride application : Higher plants generally contain appreciable amounts of calcium, usually in the range of 1–50 mg Ca g^{-1} dry matter. Calcium deficiency in plants can be minimised using calcium based fertilisers. Calcium deficiency in tomato during fertilisation can result in susceptible plants which are exposed to many calcium related disorders like blossom end rot. Calcium has been found to increase the yield of tomato when used as a fertilizer. Postharvest calcium application has also been shown to have a positive effect on many storage parameters of various fruits and vegetables. For instance, the use of calcium chloride in controlling rapid ageing in harvested fruits in general has been investigated by many authors. Postharvest calcium chloride application reduces respiration, decreases ethylene production, and delays senescence in fresh produce such as tomatoes. Exogenous application of calcium maintains cell-wall integrity and protects it from degrading enzymes, enhancing better linkages between pectic substances within the cell-wall whilst increasing the cohesion of cell-walls. This therefore gives an indication that the rate of senescence in fruits is closely related to the amount of calcium in the plant tissue and varying the calcium status affects the rate of senescence (Isaac Kojo Arah *et al*, 2015). It has been established that altering the levels of exogenous calcium application affects parameters for senescence such as protein and chlorophyll content, respiration rates, and cell membrane fluidity. Addition of calcium rigidifies cell-wall and obstructs enzymes such as polygalacturonase from reaching active sites.

Physical handling: Physical handling can have a drastic effect on the postharvest quality or life of harvested fruits. Rough handling during harvesting and after harvesting can result in mechanical injuries which affect quality. Typical industrial production systems associated with tomatoes may include mechanical harvesting, packing into crates, sorting, grading, washing, and transporting over long distances. At each of these stages there may be significant occurrence of mechanical injury which may be bruising, scarring, scuffing, cutting, or puncturing the fruits. In small-scale tomato production, mechanical injuries may result from the use of inappropriate harvesting containers and packaging materials. The effects of mechanical injuries on fruit are cumulative. Injuries which are equivalent to or greater than the bioyield point lead to a total breakdown of the structure of the affected cells which is accompanied by unwanted metabolic activities which may include increased ethylene production, accelerated respiration rates, and ripening, which results in either reduced shelf life or poor quality (Isaac Kojo Arah *et al*, 2015).. It is therefore important to handle tomato fruit with care during the harvest and postharvest activities to minimise mechanical injuries to avoid losses.

c. Postharvest handling under extreme weather conditions: The modern era of informatics and high-technology communications has allowed many growers, shippers and retailers to learn and implement practices and technologies that have had success elsewhere in the world. There have been many examples where technological protocols were "copied on the local level," and the results were similar to those previously experienced. However, we are also continually reminded that what is done at one site cannot be transmitted directly to another, even when similar cultivars are grown. The interesting point, and the good news for those working to improve postharvest quality of produce, is that a successful plan of production and handling to maintain quality may only need simple adjustments to produce good standards in other operations. It is paramount to stress the importance of environmental factors occurring during the hours before and immediately after harvest, particularly when those factors involve extreme conditions prevalent in places such as low deserts and rainy tropics. Research has shown the direct effects of environmental conditions during the hours before harvest on postharvest quality. A single stress event can produce major physical changes in the physiology and anatomy of the plant upon recovery. For example, trichomes were formed in fever few plants after recovering from a wilting event, and grafted watermelon produced adventitious roots and aerenchyma to adapt to a single flooding. Environmental factors encountered during handling of produce that can affect final quality include extremely low or high temperatures, rain and extremely low relative humidity (RH). It is possible that in the future, with more urban areas coexisting with production fields and orchards, other factors may become more important, such as environments with high emissions of CO2, as well as atmosphere with low pressure and high ethylene content (Fonseca, 2009). It could be followed the postharvest handling under extreme weather conditions as follows.

d. Postharvest handling in the tropics : Postharvest losses encountered in tropical areas with a high incidence of rainfall are likely the highest in the world. Losses in developing countries along the tropical belt are estimated to be about 50 %. Interestingly, there was no significant relationship between solar radiation and mean temperature. It could be concluded that fruit that were exposed on the day of harvest, or during the 3 days before harvest, to minimum temperatures higher than 22.4 °C showed improved tolerance to quarantine heat treatments. This induced tolerance to postharvest heat treatments has been associated with heat shock proteins. High RH and temperatures around 30 ºC can enhance wound healing of citrus fruits (Kinay *et al.* 2005). This is a major reason for the recommendation that degreening treatments occur in high RH and controlled temperatures. This beneficial effect is associated with a fast wound healing rate, triggered by the formation of lignin and the induction of antifungal compounds, such as scoparone and scopoletin. The proliferation and/or penetration in the fruit by pathogens, such as *Penicillium digitatum*, *P. italicum* and *Geotrichum candidum* is reduced when curing is induced under high RH and temperatures around 30°C. High temperatures during postharvest storage are commonly associated with high transpiration rates and subsequent degradation of quality traits. However, in the case of herbs such as marjoram even temperatures as high as 30°C do not affect oil accumulation, but rather can stimulate synthesis. Temperatures and light intensity in tropical areas fluctuate less than in hot, dry or temperate climates. Although it is not completely clear how the low daily temperature fluctuation impacts the postharvest quality of fruit, some studies suggest that shelf life of vegetative tissue may be signifi cantly impacted by the time of day of harvest (Fonseca, 2009).

In tropical countries, the production of shoot tip cutting for export to temperate countries is a major industry. It has been found that late harvests e.g. after 4 pm improve storage quality and subsequent rooting response, an effect that was attributed to higher endogenous carbohydrate status (Rapaka *et al.* 2007). On the other hand, vegetative tissue is particularly sensitive to storage temperature during the hours after harvest, which may be dependent on the time of the day or harvest if cooling facilities are not available. Chlorophyll degradation significantly increases in leafy vegetables, such as Valeriana lettuce (Ferrante and Maggiore, 2007) and fresh-cut rocket and spinach, when storage temperature increases by only 6ºC. Diurnal changes in metabolism may explain the prolonged shelf life obtained with harvest during the late hours of the day. Xyloglucan endotransglusylase hydrolase, a cell wall modifying protein, is more active during the late hours of the day, suggesting a role in preventing cell degradation during postharvest. Moreover, leaf–water relations are also known to fluctuate during the day. For example, stomata opening occurs in response to light and temperature which affects CO2 uptake productivity and assimilation of carbon productions (Rico *et al.* 2007). The "turgor" hypothesis is another factor that can partially explain the importance of weather conditions at harvest. It is agreed by many that prolonged shelf life may augment with increased turgor, which may be also improved with postharvest calcium treatments.

Physiological condition of produce at harvest determines how long the crop can be safely stored. For example, apples are picked slightly immature to ensure that they can be stored safely for several months. The onset of ripening and senescence in various fruits renders them more susceptible to infection by pathogens. On the other hand, fruits can be made less prone to decay by management of crop nutrition.

Postharvest handling operations are the main reason for *R. stolonifer* succeeds in entering and infecting most horticultural commodities. It is continuously reported that overall *R. stolonifer* infection gains better access through wounds, injuries and abrasions. A study carried out by Holmes and Stange (2002) reported that disease incidence on sweet potatoes may be associated with the type of injury. The bruise-wound type injury followed by the puncture-wound type caused the highest infection rate over the 2-year assessment period. With respect to the optimum conditions for infection to occur during postharvest handling, temperature is a very important factor. Overall temperature above 15°C accelerates infection occurring in a short period of time, but in lower temperatures, infection is only delayed. Pierson (1966) reported that in peaches the maximum growth rate of *R. stolonifer* was achieved at 25°C, while lower temperatures of 0, 4.4 and 18°C only delayed the appearance of symptoms by a few days. Spores already attached to fruits are a source of future inoculum for future infections; therefore, a source of contamination at packing plants may be the water systems used for all commodities requiring washing as a usual postharvest operation.

Major factors contributing to quality of carrots, with special focus on the role of pre- and postharvest factors and processing were updated (Randi Seljasen *et al*, 2013). The genetic factor showed the highest impact on quality variables in carrots, causing a 7–11-fold difference between varieties in content of terpenes β-carotene, magnesium, iron and phenolics as well as a 1–4-fold difference in falcarindiol, bitter taste and sweet taste. Climate-related factors may cause a difference of up to 20-fold for terpenes, 82% for total sugars and 30–40% for β-carotene, sweet taste and bitter taste. Organic farming in comparison with conventional farming had shown 70% higher levels for magnesium and 10% for iron. Low nitrogen fertilisation level might cause up to 100% increase in terpene content, minor increase in dry matter (+4 to +6%) and magnesium (+8%) and reduction in β-carotene content (−8 to −11%). Retail storage at room temperature caused the highest reduction in β-carotene (−70%) and ascorbic acid (−70%). Heat processing by boiling reduced shear force (−300 to −1000%) and crispiness (−67%) as well as content of phenolics (−150%), terpenes (−85%) and total carotenes (−20%) and increased the risk of furan accumulation. Sensory and chemical quality parameters of carrots were determined mainly by genetic and climate-related factors and to a minor extent by cultivation method. Retail temperature and storage atmosphere as well as heating procedure in processing had the highest impact in quality reduction.

During growth and development of carrots, quality can be affected from sowing to harvest by general preharvest factors such as genetic variation and climatic conditions (light, temperature, precipitation) as well as farming system-related factors that deal with organic or conventional cultivation systems, fertilisation and pest control. Postharvest factors that influence decay also include packing sanitation. It is important to maintain sanitary conditions in all areas where produce is packed. Organic matter (culls, extraneous plant parts, soil) can act as substrates for decay-causing pathogens. For example, in apple and pear packinghouses, the flumes and dump tank accumulate spores and may act as sources of contamination if steps are not taken to destroy or remove them. Chlorine readily kills microorganisms suspended in dump tanks and flumes if the amount of available chlorine is adequate. A level of 50 to 100 ppm of active chlorine provides excellent fungicidal activity. Chlorine measured as hypochlorous acid can be obtained by adding chlorine gas, sodium hypochlorite, or dry calcium hypochlorite. Although chlorine effectively kills spores in water it does not protect wounded tissue against subsequent infection from spores lodged in wounds. Organic matter in the water inactivates chlorine, and levels of chlorine must be constantly monitored. Recently, in precisely controlled tests in water or as foam, chlorine dioxide was found to be effective against common postharvest decay fungi on fruit packinghouse surfaces.

References

Agostini, J. P., Bushong, P. M., Bhatia, A., and Timmer, L. W. 2003. Influence of environmental factors on severity of citrus scab and melanose. *Plant Dis.*, **87**:1102-1106.

Alao, S.E. L.,2000. The importance of postharvest loss prevention. Paper presented at graduation ceremony of school of food storage technology. Nigerian Stored Products Research Institute, Kano p. 1-10

Amanatidou, A., Slump, R.A., Gorris, L.G.M. and Smid, E.J.,2000. High oxygen and high carbon dioxide modified atmospheres for shelf-life extension of minimally processed carrots. *J. Food Sci.*, **65**:61–66

Amiri, Ashour & Chai, Wenxuan & Schnabel, Guido. 2011. Effect of nutrient status, pH, temperature and water potential on germination and growth of *Rhizopus stolonifer* and *Gilbertella persicaria*. *J. Pl. Pathol.*, **93:** 603-612.

Ana, M. , Bocsanczy, Ute C. M., Achenbach, Arianna Mangravita-Novo, Jeanne, M. F., Yuen and David J. Norman, 2012. Comparative Effect of Low Temperature on Virulence and Twitching Motility of Ralstonia solanacearum Strains Present in Florida. *Phytopathology*, **102 :** 185-194

Anil Kumar Saini, Sushil Kumar Sharma, Anil Kumar and Pankaj Kumar, 2018. Effect of Temperature and Relative Humidity on the Development of Blue Mould Rot (*Penicillium islandicum* Sopp.) on Indian Gooseberry (*Emblica officinalis* Goerth.). *Int.J.Curr. Microbiol.App.Sci.*, **7**: 1889-1893

Anonymous ,2004. Manual for the preparation and sale of fruits and vegetables: from field to market, FAO Agri. Services Bull., No. 151.

Anonymous, 1989. Prevention of postharvest food losses fruits, vegetables and root crops, a training, FAO,UN, Rome.

Anonymous, 2001. IPCC- Working group II: impacts, adaptations and vulnerability. http://www.grida.no/

Anonymous, 2007. IPCC. In: Solomon S, Qin D, Manning M, Chen Z, Marquis M, Averyt KB, Tignor M, Miller HL (eds) The Physical Science Basis. Contribution of Working Group I to the Fourth Assessment Report of the Intergovernmental Panel on Climate Change. Cambridge University Press, Cambridge, 996p.

Anonymous,2009. Course on agribusiness management for producers' associations. Module 4 – Post- harvest and marketing. Santacoloma P, Roettger A, Tartanac F (eds) Training materials for agricultural

Arora, A., Sairam, R.K. and Srivastava, G.C.,2002. Oxidative stress and antioxidative system in plants.*Curr. Sci.*, **82**:1227–1238

Arosha Loku Umagiliyage and Ruplal Choudhary, Postharvest Ultraviolet Light Treatment of Fresh Berries for Improving Quality and Safety, *Intl.J.Food Sci.& Nutr.Engg.*, **8**: 60-71

Assad, E.D., Pinto, H.S., Zullo, J. and Jr, Ávila, A.M.H.,2004. Impacto das mudanças climáticas no zoneamento agroclimático do café no Brasil. *Pesquisa Agropecuária Brasileira,* **39**:1057–1064

Ayón-Reyna, L.E., López-Valenzuela, J.Á.and Delgado-Vargas, F., 2017. Effect of the Combination Hot Water - Calcium Chloride on the *In Vitro* Growth of *Colletotrichum gloeosporioides* and the Postharvest Quality of Infected Papaya. *Plant Pathol J.* **33**:572–581

Azeem, Muhammad Talha & Shahzad, Saleem & Sultana, Nasreen. (2016). Effect of different temperatures on postharvest rots of banana. *Intl. J. Biol.& Biotechnol.*, **13:** 417-421.

Bally, I.S.E., Hofman, P.J., Irving, D.E., Coates, L.M. and Dann, E.K. ,2009. The effects of nitrogen on postharvest disease in mango (*Mangifera indica* l. ‹Keitt›). *Acta Hortic.* **820,** 365-370

Barth, M.M., Perry, A.K., Schmidr, S.J. and Klein, B.P.,1990. Misting effects on ascorbic acid retention in broccoli during cabinet display. *J. Food Sci.,* **55**:1187–1191

Bartz, J.A. and Brecht, J.K.,2003. Postharvest physiology and pathology of vegetables. Marcel Dekker,Inc, New York, p 733

Bengtsson, G.B. and Matforsk, A.S., 2008. Storage and handling of fruit and vegetables for optimum health-related quality. In: *Tomás-Barberán, F.A., Gil, M.I., (eds) Improving the health-promoting properties of fruit and vegetable products.* Woodhead Publishing Limited/CRC Press LLC, Cambridge, England, pp 412–430

Ben-Yehoshua, S. and V. Rodov. 2003. Transpiration and water stress. *In: Postharvest Physiology and Pathology of Vegetables, J.A. Bartz and J.K. Brecht. (Eds.).* 2nd ed., Marcel Dekker, Inc., NY. Pp. 111-159.

Bhowmik, P.K., Matsui, T., Kawada, K. and Suzuki, H.,2002. Changes in storage quality and shelf life of green Asparagus.: *A Review. Asian J. Pl. Sci.*, **2**: 941-943.

Bhowmik, P.K., Matsui, T., Suzuki, H., Kosugi, Y., Alam, A.K. and Enriquez, F.G.,2003. Pre- and postharvest fi ber development in green asparagus as related to temperature during growth, storage and green asparagus over an extended harvest season. *Postharvest Biol. Technol.,* **26**:323–326

Bottcher, H., Gunther, I.and Kabelitz, L., 2003. Physiological postharvest responses of common Saint-John's wort herb (*Hypericum perforatum* L.). *Postharvest Biol. & Ecol.*, **29**: 343-35.1

Brandhoff, B., Simon, A., Dornieden, A., Schumacher, J., 2017. Regulation of conidiation in *Botrytis cinerea* involves the light-responsive transcriptional regulators bcltf3 and bcreg1. *Current Genetics*, **63**: 931–949

Burg, .S.P., 2004. Postharvest physiology and hypobaric storage of fresh produce. CABI Publishing/CAB International, Wallingford

Castejón-Muñoz,M., 2008. The effect of temperature and relative humidity on the airbone concentration of *Pyricularia oryzae* spores and the development of rice blast in southern Spain. *Spanish J. Agril. Res.*,**6** : 61-69

Clarkson, G.J.J., O'Byrne, E.E., Rothwell, S.D. and Taylor, G., 2003. Identifying traits to improve postharvest processability in baby leaf salad. *Postharvest Biol. Technol.*, **30**:287–298

Clarkson, G.J.J., Rothwell, S.D. and Taylor, G., 2005. End of day harvest extends shelf life. *Hort.Sci.*, **40**:1431–1435

Clarkson, J.P., Fawcett, L., Anthony, S.G.and Young, C.,2014. A Model for *Sclerotinia sclerotiorum* Infection and Disease Development in Lettuce, Based on the Effects of Temperature, Relative Humidity and Ascospore Density. *PLoS ONE,* **9**(4): e94049.

De Long, J. M.and Prange, R.K., 2003. Storage. In: *Thomas B, Murphy DJ, Murray BG (eds.), Encyclopedia of applied plant sciences.* Elsevier Ltd, Oxford/San Diego. doi: 10.1016/B0-12-227050-9/00346-X.

Deepshikha, Bimla Kumari , Elangbam Premabati Devi , Geeta Sharma , Shilpi Rawat and J.P. Jaiswal, 2017. Irradiation as an Alternative Method for Post–harvest Disease Management: An Overview. *Intl.J.Agric. Env. & Biotech.*, **10:** 625-633

Deng, Y., Wu, Y. and Li, Y.,2005. Changes in fi rmness, cell wall composition and cell wall hydrolases of grapes stored in high oxygen atmospheres. *Food Res. Intl.,* **38**: 769-776

Denner, F D N , Kotzé, J. M., and Putterill, J. F., 1986. The effect of temperature on spore germination, growth and appressorium formation of Colletotrichum *gloeosporioides* and *Dothiorella aromatica.* South African Avocado Growers' Association Yearbook , **9:**19-22

Dhatt, A.S. and Mahajan, B.V.C.,2007. Horticulture postharvest technology harvesting, handling and storage of horticultural crops. Punjab Horticultural Postharvest Technology Centre, Punjab Agricultural University Campus, Ludhiana.

Díaz-Pérez, Juan Muy, Dolore and Mascorro, Arturo.,2007. Fruit size and stage of ripeness affect postharvest water loss in bell pepper fruit (*Capsicum annuum* L.). *J. Sci. Food and Agric.*, **87**: 68 - 73.

Dordas, Christos, 2008. Role of Nutrients in Controlling Plant Diseases in Sustainable Agriculture: A Review. http://dx.doi.org/10.1051/agro:2007051. 28. 10.1007/978-90-481-2666-8_28.

Droby, S., Wisniewski,M. and Benkeblia, N., 2011. Postharvest biology and technology of tropical and subtropical fruits. Fundamental Issues- Woodhead Publishing Series in Food Science, Technology and Nutrition, 194-223 p

East, A.R., Tanner ,D.J,, Maguire, K.M., Mawson, A.J.,2008. The influence of breaks in storage temperature on 'Cripps Pink' (Pink Lady™) apple physiology and quality. *Hort.Sci.*, **43**:818–824

Eden, M. & Hill, R. & Beresford, Robert & Stewart, A.,2003. The influence of inoculum concentration, relative humidity, and temperature on infection of greenhouse tomatoes by *Botrytis cinerea. Plant Pathol.*, **45:** 795 - 806. Eisenback, J.D., Ma Wei and Roane, C.W., 2013. Heat Tolerance and Aging of the Anhydrobiotic Seed Gall Nematode with SEM Observations. *J.Nematol.,* **45**: 43-48.

Elazar Fallik and Zoran Ilic. 2014. Grafted vegetables – the influence of rootstock and scion on postharvest quality. *Folia Hort.* **26**: 79-90

Estrada, A.B., Dodd John and Jeffries Peter, 2001. Effect of humidity and temperature on conidial germination and appressorium development of two Philippine isolates of the mango anthracnose pathogen *Colletotrichum gloeosporioides. Plant Pathol.,* **49 :** 608 - 618.

Etienne Duveiller and Henri M.M. Maraithe, 2008. Effect of Temperature and Air Humidity on Multiplication of *Xanthomonas campestris* pv. *undulosa* and Symptom Expression in Susceptible and Field□Tolerant Wheat Genotypes. *J.Phytopathol.*, **143:** 227-232.

Everett, K. R., 2003. The effect of low temperatures on *Colletotrichum acutatum* and *Colletotrichum gloeosporioides* causing body rots of avocados in New Zealand. *Australian Pl.Path.*, **4**: 441-448

Fallik, E.,2004. Prestorage hot water treatments immersion, rinsing and brushing. *Postharvest Biol. & Technol.,* **32**: 125-134

FAO agricultural services bulletin no. 151. Food and Agriculture Organization of the United Nations, Rome

Farrar, J.J., Nunez, J.J. and Davis, R.M., 2000. Influence of Soil Saturation and Temperature on *Erwinia chrysanthemi* Soft Rot of Carrot. *Pl. Dis.*, **84:** 665-668

Felzer, B.S., Cronin, T., Reilly, J.M., Melillo, J.M. and Wang, X.,2007. Impacts of ozone on trees and crops. *C R Geosci.*, **339**:784–798

Ferrante, A. and Maggiore, T., 2007. Chlorophyll *a* fluorescence measurements to evaluate storage time and temperature of *Valeriana* leafy vegetables. *Postharvest Biol. Technol.*, **45**:73–80

Ferrante, A., Incrocci ,..L, Maggini ,R., Serra ,G. and Tognoni, F.,2004. Colour changes of fresh-cut leafy vegetables during storage. *J. Food Agric. Environ.,* **2**:40–44

Florkowski, W.J., Prussia, S.E., Shewfelt, S.L., Brueckner, B., (eds.) Postharvest handling: a systems approach, Food Science and Technology Series. Academic Press, Elsevier Inc., New York, USA, pp 57–106

Fonseca, J.M. and Saborío, D., 2001. Postharvest technology of waxed cassava (*Manihot esculenta* L.) for export markets (Spanish). National Press of Costa Rica, Costa Rica

Fonseca, J.M., 2006. Postharvest quality and microbial population of head lettuce as affected by moisture at harvest. *J. Food Sci.,* **71**: M45–M49

Fonseca, J.M., 2009. Postharvest handling under extreme weather conditions. *In: Florkowski W.J., Prussia, S.E., Shewfelt, R.L.,, Brueckner, B., (eds.) Postharvest handling: A systems approach, Food Science and Technology Series*. Academic Press, Elsevier Inc., New York, USA, pp 539–559

Fonseca, J.M., Rushing, J.W., Rajapakse, N.C., Thomas, R.L., Riley, M.B, 2005.Parthenolide and abscisic acid synthesis in fever few are associated but environmental factors affect them dissimilarly. *J. Pl. Physiol.,* **162**:485–494

Fonseca, J.M., Rushing, J.W., Thomas, R.L., Riley, M.B., Rajapakse, N.C., 2006. Postproduction stability of parthenolide in feverfew (*Tanacetum parthenium*). *J. Herbs Spices Med. Plants,* **12**:139–152

Fonseca, J.M.,2009. Postharvest handling under extreme weather conditions. In: Florkowski, W.J., Prussia, S.E., Shewfelt, R.L, Brueckner, B., (eds.) Postharvest handling: A systems approach, Food science and technology series. Academic Press, Elsevier Inc., New York, USA, pp 539–559

Forlani, M., Basile, B., Cirillo, C., Iannini, C., 2002. Effects of harvest date and fruit position along the tree canopy on peach fruit quality. *Acta. Hortic.*, **592:**459–466

Francisco Oiram Filho, Ebenézer de Oliveira Silva, Mônica Maria de Almeida Lopes, Paulo Riceli Vasconselos Ribeiro, Andréia Hansen Oster, Jhonyson Arruda Carvalho Guedes, Patrícia do Nascimento Bordallo and Guilherme Julião Zocolo, 2019. Effect of pulsed light on postharvest disease control-related metabolomic variation in melon (*Cucumis melo*) artificially inoculated with *Fusarium pallidoroseum.* bioRxiv 698407; doi: https://doi.org/10.1101/698407

Franco Nigro, Antonio Ippolito, Vincenzo Lattanzio and Di Venere, Di., 2000.Effect of ultraviolet-C light on postharvest decay of strawberry. *J.Pl.Path.*, **82**: 29-37

Friboulet, A. and Thomas, D.,2005. Systems biology – an interdisciplinary approach. *Biosens. Bioelectron.,* **20**:2404–2407

Gawai, D.U. and Mangnalikar, S.S., 2018. Effect of Temperature and pH on growth of Alternaria alternata, leaf spot pathogen of soyabean . *Bioscience Discovery*, **9**: 162-165

Gerard Morales, Concepció Moragrega, Emilio Montesinos, Isidre Llorente , 2018. Effects of leaf wetness duration and temperature on infection of *Prunus* by *Xanthomonas arboricola* pv. *pruni. PLOS,* Published: March 7, 2018

Gould, K.,2004. Nature's Swiss army knife: the diverse protective roles of anthocyanins in leaves. *J. Biomed. Biotechnol.*, **5**:314–320

Gracia-Garza,J.A. and Fravel, D.R., 1998. Effect of Relative Humidity on Sporulation of *Fusarium oxysporum i*n Various Formulations and Effect of Water on Spore Movement through Soil. *Phytopathol.*, **88**: 544-549

Gregori, Roberto & Borsetti, F & Neri, Fiorella & Mari, Marta & Bertolini, Paolo. (2008). Effects of Potassium Sorbate on Postharvest Brown Rot of Stone Fruit. *J. Food Protn.*, **71:** 1626-31.

Gruda, N.,2005. Impact of environmental factors on product quality of greenhouse vegetables for fresh consumption. *Crit. Rev. Plant Sci.*, **24**:227–247, ISSN 0735–2689

Hailu, N., Fininsa, C., Tana, T. and Mamo, G., 2017. Effects of Temperature and Moisture on Growth of Common Bean and Its Resistance Reaction against Common Bacterial Blight (*Xanthomonas axonopodis* pv. *phaseoli* strains). *J. Plant Pathol Microbiol.*, **8**: 2-6

Han, Bei-Zhong & Ma, Yong & Rombouts, Frans & Nout, M.J..2003. Effect of temperature and relative humidity on growth and enzyme production by *Actinomucor elegans* and *Rhizopus oligosporus* during sufu pehtze preparation. *Food Chemistry*, **81 :** 1

Henriquez, J. L., Sugar, D., and Spotts, R. A., 2008. Effects of environmental factors and cultural practices on bull's eye rot of pear. *Plant Dis.* **92**:421-424.

Henson, R., 2008. The rough guide to climate change, 2nd edn. Penguin Books, London, p 384

Hernández-Muñoz, P., Almenar, E., José Ocio, M. and Gavara R., 2006. Effect of calcium dips and chitosan coatings on postharvest life of strawberries (*Fragaria* × *ananassa*). *Postharvest Biol Technol*, **39** : 247-253

Hertog, M.L.A..TM., Rudell, D.R., Pedreschi ,R., Schaffer, R.J., Geeraerd, A.H., Nicolai, B.M., Ferguson, I., 2011. Where systems biology meets postharvest. *Postharvest Biol. Technol.,* **62**:223–237

Hodges ,D.W., Forney, C.F. and Wismer, W.V.,2001. Antioxidant responses in harvested leaves of two cultivars of spinach differing in senescence rates. *J. Am. Soc. Hortic. Sci.*, **126**:611–617

Hodges, D.M., Lester, G.E., Munro, K.D. and Toivonen, P.M.A,.2005. Oxidative stress: importance for postharvest quality. *Hort Sci.*, **39**: 924-929

Hodges, D.M.,2003. Postharvest oxidative stress in horticultural crops. Food Products Press,, CRC Press, ISBN 9781560229636 - CAT# HW13718

Högy, P. and Fangmeier, A.,2009. Atmospheric CO_2 enrichment affects potatoes: 2 tuber quality traits. *Eur .J. Agron.,* **30:**85–94

Houston, J.,2006. Evaporation in the Atacama Desert: an empirical study of spatiotemporal variations and their causes. *J. Hydrol.*, **330**:402–412

Huot, B., Castroverde, C.D.M. and Velásquez, A.C., 2017. Dual impact of elevated temperature on plant defence and bacterial virulence in Arabidopsis. *Nat Commun.* **8**:1808. doi:10.1038/s41467-017-01674-2

Hussain, A., Qarshi, I.A., Nazir, H. and Ullah, I.,2012. Plant Tissue Culture: Current Status and Opportunities. In: Leva A, Rinaldi LMR (eds) Recent advances in plant *in vitro* culture. InTech, Rijeka, pp 1–28, doi.org/10.5772/52760

Ibrahim, M. , Rabah, A.B. , Liman, B. and Ibrahim, N.T., 2011. Effect of Temperature and Relative Humidity on the Growth of *Helminthosporium fulvum. Nigerian J. Basic and Appl. Sci.,* **19**:127- 129.

Imahori ,Y.,2012. Postharvest stress treatments in fruits and vegetables. *In: Ahmad P, Prasad MNV (eds) Abiotic stress responses in plants: Metabolism, productivity and sustainability.* Springer, New York, pp 347–358.

Isaac Kojo Arah et.al., 2015. Preharvest and Postharvest Factors Affecting the Quality and Shelf Life of Harvested Tomatoes: A Mini Review. *Intl.J.Agron.,* Volume 2015, Article ID 478041, 6 pages

Jaganath, I.B. and Crozier, A .,2008. Overview of health-promoting compounds in fruit and vegetables. *In: Tomás-Barberán FA, Gil MI (eds) Improving the health-promoting properties of fruit and vegetable products.* Woodhead publishing series in food science, technology and nutrition no.Woodhead Publishing Limited/CRC Press LLC, Cambridge, pp 3–37

James, C.,2008. Global status of commercialized biotech/GM crops. ISAAA Brief No. 39, Ithaca, NY, p 243

January 2002, Melbourne, Australia, Sydney Postharvest Laboratory, Sydney Postharvest Laboratory Information Sheet.

Jobling, J.,2002. Postharvest management of fruit and vegetables. Good Fruit and Vegetables Magazine, January 2002, Melbourne, Australia, Sydney Postharvest Laboratory, Sydney Postharvest Laboratory Information Sheet.

John's wort herbs (*Hypericum perforatum* L.). *Postharvest Biol. Technol.,* **29**:342–350

Kader, A.A .,1995. Maturity, ripening, and quality relationships of fruit-vegetables. I*SHS Acta. Horticult*., **434:** 249–256, Strategies for Market Oriented Greenhouse Production

Kader, A.A. and Saltveit, M.E.,2003. Atmosphere modification. *In: Bartz, J.A., Brecht, J.K., (eds.) Postharvest physiology and pathology of vegetables,* 2nd edn. Marcel Dekker, Inc, New York, pp 229–246.

Kader, A.A.,1992. Postharvest biology and technology: an overview. In: Kader AA (ed) Postharvest technology of horticultural crops. UC Publication No. 3311. University of California, Division of Agriculture and Natural Resources, Oakland, pp 15–20

Kader, A.A.,1993. Postharvest handling. *In: Preece, .JE., Read, P.E., (eds.) The biology of horticulture – An An* Introductory Textbook; *Preece, J. E., Read, P. E., Eds.*; Wiley: New York, *1993*; pp 353–377

Kader, A.A. and Rolle ,R.S., 2004.'The role of postharvest management in assuring the quality and safety of horticultural produce', *Food and Agriculture Organization* (*FAO*) *of the* United Nations. Bulletin No., 152 , 125 pp

Kalnay, E., Cai, M.,. Impact of urbanization and land-use change on climate. *Nature,* **423**: 528–531.

Kanellis, A.K., Tonutti, P. and Perata, P.,2009. Biochemical and molecular aspects of modified and controlled atmospheres. In: *Yahia, E., (ed.) Modified and controlled atmospheres for storage, transportation and packaging of horticultural commodities.* CRC Press, Boca Raton, pp 553–567. ISBN 978-1-4200-6957-0

Karathanos, V.T., Bakalis, S., Kyritsi, A.and Rods, P.S.,2006. Color degradation of beans during storage. *Int. J. Food Prop.,* **9**:61–71

Kaur, L., Singh, J., Singh, N. and Ezekiel, R.,2007. Textural and pasting properties of potatoes (*Solanum tuberosum* L.) as affected by storage temperature. *J. Sci. FoodAgric.,* **87:**520–526

Kinay, P., Yildiz, F., Sen, F., Yildiz, M. and Karacali, I., 2005. Integration of pre- and postharvest treatments to minimize *Penicilium* decay of Satsuma mandarins. *Postharvest Biol. Technol.*, **37**:31–36

Kitinoja, L. and Gorny, J.,1998. Postharvest technology for fruits and vegetables Produce marketers: economic opportunities. Quality and Food Safety by, Department of Pomology, University of California, Davis. A joint publication of UC Postharvest Outreach Program and Punjab Horticultural Postharvest technology Centre, USAID/ACE

Kitinoja, L. and Kader, A.A.,2002. Small-scale postharvest handling practices: a manual for horticultural crops, 4th edn, Postharvest horticulture series no. 8E. University of California, Davis Postharvest Technology Research and Information Center, Davis

Kovacs, E. and Keresztes, A.,2002. Effect of gamma and UV-B/C radiation on plant cells. *Micron,* **33**:199–210.

Kumar Ritesh & Ghatak, Abhijeet & Balodi, Rekha & Bhagat and Arun, 2018. Decay mechanism of postharvest pathogens and their management using non-chemical and biological approaches. *J. Postharvest Technol.*, **6:** 001-011

Lamikanra, O., Imam, S. and Ukuku, D.,2005. Produce degradation: pathways and prevention. Taylor & Francis, Boca Raton, p 677

Li Hua, Chen Yong, Zhang Zhanguan, Li Bogiang, Qin Guozheng,Tian Shiping, 2018. Pathogenic mechanisms and control strategies of *Botrytis cinerea* causing postharvest decay in fruits and vegetables. *Food Quality and Safety*, **2**: 111–119

Lloyd, J. and Farquhar, G.D.,2008. Effects of rising temperatures and CO_2 on the physiology of tropical forest trees. *Philos. Trans. R. Soc. Biol. Sci.,* **363**:1811–1817

Lluis Palou, 2018. Postharvest Treatments with GRAS Salts to Control Fresh Fruit Decay *Horticulturae,***4:** 46

Lo Scalzo, R., Bianchi, G., Genna, A. and Summa, C.,2007. Antioxidant properties and lipidic profi le as quality indexes of cauliflower (*Brassica oleracea* L. var. *botrytis*) in relation to harvest time. *Food Chem.,* **100**:1019–1025

Lopez, G., Larrigaudière, C., Girona, J., Behboudian, M.H. and Marsal, J.,2011. Fruit thinning in 'conference' pear grown under defi cit irrigation: implications for fruit quality at harvest and after cold storage. *Sci Hortic.,* **129**:64–70

Ludivine Lassois, M. Haïssam Jijakli, Marc Chillet and Luc de Lapeyre de Bellaire, 2010. Crown rot of bananas: Preharvest factors involved in postharvest disease development and integrated control methods . *Pl. Dis.*, **94** : 648-658

Lurie ,S., Zhou, H.W., Lers, A., Sonego, L., Alexandrov, S., Shomer, I., 2003. Study of pectin esterase and changes in pectin methylation during normal and abnormal peach ripening. *Physiol. Plant.*, **119:**287–294

Manjunath Hubballi, Sevugapperumal Nakkeeran, Thiruvengadam Rag, 2010. Effect of Environmental Conditions on Growth of *Alternaria alternata* Causing Leaf Blight of Noni . *World J. Agril. Sci.*, **6** : 171-177

Mannaa, M. and Kim, K.D., 2018. Effect of Temperature and Relative Humidity on Growth of *Aspergillus* and *Penicillium* spp. and Biocontrol Activity of *Pseudomonas protegens* AS15 against Aflatoxigenic *Aspergillus flavus* in Stored Rice Grains. *Mycobiology*. **46** :287–295.

Manzoor Ali Abro, 2013. Nitrogen fertilization of the host plant influences susceptibility, production and aggressiveness of *Botrytis* cinerea secondary inoculum and on the efficacy of biological control. Agricultural sciences. Université d'Avignon, English. ffNNT : 2013AVIG0647ff

Mauzerall, D.L. and Wang, X.,2001. Protecting agricultural crops from the effects of tropospheric ozone exposure: reconciling science and standard setting in the United States, Europe, and Asia. *Ann. Rev Energy Environ.*, **26:**237–268

McKay, C.P., Friedmann, E.I., Gomez-Silva, B., Caceres-Villanueva, L., Andersen, D.T. and Landheim, R., 2003 . Temperature and moisture conditions for life in the extreme arid region of the Atacama Desert: four years of observations including the El Nino of 1997–1998. *Astrobiology,* **3**:393–406

Michailides, T.J. and Elmer, P.A.G., 2000. *Botrytis* gray mold of kiwifruit caused by *Botrytis cinerea* in the United States and New Zealand. *Plant Dis*, **84** : 208-223

Miedes ,E. and Lorences, E.P., 2004. Apple (*Malus domestica*) and tomato (*Lycopersicum esculentum*) fruits cellwall hemicelluloses and xyloglucan degradation during *Penicillium expansum* infection. *J. Agric/ Food Chem.*, **52:** 7957–7963

Miller, A.R.,2003. Harvest and handling injury: physiology, biochemistry, and detection. *In: Bartz, J.A., Brecht, J.K., (eds.) Postharvest physiology and pathology of vegetables*, 2nd edn. Marcel Dekker, Inc, New York, pp 177–208.

Moretti, C.L., Mattos, L.M., Calbo, A.G. and Sargent, S.A.,2010. Climate changes and potential impacts onpostharvest quality of fruit and vegetable crops: a review. *Food Res. Int.,* **43**:1824–1832.

Moscoso-Ramírez, P.A. and Palou, L. 2016. Potassium silicate: a new organic tool for the control of citrus postharvest green mold. *Acta Hortic*. **1144**: 287-292

Nag, P.K., Nag, A. and Ashtekar, S.P.,2007. Thermal limits in men in moderate to heavy work in tropical farming. *Ind. Health.,* **45:**107–117

Negi, P.S. and Handa, A.K., 2008. Structural deterioration of the produce: the breakdown of cell wall components. In: Paliyath, G., Murr, D.P., Handa, A.K., Lurie, S., (eds.) *Postharvest biology and technology of fruits, vegetables, and flowers*, vol 978-0-8138-0408-8/2008, 1st edn. Wiley- Blackwell Publishing, Ames. ISBN 978-0-8138-0408-8/2008

Pailly, O., Tison ,G. and Amouroux, A.,2004. Harvest time and storage conditions of "Star Ruby" grapefruit (*Citrus paridisi* Macf.) for short distance summer consumption. *Postharvest Biol. Technol.*, **34**:65–73

Paliyath, G. and Murr, D.P., 2008a. Common fruits, vegetables, fl owers, and their quality characteristics. In: *Paliyath, G., Murr, D.P., Handa, A.K., Lurie, S., (eds) Postharvest biology and technology of fruits, vegetables and flowers*, 1st edn. Wiley-Blackwell Publishing, Ames. ISBN 978-0-8138-0408-8/2008

Paliyath, G. and Murr, D.P., 2008b. Biochemistry of fruits. In: *Paliyath, G., Murr, D.P., Handa, A.K., Lurie, S., (eds.) Postharvest biology and technology of fruits, vegetables, and flowers*, 1st edn. Wiley- Blackwell Publishing, Ames. ISBN 978-0-8138-0408-8/2008

Paliyath, G., Murr, D.P., Handa, A.K. and Lurie, S.,2008. Postharvest biology and technology of fruits, vegetables, and flowers, 1st edn. Wiley-Blackwell Publishing, Ames. ISBN 978-0-8138-0408-8/2008

Parthasarathy, S., 2015. Effect of fertilizers on plant diseases. Agorpedia. Tamil Nadu Agricultural University, Coimbatore 641 003, Tamil Nadu, India.

Pech, J.C., Bernadac, A., Bouzayen, M. and Latche, A.,2005. Use of genetic engineering to control ripening, reduce spoilage, and maintain quality of fruits and vegetables. *In: Ben-Yehoshua, S., (ed) Environmentally friendly technologies for agricultural produce quality.* Taylor & Francis, Boca Raton, pp 397–438

Phillips, D. J., 1984. Effect of Temperature on *Monilinia fructicola* Conidia Produced on Fresh Stone Fruits. Plant Disease, **68** : 610-612

Plaza, P., Usall, J. and Texido-Vinas, T.,2003. Control of green and blue mold by curing on oranges during ambient and cold storage. *Postharvest Biol. Technol.,* **28**:195–198

Postharvest physiology and pathology of vegetables, 2nd edn. Marcel Dekker, Inc, New York, postharvest quality. *Hort. Sci.,* **39**:924–929

Prasanna, V., Prabha, T.N. and Tharanathan, R.N.,2007. Fruit ripening phenomena–an overview. *Crit. Rev., Food Sci. Nutr.,* **47:**1–19

Raftoyannis, Yannis and Dick, Michael. 2002. Effects of inoculum density, plant age and temperature on disease severity caused by Pythiaceous fungi on several plants. *Phytoparasitica.* **30** : 67-76.

Rajput, Laxman, Sinha and Puchakayala, 2017. Effect of Temperature on Rice Blast Infection Process with Emphasis on Appressoria Formation by *Magnaporthe oryzae. Intl. J. Curr. Microbiol., Appld.Sci.,* **6 :** 1931-1939.

Randi Seljasen, Hanne L Kristensen, Charlotte Lauridsen, Gabriela S Wyss, Ursula Kretzschmar, Ines Birlouez-Aragone and Johannes Kahlf, 2013. Quality of carrots as affected by pre- and postharvest factors and processing. Published online in Wiley Online Library: 6 June 2013 (wileyonlinelibrary.com) DOI 10.1002/jsfa.6189

Rapaka, V.K., Faust, J.E., Dole, J.M. and Runkle, E.S., 2007. Effect of time of harvest on postharvest leaf abscission in Lantana (*Lantana camara* L. "Dallas Red") unrooted cuttings. *Hort. Sci.,* **42**:304–308

Rathinasabapathi, B., 2004. Survival of Salmonella Montevideo on tomato leaves and mature green tomatoes. *J. Food Prot.,* **67:**2277–2279

Regina B.Bernd, Conway,W.S. and Moline, H.E., 1987. Influence of Temperature and Relative Humidity on Germinability of *Mucor piriformis* Spores. *J.Phytopathol.,* **118**: 3-8

Rico, D., Martin-Diana, A.B., Fr يas, J.M., Henehan, G.T.M. and Barry-Ryan, C.,2007. Effect of ozone and calcium lactate treatments on browning and textura properties of fresh-cut lettuce. *J. Sci. Food Agric.,* **86**:2179–2188

Robertson, A., Tirado, C., Lobstein, T., Jermini, M., Knai, C., Jensen, J., Ferro-Luzzi, A. and James, W.P.T., 2004. Food and health in Europe: a new basis for action, European series, no. 96. WHO Regional Publications, Copenhagen

Rolle, R.S.,2006. Processing of fruits and vegetables for reducing postharvest losses and adding value. *In: Postharvest management of fruit and vegetables in the Asia-Pacific Region.* Food and Agriculture Organization of the United Nations Agricultural and Food Engineering Technologies Service, Rome

Rubio, M.C., Gonzalez, M.C., Minchin, F.R., Webb, K.J., Arrese-Igor, C., Ramos, J., Becana, B.,2002. Effects of water stress on antioxidant enzymes of leaves and nodules of transgenic alfalfa over expressing superoxide dismutase. *Physiol. Plant.,* **115:**531–540

Ruoyi, K., Zhifang, Y. and Zhaoxin, L., 2005. Effect of coating and intermittent warming on enzymes, soluble pectin substances and ascorbic acid of *Prunus persica* (cv. Zhonghuashoutao) during refrigerated storage. *Food Res. Int.,* **38**:331–336

Sage, R.F. and Kubien, D.,2007. The temperature response of C 3 and C 4 photosynthesis. *Plant Cell Environ.*, **30**:1086–1106

Saini, A. K. , Chauhan, R. S., Ashwani Kumar, Narender Singh and Satish Kumar, 2019. Effect of Inoculum Load on Disease Initiation, Threshold Point and Population Build up of Postharvest Pathogen (*Penicillium islandicum* Sopp.) of Indian Gooseberry (*Emblica officinalis* Goerth). *Int.J.Curr.Microbiol.App.Sci* ., **8**: 2082-2085.

Sala, J.M. and Lafuente, M.T.,2004. Antioxidant enzymes activities and rind staining in "Navelina" oranges as affected by storage relative humidity and ethylene conditioning. *Postharvest Biol. Technol.*, **31**:277–285

Sams, C.E., Conway, W.S.,2003. Preharvest nutritional factors affecting postharvest physiology. In: *Bartz, J.A., Brecht, J.K., (eds.) Postharvest physiology and pathology of vegetables,* 2nd edn. Marcel Dekker, Inc, New York, pp 161–176. ISBN 0-8247-0687-0

Sargent, S.A., Ritenour, M.A. and Brecht, J.K.,2000. Handling, cooling, and sanitation techniques for maintaining postharvest quality. University of Florida, Coop Exten Serv, HS719.

Schick, J.L. and Toivonen, P.M.A.,2002. Reflective tarps at harvest reduce stem browning and improve fruit quality of cherries during subsequent storage. *Postharv. Biol. Technol.,* **25**:117–121

Schumacher, J., 2015.. The VELVET complex in the gray mold fungus *Botrytis cinerea*: impact of bclae1 on differentiation, secondary metabolism, and virulence. *Molecular Plant-Microbe Interactions* , **28:** 659–674

Sfakiotakis, E., Chlioumis, G. and Gerasopoulos, D.,2005. Preharvest chilling reduces low temperature breakdown incidence of kiwifruit. *Postharvest Biol. Technol.,* **38**:169–174

Shamaila, M.,2005. Water and its relation to fresh produce. In: *Lamikanra, O., Imam, S., Ukuku, D., (eds.) Produce degradation: pathways and prevention.* Taylor & Francis, Boca Raton, pp 267–291

Shao, X.F., Tu, K., Zhao, Y.Z., Che,n L., Chen, Y.Y. and Wang, H.,2007. Effects of pre-storage heat treatment on fruit ripening and decay development in different apple cultivars. *J. Hort. Sci. Biotechnol.,* **82**:297–303

Shibairo, S,I,, Upadhyayam M,K, and Toivonen, P.M.A.,2002. Changes in water potential, osmotic potential and tissue electrolyte leakage during mass loss in carrots stored under different conditions. *Sci. Hortic.*, **95**:13–21

Simkin, A.J., Zhu, C., Kuntz, M. and Sandmann, G.,2003. Light–dark regulation of carotenoids biosynthesis in pepper (*Capsicum annum*) leaves. *J. Plant Physiol.,* **160**:439–443

Simson, S.P. and Straus, M.C.,2010. Postharvest technology of horticultural crops. Oxford Book Company/Mehra Offset Press, Delhi

Sinha, N.K.,2011. Handbook of vegetables and vegetable processing. Wiley-Blackwell Publishing, Ames. ISBN 978-0-8138-1541-1

Sirivatanapa, S.,2006 Packaging and transportation of fruits and vegetables for better marketing. *In: Postharvest management of fruit and vegetables in the Asia-Pacific Region.* Food and Agriculture Organization of the United Nations Agricultural and Food Engineering Technologies Service, Rome

Smilanick ,J. L. and Mansour, M. F., 2007. Influence of temperature and humidity on survival of *Penicillium digitatum* and *Geotrichum citri-aurantii. Plant Disease*, **91**: 990-996

Smilanick, J. L., Mansour, M. F., Margosan, D. A., Mlikota Gabler, F., and Goodwine, W. R. 2005. Influence of pH and NaHCO3 on effectiveness of imazalil to inhibit germination of *Penicillium digitatum* and to control postharvest green mold on citrus fruit. *Plant Dis.*, **89**:640-648.

Smith, R.C ., 2010. Vegetable maturity dates, yields and storage. H-912 (Revised). NDSU Extension Service, North Dakota State University Fargo, ND 58108.

Solovchenko, A.E., Avertcheva, O.V. and Merzlyak, M.N.,2006. Elevated sunlight promotes ripeningassociated pigment changes in apple fruit. *Postharvest Biol. Technol.*, **40**:183–189

Song, A., Xue, G. and Cui, P., 2016. The role of silicon in enhancing resistance to bacterial blight of hydroponic- and soil-cultured rice. *Sci Rep* **6,** 24640

Steel, C. C. , Greer, L. A.., Savocchia ,S. and Samuelian, S. K., 2011. Effect of temperature on *Botrytis cinerea, Colletotrichum acutatum* and *Greeneria uvicola* mixed fungal infection of *Vitis vinifera* grape berries. *Vitis*, **50** :69–71

Stine, S.W., Song, I., Choi, C. and Gerba, C.P., 2005. Effect of relative humidity on preharvest survival of bacterial and viral pathogens on the surface of cantaloupe, lettuce and bell peppers. *J. Food Prot.,* **68**:1352–1358

Surendranathan, K.K., 2005. Postharvest biotechnology of fruits with special reference to banana – perspective and scope. *Indian J. Biotechnol.,* **4**:39–46

Terefe, H., Fininsa, C., Sahile, S. and Tesfaye, K., 2015. Effect of Temperature on Growth and Sporulation of *Botrytis fabae*, and Resistance Reactions of Faba Bean against the Pathogen. *J. Pl. Pathol Microbiol.,* **6:** 285.

Tian, S.P., Jiang, A.L., Xu, Y. and Wang, Y.S., 2004. Responses of physiology and quality of sweet cherry fruit to different atmospheres in storage. *Food Chem.,* **87**:43–49

Toivonen, P.M.A. and Hodges, D.M.,2011. Abiotic stress in harvested fruits and vegetables. In: *Shanker, A.K., Venkateswarlu, B., (eds.) Abiotic stress in plants – mechanisms and adaptations*. InTech Europe, pp 39–58.

Toivonen, P.M.A., 2005. Postharvest storage procedures and oxidative stress. *Hort. Sci.*, **39:**938–942,

Toivonen, P.M.A.,2009. Benefits of combined treatment approaches to maintaining fruit and vegetable quality. *Fresh Prod.,* **3:**58–64

Toivonen, P.M.A.,2011. Postharvest physiology of vegetables. *In: Hui, Y.H., Sinha ,N., Ahmed, J., Evranuz, E.Ö., Siddiq, M., (eds.) Handbook of vegetables and vegetable processing.* Wiley- Blackwell

Tokala, Vijay Yadav & Mahajan, B.V.C. 2018. Calcium: An indispensable element affecting postharvest life of fruits and vegetables. Emerging Postharvest Treatment of Fruits and Vegetables, Publisher: Apple Academic Press, pp.59-85

Tzortzakisa, N., Borlanda, A., Singletona, I. and Barnes, J.,2007. Impact of atmospheric ozone- enrichment on quality-related attributes of tomato fruit. *Postharvest Biol. Technol.,* **45:**317–325

Udhayakumar, R., 2018. Influence of different temperature and relative humidity levels on the mycelial growth, conidial germination and anthracnose incidence of *Colletotrichum gloeosporioides*. *Innov.Agric*., **1:** 31-33

Úrbez-Torres, J. R., Bruez, E., Hurtado, J., and Gubler, W. D., 2010. Effect of temperature on conidial germination of Botryosphaeriaceae species infecting grapevines. *Plant Dis*., **94**:1476- 1484.

Vicente, A.R., Manganaris, G.A., Sozzi, G.O. and Crisosto, C.H.,2009. Nutritional Quality of Fruits and Vegetables- Postharvest Handling: A Systems Approach. Elsevier Inc. ISBN: 978-0-12-374112-7

Wang, M., Gao, L., Dong, S., Sun, Y., She,n Q. and Guo, S., 2017. Role of Silicon on Plant-Pathogen Interactions. *Front Pl. Sci*., **8**:701.

Watkins, C.B.,2003. Postharvest physiological disorders of fresh crops. In: *Thomas, B., Murphy, D.J., Murray, B.G., (eds.) Encyclopedia of applied plant sciences*. Elsevier Ltd, San Diego.

Weerahewa, Darshani & Somapala Kosala, (2016. Role of Silicon on Enhancing Disease Resistance in Tropical Fruits and Vegetables: A Review. *OUSL J.,.* **11:** 135-162.

Whitmore ,J.S.,2000. Drought management on farmland. Kluwer Academic Publishers, Dordrecht. ISBN 0-7923-5998-4

Witkowskal Woltering, E.J ., 2010. Pre-harvest light intensity affects shelf-life of fresh-cut lettuce. *Acta Hort.,* **877**:223–227

Wszelaki ,A.L. and Mitcham, E.J., 2000. Effects of super atmospheric oxygen on strawberry fruit quality and decay. *Postharvest Biol. Technol.,* **20:**125–133

Xiangming, Xn, Lucy Guerin and Joycee robinson, 2001. Effects of temperature and relative humidity on conidial germination and viability, colonisation and sporulation of *Monilinia fructigena. Pl.Pathol.,* **50**:561-568

Yahaya, S.M. and Mardiyya, A.Y., 2019. Review of Postharvest Losses of Fruits and Vegetables. *Sci & Tech Res,* **13:** 10192-10200

Yetisir, H., Caliskan, M.E., Soylu, S. and Sakar, M.,2006. Some physiological and growth responses of watermelon [*Citrullus lanatus* (Thunb.) Matsum and Nakai] grafted onto *Lagenaria siceraria* to flooding. *Envrion. Exp. Bot.*, **58**:1–8

You, M.P., Rensing, K., Renton, M.and Barbetti, M.J., 2017. Modeling Effects of Temperature, Soil, Moisture, Nutrition and Variety as Determinants of Severity of *Pythium* Damping-Off and Root Disease in Subterranean Clover. *Front Microbiol.,***8**: 2223.

4

Storage Conditions and Food Safety Issues

Postharvest handling is the stage of crop production immediately following harvest, including cooling, cleaning, sorting and packing. The instant a crop is removed from the ground, or separated from its parent plant, it begins to deteriorate. Postharvest activities include harvesting, handling, storage, processing, packaging, transportation and marketing. Losses of horticultural produce are a major problem in the postharvest chain. The diseases which develop on harvested parts of the plants like seeds, fruits and also on vegetables are the postharvested diseases. The amount or extent of damage depends mainly on the pathogen(s) involved, on the condition of the products and the condition of storage. Major tips to avoid postharvest losses include Tips to avoid postharvest losses include assessing maturity, to check your water quality, to check water temperature, avoid injury, keep the produce cool and proper storage.

"Storage" means the phase of the postharvest system during which the products are kept in such a way as to guarantee food security other than during periods of agricultural production at the marketing level, to balance the supply and demand of agricultural products, thereby stabilizing market prices. Grain, which includes dry kitchen ingredients such as flour, rice, millet, couscous, cornmeal, and so on, can be stored in rigid sealed containers to prevent moisture contamination or insect or rodent infestation. For kitchen use, glass containers are the most traditional method. Maturity indices are the sign or indication the readiness of the commodity for harvest. It is the basis for determining harvest date. Postharvest losses occur between harvest and the moment of human consumption. They include on-farm losses, such as when grain is threshed, winnowed, and dried, as well as losses along the chain during transportation, storage, and processing.

4.1 Packing and storage environments: Harvest and handling practices have major effects on postharvest decay. Mechanical harvesting systems and time of harvest have a prominent effect on postharvest decay and mycotoxin contamination. Fruit maturity at harvest, which directly affects bruising, is a major factor affecting infection by postharvest pathogens. Harvest of fruits from the ground and contact of harvest containers with soil contaminate fruits with postharvest propagules and result in increased decay. Climatic conditions affect sources of contamination and infection of fruits: under dry, hot subtropical climates latent and quiescent infections preharvest play a significant role while under warm, humid tropical climates contamination of fruit wounds during harvest can be of great importance, affecting levels of postharvest decay (Michailides. and Manganaris, 2009). Packing directly in the field reduces production costs but increases chances for postharvest decay. Electronic noses have been constructed that can "smell" and separate decayed fruits in packinghouses and predict mycotoxin contamination. Regression models have been developed to detect the impact of mechanical damage on postharvest decay. In addition, new packaging systems have been created to minimise mechanical injuries and decrease susceptibility to decay and bruising.

Horticultural crops have high level of postharvest decay and loss due to lack of proper packaging methods and storage condition. In a study carried out to evaluate the effect of postharvest treatments and storage conditions, it was observed that products were healthy and free from defect in vegetables treated with washing or un-washing treatments and stored as covered and uncovered with polyethylene (PE) at 4°C (Koushesh Saba Mahmoud *et al*, 2014). The results showed that the weight loss and decay were different in different crop species Treatments affected weight loss and deceased yellowing and decay percent. Combination of washing and covered treatments caused less weight loss and is recommended as easy and straight methods in developing country.

The root cause of postharvest deterioration that needs to be inhibited is enhanced metabolism, whether due to natural senescence physiology or biotic or abiotic stress, with the main technological interventions involving control of temperature and humidity of the atmosphere around the produce (Mulualem Azene *et al*, 2014). It is essential to control storage temperature and relative humidity during storage as they are the main causes of fruit and vegetable deterioration during ripening and storage. Temperature of the produce and surrounding air can be reduced by forced air-cooling, hydro cooling, vacuum cooling, ice cooling, and evaporative cooling. The evaporatively cooled environment is suggested to be a good alternative for the small-scale peasant farmers, retailers, and wholesalers in Ethiopia, as it require low initial and running cost compared to other cooling methods.

Packaging and storage environments had significant ($P \leq 0.05$) interaction effects on the shelf life and most of the physiological and chemical qualities of papaya fruits. The evaporative cooler maintained temperature between 12.4 and 19.7 °C and relative humidity between 71.4 and 90.7% which is close to ranges considered to be adequate for storage of papaya. Weight loss of papaya fruits was reduced in packages compared to the control under both storage conditions. Fruits packaged with HDPE bag in the evaporatively cooled storage maintain better fresh weight during the storage period. Packaging combined with the evaporatively cooled storage best maintained the fresh weight than when combined with ambient storage. As the storage time advanced, packaged fruits stored in the evaporative cooler had shown more TSS, AA, TA, reducing and total sugar content and lower pH values. Polyethylene bag with evaporatively cooled storage were generally more effective compared to news paper and banana leaf packaging materials and the control in maintaining the quality of the fruit. The reducing and total sugar content of control fruits were higher up to day six compared to the packaged ones. However, packaged fruits maintain their sugar content better than the control towards the end of the storage period. Similarly, as the storage time advanced, fruits stored in the evaporative cooler maintained their sugar better than the control (Mulualem Azene *et al*, 2014). Over all, packaging combined with evaporative cooler storage maintained the freshness and improved the shelf life of papaya fruits. Polyethylene bag packaging with evaporative cooler storage were more effective compared to news paper, banana leaf and the control fruits in maintaining the quality as well as prolonging shelf life and marketability of papaya fruits.

To investigate the effects of packaging on the quality aspects of radish, Korean radish roots were stored at 0°C after different packaging treatments such as keeping in paper cartoon box (control), keeping in plastic crates (PC), packaging with micro perforated HDPE film in PC (HDPE + PC), curing followed by keeping in PC (Curing + PC), and curing followed by packaging with micro perforated HDPE film in PC (Curing + HDPE + PC) (Dulal Chandra *et al*, 2018). Weight losses of radish roots were remarkably lower (<3%) in both HDPE film packaged samples compared to that of control (10%) or without film (≈18%). values, whiteness index, total soluble solids, and flesh and skin firmness were better maintained in Curing + HDPE + PC treatment compared to other treatments. Lower color difference values were also found in this treatment. Both film packaged samples had lower scores of black spot, surface shrinkage, and fungal infection incidence which revealed significantly longer marketable periods. HDPE film packaged samples exhibited longer shelf life more than one and two months compared to control and unpacked samples, respectively. Results suggested that HDPE film packaging can extend postharvest life of radish while curing might have little but beneficial effects in maintaining the quality characteristics.

The optimum temperature and relative humidity for the development of carrot root rot during storage were found to be 20°C and 60%, respectively, and rot development was mostly increased as these environmental factors increased in magnitude (Brunda Devi and Bhadraiah, 2017). Both, temperature and relative humidity, were responsible for rot development in diameter rotted area as a result bio-deterioration in carrot roots inoculated with *R. arrhizus, S. Rolfsii* and *F. solani* and incubated at 3 days intervals. The protein content in inoculated carrot roots at 3 days of incubation period was low significantly compared to un-inoculated carrot roots (control). However, there was significant increase in protein content in inoculated carrot roots over control at 6 and 9 days of incubation period. Carrots inoculated with three pathogens showed gradual increase in phenol contents from 3 to 9 days incubation over control. Statistically significant increase in phenol content of carrot roots inoculated with three pathogens was found from 3 to 9 days incubation period. The amounts of total sugar and non-reducing sugars in rotted carrot roots except carrot root rot with *S. rolfsii*, were substantially decreased over healthy carrot root whereas reducing sugars in inoculated carrot roots were significantly high than un-inoculated roots of carrot. Fluctuations in total, reducing and non-reducing sugar content was observed as the storage period was prolonged from 3-9 days. In addition, the experiment on rot management and shelf-life of carrots packed at open bunch, sealed polythene bags and perforated polythene bags was done by comparing low cost zero energy cool chamber and ambient conditions. Carrot root rot caused by three pathogens was significantly controlled at cool chamber where temperature and relative humidity were maintained at 15° C – 17° C and 85 – 98 % RH, respectively. Storage life of carrots packed was significantly increased at the cool chamber over ambient temperature. Carrots stored in perforated polythene bags showed long shelf- life

The effect of harvesting stage and storage conditions on the postharvest quality of tomato cv. Roma VF during the process of ripening in storage was studid (Moneruzzaman, *et al*, 2009). Fruit of three maturity stages such as mature green (mature but green in color), half ripen (breaker stage when fruit turns to yellow) and full ripen (yellow and soft, edible stage) were kept under three different conditions; open condition (control), covering with white polythene and finally treatment by CaC_2 + polythene. High and significant variation was observed in quality characteristics under different harvesting stages, storage conditions and their combinations. The highest value for rotting and total sugar content was distinguished in full ripens tomato. The highest weight loss and shelf life was measured in mature green. The half ripen tomato showed the highest value of vitamin C and titrable acidity. The percentage of decay (rotting) and weight loss, pH, titrable acidity and total sugar were increased with gradual increasing of storage time, irrespective to maturity stages while the percentage of vitamin C was decreased with progressing time of storage. The highest value of weight loss, shelf life and titrable acidity was recorded in control treatment. The high decay (rotting percent) and total sugar content was recorded under CaC_2 + polythene covering treatment while the highest vitamin C content and pH was measured in simple polythene covering treatment at the final observation day (15th day). The values of all parameters except for vitamin C were increased gradually with the exceeding of storage time irrespective to storage condition. Tomato plants placed under CaC_2 + treatment, covered with polythene have shown the highest decay, titrable acidity and total sugar content at the final observation day.

Almost all fruits are seasonal in nature. Every year, harvesting season falls during a fixed period, say 2-3 months. This may be little early or late due to prevailing weather conditions during growing periods. The demand of any fruit beyond the harvesting season is called off-season demand. This demand can be fulfilled only if fruits are stored in the harvesting season and sold during off-season. The management of temperature, ventilation, and relative humidity are the three most important factors that affect postharvest quality and storage life of horticultural produce. Recommended storage temperature and relative humidity for cabbage, lettuce and carrots are 0 to 2 °C and 95 to 100% respectively (Anonymous, 2010). Recommended storage temperature and relative humidity for mangoes, avocados, papayas and potatoes are 13-15°C and 85-90%, respectively.

Several attempts have been made to improve shelf-life of fresh fruits and vegetables, including fruit storage under modified environment through bagging. Modified environment delays fruit peel degreening and prolongs storage life of non-climacteric fruits, retards fruit senescence, lowers respiration rates, slows down the rate of tissue softening and reduces texture loss. Nevertheless, modified environment poses challenge on getting the right gaseous composition suitable for targeted produce and such situation can lead to undesirable fermenting odour in some fruit varieties (Majubwa, *et al*, 2015). A laboratory experiment was laid out in a split plot arrangement following a complete randomized block design with three replications. Three African eggplant cultivars (Tengeru white, Manyire green and AB2) and four postharvest storage practices (perforated polyethylene bags, woven polypropene bags, closed paper boxes and on-bench storage) were used as main plot and sub-plot factors, respectively. Data were subjected to analysis of variance and means separation was conducted based on StudentNewman Keuls at $P<0.05$ using MSTAT-C statistical software. Results indicated that fruits stored in perforated polyethylene bags had significantly (p=0.001) the lowest water loss and longest shelf life though they suffered the highest incidence of decays. The decay and shelf life varied with African eggplant cultivars where cv. AB2 had significantly (P=0.001) the lowest decay incidence and longest shelf compared with cv. Tengeru white and Manyire green.

Major causes of food deterioration in storage: Major causes include the growth and activities of micro-organisms, principally bacteria, yeasts and moulds; activities of natural food enzymes; insects, parasites and rodents; temperature, both heat and cold; moisture and dryness; air and in particular oxygen; light; time.

4.2 Effect of storage environment on postharvest decay : Commercial producers and handlers modify temperature, RH and atmospheric composition during prestorage, storage, and transit to control decay. For optimum decay control, two or more factors often are modified simultaneously.

i. Temperature and RH: Proper management of temperature is so critical to postharvest disease control that all other treatments can be considered as supplements to refrigeration. Fruit rot fungi generally grow optimally at 20 to 25 °C (68 to 77 °F) and can be conveniently divided into those with a growth minimum of 5 to 10 °C (41 to 50 °F) and those with a growth minimum of -6 to 0 °C (21 to 32 °F). Fungi with a minimum growth temperature below -2 °C (28 °F) cannot be completely stopped by refrigeration without freezing the fruit. However, temperatures as low as possible are desirable because they significantly slow growth and thus reduce decay.

High temperature may be used to control postharvest decay on crops that are injured by low temperatures, such as mango, papaya, pepper, and tomato (Spotts 1984). Though hot water generally is more effective, hot air has been used to control decay in crops that are injured by hot water. Heating of pears at temperatures from 21 to 38 °C (70 to 100 °F) for 1 to 7 days reduced postharvest decay (Spotts and Chen 1987). Decay in 'Golden Delicious' apples was reduced by exposure to 38 °C (100 °F) for 4 days (Sams *et al.* 1993) and virtually eliminated when treated after inoculation (Fallik *et al.* 1995, Klien *et al.* 1997). Heat treatment eliminates incipient infections and improves coverage by fungicides (Couey 1989). The primary obstacle to the widespread use of heat to control postharvest fruit diseases or insect infestation is the sensitivity of many fruit to the temperatures required for effective treatment.

Both low and high RH have been related to postharvest decay control. Perforated polyethylene bags for fruit and vegetable storage create RH about 5 to 10% above that in storage rooms. Though shrivel and weight loss are reduced, decay may increase (Spotts, 1984). Crops with well-developed cuticle and epidermis, such as apples and pears, tolerate lower RH levels, which helps prevent storage decay (Spotts and Peters, 1981). Often fungal spore germination is inhibited at low RH, and small differences in RH can have significant effects in relation to the degree of postharvest decay.

ii. Modified or controlled atmospheres: Alterations in O_2 and CO_2 concentrations are sometimes provided around fruit and vegetables. With close control of these gases, the synthetic atmosphere is commonly called a "controlled atmosphere"; the term "modified atmosphere" is used when there is little possibility of adjusting gas composition during storage or transportation (Sommer, 1989). Because the pathogen respires as does produce, lowering O_2 or raising CO_2 above 5% can suppress pathogenic growth in the host. In crops such as stone fruits, a direct suppression occurs when fungal respiration and growth are reduced by the high CO_2 of the modified atmosphere. For example, CO_2 added to air has been widely used in the transport of 'Bing' cherries, primarily to suppress gray mold and brown rot. Low O_2 does not appreciably suppress fungal growth until the concentration falls below 2%. Important growth reductions result if O_2 is lowered to 1% or less, though there is a danger that the crop will start respiring anaerobically and develop off flavors. Other technologies that have been tested for lowering postharvest decay with limited success are storage and transport under low O_2 and the use of carbon monoxide.

Storage is an important marketing function, which involves holding and preserving goods from the time they are produced until they are needed for consumption. The storage of goods, therefore, from the time of production to the time of consumption, ensures a continuous flow of goods in the market.

Packing and packaging of fruits and vegetables: Preparation of produce for market may be done either in the field or at the packing house. This involves cleaning, sanitizing, and sorting according to quality and size, waxing and, where appropriate, treatment with an approved fungicide prior to packing into shipping containers. Packaging protects the produce from mechanical injury, and contamination during marketing. Corrugated fiberboard containers are commonly used for the packaging of produce, although reusable plastic containers can be used for that purpose. Packaging accessories such as trays, cups, wraps, liners, and pads may be used to help immobilize the produce within the packaging container while serving the purpose of facilitating moisture retention, chemical treatment and ethylene absorption (Adel A. Kader and Rosa S. Rolle, 2004). Either hand-packing or mechanical packing systems may be used. Packing and packaging methods can greatly influence air flow rates around the commodity, thereby affecting temperature and relative humidity management of produce while in storage or in transit.

4.3 Temperature and relative humidity management: Temperature is the most important environmental factor that influences the deterioration of harvested commodities. Most perishable horticultural commodities have an optimal shelf-life at temperatures of approximately 0°C. The rate of deterioration of perishables however increases two to three-fold with every 10°C increase in temperature. Temperature has a significant effect on how other internal and external factors influence the commodity, and dramatically affects spore germination and the growth of pathogens. Temperatures either above or below the optimal range for fresh produce can cause rapid deterioration due to the following disorders:

i. **Freezing :** Perishable commodities are generally high in water content, and possess large, highly vacuolate cells. The freezing point of their tissues is relatively high (ranging from -3°C to -0.5°C), and disruption caused by freezing generally results in immediate collapse of their tissues and a total loss of cellular integrity. Freezing occurs in cold storage systems either due to inadequate refrigerator design, or to thermostat failure. Freezing can also occur upon exposure to inclement weather conditions as occurs when produce is allowed to remain for even short periods of time on unprotected transportation docks during winter.

ii. **Chilling injury:** Some commodities (chiefly those native to the tropics and subtropics) respond unfavorably to storage at low temperatures which are well above their freezing points, but below a critical temperature termed their chilling threshold temperature or lowest safe temperature. Chilling injury is manifested in a variety of symptoms including surface and internal discoloration, pitting, water soaking, failure to ripen, uneven ripening, development of off flavors and heightened susceptibility to pathogen attack.

iii. **Heat injury :** High temperature conditions are also injurious to perishable crops. Transpiration is vital to maintaining optimal growth temperatures in growing plants. Organs removed from the plant, however, lack the protective effects of transpiration, and direct sources of heat, such as sunlight, can rapidly elevate the temperature of tissues to above the thermal death point of their cells, leading to localized bleaching, necrosis (sunburn or sunscald) or general collapse.

iv. **Relative humidity (RH):** It is defined as the moisture content (as water vapor) of the atmosphere, expressed as a percentage of the amount of moisture that can be retained by the atmosphere (moisture holding capacity) at a given temperature and pressure without condensation. The moisture holding capacity of air increases with temperature. Water loss is directly proportional to the vapor pressure difference (VPD) between the commodity and its environment. VPD is inversely related to the RH of the air surrounding the commodity. RH can influence water loss, decay development, the incidence of some physiological disorders, and uniformity of fruit ripening. Condensation of moisture on the commodity (sweating) over long periods of time is probably more important in enhancing decay than is the RH of ambient air. An appropriate RH range for storage of fruits is 85 to 95 percent while that for most vegetables varies between 90 and 98 percent. The optimal RH range for dry onions and pumpkins is 70 to 75 percent. Some root vegetables, such as carrot, parsnip, and radish, can best be held at 95 to 100 percent RH.

RH can be controlled by one or more of the following procedures:

- Adding moisture (water mist or spray, steam) to air with the use of humidifiers;
- Regulating air movement and ventilation in relation to the produce load in the cold storage room;
- Maintaining the temperature of the refrigeration coils in the storage room or transit vehicle to within about 1 °C of the air temperature;
- Providing moisture barriers that insulate walls of storage rooms and transit vehicles;
- Adding polyethylene liners in packing containers and using perforated polymeric films for packaging;
- Wetting floors in storage rooms;
- Adding crushed ice in shipping containers or in retail displays for commodities that are not injured by the practice;
- Sprinkling produce with sanitized, clean water during retail marketing of commodities that benefit from misting, such as leafy vegetables, cool-season root vegetables, and immature fruit vegetables (such as snap beans, peas, sweet corn, and summer squash).

Cooling methods: Temperature management is the most effective tool for extending the shelf life of fresh horticultural commodities (Adel A. Kader and Rosa S. Rolle, 2004). Packing fresh produce with crushed or flaked ice provides rapid cooling, and can provide a source of cooling and high RH during subsequent handling. The use of crushed ice is, however, limited to produce that is tolerant to direct contact with ice and packaged in moisture-resistant containers.

Clean, sanitized water is used as the cooling medium for the hydrocooling (shower or immersion systems) of commodities that tolerate water contact and are packaged in moisture resistant containers. Vacuum cooling is generally applied to leafy vegetables that release water vapor quickly, thereby allowing them to be rapidly cooled. During forced-air cooling on the other hand, refrigerated air is forced through produce packed in boxes or pallet bins. Forced air cooling is applicable to most horticultural perishables. Precise temperature and RH management are required to provide the optimum environment for fresh fruits and vegetables during cooling and storage. Precision temperature management (PTM) tools, including time temperature monitors, are increasingly being employed in cooling and storage facilities.

4.4 Refrigerated transport and storage: Cold storage facilities should be appropriately designed, of good construction, and be adequately equipped. Their insulation should include a complete vapor barrier on the warm side of the insulation; sturdy floors; adequate and well positioned doors for loading and unloading; effective distribution of refrigerated air; sensitive and properly located controls; refrigerated coil surfaces designed to adequately minimize differences between the coil and air temperatures; and adequate capacity for expected needs. Commodities should be stacked in the cold room or the refrigerated vehicle with air spaces between pallets and room walls so as to ensure proper air circulation. Storage rooms should not be loaded beyond their capacity limit if proper cooling is to be achieved. Commodity temperature rather than air temperature should be measured in these facilities.

Temperature management during transportation of fresh fruits and vegetables over long distances is critical. Loads must be stacked so as to enable proper air circulation, in order to facilitate removal of heat from the produce as well as to dissipate incoming heat from the atmosphere and off the road. Stacking of loads must also incorporate consideration for minimizing mechanical damage. Transit vehicles must be cooled prior to loading the fresh produce. Delays between cooling after harvest and loading into transit vehicles should also be avoided. Proper temperature maintenance should be ensured throughout the handling system.

As far as possible, environmental conditions (temperature; relative humidity; concentrations of oxygen, carbon dioxide, and ethylene) should be optimized in transport vehicles. Treatment with ethylene to initiate ripening during transportation is feasible, and is commercially used to a limited extent on mature green bananas and tomatoes. Produce should be cooled prior to loading and should be loaded with an air space between the palletized product and the walls of the transport vehicles in order to facilitate temperature control. Vibration during transportation should be minimized, so as to avoid damage due to bruising. Controlled-atmosphere and precision temperature management should, where possible, be observed so as to allow non-chemical insect control for markets which possess quarantine restrictions against pests endemic to exporting countries and for markets that do not want their produce exposed to chemical fumigants.

Mixing several produce items in one load is common and often compromises have to be made in selecting an optimal temperature and atmospheric composition when transporting chilling sensitive with non-chilling sensitive commodities or ethylene producing with ethylene sensitive commodities (Adel A. Kader and Rosa S. Rolle, 2004). In the latter case, ethylene scrubbers can be used to remove ethylene from the circulating air within the vehicle. Several types of insulating pallet covers are available for protecting chilling sensitive commodities when transported with non-chilling sensitive commodities at temperatures below their threshold chilling temperatures.

The cold chain and its importance: The cold chain encompasses all the critical steps and processes those foods and other perishable products must undergo in order to maintain their quality. Like any chain, the cold chain is only as strong as its weakest link. Major limitations experienced by the cold-chain include poor temperature management due to either the lack of, or limitations in, refrigeration, handling, storage, and humidity control. Investment in cold chain infrastructure ultimately leads to a reduction in the level of losses and quality degradation in fresh produce, with overall net positive economic returns

Return on investment in temperature and relative humidity management: Deficiencies in cold chain management whether due to limitations in refrigeration, improper handling and storage, or inadequate humidity control, can lead to losses in profits as well as in horticultural crops. Overcoming such deficiencies necessitates improvements in methodologies, operations and handling along the chain (Table 4). Often the level of investment required in overcoming such deficiencies is minimal in comparison to the level of losses sustained over time. A University of California study determined that a one-hour delay in cooling strawberries after harvest resulted in a 10 percent loss due to decay during marketing (Adel A. Kader and Rosa S. Rolle, 2004). The resulting economic loss exceeded the increased cost of expedited handling of the strawberries by more frequent deliveries of harvested fruit to the cooling facility and initiation of forced-air cooling. Similarly, a University of Georgia study showed that poor temperature management of lettuce resulted in a net income loss of US$172.50 per truck load of 900 cartons.

Table 4: Maintaining the cold chain for perishables

Harvest	• Minimize delays before cooling
	• Cool the product thoroughly as soon as possible
Cooling	• Store the product at optimum temperature
Temporary	• Practice first in first out rotation
Storage	• Ship to market as soon as possible
Transport to Market	• Use refrigerated loading area
	• Cool truck before loading
	• Load pallets towards the center of the truck
	• Put insulating plastic strips inside door of reefer if truck makes ultiple stops
	• Avoid delays during transport
	• Monitor product temperature during transport
Handling at destination	• Use a refrigerated unloading area
	• Measure product temperature
	• Move product quickly to the proper storage area
	• Transport to retail markets or foodservice operations in refrigerated trucks
	• Display at proper temperature range
Handling at home or foodservice outlet	• Store product at proper temperature
	• Use the product as soon as possible

Excess weight loss coupled with color deterioration resulting from delays between harvest and cooling, improper refrigeration temperature and relative humidity control during the shipping of table grapes resulted in a 15 percent loss in the value of that commodity. Resultant monetary losses were greater than the cost of improved management of temperature and RH of the grapes with perforated plastic liners in the boxes and by minimizing delays prior to cooling with humidified, forced air.

4.5 Treatments to reduce microbial contamination: Over the past few years, food safety has become and continues to be the number one concern of the fresh produce industry. A "Guide to Minimize Microbial Food Safety Hazards for Fresh Fruits and Vegetables," was published by the U.S. Food and Drug Administration in October 1998 (Adel A. Kader and Rosa S. Rolle, 2004). This guide is based on the following principles.

- Prevention of microbial contamination of fresh produce is favored over reliance on corrective actions once contamination has occurred;
- In order to minimize microbial food safety hazards in fresh produce, growers, packers, or shippers should use good agricultural and management practices in those areas over which they have control;

- Fresh produce can become microbiologically contaminated at any point along the farm-to-table food chain. Human and/or animal faeces are the source of microbial contamination of fresh produce;
- Whenever water comes in contact with produce, its quality dictates the potential for contamination. The potential of microbial contamination from water used with fresh fruits and vegetables must be minimized;
- The use of animal manure or municipal biosolid wastes as fertilizers should be closely managed in order to minimize the potential for microbial contamination of fresh produce; and
- Worker hygiene and sanitation practices during production, harvest, sorting, packing, and transport play a critical role in minimizing the potential for microbial contamination of fresh produce (Table 5).

Treatments to minimize water loss: Transpiration, or evaporation of water from the plant tissues, is one of the major causes of deterioration in fresh horticultural crops after harvest. Water loss through transpiration not only results in direct quantitative losses (loss of saleable weight), but also causes losses in appearance (wilting, shriveling), textural quality (softening, flaccidity, limpness, loss of crispness and juiciness), and nutritional quality. Transpiration can be controlled either through the direct application of postharvest treatments to the produce (surface coatings and other moisture barriers) or through manipulation of the environment (maintenance of high relative humidity).

Treatments that can be applied to minimize water loss in fruits and vegetables include:

- Curing of certain root vegetables, such as garlic, onion, potato, and sweet potato.
- Waxing and the use of other surface coatings on commodities, such as apple, citrus fruits, nectarine, peach, plum, pomegranate, and tomato.
- Packaging in polymeric films that act as moisture barriers.
- Careful handling to avoid physical injuries, which increase water loss from produce.
- Addition of water to those commodities that tolerate misting with water, such as leafy vegetables (Adel A. Kader and Rosa S. Rolle, 2004)

Table 5: Sanitizing chemicals used in produce handling

Sanitizing chemicals	Advantages	Disadvantages
Chlorine compounds Calcium hypochlorite Sodium hypochlorite Chlorine gas Chlorine dioxide	Low cost	Corrosive, irritating, trihalomethanes are by-products
Iodine compounds	Low cost, non irritating	Slightly corrosive, staining
Ozone	Faster action on microorganisms, fewer disinfection by-products than chlorine	Higher cost than chlorine
Peroxyacetic acid Hydrogen peroxide	More effective in removing and controlling microbial biofilms	Higher cost than chlorine

Treatments to reduce ethylene damage : The promotion of senescence in harvested horticultural crops by ethylene (1 ppm or higher) results in acceleration of deterioration and reduced postharvest life. Ethylene accelerates chlorophyll degradation and induces yellowing of green tissues, thus reducing the quality of leafy, floral, and immature fruit-vegetables and foliage ornamentals. Ethylene induces abscission of leaves and flowers, softening of fruits, and several physiological disorders. Ethylene may increase decay development of some fruits by accelerating their senescence and softening and by inhibiting the formation of antifungal compounds in the host tissue. In some cases, ethylene may stimulate the growth of fungi, such as *Botrytis cinerea* on strawberries and *Penicillium italicum* on oranges.

The incidence and severity of ethylene-induced deterioration symptoms is dependent upon temperature, time of exposure, and ethylene concentration. The yellowing of cucumbers can, for example, result from exposure to either 1 ppm ethylene over 2 days or to 5 ppm ethylene over 1/2 day at 10 °C. Ethylene effects are cumulative throughout the postharvest life of the commodity. Treatment of ornamental crops with 1-methylcylopropene (1-MCP), which is an ethylene action inhibitor, provides protection against ethylene damage. The commercial use of this product at concentrations of up to 1 ppm on apples, apricots, avocados, kiwifruit, mangoes, nectarines, papayas, peaches, pears, persimmons, plums, and tomatoes was approved by the United States Environmental Protection Agency. The use of 1-MCP will no doubt be extended to several other fruits and vegetables, and to use in other regions.

Treatments for decay control : A major cause of losses in perishable crops is the action of a number of microorganisms on the commodity. Fungi and bacteria may infect the plant organ at any time. "Latent" infections, in which fungi invade fruit tissues shortly after flowering, become apparent only at the onset of ripening. Postharvest rots frequently occur as a result of rough handling during the marketing process and are caused by a wide array of microorganisms (Adel A. Kader and Rosa S. Rolle, 2004). The grey mold Botrytis cineria is a very important cause of loss in many commodities (such as grapes, kiwifruit, pomegranates, raspberries, and strawberries), and is an aggressive pathogen, even at low temperatures. Virus infection frequently lowers the quality of perishable commodities, usually as a result of visual deterioration, although viruses may also affect flavor and composition.

Curing is a postharvest treatment (Table 6) that facilitates certain anatomical and physiological changes that can prolong the storage life of some root crops. It is one of the most effective and simple means of reducing water loss and decay during subsequent storage of root, tuber, and bulb crops, such as those listed in Table 6.

Table 6: Conditions for curing root, tuber, and bulb crops

Crops	Temperature (°C)	Relative humidity %	Duration (days)
Cassava root	30-40	90-95	2-5
Onion and garlic bulbs	30-45	60-75	1-4
Potato tubers	15-20	85-90	5-10
Sweet Potato roots	30-32	85-90	4-7
Yam tubers	32-40	90-100	1-4

4.6 Modified atmosphere storage : When used as supplements to keeping fresh horticultural perishables within their optimum ranges of temperature and relative humidity, controlled atmospheres (CA) or modified atmospheres (MA) can serve to extend their postharvest-life (Table 7). Optimum oxygen and carbon dioxide concentrations lower respiration and ethylene production rates, reduce ethylene action, delay ripening and senescence, retard the growth of decay-causing pathogens, and control insects. CA conditions which are not suited to a given commodity can, however, induce physiological disorders and enhance susceptibility to decay.

Several refinements in CA storage technology have been made in recent years. These include: the creation of nitrogen-on-demand by separation of nitrogen from compressed air through the use of either molecular sieve beds or membrane systems; use of low (0.7 to 1.5 percent) oxygen concentrations that can be accurately monitored and controlled; rapid establishment of CA, ethylene-free CA, programmed (or sequential) CA (such as storage in 1 percent O_2 for 2 to 6 weeks followed by storage in 2-3 percent O_2 for remainder of the storage period), and dynamic CA where levels of O_2 and CO_2 are modified as needed based on monitoring specific attributes of produce quality, such as ethanol concentration and chlorophyll fluorescence.

The use of CA in refrigerated marine containers continues to benefit from technological and scientific developments. CA transport is used to continue the CA chain for commodities (such as apples, pears, and kiwifruits) that had been stored in CA immediately after harvest. CA transport of bananas permits their harvest at a more advanced stage of maturity, resulting in the attainment of higher yields at the field level. In the case of avocados, CA transport facilitates use of a lower shipping temperature (5 °C) than if shipped in air, since CA ameliorates chilling injury symptoms. CA in combination with precision temperature management allows insect control without the use of chemicals, in commodities destined for markets that have restrictions against pests endemic to exporting countries and for markets with a preference for organic produce.

Table 7: Classification of horticultural crops according to their controlled atmosphere storage potential at optimum temperatures and relative humidities

Range of storage duration (months)	Commodity
More than 12	Almond, Brazil nut, cashew, filbert, macadamia, pecan, pistachio, walnut, dried fruits and vegetables
6-12	Some cultivars of apples and European pears
3-6	Cabbage, Chinese cabbage, kiwifruit, persimmon, pomegranate, some cultivars of Asian pears
1-3	Avocado, banana, cherry, grape (no SO_2), mango, olive, onion (sweet cultivars), some cultivars of nectarine, peach and plum, tomato (mature-green)
<1	Asparagus, broccoli, cane berries, fig, lettuce, muskmelons, papaya, pineapple, strawberry, sweet corn; fresh-cut fruits and vegetables; some cut flowers.

The use of polymeric films for packaging produce and their application in modified atmosphere packaging (MAP) systems at the pallet, shipping container (plastic liner), and consumer package levels continues to increase. MAP(usually designed to maintain 2 to 5 percent O_2 levels and 8 to 12 percent CO_2 levels) is widely applied in extending the shelf-life of fresh-cut fruits and vegetables. Use of absorbers of ethylene, carbon dioxide, oxygen, and/or water vapor as part of MAP is increasing. Although much research has been done on the use of surface coatings to modify the internal atmosphere within the commodity, commercial applications are still very limited due to inherent biological variability of commodities(Adel A. Kader and Rosa S. Rolle, 2004).

At the commercial level, CA is most widely applied during the storage and transport of apples and pears. It is also applied to a lesser extent on kiwifruits, avocados, persimmons, pomegranates, nuts and dried fruits. Atmospheric modification during long-distance transport is used for apples, avocados, bananas, blueberries, cherries, figs, kiwi-fruits, mangoes, nectarines, peaches, pears, plums, raspberries and strawberries. Technological developments geared toward providing CA during transport and storage at reasonable cost (positive benefit/cost ratio) are essential if the application of this technology to fresh fruits and vegetables is to be expanded.

Although MA and CA have both been shown to be effective in extending the postharvest life of many commodities (Table 6), their commercial application has been limited by the relatively high cost of these technologies. There are however a few cases in which a positive return on investment (cost/benefit ratio) can be demonstrated. In a comparison of losses due to decay during retail marketing of strawberries shipped in air and those shipped in an environment consisting of 15 percent CO_2-enriched air (modified atmosphere within pallet cover), the use of modified atmosphere was observed to reduce losses by 50 percent (an average of 20 percent losses was sustained in strawberries shipped in air vs 10 percent losses in those shipped by MA). The economic loss of 10 percent value (US$50-75 per pallet) was much greater than the cost of using MA (US$15-25 per pallet).

Use of controlled atmosphere (CA) during marine transportation can extend the postharvest-life of those fruits and vegetables that would normally have a short postharvest-life potential, thereby allowing the use of marine transportation instead of air transport for the shipment of such produce. In terms of cost and benefit, savings realized with the use of marine transportation are much greater than is the added cost of CA service.

4.7 Ethylene exclusion and removal: Many green vegetables and most horticultural produce are quite sensitive to ethylene damage. Their exposure to ethylene must therefore be minimised. Ethylene contamination in ripening rooms can be minimized by using ethylene levels of 100 ppm instead of the higher levels often used in commercial ripening operations, venting ripening rooms to the outside on completion of exposure to ethylene, at least once per day ventilating the area around the ripening rooms or installing an ethylene scrubber, use of battery-powered forklifts instead of engine driven units in ripening areas (Adel A. Kader and Rosa S. Rolle, 2004). Ethylene-producing commodities should not be mixed with ethylene-sensitive commodities during storage and transport. Potassium permanganate, an effective oxidizer of ethylene, is commercially used as an ethylene scrubber. Scrubbing units based on the catalytic oxidation of ethylene are used to a limited extent in some commercial storage facilities.

Return on investment in reducing ethylene damage: A University of California study showed that the use of an ethylene scrubber in storage facilities used for lettuce significantly reduced russet spotting. The difference in value of lettuce that was protected from ethylene vs that which was exposed to ethylene was estimated to be 20 to 25 percent, which was greater than the cost of the ethylene scrubber. Similar results were found with kiwi fruits, which soften very rapidly when exposed to ethylene levels as low as 50 ppb.

Treatments to enhance uniformity in fruit ripening : Ethylene treatment is commercially used to enhance the rate and uniformity of ripening of fruits such as bananas, avocados, mangoes, tomatoes, and kiwifruits. Optimal ripening conditions are as follows.

Temperature	**18 °C to 25 °C (65 °C to 77 °F)**
Relative humidity	90 to 95 percent
Ethylene concentration	10 to 100 ppm
Duration of treatment	24 to 74 hours depending on fruit type and stage of maturity
Air circulation	Sufficient to ensure distribution of ethylene within the ripening room
Ventilation	Require adequate air exchange in order to prevent accumulation of O_2 which reduces theeffectiveness of C_2H_4.

Many pathogens cause postharvest decay of fruits and vegetables grown under field and greenhouse conditions. These pathogens may infect the vegetables / fruits prior to and/or during harvest and subsequently develop during storage and shipping. Many remain undetected until environmental conditions are conducive for fungal growth. Several studies have been conducted on the effect of storage conditions viz., the range of fungi infecting fruits / vegetables sampled over a 3-year period and the potential sources of inoculum, the influence of wounding and postharvest temperatures on disease development and the composition of fruit surface mycoflora and their potential role in disease development (Punja, *et al*, 2016).

Monitoring of the temperature and humidity conditions in four shipments of tomato fruits revealed differences in the rate at which the recommended storage and shipping temperature of 13°C was reached, as well as revealing variation in the range of relative humidity levels attained. Postharvest shipment of tomato fruits at temperatures above 13°C increases the likelihood of fungal infection from resident inoculum on the fruit surface; hence, rapid cooling is recommended. At the time of arrival of the shipments at their destination, the ideal temperature range of 13–15°C had been reached, and the relative humidity was in the optimal range of 80–85%. The storage environment after unloading of the fruits was also shown to be in the optimal range. Therefore, the most significant variable in the shipments was the rate at which fruits were cooled from the initial temperature of 20–23°C at harvest time. Incubation of tomatoes at a temperature of 21°C significantly increased the extent of decay in this study compared with 13°C, highlighting the importance of cooler storage temperatures. Most postharvest infections occurred at the calyx end of the fruit, suggesting a build-up of fungal spores on the green tissue was taking place. Previous studies have not implicated the attached green calyx and stem tissue on fresh market tomatoes as a possible source of inoculum for postharvest decay. A longer duration of storage (16 days) resulted in more decay compared to 12 days. On apple fruit, prolonged periods of storage (6 months) were shown to increase the density of *Penicillium* species on the fruit surface from 10–50 spores cm−2 at 1 month to 300–400 spores cm−2. The results from this study highlight the presence of a diverse mycoflora on greenhouse tomato fruit surfaces, and the potential interactions between these populations and storage temperature and duration, as well as shipping conditions, on development of postharvest fruit decay. It was concluded that postharvest decay of tomatoes is influenced by the composition of the fruit surface mycoflora, presence of fruit injury, storage temperatures and environmental conditions during shipping.

4.8 Postharvest care

a. Handling and Packaging : Careful handling of fruit is the most important step that has to be taken at all postharvest stages. Sorting and grading of the harvested fruits is good farm house practice. Eliminating the fruits showing injury or early symptoms of a disease, revealed at this stage by careful inspection can substantially reduce the intensity of disease at later stages (Yadav, *et al*, 2013). Packing practice has to be evolved to suit the individual consignment, aiming at the maximum perfection as well as cost effectiveness (Kenneth C. Gross *et al*, 2016). Before packing it is essential to remove superficial tissues from banana bunches for reducing the incidence of crown-rot disease. Discourage the common Indian practice of using the leaves from respective fruit plants for packaging. Instead of paddy straw, shredded paper or saw dust has been suggested. In modern practice corrugated boxes/ fiber board packs/molded plastic trays/ wooden, crates are findings their places. Use impregnated paper for wrapping individual fruits such as Mango, Papaya, Banana, Guava with paper

b. Care during Transport : Horticultural produce may be transported by road, rail, sea or air, sometimes for many thousands of miles. A well planned schedule should be worked out for transportation of fruits according to the mode of journey, the nature of fruits and distance as well as duration of the journey. The fruits of guava and papaya are comparatively more prone to injury and hence need special care during transportation and if possible temperature and other ambient conditions may be controlled. High value perishables can be sent by air, either in wide bodied freights or in the cargo holds of passenger aircraft.

c. Storage : Low pressure storage, cold storage and modified atmosphere storage are aimed out preventing the perpetuation of pathogen and spread of the disease.

Microbial load : It is impossible to remove every fungal spore or bacterial cell. However, reducing the number that is present, the microbial load reduces the chance that infection will occur, even if other conditions are suitable. Reducing the microbial load generally involves pre-harvest hygiene and/or postharvest removal by washing.

Pre-harvest hygiene involves removing sources of inoculum. This may mean removing or rotary hoeing diseased crop residues, using pre-harvest fungicides and ensuring good plant health; healthy, well grown crops are less susceptible to disease infection and development than those grown under more challenging conditions.

Vigorous washing and/or brushing can reduce microbial load by 99% or more. Factors that affect how effectively the washing process removes bacteria and fungal spores include the following.

- High water temperatures can increase the effectiveness of washing. For example, short hot water rinsing and brushing treatments are used commercially to reduce postharvest disease on fruit vegetables such as capsicums and tomatoes. Treatments typically range from 50 to 60°C and last for 10 to 30 seconds. This method has been shown to remove 99.9 to 99.99% of pathogens on the product surface.
- Products with a smooth surface will be easier to clean than those with an irregular surface or complex structure, like cabbage.
- Presence and concentration of a sanitiser and pH of the water.
- Number of washes and amount of agitation: Multiple washes are more effective than one.
- Cleanliness of the water: If the water already contains a large amount of organic matter then sanitisers will be ineffective and the washing process may deposit more microbes than it removes.

4.9 Storage environment : Low temperature storage cannot prevent disease, but can limit the rate at which disease develops. The optimum temperature for spore germination of most fungal pathogens is 20–25°C. Although germination is possible at a much wider range, the further the temperature is from this optimum the slower germination will be. Cold temperatures also delay deterioration of the product, reducing susceptibility to decay. Fungi differ in their ability to grow at low temperatures; *Aspergillus niger* is unable to grow below 11°C ; *Colletotrichum* spp. cannot grow below 9°C and *Rhizopus stolonifer* cannot grow below 2°C. However, *Botrytis cinerea* can continue to grow down to –2°C and *Colletotrichum alternata* still develops at –3°C, making these diseases hard to control in storage.

High relative humidity and free moisture on produce both increase opportunities for disease development. Dry conditions mean that fungal spores are unable to germinate and grow, and bacteria cannot multiply. Even if fungal germination does occur in a small amount of moisture, low humidity can inhibit further growth. Conversely, wet conditions and high humidity allow plant pathogens to spread and multiply

i. **Storage disinfection of warehouses:** Debris from warehouses/cold storages should be removed, the walls and floors should be washed with bleaching powder, copper sulfate, etc to eradicate the pathogens which are surviving in the store structures (Kenneth C. Gross *et al.*, 2016).

ii. **Low temperature storage:** Low temperature storage is less suitable for tropical fruits like mango, banana, sapota and guava which are sensitive to chilling. Papaya fruit inoculated with *Colletotrichum gloeosporiodes* and when stored at low temperature (10 °C for 21 days) develops less anthracnose. Suitable temperature for Mango storage is more than 13 °C, temperature below 130 °C causes chilling injury, for Guava more than 15 °C, for Banana more than 10 °C and for Papaya more than 6 °C is required.

iii. **Cold storage:** Storage at low temperature is the main method for reducing deterioration of harvested fruits and vegetables. Low temperature affects both the host and the pathogen simultaneously. They prevent moisture loss from the host tissues and consequent shriveling; they retard metabolic activity and delay physiological changes that lead to ripening and senescence.

iv. **Modified atmosphere storage:** This method have found effective in extending the shelf life of harvested fruits. In storage maintaining a low temperature, an adequate humidity, low oxygen, low *ethyleneconcentration* are favorable. If atmosphere of Mango storage is modified as 2% CO2, 6% O2 and 13°C it can be stored for longer.

New directions for postharvest plant pathology: Postharvest plant pathology has changed its emphasis in the last decade. Food safety has emerged as a key element in decay control programs. Continued failure to effectively control certain postharvest diseases and the need for more environmentally friendly crop control materials require a new approach to disease control. Integrated postharvest decay control is the concept that offers the most promise for the future. Society can no longer rely on one or two control strategies but must enlist the entire spectrum of strategies to reduce postharvest losses.

4.10 Food safety issues: To ensure sufficient food availability for every inhabitant in our planet in quantity and in quality, reduction of the postharvest losses is very crucial. It will ensure that sufficient food availability and ensures about global food security. There is immense need of preservation and storage methods to extend their shelf life, for throughout the year availability either in preserved or processed form and to avoid huge losses. Although, the prime target is still to maintain the quality of the produce which is based on the various attributes such as chemical composition, nutritive values, mechanical properties like texture, firmness and sensory properties. Hence, fruits and vegetables require proper postharvest management handling and processing to maintain its quality attributes throughout the supply chain.

To ensure sufficient food availability for every inhabitant in our planet in quantity and in quality, reduction of the postharvest losses is very crucial. It will ensure that sufficient food availability and ensures about global food security. There is immense need of preservation and storage methods to extend their shelf life, for throughout the year availability either in preserved or processed form and to avoid huge losses. Although, the prime target is still to maintain the quality of the produce which is based on the various attributes such as chemical composition, nutritive values, mechanical properties like texture, firmness and sensory properties. Hence, fruits and vegetables require proper postharvest management handling and processing to maintain its quality attributes throughout the supply chain.

Fruits and vegetables start deterioration just after harvest; therefore, their proper storage is must during transportation and storage to retain maximum quality parameters and for good market value. Best storage conditions are required to prevent growth of micro flora and to maintain the nutritional values of harvested produce. Retailers and processors in every corner of world want to move toward the cheaper ways to increase the shelf life and texture of horticultural crops for better consumer preference. The purpose of this paper is to make consumers and researchers aware about different postharvest malpractices in fresh fruits and vegetables. Design/methodology/approach: Lot of chemicals like colors, artificial ripening agents, sweeteners and waxes are applied on surface of horticulture produce to siphon off money from consumers, and these have adverse health effects directly or indirectly (Anil Panghal *et al*, 2018). To ensure sufficient food availability for every inhabitant in our planet in quantity and in quality, reduction of the postharvest losses is very crucial. It will ensure that sufficient food availability and ensures about global food security. There is immense need of preservation and storage methods to extend their shelf life, for throughout the year availability either in preserved or processed form and to avoid huge losses. Although, the prime target is still to maintain the quality of the produce which is based on the various attributes such as chemical composition, nutritive values, mechanical properties like texture, firmness and sensory properties. Hence, fruits and vegetables require proper postharvest management handling and processing to maintain its quality attributes throughout the supply chain.

4.10.1 Regulation for minimizing pathogen contamination: Various food safety rules have been designed for produce quality, containment, presentation and safety. These rules either leave a gap or are not implemented strictly. Codex alimentarious standard 04.1.1 and 04.2.1 for fresh fruits and vegetables states that fresh fruit/vegetables are generally free of additives. However, these might be coated or cut or peeled for presentation to the consumer may contain additives (General Standard for Food Additives Codex Stan 192-1995, revised in 2015). Regulatory bodies, FSSAI and FDA permitted usage of carnauba wax or beeswax (yellow and white) or shellac wax as food additives and ingredients according to GMP. FSSA, 2006 and CODEX set the tolerance limits of heavy metal present in the food item. No color and mineral oil has been allowed in fresh fruits and vegetables, still these chemicals are consistently in use. Food Safety and Standards Act 2006 is not in existence practically, so the strict implementation of FSSA and Rules by the State Governments should be there. The concerned health and food safety authorities should keep a keen eye on use of hazardous postharvest treatments. There should be provision of some legal actions for violation of the provisions of act. The FDA and Codex Alimentarius should ban dyes usage in food, which serve no purpose other than coloring and cosmetic effect, but there are difficulties in law. In the meantime, companies voluntarily should replace dyes with safer, natural colorings. Food inspectors and food safety officers are empowered and educated under FSSA 2006 to inspect and examine food items intended for sale and to send them for further examination to the Food Quality Laboratory (FSSAI, under section 43[2] of FSS Act, 2006).

FSSAI started training programs for the food handlers in organized and unorganized sectors regarding food handling and safe food production. Vigilance in wholesale market should be strengthened, and officials should take steps to curb adulteration. Legal and regulatory frameworks for product chain management should be there to ensure safe and quality food. Proper training should be given to fruit traders and sellers, and there should be awareness about the possible health hazards and imbued with a sense of moral responsibility to the society. There are no simple solutions to tackle the menace of adulteration: adequate food testing facilities, complete transparency in the functioning of the department of civil and food supplies and getting samples of suspected food tested at regular intervals, are just some answers. Governments should upgrade the infrastructural facilities for proper handling, storage and marketing the fresh produce. Research should be conducted on substitutes of these entire chemical which should be environment friendly and nonhazardous cheap material to protect farmers and retailers from storage losses (Anil Panghal *et al*, 2018). This might involve quality assurance laboratories, e.g. Central Food Laboratory, Kolkata; Food Safety and Analytical Quality Control Laboratory etc. (FSSAI, 2017) and food safety authorities to reduce the risk of adulteration and to provide the produce without any hazardous treatment.

Advice for consumer food malpractices can occur at various levels such as production, entrepreneurship or sale. Consumers should preferable buy the locally grown horticultural produce because the chances of malpractices are minimal in such produce. Consumer should check ISI, GMP, FPO, Agmark, etc., labeled food products to ensure best quality of produce. There is need of awareness programs and campaigns about quality food. FSSAI has taken a step forward and is organizing training programs for all food handlers. The misbranded/mislabeled product manufacturer is liable under the Food Safety and Standards Act, 2006 to be punished with imprisonment and fined. Conclusion Malpractices such as coloring, oiling, sweeteners and hormone injections followed by retailers are mainly profit targeted. The involvement of farmers, retailers, consumers, scientists, policy makers and government agencies is highly required to solve this problem. Neither retailers nor consumers are aware about the sever health issues related to prolonged consumption of such chemicals. Industries are certified by regulatory agencies to ensure quality, but certification is highly required for farmers and vendors. Besides the certification and training, consumer should be aware about these malpractices and legal regulations. Consumer awareness is must about health hazards of chemicals and hormones used in malpractices, and consumer must report to regulatory bodies about such altered samples for further action. Extensive scientific studies are needed on development of detection methods to easily identify such maltreated samples.

The two most important causes of unsafe food are microbial toxins (Kenneth C. Gross *et al*, 2016) and contamination of horticultural products by fecal coliforms . The microbial toxins can be subdivided into bacterial toxins and toxins produced by fungi or mycotoxins. An example of a microbial toxin that is extremely toxic are the botulinum toxins produced by the anaerobic bacterium *Clostridium botulinum*. Interest in toxins produced by fungi was stimulated by the death of 100,000 turkey poults in England in 1960. Aflatoxins produced by fungi in the peanut meal used to feed the birds was the cause. Studies have since shown aflatoxins to be potent carcinogens that may occur in nuts and grain. Other toxins have been identified that are produced by the same fungi that cause postharvest decay. For example, patulin produced by *Penicillium* and *Aspergillis* spp. can be found in apple and pear products. Patulin is toxic to many biological systems but its role in causing animal and human disease is unclear .

Studies on contamination of horticultural products by fecal coliforms have increased dramatically because of documented incidences of food poisoning from apple juice and seed sprouts. Definite interactions have been shown between plant pathogens and food borne human pathogens such as *Salmonella* and Listeria. A study involving more than 400 samples each of healthy and soft rotted commodities collected in retail markets indicated that the presence of *Salmonella* on produce affected by bacterial soft rot was twice that of healthy samples (Wells and Butterfield, 1997). Controlled experiments with potato, carrot, and pepper tissues inoculated with a strain of *Salmonella* confirmed that bacterial soft rot infection increased multiplication of *Salmonella* by at least three- to ten-fold compared with multiplication on uninfected tissues. Similarly, populations of Listeria monocytogenes, inoculated into decayed apple tissue, continually increased on fruit decayed by *Glomerella cingulata* but did not survive after 5 days on fruit decayed by *Penicillium* expansum (Conway *et al.* 2000). The pH of the decayed area declined from pH 4.7 to 3.7 in the case of *P. expansum* but increased from pH 4.7 to 7.0 in the case of *G. cingulata*. This pH modification may be responsible for affecting growth of the food borne pathogen.

Contamination of produce with human pathogens is an important issue that must be addressed along with limiting decay caused by postharvest pathogens and maintaining product quality.

Minimizing pathogen contamination during harvest: During harvesting operations field personnel may contaminate fresh fruits and vegetables by simply touching them with an unclean hand or knife blade. Portable field latrines as well as hand wash stations must be available and used by all harvest crew members. Monitoring and enforcement of field worker personnel hygiene practices such as washing hands after using the latrine are a must, to reduce the risk of human pathogen contamination. Workers who are ill with hepatitis A or who have symptoms of nausea, vomiting or diarrhea should not be assigned to harvest fresh produce (Anonymous, 1989). Produce once harvested should not be placed upon bare soils before being placed in clean and sanitary field containers. Field harvesting tools and gloves should be clean, sanitary and not be placed directly in contact with soil. Field containers should be cleaned and sanitized on a regular basis as well as being free of contaminants such as mud, industrial lubricants, metal fasteners or splinters. Do not allow workers to stand in field bins during harvest to reduce pathogen spread by shoes.

Plastic field bins and totes are preferred to wooden containers since plastic surfaces are easier to clean and sanitize, which should be done after every use. If containers are not cleaned and sanitized after every use, they may become contaminated and then contaminate the next products which are placed in the container. Wooden containers or field totes are almost impossible to sanitize since they have a porous surface and wooden or metals fasteners such as nails from wooden containers may accidentally be introduced into produce. Cardboard field bins if reused should be visually inspected for cleanliness and lined with a polymeric plastic bag before reuse to prevent the risk of cross contamination. Depending upon the commodity, produce may be field packaged in containers that will go all the way to the destination market or be temporarily placed in bulk bins, baskets or bags which will be transported to a packing shed. Employees, equipment, cold storage facilities, packaging materials and any water which will be contacting the harvested produce must be kept clean and sanitary to prevent contamination (Gorny, 1999).

Minimizing pathogen contamination during postharvest handling

i. **Employee hygiene :** Gloves, hairnets and clean smocks are commonly worn by packinghouse employees in export oriented packing sheds. The cleanliness and personnel hygiene of employees handling produce at all stages of production and handling must be managed to minimize the risk of contamination. Adequate bathroom facilities and handwash stations must be provided and used properly to prevent contamination of produce by packinghouse employees. Shoe or boot cleaning stations may also be in place to reduce the amount of field dirt and contamination which enters the packing shed from field operations. Employee training regarding sanitary food handling practices should be done when an employee is hired and reviewed before they begin work each season.

ii. **Equipment :** Food contact surfaces on conveyor belts, dump tanks etc. should be cleaned and sanitized on a regular scheduled basis with food contact surface approved cleaning compounds. A 200 parts per million sodium hypochlorite (bleach) solution is an excellent example of a food contact surface sanitizer. Sanitizers should be used only after thorough cleaning with abrasion to remove organic materials such as dirt or plant materials. Use of steam to clean equipment should be avoided since steam may actually cake organic materials and form a biofilm, which renders equipment almost impossible to sanitize. Steam may also aerosolize bacteria into the air and actually spread contamination throughout the packing house facility.

iii. **Packaging materials :** All packaging materials should be made of food contact grade materials to assure that toxic compounds in the packaging materials do not leach out of the package and into the produce. Toxic chemical residues may be present in some packaging materials due to use of recycled base materials. Empty packages such as boxes and plastic bags should be stored in an enclosed storage area to protect them from insects, rodents, dust, dirt and other potential sources of contamination. These actions protect not only against the potential loss of valuable materials but protects the integrity and safety of these materials.

iv. **Wash and hydrocooling water:** All water which comes in contact with produce for washing or hydrocooling must be safe to drink. Water should contain between 100 and 150 parts per million total chlorine and have a pH of between 6 and 7.5. Chlorine use prevents the potential for cross contamination of all produce in the washing or hydrocooling system, it will not sterilize the produce. Change the water in dump tanks and hydro-coolers regularly.

v. **Ice for cooling :** Use ice made from only potable water sources.

vi. **Refrigerated transport :** Produce is best shipped in temperature controlled refrigerated trucks. Pre-cool the vehicles prior to loading. Maintaining perishables below 5°C (41° F) even while being transported to destination markets will extend shelf-life and significantly reduce the growth rate of microbes including human pathogens. Temperatures used for tranporting chilling sensitive produce will not protect against the growth of most pathogens. Trucks used during transportation should be cleaned and sanitized on a regular basis. Trucks which have been used to transport live animals, animal products or toxic materials should never be used to transport produce.

vii. **Sanitizing field containers, tools and packhouse surfaces :** High pressure wash, rinse and sanitize all crop containers, tools and packhouse surfaces prior to each day's harvest. Sanitizers should be used only after thorough cleaning with abrasion to remove organic materials such as dirt or plant materials. Most commercial sanitizers contain chlorine or quaternary ammonium compounds (QUATS, QAC, benalkonium chloride, N-alkyl dimethybenzyl ammonium chloride). Chlorine solutions prepared from chlorine gas, hypochlorites and cloramines are not compatible with

> quaternary ammonium compound sanitizers. Selection of the sanitizer to use depends upon the surface to be cleaned, hardness of the water, application equipment available, effectiveness under ambient conditions and cost. All require extreme care when handled as either compressed gas, powders or concentrated liquids. Use the following table to assist in the selection of appropriate sanitizers for your operation.

Pre-harvest food safety should focus on significant scientific advancements that are a culmination of collaborative efforts (both public health and agriculture) and significant research results. It must provide the foundation for exploring future preharvest areas and for improving and focusing on more specific intervention/control/prevention strategies. Examples include Escherichia coli and cattle, *Salmonella* and Campylobacter in poultry, and interventions and prevention and control programs. Potential emerging food safety issues and future food safety research directions are to be prioritized, viz., organic production, the FDA's Produce Rule (water and manure), genomic sequencing, antimicrobial resistance, and performance metrics. Strategic planning, coordination, epidemiology and the need for understanding food safety production as a continuum need to be emphasized (Torrence, 2016). Food safety research, whether at the pre- or postharvest level, will continue to be a fascinating complex web of food borne pathogens, risk factors, and scientific and policy interactions. Food safety priorities and research must continue to evolve with emerging global issues, emerging technologies, and methods but remain grounded in a multidisciplinary, collaborative, and systematic approach.

Microbial quality of fresh produce refers to the overall effects of microbial activity, including growth, enzymatic activity, and metabolic byproducts upon the visual and organoleptic quality of fruits and vegetable. Microbial quality is microorganism dependent, and is highly affected by chemical physical, and biological factors pertaining to the cultivar and the environment. It is highly dependent on the conditions of cultivation, harvesting, handling,transport, and postharvest storage storage, as well as marketing conditions (Shlomo Sela, Elazar Fallik 2009). Physical factors affecting growth and metabolic activity of microorganisms include temperature, pH, atmosphere and moisture content. Chemical factors include the availability of nutrients and trace elements necessary for microbial growth, while biological factors include the presence of competing flora and interactions with the plant. There are various ways in which fruits and vegetables might become colonized by microbial flora, which could cause postharvest spoilage. Perhaps the most common route is the field's colonization by soil saprophytes and plant parasites. However, microbial colonization could result from contaminated seeds, as well as via contact with other fruits and vegetables at any stage during harvest and postharvest, including handling, transportation, processing, and packaging. Microbial biota could spread not only by direct contact, but also through air and water. For example, washing produce in a water tank might share microorganisms within the entire lot.

4.10.2 Pre-harvest food safety: WHO Report: The challenge will be to further develop and implement pre-harvest food safety strategies. There is a clear need for reliable and well functioning pre-harvest food safety systems. These should include the reduction, prevention, control and surveillance of the public health risks in farm-animal production and should be guided by risk analysis (Anonymous, 2001).

a. Recommendations and general considerations: Improving public health by reducing foodborne illnesses requires a farm-to-table approach based on risk-analysis (i.e., consisting of risk assessment, risk management and risk communication) with a sharing of responsibilities between segments to reduce hazards.

- Pre-harvest food safety is an important element in sustainable national food safety policies.
- There are examples of positive correlations between the levels of foodborne pathogens in food animals and the incidence of foodborne illness in humans.
- With increasing global consumption of animal-derived foods there is an increased potential for zoonotic foodborne pathogens (including their resistance to antibiotics) to disseminate worldwide.
- Pre-harvest food safety must be primarily focused on the protection of human health and be considered in the context of a farm-to-table food safety approach.
- Pre-harvest food safety programmes are needed to fulfil public health demands caused by failure of traditional food safety systems to effectively address the new and emerging foodborne diseases/emergencies, strengthen consumer confidence and/or gain market access.
- Pre- and postharvest food safety measures should be instituted within a risk management framework to achieve the greatest public health impact in reducing foodborne illnesses. All measures available throughout the food chain and their relative contribution to reducing risks to human health should be taken into account when considering the different risk management options.

- Pre-harvest programmes have contributed to the overall control of classical foodborne zoonoses such as bovine brucellosis, tuberculosis, trichinosis, etc. in many industrialized countries.
- A reduction in the recorded incidence of foodborne human salmonellosis has been reported in some countries with extensive pre-harvest *Salmonella* control programmes.

b. Risk assessment: Data and research needs: Control of foodborne disease requires reliable and standardized epidemiological data along the food chain and targeted surveys on human health may be required, particularly in developing countries.

- An understanding of the underlying epidemiology of foodborne pathogens, particularly emerging pathogens, is required to determine the practical and effective intervention strategies that should be implemented, particularly in the pre-harvest area. The pathogenicity of foodborne pathogens for humans should be clarified by monitoring human populations and comparing pathogens from human clinical foodborne illnesses with the pathogens present in the pre-harvest area.
- Surveillance data is needed for farm-to-table risk assessment. Integrated foodborne disease monitoring and surveillance programmes should include human foodborne illnesses, foodborne pathogens in food animals and foods of animal origin.
- Risk assessments can be used to identify data gaps and research needs and focus practical risk management and communication strategies that best address important country-specific public health problems.
- Further research needs include: Identification of effective pre-harvest interventions for foodborne pathogens other than *Salmonella*, e.g. thermophilic *Campylobacter*, verotoxigenic *Escherichia coli*, *Yersinia enterocolitica, Listeria monocytogenes*; Measuring and documenting the contribution of pre-harvest interventions to the reduction of foodborne illness (e.g., through ongoing risk assessments); Determination of which pre-harvest interventions are appropriate in an extensive production system; More data are needed on the advantages or disadvantages of organic farming animal welfare measures and other alternative food animal production systems on food safety.

c. Risk management: Pre-harvest food safety strategies: Food safety programmes, including pre-harvest food safety approaches require partnership between government, industry and consumers. Pre-harvest food safety must involve producers as partners addressing their responsibilities for food safety.

- Based upon risk assessment, risk managers should institute food safety measures where they achieve the greatest public health effect, taking into account all measures available throughout the food chain. The risk management process should be outcome based, transparent and rely on sound science.
- Pre-harvest food safety strategies should focus on the risk to consumer health through the consumption of food of animal origin and be· practical, sustainable and easily implemented ; written in outcome-oriented, non-prescriptive terms to facilitate the choice of appropriate country-specific risk management options ; horizontal with a multidisease approach whenever possible ; established in a transparent manner involving all stakeholders ; developed, implemented, monitored and evaluated in close and permanent collaboration between all stakeholders.
- There are strong linkages between animal health management systems and pre-harvest food safety. Veterinary practitioners have expertise and responsibility in pre-harvest food safety (in addition to their animal health responsibilities).
- Effective risk communication is essential to the success of implementing risk management strategies.
- To be successful, pre-harvest food safety programmes should involve-National strategy development involving definition of targets and interventions determined by risk assessment; Programme development by risk managers involving all stakeholders, authorities and experts ; Progressive programme implementation of risk management and risk communication strategies ; Regular programme review ; Traceability ; Development of verification and accreditation mechanisms including third party certification processes ; Appropriate financial support from government and industry during implementation.
- Risk management strategies may include third party certification, documentation of plans, written procedures, producer organizations and education programmes in the pre-harvest sector that focus on priority public health outcomes. As the prevalence of foodborne pathogens and the level of indicator microorganisms decreases with improved food safety, more robust and sensitive monitoring tools are required for validation and evaluation.

d. Enabling developing countries to carry out pre-harvest food safety risk management interventions: The recognition of the importance of food safety by developing countries is critical to their ability to raise their domestic public health and ability to export food. Therefore developing countries should participate in the development of food safety strategies.

- Developing countries require assistance in identifying their needs, in building their capacity to assess foodborne risks and in establishing priorities for control and prevention of foodborne diseases, as well as in implementing appropriate food safety control strategies. Local and regional partnerships and coalitions should be used to enhance their ability to address food safety from the farmer to the consumer.

e. International standards setting and collaboration between international organizations

- The development of codes, guidelines and standards to address pre-harvest food safety hazards should be relevant to, and involve the full participation of, developing countries
- International organizations should work together to promote science-based approaches to pre-harvest food safety to avoid duplication and to enhance collaborative efforts to protect public (consumer) health.
- International standards setting organizations must focus on the appropriate application of risk analysis in the pre-harvest area. 5.4 International organizations should consult with all interested parties, including the private sector and nongovernmental organizations (NGOs), when developing standards, guidelines and/or recommendations to address pre-harvest food safety hazards.

The term "dry chain" was introduced to describe initial dehydration of durable commodities to levels preventing fungal growth followed by storage in moisture-proof containers (Kent J. Bradford, *et al*, 2018). This is analogous to the "cold chain" in which continuous refrigeration is used to preserve quality in the fresh produce industry. However, in the case of the dry chain, no further equipment or energy input is required to maintain product quality after initial drying as long as the integrity of the storage container is preserved. In some locations/seasons, only packaging is required to implement a "climate smart" dry chain, while in humid conditions, additional drying is required and desiccant based drying methods have unique advantages. Both climate-based and drying based approaches to implement the dry chain to minimize mycotoxin accumulation and insect infestations in dry products, reduce food loss, improve food quality, safety and security, and protect public health were proposed .

4.10.3 Future areas for improving management practices : Managing risk will continue to evolve as more agencies attempt new approaches to achieve safer food products. At the same time, there will be attempts to harmonize these (Ross C. Beier *et al*, 2004). In principle this can occur at the international level, but specific management steps will still be independent in the immediate future, depending on the problem and expertise available. Following are some areas in which improvement can be made

- Early detection and surveillance of food borne disease to speed up responses to outbreaks and improving risk assessments; e.g., next generation of FoodNet and Pulsenet and with expansion to Europe, Canada, and Mexico; and understanding antibiotic and biocide resistance patterns to predict emerging problems in food animals and humans.
- Identification and characterization of emerging pathogens; e.g., parasites, viruses, and opportunistic bacterial pathogens, as well as prions, for susceptible populations, and changing profiles of antimicrobial resistance in bacteria.
- Characterization and support of the increasing numbers of those who are susceptible to foodborne disease, including the normal aging population as well as those who have underlying conditions.
- Encouragement of research to support food safety by using both cooperative agreements and competitive grants to address data gaps in microbial food safety.
- Development of quantitative and qualitative risk assessment models for different management purposes.
- Focus on better ways to integrate risk assessment, risk management, and risk communication to produce timely recommendations for the specific problem areas.
- Creation of a better compliance environment by more effectively coordinating the various agencies (federal, state/province/local) responsible for food safety.
- Improvement of food worker behaviors and food preparation practices that directly relate to foodborne illnesses in retail food/foodservice establishments.
- Improvement of public educational programs by building on existing campaigns, more effective labeling, coordination of different communication media with the same types of messages, tailor-making the kind of information to each audience, and understanding risk perception and ways of changing opinions.
- More monitoring of imported food by targeting foods of greatest concern both in the importing and exporting countries.

- The encouragement and monitoring of new food processing and packaging technologies to reduce risks.
- The understanding and simplification of food laws and regulations both within a country and between trading nations.
- Partnering with academia in multidisciplinary projects to understand and reduce risks of foodborne disease.
- More Member Country support of international harmonization for managing risk and improving trade.
- Adapting existing and future management and assessment approaches for food safety to respond to real and potential bioterrorism attacks using food as a vehicle.

4.10.4 Future directions at National and International levels: Microbiological risk evaluation requires a more sophisticated approach than simply determining an acceptable level of risk. Many terms and concepts are currently used to describe the decision-making process of risk managers in the risk evaluation step (FAO/WHO 2002a, International Commission on Microbiological Specifications for Foods 2002, WHO 2000). These terms include: As Low As Reasonably Achievable (ALARA), Tolerable Level of Risk, Appropriate Level of (Sanitary or Phytosanitary) Protection (ALOP), and Food Safety Objectives (FSOs).

The ALARA approach, which balances risks and benefits, lies somewhere between precautionary principle and a program built entirely on the science of prevention and assumption of worst-case scenarios. ALARA divides risk into three categories, acceptable, acceptable only if a compensatory benefit is available and unacceptable, in contrast to a "bright-line" approach (above which the risk is unacceptable and below which the risk is acceptable). Tolerable Level of Risk is similar to ALARA, and some managers prefer the word tolerable to acceptable, as no risk is really acceptable but, rather, one they are willing to live with until better ways of managing risks are found (Ross C. Beier *et al*, 2004). The Proposed Draft Principles and Guidelines for the Conduct of Microbiological Risk Management under development in the Codex Committee on Food Hygiene refer to tolerable risk that is determined primarily by human health considerations (CAC 2001). This approach assumes a bright-line level of tolerable risk, unlike ALARA.

The agreement on the application of sanitary and phytosanitary measures (SPS) Agreement of the World Trade Organization refers to the ALOP, which is defined in the agreement as the level of protection deemed appropriate by the member establishing a sanitary or phytosanitary measure to protect human, animal, or plant life or health within its territory (SPS agreement; World Trade Organization 1994). This approach recognizes that there will be differences between countries but assumes there will be agreement within a country on the ALOP. It also assumes that there is a single level (a bright line) above which the risk is unacceptable. An ALOP could be reduction of an enteric disease by a certain percentage in a certain number of years; for example, 50% reduction of listeriosis cases per 100,000 people in a country from 2000 to 2010. This is measurable, but it is unclear what happens if the goal is not met. Presumably goals are adjusted along the timeframe to better reflect progress and respond to roadblocks.

FSO is a term used in Codex documents, although its final definition remains to be agreed on by members. The International Commission on Microbiological Specifications for Foods (2002) defines a FSO as a statement of the frequency or maximum concentration of a microbiological hazard in a food considered acceptable for consumer protection. An ALOP is an overall public health goal. FSOs are related to ALOPs by translating them into specific objectives that can be carried out by the food industry. FSOs are expressed as a maximum concentration or probability of occurrence of a pathogen, toxin, or metabolite in a food. The availability of quantitative information on risks (e.g., dose response data) allows the identification of a FSO. Once there is agreement what a FSO should be, either for a domestic market or for trade, criteria to achieve this can be set. Process criteria have been in place a long time; for example, for canning or pasteurization. Thus, a time/temperature process criterion could be set to achieve the desired kill step to satisfy a FSO. However, this does not give the industry any options to implement improved or less expensive approaches for control. Ideally, a FSO allows industries to use different technologies to achieve the same level of consumer protection because as long as the objective is achieved, the means of getting there is acceptable. This approach is consistent with the SPS agreement for harmonization of microbiological criteria for foods in international trade with equivalency of food standards. This is best achieved with a performance criterion, such as a reduction in the number of a pathogen per gram, or prevention of any increase in numbers. Because a FSO is set at the consumer stage, it has to allow for some consumer practices; for instance, use of an improper storage temperature or a heating step. Thus, the performance criterion will be more stringent than the FSO if growth in the stored product is expected, but less stringent if the product is to be cooked.

References

Adel A. Kader and Rosa S. Rolle, 2004. The role of postharvest management in assuring the quality and safety of horticultural produce. Food and Agriculture Organization of the United Nations, Rome, FAO Agricultural Services Bulletin 152

Anil Panghal, D.N., Yadav, Bhupender S. Khatkar, Himanshu Sharma, Vikas Kumar and Navnidhi Chhikara, 2018. "Postharvest malpractices in fresh fruits and vegetables: food safety and health issues in India", Nutrition & Food Science, https://doi.org/10.1108/NFS-09-2017-0181

Anonymous, 1989. Prevention of Postharvest Food Losses: Fruit. Vegetables and Root Crops. A Training Manual. FAO. Rome: UNFAO. 157 pp.

Anonymous, 2001. Pre-harvest food safety- Report of a WHO consultation with the participation of the Food and Agriculture Organization of the United Nations and the Office International des Epizooties., Berlin, Germany 26-28 March 2001

Anonymous, 2010. CCDC (Cold Chain Development Centre) Cold storage for fresh Horticulture produce requiring pre-cooling before storage (technical Standards number NHB -CS-Type 02-2010). National Horticulture Board (Department of Agriculture and Cooperation Ministry of Agriculture, Government of India) 85, Institutional Area, Sector 18, Gurgaon -122015, Haryana

Brunda Devi ,K. and Bhadraiah,B., 2017. Study of Carrots for their Storage Environment and its Effect on Biochemical Changes during Pathogenesis Cited by Postharvest Fungi and their Control by using Low Cost Method which Prolonged Shelf-life . *Intl. J. Sci. & Res.*, **6:** 1812-1819

Dulal Chandra, Jung-Soo Lee, Hyun Jin Choi and Ji Gang Kim, 2018. Effects of Packaging on Shelf Life and Postharvest Qualities of Radish Roots during Storage at Low Temperature for an Extended Period. *J.Food Qual.*, 2018, Article ID 3942071, 12 pages

Gorny, J.R. 1999. Food Safety for fresh horticultural produce. In : Small-Scale Postharvest Technology: Economic Opportunities, Quality and Food Safety. Postharvest Horticulture Series Kitinoja, L. and Gorny, J.R (Edrs.).No.21, Department of Pomology, University of California, Davis .

Kenneth C. Gross, Chien Yi Wang, and Mikal Saltveit (Eds.), 2016. The Commercial Storage of Fruits, Vegetables, and Florist and Nursery Stocks. USDA , Agricultural Research Service Agriculture Handbook Number 66

Kent, J. Bradford, Peetambar DahalJohan, Van Asbrouck, Keshavulu Kunusoth, Pedro Bello, James Thompson and Felicia Wu, 2018. The dry chain: Reducing postharvest losses and improving food safety inhumid climates. *Tre. Food Sci.Technol.*, **71** : 844-93

Koushesh Saba Mahmoud, Moradi Samira and Gouran Ali, 2014. The effect of postharvest treatments and storage methods on leafy vegetable decay. *J. Middle East Appl. Sci. & Technol.*, **6.** 153-156.

Majubwa, R.O. , Msogoya, T.J. and Maerere, A.P., 2015. Effects of Local Storage Practices on Deterioration of African Eggplant (*Solanum aethiopicum* L.) fruits . *Tanzania J. Agril. Sci.*, **14** : 106-111

Michailides,T.J. and Manganaris , G.A., 2009. Harvesting and handling effects on postharvest decay *Stewart Postharvest Re.*, **2:** 3

Moneruzzaman, K.M., Hossain, A.B.M S., Sani, W., Saifuddin, M.and Alenazi, M., 2009. Effect of harvesting and storage conditions on the postharvest quality of tomato (*Lycopersicon esculentum* Mill) cv. Roma VF. *Australian J. Crop Sci.*, **3:**113-121

Mulualem Azene, Tilahun Seyoum Workneh and Kebede Woldetsadik, 2014. Effect of packaging materials and storage environment on postharvest quality of papaya fruit. *J.Food Sci. Technol.*, **51**: 1041–1055.

Punja, Z.K. , Rodriguez, G., Tirajoh, A. and Formby, S., 2016. Role of fruit surface mycoflora, wounding and storage conditions on postharvest disease development on greenhouse tomatoes, *Canadian J. Pl. Pathol.*, **38:**, 448-459

Ross C. Beier, Suresh D. Pillai, Timothy D, , 2004. Preharvest and Postharvest Food Safety: Contemporary Issues and Future Directions. Blackwell Publishing and the Institute of Food Technologists, IFT Press, Institute of Food Science and Engineering Texas A&M University, 455 pp.

Shlomo Sela and Elazar Fallik 2009. Microbial Quality and Safety of Fresh Produce. In : Postharvest Handling (Second Edition), A Systems Approach, Food Science and Technology, 351-398 pp

Torrence, M.E., 2016. Introduction to Preharvest Food Safety. *Microbiol.Spectr.*, **4**: 10-28

Yadav, S. M., Patil , R. K. , Balai, L. P. and Ram Niwas, 2013. Postharvest Diseases of Horticultural Crops and Their Management . *Popular Kheti*, **1**: 20-25.

5

Quiescence in Postharvest Pathogens

In managing diseases of horticultural crops, pathogen quiescence has important implications in timing prophylaxis, the reduction of stresses that trigger the transition of quiescence to aggression, and managing the harvested crop to prolong quiescence to the point at which yield is no longer affected. The inception of pathogen quiescence and its maintenance on or within the host implies a dynamic equilibrium between the host, the pathogen, and their environment. Physiological and physical changes in the host, its environment, or both trigger changes in that equilibrium to permit the pathogen to resume aggression. Microorganisms lying on the host surface constitute a potential latent infection; as can be the case with *Botrytis cinerea* conidia that are prevented from germinating by antagonistic bacteria on leaf surfaces (William R. Jarvis, 1994). Quiescence can run the spectrum from the ungerminated spore; through symptomless, internal infections; to visible but nonexpanding lesions, such as ghost spot of tomato (*Lycopersicon esculentum*) to expanding lesions on banana caused by *Colletotrichum musae* and on avocado caused by *C. gloeosporioides.*

A latent, dormant, or quiescent parasitic relationship is a condition in which the pathogen spends long periods during the host's life in a quiescent stage until, under specific circumstances, it becomes active. Epidemiologists might refer to the period from spore landing until the parasitized tissue produces new spores as a latent period (Dov Prusky, 1996). The term "period of quiescence" is referred to differentiate the quiescent parasitic relationship from the latent period. Quiescence can be enforced during the various processes of fungal attack, which suggests that quiescence may or could occur during any of the processes leading from fungal germination to colonization.

With the exception of tomato ghost spot, most quiescent infections eventually become aggressive rots in the field and, if undetected, in stored and marketed produce. Quiescent infections are common in a wide variety of fruit, vegetable, and floral crops. They are caused by bacteria and fungi. Soft rot bacteria are common in apparently healthy tomatoes, cucumbers and potato tubers. Also in potatoes, quiescent (*Clavibacter michiganense* subsp. *sepedonicum*) is a serious impediment to export trade. The world's three most important small-berry crops, grapes, strawberries and raspberries all have quiescent infections of *Botrytis cinerea* established in their floral parts; these infections usually become aggressive in ripe fruit (Lakshiminarayana *et al.*, 1987). However, aggression may be significantly advanced in wet weather, when, for example, normally resistant green strawberries succumb to gray mold. *Botrytis cinerea* also can establish a quiescent infection in apple calyx, which becomes eye rot and in fruit wounds.

Quiescent petal flecking in many floral crops is also frequently caused by *B. cinerea* and other host-specific *Botrytis* spp. in the monocotyledonous Corolliferae and the Ranunculaceae. In the Sclerotiniaceae, *Monilinia fructicola* honey is quiescent in stone fruits and *Sclerotinia sclerotiorum* is quiescent in many vegetables including carrots. *Botrytis aclada* is quiescent in onions and is responsible for neck rot in storage. Outside the Sclerotiniaceae, *Colletotrichum* spp. are common quiescent pathogens of tropical fruit. Mechanisms of quiescence in most unripe fruit and stem and root vegetables, the mechanisms limiting pathogen aggression are associated with either preformed antimicrobial substances (phytoncides) or with phytoalexins, enzymes, or physically resistant structures. A wide variety of postulates have been, or can be, advanced to explain quiescence. Most of these postulate that aggression begins when certain physiological and physical changes occur, particularly in the host cell wall. The changes in dynamic equilibrium between climacteric metabolism and the strains induced by the stress of infection involve ethylene production. Other stresses are imposed by such factors as wounding (Bruton, 1994), an unduly wet flowering season in strawberries, harvesting in potatoes, and inappropriate storage conditions.

Implications of quiescence in disease control: Setting aside the epiphytic populations of potential pathogens on plant surfaces, which may be controlled by pesticides and biological agents, the infection process, as defined by breaching the host epidermis, may occur several days or weeks before symptoms appear. This means that it is essential to know what environmental conditions permit infection to occur. In strawberries, white flower buds and fully open flowers are the most susceptible to infection, and the calyx and corolla are more susceptible than the receptacle. Strawberries and raspberries present a difficult problem in timing prophylaxis because of the cymose inflorescences' long development sequence. This situation is further complicated by the finding that first-formed berries in the strawberry inflorescence are more susceptible to gray mold than successively later berries on the same inflorescence (William R. Jarvis, 1994). That microorganisms must be prevented from breaching the host epidermis means that the choice of pesticide is restricted to those that act on contact with the pathogen on the host surface; systemic, eradicant pesticides are less successful since they usually act imperfectly after infection.

Phonological and epidemiological studies permit formulating forecasting methods. The complex phenology of strawberries and raspberries makes forecasting difficult, but it has been done for grapes, which have a well-defined bloom period and for which the microclimates conducive to *B. cinerea* infection are known. Precise control over greenhouse microclimates permits designing effective disease prevention strategies for protected crops .

5.1 Quiescence: Biotrophs are defined by fungal structure, i.e., haustoria that facilitate an intimate parasitic relationship with living host cells. The complete absence of haustoria from the life cycle of a pathogen relegates it to necrotrophic status. Hemibiotrophs show an early asymptomatic biotrophic phase of growth but have a later stage characterized by tissue degradation and disease symptoms that is clearly necrotrophic. The transition from the biotrophic to the necrotrophic stage can be continuous, such as that described for growth of *Phytophthora infestans, Colletotrichum higginsianum*, and *C. graminicola* in the leaf. However, quiescence defines an extended sequence of time during which pathogen activity appears to be suspended and almost no growth is apparent. It is also not limited to the hemibiotrophic-necrotrophic transition. Thus, in hemibiotrophs such as *Colletotrichum*, life may begin biotrophically with infection of near-ripe fruits and continue uninterrupted to the necrotrophic stage in a manner similar to *P. infestans*. Infection by the same pathogen of green/unripe fruit leads to a suspended phase, i.e., quiescence. In contrast, the archetypical necrotroph *Botrytis cinerea* has no biotrophic stage but typically causes a quiescent infection in fruit tissues that becomes activated as the fruit ripens. Interestingly, it was reported that *B. cinerea* showed a nonpathogenic endophytic phase devoid of the necrotrophy for which it is known (Sowley *et.al.,* 2000). The generality of this finding remains to be seen. Thus, quiescence is a state that needs to be defined for each particular postharvest type of interaction.

Once the climatic and phonological conditions that permit initial infection to occur are stablished, pesticide application may be timed accurately, usually well in advance of harvesting, so that undue residues are avoided. Several studies have shown that fruit crops are better protected by applying pesticides during bloom rather than later and that vegetable crops store better when infections are controlled in the field (William R. Jarvis, 1994). Disease prevention measures can also be devised for crops in the field. Thus, green beans escape gray mold and white mold infection if the plant architecture and row orientation permits floral parts to dry rapidly. Tomatoes may escape gray mold if the correct soil N (Verhoeff, 1968) and Ca levels are maintained.

Prolonging quiescence : Even if tissues are infected quiescently, it still may be possible to obtain blemish-free floral crops or rot-free fruit and vegetables by manipulating the crop so that aggression is not triggered. How this may be done, however, is not widely understood. Water content of tissues seems to be important; for example, green strawberries are very susceptible to *B. cinerea* in wet weather, an effect simulated by supplying water directly into the fruit pedicel. The development of quiescent neck rot infections (*Botrytis aclada*) in onions depends on adequate moisture in the senescing green leaf tissue. If onions are dried adequately in windrows or cured artificially before storage, neck rot ceases to be a problem. Similarly, root crops with green leaves, like carrots, or senescent outer leaves, like cabbage should be carefully trimmed and dried before storage. Correct mineral nutrition prolongs the quiescent phase. The effect of soil N in prolonging the quiescent phase of *B. cinerea* in tomato stem tissue seems to be one of delaying tissue senescence. Biological control can be applied rationally to prolong quiescence. The temperature of insolated fruit is also important; in picked fruit left in the sun, the temperature maybe 8-10°C higher than ambient air temperature. Experience has shown that most fruits are best picked in the early morning; they have a longer shelf life if field heat is removed before shipping.

The postharvest quiescent period for pathogens was influenced by the storage temperature, the pathogen involved, and the mode of infection, ranging from 4 to 5 d at 22°C, to 14 d at 13°C, in fruit that were naturally infected with *Dothiorella* spp. and *C. gloeosporioides*, or inoculated with *D. dominicana* at the stem-end, and from 8 to 13 d at 23°C, depending upon whether the primary inoculum was soil borne or endophytic respectively (Johnson *et.al*., 1993).

The content of chlorogenic acid, (+)-catechin, (−)-epicatechin, phloretin glycosides, and quercetin glycosides in fresh and stored Golden Delicious apples (*Malus domestica* Borkh) was determined (Vincenzo Lattanzio *et.al*., 2001). The relative amount of phenolics in the peel, with the exception of chlorogenic acid and (−)-epicatechin, was higher than that in the flesh. In addition, quercetin glycosides were detected only in the skin. These compounds were tested for fungicidal activity against *Phlyctaena vagabunda* Desm., the causal agent of a postharvest rot. Chlorogenic acid only inhibited *P. vagabunda* spore germination and mycelial growth in vitro. Changes of apple phenolics and polyphenol oxidase activity during cold storage and the biological activity of these phenolics have also been analyzed with reference to the development of quiescent infections during cold storage plus shelf life at room temperature. The results obtained suggested that phloridzin and chlorogenic acid in combination with polyphenol oxidase activity could function to arrest *P. vagabunda* in quiescent infections associated with immature and ripening apple fruit.

Quiescent infection of *Colletotrichum gloeosporioides* causes anthracnose disease on mangoes and various other fruits worldwide (Pongphen Jitareerat *et.al.*, 2007) . In mangoes, it is a major postharvest disease, especially in the susceptible cultivar 'Nam Dok Mai' and it is a significant problem for Thai exports, which are well received in overseas markets. Fruit infection commonly occurs in the orchard and results in serious decay problems in transit, at the market and after sale. The fungus apparently invades the skin of young fruit and remains in a "quiescent infection" state until fruit ripening begins. Susceptibility of fruits to postharvest diseases results from physiological changes during ripening and increasing senescence that develops during storage, necessitating treatments to control diseases, such as hot water, fungicides, or their combination. Many importing countries now impose strict limits for post havest fungicidal residues because excessive use of fungicides has in general resulted in serious problems for the environment and human health and the increasing chemical resistance of pathogens.

The effects of chitosan on mycelial growth and spore germination of *Colletotrichum gloeosporioides* were investigated on potato dextrose agar (PDA) containing 0%, 0.5%, 1.0%, 1.5%, and 2.0% (w/v) chitosan dissolved in 0.5% acetic acid. Complete inhibition of mycelial growth and spore germination was observed in the medium containing only 0.5% acetic acid. Media containing chitosan were less inhibitory, but the highest concentrations of chitosan, at 1.5% and 2.0%, were more inhibitory than the lower concentrations (0.5% and 1.0%). Chitosan treatment (0.5%, 1.0%, 1.5%, and 2.0%) of mangoes previously inoculated with *C. gloeosporioides* resulted in a lower rate of disease progression compared with the controls. Chitosan concentrations of 0.5% and 1.0% had stimulatory effects on chitinase and β -1,3-glucanase activities. Additionally, chitosan coatings in mangoes delayed ripening and reduced respiration rate, ethylene production, and the loss of weight, ascorbic acid, and total titratable acidity. However, the firmness of mangoes treated at concentrations above 1 % caused a perceptible decline in sensory quality.

Preharvest treatments with chitosan have been reported to lessen the apparent severity of the quiescent infected area and the incidence of decay after harvest, whereas postharvest treatments inhibit the development of quiescent infections (Dodd *et al.* 1997). Although *C. gloesoporioides* generally infects mango fruit quiescently, its spores and mycelia may contaminate the fruit surface before harvesting and during the process of washing and storage in the packaging house. Therefore, in this work, we have examined the potential effects of the antifungal agent chitosan on in vitro and in vivo development of *C. gloeosporioides*, and also determined disease development and postharvest quality of mango fruit

The quiescence period during different stages of fungal attack of postharvest pathogens was discussed *viz.,* quiescence during spore germination and initial hyphal development, during and after appressorium formation, and quiescence of germinated appressorium and subcuticular hyphae (Dov Prusky, 1996). The different mechanisms for quiescence were reviewed: factors affecting quiescence of germinated spores, appressoria formation and germination, and fungal colonization. Special emphasis was given to mechanisms of quiescence involving fungal colonization, the pathogen's nutritional requirements, preformed antifungal compounds, the elicitation of phytoalexins and preformed compounds, and the activation of factors in fungal pathogenicity.

5.2 Quiescent stages: The quiescent stages may include quiescence during spore germination and hyphal development, appressorium formation quiescence of appressoria, germinated appressoria, and subcuticular hyphae.

In general, postharvest decay causative agents exhibit common features namely, the ability to establish in the immature fruit and to remain in a quiescent state until favorable conditions arise for the infection process to occur. Studies on quiescent infections are of key agricultural and economic importance for the development of new approaches related to the detection of quiescence and the subsequent control of postharvest diseases (Prusky *et.al.*, 2012). Amaral Daniela Dambros *et.al.,* (2017) evaluated the frequency of quiescent fungi and the effect of phosphites under modified atmosphere on *Lasiodiplodia theobromae* in papaya. The fruits were treated with a range of doses of phosphites and their actions evaluated under conditions of ambient and modified atmosphere. Of the eight fungal genera found, *Lasiodiplodia* was the most common. No interaction was observed between the evaluated factors and only atmosphere and dose were independently significant. The usage of phosphites and modified atmosphere reduced the severity of the disease, and did not affect the chemical properties of the fruits.

Insidious fungal infections by postharvest pathogens remain quiescent during fruit growth until, at a particular phase during fruit ripening and senescence, the pathogens switch to the necrotrophic lifestyle and cause decay. During ripening, fruits undergo physiological processes, such as activation of ethylene biosynthesis, cuticular changes, and cellwall loosening, changes that are accompanied by a decline of antifungal compounds, both those that are preformed and those that are inducible secondary metabolites. Pathogen infection of the unripe host fruit initiates defensive signal-transduction cascades, culminating in accumulation of antifungal proteins that limit fungal growth and development. In contrast, development of the same pathogens during fruit ripening and storage activates a substantially different signaling network, one that facilitates aggressive fungal colonization (Dov Prusky *et.al.*, 2013), who focused on responses induced by the quiescent pathogens of postharvest diseases in unripe host fruits. New genome scale experimental approaches have begun to delineate the complex and multiple networks of host and pathogen responses activated to maintain or to facilitate the transition from the quiescent to the necrotrophic lifestyle.

5.3 Quiescence and postharvest disease development : Fruits infected by postharvest fungal pathogens mostly develop disease symptoms after harvest and during storage at cold temperature, shipment, and shelf life close to consumer use. Several biotic and abiotic factors may activate postharvest development of disease symptoms. Such factors include ripening, harvesting, and wounding . Fungal pathogens exploit three main routes to penetrate the host tissue: through wounds; through natural openings, such as lenticels, stem ends, and pedicels; and by direct breaching of the host cuticle , which enables most postharvest pathogens to enter the unripe tissue and remain quiescent for months until the fruit ripens. Until then, the penetrating pathogen remains unnoticed to visual external examination, awaiting the right conditions for the infection to become active and cause the disease that impairs crop quantity, quality, and appearance (Prins *et.al.,* 2000). The period from host penetration to activation of fungal colonization and symptom appearance is described as quiescence. Several species of fungal pathogens, such as *Colletotrichum, Alternaria, Botrytis, Monilinia, Sclerotinia, Lasiodiplodia, Phomopsis,* and *Botryospheria*, have been reported to live quiescently in their hosts.

The spectrum of quiescence is host and tissue specific. The location of quiescent infections impacts their activation. Thus, germinated spores of *Colletotrichum* and *Monilinia* produce melanized appressoria that breach the fruit cuticle and remain quiescent as developing hyphae in the cuticle and the first cell layer of the fruits. In *Colletotrichum acutatum*, four types of interactions or infection strategies were described *viz.,* a predominantly biotrophic disease cycle on citrus leaves; a predominantly necrotrophic disease cycle on strawberry; a hemibiotrophic interaction comprising a combination of mostly biotrophic and necrotrophic disease cycles on blueberry fruits; and a hemibiotrophic interaction with combinations of biotrophic and mostly necrotrophic disease cycles on almond leaves and fruits The location of quiescent structures is host dependent. *Botrytis* followed different routes for penetration and establishment of quiescent infections in flowers compared with developing fruits (Dov Prusky *et.al.*, 2013). In blackcurrant, *Botrytis* conidia germinated in the stigmatic fluid, and hyphae spread endophytically throughout the stylet without eliciting visible symptoms and infected the pericarp and ovules. In pear and strawberry flowers, the hyphae grew through the vascular tissue of the upper end of the flower receptacle or of the mesocarp adjacent to the sepals.

Quiescent infections and endophytic colonization occur in a wide range of locations, e.g., cuticular wax, intercellular space, and epidermal cells, and under specific periods of host development. This suggests that during the host's transient resistance, several tissues and/or physiological mechanisms are activated to overcome the pathogen's aggressive colonization. Furthermore, different plant tissues in a given host may respond differently to invading microorganisms. Although strawberry and mango leaves showed active symptoms of infection by *B. cinerea* and *Colletotrichum gloeosporioides*, respectively, quiescence still persisted in actively developing fruits. This indicates that the transition undergone by fruits during ripening is a developmental process that coincides with increased susceptibility, and ripening fruits could exhibit physiological properties similar to those of senescing leaves. Importantly, ripening per se usually is not the only signal for termination of quiescence. Examples of *Alternaria alternata* and *C. gloeosporioides* remaining quiescent on cuticles of unripe mango and avocado fruits while being able to colonize the mesocarp of the same peeled fruits suggest that specific local changes during fruit ripening affect the transition into an active infection. Thus, fungal quiescence is a common process that affects diverse hosts and host tissues that employ specific host-pathogen interactions. Assessment of the control of quiescent infections is of key agricultural and economic importance for development.

Host regulation of quiescence: In general, it is considered that activation of the quiescent stage is the result of induced accessibility of disassembled cell wall substrates during fruit softening and ethylene induction; a decline in preformed antifungal compounds, such as polyphenols, phytoalexins, and other fungitoxic substances; the decline of inducible host-defense responses and the pH conditions in the host .

Host resistance responses during quiescence : Perception and response to fungally secreted molecules have been widely studied in leaves and roots (van Kan, 2006), but only a few reports addressed these phenomena in ripening fruits. During necrotrophic colonization, ammonia secretion by *Colletotrichum* directly activated the tomato NADPH oxidase. This resulted in production of reactive oxygen species (ROS) and induction of SA-dependent genes related to biotic stress, ion leakage, and cell death, leading to enhanced colonization. In parallel, SA repressed jasmonic acid (JA) dependent genes. This would support the hypothesis that SA induced programmed cell death (PCD) is a major process in protecting against biotrophic fungi, while at the same time inducing necrotrophic colonization. These processes activate necrotrophic development of the fungus and contribute to the termination of its quiescence, as they are also related to pH change. However, it is known that germinating *Colletotrichum* spores are copious secretors of ammonia, but in the case of germination on unripe fruit this does not lead to immediate growth but to quiescence. Thus, in unripe fruit, ammonia secretion by itself cannot trigger necrotrophic colonization. Analysis of gene expression in unripe, resistant, inoculated fruits by means of RNA seq technology has detected a large number of expressed genes, which suggests that previously described antifungal secondary metabolites represent only a portion of the antifungal factors that are activated during quiescence. Secretome analysis of *B. cinerea*–infected fruits has revealed that quiescent infections of unripe fruits activate many more plant defense related proteins than infections of red fruits, suggesting that impaired cellular responsiveness in ripe fruit could be a quality of decreased resistance.

Irrespective of whether *B. cinerea* infects unripe or ripe tomato fruits, the fungally secreted proteins detected were similar. Secretome analysis of the inoculated unripe tomato fruit showed 114 defense related proteins that were classified into four subcategories: pathogenesis related (PR) proteins (43 proteins), proteases (43 proteins), peroxidases (13 proteins), and protease inhibiting (PIN) proteins (15 proteins). Chitinases were prominent among the PR proteins 12/43. Chitinases catalyze hydrolysis of chitin and constitute the second largest group of antifungal proteins. In addition, 1,3- and 1,4-β-endoglucanases were identified. These hydrolyze the structural 1,3-β-glucan present in the fungal cell wall, leading to cell lysis and death and they can effectively inhibit fungal growth. Plant inhibitors of polysaccharide hydrolases produced by pathogens have been shown to impart reduced susceptibility to infections ; expression of a xyloglucan fungal endoglucanase-inhibitor protein has been shown to reduce susceptibility of tomato fruits to *B. cinerea*. Other proteins in the PR category include a disease-response protein, an elicitor inducible protein, a glycosyl hydrolase, three osmotin like proteins, and thaumatin.

The role of proteases can range from the elimination of pathogen effectors to the promotion of host PCD. For example, in the tomato leaf, prevention of the secretion and inactivation of the host C14 papain-like protease by *P. infestans* are crucial for its growth . In a reciprocal manner, fifteen PIN proteins were identified in unripe tomato fruits. It has been shown that protease inhibitors isolated from healthy bean and tomato plants reduced the activities of proteases from *Fusarium solani* and *Colletotrichum lindemuthianum* PIN proteins can target pathogen proteases to control PCD. Effectors are key factors in eliciting as well as subverting host-defense responses and PCD; thus, replacement of the terms quiescence and latency with establishment should be considered.

5.4 Mechanism of pathogenesis during postharvest decay

i. **Quiescent stage:** Postharvest fungal pathogens have two main modes of nutrition during the colonization on plant hosts, *viz*., biotrophy, in which the nutrients are obtained from the living host cells, and necrotrophy, in which nutrients are obtained from dead host cells killed by the fungus. Opportunistic postharvest pathogens may also reside in the matured fruit in an inactive (quiescent) mode, awaiting fruit for wounding, ripening or senescence. The length of each period may vary among pathogens, hosts, and host developmental stages (Kumar *et.al*., 2018).

ii. **Quiescent and necrotrophic stage :** Several factors at cellular levels are responsible for transition to necrotrophic from the quiescent stage and these factors are a function of nutrients and pH of the harvest. Theories regarding the mechanism for secretion of chemicals for infection are Oxalate weaken the host plants for invasion, Acids (oxalic acid and gluconic acid) loosen the plant cell wall by chelation, and Activation of Reactive Oxygen Species (ROS) generation, which leads to necrotrophic mode of development. Disease free plants have an electrochemical proton gradient across the plasma membrane. Ephemeral alkalization is responsible for plant defence response, thereby contributing to activation for development of necrotrophic mode.

The role for ammonium secretion in the induction of host ROS accumulation through the NADPH oxidase mechanism in a pH dependent manner is explained (Kumar *et.al*., 2018). ROS activation by ammonium can act as a factor for promoting cell death in tomato fruit. Ammonium can directly affect the fungus pathogenicity by activation of the expression of genes encoding pathogenicity factors, such as expression of pelB gene, which encodes pectate lyase (PL) secretion in *Colletotrichum gloeosporioides*. Finally ammonium can also indirectly affect the pathogen-host interaction by elevating pH of the infection site to create an optimal environment for production of pathogenicity factors as PL and endoglucanase. This suggests that ammonification by *Colletotrichum* at the infection site may be responsible for call death and activation of fungal pathogenicity factors. Change in pH may be responsible for transition from quiescent to necrotrophic stage in fungal pathogen in this case. Examples of pathogens which activate necrotrophic development in senescing and ripening fruits are *Alternaria, Botrytis, Colletotrichum, Monilinia* etc. *Colletotrichum* spores adhere and germinate on the plant surface, produce germ tubes, develop appresorium and penetrate by infection peg through cuticle followed by subcuticular colonization and proliferation inter- or intra-cellularly in the host; this phase is followed by necrotrophic phase of fungal growth. In contrast, *Botrytis* and *Monilinia* can penetrate through wounds and also breach the fruit cuticle by using an infection peg from an appresorium that then remains quiescent for long periods of time.

5.5 Importance of pH in quiescence: Ripening is a physiological process that involves numerous biochemical changes including change in sugar composition and change in pH of the fruit and vegetable tissues. These changes attract a variety of pathogens causing decay, thereby leading to losses up to 25-50% in harvested crops. The infection by the postharvest pathogens may remain quiescent during fruit or vegetable growth due to unfavourable growth conditions for pathogen's development to pathogenesis (Kumar *et.al.*, 2018) Owing ripening and senescence the pathogen transform to necrotrophs, activation of the pathogen is primarily due to change in pH and cause typical decay symptoms. Some pathogens have developed a mechanism for altering the pH of host environment (tissue), thereby increasing virulence. The postharvest pathogens cause acidification by producing organic acids and alkalization by ammonification of host tissues. Management of postharvest tissue decaying pathogens is usually performed by use of fungicides but due to raised environmental concerns and residual effects of fungicides on health alternative management strategies are required. This approach should be ecofriendly; the use of plant extract and biological agents are in limited use. Thus, it should be emphasized that management of postharvest food commodity (particularly those of direct consumption like fruit) pathogens can be achieved by altering the physiological environment and use of alternative management strategies to reduce the residual effect of inorganic fungicides and other chemicals.

Alkalizing fungi and acidifying fungi (e.g. *Penicillium* sp., *Botrytis cinerea*) has the ability to modify pH in the surrounding environment (the host tissue). Modification of enzyme expression in response to the ambient pH in the host by the fungus shows the importance of the specific regulatory system, which can lead to the activation of infections (Kumar *et.al.*, 2018). Postharvest pathogens secrete organic acids and cause tissue acidification for their attack e.g. *Penicillium digitatum, Botrytis cinerea* etc. In case of apple and citrus, pH change ranging from 3.95-4.31 in healthy tissue and 3.64-3.88 in the decaying tissue caused by *P. expansum* results in rapid decaying of apple in several cultivars. Similarly the pH decline has been observed in citrus after the infection of *P. digitatum* and *P. italicum*. In the necrotrophic fungus *Sclerotinia sclerotiorum*, during plant infection, a pH gradient was established in relation to oxalic acid secretion and the pH of the host reached 4.0. The production of oxalic acid during plant infection has been implicated as a primary determinant of pathogenicity in this and other phytopathogenic fungi. The size of lesions induced by different strains of *Botrytis cinerea* on grapevine and bean leaves correlated with the amount of oxalic acid and it is suggested that oxalic acid act as a co-factor in pathogenesis instead primary disease inciting agents .

The loss of PACC, pH-response transcription factor function leads to fungal misexpression of genes normally expressed only under acidic conditions. Thus, in alkaline conditions PACC is responsible for repression of genes expressed under acidic pH and for preferential induction of genes expressed under alkaline pH, including PACC itself (Caracuel *et.al.*, 2003). Early induction of PACC is critical because it activates the first wave of pathogenicity, which modulates the transformation of quiescent infections to active necrotrophic growth. The expression of pathogenicity factors, such as pectate lyase and endoglucanases, in several host-pathogen interactions is an example of PACC-controlled expression. In *C. gloeosporioides*, PELB (pectate lyase encoding gene) expression during pH alkalization followed a pattern similar to that of PAC1, suggesting the presence of a pH-regulatory mechanism for control of secreted proteins during pH increase. Similarly, for acidic environments, PG expression in *P. expansum* and F. oxysporum was controlled by acidic pH. pacC mutants of *C. gloeosporioides*, *C. acutatum, F. oxysporum*, and *S. sclerotiorum* were less virulent compared with the wild type, which suggests the activity of a complex virulence mechanism that controls pathogenicity factor expression under pH change. The activation of quiescent infection is facilitated by large gene families of cell wall– degrading enzymes. This has been observed for the endoPG families of *B. cinerea* and glucanases in *A. alternata* . Furthermore, evidence for the differential expression of each family member under varied pH conditions illustrates the ability of a fungus to finetune its pathogenicity to the prevailing environmental conditions. During the study of differential expression of endoPG by *B. cinerea* in apple and courgette, with low and neutral pHs, respectively, it was reported that Bcpg2 was negatively modulated by low ambient pH, which might account for its lack of expression in apple fruits, whereas Bcpg3 expression was induced at low pH in liquid cultures, as in apples. This fine-tuning of enzyme induction and secretion in response to the ambient pH can discriminate between hosts. It further illustrates the importance of the specific regulatory system, as controlled by environmental pH, and highlights its importance during the transformation of a quiescent to an active infection.

Postharvest pathogens can start its attack process immediately after spores land on wounded tissue, whereas other pathogens can forcibly breach the unripe fruit cuticle and then remain quiescent for months until fruit ripens and then cause major losses (Alkan *et.al.*, 2013). These fungal pH modulations of host environment regulate an arsenal of enzymes to increase fungal pathogenicity. This arsenal includes genes and processes that compromise host defenses, contribute to intracellular signaling, produce cell wall-degrading enzymes, regulate specific transporters, induce redox protectant systems, and generate factors needed by the pathogen to effectively cope with the hostile environment found within the host. Further, evidence is accumulating that the secreted molecules (organic acids and ammonia) are multifunctional and together with effect of the ambient pH, they activate virulence factors and simultaneously hijack the plant defense response and induce program cell death to further enhance their necrotrophic attack. Global studies of the effect of secreted molecules on fruit pathogen interaction, will determine the importance of these molecules on quiescence release and the initiation of fungal colonization leading to fruit and vegetable losses.

Colletotrichum gloeosporioides and *Pestalotiopsis mangiferae* are major postharvest pathogens in mango causing enormous losses. A study was conducted with these two pathogens obtained from banana and mango to find out the optimum pH level for growth and sporulation as well as the possible utilization of this knowledge in developing management strategy of quiescent infections by them (Niren Majumdar and Nakul Chandra Mandal, 2018.). Both fungi grew at Potato Dextrose Broth and PDA within the pH regime 4.0–9.0 tested at 28±1 °C, although growth rate differs significantly at different hours of incubation. The optimum pH for growth and sporulation differs significantly. The optimum pH for both growth and sporulation was 5.0 in PDA. In case of potato dextrose broth pH 6.0 was optimum for *C. gloeosporioides* and *P. mangiferae* has wide adaptability within ph range 4.0–6.0. Both pathogens sporulate in all the pH regime 4.0–9.0, however *C. gloeosporioides* best sporulate at pH 5.0 and *P. mangiferae* at pH.8.0

Growth of *Colletotrichum gloeosporioides* in pectolytic enzyme-inducing medium (PEIM) increased the pH of the medium from 3.8 to 6.5. Pectate lyase (PL) secretion was detected when the pH reached 5.8, and the level of secretion increased up to pH 6.5. PL gene (*pel*) transcript production began at pH 5.0 and increased up to pH 5.7. PL secretion was never detected when the pH of the inducing medium was lower than 5.8 or when *C. gloeosporioides* hyphae were transferred from PL-secreting conditions at pH 6.5 to pH 3.8. This behavior differed from that of polygalacturonase (PG), where *pg* transcripts and protein secretion were detected at pH 5.0 and continued up to 5.7. Under in vivo conditions, the pH of unripe pericarp of freshly harvested avocado (*Persea americana* cv. Fuerte) fruits, resistant to *C. gloeosporioides* attack, was 5.2, whereas in ripe fruits, when decay symptoms were expressed, the pericarp pH had increased to 6.3. Two avocado cultivars, Ardit and Ettinger, which are resistant to *C. gloeosporioides* attack, had pericarp pHs of less than 5.5, which did not increase during ripening. Host pH regulates the secretion of PL and may affect *C. gloeosporioides* pathogenicity (Nir Yakoby *et.al.*, 2000). The mechanism found in avocado may have equivalents in other postharvest pathosystems and suggests new approaches for breeding against and controlling postharvest diseases.

Mechanism of pH modification by pathogens: Mechanisms for tissue acidification have been proposed on the basis of several studies. Two mechanisms have been used by *Penicillium* spp. for ambient acidification *viz.,* the production of organic acids, mainly citric acid and gluconic acid and NH4+ utilization associated with H+ efflux. In decayed fruit by *P. expansum* and *P. digitatum*, both produces citric acid and gluconic acid in the decayed tissue and reduced the host pH by 0.5-1.0 (Kumar *et.al.*, 2018). Ammonium depletion from the growth medium or from the fruit tissue is directly related to acidification. Several studies on fungal pathogen demonstrate that the ambient pH conditions in host induce the expression of a particular subset of fungal genes, selected from large gene families that encode cell wall degrading enzymes. In *P. expansum*, the secretion of gluconic acid and, to a lesser extent, citric acid enhanced the expression of pectolytic enzymes and the establishment of conditions for necrotrophic development . Analysis of transcripts encoding the endo-polygalacturonase gene, pepGL, from *P. expansum* under acidic culture conditions (pH 3.5-5.0), suggests that the acidification process is a pathogenicity enhancing factor for *Penicillium* spp. This was supported by the findings that cultivars with lower pH as well as citric acid treatments that reduced tissue pH, increased *P. expansum* development. Conversely, local alkalinization with NaHCO3 reduced decay development (Prusky *et al.*, 2004). The observations demonstrate that ambient pH is a regulatory cue for processes linked to pathogenicity, and that specific genes are expressed as a result of the modified host pH created by the pathogen.

Plant pathogens also make the environment alkali for enhancing pathogenicity. In ripening and senescing fruits, the pH levels change; for example, the pH of avocado increases from 5.2-6.0 during ripening. Alkalinization of the host environment during colonization by postharvest pathogens such as *Colletotrichum* and *Alternaria* is associated with fungal secretion of ammonia. In avocado, the fruit's pericarp showed a pH of 5.2 that increased up to 7.5-8.0, upon infection caused by *C. gloeosporioides.* In tomato fruits, the initial pH ranged between 4.1 and 4.5 and decay development at the infection site increased the pH to 8.0; moreover, the ammonia accumulated to 3.6 mM compared to 0.2 mM in the healthy tissue (Kumar *et.al.*, 2018). In wide-host-range pathogens such as *A. alternata*, a threefold to tenfold increase in ammonia concentration and a pH elevation of 0.2-2.4 units were detected in several hosts like tomato, pepper, melon, cherry and persimmon. Alkalization by pathogens is achieved by active secretion of ammonia, which is produced as a result of protease activity and deamination of amino acids. The initial acidic pH has been shown to be conducive to the enhanced ammonium secretion and host tissue alkalization that facilitate the virulence of pathogen. Ammonium accumulation is a critical factor contributing to necrotrophic development in ripening fruits. Expression of the endoglucanase gene AaK1 by *A. alternata* is maximal at pH levels above 6.0 that are characteristic of decayed tissue. The AaK1 gene does not express at the lower pH values in which the pathogen is quiescent . In the pathogen *C. gloeosporioides*, the gene pelB is expressed when the pH was above 5.7 and was positively correlated with ambient alkalinization. The transcription factor pac1, which is involved in pH regulation, follows a pattern similar to that of pelB, which suggests that they are co-regulated or that pac1 regulates the expression of pelB.

Postharvest pathogens can penetrate fruit by breaching the cuticle or directly through wounds, and they show disease symptoms only long after infection. During ripening and senescence, the fruit undergo physiological processes accompanied by a decline in antifungal compounds, which allows the pathogen to activate a mechanism of secretion of small effector molecules that modulate host environmental pH (Prusky *et.al.*, 2016). These result in the activation of genes under their optimal pH conditions, enabling the fungus to use a specific group of pathogenicity factors at each particular pH. New research suggests that carbon availability in the environment is a key factor triggering the production and secretion of small pH-modulating molecules: ammonia and organic acids. Ammonia is secreted under limited carbon and gluconic acid under excess carbon. This mini review describes our most recent knowledge of the mechanism of activation of pH-secreted molecules and their contribution to colonization by postharvest pathogens to facilitate the transition from quiescence to necrotrophic lifestyle.

Colonization of nectarines and peaches fruit by *Monilinia fructicola* was accompanied by local acidification of the host tissue. The fungus acidified the host tissue in peaches and nectarines from pH 4.5 and 4.45, to pH 3.75 and 3.9, respectively. Analysis of the acidification process in colonized fruit and secondary inducing media showed that gluconic acid was the main organic acid accumulated at the infection site and under liquid-culture conditions (De cal, *et.al.*, 2013). When comparing a nectarine cv. Big Top and peach cv. Plácido with differing sensitivities to *M. fruticola*, a 250% higher accumulation of gluconic acid was observed in the susceptible peach cultivar than in the less susceptible nectarine cultivar. Under liquid conditions, at pH 3.6–3.7, the relative expression of transcripts of *mfpg 2* and *mfpg3*, encoding for two polygalacturonase genes of *M. fructicola*, increased 12-fold and 6-fold, respectively, suggesting the importance of acidification for the secretion of pathogenicity factors by *M. fructicola.* It indicated that ambient pH is a regulatory cue for processes linked to pathogenicity of postharvest pathogens, and that specific genes contributing to pathogenicity are expressed as a result of the environmental pH created by the pathogen.

Secreted molecules and quiescence release : The transition from quiescence to necrotrophic colonization probably involves modulation of fungal effectors, as reported for *C. higginsianum* and exemplified by ChNLP1 and ChToxB (Kleemann *et. al.,* 2012), or of pH modulators, exemplified by ammonia and gluconic and oxalic acid (OA) for *C. gloeosporioides, Penicillium expansum*, and *B. cinerea,* respectively. Genes regulating the expression of fungal effectors are induced during quiescence and the switch to necrotrophy, which suggests that their toxic products may contribute to the termination of the biotrophic phase in preparation for subsequent necrotrophic growth. These fungal effectors are plant induced during the switch from biotrophic to necrotrophic colonization which may indicate again that quiescence is an active process that expresses a differential set of fungal genes and host responses Although conceptually different, the transition of *P. infestans* from biotrophic to necrotrophic growth is controlled by environmental sensing that leads to differential secretion of effectors. Secreted fungal effectors play an important role in suppressing host PCD to enable biotrophic interactions. Thus, the soybean pathogen *Phytophthora sojae* effector Avr1b suppresses BAX-induced PCD, which is a positive regulator of apoptosis. The effector SNE1 from *P. infestans* suppresses PCD induced by the *P. infestans* PiNPP1.1 gene, which includes Nep1-like proteins that are highly expressed during necrotrophy. Hence, SNE1 and PiNPP1.1 are coordinately expressed during biotrophy and necrotrophy *P. infestans* utilizes the expression of these effectors to optimize its growth.

5.6 Genomic and transcriptomic studies of host factors that modulate quiescence in plant-fungal interactions: Genome-scale approaches enable molecular and genetic dissection of quiescence at unprecedented levels. However, such studies of quiescent infection, especially in fruit tissues, are rare. Early stages of *B. cinerea* infection in *Arabidopsis* leaf tissue show some similarities with quiescence and, therefore, may be used to glean data that could guide future studies of quiescence in fruits. In *Arabidopsis*, a significant proportion of inoculated plants remain quiescent despite high levels of inoculation. Also, data on the nature of the genes and pathways activated during early stages of leaf infection showed significant similarity to fruit infection (Abu Qamar *et.al.,* 2006). Global expression profiling identified diverse early induced genes that contributed directly to resistance , whereas, in contrast, late stages of infection caused perturbation as an indication of general stress. More recently, very extensive and detailed analysis of gene expression in *B. cinerea–*

inoculated *Arabidopsis* tissues documented the chronology and regulation of the transcriptome; approximately one-third of the *Arabidopsis* genome was differentially expressed during the first 48 h after infection, with the majority of changes in gene expression occurring before significant lesion development. This study also documented the times at which signaling and metabolic processes changed, and it identified transcription factor families operating at different stages of infection and regulatory networks impacting resistance to necrotrophic infections. Overall, the data from early stages of infection, prior to lesion development, may be informative on the nature of regulatory networks important for quiescence. Global analyses of pathogen-induced transcriptomes also identified important early regulators of the host responses to *B. cinerea* and other pathogens; components of pathogen perception and signaling, as well as regulators of gene expression, were identified.

The *Arabidopsis* BIK1 and the plant-specific transcription factor WRKY33 were isolated during such studies: BIK1 is a central regulator of PAMP-triggered immunity (PTI) in diverse pathogens, and its tomato homolog performs similar functions through its contribution to ethylene signaling. PTI is a form of quantitative resistance triggered by host recognition of pathogen-derived molecules. In the case of WRKY33, extensive data support its critical role in resistance to necrotrophic infections. Global expression profiling uncovered massive transcriptional reprogramming as a result of *B. cinerea* infection, and the loss of WRKY33 caused inappropriate activation of SA-dependent responses and caused elevated SA levels accompanied by suppression of JA-associated responses. Genes involved in redox homeostasis, SA signaling, ethylene- and jasmonate-mediated cross talk, and camalexin biosynthesis were identified as direct targets of WRKY33, indicating the pervasive role of this protein. WRKY33 also regulated the machinery of plant autophagy, a process that may be important for defense and senescence, including fruit ripening). The roles of such critical regulators should be evaluated for their impact on quiescence. In light of the genetic data that demonstrate accelerated disease development in mutants of these genes in vegetative tissue, it is tempting to speculate that such factors may delay the necrotrophic phase, even in fruit tissues, but the details of how they are regulated and what their interactions are with the ripening process, including changes in the plant cell wall, need to be examined.

One of the studies that examined the defense-related transcriptome of infected fruit tissue was by Cantu *et.al.,* (2008) , who compared changes in gene expression in mature green and red ripe tomato fruit tissues after a *B. cinerea* challenge. *B. cinerea*–induced genes in tissues of red ripe and mature green tomato fruits showed significant overlap: 36% of *B. cinerea*–induced genes in mature green fruits were induced in red ripe fruits also, and 12% of genes down regulated by *B. cinerea* in mature green fruits were down regulated by infection in red ripe fruits also. Transcriptional changes observed in both mature green and red ripe fruit in response to *B. cinerea* suggested that the pathways were activated by *B. cinerea*, regardless of the ripening stage of the fruit. Interestingly, in mature green tomato fruits, *B. cinerea* induced genes normally associated with ripening, which indicates pathogen driven modification of physiological processes that favor necrotrophy. Overall, the kinetics and nature of gene expression found in fruit tissues share significant overlap with those found in comparable experiments in *B. cinerea* infected leaf tissues.

Recently, next generation sequencing of the transcriptome of *C. gloeosporioides* infected, unharvested and harvested avocado fruits was applied to avocado fruit tissues to determine their early and late host responses. A total of 639 genes were identified that encoded proteins functioning in secondary metabolism, signal transduction, transcriptional control, defense, stress-response, and cellular transportation processes. Such studies may uncover the biological bases of the fruit responses to *C. gloeosporioides* infection in such a complex system, provided follow-up studies on specific genes are completed. Further studies are required to elucidate major host components and those that can be used as markers of quiescent or active infection, and thereby to enable monitoring of pathogen progress in fruits. Biological processes that modulate the state of infection in fruit as well as vegetative tissue should be further investigated.

Fungal pathogens exploit different routes to penetrate the unripe fruit tissue but remain quiescent for months until the fruit ripens, awaiting the right conditions for the infection to become active and cause disease. The period from host penetration to activation of fungal colonization is described as quiescence.

In hemibiotrophs, quiescence may last through the duration of the biotrophic phase until the pathogen switches to the necrotrophic phase. As quiescent fungi develop, they elicit biotrophic interactions. The fungi manipulate host responses by secretion of effector proteins during quiescence, emphasizing that quiescence is not a passive stage.The release from quiescence is the result of induced accessibility of disassembled cell wall substrates and ethylene induction, of decline in preformed antifungal compounds, and of decline of inducible host-defense responses and pH conditions in the host. Enzymes involved in host cell wall metabolism play a central role in modulating the transformation from quiescence to active infection, not only as physical barriers but also as factors involved in pathogen recognition and in activation and deployment of plant responses.

Postharvest pathogens regulate the duration of and exit from quiescence by induction of fruit-senescing processes through actively secreting plant hormones, effectors, and pH modulators that induce host ripening, cuticular permeability, NADPH oxidase, and SA pathway. Pre-quiescence stage pathogens activate the unripe fruit defense-related proteins, including PR proteins, proteases, peroxidases, and PIN proteins. The transition from quiescence to necrotrophic colonization involves fungal pH modulators, exemplified by ammonia and organic acids. These factors are part of the first and subsequent waves of pathogenic attack and determine the capability of the pathogen to successfully colonize, invade, and kill the targeted host cell.

Genome scale analysis of gene expression in leaf tissue inoculated by a postharvest pathogen (*B. cinerea*) documented that one-third of the leaf genome was differentially expressed during the first 48 h after infection, with the majority of changes in gene expression occurring before significant lesion development. The kinetics and nature of gene expression found in leaf tissues suggest that they share significant overlap with what occurs in fruit tissues.

5.7 Managing quiescent infections : The phenomenon of latent or quiescent infection in *Colletotrichum* was defined as 'the inhibition of the pathogen through physiological conditions imposed by the host until some stage of maturation has been accomplished (Parthasarathy *et. al.,* 2017)'. Latency is defined as a quiescent or dormant parasitic relationship, which after a time changes to an active one. Molecular and physiological basis of quiescent infection during transition from immature to ripening stage was clearly described. Latent infections have been reported in stone fruits, apples, avocados, mangoes, papayas, citrus fruits, grapes and strawberries. In strawberries, *Botrytis* rots start on infected, senescent floral parts that become latent infections which develop into active rots on ripe fruits. The yeast-like fungus *Aureobasidium pullulans* and the yeast *Candida oleophila* was effective against *B. cinerea* storage rots on strawberries when applied in the field at bloom compared to application immediately after harvesting. Also, the application of *A. pullulans* and *Epicoccum purpurascens* to sweet cherry blossoms reduced the number of latent infections by *Monilinia laxa* in green strawberry fruits. The importance of establishment of a biocontrol agent before pathogen arrival especially, when shortlived colonizable substrates (e.g. petals, stamens) act as infection foci.

Repeated applications of antagonists was required to control postharvest decays originating from latent infections in the orchard as demonstrated by three sprays of *Bacillus subtilis* to control avocado anthracnose. In addition, continuous application of *B. subtilis* over a four-year period resulted in a gradual buildup of the antagonist population, reduction in the level of pathogen inoculum and more effective disease control. The population dynamics of *A. pullullans, Rhodotorula glutinis* and *B. subtilis* on apple fruits, for which the antagonists were applied to trees in the field late in the growing season. After transfer from field to cold storage, the population of the introduced antagonists remained higher than in control treatments, except for *B. subtilis* which declined. The efficacy of preharvest application of *A. pullulans* against *Botrytis* rot of sweet cherry and table grapes and *R. glutinis* against *Botrytis* rot and blue mold of pear were also reported. Field sprays of *C. guilliermondii* A42 and *Acremonium cephalosporium* B11 two to five times at 7-10 days interval on table and wine grapes could reduce decay caused by *B. cinerea* and *A. niger.*

Novel postharvest treatments of strawberries have been suggested to prevent *B. cinerea* infections during storage. Examples are edible fruit coatings of chitosan, silk fibroin or methylcellulose that prevent water loss and can include antifungal compounds (Stefan Petrasch, *et.al.*, 2019). MeJA treatment to induce fruit defence mechanisms, ultraviolet and visual light treatment, enrichment of storage atmosphere with chlorine or ozone, and soft mechanical stimulation have also been tested as alternative treatments. Most of these approaches are still in an experimental stage or not yet adaptable to commercial settings. It is common practice to handpick strawberries and place them into clamshells in the field, in order to reduce wounding and bruising of the fruit. Rapid and constant cooling of strawberries at temperatures below 2.5 °C is another critical strategy to reduce or inhibit reactivation of *B. cinerea* quiescent infections. Often, strawberries are also stored in modified atmospheres, which are generally low in oxygen and high in carbon dioxide to slow down metabolic processes, senescence and fungal decay. Relative humidity during storage is usually kept around 85%–90% to prevent dehydration of fruit, but limit fungal growth

For mango anthracnose management, use of resistant cultivars is an ideal, simplest and cheapest method for the control of plant disease. Heat treatment reduced disease incidence in plantain banana, mango, lychee and longan (Md. Nasir Uddin *et al.*, 2018)]. Anthracnose is one the most important diseases of mango fruit that affects pre-harvest and postharvest quality of fruits. It also affects the flowers, leaves and inoculum year-round throughout the canopy. Anthracnose disease is prevalent but well controlled and non-significant in dry regions. Management of these diseases is essentially needed to produce high yield and quality fruit. Management of disease requires an awareness of this everpresent threat and the weather conditions that increase infection and disease development. Optimum control relies on Disease management. The following postharvest treatments were proposed, *viz.*, Scrubbing with 1% NaOCl; Hot water dip (50-55°C for 3-10 minutes); Hot benomyl dip

(500-1000 ppm); Hot/cold prochloraz dip (400-l000 ppm); Hot imazalil (1000 ppm); Hot water + 20 k RAD irradiation ; Hot water + 75 k RAD irradiation 8. Hot benlatc/iprodionc (1000 ppm)+ 75 k RAD irradiation + waxing.

In laboratory studies and experimental packingline trials that closely simulated fungicide treatments under commercial conditions, the “reduced-risk” fungicides fenhexamid and fludioxonil were highly effective in managing postharvest brown rot and gray mold decays of peach, nectarine, and plum caused by *Monilinia fructicola* and *Botrytis cinerea*, respectively (Förster *et.al.,* 2007). Tebuconazole was more effective against brown rot than against gray mold. Additionally, fludioxonil and tebuconazole significantly reduced the incidence of *Rhizopus* rot caused by *Rhizopus stolonifer*. In laboratory studies, the QoI fungicide azoxystrobin was not consistent in its decay control activity. The high efficacy of fenhexamid and fludioxonil against brown rot and gray mold was substantiated by low effective concentrations necessary (≤0.063 mg/liter)for 50% inhibition of mycelial growth in vitro. In general, fungicides applied 14 to 16 h after wound inoculation were significantly more effective than those applied before inoculation. These results indicate that the fungicides act mainly as protectants that do not penetrate deeply enough into the fruit to prevent decay from wounds that extend below the fruit epidermis. Laboratory treatments of plum fruit generally were less effective than those of peach or nectarine fruit. Comparative studies using either low volume spray or high-volume, in-line drench applications over a roller bed demonstrated that drench applications were significantly more effective in reducing postharvest decays of inoculated plum fruit. Decay incidence using fenhexamid or fludioxonil was ≤1.1%, whereas incidence after the spray applications was between 25.2 and 40.4% for brown rot, between 12.0 and 24.3% for gray mold, or 62.6% for *Rhizopus* rot (fludioxonil only).

The last decade was accompanied with major advantages in high throughput sequencing of DNA and RNA and computational mapping. Those methods could be applied to fast sequencing of fungal genomes, fruit pathogen interaction in the transcriptional level and study the dynamics of microorganism in the fruit stem end. Today, the genomes of most of the hosts and SER causing pathogens are available, which will enable transcriptome analysis that could open new insights for better understanding the host effective defense response and the switch of pathogenic fungi from endophytic to pathogenic stage (Ortal Galsurker *et.al.,* 2018). This transcriptome analysis could lead to the development of new control methods. The microbiome consists of all of the microorganisms (fungi and bacteria) inhabiting the plant tissue. The advances in high throughput sequencing of DNA have been applied to study the plant microbiome. In apples, variety and rootstock modulate the endophytic microbiome, suggesting coevolution of a specific genotype with its microbiome. When these methods were used to study the mango stem end microbiome during postharvest storage and disease development, healthier stem ends (with a lower incidence of SER) showed more diverse microbial communities. In contrast, an increase in SER incidence was correlated with a reduction in microbiome diversity and an expansion of one or several fungal pathogen families, such as *Pleosporaceae* and *Botryosphaeriaceae*. In addition, the increase in fungal abundance and SER was correlated with an increase in the chitin degrading *Chitinophagaceae* bacteria and a reduction in biocontrol agents]. This implies that the stem end microbiota is a dynamic system that can be modified to control SER causing pathogens.

References

Abu Qamar, S., Chen, X., Dhawan, R., Bluhm, B. and Salmeron, J., 2006. Expression profiling and mutant analysis reveals complex regulatory networks involved in *Arabidopsis* response to *Botrytis* infection. *Plant J.* **48:**28-44.

Alkan,N., Espeso,E.A. and Prusky,D., 2013. Virulence regulation of phytopathogenic fungi by pH. *Antioxid Redox.Signal,* **19**: 1012-1025.

Amaral, Daniela Dambros, Monteiro, Ana Leticia Rocha, Silva, Elias Inácio Da, Lins, Severina Rodrigues De Oliveira, & Oliveira, Sonia Maria Alves De., 2017. Frequency of Quiescent Fungi and Postharvest Alternative Management of Stem End Rot in Papaya. *Revista Caatinga*, **30:** 786-793.

Antonio Ippolito and Mario Salerno. 2001. Low Temperature Metabolism of Apple Phenolics and Quiescence of *Phlyctaena vagabunda. J. Agric. Food Chem.* **4 9**: 5817-5821*Appl. & Env.Microbiol.*, **66** : 1026-1030.

Bruton, B.D., 1994. Mechanical injury and latent infections leading to postharvest decay. *Hort. Sci*., **29:** 747-749.

Caracuel, Z., Roncero, M.I.G., Espeso, E.A., Gonzalez-Verdejo, C.I., Garcia-Maceira, F.I.and Di Pietro, A.,. 2003. The pH signalling transcription factor PacC controls virulence in the plant pathogen *Fusarium oxysporum. Mol. Microbiol.*, **48:**765–79

De cal, A.., Sandin-Espana, P., Martinex, F., Eguen,B., Chien-Ming,C., Lee, M.H., Melgarejo, P. and Prusky, D., 2013. Role of gluconic acid and pH modulation in virulence of *Monilinia fructicola* on peach fruit. *Postharvest Bio. & Tecnol.*, **86**: 418-442

Dodd, J.C., Prusky, D. and Jefries, P., 1997. Disease of mango fruits. *In:.* The mango: botany, production and uses. Litz, R.E., (Ed.), Wallingford, United Kingdom, CAB International. Pp. 257-280.

Dov Prusky , 1996. Pathogen Quiescence in Postharvest Diseases . *Ann.. Rev. Phytopathol..* **34**: 413–34

Dov Prusky, Noam Alkan, Tesfaye Mengiste, and Robert Fluhr, 2013. Quiescent and Necrotrophic Lifestyle Choice during Postharvest Disease Development. *Annu. Rev. Phytopathol..* **51**:155–76

Förster, H., Driever, G. F., Thompson, D. C., and Adaskaveg, J. E. 2007. Postharvest decay management for stone fruit crops in California using the "reduced-risk" fungicides fludioxonil and fenhexamid. *Plant Dis*., **91**:209-215.

Johnson, G., Cooke, T. and Mead, A., 1993. Infection and Quiescence of Mango Stem-End Rot Pathogens. *Acta Hortic*. **341** : 329-336

Kleemann, J., Rincon-Rivera, L.J., Takahara, H., Neumann, U., van Themaat, E.V.L., 2012. Sequential delivery of host-induced virulence effectors by appressoria and intracellular hyphae of the phytopathogen *Colletotrichum higginsianum*. *PLoS Pathog*., 8:e1002643

Kumar, R., Ghatak, A., Balodi, R. and Bhagat, A.P. 2018. Decay mechanism of postharvest pathogens and their management using non-chemical and biological approaches. *J. Postharvest Technol*., **6** :001-011.

Lakshiminarayana, S.,. Sommer, N.F., Polito, V. and Fortlage, R.J., 1987. Development of resistance to infection by *Botrytis cinerea* and *Penicillium expansum* in wounds of mature apple fruits. *Phytopathology,* **77:**1674 1678.

Md. Nasir Uddin, Shoaibe Hossain Talukder Shefat, Mansura Afroz and Nusrat Jahan Moon, 2018. "Management of Anthracnose Disease of Mango Caused by *Colletotrichum gloeosporioides*: A Review". *Acta Sci.Agri*., **2:** 169-177.

Nir Yakoby, Ilana Kobiler, Amos Dinoor, Dov Prusky, 2000. pH Regulation of Pectate Lyase Secretion Modulates the Attack of *Colletotrichum gloeosporioides* on Avocado Fruits.

Niren Majumdar and Nakul Chandra Mandal, 2018. Effect of pH on Mycelial Growth and Sporulation of Postharvest Pathogen *Colletotrichum gloeosporioides* (Penz.) Penz & Sacc. and *Pestalotiopsis mangiferae* (Henn.) Steyaert. *Intl.J. Biores.& Stress Mnt*., **9 :** 416 – 420.

Ortal Galsurker , Sonia Diskin , Dalia Maurer, Oleg Feygenberg and Noam Alkan, 2018. Fruit Stem-End Rot. *Horticulturae,* **4**:1 -16.

Parthasarathy, S., Rajalakshmi, J., Narayanan, P., Arunkumar, K and Prabakar, K., 2017. Bio-Control Potential of Microbial Antagonists Against Postharvest Diseases of Fruit Crops: A Review . *Research & Reviews: J. Bot. Sci.,* **6**: 17-23.

Pongphen Jitareerat and Sudkanueng Paumchai , 2007. Sirichai Kanlayanarat and Somsiri Sangchote, Effect of chitosan on ripening, enzymatic activity, and disease development in mango (*Mangifera indica*) fruit. *New Zealand J. Crop & Hort. Sci*., **35**: 211—218

Prins, T.W., Tudzynski, P., von Tiedemann, A., Tudzynski, B. and ten Have, A., 2000. Infection strategies of *Botrytis cinerea* and related necrotrophic pathogens. *In : Fungal Pathology*, J.W. Kronstad (Ed.), pp. 33–65. Dordrecht, Neth.: Kluwer Acad.

Prusky, D. and Kobiler, I., 2012. Mechanisms modulating the activation of quiescent infections by postharvest pathogens. *In: Avanços tecnológicos na patologia pós-colheita, Oliveira, S. M. A.; Lins, S. R. O.; Santos, A. M. G. (Eds.).* Recife: Editora UFRPE, **1** : 275-296

Prusky, D., McEvoy, J.L., Saftner, R., Conway, W.S. and Jones, R. 2004. The relationship between host acidification and virulence of *Penicillium* spp. on apple and citrus fruit. *Phytopathology*, **94**: 44–51.

Prusky, D.B., Moral, B. F. and Barad, J,S, 2016. How Does Host Carbon Concentration Modulate the Lifestyle of Postharvest Pathogens during Colonization? *Frontiers Pl. Sci*.**,** **7**:1306.

Roberts, R.G., 1994. Integrating biological control into postharvest disease management strategies. *Hort Sci*., **29**:758–762.

Sowley, E.N.K., Dewey, F.M. and Shaw, M.W., 2010. Persistent, symptomless, systemic, and seed-borne infection of lettuce by *Botrytis cinerea*. *Eur. J. Plant Pathol*., **126**:61–71

Stefan Petrasch, Steven J. Knapp, Jan A. L. van Kan and Barbara Blanco-Ulate. 2019. Grey mould of strawberry, a devastating disease caused by the ubiquitous necrotrophic fungal pathogen *Botrytis cinerea. Mol Plant Pathol.,* **20** : 877–892.

van Kan, J.A.L., 2006. Licensed to kill: the lifestyle of a necrotrophic plant pathogen. *Trends Plant Sci.* **11**:247–53.

Vincenzo Lattanzio, Vincenzo Lattanzio, Donato Di Venere, Vito Linsalata, Paolo Bertolini and William R. Jarvis, 1994. Latent Infections in the Pre- and Postharvest Environment. *Hort Sci*., **29** : 762–768.

6

Mycotoxins

Mycotoxins are naturally occurring toxins produced by certain molds and can be found in food. The molds grow on a variety of different crops and foodstuffs including cereals, nuts, spices, dried fruits, apples and coffee beans, often under warm and humid conditions. Mycotoxins can cause a variety of adverse health effects and pose a serious health threat to both humans and livestock. The adverse health effects of mycotoxins range from acute poisoning to long-term effects such as immune deficiency and cancer. A scientific expert committee jointly convened by WHO and the Food and Agriculture Organization of the United Nations (FAO), called JECFA, is the international body responsible for evaluating the health risk from natural toxins including mycotoxins. International standards and codes of practice to limit exposure to mycotoxins from certain foods are established by the Codex Alimentarius Commission based on JECFA assessments.

Mycotoxins are toxic compounds that are naturally produced by certain types of molds (fungi). Molds that can produce mycotoxins grow on numerous foodstuffs such as cereals, dried fruits, nuts and spices. Mold growth can occur either before harvest or after harvest, during storage, on/in the food itself often under warm, damp and humid conditions. Most mycotoxins are chemically stable and survive food processing. Several hundred different mycotoxins have been identified, but the most commonly observed mycotoxins that present a concern to human health and livestock include aflatoxins, ochratoxin A, patulin, fumonisins, zearalenone and nivalenol/deoxynivalenol. Mycotoxins appear in the food chain as a result of mold infection of crops both before and after harvest. Exposure to mycotoxins can happen either directly by eating infected food or indirectly from animals that are fed contaminated feed, in particular from milk.

All mycotoxins are low-molecular-weight natural products (i.e., small molecules) produced as secondary metabolites by filamentous fungi. These metabolites constitute toxigenically and chemically heterogeneous assemblage that are grouped together only because the members can cause disease and death in human beings and other vertebrates. Not surprisingly, many mycotoxins display overlapping toxicities to invertebrates, plants, and microorganisms (Ida *et al.*,2014). The term mycotoxin was coined in 1962 in the aftermath of an unusual veterinary crisis near London, England, during which approximately 100,000 turkey poults died). When this mysterious turkey X disease was linked to a peanut (groundnut) meal contaminated with secondary metabolites from *Aspergillus flavus* (aflatoxins), it sensitized scientists to the possibility that other occult mold metabolites might be deadly. Soon, the mycotoxin rubric was extended to include a number of previously known fungal toxins (the ergot alkaloids), some compounds that had originally been isolated as antibiotics (patulin), and a number of new secondary metabolites revealed in screens targeted at mycotoxin discovery (ochratoxin A). The period between 1960 and 1975 has been termed the mycotoxin gold rush because so many scientists joined the well-funded search for these toxigenic agents. Depending on the definition used, and recognizing that most fungal toxins occur in families of chemically related metabolites,) some 300 to 400 compounds are now recognized as mycotoxins, of which approximately a dozen groups regularly receive attention as threats to human and animal health. Mycotoxicoses are the animal diseases caused by mycotoxins; mycotoxicology is the study of mycotoxins.

While all mycotoxins are of fungal origin, not all toxic compounds produced by fungi are called mycotoxins. The target and the concentration of the metabolite are both important. Fungal products that are mainly toxic to bacteria (such as penicillin) are usually called antibiotics. Fungal products that are toxic to plants are called phytotoxins by plant pathologists (confusingly, the term phytotoxin can also refer to toxins made by plants.. Mycotoxins are made by fungi and are toxic to vertebrates and other animal groups in low concentrations. Other low-molecular-weight fungal metabolites such as ethanol that are toxic only in high concentrations are not considered mycotoxins. Finally, although mushroom poisons are definitely fungal metabolites that can cause disease and death in humans and other animals, they are rather arbitrarily excluded from discussions of mycotoxicology. Molds (microfungi) make mycotoxins; mushrooms and other macroscopic fungi make mushroom poisons. The distinction between a mycotoxin and a mushroom poison is based not only on the size of the producing fungus, but also on human intention. Mycotoxin exposure is almost always accidental. In contrast, with the exception of the victims of a few mycologically accomplished murderers, mushroom poisons are usually ingested by amateur mushroom hunters who have collected cooked and eaten what was misidentified as a delectable species.

6.1 Major mycotoxins: Fungi are the major source of mycotoxins in foods. *Aspergillus*, *Penicillium* and *Fusarium* are the fungi of greatest importance to humans with respect to food poison by mycotoxins. Aflatoxins are known as the most important fungal metabolites of a direct health hazard to humans. *Aspergillus flavus* is widely distributed in soil, air, and rotten plant residues. Aflatoxin production and contamination can take place in the field, in transit, or in storage. However, most infection takes places in the field while aflatoxin production takes place at any point under favourable conditions (Ashiq, 2015**)**. Mycotoxins are produced by a large number of fungal species and some are able to produce more than one mycotoxin (Table 8). Also, some mycotoxins can be produced by more than one species (Hunssein and Brasel, 2001). The effects of some food-borne mycotoxins are acute with symptoms of severe illness appearing quickly after consumption of food products contaminated with mycotoxins. Other mycotoxins occurring in food have been linked to long-term effects on health, including the induction of cancers and immune deficiency. Of the several hundred mycotoxins identified so far, about a dozen have gained the most attention due to their severe effects on human health and their occurrences in food. Most mycotoxins are chemically and thermally stable during food processing, including cooking, boiling, baking, frying, roasting, and pasteurization. Mycotoxins can also come to the human plate via animal products such as meat, eggs, milk as the result of the animal eating contaminated feed.

There are over 200,000 species of fungi, including mold, yeast, and mushrooms. More than 100,000 mold species have been identified. Among the mycotoxins fitting into this major group would be the aflatoxins, deoxynivalenol, fumonisins, zearalenone, T-2 toxin, ochratoxin and certain ergot alkaloids. The diseases (mycotoxicoses) caused by these mycotoxins are quite varied and involve a wide range of susceptible animal species including humans. Most of these diseases occur after consumption of mycotoxin contaminated grain or products made from such grains but other routes of exposure exist. The diagnosis of mycotoxicoses may prove to be difficult because of the similarity of signs of disease to those caused by other agents. Therefore, diagnosis of a mycotoxicoses is dependent upon adequate testing for mycotoxins involving sampling, sample preparation and analysis.

Modern mycotoxicology began with the discovery of aflatoxins after losses of a large number of animals in England in 1961 was attributed to consumption of peanut meal incorporated in the diets. Subsequent to the finding of the carcinogenicity of the aflatoxins, the immunosuppressive nature of the aflatoxins and the discovery that the aflatoxins were not only a storage problem in grains but actually contaminated certain crops pre-harvest, the total mycotoxin problem became a multidisciplinary issue involving analytical chemists, microbiologists, agronomists, agricultural engineers, entomologists, plantpathologists, crop breeders, geneticists, medical and veterinary practitioners and producers (farmers/ranchers) (John L. Richard, 2007). The combined efforts of interactions of these groups led to the discovery of many new mycotoxins and their involvement in diseases of animals.

Table 8: Major mycotoxins

Major genera of mycotoxin producing fungi	**Species**	**Mycotoxins**
Aspergillus	*A.flavus* *A.parasiticus* *A.nomius* *A. pseudotamarii*	Aflatoxin (B_1, B_2, G_1, G_2)
	A. ochraceus	Ochratoxin (Ochratoxin A)
	A.clavatus *A. terreus*	Patulin
	A.flavus *A. versicolor*	Cyclopiazonic acid (CPA)
Claviceps	*C.purpurea* *C.fusiformis* *C.paspali* *C. africana*	Penitrem A **Ergot alkaloids:** Clavines (Argroclavine) Lysergic acids Lysergic acid amides (Ergine) Ergopeptines (Ergotamine, Ergovaline)
Fusarium	*F. verticillioides* *(syn. F. moniliforme)* *F. proliferatum*	Fumonisin (B1, B2, B3) Fusaric acid
	F. graminearum *F. avenaceum* *F. culmorum*	**Type A Trichothecenes:** T-2 toxin, HT-2 toxin, diacetoxyscirpenol
	F. poae *F. equiseti* *F. crookwellense* *F. acuminatum* *F. sambucinum* *F. sporotrichioides*	**Type B Trichothecenes:** Nivalenol, Deoxynivalenol, Fusarenon-X
	F. graminearum *F. culmorum* *F. sporotrichioides*	Zearalenone
Penicillum	*P. verrucosum* *P. viridicatum*	Ochratoxin (Ochratoxin A)

	P. citrinum *P. verrucosum*	Citrinin
	P. roqueforti	Roquefortine C PR toxin Penitrem A
	P. cyclopium *P. camemberti*	Cyclopiazonic acid (CPA) Penitrem A
	P. expansum *P. claviforme* *P. roquefortii*	Patulin
Neotyphodium	*N. coenophialum*	Tall fescue toxins: Ergot alkaloids, Lolines, Peramine
	N. lolii	Ryegrass toxins: Lolitrems, Peramine, Ergot alkaloids (e.g. Ergovaline)
Pithomyces	*P. chartarum*	Sporidesmin
Stochybotrys	*S.chatarum*	Satratoxins
Alternaria	*A. alternata*	Tenuazonic acid
Monographella	*M. nivalis*	Bentenolide
Chaetomium	*C. globosum*	Chaetoglobosin
Acremonium	*A. crotocinigenum*	Crotocin
Rhizoctonia	*R.leguminicola*	Slaframine
Trichoderma	*T.viride*	Trichodermin
Trichothecium	*T.roseum*	Trichothecin
Eurotium	*E.chevalieri*	Xanthocillin
Others	*Myrothecium roridum,* *M. verrucaria, Dendrodochium* spp., *Cylindrocarpon spp. , Stachybotrys* spp.	Roridin E
	Myrothecium verrucaria, *Dendrodochium* spp., *Stachybotrys chartarum*	Verrucarin
	Trichophyton spp. *Penicillium viridicatum*	Viopurpurin

Primary mechanism of action of main mycotoxin groups:

Aflatoxin: Binds to guanine (DNA-adduct) after metabolic activation in the liver

Trichothecenes: Inhibition of protein synthesis

Zearalenone :Binds to mammalian estrogen receptor

Ochratoxins : Blocks protein synthesis

Ergot alkaloids : Binding to adrenergic, dopaminergic and serotonin receptors

Fumonisins : Inhibit ceramide synthase (sphingolipid biosynthesis)

Mycotoxins are capable of direct-target toxicity towards certain organs such as the liver, nervous system, kidney, skin, cardiovascular, reproductive and immune systems. Non-direct-target effects include carcinogenicity, teratogenicity and mutagenesis (Bryden, 2012). Mycotoxins are absorbed through the gastro-intestinal tract (GIT), the lungs, the skin and other organs like the eyes.In the GIT, mycotoxins can be absorbed in the mouth and esophagus (minimal absorption), in the stomach and in the small intestine. In the stomach, absorption is generally correlated with the fraction of non-dissociated contaminants like weak acids.The small intestine, is where the maximal absorption takes place. A fraction of mycotoxins can also be absorbed in the colon.

In the lungs, mycotoxins that are generally carried by dust are absorbed in the alveoli (i.e. ochratoxin A).Once in the body, mycotoxins can be distributed through different pathways. Some mycotoxins can bind to plasma proteins and be transported into the blood plasma (e.g. ochratoxin A). Mycotoxins can also be lipophilic and accumulate in the fat tissues. Lipophilic compounds can easily penetrate the blood-brain barrier and the placental barrier. The half-life elimination time, or the time required to reduce the initial plasma concentration of the toxin, can be very long in the case of chronic exposure (ochratoxin A half-life elimination time in humans is longer than 560 hours. Mycotoxins can be partially excreted through several pathways. Usually polar and hydrophilic substances are excreted via the kidneys with the urine. Accumulation of mycotoxins in kidneys has been observed and can produce toxic effects (i.e. ochratoxin A). Compounds that are characterized by high molecular weight are usually excreted via the bile. Other pathways of excretion include transfer to milk (aflatoxin B_1) and elimination through sweat and saliva. Unabsorbed mycotoxins may be excreted in feces but could still have had an effect on the gut wall during passage.

Toxic effects of mycotoxins can be reversible and irreversible. Reversible effects include minor damages that can heal such as skin irritation. Irreversible effects involve permanent damage to health. One example, is vasoconstriction (blood vessel constriction) caused by ergot alkaloids, leading to necrosis of extremities (e.g. toes, ears or tail). The main toxic effects of mycotoxins are carcinogenicity, genotoxicity, nephrotoxicity, hepatotoxicity, estrogenicity reproductive and digestive disorders, immunosuppression and dermal effects. There are several factors which influence symptoms, *viz.*, type of mycotoxins consumed, intake level and duration of exposure, animal species, sex, breed age, general health, immune status, farm management: hygiene, temperature production density and possible synergism between mycotoxins simultaneously present in feeds.

6.1.1. Aflatoxins : The aflatoxins were isolated and characterized after the death of more than 100,000 turkey poults (turkey X disease) was traced to the consumption of a mold-contaminated peanut meal (Blanc, *et al.*, 1995). The four major aflatoxins are called B_1, B_2, G_1, and G_2 based on their fluorescence under UV light (blue or green) and relative chromatographic mobility during thin-layer chromatography. Chemically it is (6a*R*,9a*S*)-2,3,6a,9a-Tetrahydro-4-methoxy-1*H*,11*H*-cyclopenta[c]furo[3',2':4,5]furo[2,3-h][1]benzopyran-1,11-dione. Aflatoxin B_1 (Fig. 31) is the most potent natural carcinogen known and is usually the major aflatoxin produced by toxigenic strains. It is also the best studied: in a large percentage of the papers published, the term aflatoxin can be construed to mean aflatoxin B_1. However, well over a dozen other aflatoxins (e.g., P_1. Q_1, B_{2a}, and G_{2a}) have been described, especially as mammalian biotransformation products of the major metabolites.

Major types : Aflatoxin B1 is considered the most toxic and is produced by both *Aspergillus flavus* and *A. parasiticus.* Aflatoxin M1 is present in the fermentation broth of *A. parasiticus*, but it and aflatoxin M_2 are also produced when an infected liver metabolizes aflatoxin B_1 and B_2. Aflatoxin B_1 and B_2 (AFB), produced by *A. flavus* and *A. parasiticus.* Aflatoxin G_1 and G_2 (AFG), produced by some Group II *A. flavus* and *A. parasiticus.* Aflatoxin M_1, metabolite of aflatoxin B_1 in humans and animals (exposure in ng levels may come from a mother's milk) (Geiser *et al.*, 2000). Aflatoxin M_2, metabolite of aflatoxin B_2 in milk of cattle fed on contaminated foods. Aflatoxicol (AFL): metabolite produced by breaking down the lactone ring. Aflatoxin Q_1 (AFQ_1), major metabolite of AFB_1 in *in vitro* liver preparations of other higher vertebrates: FM, AFQ, and AFL retain the possibly to become an epoxide. Nevertheless, they appear much less capable of causing mutagenesis than the unmetabolized toxin.

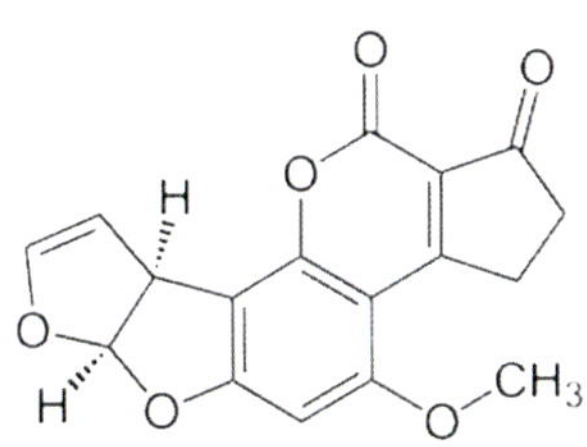

Aflatoxin B1

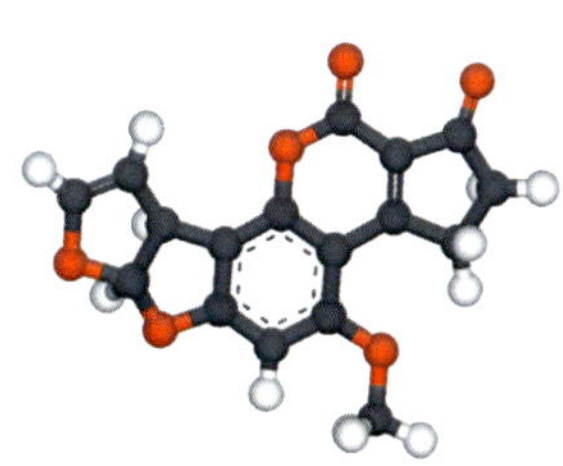

3D structure of aflatoxin B1

Fig. 31: Aflatoxin B1 3D structure of aflatoxin B1

Aspergillus flavus **:** *Aspergillus flavus* is a fungus of the Trichocomaceae family with a worldwide distribution. The mold lives in soil, surviving off dead plant and animal matter, but spreads through the air via airborne conidia. This fungus grows in long branched hyphae and is capable of surviving on numerous food sources including corn and peanuts.The fungus and its products are pathogenic to a number of species, including humans.While toxicity of its products, aflatoxins, are explored throughout this article, *Aspergillus flavus* itself also exerts pathogenic effects through aspergillosis, or infection with the mold (Hedayati *et al.,* 2007). This infection largely occurs in the lungs of immune compromised patients but infection may also occur in the skin or other organs. Unlike many mold species, *Aspergillus flavus* prefers hot and dry conditions. Its optimal growth at 37 °C (99 °F) contributes to its pathogenicity in humans.

Aflatoxins are difuranocoumarin derivatives produced by a polyketide pathway by many strains of *Aspergillus flavus* and *Aspergillus parasiticus*; in particular, *Aspergillus flavus* is a common contaminant in agriculture. *Aspergillus bombycis, Aspergillus ochraceoroseus*, *Aspergillus nomius*, and *Aspergillus pseudotamari* are also aflatoxin-producing species, but they are encountered less frequently (Peterson *et al.,* 2001). From the mycological perspective, there are great qualitative and quantitative differences in the toxigenic abilities displayed by different strains within each aflatoxigenic species. For example, only about half of *Aspergillus flavus* strains produce aflatoxins, while those that do may produce more than 10^6 µg/kg.

Many substrates support growth and aflatoxin production by aflatoxigenic molds. Natural contamination of cereals, figs, oilseeds, nuts, tobacco, and a long list of other commodities is a common occurrence. Like the genetic ability to make aflatoxin, contamination is highly variable. Sometimes crops become contaminated with aflatoxin in the field before harvest, where it is usually associated with drought stress; even more problematic is the fate of crops stored under conditions that favor mold growth. In storage, usually the most important variables are the moisture content of the substrate and the relative humidity of the surroundings. Aflatoxin contamination has been linked to increased mortality in farm animals and thus significantly lowers the value of grains as an animal feed and as an export commodity. Milk products can also serve as an indirect source of aflatoxin. When cows consume aflatoxin-contaminated feeds, they metabolically biotransform aflatoxin B_1 into a hydroxylated form called aflatoxin M_1. Aflatoxin is associated with both toxicity and carcinogenicity in human and animal populations (Eaton and Groopman, 1994). The diseases caused by aflatoxin consumption are loosely called aflatoxicoses. Acute aflatoxicosis results in death; chronic aflatoxicosis results in cancer, immune suppression, and other "slow" pathological conditions. The liver is the primary target organ, with liver damage occurring when poultry, fish, rodents, and nonhuman primates are fed aflatoxin B_1. There are substantial differences in species susceptibility. Moreover, within a given species, the magnitude of the response is influenced by age, sex, weight, diet, exposure to infectious agents, and the presence of other mycotoxins and pharmacologically active substances. Thousands of studies on aflatoxin toxicity have been conducted, mostly concerning laboratory models or agriculturally important species.

Cytochrome P450 enzymes convert aflatoxins to the reactive 8,9-epoxide form (also referred to as aflatoxin-2,3 epoxide in the older literature), which is capable of binding to both DNA and proteins. Mechanistically, it is known that the reactive aflatoxin epoxide binds to the N^7 position of guanines. Moreover, aflatoxin B_1-DNA adducts can result in GC to TA transversions. A reactive glutathione *S*-transferase system found in the cytosol and microsomes catalyzes the conjugation of activated aflatoxins with reduced glutathione, leading to the excretion of aflatoxin. Variation in the level of the glutathione transferase system as well as variations in the cytochrome P450 system is thought to contribute to the differences observed in interspecies aflatoxin susceptibility.

Because of the differences in aflatoxin susceptibility in test animals, it has been difficult to extrapolate the possible effects of aflatoxin to humans, but acute toxicity of aflatoxins in *Homo sapiens* has not been observed very often. It is believed that a 1974 Indian outbreak of hepatitis in which 100 people died may have been due to the consumption of maize that was heavily contaminated with aflatoxin. Some adults may have eaten 2 to 6 mg of aflatoxin in a single day. Subsequently, it was calculated that the acute lethal dose for adults is approximately 10 to 20 mg of aflatoxins (Pitt, 2000). One anecdotal report refutes this estimate. A woman who had ingested over 40 mg of purified aflatoxin in a suicide attempt was still alive 14 years later. Multiple laboratory tests of her urine and blood, and X-ray, ultrasound, and computerized axial tomography analyses of her abdomen, liver, and spleen all gave normal results.

It has been hypothesized that kwashiorkor, a severe malnutrition disease, may be a form of pediatric aflatoxicosis. Further early speculations that aflatoxin might be involved in Reye's syndrome, an encephalopathy, and fatty degeneration of the viscera in children and adolescents have not been substantiated. Nevertheless, aflatoxin has achieved some notoriety as a poison. Nevertheless, aflatoxin's reputation has a potent poison may explain why it has been adopted for use in bioterrorism. There is substantial evidence that Iraq stockpiled aflatoxin to be delivered in missiles (see section on bioterrorism below). The data on aflatoxin as a human carcinogen are far more damning than the data implicating it in acute human toxicities. Exposure to aflatoxins in the diet is considered an important risk factor for the development of primary hepatocellular carcinoma, particularly in individuals already exposed to hepatitis B. In classical epidemiology, several studies have linked liver cancer incidence to estimated aflatoxin consumption in the diet). The results of these studies have not been entirely consistent, and quantification of lifetime individual exposure to aflatoxin is extremely difficult. The incidence of liver cancer varies widely from country to country, but it is one of the most common cancers in China, the Philippines, Thailand, and many African countries. The presence of hepatitis B virus infection, an important risk factor for primary liver cancer, complicates many of the epidemiological studies. In one case-control study involving more than 18,000 urine samples collected over 3.5 years in Shanghai, China, aflatoxin exposure alone yielded a relative risk of about 2; hepatitis B virus antigen alone yielded a relative risk of about 5; combined exposure to aflatoxin and hepatitis B yielded a relative risk of about 60. Vaccination against hepatitis B virus is recommended as a more realistic and cost-effective strategy for lowering liver cancer incidence than removing aflatoxin from the diet.

In molecular epidemiology, it is possible to demonstrate with more certainty the association between putative carcinogens and specific cancers. Biomonitoring of aflatoxins can be done by analyzing for the presence of aflatoxin metabolites in blood, milk, and urine; moreover, excreted DNA adducts and blood protein adducts can also be monitored (Sabbioni and Sepai, 1994). The aflatoxin B_1-N^7-guanine adduct represents the most reliable urinary biomarker for aflatoxin exposure but reflects only recent exposure. Numerous studies have shown that carcinogenic potency is highly correlated with the extent of total DNA adducts formed *in vivo*. Inactivation of the p53 tumor suppressor gene may be important in the development of primary hepatocellular carcinoma. Studies of liver cancer patients in Africa and China have shown that a mutation in the p53 tumor suppressor gene at codon 249 is associated with a G-to-T transversion. Mechanistically, it is known that the reactive aflatoxin epoxide binds to the N^7 position of guanines. Moreover, aflatoxin B_1-DNA adducts can result in GC to TA transversions. The specific mutation in codon 249 of the p53 gene has been called the first example of a "carcinogen-specific" biomarker that remains fixed in the tumor tissue.

There is also considerable evidence associating aflatoxin with neoplasms in extrahepatic tissues, particularly the lungs. For example, one early epidemiological study of Dutch peanut processing workers exposed to dust contaminated with aflatoxin B_1 showed a correlation between both respiratory cancer and total cancer in the exposed group compared with unexposed cohorts. Anecdotal and circumstantial evidence of air-borne aflatoxin exposure leading to cancer has been reported. To recapitulate, there is no other natural product for which the data on human carcinogenicity are so compelling. The International Agency for Research on Cancer has classified aflatoxin B_1 as a group I carcinogen.

In developed countries, sufficient amounts of food combined with regulations that monitor aflatoxin levels in these foods protect human populations from significant aflatoxin ingestion. However, in countries where populations are facing starvation or where regulations are either not enforced or nonexistent, routine ingestion of aflatoxin may occur. Worldwide, liver cancer incidence rates are 2 to 10 times higher in developing countries than in developed countries. Unfortunately, strict limitation of aflatoxin-contaminated food is not always an option. A joint Food and Agriculture Organization/World Health Organization/United Nations Environment Programme Conference report stated that "in developing countries, where food supplies are already limited, drastic legal measure may lead to lack of food and to excessive prices

There has been considerable fundamental work on the aflatoxins, especially on their biosynthesis and molecular biology. The first stable step in the biosynthetic pathway is the production of norsolorinic acid, an anthraquinone precursor, by a type II polyketide synthase. An elaborate series of at least 15 post-polyketide synthase steps follows, yielding a series of increasingly toxigenic metabolites. Sterigmatocystin, a related dihydrofuran toxin, is a late metabolite in the aflatoxin pathway and is also produced as a final biosynthetic product by a number of species such as *Aspergillus versicolor* and *Aspergillus nidulans*. Sterigmatocystin is both mutagenic and tumorigenic but is less potent than aflatoxin Analysis of the molecular genetics of sterigmatocystin biosynthesis in the genetically tractable species *Aspergillus nidulans* has provided a useful model system. The genes for the sterigmatocystin gene cluster from *Aspergillus nidulans* have been cloned and sequenced). Cognate genes for aflatoxin pathway enzymes from *Aspergillus flavus* and *Aspergillus parasiticus* show high sequence similarity to the sterigmatocystin pathway genes. The organization of genes in the *Aspergillus flavus, Aspergillus nidulans*, and *Aspergillus parasiticus* sterigmatocystin-aflatoxin pathway has been compared (Hicks *et al*. 2002).

Aspergillus oryzae and *Aspergillus sojae*, species that are widely used in Asian food fermentations such as soy sauce, miso, and sake, are closely related to the aflatoxigenic species *Aspergillus flavus* and *Aspergillus parasiticus*. Although these food fungi have never been shown to produce aflatoxin, they contain homologues of several aflatoxin biosynthesis pathway genes. Deletions and other genetic defects have led to silencing of the aflatoxin pathway in both *Aspergillus oryzae* and *Aspergillus sojae.*

6.1.2 Citrinin : Citrinin was first isolated from *Penicillium citrinum* prior to World War II **(**Manabe,2001**)** subsequently, it was identified in over a dozen species of *Penicillium* and several species of *Aspergillus* like *Aspergillus terreus* and *Aspergillus niveus*, including certain strains of *Penicillium camemberti* (used to produce cheese) and *Aspergillus oryzae* (used to produce sake, miso, and soy sauce) More recently, citrinin has also been isolated from *Monascus ruber* and *Monascus purpureus,* industrial species used to produce red pigments.

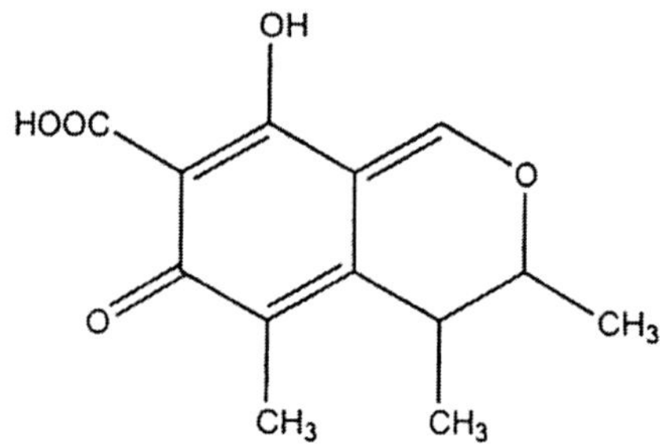

Fig. 32 : Citrinin

Citrinin has been associated with yellow rice disease in Japan. It has also been implicated as a contributor to porcine nephropathy. Citrinin acts as a nephrotoxin in all animal species tested, but its acute toxicity varies in different species. The 50% lethal dose for ducks is 57 mg/kg; for chickens it is 95 mg/kg; and for rabbits it is 134 mg/kg. Citrinin can act synergistically with ochratoxin A to depress RNA synthesis in murine kidneys).Wheat, oats, rye, corn, barley, and rice have all been reported to contain citrinin. With immunoassays, citrinin was detected in certain vegetarian foods colored with *Monascus* pigments. Citrinin has also been found in naturally fermented sausages from Italy. Although citrinin is regularly associated with human foods, its significance for human health is unknown.

Citrinin is a polyketide mycotoxin, which is a secondary metabolite of some fungi species. Its chemical name is (3R,4S)-4,6-dihydro-8-hydroxy-3,4,5-trimethyl-6-oxo-3H-2-benzopyran-7-carboxylic acid and the molecular formula is $C_{13}H_{14}O_5$ (Fig.32**)**. Citrinin has a molecular weight of 250.25 g/mol. It forms disordered yellow crystals which melt at 175 °C. Citrinin is a planar molecule which contains conjugated bonds. As a result of these conjugated bonds citrinin is autofluorescent. Citrinin crystals can hardly be dissolved in cold water, however in polar organic solvents and aqueous sodium hydroxide, sodium carbonate and sodium acetate dissolving is possible.

Citrinin decomposes at temperatures higher than 175 °C, providing that it is under dry conditions. When water is present, the decomposition temperature is around 100 °C. Several decomposition products of citrinin are known, including phenol A, citrinin H1, citrinin H2 and dicitrinin A. Citrinin H1 is produced out of two citrinin molecules and its toxicity is increased compared to the original toxicity of citrinin. Citrinin H2, a formylated derivative of phenol A, is less toxic than citrinin. Phenol A seems to be produced mainly under acidic conditions. Dicitrinin A is a dimer of citrinin molecules which is mainly formed during decomposition in a neutral environment, when a high concentration of citrinin is present. Citrinin often occurs together with other mycotoxins like ochratoxinA or aflatoxin B1 , because they are produced by the same fungi species. The combination which is observed most often is citrinin with ochratoxin A and this is also the most studied combination. The effects of co-occurrence of these mycotoxins are either additive or synergistic. The nephrotoxic effects of ochratoxin A and citrinin, for example, are increased synergistic when exposure to both takes place (Anonymous, 2012.). Next to that, the co-exposure of these compounds is expected to be involved in the pathogenese of a human kidney disease, called Balkan Endemic Nephropathy. The interaction of both substances might also influence apoptosis and necrosis in hepatocytes.

Mode of action: Various *in vitro* studies have revealed the involvement of citrinin toxicity in reduced cytokine production, inhibition of RNA and DNA synthesis, induction of oxidative stress, inhibition of nitride oxide gene expression, increase in ROS production and activation of apoptotic cell death via signal transduction pathways and the caspase-cascade system.

6.1.3 Deoxynivalenol (Vomitoxin): It is a type B trichothecene, an epoxy-sesquiterpenoid. This mycotoxins occurs predominantly in grains such as wheat, oats, barley, rye and corn and less often in rice, sorghum and triticale. The occurrence of deoxynivalenol is associated primarily with *Fusarium graminearum* (*Gibberella zeae*) and *F. culmorum* both of which are important plant pathogens which cause fusarium head blight in wheat and gibberella or ear blight in corn. A direct relationship between the incidence of fusarium head blight and contamination of wheat with deoxynivalenol has been established. The incidence of fusarium head blight is strongly associated with moisture at the time of flowering (anthesis), and the timing of rainfall, rather than the amount, is the most critical factor. However, increased amount of moisture towards harvest time has been associated with lower amount of vomitoxin in wheat grain due to leaching of toxins. Furthermore, deoxynivalenol contents are significantly affected by the susceptibility of cultivars towards *Fusarium* species, previous crop, tillage practices, and fungicide use (Beyer *et al.*, 2006).

F. graminearum grows optimally at a temperature of 25 °C and at a water activity above 0.88. *F. culmorum* grows optimally at 21 °C and at a water activity above 0.87. The geographical distribution of the two species appears to be related to temperature, *F. graminearum* being the more common species occurring in warmer climates. Deoxynivalenol has been implicated in incidents of mycotoxicoses in both humans and farm animals. Chemically it is (3a,7a)-3,7,15-trihydroxy-12, 13-epoxytrichothec-9-en-8-one (Fig. 33).

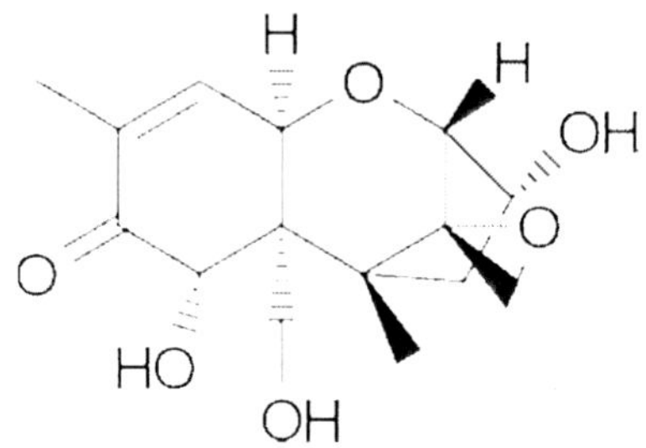

Fig. 33: Deoxynivalenol

Mode of action: Vomitoxin belongs to a class of mycotoxins which are strong inhibitors of protein synthesis, exposure to vomitoxin causes the brain to decrease its uptake of the amino acid tryptophan and, in turn, its synthesis of erotonin. Reduced levels of serotonin are believed to be responsible for the anorexic effects of DON and other trichothecenes. Irritation of the gastrointestinal tract may also play a role in reducing feed intake, and may also partially explain the high incidence of paraesophageal stomach ulcers observed in sows during feed refusal. In humans DON is extensively glucuronidated and excreted via urine (Pestka James,2010).

6.1.4 Ergot alkaloids: The ergot alkaloids are among the most fascinating of fungal metabolites. They are classified as indole alkaloids and are derived from a tetracyclic ergoline ring system. Lysergic acid, a structure common to all ergot alkaloids, was first isolated in 1934. The clavines have ergoline as a basic structure but lack peptide components; the lysergic acid alkaloids include ergotamine and lysergic acid amide (ergine) (Bennett and Bentley,1999)). The structure of ergotamine is shown in Fig. 34.

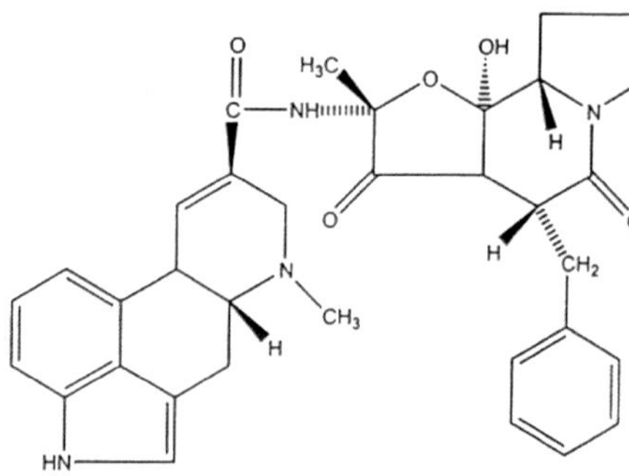

Fig. 34: Ergotamine

These compounds are produced as a toxic cocktail of alkaloids in the sclerotia of species of *Claviceps*, which are common pathogens of various grass species. The ingestion of these sclerotia, or ergots, has been associated with diseases since antiquity. An Assyrian tablet dated to 600 B.C.E., referring to a "noxious pustule in the ear of grain," is believed to be an early reference to ergot. The human disease acquired by eating cereals infected with ergot sclerotia, usually in the form of bread made from contaminated flour, is called ergotism or St. Anthony's fire. Two forms of ergotism are usually recognized, gangrenous and convulsive. The gangrenous form affects the blood supply to the extremities, while convulsive ergotism affects the central nervous system.

The most prominent member of this group is *Claviceps pururea* ("rye ergot fungus"). This fungus grows on rye and related plants, and produces alkaloids that can cause ergotism in humans and other mammals who consume grains contaminated with its fruiting structure. *Claviceps* includes about 50 known species, mostly in the tropical regions. Economically significant species include *C. purpurea* (parasitic on grasses and cereals), *C. fusiformis* (on pearl millet, buffel grass), *C. paspali* (on dallis grass), *C. africana* (on sorghum), and *C. lutea* (on paspalum) *C. purpurea* most commonly affects outcrossing species such as rye (its most common host), as well as triticale, wheat, and barley. It affects oats only rarely. *C. purpurea* has at least three races or varieties, which differ in their host specificity. G1 - and grasses of open meadows and fields; G2-grasses from moist, forest, and mountain habitats; G3 (*C. purpurea* var. *spartinae*) - salt marsh grasses (*Spartina*, *Distichlis*).(Schardl *et al.*, 2006).

The ergot sclerotium contains high concentrations (up to 2% of dry mass) of the alkaloid ergotamine, a complex molecule consisting of a tripeptide-derived cyclol-lactam ring connected via amide linkage to a lysergic acid (ergoline) moiety, and other alkaloids of the ergoline group that are biosynthesized by the fungus. Ergot alkaloids have a wide range of biological activities including effects on circulation and neurotransmission.

Ergot alkaloids are classified asderivatives of 6,8-dimethylergoline and lysergic acid derivative. There are three main classes of EA, short chain substituted amides of lysergic acid, clavine alkaloids (relatively insignificant substituted 6,8-dimethylergolines), and ergopeptines, which are peptide EA comprising (+)-lysergic acid and a tripeptide system containing L-proline. *Claviceps purpurea* produces lactam ergot alkaloids (ergopeptams) containing isoleucine as a second amino acid, which have been found to predominate in some infected wild grasses from Norway (Uhlig and Petersen, 2008). Common 5R,8R-nonpeptidic amides found in ergot include ergonovine (ergometrine, ergobasine), lysergic acid 2-hydroxyethylamide, lysergic acid amide (ergine), and paspalic acid. Lysergic Acid Diethylamide - LSD. LSD is one of the most powerful hallucinogenic drugs known (Fig. 35). It was invented in 1938 by the Swiss chemist, Albert Hoffman, who was interested in developing medicines from compounds in ergot, a fungus that attacks rye.

Human ergotism was common in Europe in the Middle Ages. For example, a three-volume work entitled *Handbook of Geographical and Historical Pathology* published in London by August Hirch between 1883 and 1886 recorded 132 epidemics of European ergotism between the 6th and 18th centuries. The "slow nervous fever" described by the 18th century English physician Jon Huxham may be another example of human ergotism. Slow nervous fever usually occurred in the summer and fall after a severe winter; Huxham suspected "bad food" as the source of the trouble. Ergot alkaloids may have had a strong influence on fertility trends in England and other European countries during the 17th and 18th centuries

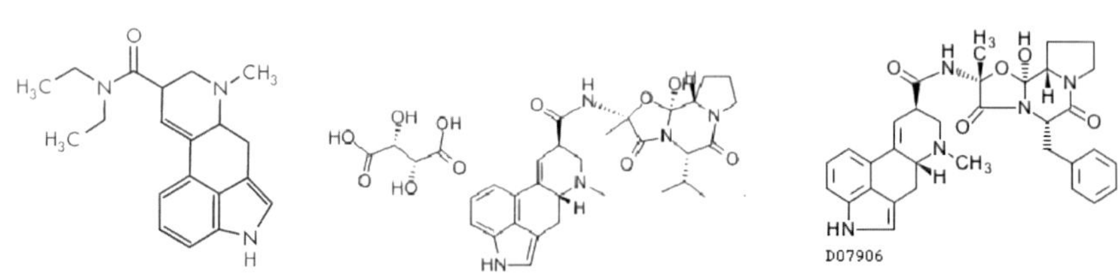

LSD **Ergovaline** **Ergotamine**

Fig. 35: LSD Ergovaline Ergotamine

Ergotism is still an important veterinarian problem. The principal animals at risk are cattle, sheep, pigs, and chickens. Clinical symptoms of ergotism in animals include gangrene, abortion, convulsions, suppression of lactation, hypersensitivity, and ataxia. Sometimes the line between toxin and drug is defined with the shift of a decimal point or a change in a small chemical moiety. The ergot alkaloids are a case in point. Their myriad actions have long engaged the interest of physicians and pharmacologists. Several ergot alkaloids induce smooth muscle contractions. For centuries it had been observed that grazing on grass infected with ergot caused abortion in pregnant farm animals, so it is not surprising that midwives and others adopted ergot as a folk medicine, using it as both an abortifacient and a drug to accelerate to uterine contractions for women in labor. During the 20th century, the famous hallucinogen lysergic acid diethylamide (LSD) was discovered as the result of research with ergot alkaloids conducted at the Sandoz Laboratories in Basel, Switzerland. A chemist named Hofmann combined different amines in peptide linkage with lysergic acid to produce ergobasine (also called ergometrine and ergonovine), the first semisynthetic ergot alkaloid. Then, by varying the amino alcohol constituent, he obtained Methergine, a compound prescribed widely for decades to control hemorrhage after childbirth. Hofmann continued to synthesize new lysergic acid derivatives; the 25th substance in his series was Δ-lysergic acid diethylamide (LSD-25).

In 1943, after accidentally ingesting some of the compound, he discovered the hallucinogenic properties of this semisynthetic derivative. For a while, Sandoz marketed LSD to psychiatrists under the trademark Delysid. It was used unsuccessfully to treat schizophrenia. In a bizarre Chapter of American history, the Central Intelligence Agency, under the code name MK- ULTRA, used LSD as a truth serum for interrogating suspected communists (Ulrich and Patten, 1991). More recently, pure ergotamine has been used for the treatment of migraine headaches. Other ergot derivatives are used as prolactin inhibitors, in the treatment of Parkinsonism, and in cases of cerebrovascular insufficiency. The therapeutic administration of ergot alkaloids may cause sporadic cases of human ergotism.

It has been hypothesized that the Salem witchcraft affair may have been a form of convulsive ergotism related to consumption of rye infected with *Claviceps* sclerotia. Although some historians dispute this hypothesis, the epidemiological and clinical data are quite provocative. Robin Cook, the author of a number of bestsellers, used the Salem-ergot hypothesis as the basis for a novel in which a young physician isolates *Claviceps* spores from a damp New England cellar, cultures the fungus, and then discovers a new alkaloid with mind- and energy-enhancing properties. Taking a cue from the Central Intelligence Agency, he names his drug Ultra and founds a biotechnology company in order to reap profits from the would-be "billion-dollar molecule." In the end it is discovered that the euphoria-inducing fictional alkaloid has horrifying side effects. Cook's novel is our personal favorite example of mycotoxins in popular culture.

Mode of action: Ergot alkaloids have structures similar to the biogenic amines norepinephrine, serotonin, and dopamine. Vasoconstriction is produced by an agonist activity and this effect varies with different vascular beds. Hyperthermia and uterine stimulation are other effects. The alkaloids are antagonistic to dopamine at D_1 vasodilatory receptors. Dopaminergic activity at D_2 receptors causes inhibition of proclatin secretion (Goldstein *et al.*, 1980). The dopamine antagonist domperidone is an effective treatment in mares suffering agalactia and uterine effects induced by ergot alkaloids in endophyte infected tall fescue grass (Walkembach *et al.*, 2005). The mechanism of action of ergotamine is complex. The molecule shares structural similarity with neurotransmitters such as serotonin, dopamine, and epinephrine and can thus bind to several receptors acting as an agonist. The anti-migraine effect is due to constriction of the intracranial extracerebral blood vessels through the 5-HT1D receptors, and by inhibiting trigeminal neurotransmission by the 5-HT1D receptors. Ergotamine also has effects on the dopamine and norepinephrine receptors.

6.1.5 Fumonisins: Fumonisins were first described and characterized in 1988. The most abundantly produced member of the family is fumonisin B_1 (Fig. 36). They are thought to be synthesized by condensation of the amino acid alanine into an acetate-derived precursor. Fumonisins are produced by a number of *Fusarium* species, notably *Fusarium verticillioides* (formerly *Fusarium moniliforme* = *Gibberella fujikuroi*), *Fusarium proliferatum,* and *Fusarium nygamai,* as well as *Alternaria alternata* f. sp. *lycopersic*. These fungi are taxonomically challenging with a complex and rapidly changing nomenclature that has perplexed many nonmycologists (and some mycologists, too) (Marasas *et al.,* 1984). The major species of economic importance is *Fusarium verticillioides*, which grows as a corn endophyte in both vegetative and reproductive tissues, often without causing disease symptoms in the plant. However, when weather conditions, insect damage, and the appropriate fungal and plant genotype are present, it can cause seedling blight, stalk rot, and ear rot. *Fusarium verticillioides* is present in virtually all corn samples. Most strains do not produce the toxin, so the presence of the fungus does not necessarily mean that fumonisin is also present. Although it is phytotoxic, fumonisin B_1 is not required for plant pathogenesis.

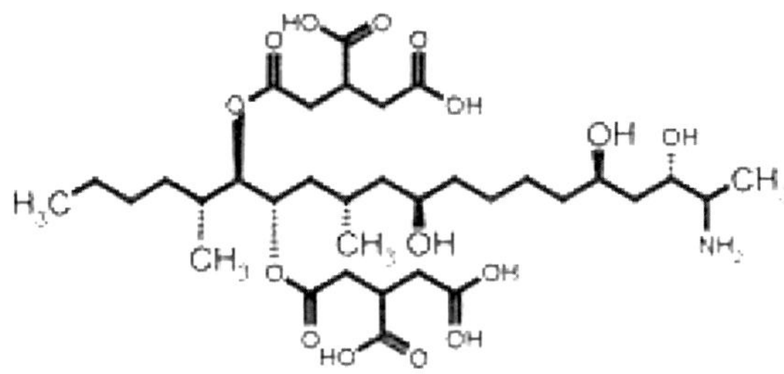

Fig. 36 : Fumonisin B_1

The maize and maize-based products are most commonly infected with fumonisins besides their presence in several other grains (rice, wheat, barley, maize, rye, oat, and millet) and grain products (tortillas, corn flask, chips)] which have major influences on health. More than 15 fumonisin homologues have been known and characterized as fumonisin A, B, C, and P. Further among fumonisin B, FB1, FB2, and FB3 are most abundant with FB1 being the most toxic form that can co-exists with other forms of fumonisin, i.e., FB2 and FB3. These (FB1, FB2, and FB3) forms are the main food contaminants. FB1 consists of a diester with propane-1,2,3-tricarboxylic acid (TCA) and 2-amino-12,16-dime thyl-3,5,10,14,15-pentahydroxyleicosane where hydroxyl (OH-) groups at the C-14 and C-15 positions involved with the carboxyl groups (-COOH) of TCA to form an ester. On the other hand, FB2 and FB3 are actually the C-5 and C-10 dehydroxy analogues of FB1 (Damiani *et al.*, 2019).

Fumonisins affect animals in different ways by interfering with sphingolipid metabolism. They cause leukoencephalomalacia (hole in the head syndrome) in equines and rabbits); pulmonary edema and hydrothorax in swine and hepatotoxic and carcinogenic effects and apoptosis in the liver of rats). In humans, there is a probable link with esophageal cancer. The occurrence of fumonisin B_1 is correlated with the occurrence of a higher incidence of esophageal cancer in regions of Transkei (South Africa), China, and northeast Italy. It has been isolated at high levels in corn meal and corn grits, including seven samples from a supermarket in Charleston, S.C., a city which has the highest incidence of esophageal cancer among African-Americans in the United States. A possible case of acute exposure to fumonisin B_1 involved 27 villages in India, where consumption of unleavened bread made from moldy sorghum or corn caused transient abdominal pain, borborygmus, and diarrhea. All those affected recovered fully. Finally, fumonisins can cause neural tube defects in experimental animals and thus may also have a role in human cases. It has been hypothesized that a cluster of anencephaly and spina bifida cases in southern Texas may have been related to fumonisins in corn products. Unlike most known mycotoxins, which are soluble in organic solvents, fumonisins are hydrophilic. This makes them difficult to study. Usually they are extracted in aqueous methanol or aqueous acetonitrile. High-performance liquid chromatography with fluorescent detection is the most widely used analytic method. The fumonisin story raises the specter that there may be many other occult but toxic products of fungal metabolism that have not yet been discovered because of their hydrophilic nature.

Mode of action: Fumonisin B_1 inhibits the enzyme ceramide synthase (sphingosine N-acyltransferase), which acylates sphingoid bases. This blocks the formation of ceramide via two pathways. It inhibits de formation via de novo sphinganine and fatty acyl-CoA and via sphingosine produced by the breakdown of ceramide by ceramidase.

6.1.6 Ochratoxin : Ochratoxin A (Fig.37) was discovered as a metabolite of *Aspergillus ochraceus* in 1965 during a large screen of fungal metabolites that was designed specifically to identify new mycotoxins It is a phenylalanine derivative resulting from the formal condensation of the amino group of L-phenylalanine with the carboxy group of (3R)-5-chloro-8-hydroxy-3-methyl-1-oxo-3,4-dihydro-1H-2-benzopyran-7-carboxylic acid (ochratoxin alpha). Shortly thereafter, it was isolated from a commercial corn sample in the United States and recognized as a potent nephrotoxin. Members of the ochratoxin family have been found as metabolites of many different species of *Aspergillus*, including *Aspergillus alliaceus*, *Aspergillus auricomus*, *Aspergillus carbonarius*, *Aspergillus glaucus*, *Aspergillus melleus*, and *Aspergillus niger*. Because *Aspergillus niger* is used widely in the production of enzymes and citric acid for human consumption, it is important to ensure that industrial strains are nonproducers. Although some early reports implicated several *Penicillium* species, it is now thought that *Penicillium verrucosum*, a common contaminant of barley, is the only confirmed ochratoxin producer in this genus. Nevertheless, many mycotoxin reviews reiterate erroneous species lists.

Ochratoxin B is a phenylalanine derivative resulting from the formal condensation of the amino group of L-phenylalanine with the carboxy group of (3R)-8-hydroxy-3-methyl-1-oxo-3,4-dihydro-1H-2-benzopyran-7-carboxylic acid. Ochratoxin C is N-[[(3R)-5-chloro-3,4-dihydro-8-hydroxy-3-methyl-1-oxo-1H-2-benzopyran-7-yl]carbonyl]-L-phenylalanine, ethyl ester. Ochratoxin TA chemically is l-phenylalanine-N-[(5-chloro-3,4-dihydro-8-hydroxy-3-methyl-1-oxo-1H-2-benzopyrane-7-yl)carbonyl]-(R)-isocoumarin. OTA is a potent nephrotoxic mycotoxin in nature and also displays other adverse effects such as hepatotoxicity, teratogenicity, and immunosuppression

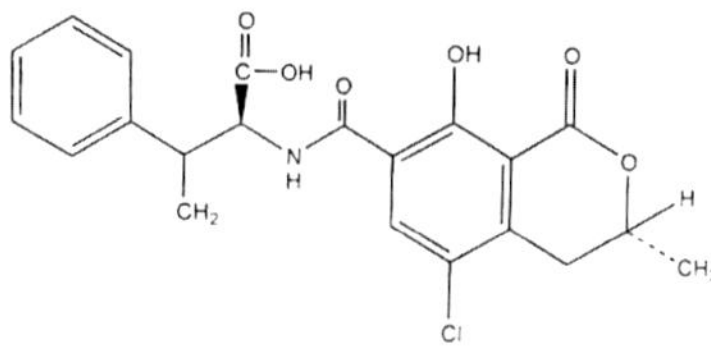

Ochratoxin A

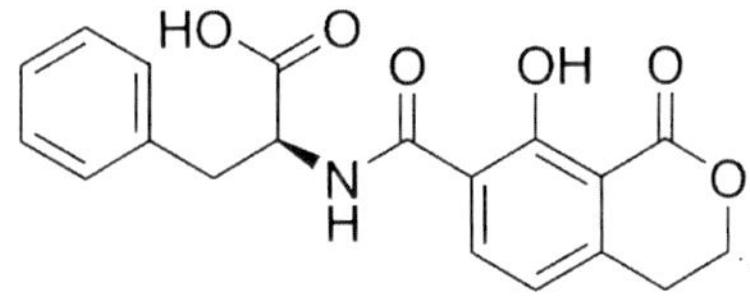

Ochratoxin B

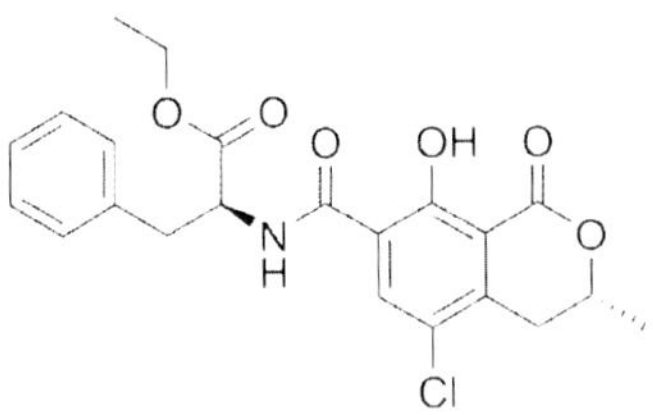

Ochratoxin C

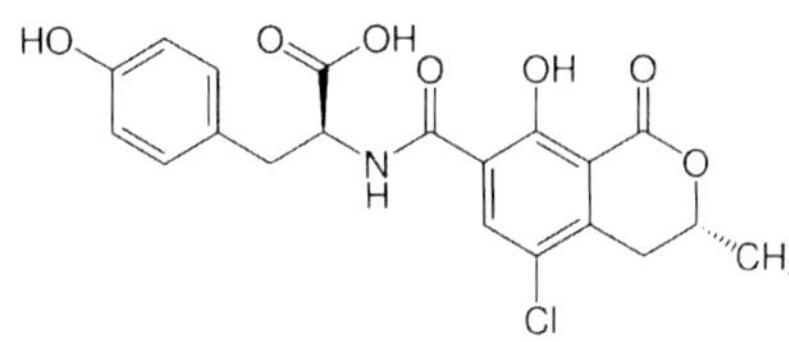

Ochratoxin TA

Fig. 37: Ochratoxin A Ochratoxin B Ochratoxin C Ochratoxin TA

As with other mycotoxins, the substrate on which the molds grow as well as the moisture level, temperature, and presence of competitive microflora interact to influence the level of toxin produced. Ochratoxin A has been found in barley, oats, rye, wheat, coffee beans, and other plant products, with barley having a particularly high likelihood of contamination. There is also concern that ochratoxin may be present in certain wines, especially those from grapes contaminated with *Aspergillus carbonarius*. Of the *Aspergillus* toxins, only ochratoxin is potentially as important as the aflatoxins. The kidney is the primary target organ. Ochratoxin A is a nephrotoxin to all animal species studied to date and is most likely toxic to humans, who have the longest half-life for its elimination of any of the species examined. In addition to being a nephrotoxin, animal studies indicate that ochratoxin A is a liver toxin, an immune suppressant, a potent teratogen, and a carcinogen. Ochratoxin A disturbs cellular physiology in multiple ways, but it seems that the primary effects are associated with the enzymes involved in phenylalanine metabolism, mostly by inhibiting the enzyme involved in the synthesis of the phenylalanine-tRNA complex. In addition, it inhibits mitochondrial ATP production and stimulates lipid peroxidation.

Ochratoxin has been detected in blood and other animal tissues and in milk, including human milk. It is frequently found in pork intended for human consumption. Ochratoxin is believed to be responsible for a porcine nephropathy that has been studied intensively in the Scandinavian countries. The disease is endemic in Denmark, where rates of porcine nephropathy and ochratoxin contamination in pig feed are highly correlated In addition, ochratoxin is associated with disease and death in poultry. There has been speculation that ochratoxins are involved in a human disease called endemic Balkan nephropathy. This condition is a progressive chronic nephritis that occurs in populations who live in areas bordering the Danube River in parts of Romania, Bulgaria, and the former Yugoslavia. In one Bulgarian study, ochratoxin contamination of food and the presence of ochratoxin in human serum were more common in families with endemic Balkan nephropathy and urinary tract tumors than in unaffected families. In addition to ochratoxin poisoning, this curious disease has been attributed to genetic factors, heavy metals, and possible occult infectious agents. The current consensus is that endemic Balkan nephropathy is of unknown etiology, but many mycotoxin reviews list it, without caveat, as an ochratoxicosis. It has also been hypothesized that the gene for phenylketonuria might occur in relatively high frequency because of a heterozygous advantage against ochratoxin poisoning and that ochratoxin might be a risk factor for testicular cancer.

Ochratoxin has been detected in human blood and serum. Analyses of urine from children in Sierra Leone detected both ochratoxin and aflatoxin throughout the year. Several detailed risk assessments have been conducted for ochratoxin A. Given the known human exposure and the abundance of toxicological data from animal studies, the European Union Scientific Committee has recommended that ochratoxin A levels be reduced to below 5 ng/kg of body weight per day. In addition, several European countries have proposed individual regulations, with maximum tolerated concentrations varying greatly from country to country. The International Agency for Research on Cancer has rated ochratoxin as a possible human carcinogen (category 2B).

It would be prudent for the medical community to pay closer attention to the possibility of ochratoxin toxicity in patients with symptoms of renal pathology. Although the role of ochratoxin A in human disease is still speculative, its acute nephrotoxicity, immunosuppressive actions, and teratogenic effects in animal models, coupled with its ability to be carried through the food chain, merit concern.

Mode of action: Ochratoxin absorbed into the body is distributed at a high concentration in the kidney. It inhibits the synthesis of proteins, DNA and RNA in the cell. It shows renal toxicity by inhibiting various enzyme activities in the kidney. Ochratoxin A has a number of toxic effects in mammals, the most notable of which is nephrotoxicity. It is also immunosuppressive, teratogenic and carcinogenic (Dirheimer and Creppy, 1991).

6.1.7 Patulin : Patulin, 4-hydroxy-4H-furo[3,2c]pyran-2(6H)-one, is produced by many different molds but was first isolated as an antimicrobial active principle during the 1940s from *Penicillium patulum* (later called *Penicillium urticae,* now *Penicillium griseofulvum*). The same metabolite was also isolated from other species and given the names clavacin, claviformin, expansin, mycoin c, and penicidin. The chemical structure is shown in Fig.38. A number of early studies were directed towards harnessing its antibiotic activity. For example, it was tested as both a nose and throat spray for treating the common cold and as an ointment for treating fungal skin infections. However, during the 1950s and 1960s, it became apparent that, in addition to its antibacterial, antiviral, and antiprotozoal activity, patulin was toxic to both plants and animals, precluding its clinical use as an antibiotic(Llewellyn, *et al*., 1998). During the 1960s, patulin was reclassified as a mycotoxin. It is an organic compound that is classified as a polyketide. It is a white powder that is soluble in acidic water as well as in organic solvents . It is lactone that is heat-stable, so it is not destroyed by pasteurization or thermal denaturation

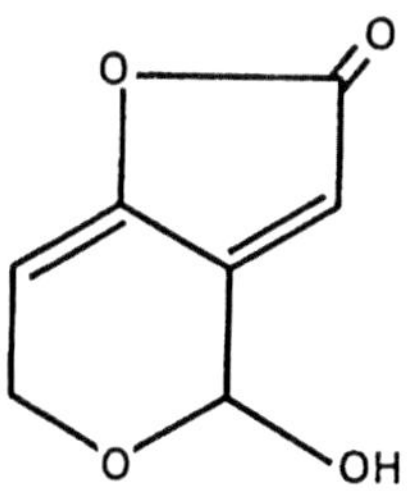

Fig. 38 : Patulin

Penicillium expansum, the blue mold that causes soft rot of apples, pears, cherries, and other fruits, is recognized as one of the most common offenders in patulin contamination. Patulin is produced by numerous *Penicillium* and *Aspergillus* species and *Byssochlamys nivea. Penicillium expansum*, which commonly occurs in rotting apples, produces patulin. Patulin is of some public health concern because of its potential carcinogenic properties, and because it has been found in commercial apple juice and other apple products. Patulin appears to be unstable in grains, cured meats, and cheese. When administered orally to rats, patulin showed no toxicity or carcinogenicity. Patulin is regularly found in unfermented apple juice, although it does not survive the fermentation into cider products. Patulin is toxic at high concentration in laboratory settings, but evidence for natural poisoning is indirect and inconclusive. Nevertheless, the Joint Food and Agriculture Organization-World Health Organization Expert Committee on Food Additives has established a provisional maximum tolerable daily intake for patulin of 0.4 mg/kg of body weight per day.

Mode of action: Patulin alters the barrier function of the intestinal epithelium: mechanism of action of the toxin and protective effects of glutathione.. Since TER reflects the organization of tight junctions, these data indicate that patulin affected the barrier function of the intestinal epithelium (Mahfoud *et al*., 2002). Patulin has also played an important role in the study of the classical biochemistry of polyketide biosynthesis. The first cell extract for a fungal polyketide synthase involved studies on a 6-methylsalicylic acid synthase from the species then called *Penicillium urticae* (now *Penicillium griseofulvum).*

6.1.8 Trichothecenes: The trichothecenes constitute a large family of more than sixty sesquiterpenoid metabolites produced by a number of fungal genera, including *Fusarium, Myrothecium, Phomopsis, Stachybotrys, Trichoderma, Trichothecium* and others. The term trichothecene is derived from trichothecin, which was the one of the first members of the family identified. All trichothecenes contain a common 12,13-epoxytrichothene skeleton and an olefinic bond with various side chain substitutions and are designated as 12,13-epoxytrichotec-9-enes because of their double bond between C-9 and C-10 and the epoxide at C-12 and C-13. The positions C-3, C-4, C-7, C-8, and C-15 of the skeleton can be oxygenated and/or esterified.. They are commonly found as food and feed contaminants, and consumption of these mycotoxins can result in alimentary hemorrhage and vomiting; direct contact causes dermatitis. The most important structural features causing the biological activities of trichothecenes are the 12,13-epoxy ring, the presence of hydroxyl or acetyl groups at appropriate positions on the trichothecene nucleus, and the structure and position of the side-chain. They are produced on many different grains like wheat, oats or maize by various *Fusarium* species such as *F. graminearum*, *F. sporotrichioides*, *F. poae* and *F. equiseti*. Some molds that produce trichothecene mycotoxins, such as *Stachybotrys chartarum*, can grow in damp indoor environments (Etzel *et al.*, 2002).. It has been found that macrocyclic trichothecenes produced by *S. chartarum* can become airborne and thus contribute to health problems among building occupants. A poisonous mushroom in Japan and China, *Podostroma cornu-damae*, contains six trichothecenes, including satratoxin H, rorodin E and verrucarin.

Trichothecenes are classified as macrocylic or nonmacrocyclic, depending on the presence of a macrocylic ester or an ester-ether bridge between C-4 and C-15. The nonmacrocylic trichothecenes in turn can be subclassified into two groups: type A, which have a hydrogen or ester type side chain at the C-8 position, and include T-2 toxin (Fig. 39), neosolaniol, and diacetoxyscirpenol, while the type B group contain a ketone and include fusarenon-x, nivalenol, and deoxynivalenol. *Fusarium* is the major genus implicated in producing the nonmacrocylic trichothecenes. Many members of this genus are significant plant pathogens; their convoluted taxonomy has already been mentioned.

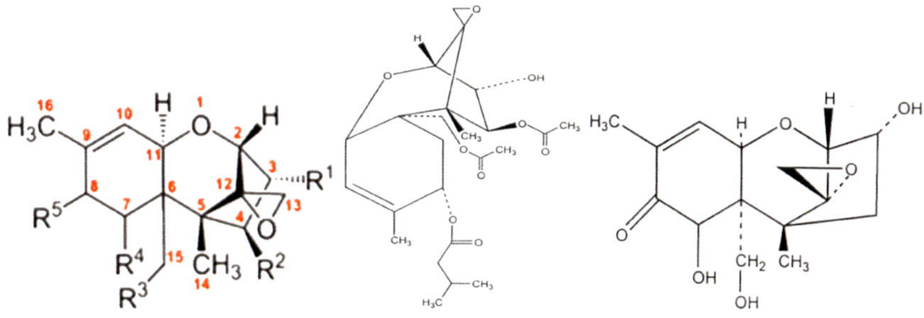

Fig. 39: Trichothecene **T-2 toxin** **Deoxynivalenol**

The trichothecenes are extremely potent inhibitors of eukaryotic protein synthesis; different trichothecenes interfere with initiation, elongation, and termination stages. Trichodermin was the first trichothecene shown to inhibit peptidyl transferase activity. Subsequently, it would appear that while all trichothecenes inhibit peptidyl transferase by binding to the same ribosome-binding site, they exert different effects which can be correlated with different functional groups (McCormick *et al.,* 2011). The 12,13-epoxide group is essential for inhibition of protein synthesis; reduction of the 9,10 double bond reduces toxicity. There is a long history of moldy grain "intoxications" in Japan, where disease in both human beings and farm animals has been attributed to *Fusarium* mycotoxicoses. *Fusarium graminearum* (teleomorph *Gibberella zeae*), regularly found on barley, oats, rye, and wheat, is considered the most important plant pathogen in Japan and is believed to be the cause of red mold disease (Akakabi toxicosis). As with all mycotoxins, depending on weather conditions, the growth of trichothecene-producing fungi and subsequent production of toxins vary considerably from year to year and from place to place. Trichothecenes are a group of over 150 chemically related mycotoxins.

Diacetoxyscirpenol, deoxynivalenol, and T-2 are the best studied of the trichothecenes produced by *Fusarium* species. Deoxynivalenol is one of the most common mycotoxins found in grains. When ingested in high doses by agricultural animals, it causes nausea, vomiting, and diarrhea; at lower doses, pigs and other farm animals exhibit weight loss and food refusal. For this reason, deoxynivalenol is sometimes called vomitoxin or food refusal factor. Although less toxic than many other major trichothecenes, it is the most prevalent and is commonly found in barley, corn, rye, safflower seeds, wheat, and mixed feeds. The symptoms produced by various trichothecenes include effects on almost every major system of the vertebrate body; many of these effects are due to secondary processes that are initiated by often poorly understood metabolic mechanisms related to the inhibition of protein synthesis. Of the naturally occurring trichothecenes, T-2 and diacetoxyscirpenol appear to be the most potent in animal studies. In addition to their cytotoxic activity, they have an immunosuppressive effect that results in decreased resistance to infectious microbes .

It has been hypothesized that T-2 and diacetoxyscirpenol are associated with a human disease called alimentary toxic aleukia. The symptoms of the disease include inflammation of the skin, vomiting, and damage to hematopoietic tissues. The acute phase is accompanied by necrosis in the oral cavity, bleeding from the nose, mouth, and vagina, and central nervous system disorders. It is possible that alimentary toxic aleukia may sometimes have been misdiagnosed as diphtheria or scurvy. The Soviet literature contains many reports of disease outbreaks associated with eating moldy grain, some dating back to the 19th century. Alimentary toxic aleukia affected a large population in the Orenburg district of the former U.S.S.R. during World War II. The sick people had eaten overwintered grain colonized with *Fusarium sporotrichioides* and *Fusarium poae*. The macrocyclic trichothecenes are produced largely by *Myrothecium, Stachybotrys*, and *Trichothecium*species. Glutinosin, a mixture of the macrocyclic trichothecenes verrucarin A and B, was originally identified as an antimicrobial agent. Recently, the trichothecenes produced by *Stachybotrys atra*(*Stachybotrys chartarum*) have received the most attention. They include satratoxins (Fig. 40), roridins, verrucarins, and atranones). The macrocyclic trichothecenes are a group of potent protein synthesis inhibitors that have been encountered in indoor air and food as a result of infestation by the fungus *Stachybotrys* (Mi-Gyung Lee *et al.*, 1999).

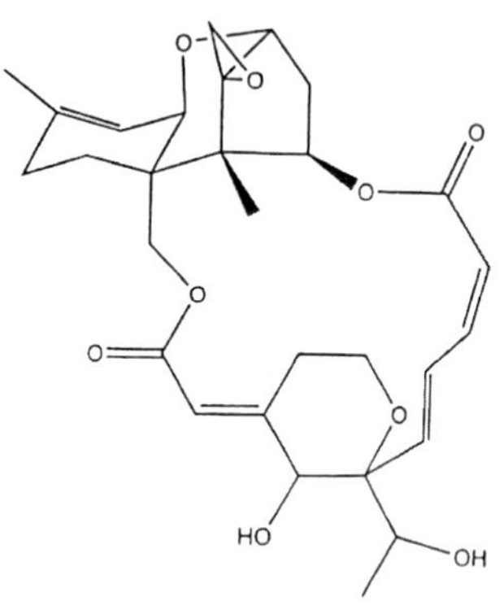

Fig. 40: Satratoxin H

Stachybotryotoxicosis was first described as an equine disease of high mortality associated with moldy straw and hay. Until recently, human stachybotryotoxicosis was considered a rare occupational disease limited largely to farm workers who handle moldy hay. However, it has become apparent that *Stachybotrys* grows well on all sorts of wet building materials with high cellulose content, for example, water-damaged gypsum board, ceiling tiles, wood fiber boards, and even dust-lined air conditioning ducts. The presence of *Stachybotrys* has been associated with pulmonary bleeding in infants. Although toxic mold was found in the homes of the children with pulmonary bleeding, a cause-and-effect relationship has been difficult to prove). Nevertheless, this incident and a few other highly publicized mold cases seem to have precipitated a series of multimillion-dollar law suits against contractors, real estate developers, and insurance companies as well as a great deal of public apprehension about "toxic molds." See the section on sick building syndrome below for more on this aspect of mycotoxicology.

It was suggested that the 10th plague visited on Egypt was a *Stachybotrys* infestation, but since the plague infected grain, not hay, it is more likely that *Aspergillus* or *Penicillium* was the culprit. Many of the genes involved in trichothecene biosynthesis in *Fusarium sporotrichioides* and in *Myrothecium roridum* have been cloned and sequenced and there is also considerable progress in understanding the regulation of the trichothecene gene cluster. Homologues of these trichothecene pathway genes have also been reported for *Fusarium graminearum*.

Mode of action: The toxicity of tricothecenes is primarily the result of their widely cited action as protein synthesis inhibitors. This inhibition occurs at ribosomes during all three stages of protein synthesis: initiation, elongation, and termination. During intitiation, trichothecenes can either inhibit the association of the two ribosomal subunits, or inhibit the function of the mature ribosome by preventing the association of the first tRNA with the start codon (Kiessling,1986). Inhibition at elongation most likely occurs due to trichothecenes preventing the function peptidyl transferase the enzyme which catalyzes the formation of new peptide bonds on the 60s ribosomal subunit. Inhibition during termination can also be the result of peptidyl transferase inhibition or the ability of trichothecenes to prevent the hydrolysis required at this final step.It is interesting to note that the substitution pattern of the ring core of trichothecenes influences the toxin's action as either an inhibitor of initiation or as an inhibitor of elongation/termination. Trichothecenes also have the ability to affect general cellular enzyme function due to the tendency of active site thiol groups to attack the 12,13 carbon epoxide ring. These inhibitory effects are seen most dramatically in actively proliferating cells such as in the gastrointestinal tract or the bone marrow.

6.1.9 Zearalenone: Zearalenone (6-[10-hydroxy-6- oxo-*trans*-1-undecenyl]-B-resorcyclic acid lactone), a secondary metabolite from *Fusarium graminearum* (teleomorph *Gibberella zeae*) was given the trivial name zearalenone as a combination of *G. zeae*, resorcylic acid lactone, -ene (for the presence of the C-1′ to C-2 double bond), and -one, for the C-6′ ketone. Chemically it is (3*S*,11*E*)-14,16-Dihydroxy-3-methyl-3,4,5,6,9,10-hexahydro-1*H*-2-benzoxacyclotetradecine-1,7(8*H*)-dione (Kuiper-Goodman, *et al.*, 1987). Almost simultaneously, a second group isolated, crystallized, and studied the metabolic properties of the same compound and named it F-2. Much of the early literature uses zearalenone and F-2 as synonyms; the family of analogues are known as zearalenones and F-2 toxins, respectively. The structure of zearalenone is given in Fig. 41. Perhaps because the original work on these fungal macrolides coincided with the discovery of aflatoxins, Chapters on zearalenone have become a regular fixture in monographs on mycotoxins Nevertheless, the word toxin is almost certainly a misnomer because zearalenone, while biologically potent, is hardly toxic; rather, it sufficiently resembles 17β-estradiol, the principal hormone produced by the human ovary, to allow it to bind to estrogen receptors in mammalian target cells Zearalenone is better classified as a nonsteroidal estrogen or mycoestrogen. Sometimes it is called a phytoestrogen. Zearalenone is the primary toxin, causing infertility, abortion or other breeding problems, especially in swine. Zearalenone is heat-stable and is found worldwide in a number of cereal crops, such as maize, barley, oats, rice, wheat and sorghum. In addition to its actions on the classical estrogen receptors , zearalenone has been found to act as an agonist of the GPER (GPR30) (Prossnitz, *et al.*, 2014).

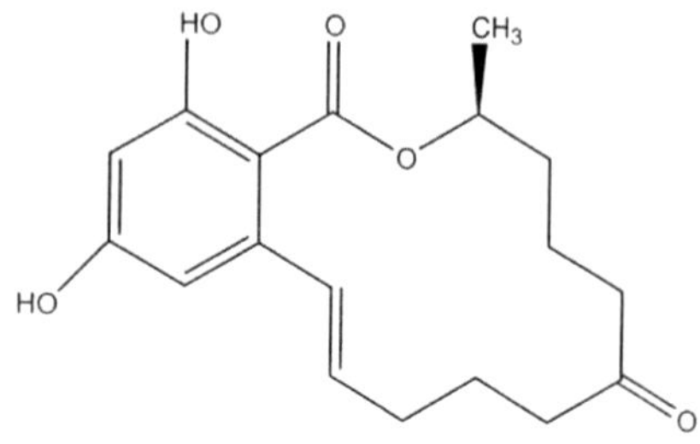

Fig. 41: Zearalenone

The zearalenones are biosynthesized through a polyketide pathway by *Fusarium graminearum*, *Fusarium culmorum*, *Fusarium equiseti*, and *Fusarium crookwellense*. All these species are regular contaminants of cereal crops worldwide. An association between moldy grain consumption and hyperestrogenism in swine has been observed since the 1920s; modern work shows that dietary concentrations of zearalenone as low as 1.0 ppm may lead to hyperestrogenic syndromes in pigs; higher concentrations can lead to disrupted conception, abortion, and other problems. Reproductive problems have also been observed in cattle and sheep.

The reduced form of zearalenone, zearalenol, has increased estrogenic activity. A synthetic commercial formulation called zeranol (Ralgro) has been marketed successfully for use as an anabolic agent for both sheep and cattle. In 1989, zeranol was banned by the European Union, but it is still used in other parts of the world. Zearalenone has also been used to treat postmenopausal symptoms in women (Utian,1973.), and both zearelanol and zearalenone have been patented as oral contraceptives. It has been claimed that the high frequency of early menarche in Puerto Rico might be due to zearalenone and related compounds in the human diet; however, studies by the Food and Drug Administration do not support this.

Recently, endocrine (hormone) disrupters have received a lot of public attention and are widely believed to reduce male fertility in human and wildlife populations but it is not clear how much the zearalenones contribute to the total environmental load of xenoestrogens. Sometimes, hormone disrupters are labeled environmental toxicants, further muddying the distinction between a compound that can cause death (toxin) and a compound that has other pharmacological activities. The zearalenone family of metabolites illustrates some of the limitations of scientific language. The biological potency of these compounds is high, but the actual toxicity is low. The 50% lethal dose in female rats is greater than 10,000 mg/kg; in female guinea pigs it is 5,000 mg/kg, while as little as one µg/kg may create a detectable uterogenic response in female swine. Thus, mycoestrogen is a more appropriate rubric than mycotoxin. Moreover, like the ergot alkaloids, in certain formulations some analogues of these macrolides can be called drugs.

Mode of action: Zearalenone and its analogs could albeit poorly bind to the rat uterine and brain ERs *in vivo* and *in vitro*, and their binding ability was of following order: α-zearalanol (α- ZAL)>α-zearalenol (α-ZEL)>β-ZAL>ZEN>β-ZEL. RNA polymerase activities in the uterine nuclei isolated from ZEN-treated rats were enhanced 4-fold of the vehicle control. Furthermore, a 52-kDa protein was specifically induced immediately after the addition of ZEN and its derivatives such as α-ZEL and α-ZAL to the immature rat uteri *in vivo* and *in vitro*. On the other hand, the induction of 52-kDa protein was hardly detected in the presence of RNA polymerase inhibitors including α-amanitin and actinomycin D, suggesting that ZEN and its analogs-ER complexes might act as a transcriptional activating factor. These findings clearly indicate that ZEN and its analogs, despite their non-steroidal structure, exhibit an estrogenic activity toward target tissues in a similar manner to that of natural estrogens. In addition, the binding ability of ZEN to the rat brain ER could assume the possibility that ZEN affects the estrogen feedback system in the rat brain to result in the shrinkage of ovary and testis and participates in behavioral abnormalities (Fumio Tashiro, 2000).

6.1.10. Cyclopiazonic acid : Cyclopiazonic acid (CPA) is a toxic fungal secondary metabolite (Fig. 42). Chemically, CPA is related to ergoline alkaloids. CPA was originally isolated from *Penicillium cyclopium* and subsequently from other fungi including *P. griseofulvum, P. camemberti, P. commune, A.flavus,* and *A. vericolor*. It is an indole-tetramic acid mycotoxin produced by the

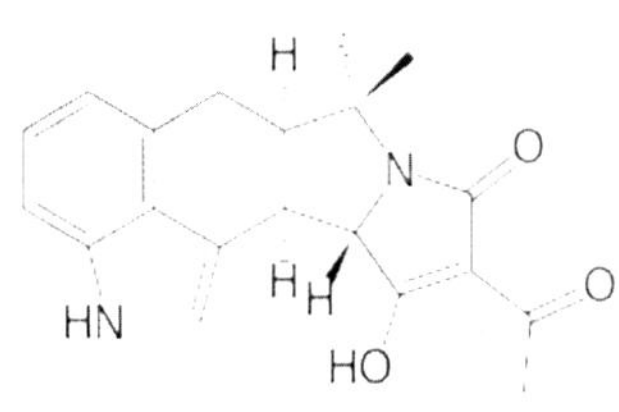

Fig. 42: Cyclopiazonic acid

ubiquitous genera of molds *Aspergillus* and *Penicillium.* Beside colonizing various grains and seeds, these molds can grow on many food substrates, such as cheese and meat products. Therefore, CPA can contaminate a number of agricultural commodities, animal feeds, and food sources (Abbas *et al.*, 2008).. This toxin has been found in edible tissue in poultry, milk, and eggs presumptively due to animals' consumption of contaminated feeds. Despite the wide presence of CPA, few incidents of animal mycotoxicosis and no confirmed incident of human poisoning have been attributed to CPA. Despite its early discovery, the benign nature of CPA has rendered it to receive much less attention of the mycotoxin research community than its counterparts such as aflatoxins, trichothecenes, fumonisins and ochratoxins in the past two decades.

CPA only appears to be toxic in high concentrations. Ingestion of CPA causes one to be an anorexic, to be dehydrated, to lose weight, to be immobile, and has some sign of spasm when near death. They can be found in molds, corns, peanuts, and other fermented products, such as cheese and sausages. Biologically, CPA is a specific inhibitor of SERCA ATPase in intracellular Ca^{2+} storage sites.CPA inhibits SERCA ATPase by keeping it in one specific confirmation, thus, preventing it from forming another. CPA also binds to SERCA ATPase at the same site as another inhibitor, thapsigargin (TG) (Ma *et al.*, 1999). In this way, CPA lowers the ability of SERCA ATPase to bind an ATP molecule. Its chemical name is (6a*R*,11a*S*,11b*R*)-10-Acetyl-11-hydroxy-7,7-dimethyl-2,6,6a,7,11a,11b-hexahydro-9H-pyrrolo[1',2':2,3]isoindolo[4,5,6-cd]indol-9-one.

Mode of action: It is a highly specific inhibitor of calcium ATPase of sarcoplasmic reticulum, altering calcium homeostasis and ATP-dependent calcium transport and resulting in the release of intracellular stored Ca2+ without increasing IP3. The sarcoplasmic reticulum Ca2-ATPase is essential for calcium reuptake in the muscle contraction-relaxation cycle. Karine Moncoq *et al.*, (2007) presented structures of a calcium-free state with bound cyclopiazonic acid (CPA) and magnesium fluoride at 2.65 A° resolution and a calcium-free state with bound CPA and ADP at 3.4 A° resolution. In both structures, CPA occupies the calcium access channel delimited by transmembrane segments M1–M4. Inhibition of Ca2-ATPase is stabilized by a polar pocket that surrounds the tetramic acid of CPA and a hydrophobic platform that cradles the inhibitor. The calcium pump residues involved include Gln56, Leu61, Val62, and Asn101. We conclude that CPA inhibits the calcium pump by blocking the calcium access channel and immobilizing a subset of transmembrane helices. In the E2(CPA) structure, ADP is bound in a distinct orientation within the nucleotide binding pocket. The adenine ring is sandwiched between Arg489 of the nucleotide-binding domain and Arg678 of the phosphorylation domain. This mode of binding conforms to an adenine recognition motif commonly found in ATP-dependent proteins.

6.1.11. Penitrem : Penitrem is an indole-diterpenoid mycotoxins produced by certain species of *Aspergillus, Claviceps* and *Penicillium,* which can be found growing on various plant species such as rye grass. Penitrem A is one of many secondary metabolites following the synthesis of paxilline in *P. crostosum*. Penitrem A poisoning in humans and animals usually occurs through the consumption of contaminated foods by mycotoxin-producing species, which is then distributed through the body by the bloodstream. It bypasses the blood-brain barrier to exert its toxicological effects on the central nervous system. In humans, penitrem A poisoning has been associated with severe tremors, hyperthermia, nausea / vomiting, diplopia, and bloody diarrhea (Berntsen *et al.,* 2007). In animals, symptoms of penitrem A poisoning has been associated with symptoms ranging from tremors, seizures, and hyperthermia to ataxia and nystagmus Chemically it is (2R,3S,3aS,4aR,4bS,6aR,7S,8S,9aR,14bS,14cR,16aS)-12-Chloro-2-isopropenyl-14b,14c,17,17-tetramethyl-10-methylene-3,3a,6,6a,7,8,9,9,10,11,14,14b,14c,15,16,16a-hexadecahydro-2H,4bH-7,8-(epoxymethano)cyclobuta[5,6]benzo[1,2-e]oxireno[4',4a']chromeno[5',6':6,7]indeno[1,2-b]indole-3,4b,7d(5H)-triol (Fig. 42).

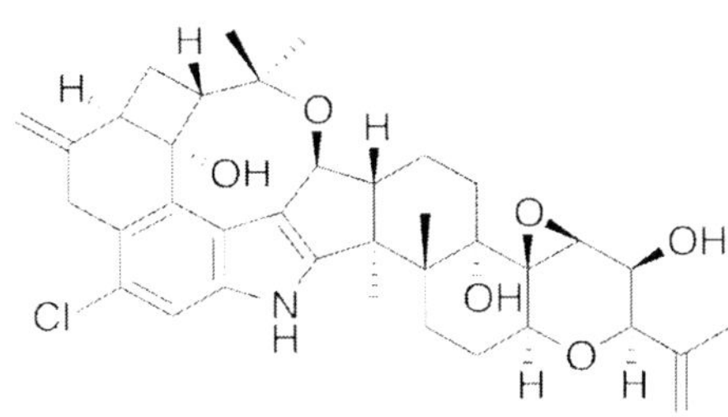

Fig. 42: Penitrem

Mode of action: Penitrem A impairs GABAergic amino acid neurotransmission and antagonizes high-conductance Ca^{2+}-activated potassium channels in both humans and animals. (Moldes-Anaya, *et al.*, 2011). Impairment of the GABAergic amino acid neurotransmission comes with the spontaneous release of the excitatory amino acids glutamate and aspirate as well as the inhibitory neurotransmitter y-aminobutyric acid (GABA). The sudden release of these neurotransmitters results in imbalanced GABAergic signalling, which gives rise to neurological disorders such as the tremors associated with penitrem A poisoning. Penitrem A also induces the production of reactive oxygen species (ROS) in the neutrophil granulocytes of humans and animals. Increased ROS production results in tissue damage in the brain and other afflicted organs as well as hemorrhages in acute poisonings.

6.1.12. T-2 toxin : T-2 Mycotoxin is a trichothecene mycotoxin. It is a naturally occurring mold byproduct of fungus which is toxic to humans and animals. The T-2 mycotoxin consiss of over 40 different toxic compounds produced by fungi of the genera *Stachybotrys* and *Fusarium. The* fungus responsible for producing the T2 toxin is commonly found in grain. Farmers may become exposed via contact with contaminated hay and grain dust.The clinical condition it causes is *alimentary toxic aaleukia* and a host of symptoms related to organs as diverse as the skin, airway, and stomach. Ingestion may come from consumption of moldy whole grains. T-2 can be absorbed through human skin. Although no significant systemic effects are expected after dermal contact in normal agricultural or residential environments, local skin effects cannot be excluded. Hence, skin contact with T-2 should be limited (Boonen, *et al.*, 2012). Although it is still controversial, it is suspected that T-2 mycotoxin has been used as a chemical warfare agent from the 1970s till the 1990s. Based on the descriptions of eyewitnesses and victims, T-2 mycotoxin was mostly delivered by low-flying aircraft that released a yellow oily liquid. Hence, this phenomenon is also named “yellow rain”.

This compound has a tetracyclic sesquiterpenoid 12,13-epoxytrichothene ring system, which relates it to the trichothecenes ((2α,3α,4β,8α)-4,15-bis(acetyloxy)-3-hydroxy-12,13-epoxytrichothec-9-en-8-yl 3-methylbutanoate) (Fig. 44). These compounds are generally very stable and are not degraded during storage/milling and cooking/processing of food. They do not degrade at high temperatures either. This compound has an epoxide ring, and several acetyl and hydroxyl groups on its side chains. These features are mainly responsible for the biological activity of the compound and make it highly toxic. T-2 mycotoxin is able to inhibit DNA and RNA synthesis *in vitro* and *in vivo* and can induce apoptosis. However, *in vivo* the compound rapidly metabolizes to HT-2 mycotoxin.

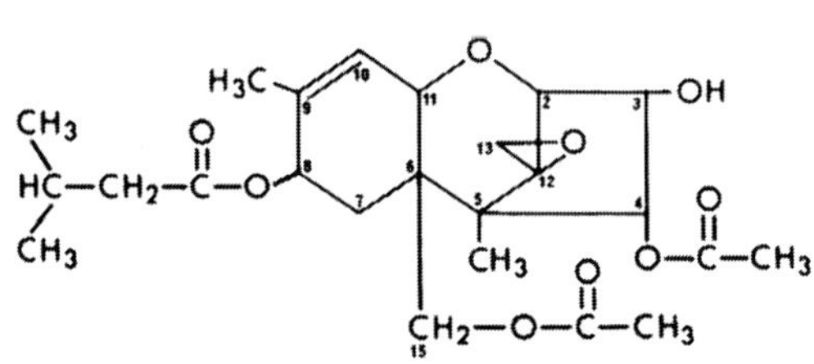

Fig. 44: T-2 toxin

Mode of action: The toxicity of T-2 toxin is due to its 12,13-epoxy ring. Epoxides are in general toxic compounds; these react with nucleophilcs and then undergo further enzymatic reactions. The reactivity of epoxides can lead to reactions with endogenous compounds and cellular constituents like DNA bases and proteins (Li *et al.*, 2011) These reactions could be the reason for the noticed actions and effects of T-2 mycotoxin. The toxic compound influences the metabolism of membrane phospholipids, leads to an increase of liver lipid peroxidases and has an inhibiting effect on DNA and RNA synthesis. In addition it can bind to an integral part of the 60s ribosomal subunit, peptidyltransferase, thereby inhibiting protein synthesis. These effects are thought to be the explanation for T-2 toxin inducing apoptosis (cell death) in different tissues as the immune system, the gastrointestinal tissue and also fetal tissue. With regard to apoptosis there has been noticed that the level of the pro-apoptotic factor Bas (Bcl-2-associated X protein) was increased and the level of Bcl-xl, an anti-apoptotic factor, was decreased in human chrondocytes when exposed to T-2 mycotoxin. Furthermore, the level of Fas, an apoptosis-related cell-surface antigen and p53, a protein regulating the cell cycle, were increased.

The identification of detoxification agents against T-2 mycotoxicosis is more important, and its side effects need to be assessed. Future potential applications of the detoxification of mycotoxin by microorganisms, enzymes and genes could prove to be beneficial. These detoxification enzymes should have pronounced capabilities to eliminate mycotoxins from human and animal systems and from foods and feedstocks. Beneficial detoxification genes used could be cloned and expressed in microorganisms to develop recombinant microorganisms that are biologically and ethically appropriate for industrial-scale enzyme production and purification (Manish Adhikari *et al.,* 2017). The development of host resistance strategies and characterization studies can also be employed to realize efficient methods which target T-2 toxin decontamination. RNAi gene silencing and gene mapping can prove to play a pivotal role in building strategies to increase the contribution of selected resistance-associated proteins in seed/crop resistance to T-2 contamination.

Plausible therapeutic uses *:* T-2 mycotoxin and their derivatives have been attributed a diverse range of abilities, including their application as growth promoters, antibiotics, and a range of other drugs. The trichothecene family has been credited with numerous biological properties consisting of antiviral abilities (chiefly as Herpes replication inhibitors), immunotoxic activities, antileukemic and antimalarial capabilities. This vast amount of the aggregated understanding of mycotoxins has opened a new era of applications utilizing an amalgamation of toxin dexterity factors with progress made with scientific techniques in various fields such as immunology, biotechnology, molecular biology, cell biology and nanotechnology in order to develop target-specific strategies that can adapt a fatal toxin into a potential therapeutic agent.

Fungal secondary metabolites such as the T-2 toxin have had severe adverse effects and continue to poison farm animals worldwide. The T-2 toxin and its metabolites exist in various countries but are mainly found in tropical and subtropical regions, such as South Korea. These toxins contaminate animal feedstuff and their production is aided by the grain moisture content and weather conditions in the affected areas. The symptoms of toxicity are diverse, affecting the GI tract mucosa and digestion process and causing skin blistering, edema, irritation, necrosis and apoptosis. These all present bio-related threats to humans and can induce oxidative stress, causing DNA damage, inhibiting protein synthesis, and damaging lipids (Manish Adhikari *et al.,* 2017).These lethal properties of the toxin support its candidature as a fatal biological warfare agent. Despite various guidelines and regulations established regarding its usage and detection strategies pertaining to maximum permissible limits in feed and food stocks, its presence can prove to be toxic. Presently, T-2 toxin treatments of induced damage emphasize mainly the use of natural substances, probiotics, and amino acids, and the quest for a precise antidote against the toxin continues to date. Therefore, stringent regulations must be established and quarantine activities need to be undertaken in order to prevent planned/unplanned exposure on a large scale.

6.1.13 Other mycotoxins and purported mycotoxicoses: *Penicillium roqueforti* and *Penicillium camemberti* (=*Penicillium caseicola*), used to manufacture mold-ripened cheeses, produce a number of toxic metabolites, including penicillin acid, roquefortine, isoflumigaclavines A and B and PR toxin. Several mycotoxins induce tremors as a neurological response in farm animals , most of these fungal tremorgens contain a modified indole moiety and are produced by certain species of *Aspergillus*, *Penicillium*, and *Claviceps*. The tremorgenic mycotoxins include the penitrems, janthitrems, lolitrems, aflatrem, paxilline, paspaline, paspalicine, paspalinine, and paspalitrem A and B. *Penicillium crustosum* produces penitrem A, a compound implicated in several cases of canine intoxication and one case of human tremor, vomiting, and bloody diarrhea.

Penitrem A: Penitrem A (tremortin) is an indole-diterpenoid mycotoxins produced by certain species of *Aspergillus*, *Claviceps* and *Penicillium*, which can be found growing on various plant species such as rye grass. Penitrem A is one of many secondary metabolites following the synthesis of paxilline in *Penicullium crostosum*. Penitrem A poisoning in humans and animals usually occurs through the consumption of contaminated foods by mycotoxin-producing species, which is then distributed through the body by the bloodstream. It bypasses the blood-brain barrier to exert its toxicological effects on the central nervous system. In humans, penitrem A poisoning has been associated with severe tremors

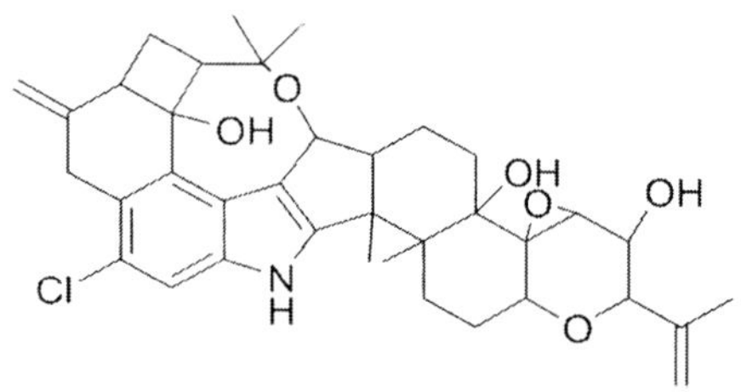

Fig. 45: Penitrem A

hyperthermia, nausea / vomiting, diplopia and bloody diarrhea. In animals, symptoms of penitrem A poisoning has been associated with symptoms ranging from tremors, seizures and hyperthermia to ataxia and nystagmus. Chemically it is $C_{37}H_{44}ClNO_6$ (Fig. 45).

Mode of action: Penitrem A impairs GABAergic amino acid neurotransmission and antagonizes high-conductance Ca^{2+} activated potassium channels in both humans and animals (Moldes-Anaya *et al.,* 2001). Impairment of the GABAergic amino acid neurotransmission comes with the spontaneous release of the excitatory amino acids glutamate and aspartate as well as the inhibitory neurotransmitter Gama aminobutyric acid (GABA). The sudden release of these neurotransmitters results in imbalanced GABAergic signalling, which gives rise to neurological disorders such as the tremors associated with penitrem A poisoning. Penitrem A also induces the production of reactive oxygen species (ROS) in the neutrophil granulocytes of humans and animals. Increased ROS production results in tissue damage in the brain and other afflicted organs as well as hemorrhages in acute poisonings.

6.2 Mycoses and Mycotoxicoses : Mycosis is tissue invasion by a fungi and mycotoxicosis is toxin production by a fungi growing on foodstuffs which is then ingested causing disease. Mycotoxicoses are acute and chronic toxic diseases caused by mycotoxins. Their clinical features, target organs, and outcome depend on the intrinsic toxic features of the mycotoxin and the quantity and length of exposure, as well as the health status of the exposed person. Historical outbreaks of mycotoxicoses occurred frequently in mild zones and today they are present mainly in tropical zones. International food trade from tropical countries increases the risk of mycotoxicoses all around the world. Although it is not realistic that medical doctors recognize mycotoxicosis in a single person, they should at least have an idea of the clinical symptoms of the most common mycotoxicoses.

The fungi that cause mycoses can be divided into two categories.

- **Primary pathogens** (e.g., *Coccidioides immitis* and *Histoplasma capsulatum*). Primary pathogens affect otherwise healthy individuals with normal immune systems.
- **Opportunistic pathogens** (e.g., *Aspergillus fumigatus* and *Candida albicans*). Opportunistic pathogens produce illness by taking advantage of debilitated or immunocompromised hosts. The majority of human mycoses are caused by opportunistic fungi.

The mechanisms of pathogenesis of both primary and opportunistic fungi are complex, and medical mycologists have devoted considerable research energy trying to identify the factors that distinguish fungal pathogens from other species. Some infections remain localized, while others progress to systemic infection. For many mycoses, the ordinary portal of entry is through the pulmonary tract, but direct inoculation through skin contact is not uncommon.

In contrast to mycoses, mycotoxicoses are examples of "poisoning by natural means" and thus are analogous to the pathologies caused by exposure to pesticides or heavy metal residues. The symptoms of a mycotoxicosis depend on the type of mycotoxin; the amount and duration of the exposure; the age, health, and sex of the exposed individual; and many poorly understood synergistic effects involving genetics, dietary status, and interactions with other toxic insults. Thus, the severity of mycotoxin poisoning can be compounded by factors such as vitamin deficiency, caloric deprivation, alcohol abuse, and infectious disease status. In turn, mycotoxicoses can heighten vulnerability to microbial diseases, worsen the effects of malnutrition, and interact synergistically with other toxins.

Effects on animals : There are six major types of mycotoxins that affect animals and their effects on animals is summarized below (Binder Eva , 2007)

Mycotoxin	Fungi	Effect on animals
Aflatoxin	*Aspergillus flavus*, *A. parasiticus*	Liver disease, carcinogenic and teratogenic effects
Trichothecenes	*Fusarium graminearum*, *F. sporotrichioides*	Immunologic effects, hematological changes, digestive disorders, edema
Zearalenone	*Fusarium graminearum*	Estrogenic effects, atrophy of ovaries and testicles, abortion
Ochratoxin	*Aspergillus ochraceus* *Penicillium verrucosum*	Nephrotoxicity, mild liver damage, immune suppression
Ergot alkaloid	*Claviceps purpurea*, *C. paspaspali*	Nervous or gangrenous syndromes
Fumonisin	*Fusarium verticillioides*, *F.proliferatum*	Pulmonary edema leukoencephalomalacia, nephrotoxicity, hepatotoxicity

Effect on feed production: The most common mycotoxin is aflatoxin. It can be very carcinogenic to both humans and animals. It is produced by two species of *Aspergillus*, *A. flavus* and *A. parasiticus*, which are known to affect plants including cereal grains, figs, nuts, and tobacco (Binder Eva, 2007). Cereal grains are one of the main ingredient in animal feed. The animals most at risk of having serious problems with aflatoxins are trout, ducklings, and pigs, while cattle are less at risk. Another animal feed product is genetically altered grass and animals including cattle, sheep, and horses eat tons of it. Ergot alkaloids are associated with grasses that are produced in a structure of *Claviceps* called the sclerotia. Some of the conditions that result from ergot ingestion in animals include gangrene, abortion, convulsions, hypersensitivity, and ataxia.

Fumonisins were the most recent mycotoxin found to affect humans and animals negatively. The most produced toxin for this group of fungi is fumonisin B1. Studies have shown that it can cause diseases such as equine leukoencephalomalacia in horses, hydrothorax and porcine pulmonary edema in swine, and it can negatively affect the immune system.

Turkey X disease: This disease was the turning point for the use of the term mycotoxin. In the 1960s, about 100,000 turkey poults died near London, England due to peanut meal that was contaminated by Mycotoxins produced by *Aspergillus flavus*. Studies showed that the age group that was most affected was turkeys from two to twenty weeks old. Some of the first signs of Turkey X were neurological symptoms and coma, which would result in death.

6.2.1 Mycotoxins and human health: The chemical nature of most of the mycotoxins makes them highly liposoluble compounds that can be absorbed from the site of exposure such as from the gastrointestinal and respiratory tract to the blood stream where it can be dissimilated throughout the body and reach different organs such as the liver and kidneys. Mycotoxins have a strong tendency and ability to penetrate the human and animal cells and reach the cellular genome where it causes a major mutagenic change in the nucleotide sequence which leads to strong and permanent defects in the genome. This defect will eventually be transcribed, translated and lead to the development of cancer (Ahmed Adam *et al.*, 2017). The chemical and physical nature of mycotoxins, the action of mycotoxins on the cellular genome and its effect on humans, mycotoxins and their carcinogenicity and mycotoxins research gaps were discussed and new research areas were suggested. The research review posed various questions, like what are the different mycotoxins that can cause cancer, what is the role of mycotoxins in causing cancer and what types of cancers can be caused by mycotoxins? These questions have been selected due to the significant increase in the mycotoxin contamination and the cancer incidence rate in the contemporary world. By revealing and understanding the role of mycotoxins in developing cancer, measures to reduce the risks and incidents of cancer could be taken.

The most important mycotoxins are Citrine, Aflatoxin B1, Ergot, fumonisins akaloids, Patulin, Trichothecene, Zearalenone and Ochratoxin A and the reason behind their importance is the fact that they are involved in many human and animal diseases which sometimes leads to death. The different mycotoxins that can cause cancer are Aflatoxin B1, Citrinin, Fumonisins, Patulin, Trichothecene and Zearalenone. The role of mycotoxins in causing cancer will vary with the type of mycotoxin and it will depend on the mode of action within the biological system.

Aflatoxin B1 : It can penetrate the cell membrane and attach to its DNA where it causes irreversible mutations. As soon as aflatoxin B1 enters the cells, they are metabolized by cytochrome P450, to aflatoxin-8, 9-epoxide which is highly reactive and unstable and requires the binding to a DNA or to the protein molecule in order to become more stable and it will form aflatoxin-N7-guanine that cause GC to TA transversion mutations which will directly affect the cell cycle by affecting the P53 gene which encodes for tumor suppressor protein that inhibits the development of cancer.

Ochratoxin A: It can cause an alteration of gene expression that results in affecting calcium homeostasis which will disrupt the different pathways regulated by the HNF4α and Nrf2 in the kidney, influence the expression of genes that are involved in DNA synthesis and repair, in response to DNA damage and it has a very small influence on the genes that are responsible for apoptosis. Ochratoxin A can cause genotoxicity by forming a DNA adduct and single cell death. Furthermore, Ochratoxin A can increase phosphorylation of a typical-PKC which in turn will lead to a downstream activation of ERK1 and ERK 2 and their substrate ELK1, ELK2 and p90RSK where PKC and the MEK-ERK MAP-kinase pathways activated by Ochratoxin A are involved in cell proliferation, cell survival, anti-apoptotic activity and most importantly renal cancer development. Ochratoxin A is suggested to have a renal carcinogenesis effect. Its mode of action is by means of direct genotoxicity through covalent DNA adduct formation. The effects of aflatoxin B1 and Ochratoxin A in causing breast cancer and lung cancer are demonstrated in only a few studies and additional studies need to be conducted in order to confirm the course of action which will help in controlling and reducing the rates of these cancers which are considered as the most common cancers in the world.

Citrinin : It produced micronuclei and DNA migration and it has the potential of inducing micronuclei. Citrinin would cause DNA damage which would increase the risk of cancer in humans. Citrinin can induce chromosomal aberrations in both human and rat livers and it can cause complex translocations within the cellular DNA at a high frequency and it is capable of inducing defects in chromosomal coiling which may lead to the formation of DNA adducts. Apart from this, it can cause DNA fragmentation and apoptosis and finally, it can cause a mitotic arrest and CREST-positive micronuclei where it will induce the formation of micronuclei containing a whole chromosome or chromatids and will eventually lead to cancer development.

Fumonisins : These are capable of inhibiting the biosynthesis of sphingosine and the complex sphingolipids by inhibiting the enzyme ceramide synthase. That, in turn, will lead to the increase in free sphinganine and the free sphingoid bases which are toxic in cells, and that will inhibit the growth, increase cell death, and initiate involvement in mitogenicity that would lead to the development of cancer.

Patulin : This will cause an inactivation of the active site of protein tyrosine phosphatase (PTP) and this protein is a key regulator of intestinal epithelial barrier function. The active site of PTP contains a cysteine residue (Cys215) which is important and required for phosphatase activity and sulfhydryl-reacting compounds such as acetaldehyde. Acetaldehyde decreases the transepithelial resistance (TER) through a covalent modification of cysteine residue Cys215 of the protein tyrosine phosphatase and this toxicity can lead to damage in the intestinal cells that may eventually lead to intestinal and stomach cancer.

T-2 : This is a trichothecene mycotoxin is capable of inducing breaks in the DNA of lymphocytes when administered *in vivo* and *in vitro* once the fibroblast cells are treated with this mycotoxin combined with ^{3}H thymidine. In addition, Nivalenol, a trichothecene mycotoxin is also involved in causing chromatid exchange in cells and it can cause DNA damage and that makes it a potential genotoxic agent. Finally, trichothecene mycotoxin can inhibit the proliferation and Ig production in the mitogen-stimulated human lymphocytes and it can induce apoptosis, the migration of sister chromatids as well as cause protein synthesis inhibition. The action of T-2 and Nivalenol on a cellular level can be responsible for the development of tumors and cancer.

Zearalenone : It shows hormonal activity in many animal species which makes it genotoxic, carcinogenic or both at the same time, and it can impose changes in the reproductive organs and systems of many laboratory animals such as mice and rats and different domestic animals such as horses. Thus, Zearalenone is capable of causing reproductive organ cancer in both man and animal.

The cancers that can be caused by mycotoxins include liver cancer, kidney cancer, esophageal cancer, lung cancer, renal cancer, breast cancer, testicular cancer, stomach cancer and reproductive organ cancer. Considering the danger caused by mycotoxins, additional investigations need to be conducted to confirm the ability of each mycotoxin in inducing cancer, as stated above.

Hundreds of mycotoxins have been identified thus far, with some, such as aflatoxins, ochratoxins, trichothecenes, zearalenone, fumonisins, and patulin, considered agro-economically important. Several factors contribute to the presence of mycotoxins in food, such as climatic conditions, pest infestation, and poor harvest and storage practices. Exposure to mycotoxins, which occurs mostly by ingestion, leads to various diseases, such as mycotoxicoses and mycoses that may eventually result in death . In light of this, Oluwadara Pelumi Omotayo *et al.*, (2019) reviewed the relevant literature on mycotoxin contamination, as well as various methods for the prevention and control of their prevalence, to avert its debilitating consequences on human health. Clear evidence of mycotoxin contamination was present in Africa, and it was therefore recommended that adequate prevention and control of these toxic substances in food system should be encouraged and that appropriate measures must be taken to ensure food safety as well as the enhanced or long-lifespan of the African populace. Governments, research institutions, and non-governmental organizations should tailor the limited resources available to tackle mycotoxin prevalence, as these will offer the best prospects for successful development of a sustainable food system in Africa.

Acute mycotoxicoses can cause serious and sometimes fatal diseases. The possibility of mycotoxin intoxication should be considered when an acute disease occurs in several persons when there is no evidence of infection with a known etiological agent, and no improvement in the clinical picture following treatment. Most of the outbreaks of mycotoxicoses described are a consequence of the ingestion of food that is contaminated with mycotoxins. The strict control of food quality, in both industrialized and developing countries, is therefore necessary to avoid such outbreaks.

6.2.2 Fungalbionics : The term 'Fungalbionics' was created in an attempt to describe one of the most dynamic microbial chemical factories ever encountered in the history of scientific exploration, the fungus. Fungi are masters at producing a wide array of biologically active substances which serve the producing fungus extremely well. These biological metabolites are anti-predatory, *i.e.* territory-protective, and exist to ensure that the fungus will survive as long as possible in this quite hostile world. These metabolites are anti-viral, anti-bacterial. anti-protozoan, anti-insect, anti-animal and, of course, anti-human. These metabolites are referred to as mycotoxins. The term is derived from the Greek mykes, meaning fungus, and toxicum, meaning toxin or poison.

Fungalbionics implicates fungi and fungal metabolites as the cause of gout/hyperuricemia. The mechanisms are not the usual patterns of invasive type mycoses nor of mycotoxicoses, but rather incorporate occult features of both mechanisms (Costantini, 1988). Gout and/or hyperuricemia can be induced in fowl by the mycotoxins oosporein, ochratoxin and by oosporein producing fungi. Gouty tophi have been induced in primates by aflatoxin. Fungi produce preformed uric acid, preformed urate crystals, lipoproteins glycosaminoglycans and glutamates, excess of which are found in gout. Gouty tophi are granulomatous and possess all of the features of delayed hypersensitivity. Giant cells in avian and human gouty tophi contain asteroid bodies which are fungal in origin. Fungal-like spherules have been found in avian gouty lesions; cultural and immunological electron microscopy studies are in progress. The findings in acute gout are those of an acute infection. All drugs used in treating gout/hyperuricemia are antifungal. Griseofulvin, an antifungal antibiotic, is as effective as colchicine in gout. Both are antitubulins and arrest fungal cell division. Probenecid, allopurinol, corticosteroids, NSAIDS, possess antifungal activity. The fungalbionic concept gives a unitarian explanation of gout, hyperuricemia and related diseases and findings.

Frank growth of fungi on animal hosts produces the diseases collectively called mycoses, while dietary, respiratory, dermal, and other exposures to toxic fungal metabolites produce the diseases collectively called mycotoxicoses. Mycoses range from merely annoying (e.g., athlete's foot) to life-threatening (e.g., invasive aspergillosis). The fungi that cause mycoses can be divided into two categories, primary pathogens (e.g., *Coccidioides immitis* and *Histoplasma capsulatum*) and opportunistic pathogens (e.g., *Aspergillus fumigatus* and *Candida albicans*). Primary pathogens affect otherwise healthy individuals with normal immune systems. Opportunistic pathogens produce illness by taking advantage of debilitated or immunocompromised hosts. The majority of human mycoses are caused by opportunistic fungi (Bennett and Klich, 2003). The mechanisms of pathogenesis of both primary and opportunistic fungi are complex, and medical mycologists have devoted considerable research energy trying to identify the factors that distinguish fungal pathogens from saprophytic and commensal species). Some infections remain localized, while others progress to systemic infection. For many mycoses, the ordinary portal of entry is through the pulmonary tract, but direct inoculation through skin contact is not uncommon.

In contrast to mycoses, mycotoxicoses are examples of "poisoning by natural means" and thus are analogous to the pathologies caused by exposure to pesticides or heavy metal residues. The symptoms of a mycotoxicosis depend on the type of mycotoxin; the amount and duration of the exposure; the age, health, and sex of the exposed individual; and many poorly understood synergistic effects involving genetics, dietary status, and interactions with other toxic insults. Thus, the severity of mycotoxin poisoning can be compounded by factors such as vitamin deficiency, caloric deprivation, alcohol abuse, and infectious disease status. In turn, mycotoxicoses can heighten vulnerability to microbial diseases, worsen the effects of malnutrition, and interact synergistically with other toxins.

The number of people affected by mycoses and mycotoxicoses is unknown. Although the total number affected is believed to be smaller than the number afflicted with bacterial, protozoan, and viral infections, fungal diseases are nevertheless a serious international health problem. Mycoses caused by opportunistic pathogens are largely diseases of the developed world, usually occurring in patients whose immune systems have been compromised by advanced medical treatment. Mycotoxicoses, in contrast, are more common in underdeveloped nations. One of the characteristics shared by mycoses and mycotoxicoses is that neither category of illness is generally communicable from person to person. Mycoses are frequently acquired via inhalation of spores from an environmental reservoir or by unusual growth of a commensal species that is normally resident on human skin or the gastrointestinal tract. These commensal species become pathogenic in the presence of antibacterial, chemotherapeutic, or immunosuppressant drugs, human immunodeficiency virus infection, in-dwelling catheters, and other predisposing factors. The majority of mycotoxicoses, on the other hand, result from eating contaminated foods. Skin contact with mold-infested substrates and inhalation of spore-borne toxins are also important sources of exposure. Except for supportive therapy (e.g., diet, hydration), there are almost no treatments for mycotoxin exposure, although Fink-Gremmels described a few methods for veterinary management of mycotoxicoses, and there is some evidence that some strains of *Lactobacillus* effectively bind dietary mycotoxins. Oltipraz, a drug originally used to treat schistosomiasis, has been tested in Chinese populations environmentally exposed to aflatoxin.

6.3 Bioterrorism : Mycotoxins are toxic metabolites produced by a diverse groups of fungi, namely by those of ubiquitous *Aspergillus, Penicillium, Fusarium, Alternaria, Claviceps* genera. Some mycotoxins could be used as bioterrorist weapons. The risk of such an attack is now considered to be real because the exposition to mycotoxins can have a wide range of adverse biological effects including hemorrhagic, hepatotoxic, nephrotoxic, neurotoxic, estrogenic, teratogenic, mutagenic and carcinogenic ones. Mycotoxins would be suitable as chemical warfare agents for the actions of sabotage, because their use is not associated with any primary alarming signs such as blast, fire or suspicious odors and because there are no real-time detection and alarm systems for these toxins (Malíř František *et al.*, 2007). Therefore, there would be no reason to escape and then it would be too late to do so after the signs of intoxication become evident. The abuse of mycotoxins for purposes of the biological warfare does not seem to be a purely theoric threat. There is some evidence that Iraqi scientists were developping aflatoxins as part of Iraqi biowarfare programme during the 1980s. The potent acute toxicity and chemical stability of mycotoxins Aflatoxin B1 (AFB1), fumonisin B1 (FB1), ochratoxin A (OTA), T-2 (Tricothecene) toxin and deoxynivalenol (DON) make them apt to be weaponized for bioterrorism. The prospect of their abuse is all the more higher that publicly available literature includes information on the production of these mycotoxins by fermentation and on their extraction and purification. T-2 mycotoxins are nonvolatile compounds produced by filamentous fungi primarily of the genera *Fusarium*, *Myrotecium*, *Trichoderma* and *Stachybotrys*.

The potential for biological weapons to be used in terrorism is a real possibility. Biological weapons include infectious agents and toxins (Anderson, 2012). Toxins are poisons produced by living organisms. Toxins relevant to bioterrorism include ricin, botulinum, *Clostridium perfrigens* epsilson toxin, conotoxins, shigatoxins, saxitoxins, tetrodotoxins, mycotoxins, and nicotine. Toxins have properties of biological and chemical weapons. Unlike pathogens, toxins do not produce an infection. Ricin causes multi organ toxicity by blocking protein synthesis. Botulinum blocks acetylcholine in the peripheral nervous system leading to muscle paralysis. Epsilon toxin damages cell membranes. Conotoxins block potassium and sodium channels in neurons. Shigatoxins inhibit protein synthesis and induce apoptosis. Saxitoxin and tetrodotoxin inhibit sodium channels in neurons. Mycotoxins include aflatoxins and trichothecenes. Aflatoxins are carcinogens. Trichothecenes inhibit protein and nucleic acid synthesis. Nicotine produces numerous nicotinic effects in the nervous system.

Mycotoxins can be used as chemical warfare agents (Chu, 1997). There is considerable evidence that Iraqi scientists developed aflatoxins as part of their bioweapons program during the 1980s. Toxigenic strains of *Aspergillus flavus* and *Aspergillus parasiticus* were cultured, and aflatoxins were extracted to produce over 2,300 liters of concentrated toxin. The majority of this aflatoxin was used to fill warheads; the remainder was stockpiled. Aflatoxins seem a curious choice for chemical warfare because the induction of liver cancer is "hardly a knockout punch on the battlefield" (Stone, 2001). Even so, the repugnance caused by the use of chemical and biological weapons is the kind of emotional response that terrorists seek to elicit. Furthermore, if used against ethnic groups such as the Kurds, the long-term physical and psychological results could be devastating. Finally, some experts think aflatoxin might have been selected simply because it was the "pet" toxin of an influential Iraqi scientist.

Unlike the aflatoxins, trichothecenes can act immediately upon contact, and exposure to a few milligrams of T-2 is potentially lethal. In 1981, then

Secretary of State Alexander Haig of the United States accused the Soviet Union of attacking Hmong tribesman in Laos and Kampuchea with a mysterious new chemical warfare agent, thereby violating the 1972 Biological Weapons Convention. The symptoms exhibited by purported victims included internal hemorrhaging, blistering of the skin, and other clinical responses that are caused by exposure to trichothecenes. Leaf samples from Kampuchea were analyzed by Chester Mirocha of the University of Minnesota, who found nivalenon, deoxynivalenon, and T-2. Furthermore, Mirocha stated that the quantity and proportions of the toxins were unknown in nature and that, based on reports in the scientific literature, these mycotoxins did not occur naturally in Southeast Asia. (We now know that trichothecene-producing fungi do exist in Southeast Asia.) The purported chemical warfare agent came to be known as yellow rain (Marshall, 1982).

In a bizarre turn of events, Matthew Meselson of Harvard University produced compelling evidence that yellow rain might have a benign natural explanation. Specks of yellow material found on foliage and other samples of the purported warfare agents from Southeast Asia were composed of plant pollen. The pollen grains were the result of so-called "cleansing flights" (mass defecations) dropped by swarms of wild Asian honeybees. Although many government officials tried to dismiss Meselson's explanation as a "great bee caper," the pollen hypothesis was widely accepted. An editorial by John Maddox stated, for example, that the pollen work created "a serious challenge for those who have been arguing that visual observations of 'yellow rain' can be counted as evidence of the use of chemical weapons against embattled populations" (Maddox, 1984).

At the time, several well-written and accessible summaries of the yellow rain controversy appeared in the popular press. Two books were also published on the subject. The government mycotoxin theory was promulgated in *Yellow Rain* , while the opposing position was addressed in *The Yellow Rainmakers*: *Are Chemical Weapons Being Used in Southeast Asia ?* (Evans, 1983.). Although analytical methods were in their infancy during the controversy, they were still sensitive enough to pick up low levels of environmental trichothecene contamination. The government research highlighted, if nothing else, that natural mycotoxicoses were an important health hazard in Southeast Asia.

The threat of biological warfare is an increasing concern in recent years. Advances in microbiology and genetic engineering have made it possible to create extremely dangerous toxigenic and pathogenic microorganisms (Venkataramana *et al.,* 2014). Accurately detecting and identifying biological warfare agents is the focal point for countering bioterrorism, methods for detection of these agents and treatments for infection are currently limited, and the development of highly rapid and sensitive field-deployable methods for early detection of biological warfare agents is crucial. There has been a remarkable progress in the detection, protection, and decontamination of biological warfare agents as many instrumentation platforms and detection methodologies are developed and commissioned. Furthermore in the past decade, there have been threats due to the emerging new diseases and also the reemergence of old diseases and development of antimicrobial resistance and spread to new geographic regions. The preparedness against these agents needs complete knowledge about the disease, better research and training facilities and diagnostic facilities, and improved public health system. Even then, the threat of biological warfare agents and their use in bioterrorist attacks still remain a leading cause of global concern.

6.4. Symptoms and diagnosis : Trichothecene mycotoxins, biological toxins, are a group of 40 compounds produced by a common grain mold. Biological toxins are harmful substances that are produced by living organisms such as fungi. Trichothecene mycotoxins are of more concern, especially T-2, as a biological weapon because they are very stable, resistant to disinfectants, easy to produce in large quantities, and can be dispersed through a number of different ways (Anonymous, 2004.). Additionally, there is strong evidence to suggest that they have been used as biological warfare agents in the past. can be inhaled, ingested, or absorbed through the skin. The T-2 mycotoxins are the potential biological agent that can adhere to and penetrate intact skin. Trichothecene mycotoxins cannot be transmitted person to person. This usually occurs when contaminated foods, such as moldy grain, are eaten.

Mycotoxins act by inhibition of protein synthesis. Symptoms start within minutes to hours after exposure, and involve eyes, skin, respiratory and gastrointestinal tracts. After exposure early symptoms begin within 5 to 60 minutes. Symptoms are dependent on the route of exposure. Inhalational exposure results in nasal itching, pain, sneezing, bloody and runny nose, difficulty breathing, wheezing, cough, and blood-tinged saliva and sputum. Exposure through ingestion causes loss of appetite, nausea and vomiting, stomach cramping, and watery and/or bloody diarrhea. Skin symptoms include burning, tender and reddened skin, swelling, and blistering progressing to death of skin tissues and, in lethal cases, sloughing of large areas of skin. After exposure to the eyes, pain, tearing, redness, and blurred vision occur. Exposure through any route can lead to full body illness, the symptoms of which include weakness, fatigue, dizziness, lack of muscular coordination, irregular heartbeat, hyperthermia or hypothermia, extensive bleeding, and low blood pressure.

A diagnosis is generally made based on the symptoms the patient is experiencing and the results from blood and urine tests. Environmental tests may also be done to help support the diagnosis. Additionally, if the exposure were the result of a bioterrorism event, many patients may report seeing a "yellow rain" or a smoke attack. There is no antitoxin for the trichothecene mycotoxins therefore, the only thing that can be done is to provide supportive care and treat the symptoms that occur. If given early, superactivated charcoal can be useful in treating patients who have ingested trichothecene mycotoxins. People who are known have been exposed to trichothecene mycotoxins should seek immediate medical aid. Exposed skin should be washed thoroughly with soap and water and eyes should be flooded with saline or water. The only way to prevent exposure is to avoid contact with the trichothecene mycotoxins. The can be done by wearing protective clothing and a mask.

6.5 Management of mycotoxins : Fungal secondary metabolites (mycotoxins) produced as contaminants on food and feed commodities are considered to be economically and toxicologically important world wide. Aflatoxins, orchratoxins, citrinin, fumonisins, zearalenone and trichothecenes are important mycotoxins which have been analysed as natural contaminants in various agricultural commodities. Many developed countries have laid down specific regulations for import and export of those items in terms of economic implications. In the light of present status different measures (prevention, elimination, and decontamination/inactivation) have been discussed in this review to minimize the risk of mycotoxin contamination in preharvest, harvest and post harvest conditions (Ajoy Kumar Choudhary and Priyanka Kumari, 2010).

Integrated mycotoxin management programmes: Strategies to address the food safety and economic issues employ both preharvest and postharvest measures to reduce the risk of mycotoxin contamination in food and feed. Postharvest measures, such as adequate storage, detection and decontamination or disposal as well as continuous monitoring of potential contamination during processing and marketing of agricultural commodities, have proved to be critical and indispensable in ensuring food and feed safety. However, the postharvest contamination is usually the result of preharvest presence of fungal contamination. Preharvest control includes good cultural practices, biocontrol and development of resistant varieties of crops through new biotechnologies. Processed food cannot be safe if prevention, control, good manufacturing practices and quality control are not used at all stages of production. The Hazard Analysis and Critical Control Point (HACCP) approach to processing mycotoxin contaminated commodities should be considered. Selected aspects of an integrated mycotoxin control programme should involve different phases such as ones outlined below:

Preharvest procedures: Significant levels of mycotoxins can occur in the food crops in the fields. Some of the strategies for prevention of mycotoxins in the field are reduction in plant stress through irrigation, mineral nutrition, protection from insect damage, avoidance of environmental conditions that favour infection in the field eg. Drought, insect infestation, primary inoculum, delayed harvesting etc.

Potentially effective methods: Good cultural practices viz. crop rotation, cropping pattern, irrigation, timely planting and harvesting and use of biopesticides have protective actions that reduce mycotoxin contamination of field crops, breeding of cultivars resistant to fungal infection, use of crop protection chemicals that are antifungal agent, identification of plant constituents that disrupt aflatoxin biosynthesis or fungal growth and their use in new biochemical marker-based breeding strategies to enhance resistance in crops, development of transgenic plants resistant to fungal infection, development of transgenic cultivars capable of catabolism/interference with toxin production, development of crops genetically engineered to resist damage, development of crop seeds containing endophytic bacteria that exclude toxigenic fungi, exclusion of toxigenic fungi by pre infection of plants with biocompetitive non toxigenic fungal strains.

The fungal genome of *A.flavus* has been sequenced to understand the regulation of aflatoxin formation by environmental factors (Ajoy Kumar Choudhary and Priyanka Kumari, 2010). This information can be used in development of host resistance against aflatoxin contamination by studying the effects of various physiological parameters eg - drought stress on gene expression in toxigenic fungi.

Harvesting procedure : Mechanical damage of seeds may incur during harvesting. When damage is kept to a minimum during this phase, subsequent contamination is significantly reduced. Field crops should be harvested in timely manner to reduce moisture or water activity (Aw) level to a point where mycotoxin formation will not occur.

Postharvest procedures : Even if the contamination occurs or persists after this phase, the hazards associated with toxin must be managed through postharvest procedures, if the product is to be used for food and feed purposes. Storage and processing are the major areas where contamination can be prevented.

Postharvest strategy

i. Removal of damaged grain and drying of grain to the minimal moisture level.

ii. Control of insect and rodent activity and maintenance of appropriate moisture levels and temperature.

iii. Appropriate packaging is often successful way of excluding insects and moulds.

iv. Frequent cleaning of food/feed delivery systems and short term storage. Use of antifungal agents such as propionic acid and acetic acid. Thermal inactivation is also one of the alternatives for products that are usually heat processed. Fumonisins and ochratoxin levels have been shown to be lower in thermally processed maize and wheat products.

Regulations: Regulatory limits are law, violation of which has legal consequences. Regulatory limits or standard provide a benchmark against which effectiveness of food safety programme can be tested. Adverse health effects from the ingestion of these toxins have caused regulatory agencies throughout the world to limit the amount of aflatoxins that are permitted in food or feed that is available for sale. Currently, 77 countries have imposed regulatory limits for mycotoxins. The establishment of regulatory limits may vary in each country depending on level of exposure and sociological, political and economic factors. This results in undue economic burden on growers (Ajoy Kumar Choudhary and Priyanka Kumari, 2010). According to FAO estimates, world losses of food stuffs due to mycotoxins are in the range of 1000 million tons a year. Since aflatoxins, ochratoxin A, deoxynivalenol, zearalenone, T-2 toxin and fumonisins are the mycotoxins of most concern thereby, stringent mandatory standard, voluntary regulations or guidelines, tolerance level have been laid down for these mycotoxins by many countries.

At the international level, CAC (Codex Alimentarius Commission) through its committee on Food Additives and Contaminants (CCFAC) and other relevant commodity Committee is considering the establishment of international guidelines for levels of mycotoxins in food. An updated status of existing regulations and also the role of CAC in providing a platform to reach a consensus on the maximum permitted levels in food have been reviewed recently (Vishwanath, 2004). Since mycotoxins can occur both in raw products and finished byproducts, this has necessitated the testing of mycotoxins at various stages in food-chain. Important commodities under regulation include - food and feed products: cereal grains and cereal based products, peanuts and their products, nuts and other oil-seeds, milk and milk products, infant and children food, fruit and fruit juices, miscellaneous food products. The establishment of regulatory limits may vary in each country depending on level of exposure and sociological, political and economic.

A good example of management through a monitoring programme is the aflatoxin control programme used by the state of Arizona. During the year 1978, almost 910,000 pounds of milk were dumped with contamination levels as high as 10 ppb aflatoxin M1. As a result of this huge commercial loss and the need to establish an effective aflatoxin control programme, the state instituted a programme to monitor aflatoxin levels in whole cottonseed and cottonseed products, the major source of aflatoxin contamination in feed. Once the regulatory limit has been set, sampling plays an important role in determining the fate of particular product. Adequate random sampling techniques should be used at each point of analysis. It is important to consider the existence of 'hot spot' (or highly contaminated portions of the product). If these highly contaminated portions are not detected, the toxin would then be distributed throughout the lot, rendering a highly contaminated product. Once the level of contamination has been determined through the monitoring programme, the final use of the commodity can be determined.

Control of mycotoxins :As mycotoxins are potential risk to human and animal health and thereby much time and effort has been expended on seeking methods for removal of or destruction of mycotoxins in contaminated products (Ajoy Kumar Choudhary and Priyanka Kumari, 2010). The best approach to the control of mycotoxin is prevention. However, contamination of mycotoxin is sometimes unavoidable, and if prevention fails, other alternative must be considered. The postharvest contamination is the result of preharvest presence of fungal contamination. Therefore, research focus in the past decades has shifted from postharvest control to a more preventive approach employing various preharvest control measures.

6.5.1. Pre harvest management of mycotoxins: Ecophysiological conditions are substantially different during preharvest and postharvest stages of agricultural crops. The mould invasion, infestation and mycotoxin elaboration to a great extent depend upon various environmental factors prevailing in the vicinity. Preharvest period begins with the emergence of seedling and continues upto maturity, finally ending with the harvesting of the crops. It is now established that some of the storage fungi (responsible for postharvest mycotoxin production) get associated with food grains from the field (preharvest stage). However, fungi that are associated with standing crops may or may not be able to grow in fields because conditions required for successful mould growth are less favourable in fields than in warehouses and godowns.

Moreover, only such moulds are able to grow under fields that exhibit certain degree of parasitism for invading receptive parts of the susceptible varieties of plants. Some preconditions are absolutely essential for preharvest production of mycotoxins. The foremost are availability of toxigenic fungal strain, susceptible host and favourable agro climatic niche (Ajoy Kumar Choudhary and Priyanka Kumari, 2010)..

In standing crops, host fungus environment interaction is critical in predisposition of mycotoxin contamination. Preharvest control includes good cultural practices such as insect control, irrigation during drought conditions, planting and harvesting dates, cropping patterns etc. Pre harvest prevention especially through host resistance is probably the best and widely explored strategy for control of mycotoxins. Currently new biotechnological approaches are employed for preharvest control of mycotoxins. Enhancement of host resistance Breeding Host resistance enhancement can be achieved through identification of germplasm resistance to aflatoxin and also identification of natural resistance mechanisms and traits. Aflatoxin contamination involves multiple chromosome regions and several genes (Davis and William, 1999). Therefore, attempt to select resistant trait, while maintaining desirable agronomic characteristics, have slowed down due to failure to identify expressed genes and proteins involved in resistance. Present approach is to elucidate the biochemical mechanisms that confer resistance in corn kernels that are vulnerable to aflatoxin contamination. These mechanisms could be used to enhance germplasm through marker assisted breeding and/or genetic engineering.

Genetically engineered A.flavus with gene construct consisting of GUS (β-glucuronidase) or Green Fluorescent Protein (GFP) reporter gene linked to an A. flavus α- tubulin gene promoter for monitoring fungal growth or with the reporter gene linked to an aflatoxin biosynthetic pathway gene, could provide a quick and economical way to indirectly measure aflatoxin levels. It is now possible to accurately assess fungal infection levels and to predict the corresponding aflatoxin levels in the same kernels as a result of fungal infection. In this way, it is now possible to determine whether kernel resistance mechanisms are affecting fungal growth or aflatoxin biosynthesis. Chromosomal regions with resistance to *A. flavus* and inhibition of aflatoxin production in corn have been identified through Restriction Fragment Length Polymorphism (RFLP) (White *et al*, 1998). However, limiting the growth of aflatoxigenic fungi might at times not be enough to maintain aflatoxins "at acceptable" levels in corn crops.

Therefore, identification of compounds (or plant constituents) that disrupt aflatoxin biosynthsis or fungal growth and their use in new biochemical marker based breeding strategies to enhance resistance in crops to aflatoxins, could potentially save the agricultural industries thousands of millions of dollars. Studies of the genetics of aflatoxin biosynthetic pathway for understanding how and why this fungus makes aflatoxin have enabled scientists to examine strategies to interrupt aflatoxin synthesis, thereby, preventing aflatoxin contamination of crops. The fungal genome of *A.flavus* has been sequenced to understand the regulation of aflatoxin formation by environmental factors. This information is being used to assist in the development of host resistance against aflatoxin contamination by studying the effect of various physiological parameters eg- drought stress on gene expression in toxigenic fungi.

i. Genetic modification : Genetically modified crops are plants of which the DNA has been altered through the introduction of a foreign gene to express a trait not inherent to the modified plant. Three transgene-mediated strategies have been proposed for the management of mycotoxigenic fungi and mycotoxins in maize (Lindy J. Rose, *et al.*, 2018). These include the reduction of fungal infection, (the degradation of mycotoxins and interfering with the mycotoxin biosynthetic pathway. To reduce infection by the fungus, the incorporation of antifungal and/or resistance genes, as well as the overexpression of defence-related genes, is required. Catabolic enzymes from microbes have been used to detoxify certain mycotoxins both *in vitro* and *in situ*, before they accumulate in the plant. Fumonisin esterase and amine oxidase genes encoding FUM-degrading enzymes have been identified in *Exophiala spinifera* . None of these genes have, however, been successfully introduced into maize. Maize plants have, however, been genetically engineered to interfere with the biosynthesis of AF and TCT. The best-known example of using genetically modified maize for reducing FER and FUM contamination of grain is *Bt* maize. This is due to the close association between kernel damage by insects and infection by *F. verticillioides*. *Bt* maize plants that prevent insect damage, therefore, also reduce FUM contamination of maize grain. Genetically modified maize is not authorised in all countries and, consequently, conventional breeding efforts are still commonly used.

ii.Mutation breeding : Exposure of seeds or other heritable materials to chemicals or radiation with the purpose to induce DNA changes (mutations) is known as mutation breeding. Nuclear technology for crop improvement makes use of ionising radiation, which causes induced mutations with a high mutation frequency in plants. These mutations might be beneficial and alter physiological characters of plants, including plant height, ear height and improved root architecture (Borzouei *et al.*, 2010). The radiation of seeds may also cause genetic variability that enables breeders to select new genotypes with improved grain yield and quality. Mutation breeding has been successfully used to generate genetic variation in cereal crops, including maize, for a number of aspects including enhanced yield and productivity, altered ear length, drought tolerance and enhanced stem structure. It can thus potentially provide an attractive means for generating tolerance to mycotoxigenic fungi and their mycotoxins.

iii. Host-plant resistance : The planting of disease-resistant plants is an effective, affordable and environmentally sound strategy to control ear rot diseases and mycotoxin accumulation. Commercial hybrids differ in their ability to accumulate mycotoxins (Lindy J. Rose *et al.*, 2018), while hybrids grown outside of their adapted range are more susceptible to mycotoxins than those grown within their adapted range. Determining host-plant resistance to mycotoxigenic fungi and mycotoxin accumulation is a fundamental step towards developing commercially tolerant plant varieties. Several factors require careful consideration when screening materials for resistance to mycotoxigenic fungi and their mycotoxins. Inoculation technique significantly contributes to the efficacy of the screening protocol and should, therefore, be appropriate, produce consistent results and consider the disease cycle of the pathogen. Numerous studies relating to different crops report on the importance of screening for resistance under variable environmental conditions since genotype by environment interactions (GEI) plays such a vital role in disease development and mycotoxin contamination. Furthermore, GEI and stability indicators provide for the selection of material tolerant across a broad range of environments or alternatively exhibiting tolerance in specific environments.

Various countries have reported on the tolerance levels of maize and wheat cultivars to mycotoxigenic fungi and associated mycotoxins (Janse van Rensburg *et al.,* 2015). However, focus has been placed on the characterisation of inbred lines for the identification of appropriate breeding material towards resistance to mycotoxigenic fungi and their toxins. Genetically modified maize, expressing *Bacillus thuringiensis* genes (BT maize), has been found to accumulate less FUM than its non-modified isolines.

iv. Resistance Associated Protein (RAP) : Two kernel proteins, one 28 KDa and inhibitory to *A. flavus* growth, the other over 100 KDa in size and primarily inhibitory to toxin formation were identified from resistant corn inbred. Chen *et al.* (1998) observed kernel protein profiles of 13 corn genotype which revealed a constitutively expressed 14 KDa Trypsin Inhibitor Protein (TI), present in relatively high concentrations in seven resistant corn lines, but at low concentration or absent in six susceptible lines. The mode of action of TI against fungal growth may be partially due to inhibition of fungal amylase, limiting A.flavus growth but also for toxin production. TI also demonstrated antifungal activity against other mycotoxigenic species. The identification of these proteins and their corresponding genes may provide markers for plant breeders and facilitate the introduction of antifungal genes through genetic engineering into other aflatoxin susceptible crops. Using 2-D gel electrophoresis, several resistant and susceptible genotypes were compared and over a dozen such protein spots, either unique or 5 fold upregulated in resistant lines, were identified and isolated (Ajoy Kumar Choudhary and Priyanka Kumari, 2010).

v. Other methods : Green Fluorescent Protein (GFP) is being utilized to monitor gene expression and food colonization by *Aspergillus flavus.* These studies demonstrated that the modified GFP encoded by pNucEm2 and pGAP33 was highly expressed in *A.flavus*. The intensity of fluorescence is sufficient to allow the visualization of a GFP- containing strain under a standard laboratory UV-light. The use of these constructs in strains of *A.flavus* could facilitate the detection of the fungus in substrates such as soils or foods and can be useful in screening corn genotype for resistance to aflatoxin accumulation and making screening faster and more commercial. Aqueous plant extracts of cinnamon, pippermint, basil, origanum, epizote, cloveand thyme caused total inhibition of fungal development on maize kernels and optimal dosage varies from 3 to 8 per cent. *O.methoxycynamaldehyde* from cinnamon has been reported to be highly effective against *A.flavus* and *A.parasiticus*. Inhibition of mycelial growth and aflatoxin production on maize grains when treated with onion, garlic and eugenol. Has been reported (Ajoy Kumar Choudhary and Priyanka Kumari, 2010). Caffeine appears to inhibit aflatoxin synthesis by restricting the uptake of carbohydrate, which are ultimately used by the mould to synthesize aflatoxins. Capsanthin of *Capsicum annum* was also reported to check growth and toxin production at all tested concentrations. Phenolics like tannic acid, caffeic acid and phloroglucinol at 0.01 M concentration prevented aflatoxin production by more than 55%. Antifungal efficacy of lemongrass (*Cymbopogon citrates*) powder and essential oil on *A.flavus* growth and aflatoxin contamination have been studied. Recently, components of *Eucalyptus* oil limonene andgeranial have confirmed the antifungal activity.

6.5.2 Postharvest management of mycotoxins: Once crop becomes infected under field conditions, the fungal growth continues usually with increasing vigour at postharvest stage and in storage. Most fungi cannot grow below a critical moisture level. *A.flavus* cannot grow in corn with moisture content less than 12 to 13 per cent. If corn is dried below this level no additional growth of fungus or production of aflatoxin will occur if proper storage practices are followed. In addition, fungus can survive in residues left in storage and feeding facilities and thereby, produce mycotoxins under such conditions. Food and feed residues should be discarded soon and storage and feeding facilities should be decontaminated. Periodical evaluation of storage suitability be monitored with the help of CO_2 sensor (Ajoy Kumar Choudhary and Priyanka Kumari, 2010). However, in most of the cases preharvest and postharvest preventive strategies have to be supplemented with control strategies viz Elimination and/ or Detoxification (inactivation) of mycotoxins.

i. CO_2 Sensor: It is essential for grain industry to have effective management programmes to protect against economic loss due to contamination from stored-product insects, moulds and mycotoxins. Manual grain inspection (human sensory exposure) and measuring grain temperature are the main tools used by the farmers and grain industry for monitoring proper storage conditions (Bortosik *et al*, 2008). Human sensory exposure literally means having personnel "walk" to the grain mass, smell the grain aeration discharge stream and look at the grain. Human sensory exposure for mould spoilage and other quality parameters could be biased and it varies from person to person. These cables are routinely placed in modern grain bin. Unfortunately, a temperature cable will not detect the fungal growth several feet away from the cable until the size of the spoiling grain mass is large enough to raise the temperature around the volume of the temperature cable. These limitations are overcome with CO_2 sensors.

CO_2 sensors can be effectively used to monitor early detection of spoilage during storage (Maier *et al*, 2010). CO_2 sensors can effectively be used to detect stored product insect infestation and grain spoilage due to mould infections well before spoilage detection by traditional methods such as visual inspection, smell and temperature cables. Such an early warning system would provide more timely information to farmers to make correct management decision to avoid the cost of spoilage mitigation measures such as turning, aeration and fumigation. This would help in continuing to store grain or market it early to avoid further quality deterioration.

ii. Elimination of mycotoxins from contaminated food/feed : Mycotoxins in contaminated commodities usually reside in relatively small number of seeds or kernels. Those seeds or kernels can be removed following one of the following methods.

a. Physical separation : The principle of this method is based on the identification of damaged kernels in the seed lots because of variations in size, shape, colour and also visible mould growth. Aflatoxin contaminated kernels are usually damaged, shriveled or discoloured (Ajoy Kumar Choudhary and Priyanka Kumari, 2010). Significant amount of aflatoxins in shelled peanuts can be removed by electronic sorting and hand picking However, complete removal of all contaminated particles or aflatoxin cannot be expected with physical methods of separation. Since the toxin can diffuse into the interior of the kernel, residual contamination may be present at very low levels in the final product. If there is a high level of residual contamination, other procedures must be used to manage the residual contamination in final product. Similar process based on fluorescence has been suggested. There appears to be good correlation between the presence of aflatoxins and the occurrence of a Greenish yellowish fluorescence (GYF) in thecontaminated seeds under UV-light. However, this type of separation was not successful in pecans because of inherent intense fluorescence in the kernels (Escher, 1974). Another approach is through floating and density segregation.

It was observed that 95% of the aflatoxin in 21 of 29 samples of peanuts was contained in kernels that floated in tap water (Ajoy Kumar Choudhary and Priyanka Kumari, 2010). Mean aflatoxin level was decreased from 301 ppb to 20 ppb using floating as a separation mechanism. There is still aflatoxin residue though separation has shown to significantly reduce aflatoxin contamination and thereby, can be considered as an effective first line defense for certain products. Corn screening (based on size and broken corn kernels) usually contains about 10 times the fumonisin content compared to intact corn. Significant amount (Ca 25%) of Deoxynivalenol (DON) can also be removed by cleaning andpolishing. Apples rotten with *Pencillium expansum* contain patulin, is mostly used for commercial juice production. It has been reported that trimming of apples to remove the rot reduces patulin content by 93% to 99%.. Physical separation or trimming of apples before processing is the best method to reduce patulin contamination.

b. Filtration : Aflatoxin was removed up to 90% through single filtration, but in recirculation of oil, they could achieve even up to 100% removal since aflatoxin in crude peanut oil remains in finely suspended form and can easily be separated by filtration. Basappa and Sreenivasamurthy (1979) at CFTRI, India have developed a special filter pad system which can easily be adopted in oil mills to remove aflatoxin from crude oil. This filter pad can be prepared by impregnating fuller'searth salt slurry in between two filter cloth layers and dried completely at 100°C for eight hours. The pads can be stored under desiccated conditions up to 2 months without loss in activity (Ajoy Kumar Choudhary and Priyanka Kumari, 2010)..

c. Milling ; In the processed products the levels of aflatoxin vary with the nature of processing, food materials and the affinity or solubility of toxin in the products. Laboratory studies have reported that during wet milling of inoculated corn, aflatoxin B1 was distributed in the milling fractions viz- it went primarily in steep water (39% to 42%), fibre (30% to 38%), with the remainder found in gluten (13% to 17%), germ (6% to 10%)and starch (only 1%). In a good risk management plan, individual fraction should be considered for further utilization (Basappa & Sreenivasamurthy, 1974). In an another investigation (CFTRI, India) it was found that 85% of the aflatoxins present in groundnut seeds goes into the cake after crushing in expeller oil mill or hydraulic press, and only 15% remained in oil.

Other mycotoxins (zearalenone, DON, nivalenol) are also distributed during in wet milling of corn kernels. Recent wet milling laboratory scale studies on naturally contaminated corn have shown that the toxin remains mostly in gluten while the germ remains free of the toxin. The starch fractions which account for 65% to 71% of milled products were free of detectable zearalenone. Additional chemicals can be added to the steeping solution so that the modified process would be used to ensure the safety of the final product (Ajoy Kumar Choudhary and Priyanka Kumari, 2010).. The concentration of toxin in a particular fraction simplifies the process of risk management. In dry milling of naturally contaminated corn, the highest levels of aflatoxin B1 occurred in germ and hull fractions, but distribution varied with the contamination level. In artificially contaminated rice, aflatoxin was greatly reduced by milling, with more than 95% in bran and polishfraction. In drumwheat, aflatoxin B1 was determined to be in peripheral parts of the kernels. Upon milling, the aflatoxin concentration in the flour varied according to the quality of the final product. Dry cleaning of the zearalenone naturally contaminated corn does not significantly reduce toxin levels in corn lot. On industrial perspective, combination of cleaning, sorting and milling can significantly reduce the risk of mycotoxins.

Solvent extraction procedure : Extraction of aflatoxin with solvent offers several advantages like the aflatoxin can completely be removed under suitable conditions ; there is little likelihood of forming from the aflatoxins other products having adverse physiological activity ;extraction can be carried out and the solvent can be recovered without nutritional loss in many cases ; the possible introduction of offflavour Solvent able to fit all the criteria, however, is not an easy task. Other factors become involved like cost of solvent, percentage of recovery, toxic affect of solvent and its residues, and from safety point of view the inflammability, explosiveness and boiling point, may also be taken into considerations (Ajoy Kumar Choudhary and Priyanka Kumari, 2010). The disadvantages are special solvent extraction equipment may be required , the extraction of some of the soluble components with aflatoxins and the added cost of traditional processing.

Aflatoxin is soluble in polar solvent such as methanol and is insoluble in water and petroleum hydrocarbons. The extraction methods used in analysis of aflatoxins employ chloroform/water and acetone/water to remove the mycotoxins. Aqueous isopropanol has been found to be an effective solvent for removal of aflatoxins from both contaminated cottonseed and groundnuts. Six extractions with 80% aqueous isopropanol at 60°C resulted in complete removal of aflatoxin in both meals. A binary solvent system of 90% acetone and 10% water (by weight) reduced the aflatoxin content of pre pressed cottonseed and groundnut meal to less than 10μg/kg insmall scale batch extractions and less than 40 μg/kg in continuous plant extractions.

Removal of all aflatoxins from peanut meal by an aqueous solution of calcium chloride has also been reported (Ajoy Kumar Choudhary and Priyanka Kumari, 2010). There are, however, many disadvantages inherent with the use of these solvent systems. Aqueous alcohol and aqueous acetone removes 2.8 to 10 % of soluble carbohydrate from groundnut products making such processes uneconomical. There is also a problem of disposal of the toxic extracts. Acetone has also been reported to produce undesirable off-flavours due to mesityl oxide on the detoxified product. High concentrations of iso-propanol are also unsuitable as they cause formation of protein gels.

6.5.3 Inactivation of mycotoxins : When the risk management of aflatoxin through elimination is not possible/ successful or as an alternative, one can apply inactivation or detoxification of mycotoxin either through physical or chemical method, which becomes inevitable.

Physical methods

a. Thermal inactivation : Thermal inactivation of aflatoxin is achieved by cooking, roasting, frying, spray drying, baking etc. Different rice cooking methods used by Indians were compared and its was found that pressure cooking at 15 psi for 5 minutes gave maximum destruction (72%) of aflatoxin to the method of ordinary cooking (50%) and cooking with excess water (50%) (Ajoy Kumar Choudhary and Priyanka Kumari, 2010). They recommended pressure cooking of rice not only because of destroying maximum amount of aflatoxins but also for preserving nutrients of rice. Microwave roasting destroy aflatoxins almost completely. Unfortunately, this process would involve an increase in processing cost, making it commercially impractical. Thereby, microwave roasting cannot be a good solution.

Cold Atmospheric pressure Plasma (CAP) represents a promising, low-cost, and environmentally friendly means to degrade mycotoxins with negligible effect on the quality of food products. Despite this promise, the study of CAP-mediated mycotoxin degradation has been limited to a small subset of the vast number of mycotoxins that plague the food supply chain. This study explored the degradation of aflatoxins, trichothecenes, fumonisins, and zearalenone using CAP generated in ambient air (Nataša Hojnik *et al,* 2019). CAP treatment was found to reduce aflatoxins by 93%, trichothecenes by 90%, fumonisins by 93%, and zearalenone by 100% after 8 minutes exposure. To demonstrate the potential of CAP-mediated mycotoxin degradation against more conventional methods, its efficiency was compared against ultraviolet C (UVC) light irradiation. In all cases, CAP was found to be considerably more efficient than UVC, with aflatoxin G1 and zearalenone being completely degraded, levels that oculd not be achieved using UVC irradiation.

b. Inactivation through light / irradiation : Aflatoxin destroy in chloroform solution to the extent of 99% when exposed to UV–light. Before the degradation of the aflatoxin B1 molecule, photodiamerization of the coumarin moiety takes place. Exposure to sunlight of aflatoxin contaminated vegetable oils, revealed encouraging results. Unrefined groundnut oil containing aflatoxin (> 100 mg/kg) was exposed to bright sunlight, gas filled tungsten lamp or longwave UV light. Sunlight destroyed 99% of the aflatoxin present in 15 min, whereas, tungsten lamp light and ultraviolet light destroyed 82% to 85% of aflatoxin in 18 hours and 30% to 40% of aflatoxin in 2 hours, respectively.

6.5.4 Chemical methods: Various chemicals *viz.* chlorine, ozone, hydrochloric acid, benzoyl peroxide, ammonia, sodium hypochloride and ethanolamine have shown to destroy pure aflatoxin or aflatoxin in contaminated materials. However, the detoxification process must satisfy criteria *viz.*, it must destroy or detoxify the aflatoxin, must not produce or leave any toxic or carcinogenic residues in the final product, should destroy fungal spores and mycelium as under suitable conditions they could grow and recontaminate the product., should preserve, as far as possible, the nutritive value and palatability of the starting material and should not alter significantly the important technological properties of the starting material (Ajoy Kumar Choudhary and Priyanka Kumari, 2010).

Unfortunately, many of the chemicals either decrease significantly the nutritive value or produce toxic products or have undesirable side effects. Ammoniation of corn, peanuts, cottonseed and meals to alter the toxic and carcinogenic effects of aflatoxin contamination has been intensively investigated. Efficacy and safety of ammoniation as a practical solution toaflatoxin decontamination in animal feeds have been demonstrated. Use of ammonia treatment to reduce risks posed by aflatoxin contamination, particularly in the high temperature / high pressure treatment has been recommended. With the ambient temperature/

atmospheric pressure method, caution must be exercised to assure even distribution of ammonia gas / water or aqueous ammonia solution in the product. This process is limited to whole kernel seed / nut products and has not been successful for ground or meal products due to inability of ammonia to penetrate commodity. For another toxin, fumonisins. Atmospheric pressure/ ambient temperature ammoniation reduced fumonisin content of *F.moniliforme* but did not reduce the toxicity of material when fed to animals.

Traditional alkaline heat treatment (Nixtamalization) of corn used in the manufacture of tortillas reduces significantly the levels of aflatoxin. Nixtamalization greatly reduces FB1 content. However, when nixtamalized corn was fed to weaning rats, the toxicity of the inoculated corn was not reduced (Ajoy Kumar Choudhary and Priyanka Kumari, 2010).

Modified 'nixtamalization' procedure involves the addition of hydrogen peroxide and sodium bicarbonate to traditional process. Results from this study show 40% reduction of brine shrimp mortality compared to traditional nixtamalization. In oil seed meal oxidative destruction of aflatoxin by ozone has also been considered as a practical method. Total aflatoxin B1 was found to be inactivated after incubation with ozone for two hours at 100°C in cottonseed and peanut meals having 22 per cent and 30 per cent moisture. Hydrogen peroxide in different concentrations (3%, 5% and 10%) has its efficacy in destruction of zearalenone in contaminated corn However, thc percentage of destruction was dependent on the concentration of hydrogen peroxide, temperature, period of exposure, maximum destruction was recorded when the contaminated corn was treated with 10% hydrogen peroxide at 80°C for 16 hours. Calcium hydroxide monomethylamine effectively decontaminated feed containing T2 toxin and zearalenone. Patulin andpenicillic acid are moderately stable in apple juice with a half-life of several weeks (Ajoy Kumar Choudhary and Priyanka Kumari, 2010).

Concern about chemicals and toxic residues in the food chain is a matter of great concern. Processing of food not only alters the food but also adds new conditions and ingredients. For risk management there is a need to explore the application of new food additives to control mycotoxins as an integral part of the process.

6.5.5 Biomanagement: Biological management using microbial antagonists either alone or as part of an integrated control strategy to reduce pesticide inputs, has emerged as a promising approach for control of mycotoxins in crops both pre- and postharvest. Several organisms including atoxigenic Aspergilli yeasts, bacteria and fungi have been tested for their ability to reduce both fungal infection and mycotoxin contamination. For instance, atoxigenic fungal strains are being used widely to prevent pre-harvest aflatoxin contamination of crops such as peanuts, pistachios, maize, and cottonseed in several parts of the world including the Mediterranean area (Dimitrios *et al.*, 2012). Recent advancements in the use of biocontrol strategies have led to registration of commercial products with increased practical applications for the benefit of growers in several countries.

The most prevalent toxigenic fungi belong to the genera *Aspergillus, Fusarium, Penicillium* and *Alternaria* (Table 9). These fungi pose serious phytopathological and mycotoxicological risks at preharvest and postharvest stages, as well as in processed food products (Castoria *et al.*, 2008). *Fusarium* and *Alternaria* species are often classified as 'field fungi', because they require high moisture content in the substrate for growth and mycotoxin synthesis (>20%) whereas the 'storage fungi', mainly species of *Aspergillus* and *Penicillium*, can also grow well at lower moisture contents. Thus, Fusarium and Alternaria usually pose a high mycotoxicological risk at pre-harvest level or in freshly harvested products that are drying, whereas toxigenic species of *Aspergillus* and *Penicillium* represent a higher risk for food and feed products in storage or other kind of processing.

Table 9: Major biocontrol agents

Targeted mycotoxigenic fungus	Targeted mycotoxin	Biocontrol agent
A. flavus, A. parasiticus	Aflatoxin	Atoxigenic *Aspergillus* strains
Aspergillus spp.	Aflatoxin	Lactic acid bacteria
Aspergillus spp.	Aflatoxin	Fungal strains that detoxify AF
A. carbonarius, A. niger	Ochratoxin A	Phyllosphere yeasts (Ex:. *Aureobasidium pullulans*)
F. verticillioides	Fumonisin	*Bacillus subtilis, B. amyloliquefaciens, Kluyveromyces* sp.
F. graminerarum, F. culmorum	DON	*Fusarium equiseti,* yeasts, *Bacillus* spp., *Pseudomonas* spp.
Penicillium expansum	Patulin	*A. pullulans, P. syringae, Candida sake, Pantoea agglomerans*

Che-Hsing *et al.*, (2019) investigated the ability of MSMEG_5998, an aflatoxin-degrading $F_{420}H_2$-dependent reductase from *Mycobacterium smegmatis*, to degrade aflatoxin B1 (AFB1) and reduce AFB1-caused damage in HepG2 cell culture model and whether a thioredoxin (Trx) linkage of MSMEG_5998 enhanced the enzyme activity. They showed that Trx-linked MSMEG_5998 degraded 63% AFB1 and native MSMEG_5998 degraded 31% after 4 h at 22 °C, indicating that the Trx-linked enzyme had a better AFB1-degrading ability. In a HepG2 cell culture model, Trx-linked MSMEG_5998 reduced DNA damage and p53-mediated apoptosis caused by AFB1 to a greater extent than the native enzyme. These findings suggested that Trx-linked MSMEG_5998 could potentially be developed to protect the liver from AFB1 damage, or as a candidate protein to reduce AFB1-related toxicity in animals.

Applying postharvest control measures such as adding mycotoxin detoxifying agents is a frequently-used mitigation strategy for mycotoxins (Marianne Lauwers *et al*., 2019). EFSA states that the efficacy of these detoxifiers needs to be tested using specific biomarkers for exposure. the biomarkers for exposure are not further optimized for specific target species. A study was taken up to evaluate the most suitable biomarkers for deoxynivalenol (DON) and zearalenone (ZEN) in porcine plasma, urine and feces; and DON, aflatoxin B1 (AFB1) and ochratoxin A (OTA) in plasma and excreta of broiler chickens and to determine the efficacy of a candidate detoxifier, as a proof-of-concept study. Therefore, a mixture of mycotoxins was administered as a single oral bolus with or without detoxifying agent. In accordance with literature AFB1, OTA, and DON-sulphate (DON-S) proved optimal biomarkers in broilers plasma and excreta whereas, in pigs DON-glucuronide (DON-GlcA) and ZEN-glucuronide (ZEN-GlcA) proved the optimal biomarkers in plasma, DON and ZEN-GlcA in urine and, ZEN in feces. A statistically significant reduction was seen between control and treatment group for both AFB1 and DON in broiler plasma, under administration of the mycotoxins blend and detoxifier dose studied suggesting thus, beneficial bioactivity.

Many additives can enhance the biological control activity of antagonists and salts usually are the most related ones. Thus, the combination of 2% sodium bicarbonate with the yeast antagonist *Candida oleophila* enhanced significantly its biocontrol performance (curative and protective effect) against fungi rot diseases caused by *Botrytis, Penicillium, Monilinia* and *Rhizopus* in apple and peach (Flávio Henrique Vasconcelos de Medeiros *et al*., 2012). Similarly, a sugar analogue (2-deoxy-dscp-glucose) has also demonstrated the capacity to increase efficacy. The bioagent *Candida saitoana* supplemented with 0.2% 2-deoxy-D-glucose was more effective in controlling decay of apple, orange, and lemon caused by *Botrytis cinerea, Penicillium expansum*, and *P. digitatum* than either *C. saitoana* or the application of a 0.2% solution of 2-deoxy-D-glucose alone. Other additives such as ethanol, silicon, nisin and chitosan, have been reported to enhance the quality and control efficacy of different biocontrol agents (Janisiewicz Conway, 2010).

Aflatoxins produced primarily by two closely related fungi, *Aspergillus flavus* and *Aspergillus parasiticus*, are mutagenic and carcinogenic in animals and humans. Of many approaches investigated to manage aflatoxin contamination, biological control method has shown great promise. Numerous organisms, including bacteria, yeasts and nontoxigenic fungal strains of *A. flavus* and *A. parasiticus*, have been tested for their ability in controlling aflatoxin contamination. Great successes in reducing aflatoxin contamination have been achieved by application of nontoxigenic strains of *A. flavus* and *A. parasiticus* in fields of cotton, peanut, maize and pistachio. The nontoxigenic strains applied to soil occupy the same niches as the natural occurring toxigenic strains. They, therefore, are capable of competing and displacing toxigenic strains (Yan-ni Yin *et al.,* 2008).

Aflatoxins produced by *Aspergillus flavus* and *A. parasiticus* are common contaminants of peanut and a major threat to consumers, particularly in Sub-Saharan Africa (Waliyar, *et al.,* 2008). Aflatoxin contamination is a serious concern given their hepatotoxic properties and their widespread occurrence during cultivation, harvest, drying, storage, transit and distribution. Preharvest infection by *A. flavus* is the major cause of aflatoxin contamination in peanut. Its prevention is a complicated task that requires a series of intervention strategies to be merged with traditional farming practices. The International Crops Research Institute for the Semi-Arid Tropics (ICRISAT) and its partners have developed an integrated approach to mitigate *A. flavus* infestation and aflatoxin contamination by combining: host plant resistance, soil amendments with lime and organic supplements to enhance water holding capacity, plant vigor and seed health, timely harvesting and postharvest drying methods, the use of antagonistic biocontrol agents, such as *Trichoderma* and Pseudomonads and awareness campaigns and training courses to disseminate technology to the end-users.

References

Abbas, H.K., Accinelli C., Zablotowicz, R.M., Abel, C.A., Bruns, H.A., Dong, Y. and Shier, W.T., 2008. Dynamics of mycotoxin and *Aspergillus flavus* levels in aging Bt and non-Bt corn residues under Mississippi no-till conditions. *J. Agric. Food Chem.*, **56**: 7578–7585

Ahmed Adam, M.A., Tabana, Y.m., Musa, K B. and sandai, D.A., 2017. Effects of different mycotoxins on humans, cell genome and their involvement in cancer (Review).*Oncol. Rep.*, **37**: 1321-1336

Ajoy Kumar Choudhary and Priyanka Kumari, 2010, Review Article Management of Mycotoxin Contamination In Preharvest And Postharvest Crops: Present Status And Future Prospects. *J. Phytol.,* **2**: 37-52

Anderson, P.D., 2012. Bioterrorism: toxins as weapons. *J. Pharm. Pract.*, **25**:121-129.

Anonymous, 2004. Arizona Department of Health Services, Division of Public Health Services, Office of Public Health Emergency Preparedness and Response.

Anonymous, 2012. Scientific Opinion on the risks for public and animal health related to the presence of citrinin in food and feed. European Food Safety Authority. *EFSA Journal*. **10** (3).

Ashiq, S., 2015. Natural occurrence of mycotoxins in food and feed: Pakistan perspective. *Comprehensive Reviews in Food Science and Food Safety* **14** :159–175.

Basappa, S.C. and Sreenivasamurthy, V. 1974. Partitionof aflatoxin during separation of different constituents of groundnut kernel. *J. Food Sci Technol.,* **11**: 137.

Basappa, S.C. and Sreenivasamurthy, V. 1979. Decontamination of groundnut oil from aflatoxin. *Indian J. Technol.,* **17**: 440.

Binder, Eva M., 2007. "Managing the risk of Mycotoxins in modern feed production". *Animal Feed Sci. & Technol.,* **133** : 149–166

Bennett, J. W., and R. Bentley. 1999. Pride and prejudice: the story of ergot. *Perspect. Biol. Med,* .**42:**333-355

Bennett, J.W.and Klich, M., 2003. Mycotoxins . *Clin.Microb.Rev.,* **16**: 497–516

Berntsen, H.F., Bogen, I.L.,Wigestrand, M.B., Fonnum, F., Walaas, S.I.and Moldes-Anaya, A ., 2017. The fungal neurotoxin penitrem A induces the production of reactive oxygen species in human neutrophils at submicromolar concentrations. *Toxicology,* **392**: 64–70

Beyer, M., Klix, M.B., Klink, H. and Verreet, J.A .,2006. Quantifying the effects of previous crop, tillage, cultivar and triazole fungicides on the deoxynivalenol content of wheat grain – a review. *J. Pl. Dis.& Protn.,* **113:** 241–246.

Bezuidenhout, S. C., Gelderblom, W. C. A., Gorst-Allman, C. P., Horak, R. M., Marasas, W. F. O., Spiteller, G. and Vleggaar, R., 1988. Structure elucidation of the fumonisins, mycotoxins from *Fusarium moniliforme. J. Chem. Soc. Chem. Commun.,* **743** : 2-5.

Blanc, P. J.,. Loret, M. O . and Goma ,G., 1995. Production of citrinin by various species of *Monascus*. *Biotechnol. Lett.* **17:**291-294.

Boonen Jente, Malysheva Svetlana, V., Taevernier Lien, Diana Di Mavungu, José De Saeger and Sarah De Spiegeleer Bart, 2012. "Human skin penetration of selected model mycotoxins". *Toxicology*. **301** : 21–32.

Bortosik, R., Cardoso, L. and Rodriguez, J., 2008. Early detection of spoiled grain stored in hermetic plastic bags (Silo bags) using CO2 monitoring. In: Guo, D.L., Navarro, S., Jian, Y., Cheng, T., Zuxun, J., Yue, L., Yang L. and Haipeng, W. (Eds.) Proceedings of 8th International conference, 15-18 July 2008, Chicago, IL, USA

Borzouei, A., Kafi, M., Khazaei, H., Naseriyan, B., Majdabadi, A.., 2010. Effects of gamma radiation germination and physiological aspects of wheat (*Triticum aestivum* L.) seedlings. *Pakistan J. Bot.,* **42**:281-290

Bryden,W.L., 2012. Food and feed, mycotoxins and the perpetual pentagram in a changing animal production environment. *Ani. Prod. Sci*., **52**: 383-397

Castoria, R., S.A.I. Wright and Droby, S.., 2008. Biological control of mycotoxigenic fungi in fruits. *In: Mycotoxins in Fruits and Vegetables*. R. Barkai-Golan, N. Paster (Edrs.), 311–333

Che-Hsing Li ,Wei-Yang Li ,I-Ning Hsu ,Yung-Yu Liao ,Chi-Ya Yang ,Matthew C. Taylor ,Yu-Fan Liu ,Wei-Hao Huang ,Hsiang-Hua Chang,Ho-Lo Huang,Shao-Chi Lo ,Ting-Yu Lin ,Wei-Che Sun ,Ya-Yi Chuang,Yu-Chieh Yang ,Ru-Huei Fu and Rong-Tzong Tsai, 2019. Recombinant Aflatoxin-Degrading $F_{420}H_2$-Dependent Reductase from *Mycobacterium smegmatis* Protects Mammalian Cells from Aflatoxin Toxicity. *Toxins* , **11 :** 259

Chu, F. S. 1997. Trichothecene mycotoxicoses. *Encyclopedia Hum. Biol.,* **8** :511-522

Costantini, A., 1988. 24 Fungalbionics-A New Concept of the Etiology of Gout and Hyperuricemia. *Pediatr. Res.,* **24,** 115

Damiani, T., Righetti, L., Suman, M., Galaverna, G. and Dall'Asta, C., 2019. Analytical issue related to fumonisins: A matter of sample comminution? *Food Control*, **95**:1–5

Davis, G.L. and William, W.P. 1999. QTL for aflatoxin reduction in maize. *Maize Genetic Conference*, **41**: 22.

Dimitrios I., Tsitsigiannis, Myrto Dimakopoulou, Polymnia, P., Antoniou and Eleftherios, C.and Tjamos, 2012. Review Biological control strategies of mycotoxigenic fungi and associated mycotoxins in Mediterranean basin crops. *Phytopathologia Mediterranea* , **51** :158–174

Dirheimer,G. and Creppy,E.E., 1991. Mechanism of action of ochratoxin A. *IARC Sci Publ.* **1**;171-86.

Eaton, D. L., and Groopman J. D., (ed.). 1994. The toxicology of aflatoxins: human health, veterinary, and agricultural significance. Academic Press, San Diego, Calif.

Evans, G. 1983. The yellow rainmakers. Are chemical weapons being used in Southeast Asia? Verso/Schocken, London, United Kingdom.

Flávio Henrique Vasconcelos de Medeiros, Samuel Julio Martins, Tiago Domingues Zucchi; Itamar Soares de Melo, Luis Roberto Batista, José da Cruz Machado, 2012. Biological control of mycotoxin-producing molds. *Ciênc. agrotec.* vol.36 no.5

Fumio Tashiro, Mode of action of zearalenone, an estrogenic mycotoxins. 2000 Issue, *JSM Mycotoxins.* **50** : 105-110

Geiser, D.M., Dorner, J.W., Horn, B.W. and Taylor, J.W., 2000. The phylogenetics of mycotoxin and sclerotium production in *Aspergillus flavus* and *Aspergillus oryzae . Fungal Gen. & Biol.,.* **31**: 169–79.

Hedayati, M. T., Pasqualotto, A. C., Warn, P. A., Bowyer, P.and Denning, D. W., 2007. *Aspergillus flavus:* human pathogen, allergen and mycotoxin producer. *Microbiology*, **153**: 1677–1692

Hicks, J. K., K. Shimizu, and N. P. Keller. 2002. Genetics and biosynthesis of aflatoxins and sterigmatocystin, p. 55-69. *In F. Kempken (ed.), The mycota, vol. XI. Agricultural applications*. Springer-Verlag, Berlin, Germany.

Ida ,M.L., Drejer Strom, Rie Romme Rasmussen and Peter Have Rasmussen, 2014. Occurrence of Pre- and Postharvest Mycotoxins and Other Secondary Metabolites in Danish Maize Silage. *Toxins* , **6**: 2256–2269.

Janse van Rensburg, B., Mclaren, N.W., Flett, B.C. and Schoeman, A., 2015. Fumonisin producing *Fusarium* spp. and fumonisin contamination in commercial South African maize. *European J. Pl. Pathol.*,**141**: 491-504

John L. Richard, 2007. Some major mycotoxins and their mycotoxicoses. *Intl.J. Food Microbiol.*,**119**: 3-10

Karine Moncoq, Catharine A. Trieber and Howard S. Young. 2007. The Molecular Basis for Cyclopiazonic Acid Inhibition of the Sarcoplasmic Reticulum Calcium Pump. *J. Biol. Chem..* **282** : 9748 –9757,

Kiessling, K.,1986. "Biochemical mechanism of action of mycotoxins" (PDF). *Pure & Applied Chem.,* **58** : 327–338.

Kuiper-Goodman, T., Scott, P. M. and Watanabe, H., 1987. "Risk Assessment of the Mycotoxin Zearalenone". *Regulatory Toxicol. & Pharmacol.,* **7** : 253–306

Li, Y., Wang, Z., Beier, R. C., Shen, J., De Smet, D., De Saeger, S. and Zhang, S., 2011. T-2 toxin, a Trichothecene Mycotoxin: Review of toxicity, metabolism and analytical methods. *J. Agril. & Food chem..*, **59 :** 3441-53

Lindy J. Rose, Sheila Okoth, Bradley C. Flett, Belinda Janse van Rensburg and Altus Viljoen, 2018 Preharvest Management Strategies and Their Impact on Mycotoxigenic Fungi and Associated Mycotoxins, *Intech Open*, DOI: 10.5772/intechopen.76808.

Llewellyn, G.C., McCay, J.A., Brown, R.D., Musgrove, D.L., Butterworth, L.F., Munson, A.E. and White, K.L., Jr. 1998. Immunological evaluation of the mycotoxin patulin in female B6C3F1 mice. *Food Chem. Toxicol.*, **36 :** 1107–1111

Ma Hailun, Zhong Lilin, Inesi Giuseppe, Fortea Isabel, Soler Fernando, Fernandez-Belda and Francisco , 1999. Overlapping Effects of S3 Stalk Segment Mutations on the Affinity of Ca^{2+}-ATPase (SERCA) for Thapsigargin and Cyclopiazonic Acid. *Biochemistry*, **38** : 15522–15527

Mahfoud, R., Maresca, M., Garmy, N. and Fantini, J., 2002. The mycotoxin patulin alters the barrier function of the intestinal epithelium: mechanism of action of the toxin and protective effects of glutathione. *Toxicol Appl Pharmacol.*, **181**:209-18.

Malíř František, Roubal Tomas , Ostry Vladimir, Tuma Jiri and Procházková, Hana & Říčařová, Blanka & Marešová, Hana & Rolečková, Eva, 2007. *Mycotoxins & bioterrorism Chemicke List,* . **101:** 119-121.

Manabe, M. 2001. Fermented foods and mycotoxins. *Mycotoxins,* **51**:25-28

Manish Adhikari, Bhawana Negi, Neha Kaushik, Anupriya Adhikari, Abdulaziz A. Al-Khedhairy, Nagendra Kumar Kaushik and Eun Ha Choi , 2017. T-2 mycotoxin: toxicological effects and decontamination strategies. *Oncotarget.* **8**: 33933–33952.

Marasas, W. F. O., Nelson, P. E. and Toussoun,. T. A., 1984. Toxigenic *Fusarium* species: identity and mycotoxicology. The Pennsylvania State University Press, University Park, Pa.

Maria L.Fernandez-Cruz, Marcia L Mansilla and Jose L. Tadeo, 2010. Mycotoxins in fruits and their processed products: Analysis, occurrence and health implications. *J.Adv.Res.*, **1**: 113-122

Marianne Lauwers ,,Siska Croubels ,,mBen Letor ,Christos Gougoulias and Mathias Devreese , 2019. Biomarkers for Exposure as a Tool for Efficacy Testing of a Mycotoxin Detoxifier in Broiler Chickens and Pigs. *Toxins* , **11 :** 187

Marshall, E., 1982. Yellow rain: filling in the gaps. *Science,* **217:** 31-34

McCormick, S.P., Stanley, A.M., Stover, N.A.and Alexander, N.J., 2011. "Trichothecenes: from simple to complex mycotoxins". *Toxins,* **3** : 802–14.

Mi-Gyung Lee, Shiguang Li, Bruce B.Jarvis,James J.Pestka, 1999. Effects of satratoxins and other macrocyclic trichothecenes on il-2 production and viability of el-4 thymoma cells. *J. Toxicology & Env.Health*-Part A-Current Issues, **57:** 459-474

Moldes-Anaya, Angel S, Fonnum, Frode, Eriksen, Gunnar S, Rundberget, Thomas, Walaas, S. Ivar and Wigestrand, Mattis B., 2011. *In vitro* neuropharmacological evaluation of penitrem-induced tremorgenic syndromes: Importance of the GABAergic system. *Neurochemistry International*, **59**: 1074–1081

Nataša Hojnik ,Martina Modic,Gabrijela Tavčar-Kalcher ,Janja Babič ,James L. Walsh and Uroš Cvelbar, 2019, Mycotoxin Decontamination Efficacy of Atmospheric Pressure Air Plasma. *Toxins* **11 :** 219;

Oluwadara Pelumi Omotayo, Abiodun Olusola Omotayo, Mulunda Mwanza and Olubukola Oluranti Babalola , 2019. Prevalence of mycotoxins and their consequences on human health. *Toxicological Res.*, **35:** 1-7

Pestka, James, J., 2010. Deosynivalenol: Mechanisms of action, human exposure and toxixological relevance. *Archives Toxicol.,* **84:** 663-679

Pitt, J. I., 2000. Toxigenic fungi: which are important? *Med. Mycol.*, **38 :**17-22.

Prossnitz Eric R. and Barton Matthias, 2014. Estrogen biology: New insights into GPER function and clinical opportunities. *Mol. & Cellular Endocrinol.,* **389** : 71–83.

Sabbioni, G. and Sepai, O.,1994. Determination of human exposure to aflatoxins, p. 183-226. *In* K. K. *Sinha and K. Bhatnagar (ed.), Mycotoxins in agriculture and food safety.* Marcel Dekker, Inc., New York, N.Y.

Schardl, C.L., Panaccione, D.G. and Tudzynski, P., 2006. *Ergot alkaloids – biology and molecular biology*. The Alkaloids. *Chem. & Biol.*, **63**:. 45–86

Stone, R. 2001. Down to the wire on bioweapons talks. *Science,* **293:**414-416

Uhlig, S. and Petersen, D., 2008. Lactam ergot alkaloids (ergopeptams) as predominant alkaloids in sclerotia of *Claviceps purpurea* from Norwegian wild grasses. *Toxicon.,***52**: 175–185.

Ulrich, R. F. and Patten, B. M., 1991. The rise, decline, and fall of LSD. *Perspect. Biol. Med.*, **34:**561-578.

Utian, W. H., 1973. Comparative trial of P1496, a new non-steroidal oestrogen analogue. *Br. Medical J.*, **1:** 579-581

Venkataramana, M., Chandranayaka, S..,. Prakash, H. S. and Niranjana, S. R., 2014. Mycotoxins Relevant to Biowarfare and Their Detection. *Biol. Toxins & Bioterrorism,* pp 295-319

Vishwanath, Prema, 2004. Mycotoxin regulations for foods: A global update. *J.Food Sci. Technol.*, **41**: 115-123.

Waliyar, F., Kumar, L., Traoré, A., Ntare, B.R., Diarra, B. and Kodio, O., 2008. Pre-and postharvest management of aflatoxin contamination in peanuts. *In : J.F. Leslie, R. Bandyopadhyay and A. Viscont, Mycotoxins: detection methods, management, public health and agricultural trade.* (p. 209-218). Wallingford: CAB International.

Walkembach, J., Brüss, M., Urban, B.W., Barann, M., 2005. Interactions of metoclopramide and ergotamine with human 5-HT3A receptors and human 5-HT reuptake carriers . *Br. J. Pharmacol.*, **146** : 543–52

Yan-ni Yin, Lei-yan Yan, Jin-hua Jiang and Zhong-hua Ma, 2008. Biological control of aflatoxin contamination of crops. *J Zhejiang Univ Sci B.*, **9** :787–792.

7

Postharvest Disease Management

The potential for significant economic losses from decay caused by postharvest pathogens is greater than is often realized. Harvesting and handling, and the attendant costs for labor and postharvest chemical treatments, can increase the cost of fresh fruits and vegetables several fold from the field to the consumer. Therefore, any produce lost in storage is worth more economically than an equivalent amount lost in the field . Considering economics and the problem of world hunger, it is imperative to understand the importance of postharvest disease and to continually strive to reduce postharvest losses from pathogens. In recent years, many international organizations have identified the necessity of reducing postharvest food losses as an important worldwide goal. While postharvest losses vary from season to season, an average loss of 10% to 30% after harvest is an average estimate. Significantly higher losses are not uncommon in developing countries. The extent of these losses justifies increased research efforts in postharvest pathology (Carl E. Sams, 1994). Chemical treatments and irradiation have effectively reduced the number of microorganisms on various commodities. However, consumer fears associated with chemical residues and the effects of ionizing radiation on fresh produce are forcing the development of alternative protection methods. Also, many pathogens have developed resistance to several commonly used chemicals, while other decaycausing microorganisms currently cannot be controlled by chemicals. Therefore it is better to seek control strategies that will supplement or replace current disease control methods.

Postharvest diseases are the result of two types of infections ; those resulting from injuries incurred during harvesting and handling, or latent or quiescent infections initiated during the growing season. In both wound and latent infections, natural antifungal metabolizes produced by the host play a role in host defense against decay. However, as fruit ripens, the concentrations of natural antifungal inhibitors drop to ineffective levels as ripening fruit tissue senesces. The onset of ripening and senescence in various fruits and vegetables renders them more susceptible to infection by pathogens. Therefore, delaying any change that is associated with ripening of the stored product can also delay the onset of susceptibility of fruit to decay and prolong storage life. Refrigerated storage has been reasonably effective in delaying senescence and reducing losses due to postharvest decay. Combining refrigeration with various types of controlled-atmosphere storage has proven even more effective in reducing postharvest losses. Controlled-atmosphere storage delays senescence of the commodity and slows the growth and development of many pathogens.

Many postharvest pathogens can degrade the polysaccharides of the cell wall by using pectolytic enzymes. Many of these enzymes are produced inductively, and the cell walls of immature fruits and vegetables may not yet have the substrates necessary to induce these fungal enzymes. Any factor that can delay the production or reduce the activity of these pectolytic enzymes can therefore reduce decay. The presence of proteinaceous inhibitors of pectolytic enzymes has been documented in some fruits and vegetables.

7.1 Principles of Postharvest Disease Management: With increased global trade, fresh produce is increasingly being moved over vast distances to foreign markets. This requires effective cold chain and product management practices to ensure premium quality fresh produce with increased shelf life. Product safety has now also become an additional minimum requirement for market entry. Producers should therefore take cognisance of the impact postharvest treatments and practices have on product quality and safety in a more holistic approach. Traditionally, exporters relied heavily on agricultural chemicals to provide effective crop protection. However, with global chemophobia, producers have to be more resilient in terms of alternative postharvest treatment. With increased global trade, fresh produce is increasingly being moved over vast distances to foreign markets. This requires effective cold chain and product management practices to ensure premium quality fresh produce with increased shelf life. Product safety has now also become an additional minimum requirement for market entry. Producers should therefore take cognisance of the impact postharvest treatments and practices have on product quality and safety in a more holistic approach. Traditionally, exporters relied heavily on agricultural chemicals to provide effective crop protection (Lise Korsten, 2006). However, with global chemophobia, producers have to be more resilient in terms of alternative postharvest treatment.

With increased global trade, fresh produce is increasingly being moved over vast distances to foreign markets. This requires effective cold chain and product management practices to ensure premium quality fresh produce with increased shelf life. Product safety has now also become an additional minimum requirement for market entry. Producers should therefore take cognisance of the impact postharvest treatments and practices have on product quality and safety in a more holistic approach. Traditionally exporters relied heavily on agricultural chemicals to provide effective crop protection. However, with global chemophobia, producers have to be more resilient in terms of alternative postharvest treatmentWith increased global trade, fresh produce is increasingly being moved over vast distances to foreign markets. This requires effective cold chain and product management practices to ensure premium quality fresh produce with increased shelf life. Product safety has now also become an additional minimum requirement for market entry. Producers should therefore take cognisance of the impact postharvest treatments and practices have on product quality and safety in a more holistic approach. Traditionally, exporters relied heavily on agricultural chemicals to provide effective crop protection.

However, with global chemophobia, producers have to be more resilient in terms of alternative postharvest treatments. With increased global trade, fresh produce is increasingly being moved over vast distances to foreign markets. This requires effective cold chain. With increased global trade, fresh produce is increasingly being moved over vast distances to foreign markets. This requires effective cold chain and product management practices to ensure premium quality fresh produce ith increased shelf life. Product safety has now also become an additional minimum requirement for market entry. Producers should therefore take cognizance of the impact postharvest treatments and practices have on product quality and safety in a more holistic approach. Traditionally, exporters relied heavily on agricultural chemicals to provide effective crop protection. However, with global chemophobia, producers have to be more resilient in terms of alternative postharvest treatments.

Sustainable agriculture is a core part of the concept of sustainable development. Given the forecast in population increase, sustainable agriculture has to achieve food security in combination with economic viability, social responsibility and have as little effect on biodiversity and natural ecosystems as possible. Based on Agenda signed at the world summit in Rio de Janeiro 1992, sustainable agriculture takes a truly global perspective. This concept requires a thorough understanding of agro-ecosystem functions. The protection of soil and water is one necessary prerequisite as well as the effi cient use of mineral and organic fertilizers. This might be achieved by means of improved technology and better understanding of the basic processes in soils. Solving the persistent hunger problem is not simply a matter of developing new agricultural technologies and practices. Most poor producers cannot afford expensive technologies. They will have to find new types of solutions based on locally-available and cheap technologies combined with making the best of natural and human resources. Sustainable intensification is the use of the best available technologies and inputs such as best genotypes, best agronomic management practices and best postharvest technologies to maximize yields, while at the same time minimizing or eliminating harm to the environment. Clearly, over the next 50 years we will need to learn to do just this. Therefore, this review will be focused on the postharvest physiology and management including harvesting, handling, packing, storage and hygiene of fruits and vegetables to enhance using of new postharvest biotechnology. The postharvest biology including biochemical parameters of horticultural crops quality, postharvest handling under extreme weather conditions, potential impacts of climate changes on vegetable postharvest and postharvest biotechnology have been highlighted.

In postharvest, fresh harvested food crops can be considered isolated small scale systems. Postharvest research aims to understand the quality of these 'systems' as influenced by postharvest conditions. The phenotypic quality of horticultural produce is based on genetic traits that are expressed through a cascade of reactions subject to complex regulatory mechanisms and diverse environmental conditions.Ultimately, to fully understand postharvest phenomena, a systemic approach that links genetic and environmental responses and identifies the underlying biological networks is required. Thanks to the development of high throughput omics techniques such system-wide approaches have become a viable option to support traditional postharvest research (Hertog *et al.* 2011). The structure of a biological system is defi ned by its physical parts e.g., tissues,cells, organelles and their composition e.g., DNA, proteins, metabolites, lipids. Their behaviour involves inputs e.g., external stimuli such as light, temperature, atmospheric composition, pH and nutrient levels and internal stimuli such as proteins, metabolites, hormones and various other compounds that might act as signal molecules, processing e.g., via catabolic and anabolic pathways, or processes such as gene expression, differentiation and cell division and outputs of material e.g., proteins, metabolites, information e.g., transcripts or energy e.g., heat, ATP, movement. Finally, a biological system is characterized by a high degree of interconnectivity between its various parts, showing both functional relationships e.g., through metabolic, signaling and gene regulation pathways and structural relationships between each other e.g., through compart-mentalization, receptor molecules, membrane transporters, plasmodesmata, vascular tissue, and the cytoskeleton. Systems biology relies on a multidisciplinary approach to integrate data from various disciplines of biology bringing together molecular disciplines e.g., genetics, biochemistry, molecular biology with those involving more complex systems e.g., cell biology, microbiology, plant or human physiology. The aim of systems biology is to link the quantitative data in a mathematically defined sense across the different scales of biological organization from DNA, RNA, protein to cell, tissue, organs. Mathematical modelling is used to drive integration with an aim of reaching a unified understanding of biological systems.

7.2 Management strategies: The main objective of postharvest disease management is to keep the disease free or symptom-free until it is marketed or consumed. Hence, the management strategies should aim at prevention, eradication and delaying the symptoms of diseases during transit and storage of fruits and vegetables. To manage postharvest diseases, the treatments are broadly divided into five groups *i.e.,* cultural, physical, chemical, biological and resistance approaches. The effectiveness of treatment depends on the ability of the treatment or agent to reach the pathogen, the level and sensitivity of the infection and the sensitivity of the host produce (Nabi, *et al*., 2017). The various methods of postharvest disease control of fruits as described briefly hereunder:

a. Cultural approaches: Fruits and vegetables that are injured during harvesting, sorting, packaging or transportation, and have succeeded in avoiding infection by wound pathogens, are still liable to come into contact with pathogens during packing or storage. Since disease development requires the presence of a given pathogen along with an available wound for penetration, a reduction in either of these factors will lead to the suppression of disease development. Wounding to the produce can be minimized by careful harvesting, sorting, packaging and transportation, including preventing the fruit from falling at all stages. Regarding the avoidance of wounds one should remember that physiological injuries are also caused by cold, heat, oxygen deficiency, and other environmental stresses, which predisposes the commodity to attack by wound pathogens. A general reduction in wounds also reduces the chances of infection of fruits and vegetables by pathogens during transit and storage, such factors should also be taken into consideration while packaging or storing the fruits. The source of pathogen should be immediately be removed either by disposal of rotten fruits or immersing it in a disinfectants solution like formaldehyde, isopropyl, alcohol, ammonium compounds sodium or calcium hypochlorite, in a special container.

b. Physical approaches: The postharvest diseases may be controlled by various physical treatments, such as, low temperature storage, high temperature treatments, magnetic fields and radiation. The various radiations include sound, ultrasound, radio, microwave, infrared, visible light, ultraviolet X-rays, gamma rays and cathode ray spectra. Some are highly fungicidal while others are less effective. Among these, a few have been used potentially as postharvest treatments of fruits as described briefly hereunder.

i. Use of gamma irradiation : Gamma irradiation can penetrate the produce and inactivate the deep seated pathogens. Mature fruits are relatively resistant to radiation damage because cell division rarely occurs in immature tissues. Doses required to eradicate infections range from 2,000 – 3,000Gy, in some cases as low as 1,000 and in others as high as 6,000Gy, which is far higher than the dose required for disinfection (75 – 300Gy) In crop like apple minimum dose required is 150Gy and in apricot and peaches 200Gy is required to eradicate the postharvest pathogens.

ii. Use of low temperature : Use of low temperature is considered very important in controlling decay in several fruits and vegetables. Low temperature may slow down the growth of the pathogens, but it also slows down fruit ripening process. Temperature management is important in reducing physiological deterioration and preventing moisture loss and shriveling as well as reducing disease incidence. For this reason, with many commodities refrigeration can be considered supplements to fungicidal treatments in several fruits. In general, it is recommended to store fruits at the lowest possible temperature that does not harm the host. With many fruits, the lowest desirable temperature is just above the freezing temperature. Certain varieties of apple, pears, plums, peaches can thus be stored between 0 and -2 °C. It is commonly observed that apples and pears stored at slightly below 0 °C are attacked by *Botrytis cinerea, Penicillium expansum* and *Cladosporium*. The pathogenic growth of most fungi, however is completely stopped at temperature near 0 °C.

iii. Use of high temperature : Heat may be applied to fruits in several ways such as hot water dips, vapour heat, hot dry air or hot water rinsing and brushing. However, the major factors to be considered while developing postharvest heat treatments include thermal sensitivity of target organism location of the target organism in or on the fruit and thermal sensitivity of the fruit.

These factors largely determine temperature, duration and type of the heat treatment required. Heat treatment in the form of either moist hot air or hot water dips has been commercialized for the control of postharvest diseases in several fruits such as stone fruits. This eco- friendly technique has been used to control postharvest diseases in many fruits. Hot water treatment reduces severity of various fruit rots. The advantage of hot water dipping is that it controls surface infections as well as infections that have penetrated deep in the peel, and it leaves no chemical residues in the produce. Hot water dips must be precisely administered as the range of temperature necessary to control disease (50–55^0 C) can damage the produce.

c. Chemical approaches: The success of a chemical treatment for disease control depends on the initial pathogen load, the depth of the infection within the host tissues, the growth rate of the infection, the temperature and humidity and the depth to which the chemical can penetrate the host tissues. Moreover, the applied chemical must not be phytotoxic and must fall within the ambit of the local food additive laws. For controlling postharvest diseases of fruits fungicides can be applied in different ways as under:

i. Pre-harvest chemical treatments: In most cases, control of postharvest diseases should start before harvest in the field of orchard itself. The possibility of controlling well-established pathogens by postharvest disinfection is very less since most fungicides are unable to penetrate deeply into the tissues and effective concentrations of the fungicide would not reach deep-seated infections. Hence, the effective way to reduce infections initiated in the field, including, quiescent infections is the application of broad-spectrum protective fungicides to the fruits on the plant itself. The developing fruits are sprayed to prevent spore germination and subsequent formation of appressoria and infective hyphae, which are quiescent stages of the fungus.

ii. Postharvest chemical treatments : Injuries to the produce during harvesting, handling and packaging, are the major sites of invasion by postharvest wound pathogens, the protection of wounds by chemicals will considerably decrease decay in storage. Other potential sites of infection are the natural openings in the host surface, such as lenticels and stomata, whose sensitivity to infection is increased by wounding or after washing the commodity in water. An efficient disinfection process should reach the pathogenic microorganisms accumulated in all those sites. Different chemicals/fungicides which are being used to control postharvest decay in developed fruits are as under. .

Methods of application : There are several methods by which the fungicides can be applied to the fruits for controlling postharvest diseases (Nabi *et al.*, 2017). These methods include dipping, electrostatic sprays, dusting, fumigation and use of chemical pads described briefly hereunder.

i. Dipping : In this method, fruits are immersed in water containing an appropriate concentration of a chemical, which is toxic to the disease causing fungi. The produce may be passed below a shower of the diluted chemical, which is called as „cascade application. For improving the effectiveness of dips, additives may be included in the formulation. These include wetting agents (e.g. teepol or Triton-x-100), which reduce the surface tension and allow a better coating of the chemical on the produce, and acids such as citric acid, which lowers the pH of the fungicide and can, make it more effective.

ii. Electrostatic sprays : Breaking up the pesticide solution into fine droplets and then giving them an effective electric charge for field sprays. The main advantage of this system is the increased uniformity of application of spraying materials. Similarly, there is no loss of biological activity of the materials with such sprays. The principle on which they work is that all particles have the same electrical charge and thus they repel each other. They are attracted towards crop, and then form a thin even layer/ coat. This method is being used in several fruits in many advanced countries.

iii. Fumigation : Fumigation is also considered as an effective method of chemical application in some fruits. It has several promising applications. It can be carried out immediately after harvest to prevent infection of injuries on the fruit to be transported to long distances, degreened before processing or to be held for several days before processing.

iv. Chemical pads : Paper pads impregnated with fungicidal chemicals first time were developed to prevent banana infections during transportation. These are also known as crown pads, and are used to prevent the fungal infections on the cut crowns of fruits. The pads are made from several layers of soft paper previously soaked in a fungicide (often thiobendazole) and then dried.

d. Biological approach: In recent years, there has been considerable interest in the use of antagonistic microorganisms for the control of postharvest diseases. The global trend appears to be shifting towards reduced use of fungicides on produce and hence, there is a strong public and scientific desire to seek safer and eco-friendly alternatives for reducing the decay loss in the harvested commodities. Several postharvest diseases can now be controlled by microbial antagonists. The various mechanism(s) most widely accepted by which microbial antagonists suppress the postharvest pathogens is competition for nutrients and space. In addition, production of antibiotics, direct parasitism, and possibly induced resistance in the harvested commodity are other modes of their actions by which they suppress the activity of postharvest pathogens in fruits. Microbial antagonists are applied either pre or postharvest on produce, but latter applications are more effective than former. Mixed cultures of the microbial antagonists appear to provide better control of postharvest diseases over individual cultures or strains.

The postharvest environment provides the advantages for biological control measures, *viz.*, the partially controlled environment in storage may result in a shift in the balance of interactions between host, pathogen, and antagonistic microbe in favour of antagonist; the efficacy of antagonist may be enhanced because the biocontrol product can be applied directly onto the site where needed in the harvested product; the harvested commodity may be protected relatively free of potential interfering factors; protection is needed for a relatively short period as compared with period of protection required for field crops, and the harvested fruits have high market value, use of a relatively high-cost biocontrol product may be justified (Nabi, *et al.*, 2017).

On the other hand, certain obstacles have to be overcome in the attempts to protect the harvested commodity by using microbial agents like the level of control to be achieved is extremely high (95–98%), since the market value of commodity depends on the blemish-free, attractive appearance; very strict food safety considerations demand careful examination of products for the possible presence of toxic or unacceptable substances produced by biocontrol agents; the potential market for the use of bioagents against postharvest diseases (bio fungicide) is relatively less as compared to control of soil borne field crop diseases.

i. Microbial antagonists : Several microbial antagonists have been reported to control postharvest diseases of fruits successfully. For their use, two basic approaches have to be employed. First approach is to promote and manage those antagonistic microbes, which already exist on the fruits itself, and the second approach should be to artificially induce the desirable microbial antagonists against postharvest pathogens. Both these approaches have been discussed briefly hereunder. Many researchers have tried different bio agents for controlling postharvest diseases of temperate fruits.

ii. Mode of action of microbial antagonists: There are two reasons to understand about the mode of action of the microbial antagonists, it will help in developing more reliable procedure for better results from the known antagonists, and it would help in providing more rationale for selecting more effective antagonists. The available information indicates that antibiotic production is the major mode of action of antagonists, which are used for controlling postharvest diseases. In addition, nutrient competition, direct parasitism, and possibly induced resistance are other actions of antagonists, which suppress the activity of pathogens in fruits during postharvest handling.

e. Resistance approach: Development and use of resistant varieties against pathogens is considered as the most reliable method of disease management. Unfortunately, a little attention has been paid to develop resistant varieties against postharvest pathogens in horticultural crops. In general, those varieties are preferred, which have thin peel, have low tannin content and high sugar content, and unfortunately, all these factors favour susceptibility to postharvest pathogens. Plant breeder needs to recognize the resistance to postharvest diseases, which is different from the field resistance and hence, breeding programme should be developed to use only this type of resistance.

Plants possess various biochemical and structural defence mechanisms which protect them against infection. Some of these mechanisms are in place before arrival of the pathogen (i.e. constitutive resistance), while others are only activated in response to infection (i.e. induced resistance). Compared to intact plants, relatively little is known about host defence responses in harvested commodities, although there is growing interest in this area. In avocado for example, research in Israel and Sri Lanka is uncovering the biochemical mechanisms associated with host resistance to anthracnose. Preformed antifungal diene compounds present in fungitoxic concentrations in unripe fruit decline to sub-lethal concentrations in ripe fruit allowing the pathogen to develop once fruit ripening commences. Levels of the diene compounds can be increased by applying various treatments such as challenge inoculation with either pathogenic or non-pathogenic strains of *Colletotrichum* or by treatment with certain antioxidants or high concentrations of CO_2. In contrast to preformed antimicrobial compounds, phytoalexins are only produced in response to pathogen invasion, although in some cases they can be elicited by certain chemical and physical treatments. For example, non-ionising ultraviolet-C radiation is known to induce production of phytoalexins in various crops. UV treatment of carrot slices induces production of 6-methoxymellen which is inhibitory to *B. cinerea* and *Sclerotinia sclerotiorum*.

f. Natural fungicides: Many compounds produced naturally by microorganisms and plants have fungicidal properties. Chitosan, for example, is not only an elicitor of host defence responses but also has direct fungicidal action against a range of postharvest pathogens. Antibiotics produced by various species of *Trichoderma* have potent antifungal activity against *B. cinerea, S. sclerotiorum, Corticium rolfsii* and other important plant pathogens. There are many other natural compounds which have been isolated and shown to possess considerable antifungal activity. Although these compounds may be more desirable than synthetic chemicals from a consumer viewpoint, their potential toxicity to humans needs to be evaluated before useable products are developed. On a positive note, the high specificity of many of these compounds also means that they biodegrade readily.

To conclude, a wide variety of fungal and bacterial pathogens cause postharvest disease in fruit and vegetables. Some of these infect produce before harvest and then remain quiescent until conditions are more favourable for disease development after harvest. Other pathogens infect produce during and after harvest through surface injuries. In the development of strategies for postharvest disease control, it is imperative to take a step back and consider the production and postharvest handling systems in their entirety. Many preharvest factors directly and indirectly influence the development of postharvest disease, even in the case of infections initiated after harvest. Traditionally fungicides have played a central role in postharvest disease control. However, trends towards reduced chemical usage in horticulture are forcing the development of new strategies.

7.3 Postharvest treatments designed to minimize produce contamination and to maximize quality: Treatments to reduce microbial contamination Over the past few years, food safety has become and continues to be the number one concern of the fresh produce industry. A "Guide to Minimize Microbial Food Safety Hazards for Fresh Fruits and Vegetables," was published by the U.S. Food and Drug Administration in October 1998. This guide is based on the principles viz., prevention of microbial contamination of fresh produce is favored over reliance on corrective actions once contamination has occurred (Adel A. Kader and Rosa S. Rolle, 2004) ; In order to minimize microbial food safety hazards in fresh produce, growers, packers, or shippers should use good agricultural and management practices in those areas over which they have control; Fresh produce can become microbiologically contaminated at any point along the farm-to-table food chain. Human and/or animal faeces are the source of microbial contamination of fresh produce; Whenever water comes in contact with produce, its quality dictates the potential for contamination. The potential of microbial contamination from water used with fresh fruits and vegetables must be minimized; The use of animal manure or municipal biosolid wastes as fertilizers should be closely managed in order to minimize the potential for microbial contamination of fresh produce; and Worker hygiene and sanitation practices during production, harvest, sorting, packing, and transport play a critical role in minimizing the potential for microbial contamination of fresh produce.

Clean water containing an appropriate concentration of sanitizers is required in order to minimize the potential transmission of pathogens from water to produce, from healthy to infected produce within a single lot, and from one lot of produce to another, over time. Waterborne microorganisms, including postharvest plant pathogens and agents of human illness, can be rapidly acquired and taken up on plant surfaces. Natural plant surface contours, natural openings, harvest and trimming wounds and scuffing can provide points of entry as well as safe harbor for microbes. When located in these protected sites, microbes are largely unaffected by common or permitted doses of postharvest water sanitizing treatments. It is therefore essential that the sanitizer concentration is sufficient to kill microbes before they attach or become internalized in produce. The concentration of sanitizer is important in some pre-harvest water uses (such as spraying pesticides or growth regulators) and in all postharvest procedures involving water, including washing, cooling, water-mediated transport (flumes), and postharvest drenching with calcium chloride or other chemicals.

i. Treatments to minimize water loss: Transpiration, or evaporation of water from the plant tissues, is one of the major causes of deterioration in fresh horticultural crops after harvest. Water loss through transpiration not only results in direct quantitative losses (loss of saleable weight), but also causes losses in appearance (wilting, shriveling), textural quality (softening flaccidity, limpness, loss of crispness and juiciness), and nutritional quality. Transpiration can be controlled either through the direct application of postharvest treatments to the produce (surface coatings and other moisture barriers) or through manipulation of the environment (maintenance of high relative humidity). Treatments that can be applied to minimize water loss in fruits and vegetables include: *a.* Curing of certain root vegetables, such as garlic, onion, potato, and sweet potato. *b.* Waxing and the use of other surface coatings on commodities, such as apple, citrus fruits, nectarine, peach, plum, pomegranate, and tomato. *c.* Packaging in polymeric films that act as moisture barriers. *d.* Careful handling to avoid physical injuries, which increase water loss from produce. *e.* Addition of water to those commodities that tolerate misting with water, such as leafy vegetables

ii.Treatments to reduce ethylene damage The promotion of senescence in harvested horticultural crops by ethylene (1 ppm or higher) results in acceleration of deterioration and reduced postharvest life. Ethylene accelerates chlorophyll degradation and induces yellowing of green tissues, thus reducing the quality of leafy, floral, and immature fruit-vegetables and foliage ornamentals. Ethylene induces abscission of leaves and flowers, softening of fruits, and several physiological disorders. Ethylene may increase decay development of some fruits by accelerating their senescence and softening and by inhibiting the formation of antifungal compounds in the host tissue (Adel A. Kader and Rosa S. Rolle, 2004). In some cases, ethylene may stimulate the growth of fungi, such as Botrytis cineria on strawberries and Penicillium italicum on oranges. The incidence and severity of ethylene-induced deterioration symptoms is dependent upon temperature, time of exposure, and ethylene concentration. The yellowing of cucumbers can, for example, result from exposure to either 1 ppm ethylene over 2 days or to 5 ppm ethylene over 1/2 day at 10 °C. Ethylene effects are cumulative throughout the postharvest life of the commodity. Treatment of ornamental crops with 1-methylcylopropene (1-MCP), which is an ethylene action inhibitor, provides protection against ethylene damage. The commercial use of this product at concentrations of up to 1 ppm on apples, apricots, avocados, kiwifruit, mangoes, nectarines, papayas, peaches, pears, persimmons, plums, and tomatoes was approved by the United States Environmental Protection Agency. The use of 1-MCP will no doubt be extended to several other fruits and vegetables, and to use in other regions.

The management of postharvest diseases is a not an easy task and there is intensive need to focus upon by scientists, administrators and policy makers. The future investigations should be focused on, Proper diagnosis and detection of postharvest diseases and pathogens, exploring the possibility and potentiality of use of non-pathogenic or attenuated strains of postharvest pathogens, Identification of compounds, which when injected into the trunk or sprayed on foliage long before harvest, are translocated to the fruit, making it resistant to infection by postharvest pathogen, use of biotechnological tools for identification of genes that promote epiphytic antagonists and producing resistant fruits with good quality characters.

It is essential to ensure that close coordination, constant surveillance, and efficient technical support for rapid detection and precise identification of microbial pathogens, and feedback on the effectiveness of corrective measures taken to restrict the incidence and subsequent spread of diseases, are available. Avoidance of wounds to the harvested produce is the basic precaution, to be strictly enforced in all cases. Though these methods are nonspecific, they will effectively reduce the chances of infection by microbial pathogens. The strategies that can directly act on the microbial pathogens may be integrated broadly into four combinations, *viz*., physical + chemical, physical + biocontrol, biocontrol + chemical, and resistance + biocontrol + physical + chemical methods. Various strategies for the management of postharvest diseases have been found to be effective to varying levels under wide range of conditions which interact with each other. The possibility of integrating the different effective strategies to achieve higher level of control of postharvest pathogens and to minimize or replace the use of synthetic fungicides has to be explored in certain host–pathogen systems. The usefulness of integrating different strategies to provide better control of diseases and to obtain safe, disease- and residue-free food products is the need of hour.

i. **Preventative :** Major preventive measures include -Avoidance of the pathogen (Cultural practices), Host resistance (Resistant varieties), Exclusion (Quarantines and Sorting/Grading), Eradication (Eliminating or reducing inoculum - Sanitation), Protection/Prevention (Chemical or biological or physical treatments – Cold temperature).

ii. **Curative** : Therapy (Physical or chemical treatments)

iii. **Eradication:** Sanitation washes using oxidizing materials (chlorine, ozone, peroxide, etc.), disinfestation of non-injured commodity surfaces and of micro-organisms in water ; Strategies of postharvest decay control for protection, suppression, or eradication of decay ; Altering the micro-environment (Treatments with indirect effects on pathogen-Change in pH); Treatments with direct effects on pathogen: Biocontrols, Fungicides.

iv. **Altering the host physiology and susceptibility :** Indirect effect on pathogen; Plant growth regulators (Effective against weak pathogens).

v. **Altering the micro-environment:** Treatments with indirect effects on the pathogen (Change in pH , Alkaline solutions of borax, sodium carbonate (soda ash), and sodium bicarbonate); Accumulation of acid in potential infection sites, (e.g. SO_2)

vi. **Treatments with direct effects on pathogen:** Biocontrol: (Competition, antibiosis, parasitism) ; Fungicides (Direct toxicity).

vii. **Borax, sodium carbonate (soda ash), and sodium bicarbonate:** Change in pH (Accumulation of alkali in potential infection sites on fruit surface); Germination of pathogen spores is inhibited (fungistatic action), Heated solutions are more toxic

viii. **Biocontrol method**: Competition, antibiosis, parasitism. Development is driven by safety concerns ; Activity from laboratory experiments is difficult to transfer into a commercial scale ; No activity against existing infections (infections that occur at harvest); Efficacy is generally inconsistent and never complete

ix. **Integrated approach :** Hormoneous integration of appropriate methods.

Postharvest management : Postharvest management is a set of post-production practices that includes: cleaning, washing, selection, grading, disinfection, drying, packing and storage. These eliminate undesirable elements and improve product appearance, as well as ensuring that the product complies with established quality standards for fresh and processed products. Postharvest practices include the management and control of variables such as temperature and relative humidity, the selection and use of packaging, and the application of such supplementary treatments as fungicides. After they are harvested, the value of fruits and vegetables is added in successive stages up to the point when someone eats them. The aim of postharvest management is to maximize this added value (El-Ramady *et al.*, 2015). This ultimately should benefit the whole community, whether through increased export earnings or extending the availability of fresh produce through the year. Conversely losses hurt everyone. Postharvest management can influence all them, with the two most important areas being temperature management and packaging. Another point to remember is that the loss of value of a downgraded product is likely to be substantially greater for highly differentiated branded products which sell at a premium in the market. All the hard work that has gone into promoting and raising the profile of a branded product can be quickly eroded if there are postharvest quality problems with some lines of that product.

7.4 Postharvest care

i. **Handling and packaging** : Careful handling is the most important step that has to be taken at all postharvest stages. Sorting and grading of the harvested products is good farm house practice. Eliminating the products showing injury or early symptoms of a disease, revealed at this stage by careful inspection can substantially reduce the intensity of disease at later stages. Packing practice has to be evolved to suit the individual consignment, aiming at the maximum perfection as well as cost effectiveness. Before packing remove superficial tissues from products like banana for reducing the incidence of crown-rot disease. In modern practice corrugated boxes/ fiber board packs/molded plastic trays/ wooden, crates are findings their places. Use impregnated paper for wrapping individual fruits such as mango, papaya, banana, guava with paper.

ii. **Care during transport** : Horticultural produce may be transported by road, rail, sea or air, sometimes for many thousands of miles. A well planned schedule should be worked out for transportation of fruits according to the mode of journey, the nature of fruits and distance as well as duration of the journey. The fruits of guava and papaya are comparatively more prone to injury and hence need special care during transportation and if possible temperature and other ambient conditions may be controlled. High value perishables can be sent by air, either in wide bodied freights or in the cargo holds of passenger aircraft (Yadav *et al.*, 2013).

iii. **Storage :** Low pressure storage, cold storage and modified atmosphere storage are aimed out preventing the perpetuation of pathogen and spread of the disease. Storage disinfection of warehouses: Debris from warehouses/cold storages should be removed, the walls and floors should be washed with bleaching powder, copper sulfate, etc to eradicate the pathogens which are surviving in the store structures.

a. **Low temperature storage**: Low temperature storage is less suitable for tropical fruits like mango, banana, sapota and guava which are sensitive to chilling. Papaya fruit inoculated with *Colletotrichum gloeosporiodes* and when stored at low temperature (10° C for 21 days) develops less anthracnose. Suitable temperature for mango storage is more than 13°C, temperature below 13° C causes chilling injury, for guava more than 15° C, for banana more than 10° C and for papaya more than 6° C is required.

b. **Cold storage**: Storage at low temperature is the main method for reducing deterioration of harvested fruits and vegetables. Low temperature affects both the host and the pathogen simultaneously. They prevent moisture loss from the host tissues and consequent shriveling; they retard metabolic activity and delay physiological changes that lead to ripening and senescence.

c. **Modified atmosphere storage**: This method have found effective in extending the shelf life of harvested fruits. In storage maintaining a low temperature, an adequate humidity, low oxygen, low ethylene concentration are favorable. If atmosphere of mango storage is modified as 2% CO_2, 6% O_2 and 13° C it can be stored for longer.

7.5 The nature of postharvest management : The horticultural produce includes fruits, vegetables, flowers and other ornamental plants, plantation crops, aromatic and medicinal plants and spices. According to Oxford English Dictionary, fruit can be defined as '*the edible product of a plant or tree, consisting of seed and its envelope, especially the latter when it is juicy or pulpy*'. The consumer definition of fruit would be 'plant products with aromatic flavors, which are either naturally sweet or normally sweetened before eating. The classification of fruits and vegetables is arbitrary and according to usage. Botanically many crops, defined as vegetables, are fruits e.g., tomato, capsicum, melons etc. Morphologically and physiologically the fruits and vegetables are highly variable, may come from a root, stem, leaf, immature or fully mature and ripe fruits. They have variable shelf life and require different suitable conditions during marketing. All fresh horticultural crops are high in water content and are subjected to desiccation (wilting, shriveling) and to mechanical injury. Various authorities have estimated that 20–30% of fresh horticultural produce is lost after harvest and these losses can assume considerable economic and social importance. That is why, these perishable commodities need very careful handling at every stage so that deterioration of produce is restricted as much as possible during the period between harvest and consumption (El-Ramady *et al.*, 2015)

Horticultural produce is alive and has to stay alive long after harvest. Like other living material it uses up oxygen and gives out carbon dioxide. It also means that it has to receive intensive care. For a plant, harvesting is a kind of amputation. In the field it is connected to roots that give it water and leaves which provide it with the food energy it needs to live. Once harvested and separated from its sources of water and nourishment it must inevitably die.

The role of postharvest handling is to delay that death for as long as possible. Horticultural managers must possess many skills to succeed in this. They need a keen appreciation of horticultural diversity. For example, spinach and apples, bananas and potatoes each have their own requirements. The optimum postharvest management of horticultural products is not the same for all products. Growers, wholesalers, exporters and retailers must all be aware of the specific needs of a product if the postharvest shelf life and quality is to be maximized .

7.6 Postharvest treatments : Fresh fruits are living tissues subject to continuous change after harvest. Some changes are desirable from consumer point of view but most are not. Postharvest changes in fresh fruit cannot be stopped, but these can be slowed down within certain limits to enhance the shelf life of fruits. The postharvest treatments play an important role in extending the storage and marketable life of horticultural perishables (El-Ramady *et al.*, 2015). The most important postharvest treatments include the following:

7.6.1. Chlorination: The primary applications are intended to minimize the redistribution and transmission of pathogens from adhering soil, infested fruit or vegetable surfaces or debris to non-infested surfaces such as harvest and trimming cuts, or natural plant surface openings. Another critical use continues to be in water disinfection, primarily for harvest and postharvest handling and cooling. Chlorine treatment (100–150 ppm available chlorine) can be used in wash water to help control inoculum build up during packing operations. Maintain pH of wash water between 6.5 and 7.5 for best results.

Chlorine is a very irritating, heavy, greenish yellow gas with a strong, pungent odor. Free chlorine is very reactive, combining with any chemical that will react with oxygen, and is never found uncombined in nature. Chlorine in the gaseous form is a very potent disinfectant, although it is seldom used in that form. It is much safer and easier to use when dissolved in water. Disinfection of produce using chlorine or some other chemical is nearly always done during hydrocooling or during the process of washing the produce to remove soil. Chlorine for disinfection may be obtained from one of three sources: pressurized chlorine gas, calcium hypochlorite (a soluble solid), or a solution of sodium hypochlorite. Chlorination is a dynamic chemical process. Its effectiveness is influenced by a number of factors (Mike Boyette, *et al.*, 1993). Proper chlorination requires frequent monitoring of the solution and a thorough understanding of the factors involved. These factors include the pH of the solution, chlorine concentration, water temperature, amount of organic matter present, exposure time, and the growth stage of the pathogens present.

a. Solution pH : The pH of a solution is a measure of its acidity or alkalinity. A solution that is neutral (neither acid nor alkaline) has a pH of 7.0. Solutions with pH numbers less than 7.0 are acid; the lower the number, the greater the acidity. On the other hand, the greater the number above 7.0, the more alkaline, the solution. A change of one pH unit indicates a ten-fold change in acidity or alkalinity. The pH of the solution has a significant effect on the level of chlorination activity. When chlorine gas or one of the hypochlorite salts is added to water, each will generate chlorine gas (Cl_2), hypochlorous acid (HOCl), or hypochlorite ions (OCl) in various proportions, depending on the pH of the solution. The form desired for chlorination is hypochlorous acid (HOCl). Hypochlorite ions are relatively inactive, and chlorine gas quickly bubbles out of the solution, causing worker discomfort and serving no useful purpose.

At a pH slightly above neutral, half of the chlorine will be in the form of hypochlorous acid and the other half in the form of hypochlorite ions. Very little will be in the gaseous form. Solutions that are more acid have a higher percentage of hypochlorous acid but are very unstable, allowing more of the chlorine to escape from the solution as a gas. To maximize the proportion of hypochlorous acid and hence the effectiveness of the solution, the pH should be kept in the practical range between 6.5 and 7.5. Because well water in North Carolina varies from moderately acid to moderately alkaline, the pH should be checked with a pH meter or test papers before and after the chemicals are added and frequently during operation. Furthermore, even if the water initially has a near-neutral pH, the addition of hypochlorites will change the pH.

Different sources of chlorine have different effects on pH, viz., chlorine gas decreases pH , sodium hypochlorite increases pH and calcium hypochlorite increases pH slightly. It may be necessary to add a ommon acid (like vinegar) to lower the pH. Small amounts of sodium hydroxide (lye) may be used to raise the pH. Inexpensive test papers for checking both the chlorine level and pH may be obtained from most swimming pool and chemical supply houses.

Chlorine concentration : The concentration of a small amount of chemical in a solution is measured in units of parts per million (ppm). In the case of chlorination, this unit of measure indicates the number of parts of available chlorine, by weight, that there are in a million parts of solution. The quantity of calcium or sodium hypochlorite that must be added to a certain quantity of water to obtain a given concentration depends on the available chlorine content of the compound, the concentration of the compound and the volume of water to be treated.

Temperature : The activity of chlorine increases with the temperature of the solution. Unnecessarily warm solutions should be avoided, however, because the chlorine escapes into the air more rapidly as the temperature increases. On the other hand, in hydrocooling the combined effects of low temperature and high pH values reduce chlorination efficiency.

Organic matter : Chlorine has a particular affinity for soil particles and organic matter. Chlorinating dirty produce therefore depletes the chlorine supply much faster than relatively clean produce. The amount of chlorine constantly decreases with chlorination reactions. The more organic matter (such as fruit, leaves, or soil) in the tank, the more chlorine will be lost. As a result, the chlorine level should be checked and adjusted hourly, especially when large loads of produce are being processed. Extremely dirty produce (such as sweetpotatoes) is commonly washed with clean water before it is placed into the chlorination tank.

Exposure time : The effectiveness of chlorination depends greatly on the length of time the produce is exposed to the chlorine solution. Quick dips are much less effective than longer exposures. However, most of the sanitizing action of the chlorine is accomplished within the first several minutes of exposure. Prolonged exposure to strong chlorine solutions has been known to cause surface bleaching. Experience is the best guide to the correct combination of treatment time and chlorine concentration for the crop being processed.

Growth stage of the pathogen : Disease organisms may be either in the active vegetative form or in the form of spores. Chlorine will readily kill the vegetative form, but fungal spores are 10 to 1,000 times more difficult to kill. Therefore, chlorine treatment rarely eliminates all pathogens and sterilizes the surface of the produce. Many spores may remain on the surface to develop later should the opportunity arise. Further, chlorine kills only on contact, not systemically, and is effective only on exposed pathogens such as those suspended in water or those on the surface of produce; chlorine does not kill pathogens below the skin because it cannot contact them. Chlorination leaves no residual effect. Therefore, produce exposed to pathogens after treatment is susceptible to reinfection.

Wastewater disposal : Chlorinated wastewater is customarily drained at the end of each work day or more often if circumstances dictate. This wastewater often contains sediment, pesticides, and other suspended matter. If it is discharged to a municipal wastewater treatment plant or to surface waters (canals, creeks, or ponds), regulatory agencies may consider it to be industrial wastewater. Land application of this material is normally permitted, but a nondischarge permit may be required (Trevor Suslow , 2015). Operators may be required to obtain wastewater discharge permits. Illegal disposal of hydrocooler wastewater may result in a substantial penalty.

The Environmental Protection Agency (EPA) regulation of May 8, 1991, (40 CFR part 180) exempts calcium hypochlorite and chlorine gas from residue tolerance requirements when it is used before or after harvest on any raw agricultural commodity. Any amount can legally be used; however, excessive chlorination can damage equipment, injure the surface of fruits and vegetables, and waste money. In addition, the use of excessive amounts of chlorine may pose a worker health and safety hazard that is regulated by the North Carolina Department of Labor.

Practical rules for successful chlorination: If water is not necessary in the packing process, do not use it. Wetting the produce greatly increases the likelihood of damage by postharvest diseases. If the produce must be washed to remove soil, there is no alternative to wetting. Hydrocooling also necessitates wetting the produce, although other methods, such as forced-air cooling, may be a viable option in some cases. When water is necessary in packing lines (for example, in dumping tanks, flumes, or a hydrocooler), always treat it to reduce the risk of disease (Mike Boyette, *et al.*, 1993).

i. Monitor the chlorine concentration and the condition of the water. Check the chlorine concentration and pH frequently using test papers or electronic equipment. Automatic chlorination equipment is available that will continually monitor the condition of the solution, add chlorine, and correct the pH. Also, monitor the water temperature.

ii. Avoid overexposure. Do not allow the produce to remain in contact with the solution longer than necessary. Check circulation patterns in chlorination tanks to eliminate dead spots.

iii. Change the water frequently. Chlorination efficiency is poor in dirty water. If necessary, wash very dirty produce with clean water before it comes into contact with the chlorinated water.

iv. Dispose of wastewater properly. Before installing chlorination equipment, plan how you will dispose of the wastewater. Land application of wastewater is normally allowed, but check to see if a permit is needed. Illegal disposal of chlorinated water could result in a substantial fine.

v. Practice good sanitation. Hose off the packing equipment and floors daily; remove any dirt and trash that has settled in the chlorination tank. Sanitize the equipment with a spray solution composed of four pints of

5.25 percent sodium hypochlorite solution in 10 gallons of water. As an alternative, steam clean the equipment with an approved detergent. Do not allow culls or decayed produce to remain in or around the packing house.

vi. Protect workers. For their safety and comfort, workers must be protected from the chlorine fumes associated with excessively high levels of chlorine. If the amount of chlorine gas in the work area is great enough to cause worker discomfort, the amount of chlorine being used is well above that required for proper postharvest sanitation. If air monitoring equipment is not available, chlorine concentrations can be checked by asking a person who has not been desensitized by the odor to enter the work area. If he or she can smell the chlorine, the level is probably adequate. The concentration is too high if workers are continually irritated by the odor.

vii. Remember that chlorination will not solve all problems. Even the best chlorination program may not be sufficient to prevent all postharvest decay. Prompt handling, proper sanitation, and rapid cooling should all be part of postharvest disease management program. Produce infected in the field or otherwise damaged cannot be saved by chlorination.

7.6.2 Ethylene inhibitors/Growth regulator/fungicide treatments: 1-MCP (1-methyl cyclopropene), AVG (Amenoethoxyvinyl gycine), silver nitrate, silver thiosulfate, cycloheximide, benzothiadiazole etc. are some of the chemicals which inhibit ethylene production and/or action during ripening and storage of fruits. The growth regulators or fungicidal application such as GA 3 can be effectively used to extend/enhance the shelf life of fruits.

The plant hormone ethylene has become the focus of plant biology over the last 100 years. It is a gaseous plant hormone that is responsible for fruit ripening, growth inhibition, leaf abscission, aging and a wide range of other plant processes. Extensive reviews of ethylene biosynthesis have been described by researchers and the effects of ethylene in plants are well established. Most studies on fruit ripening and ethylene biosynthesis have been confined to climacteric fruits such as apple, tomato, avocado, banana and to a lesser extent in papaya, kiwi fruit and breadfruit. Ethylene is also at the centre of postharvest technology acting as a key in the extension of shelf life and fruit quality during storage. The use of postharvest technology is being examined to preserve the quality and shelf life of produce through the removal/inhibition of ethylene from storage systems and describe systems that primarily employ ethylene inhibition/removal as its core (Golden, *et al.*, 2014).

As inhibitors of ethylene biosynthesis and action, good results have been found with polyamines and 1-methylcyclopropene (1-MCP) in terms of maintenance of fruit and vegetable quality and extension of postharvest shelf-life. As ethylene scavengers, the best results can be achieved by adsorbers combined with catalysts, either chemical or biological (biofilters) (Domingo Martínez-Romero *et al.*, 2007).

Inhibition of ethylene production: Low temperature, Low O_2, Chemical inhibitors of enzymes, AVG / AOA/ Others, Antisense technology, Gene/ DNA inserted reverse order, nonfunctional and natural mutants.

Inhibition of ethylene action: Low temperature, High CO_2, Low O_2 STS, Silver thiosulphate, (Cut nonedible flowers), Cyclopropenes, 1-MCP (Smartfresh™; Ethylbloc™).

Calcium application: The postharvest application of $CaCl_2$ or Ca $(NO_3)_2$ play an important role in enhancing the storage and marketable life of fruits by maintaining their firmness and quality. Calcium application delays aging or ripening, reduces postharvest decay, controls the development of many physiological disorders and increases the calcium content, thus improving their nutritional value. The postharvest application of $CaCl_2$ (2–4 %) or Ca $(NO_3)_2$ for 5–10 min dip extend the storage life of pear up to 2 months, plum up to 4 weeks and apple up to 6 months at 0–2 °C with excellent color and quality. Calcium infiltration reduces chilling injury and increase disease resistance in stored fruit.

Calcium is a safe and viable practice for postharvest disease control in papaya. Calcium salts may also have a significant role when fruit are subjected to long distance transportation. Moreover, the work will continue investigating pre-harvest calcium application at 1.5 or 2.0% concentrations combined with postharvest dipping and infiltration of calcium to control anthracnose disease. pre-harvest calcium chloride sprays at 1.5-2.0% concentration increased significantly fruit calcium content, reduced significantly spore germination and the anthracnose incidence on fruits during five weeks of storage (Madani *et al.*, 2014). After 35 days of storage, the anthracnose incidence was 100%, 62% and 46% in untreated (control), 1.5% calcium chloride treated, 2.0% calcium chloride treated fruit, respectively.

Calcium is a key plant nutrient that has a significant role in cell functions, including reducing softening and senescence of fruits. Many disorders of crops such as bitter pit in apple, cork spot in pear, and blossom end rot in tomato that are caused by calcium deficiency could be reduced by calcium spraying. The most common form of calcium supplement is calcium chloride, however many proprietary products containing calcium in other forms, or combined with other nutrients are available. Several studies have shown that calcium salts may have potential as environmentally compatible, nontoxic fungicides for controlling postharvest pathogens. Calcium chloride and calcium hydroxide restrict *in vitro* conidial germination and germ tube growth of *C. acutatum, C. gloeosporioides, A. alternata*, and *P. expansum*.

Synthetic fungicides have been routinely used to control postharvest diseases. However, the development of fungicide resistance and an increasing environmental concern over fungicide residues in food have stimulated to find alternative means for controlling postharvest decay (Stefan Stošić *et al.*, 2014). Among alternatives to synthetic chemicals, several inorganic salts and organic, lipophilic acids and their salts have shown increasing evidence of efficacy in controlling plant pathogens. Many of these salts are widely used in industry as preservatives and antimicrobials and have shown some advantages for utilization as postharvest chemicals, such as a broad-spectrum antimicrobial activity with low mammalian toxicity and biocompatibility. Several studies have been undertaken in recent years identifying the fungicidal properties of many different antimicrobial salts. Sodium carbonate, sodium bicarbonate, potassium carbonate, potassium bicarbonate, ammonium bicarbonate, and potassium silicate have been tested for inhibition of fungal pathogens on fruits, vegetables, field crops and ornamentals. These salts demonstrated in vitro and/or in vivo inhibition of *Fusarium tricinctum, F. graminearum, F. sporotrichioides, Aspergillus ochraceus, A. flavus, A. niger, P. griseofulvum, P. notatum, P. digitatum, P. expansum, Botrytis cinerea, Helminthosporium solani, C. acutatum*, and *C. gloeosporioides*. Sorbic and propionic acids and their salts are used as preserving additives in milled corn and have been tested for the suppression of mycotoxins and storage molds.

High concentration of extracellular calcium may increase calcium in the cytosol to toxic levels. With a minimum level of calcium ion concentration being necessary for normal cell growth, any limitation in regulation of intracellular calcium level may result in reduced organism development (Miceli *et al.*, 1999). Calcium ions may reduce the incidence of fungal infection by directly inhibiting fungal growth and by inhibiting cell wall–degrading enzymes produced by the pathogens. These effects were probably due the toxicity of higher concentration of calcium on pathogens by affecting the osmotic balance in the fungal cells and inhibition of pectinolytic enzymes.

Thermal treatments : Heat treatments have been investigated for use in many aspects of postharvest storage. They have been developed for insect control, prevention of fungal development and prevention of postharvest storage disorders including chilling injury. The treatment times and temperature range vary widely, from days at 35 °C to 39 °C in hot air, to up to 63 °C for less than a minute in hot water. Much of the research has been performed to develop solutions to a particular problem, and less investigation has been conducted on the responses of the commodity to the treatment (Susan Lurie and Romania Pedreschi, 2014). However, since the turn of the century, a number of groups have been active in examining the molecular responses and changes that occur in commodities during and after the heat treatment.

Heat stress affects the stability of various proteins, membranes, RNA species and cytoskeleton structures, and alters the efficiency of enzymatic reactions in the cell, causing metabolic imbalance. Disrupting the steady-state flux of metabolites can cause the accumulation of toxic compounds, such as reactive oxygen species (ROS). In fact, ROS is generally a component of heat stress and increases in antioxidant processes are part of the heat stress response. As a response to heat stress, plants reprogram their transcriptone, proteome, metabolome and lipidome. These changes establish a new steady-state balance of metabolic processes that can enable the organism to function at the new temperature. This new balance may also be helpful in responding to other stresses, such as the low temperatures of storage

A high temperature will affect most macromolecules and will cause changes such as increased membrane fluidity, partial melting of DNA and RNA strands, protein subunit disassociation and exposure of hydrophobic cores. The specific change which might be a sensor of high temperature stress is unknown, and may be different in different plants. The classical model for heat stress activation involves HSFs. Heat-stressed unfolded proteins in the cell cause molecular chaperones to be released from their constitutive inhibitory association with HSF monomers and to bind the unfolded proteins, while the free HSF subunits trimerize, undergo phosphorylation and bind to heat shock response promoters in the genome to activate heat stress responses. However, mild heat stresses may not lead to protein unfolding and yet the plant shows a stress response. Other models which appear to be dominant in some situations include; a specific calcium channel opening in the plasma membrane due to heat-induced changes in membrane fluidity, lipid signaling due to membrane fluidity changes and ROS signaling, particularly the respiratory burst mediated by a plasma membrane NADP oxidase. It is noteworthy that these models all include some action at the plasma membrane, which is also the suggested site of response to cold stress. Recently, there has been some evidence that smRNAs may also be involved in responses to heat stress, although so far only as response to high temperature and not involved in acquired thermotolerance (Susan Lurie and Romania Pedreschi, 2014).

Thermal treatments includes

a. Hot water treatment: Fruits may be dipped in hot water before marketing or storage to control various postharvest diseases and improving peel color of the fruit (Table 10).

b. Vapor heat treatment (VHT): This treatment proved very effective in controlling infection of fruit flies in fruits after harvest. The boxes are stacked in a room, which are heated and humidified by injection of steam.

Table 10: Hot water treatments for different fruits

Commodity	Pathogens	Temperature (°C) & Time (min.)
Apple	*Gloeosporium* sp. *Penicillium expansum*	45 10
Grapefruit	*Phytophthora citrophthora*	48 3
Lemon	*Penicillium digitatum Phytophthora* sp.	52 5–10
Mango	*Collectotrichum gloeosporioides*	52 5
Orange	*Diplodia* sp. *Phomopsis* sp. *Phytophthora* sp.	53 5
Papaya	Fungi	48 20
Peach	*Monolinia fructicola Rhizopus stolonifer*	52 2.5

Evaluating effect of hot water treatment on development of anthracnose and quality of mango fruit is imperative. A total of three hot water levels 48, 52 and 56 °C at two time interval (5 and 10 min) were tested with factorial arrangement in completely randomised design. The study indicated that hot water treatment at different temperatures and time interval significantly ($p < 0.001$) affects disease development and shelf life and postharvest quality of mango fruits. Hot water treatments reduced the incidence and severity of anthracnose disease significantly ($p < 0.001$) in mango fruits as compared to control. There was a highly significant difference ($p < 0.0001$) on weight loss, total soluble solids, titratable acidity and fruit firmness of mango fruits due to treatment. Hot water treatment has a potential in reducing the postharvest loss due to anthracnose and improving the shelf life and quality of mango fruits. However, the reduction of disease pressure on fruits was not at applicable level, which call ups future effort on developing on integrated disease management strategies for reduction of postharvest loss of mango fruits (Agurash Seid, *et al.,* 2017).

The mango fruits were harvested at proper stage of maturity and given hot water treatments at 52°C for 10 minutes alone and in combination with fungicidal dips *viz.*, Bavistin (0.1%) and Captan (0.2%) (Waskar and Gaikwad, 2005). The mango fruits were then packed in corrugated fibre board box and then stored in two storage environments *viz.*, at room temperature (28.12 to 36.18°C temperature and 46.18 to 71.25% RH) and in cool chamber (21.47 to 27.10°C temperature and 91 to 95% RH). It was observed that hot water treatment combined with Bavistin (0.1%) was found to be the best in controlling the incidence of anthracnose and stem-end rot. It was observed that shelf-life of mango fruits could be extended for more than 28 days when given hot water treatment coupled with fungicides and stored in cool chamber. On the contrary, the shelf life of fruit was found be 21 days when given same set of treatments and hardly 17 days when untreated and stored at room temperature. It was also observed that hot water treatment coupled with fungicide to mango fruits recorded lower physiological loss in weight and high organoleptic score when stored in cool chamber as compared to room temperature storage. The untreated (control) fruits were found to have infected with *Colletotrichum gloeosporioides* and *Diploidia natalensis*.

7.6.3 Fumigation : The fumigation of SO_2 is successfully used for controlling post- harvest diseases of grapes. This is achieved by placing the boxes of fruit in a gas tight room and introducing the gas from a cylinder to the appropriate concentration. However, special sodium metabisulphite pads are also available which can be packed into individual boxes of a fruit to give a slow release of SO_2 . The primary function of treatment is to control the *Botrytis cinerea* . The SO_2 fumigation is also used to prevent discoloration of skin of litchis.

The effect of chlorine dioxide (ClO_2) fumigation on postharvest diseases of longan fruit 'Daw' during storage at room temperature was investigated. Fresh longan fruit were fumigated with (control), 2.5, 5,10 and 25 mg/L ClO_2 for 10 min, then packed in cardboard boxes and stored at 25±1°C for seven days. ClO_2 significantly reduced and delayed the occurrence of disease and total microbial load in all treatments. ClO_2 at the concentrations of 10 and 25 mg/L were most effective with no difference in both disease incidence and total microorganisms. On day 7 of storage, treatment with 10 and 25 mg/L ClO_2 reduced the total microorganisms by 81.5 and 83.2%, respectively (Chumyam *et al*., 2015). In the second experiment, the effect of ClO_2 fumigation with0 (control) and 10 mg/L ClO_2 for10 min and kept at room temperature for seven days on the growth of *Cladosporium* sp., *Lasiodiplodia* sp. and *Fusarium* sp. isolated from longan fruit and cultured on potato dextrose agar were determined. It was shown that ClO_2 treatment significantly reduced growth of the three fungi. On day 7, ClO_2 fumigation inhibited growth of *Cladosporium* sp., *Lasiodiplodia* sp. and *Fusarium* sp. by 96.7, 87.4 and 31.2% respectively.

A variety of volatile organic compounds (VOCs) produced by *Ceratocystis fimbriata* have strong bioactivity against a wide range of fungi, bacteria and Oomycetes (Li *et al*., 2015). Mycelial growth, conidial production, and spore germination of fungi and oomycetes were significantly inhibited after exposure to cultures of *C. fimbriata*, and colony formation of bacteria was also inhibited. Two postharvest diseases, peach brown rot caused by *Monilinia fructicola* and citrus green mold caused by *Penicillium digitatum*, were controlled during a 4-day storage by enclosing wound-inoculated fruits with 10 standard diameter Petri plate cultures of *C. fimbriata* in a 15 L box. The fruits were freshly inoculated at onset of storage and the cultures of *C. fimbriata* were 6 days old. Percentage of control was 92 and 97%, respectively. After exposure to *C. fimbriata* VOCs, severely misshapen hyphae and conidia of these two postharvest pathogens were observed by scanning electron microscopy, and their pathogenicity was lost or greatly reduced.

Cucumber plants subjected to consecutive monoculture for 9 years were found to suffer from severe Fusarium wilt disease, caused by the soil-borne fungus *Fusarium oxysporum* f. sp. *cucumerinum*. Greenhouse experiments were conducted to evaluate the influence of ammonia gas fumigation on *Fusarium* wilt suppression, fungal abundance and fungal community composition (Jun Zhao, *et al,* 2017). Results showed that ammonia gas fumigation remarkably reduced disease incidence from 80% to 27%, resulting in a four-fold increase in yield, compared to the control. Total fungal abundance declined dramatically after fumigation and reached the lowest level at day 32, at 243 times lower than the control. Moreover, fumigation significantly increased soil fungal diversity, though it also decreased considerably coinciding with cucumber growth. Fumigation also significantly altered soil fungal community composition, relative to the control. *Fusarium* was strongly inhibited by fumigation in both relative abundance (3.8 times lower) and targeted quantification (a decrease of 167 fold). Collectively, the application of ammonia gas fumigation to control *Fusarium* wilt of cucumber resulted in a re-assembly of the fungal community to resemble that of a non-disease conducive consortium. Additional strategies, such as bioorganic fertilizer application, may still be required to develop sustainable disease suppression following fumigation.

Postharvest green mold and blue mold, caused by *Penicillium digitatum* and *Penicillium italicum*, respectively, were effectively controlled by fumigation of lemons and oranges for 6 h at 220 C with two applied dosages of 3000 L L−1 of ammonia that was injected initially and again 2 h later (Clara Montesinos-Herreroa, *et al*., 2011). This treatment did not injure oranges, however, it caused the tissue within previously injured areas on the rind of lemons to become darker in color. Fumigation of lemons with 6000 L L−1 of ammonia slightly accelerated the natural transition of rind color from green to yellow. The germination of conidia of *P. italicum* was more sensitive to ammonia than those of *P. digitatum*, although many survived fumigation. About 30% of the conidia of *P. digitatum* and 10% of those of *P. italicum* could germinate after a 6 h fumigation where two injections of 6000 L L−1 of ammonia were applied, one initially and a second 2 h later. Ammonia fumigation controlled an isolate of *P. digitatum* with a high level of resistance to imazalil (IMZ). The influence of ammonia fumigation on the effectiveness of this common postharvest fungicide was examined. When fruit were first immersed in 10 or 30 mg L−1 IMZ (about 10% of typical commercial rates) before ammonia fumigation, a single fumigation with 1500 L L−1 of ammonia was adequate to control both diseases and the increase in effectiveness was usually additive and sometimes synergistic. This effect was probably due in part to the influence of pH on IMZ activity, because the neutral form of IMZ increases with increasing pH and it has markedly higher antifungal activity than the ionized molecule. Fumigation with 1500, 3000, or 6000 L of ammonia per liter increased the pH (±SD) of albedo tissue of pre-existing wounds on oranges and lemons from 5.9 (±0.2) before fumigation by 0.6 (±0.3), 0.9 (±0.4), or 1.3 (±0.3) units, respectively. IMZ can be applied immediately after harvest by drenching fruit within harvest bins with aqueous IMZ solutions. Subsequent ammonia fumigation on their arrival to packinghouses may be a feasible practice, since it could employ the existent ethylene degreening chambers present at all packinghouses, if these were modified to be gas tight. Ammonia could replace synthetic fungicides or augment IMZ performance in citrus postharvest decay management. Its capacity to control IMZ resistant isolates of *P. digitatum*, common in citrus packinghouses, is particularly valuable.

7.6.4 Irradiation: Considering the high economic worldwide loss due to the postharvest decay of fruit and vegetables, and the frequent development of pathogen isolates that are resistant to synthetic fungicides, there is a need to develop more eco-friendly alternatives for controlling postharvest diseases. Irradiation (Gamma and UV-C) is one such physical method which is promising, but when used separately is not as effective as fungicides (Deepshikha, *et al.*, 2017). Therefore, to improve its effectiveness and persistence, irradiation method can be combined with other complementary control measures such as sodium carbonate, heat treatment, chemicals, cold storage and bio-control agents as apart of integrated disease management strategies. Low doses of UV radiation, particularly UV-C have proven to be effective in delaying ripening and senescence, diminishing decay and even in increasing the content of beneficial compounds owing to its germicidal properties. It mainly inactivates microorganisms, especially pathogens by creating damage in the genetic material (DNA) of the cell. These methods can be efficiently used to delay the ripening of fruits, to inhibit germination, improve nutritional quality, minimize insect infestation and deactivate viruses. Short shelf life of mushroom can be extended by inhibiting cap opening and browning, stalk elongation, reducing the level of microbial contamination and finally by increasing the concentration of vitamin D2 significantly, without causing any adverse effect on its taste.

Irradiation is a process of controlled application of energy from ionizing and non-ionizing radiations to increase the safety and shelf-life of food. Ionizing and non-ionizing radiations are physical treatments that can be used for the control of postharvest diseases. Irradiation has been used to delay ripening-associated processes and to control pathogens and insects. Different types of radiation. have been tested for fruits. The primary mode of action of many physical treatments acts as a disinfection of the commodity. Thus, fungal spores and mycelial infections both on and in the outer cell layer of fruits or vegetables are either removed or destroyed. A complex defense response against the pathogens arise in the irradiated tissue, which includes the induction of a wide range of genes, the modification of enzyme activities, as well as de novo synthesis of pathogenesis related proteins (PR) and the increase of secondary metabolites (Civello *et al.* 2014). Ionizing radiation can be applied to fresh fruits and vegetables to control micro-organisms and inhibit or prevent cell reproduction and some chemical changes. It can be applied by exposing the crop to radiations from radioisotopes (normally in the form of gamma-rays measured in Grays (Gy), where 1 Gy = 100 rads.

The effect of UV radiation in reducing green mould in grapefruit is mediated through the host response, rather than being the only result of the germicidal action of the UV treatment. an increase in grapefruit resistance to *P. digitatum* was attributed to chitinase and β-1, 3- endoglucanase induction in the fruit skin. The treatment can also extend the postharvest life of fruits by delaying ripening. The occurrence of *Alternaria alternata, Botrytis cinerea* and *Rhizopus stolonifer* in tomato were reduced with different UV-C dosages, and these fruits were firmer in texture and less red in colour than those used as controls . UV-C treatment significantly reduced *Botrytis storage* rot (*Botrytis cinerea*) in cv. 'Pajaro' of strawberries at UV has a dosage of about 0.50 and 1.00 kJ/m_2 while the single table grape berries (cv. Italia) at 0.125 to 0.5 kJ/m_2 in comparison with the un-irradiated control. The effect of different types of UV radiations on *in vitro* and *in vivo* growth of *Colletotrichum musae*, causing anthracnose disease of banana were investigated and UV-C irradiation was found to be the most effective in inhibiting the fungal growth. The same fact was observed for citrus black spot, caused by *Guignardia citricarpa*. UV-C light (254 nm) at 79.2 kJm-2 and 105.6 kJm-2 reduced the mycelial growth *Fusarium solani* f.sp. *eumartii*, causal agent of potato dry rot (Deepshikha, *et al*., 2017). The possibility of inducing resistance to bitter rot (*Colletotrichum gloeosporioides*), brown rot (*Monilinia fructicola*), and green mould

(*Penicillium digitatum*) in 'Golden delicious' apples, 'Elberta' peaches, and 'Dancy' tangerines, respectively, by treating them with ultraviolet light-C at the stem end in a stationary position without rotation were compared with the conventional procedure where fruits were rotated four times, thereby exposing the entire surface area to the full effects of the UV-C light.

In a preliminary study, Rae-Dong Jeong, *et al.*, (2015) evaluated ionizing radiations (gamma, X-ray, or e-beam irradiation) for their antifungal activity against *Botrytis cinerea*, *Penicillium expansum*, and *Rhizopus stolonifer* through mycelial growth, spore germination, and morphological analysis under various conditions. Different fungi exhibited different radiosensitivity (Figs. 46 & 47). The inhibition of fungal growth showed in a dose-dependent manner. Three fungal pathogens have greater sensitivity to the e-beam treatment compared to gamma or X-ray irradiations. The inactivation of individual fungal viability to different irradiations can be considered between 3–4 kGy for *B. cinerea* and 1–2 kGy for *P. expansum* and *R. stolonifer* based on the radiosensitive and radio-resistant species, respectively.

Gamma irradiation alone or in combination with fumigation was evaluated to manage the postharvest decay caused by *Colletotrichum gloeosporioides* and *Penicillium expansum* and also to extend the shelf life of apples in South Korea (Wonsu Cheon *et al.*, 2016). An irradiation dose of 2.0 kGy resulted in the maximum inhibition of *C. gloeosporioides* and *P. expansum* spore germination. The gamma irradiation dose required to reduce the spore germination by 90% was 0.22 and 0.35 kGy for *C. gloeosporioides* and *P. expansum*, respectively. Microscopic observations revealed that when the fungal spores were treated with gamma irradiation (4.0 kGy), conidial germination was stopped completely resulting in no germ tube formation in *C. gloeosporioides*. Treatment with the eco-friendly fumigant ethanedinitrile had a greater antifungal activity against *C. gloeosporioides* and *P. expansum* in comparison with the non-treated control under *in vitro* conditions. The *in vitro* antifungal effects of the gamma irradiation and fumigation treatments allowed us to further study the effects of the combined treatments to control postharvest decay on stored apples. Interestingly, when apples were treated with gamma irradiation combined with fumigation, disease inhibition increased more at lower (< 0.4 kGy) than at higher doses of irradiation, suggesting that combined treatments reduced the necessary irradiation dose in phytosanitary irradiation processing under storage conditions.

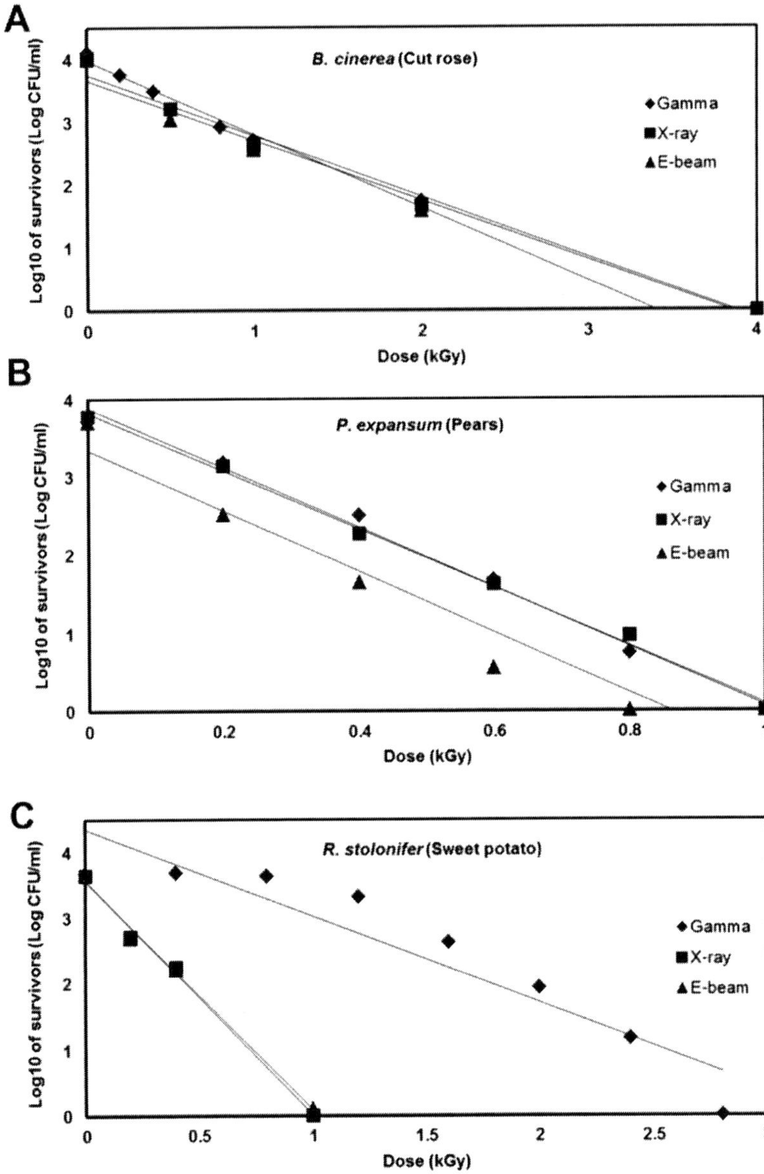

Fig. 46 : Sensitivity of *B. cinerea* (A), *P. expansum* (B), and *R. stolonifer* (C) following gamma-rays, X-rays, or e-beam treatment. The D_{10} values for three fungal pathogens were determined by calculating the negative reciprocal of the survival slope. Bars represent the means±SD. Experiments with three fungi were conducted independently three times

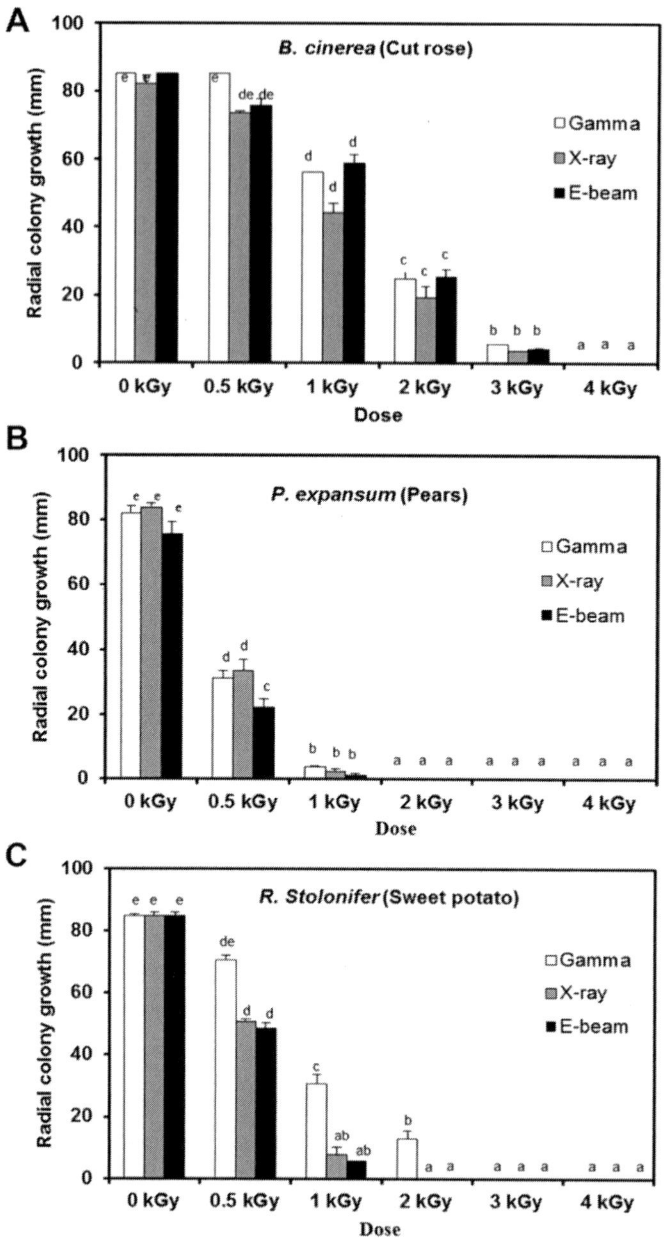

Fig. 47: Radial colony growth of *B. cinerea* (A), *P. expansum* (B), and *R. stolonifer* (C) following gamma-rays, X-rays, or e-beam treatment. Bars represent the means±SD. The value followed by the same letters are not significantly different ($p<0.05$) based on Duncan's multiple range test (n = 3)

With increasing demand for the use of non-chemicals, ionizing radiation has been investigated for application in many aspects of postharvest treatment, especially as an approach with significant potential for the control of postharvest disease. Recently, a number of more or less technologically advanced methodologies, e.g. irradiation combined with other types of treatments and induced disease-resistance, have been developed to control postharvest diseases as well as to increase the quality and storage life of fresh commodities (Jeong and Jeong, 2018). Challenges for future application of irradiation in fresh produce are the cost, lack of irradiation facilities, lack of knowledge about optimal conditions for different commodities, and lack of acceptance of irradiated fresh produce. This review aims to document the advances in understanding of the effects of irradiation on postharvest disease, the possible modes of action, and the perspectives in commercial use.

Postharvest ionizing radiation has been studied extensively since 1950 and used successfully in more than 50 countries to prevent decay of over 60 products, despite the safety issues raised by consumers. Postharvest ionizing radiation preserves fresh produce by inactivating the pathogens, as well as by inhibiting sprouting of potatoes and onions. Additionally, ionizing radiation is used worldwide as a promising phytosanitary treatment to disinfect agricultural commodities of quarantine pests. The ability of ionizing radiation to penetrate fresh produce renders irradiation a more effective means for controlling postharvest diseases. However, the use of high doses of ionizing radiation is limited because of the susceptibility of the host tissues, leading to changes in colour, texture, flavour or aroma. Thus, the employment of ionizing radiation for decay control should be considered while keeping in mind the balance between pathogen sensitivity and host resistance. The optimum radiation doses and type of radiation, such as gamma rays, X-rays and electron-beam, for fresh produce are usually decided empirically. Currently, the USA, EU and Asian countries allow irradiation doses of 1 kGy or less for fruits and vegetables. Many studies have investigated physiological responses of postharvest pathogens to ionizing radiation alone or combined with other treatments to assess the optimal doses or conditions for their inactivation. Nevertheless, postharvest ionizing radiation is not used as much as other conventional technologies due to the high costs of irradiation units and the negative perception of consumers about its safety.

Synergistic effects of gamma radiation combined with heat treatment (38 °C for 4 days) contributed to the 5- to 10- fold increase in the inactivation of spores of postharvest pathogens (Temur & Tiryaki, 2013). Heat treatment may also affect the susceptibility of the host to pathogens by triggering the synthesis of an inhibitory substance in the peel. The combination of hot water and gamma irradiation synergistically reduced fungal development in tomato fruits, resulting in 1.7% and 10.0% infection rates by *B. cinerea* and *Rhizopus stolonifer*, respectively. Moreover, a hot water dip followed by irradiation at 0.5 kGy totally eliminated the decay caused by *Alternaria alternata* for 8 days at 23 °C in mango fruits. However, the combination of hot water with irradiation is not commonly used on citrus due to the detrimental effect it has on the quality of treated clementines and the contradictory results presented in different studies. Another study suggested that a combined treatment of 0.5 kGy of gamma irradiation with hot water (47 °C for 7 min) inactivates *P. expansum* but does not prevent the growth of *B. cinerea* and *Alternaria tenuis*. In addition, the fungal population on mangoes was reduced by treatment with hot water (55 °C for 5 min) and 1 kGy of gamma irradiation (A similar study on tomato found that the *B. cinerea, R. stolonifer* and *A. alternata* decay is reduced with a combined treatment of 1 kGy of gamma irradiation and a hot water dip at 50 °C for 2 min

Mechanism of pathogen inactivation by ionizing radiation: An understanding of the physiological response of pathogens to ionizing radiation is critical for their postharvest control in produce. It helps to explain the biological meaning of the kinetics of inactivation and predict the pathogen inactivation under various irradiation conditions. Thus, a better knowledge of the mechanism of inactivation of pathogens by irradiation gives rise to effective irradiation processes.

Evidence of the antimicrobial activity of ionizing radiation is the disruption of the pathogen's cell membrane, resulting in a loss of cytoplasmic nutrients from the cell. In addition, ionizing radiation can cause either direct or indirect damage. The main direct effect of ionizing radiation on a pathogen is damage to the genetic material, either DNA or RNA, described above. Single-strand breaks may not be lethal or may lead to a mutation, but multiple breaks render the pathogen non-viable. The indirect effect of ionizing radiation on pathogens involves the interaction between the radiation and other atoms or molecules within the pathogens. Thus, when water molecules are irradiated, they lose an electron, $H_2O \rightarrow H_2O^+ + e^-$. These products combine with each other or other water molecules to form molecular hydrogen and oxygen, hydrogen peroxide (H_2O_2), hydroxyl radicals (^-OH), hydrogen radicals ($H^\cdot$), and hydroperoxyl radicals ($HO_2^\cdot$). Of these compounds, hydroxyl radicals and hydrogen peroxide are important because they affect the nucleic acids and break the bonds that maintain their structure (Shama and Alderson, 2005). Moreover, ionizing radiation can also affect enzymes, plasmids or proteins that are essential for the growth of pathogens.

Susceptibility or sensitivity of postharvest pathogens to ionizing radiation: The control of postharvest diseases by irradiation depends on the radiation sensitivity of the pathogens. The sensitivity of an organism is correlated with its complexity and the repair mechanisms by which many cells can reverse the damage caused by radiation. Thus, viruses show the highest resistance to inactivation by irradiation, whereas insects and parasites are the most sensitive. Spores and cysts are resistant to ionizing radiation because of the small amount of DNA they contain. The effectiveness of irradiation for control also depends on the type and growth stages of the pathogen, the number of viable cells on or within the tissue, treatment dose, moisture condition, composition of products, and storage conditions. However, the effects of the inevitable mutations caused by irradiation should be considered, as they can lead to greater, lower, or similar levels of pathogenicity. Moreover, repeated cycles of irradiation give rise to radiation resistant microorganisms, and therefore, safe irradiation doses should be employed to completely kill the pathogens. The lethal effect of irradiation on microbes is measured by the D_{10} value, which is the dose required to inactivate 90% of a population; for example, the D_{10}value for *Penicillium expansum* is 0.23 kGy, which means that irradiation with 0.23 kGy will kill 90% of a population of this pathogen.. Due to the deeper penetration, a gamma ray is the method most used for elimination of pathogens in industrial processes. However, irradiation of fruits and vegetables is limited to doses below 1 kGy, owing to nutritional losses and chemical changes in irradiated products. In addition, the use of gamma ray, electron beam, and X-ray irradiation as a phytosanitary treatment against insects has increased dramatically in recent years. This application of irradiation technology is critical owing to uncertainties in the future and restrictions imposed under the Montreal Protocol on the use of methyl bromide, which is still widely used for quarantine purposes.

The inhibition of spore germination and morphology of fungal spores subsequent to treatment are directly related to the strength of the radiation dose. Increasing doses inhibit spore germination and crush the spore surface (Jeong *et al.*, 2015). In addition, lower doses of gamma radiation increase the total protein or total lipid production in *Fusarium solani* and *P. notatum*, suggesting that proteins or lipids may play a role in protection against the harmful effects of radiation. Identification of these proteins and lipids would elucidate their roles as radioprotectors. The radiosensitivity of a specific fungal species may depend on various factors including the moisture content of spores or commodities, the age of spores, and the time of storage before or after ionizing radiation. Dried spores show more resistance to radiation. Postharvest pathogens that exist in products with a high moisture content will easily recover after irradiation if inactivation is not complete.

Advantages and disadvantages of irradiation : Irradiation is a physical treatment used to control postharvest diseases and appears to offer one of the most viable alternatives to the use of chemicals. The penetrating power and uniformity of irradiation and its derived effectiveness on the inactivation of pathogens is better than that of fungicides. Irradiation is essentially a nonthermal process, with minimal to no heating of fresh produce, thus eliciting negligible sensory changes in the produce. Ionizing radiation does not render food radioactive because it is not strong enough to affect the neutrons in the nucleus. Therefore, ionizing radiation is considered a safe, energy efficient and environmentally safe process.

While irradiated products are still not widely popular in developed countries, ionizing irradiation of fresh produce is used to control quarantine pests rather than postharvest diseases. The maximum US FDA permitted dose of 1 kGy to treat fruits and vegetables is too low for effective fungal pathogen control, whereas higher effective doses are often phytotoxic to treated fresh produce. In addition to safety concerns, irradiated products have other issues such as resistant strains, cost and nutrient content of treated produce. One common concern about irradiating fresh produce is that long-term use of irradiation promotes development of radiation resistant strains of fungi and bacteria. In addition, some studies have shown that irradiation increases mycotoxin levels because of the absence of competitive inhibition by mould growth, leading to toxin-producing strains eventually dominating the mycoflora. Another disadvantage is that irradiated fresh produce can be more expensive owing to the upfront costs of an irradiation facility (Mostafavi *et al*., 2012). Irradiation can also have a negative effect on the physical and/or physiological conditions of the fruit, for example, causing softening, discolouration, and the development of 'off-flavours'. Moreover, irradiation can deplete fruits and vegetables of nutrition such as vitamins, proteins, carbohydrates, lipids and undesirably changes the taste at high dosages.

Safety issues of irradiated fresh produce: Food irradiation remains unpopular compared to conventional technologies. Nuclear disasters and their derived radioactive contamination have given consumers a negative perception of irradiated fresh produce. For this reason, more than 50 years of extensive research has been devoted to the safety and effectiveness of irradiation as a food sterilization technique to kill pathogens in food and extend its shelf life. There are three primary categories of safety that need to be considered: radiological safety (existence of radioactivity in the irradiated food), microbiological safety (occurrence of radiation-resistant microbial populations), and toxicological safety (effect of irradiated food on humans). The safety and efficacy of irradiation has been proved and accepted by the World Health Organization

(WHO), the FAO, the International Atomic Energy Agency (IAEA), Codex Alimentarius, and the governments of different countries commending the process. It was concluded that there are no microbiological risks for consumers, and that chemical substances generated during irradiation in fresh produce are not significantly higher than those induced by heat treatment. These documents have been widely adopted internationally, and specific applications are being applied in over 55 countries worldwide

Irradiation of mushrooms through UV-C: UV-C radiation could potentially be used for sanitizing fresh mushrooms and may act as a useful non-chemical way of maintaining mushroom's quality and extending their postharvest shelf-life. The effects of UV-C light (0.45–3.15 kJm-2) was investigated in the inactivation of *Escherichia coli* O157:H7 and the inhibition of lesion development on the mushroom surface (Guan *et al.* 2012). UV-C treatment (4 kJm-2) enhanced antioxidant enzyme activities (catalase, superoxide dismutase, ascorbate peroxidase and glutathione reductase) in shiitake mushrooms (*Lentinus edodes*) (Jiang *et al.* 2010). The result of some recent works on the UV application of the postharvested fruits and vegetables, its direct germicidal activity and its hormectic effects were reviewed (Deepshikha, *et al.*, 2017). Pulsed UV light (PUV) can be very effective in rapidly increasing the vitamin D2 content of fresh *Agaricus bisporus* and adenosine, cordycepin, polysaccharides and antioxidant activity in *Cordyceps militars*. A comprehensive study was done (Fernandes *et al.* 2012) on irradiations effect on physio-chemical and bio-chemical parameters of different species of mushrooms.

7.6.5 Nanopartilces and salts: Synthetic fungicides are the first choice worldwide to control postharvest diseases of fruit and vegetables. However, several problems and constraints related to their use have forced scientists to develop alternatives control means to prevent postharvest diseases. Physical and biological means, resistance inducers, and GRAS (generally recognized as safe) compounds are the most important alternatives used during the last 20 years. Recently, nanomaterial treatments have demonstrated promising results and they are being investigated to reduce the utilization of synthetic fungicides to control postharvest rot in fruit and vegetables. In the field of postharvest technology of fruits and vegetables, the nanotechnology approach is helpful for controlling postharvest diseases; introducing a new invention for packaging films; preventing impact of gases and unsafe rays; improving packaging appearance; and helping for labeling fresh products using the multiple chips (nanobiosensors) (Sergio Ruffo Roberto *et al.*, 2019). Nanotechnology was used to produce nanomaterials for use as antifungal agents in several commodities, including fruits and vegetables. Several nanomaterials have shown potential in postharvest management and have been developed to control diseases in citrus, grapes banana, apple, mango, peach, and nectarine.

Clay-chitosan nanocomposite (CCNC) was prepared at different ratios of clay/chitosan and was tested against *P. digitatum* on 'Valencia Late' sweet orange (*Citrus sinensis* L. Osb.). In in vitro tests, a complete inhibition of the pathogen was achieved at 20 $\mu g \cdot mL^{-1}$ for clay/chitosan (1:0.5), clay/chitosan (1:1), and clay/chitosan (1:2). In in vivo trials, a complete inhibition of the disease was obtained, and a high reduction (around 70%) of lesion development was reported for clay/chitosan at 1:2 ratio at 20 $\mu g \cdot mL^{-1}$. Scanning electron microscopy showed that treatment with this nanocomposite caused severe collapse and irregular branching of hyphae in the apical part. In addition, CCNC caused a degradation of DNA of the tested pathogen at 20 $\mu g \cdot mL^{-1}$. Silver nanoparticles, with an average diameter of 10 ± 5 nm were tested in vitro against *Alternaria alternata*, *P. digitatum* and *Alternaria citri* isolated from citrus fruits]. This nanoparticle at 150 $mg \cdot L^{-1}$ showed a good antifungal efficacy against the three tested pathogenic fungi. The results of this study confirmed silver nanoparticles as promising nanomaterials, even if compared to iprodione or difenoconazole fungicides at the same concentration (Sergio Ruffo Roberto *et al.*, 2019). To understand the mechanism in which silver nanoparticles works, it has been mentioned that it may cooperate with phosphorus and sulphur and their interaction may damage the DNA and proteins, and be consequential in cell death.

In 'Valencia Late' sweet orange (*C. sinensis* L. Osb.), copper nanoparticles (48 nm) were tested against *P. digitatum* and *Fusarium solani.* The in vitro tests showed that a complete inhibition of *P. digitatum* was recorded at 20 $\mu g \cdot mL^{-1}$ for copper nanoparticles. The percentage of reduction in colony diameter of *P. digitatum* was 43%, 60%, and 100% for copper nanoparticles, respectively. A complete inhibition of *F. solani* was obtained at 60 $\mu g \cdot mL^{-1}$ copper nanoparticles. The *in vivo* experiments demonstrated that copper nanoparticles completely inhibited green mold, in a direct action. In addition, when copper nanoparticles and *P. digitatum* were added into separate wounds, the reduction was 88.5% for copper nanoparticles. For fusarium rot, in direct action, copper nanoparticles completely inhibited the rot development. The mechanism by which copper nanoparticles acted against the pathogen was explained by degradation of fungal DNA. In fact, earlier investigations related to the mechanism of copper nanoparticles on plant pathogenic fungi have not been enough to explain the precise action. Indeed, it gives the impression to suppose that the small-size range of copper nanoparticles may possibly add to its antifungal effect, given that it can simply penetrate pathogen cell membrane and lead to cell death.

Alternaria alternata can attack a wide range of agricultural products causing black rot of mandarins and brown spot on citrus fruits. The ZnO nanoparticles (<50 nm) at concentrations from 0 mM to 15 mM were tested against a number of fungal contaminants, including *A. alternata* . The pathogen was inhibited by ZnO nanoparticles at concentrations higher than 6 mM. In the presence of ZnO nanoparticles, clear morphological aberrations in the fungal structures were detected. (Sergio Ruffo Roberto *et al.*, 2019). On lemon fruits, TiO_2 nanoparticles (7 nm) applied on polypropylene film were prepared to control *Penicillium expansum*. The fruits were enveloped in the prepared film and treated with irradiation by UVA during two-weeks storage period at 25 °C. The authors confirmed a noteworthy decrease of decay in lemons wrapped in prepared coated film as compared to the control. They also referred this efficacy to the accumulation of ROS from the UVA-irradiated TiO_2 surfaces that suppress fungus development

The antifungal efficacy in SiO_2/Ag_2S nanocomposites (300 nm) was reported for the first time against *A. niger*. The results showed that SiO_2/Ag_2S nanocomposites suppress the fungal development and sporulation. The authors refereed this effect to the presence of Ag_2S nanophases, and they observed a synergistic effect due to Ag_2S antifungal centres and the SiO_2 surfaces in supporting the adsorption of the fungus (Fateixa *et al.*, 2009)]. It has been recommended that for metallic Ag, there is interaction of silver species with the sulphydryl groups of respiratory enzymes in the plasma membranes of susceptible pathogens, causing changes in the membrane permeability. SEM showed that *A. niger* hyphae become covered by SiO_2/Ag_2S nanomaterials when the fungus was exposed to the compound. Another hypothesis to explain this mechanism was the binding of Ag^+ to microbial genetic material.

Silver nanoparticles (38 nm), copper nanoparticles (20 nm), and silver-copper nanoparticles were examined at different concentrations (0, 1, 5, 10 and 15 $mg \cdot L^{-1}$) against *A. alternata* and *B. cinerea*. Fifteen $mg \cdot L^{-1}$ resulted in being the best concentration to inhibit both pathogens. Using SEM the authors demonstrated that the nanoparticles obviously damaged the hyphae and conidia of *A. alternata*. In *B. cinerea*, damage to the fungal hyphae surface was noted resulting in the discharge of internal cellular materials with shrinkage of pathogen hyphae. Regarding the cell wall components of *P. expansum*, a huge reduction in total sugar and protein was recorded while a little reduction in lipids was observed due to the application of silver nanoparticles. In case of *B. cinerea*, a highly decrease in total protein of culture filtrate and total protein and N-acetyl glucoseamine of cell wall was noted, while an increasing in total lipids for culture filtrate and cell wall was observed following silver nanoparticles treatments

Many scientists have investigated the development or preparation of nanomaterials coating formulations to extend postharvest storage period. Regarding the packaging of red table grapes, a chitosan-TiO_2 composite film (70 μm thickness,) was prepared and evaluated against *Escherichia coli, Staphylococcus aureus, Candida albicans,* and *A. niger*. The chitosan-TiO_2 film (50–80 nm) was examined as packaging table grapes in order to avoid postharvest infections and expand the shelf life period. In this study, table grape was preserved for 22 days with composite film compared to two weeks with bare chitosan film before mildew occurred (Zhang *et al.,* 2017). Nanocomposites of chitosan/gelatin and silver nanoparticles (25–45 nm) were successfully prepared using the solution casting technique. The grapes wrapped with the nanocomposites hybrid film showed a fresh appearance without putridity, and the surface was kept smooth without any leakage of the grape juice. This nanocomposite led to extending the shelf life period for extra two weeks.

Banana anthracnose caused by *Colletotrichum musae* is the most important disease attacking banana fruits. The effectiveness of ajwain (*Trachyspermum ammi*) and neem (*Azadirachta indica*) leaf extracts of copper, silver, nickel and magnesium (68 nm) were tested in vitro against the causal agent of banana anthracnose. The results showed that ajwain-magnesium NPs, ajwain nickel NPs at 0.2% and neem silver NPs at 0.1% and 0.2% completely inhibited the spore germination of the pathogen. Silver neem nanoparticles were examined in vivo against banana anthracnose disease at different concentrations, and spraying this nanomaterial at 0.2% showed the lowest anthracnose severity.

A novel nanopackaging was developed by coating polyvinyl chloride film with ZnO NPs (200–400 nm diameter) and its effect on fresh cut 'Fuji' apple (*M. domestica* Bork.) preservation quality was tested. Fresh cut apple stored with ZnO NPs packaging had lower fruit decay percentage (21.5%) as compared to control (42.4%). Also, nanopackaging reduced the accumulation of malondialdehyde and decreased polyphenoloxidase and pyrogallol peroxidase activities as compared to control fruit. The primary appearance of apple slices was maintained and the browning index was avoided in nanopackaging 'Fuji' apple fruits stored at 4 °C for 12 days. The nanopackaging was also able to maintain the level of soluble solids and titratable acidit.

For 'Red Delicious' apple (*M. domestica* Bork.), nano-calcium and calcium chloride were applied at different concentrations in the field five times, starting 70 days following full-bloom untill one month before apple harvesting. Firmness, titratable acidity, total phenolic content, total antioxidant activity, and fiber content increased in apple fruits treated with nano-calcium and calcium chloride as compared to untreated fruit. The quality of apples treated with nano-calcium was better than apple fruits treated with calcium chloride in all investigated parameters during the storage period. The authors proved that nano-calcium decreased cell wall enzyme activities including polygalacturonase, pectin methylesterase and β-galactosidase].

ZnO NPs (70 ± 15 nm) were tested against *P. expansum*, the causal agent of postharvest blue mold of apple, at different concentrations (0, 3, 6 and 12 $mmol \cdot L^{-1}$). The pathogen growth was inhibited at concentrations higher than 3 $mmol \cdot L^{-1}$. The mechanism of ZnO NPs on the pathogen proved to be related to the prevention of growth of conidiophores and conidia and death of hyphae. The antifungal activity of ZnO NPs (sizes < 50 nm) against *P. expansum* was also investigated by (Sardella, *et al.*, 2018). The results showed that the minimum inhibitory concentration and non-inhibitory concentration were 9.8 and 1.8 mM, respectively.

The mechanism of salts against gray mold on different hosts, showed that potassium bicarbonate, sodium silicate, and calcium chelate can change the morphology and function of mitochondria in the fungus. The aforementioned salts caused accumulation of ROS, reduced adenosine triphosphate content and decreased spore mitochondrial membrane potential, resulting in the loss of mitochondria role. In addition to the direct mechanism, induction of resistance in the fruit and vegetable was also proposed. In nature, plants are frequently challenged by a diverse array of pathogenic microorganisms, but they are able to defend themselves against almost all of the attacks. In many cases, their protective mechanisms involve inducible defense reactions, such as the accumulation of hydrolytic enzymes, phytoalexins, or the production of structural barriers. The failure of plants to invoke such defense reactions may result in susceptibility to disease. Induced disease resistance has been defined as "the process of active resistance dependent on the host plant physical or chemical barriers, activated by biotic or abiotic agents"]. It is well documented that treating plants with various agents (e.g. virulent or avirulent pathogens, non-pathogens, cell wall fragments, plant extracts, and synthetic chemicals) can lead to the induction of resistance, both locally and systemically, to subsequent pathogen attacks. Hence, there is a huge interest in exploring new strategies based on the activation of the plant defense mechanisms] as an alternative to traditional means for controlling diseases.

The induction of plant defense reactions is presumed to be mediated by an initial natural recognition process between plants and pathogens, which involves the detection of certain structural features of incompatible pathogens by plant receptor-like molecules. This hypothesis is supported by the finding that certain compounds, known as elicitors, and isolated from pathogenic microorganisms, induce biochemical events characteristic of the resistance responses in plants. In 'Red Top' and 'Anjiry' peaches (*P. persica* L.), and 'Red Gold', 'Songlu' and 'Independence' nectarines (*P. persica* L.), potassium permanganate coated with nanozeolites (nano-absorbent granules) was prepared to check its role on quality properties and shelf life after removing ethylene from the storage environment. Treatment with zeolite nanoparticles coated with potassium permanganate in a machine circulating the ethylene-containing air, avoided mass loss, reduced the firmness and contamination, and increased shelf life and the quality of peach and nectarine samples during storage].

Mango fruits were coated and stored for nine days at room temperature and quality parameters were evaluated. The overall results showed that increasing the concentration of ZnO NPs led to a reduction of mass loss (%), increasing to titratable acidity, increasing of ascorbic acid content in fruit, maintaining a low soluble solids content, decreasing the pH, increasing thickness and increasing the transmittance and elongation (%) of the formulated films. Copper oxychloride-conjugated silver NPs were synthesized using neem extract and tested for their activity against the causal agent of mango anthracnose (*C. gloeosporioides*) by Raghavendra *et al.,* (2019). The resulted nonmaterial had spherical shape and an average particle size of 21–25 nm. The nanomaterials displayed the highest growth inhibition of the pathogen (~187%) as compared to copper oxychloride. The obtained results confirmed that this nanomaterial can be successfully used as alternative control means for controlling anthracnose disease in mango.

Some metal nanomaterials such as MgO and ZnO composites (52–219 nm) were tested against *C. gloeosporioides*, the causal agent of anthracnose on papaya and avocado tropical fruits. All nanomaterials significantly inhibited the germination of conidia and the action proved to be fungicidal since the minimum inhibitory concentrations were equal to the minimum fungicidal concentration. Regarding the radial growth, *C. gloeosporioides* from papaya was more resistant to benomyl than *C. gloeosporioides* from avocado, and had minimum inhibitory concentrations of 0.625 and 0.312 $\mu g\ mL^{-1}$, respectively. The authors summarized that the tested nanomaterials can be used as alternatives to control anthracnose.

Nanoemulsions (NE) are prepared using aqueous phase, oils, surfactants, and co-surfactants. It was demonstrated that chitosan in the form of nanoemulsions exhibited better antifungal efficacy when compared to its raw form. The effect of chitosan in the form of nanoemulsions (200–1000 nm) against *C. musae* and *C. gloeosporioides*, the causal agents of anthracnose of tropical fruits including banana, papaya and dragon fruits was investigated. The fruits treated with conventional chitosan showed higher disease incidence and severity when compared to fruits treated with chitosan nanoemulsions. Nevertheless, in banana fruits, the start of anthracnose symptoms was delayed and the symptoms appeared only after two weeks of cold storage period (Zahid *et al*., 2012). While in papaya and dragon fruits, the symptoms appeared after seven days of cold storage. Also, it is worth mentioning that chitosan-loaded nanoemulsion treatments did not demonstrate any negative effects on the quality of tested fruits.

There are several and easy postharvest practices such as suitable harvest timing, fruit washing and disinfection, categorization, wrapping, pre-cooling and suitable transportation and distribution, etc. to reduce postharvest losses. However, those practices are not always efficient at maintaining fruit quality after harvest. Fungicide application is still one of the most important means to control postharvest disease of fresh fruits to extend their shelf-life (Sergio Ruffo Roberto, *et al*., 2019). One of the most important factors for successful management of postharvest diseases is the synthesis of low-cost nanomaterials with high efficiency for better storage conditions with maintaining the fruit quality. For this reason, a lot of effort is needed to enhance using nanomaterials as alternative control means for postharvest disease management. Nanomaterials have been exploited as the novel generation of a revolution in the field of postharvest technology in general and particularly disease management. A wise application of nanomaterials is needed since their risks and hazards are still unknown and not fully studied. Future investigations should look at possible human health effects since some materials, such as TiO_2, have been considered to produce colon cancer. The size of the nanomaterials, as an important characteristic, was missing in some studies. Accordingly, we suggest that researchers provide the size in the future studies to enable other scientists to properly compare the nanoparticles between themselves and then plan the most efficient new design. Several factors including fruit species, dose and kind of nanomaterials, treatment timing, fruit-ripening stage and others can affect the efficacy of nanomaterials.

7.6.6 Waxing: Waxing of fruits or vegetables is a common postharvest practice. Food grade waxes are used to replace some of the natural waxes removed during harvesting and sorting operations and can help reduce water loss during handling and marketing. It also helps in sealing tiny injuries and scratches on surface of fruits and vegetables. It improves cosmetic appearance and prolongs the storage life of fruits and vegetables. The wax coating must be allowed to dry thoroughly before packing (El-Ramady *et al*., 2015). Waxes are generally applied by foaming, spraying and brushing of which foaming is the best, since it leaves a very thin coating. Some of the common coating materials are semperfresh, prolong and waxol. Vegetables like tomato, brinjal, sweet pepper, cucumber, muskmelon, carrot are often waxed (Mudasir Magray, *et al*., 2017).

Pineapple fruits ripen and soften rapidly after harvest. A study was conducted on examining the effect of wax application on the ripening and storage quality of the fruit (Li Xueping *et al.,* 2018). An optimized waxing treatment (65 g L^{-1} for 1 min) could effectively control fruit ripening by delaying fruit color change, decreasing the respiration rate and ethylene production, decreasing content of organic acids and relieving the symptoms of internal browning of pineapple fruit. Waxing enhanced the relative level of pentose phosphate pathway of respiration and affected the enzymes involved in organic acid metabolism. Waxing also reinforced the antioxidant system and enhanced the expression levels of genes related to defense, such as PGIP. These results indicated that wax treatment could effectively improve the fruit quality, mainly through the reduction of organic acid content.

Residue loading of pyrimethanil (PYR) with application methods typically used in South African packhouses and green mould control was studied. PYR was applied curatively (after inoculation) and protectively (before inoculation) in dip, drench and wax coating treatments and fruit were inoculated with an imazalil (IMZ)-sensitive or an IMZ-resistant isolate of *Penicillium digitatum*, the causal agent of citrus green mould (Kellerman Mareli *et al.,* 2018).The dip treatments consisted of PYR concentrations 0, 50, 100, 250, 500 and 1000 mg L^{-1} ; fruit were dipped for 60 s at 18 °C at pH 7. Residues loaded differed between fruit type and batch. At the registered concentration of 1000 mg L^{-1} , an average of 1.96 mg kg^{-1} PYR (range 1.40-2.95 mg kg^{-1}) was loaded. Increased residue levels generally resulted in improved control, but better curative than protective control was observed with effective residue levels of 0.27 and 0.91 mg kg^{-1} for 50% and 75% curative control of green mould, respectively. Effective residue levels for 50% protective control were 0.97-1.43 mg kg^{-1}. In conclusion, PYR provided very good curative control in dip applications, but gave poor results in drench and wax applications. Green mould control following PYR drench treatment improved on fruit inoculated with smaller spore loads, or when PYR was mixed with other fungicides.

The horticultural postharvest system comprises multiple and interconnected activities from harvest through storage, from marketing to the consumer's home. Produce losses, in quantity and quality, can occur at any step during the postharvest chain, with important economic impact. Postharvest losses of fruits and vegetables are di□cult to predict; the major agents producing deterioration are those attributed to fungal pathogens such as those of the genera *Penicillium, Botrytis, Monilinia, Rhizopus, Alternaria*, *Aspergillus Fusarium, Geotrichum, Gloeosporium* and *Mucor* or physiological disorders, above all chilling injuries. However, there are few peer reviewed published studies that estimate the total amount of food losses in developed countries and even fewer attempts to estimate the monetary value. For example, considering European stone fruit production of 8.9 MT in 2011 (Rivka Barkai-Gorivka Barkai-Golanlan, 2001) and by calculating an average percentage of losses of at least of 10% due to *Monilinia* rots, the economic damage can account for 850-900 M€ /year.

Since the beginning of the 1960s, treatments based on chemical fungicides as the main method of reducing postharvest fruit losses have obtained satisfactory results. However, in the last few decades, stricter regulatory policies have started to be imposed on their use. The reasons for these restrictions concern human safety, the environment and the development of pathogen strains resistant to the common fungicides. Therefore, the fungicide option for managing postharvest decays is increasingly limited; in addition, the use of fungicides on fruit after harvest is regulated by deferent food protection agencies. The number and type of registered fungicides can change depending on the countries considered, creating confusion in agricultural commodity exchanges. For example, blue mould of pome fruits caused by *P. expansum* was generally controlled until the mid-1990s by drenching fruit after harvest with benzimidazole fungicides. The appearance of strains resistant to the active ingredient required the development of new fungicides. In the USA, there are four fungicides currently labelled for controlling *P. expansum* with postharvest application: captan, thiabendazole (TBZ), fludioxonil and pyrimethanil. In China only TBZ is used while in Europe none of them are allowed for postharvest fruit treatment.A new fungicide, Philabuster, containing imazalil and pyrimethanil, has recently been registered for postharvest treatments of pome fruits .

In the attempt to mitigate consumer concerns regarding the pesticide residues in fresh produce, many retailers pursue a policy of residue reduction to 3-4 active ingredients or their complete elimination from fruit, partially closing the door to the use of fungicides in the postharvest phase.

7.6. Biofungicides : Although the use of synthetic chemical fungicides remains a primary method of controlling postharvest diseases, the global trend appears to be shifting towards a reduced use of fungicides on produce and hence there is a strong public and scientific desire to seek safer and eco-friendly alternatives to reduce postharvest losses. Biological control using microbial antagonists such as bacteria and yeasts has emerged as one of the most promising of these alternatives and after three decades of study and promising results, it is time to reflect on the direction of the current research and on the foreseeable expectations. The greate ort made by researchers and documented by many peer reviewed publications has been only partially compensated for by the appearance on the market of biological agents available for industrial application to control postharvest decays (Table 11). In 1995 two biocontrol agents (BCAs): a yeast, *Candida oleophila*, used in Aspire and a saprophytic strain of *Pseudomonas syringae* used in Biosave obtained registration from the EPA (United States Environmental Protection Agency).

Other products, such as Yeld Plus (*Cryptococcus albidus*), Avogreen (*Bacillus subtilis*) and Shemer (*Metschnikowia fructico*-remains limited: Aspire was taken off the market 3 years after its large scale commercial introduction; Avogreen, registered for the control of avocado postharvest diseases, did not obtain widespread diffusion, due probably to its inconsistent results and none of these are registered in Europe for fruit treatments after harvest. Three new products will soon be available: Candifruit based on Candida sake, developed in Spain; Boni-Protect, based on *Aureobasidium pullulans* and developed in Germany, and NEXY, based on *C. oleophila* and developed in Belgium, showing the interest for biological control against postharvest diseases (Table 10), are also on the market in various countries. There are several reasons for this limited diffusion, some relate to an intrinsic ineffectiveness of BCAs such as their inconsistency, and the variability of the control under commercial conditions. Some relate to their formulation that reduces the activity of antagonists with respect to the fresh cells, the high cost of the production, and the regulatory barriers to BCA registration in deferent countries that do not encourage their dissemination.

Table 11: Microbial antagonists registered or in progress of registration for postharvest phase

Microorganism	Product name	Pathogen target	Country
Bacillus subtilis	Avogreen	*Colletotrichum gloeosporioides*	South Africa
Aerobasidium pullulans	Boniprotect	*Penicillium expansum, Botrytis cinerea, Monilinia* sp.	Germany
Candida sake	Candifruit	*P. expansum, B. cinerea, Rhizopus* sp.	Spain
Pantoea agglomerans	Pantovital	*P. expansum, B. cinerea, Monilinia* sp.	Spain
Pseudomonas syringae	Biosave	*Penicillium* sp., *B. cinerea, Mucor* sp.	USA
Cryptococcus albidus	Yield Plus	*B. cinerea,, Penicillium, Mucor*	South Africa
Metschnikowia fructicola	Shemer	*B. cinerea, Penicillium* sp.,*Rhizopus* sp. *Aspergillus* sp.	Israel
Candida oleophila	Nexy	*B. cinerea, Penicillium* sp.	Belgium

Synthetic fungicides are primarily used to control postharvest decay loss. However, the recent trend is shifting toward safer and more eco-friendly alternatives for the control of postharvest decays. Of various biological approaches, the use of antagonistic microorganisms is becoming popular throughout the world. Several postharvest diseases can now be controlled by microbial antagonists. Although the mechanism(s) by which microbial antagonists suppress the postharvest diseases is still unknown, competition for nutrients and space is most widely accepted mechanism of their action. In addition, production of antibiotics, direct parasitism, and possibly induced resistance in the harvested commodity are other modes of their actions by which they suppress the activity of postharvest pathogens in fruits and vegetables. Microbial antagonists are applied either before or after harvest, but postharvest applications are more effective than preharvest applications. Mixed cultures of the microbial antagonists appear to provide better control of postharvest diseases over individual cultures or strains. Similarly, the efficacy of the microbial antagonist(s) can be enhanced if they are used with low doses of fungicides, salt additives, and physical treatments like hot water dips, irradiation with ultraviolet light *etc* (Sharma *et al*., 2009). At the international level, different microbial antagonists like *Debaryomyces hansenii*, *Cryptococcus laurentii*, *Bacillus subtilis* and *Trichoderma harzianum* are being used. Biocontrol products like Aspire, BioSave, and Shemer etc., have also been developed and registered.

The birth of the field of biological control of postharvest diseases can be traced back to 1984 when a researcher testing an antagonist (*Bacillus subtilis*) in the field to control brown rot of peaches (caused by *Monilinia fructicola*) decided to apply the antagonist directly to the peach to control brown rot. The results were striking. It became apparent from this simple experiment that postharvest diseases and biocontrol were meant for each other. The advantages of applying biological control practices in the packinghouse vs. the field became clear. Biological control holds great promise of providing an alternative to present postharvest chemical fungicides for the control of postharvest diseases and reducing the health and environmental hazards presented by synthetic fungicides (Wilson *et al*., 2011). Researchers around the world have been pursuing the use of antagonistic yeasts and bacteria as such alternatives and registered products are now on the market. The major shortcoming of these first products has been their lack of an eradicant action. In other words once a postharvest infection occurs the antagonist is unable to arrest it. This has required the addition of low dosages of synthetic fungicides with the antagonist in order to reach levels of disease control comparable to the synthetic fungicide alone.

The early concept of postharvest biocontrol as being a one on one relationship between an antagonist and a pathogen limited approaches to the development of this technology. Using an expanded concept, combinations of antagonists with natural fungicides and physical treatments that induce host resistance are yielding more effective control and enhancing the potential of postharvest biocontrol. Advances in the fields of molecular biology and microecology hold promise for a better understanding of postharvest biocontrol systems and expanding the application of postharvest biocontrol strategies. There is the potential of identifying and manipulating "biocontrol genes" in antagonists to produce more effective antagonists. Genes in antagonists responsible for suppression of pathogens have also been used to screen for more effective antagonists. There is evidence that plants can genetically control epiphytes on their surfaces that contribute to biocontrol. This presents the prospect that "biocontrol genes" can be identified in postharvest commodities that will promote suppressive epiphytes and thus reduce the development of postharvest diseases.

In recent years, there has been considerable interest in the use of antagonistic microorganisms for the control of postharvest diseases. Such organisms can be isolated from a variety of sources including fermented food products and the surfaces of leaves, fruit and vegetables. Once isolated, organisms (whether they be bacteria, yeasts or filamentous fungi) can be screened in various ways for inhibition of selected pathogens. In most reported cases, pathogen inhibition is greater when the antagonist is applied prior to infection taking place. For this reason, control of quiescent field infections (e.g. *Colletotrichum* sp.) using postharvest applications of antagonists is often more difficult to achieve than control of infections occurring after harvest (e.g. *Penicillium* spp.). Unless an antagonist has eradicant activity or has some effect on host defence responses, field applications are often necessary to achieve control of quiescent infections. In South Africa, both field and postharvest applications of *Bacillus subtilis* and *B. licheniforrnis* suppress anthracnose (caused by *Colletotrichum gloesponoides*) and stem end rot (caused by *Dothiorella* spp. and other fungi) development in avocado.

A number of modes of action are involved in these pathogen-antagonist interactions, including site exclusion, nutrient and space competition and antibiotic production. In Australia however, only field applications of a non-antibiotic producing strain of *Bacilllus* sp. have shown potential for the control of anthracnose in avocado. There are a number of reports in the literature concerning the biological control of wound pathogens in various fruit and vegetables. To be effective against wound pathogens, an antagonist must be able to successfully colonise wound sites to the exclusion of the pathogen.

Antagonists which act against postharvest pathogens by competitive inhibition at wound sites include the yeasts. While the potential for biological control of postharvest diseases clearly exists, future success relies on the ability to achieve consistent results in the field and after harvest. It will be necessary to enhance the efficacy of biological control agents against postharvest disease and commercialise the technology involved.

Ideally, an antagonist should be effective against a broad spectrum of pathogens on a wide range of fruit and vegetables, it should be unique and be able to be produced on inexpensive growth media. Formulations incorporating such antagonists should have a long shelf-life and be able to be manufactured at low cost. Most antagonists do not satisfy all of these criteria. Many are quite specific in their activity against pathogens and for this reason may not be particularly appealing to prospective investors.

Basic approaches for using the microbial antagonists: There are two basic approaches for using the microbial antagonists for controlling the postharvest diseases of fruits and vegetables:, *viz*., use of microorganisms which already exist on the produce itself, which can be promoted and managed, or those that can be artificially introduced against postharvest pathogens (Sharma *et al.* 2009).

Considerations for biocontrol of postharvest diseases: Postharvest application of biocontrol agents has acquired much importance in the recent years. The confined environment for the storage of harvested perishable commodities provides a great opportunity for applying biocontrol agents since temperature and humidity are stable and controllable (Seethapathy Parthasarathy *et al*., 2017). In addition, biocontrol agents are easy to apply they act on more concentrated plant parts, and the value of the harvested crops is great enough to justify expensive control procedures. But relatively few studies have been carried out on the control of postharvest diseases by pre-harvest application of microbial antagonists. The major aspects taken care of by pre-harvest application of biocontrol agents are, management of latent or quiescent infections, controlling wound infections during harvest and postharvest handling and suppression of pathogen inoculum.

Bioagents for the management: Biocontrol agents are harmless; cheapen than pesticides, highly effective throughout the crop growth period with high rhizosphere competency and competitive saprophytic ability. The other avenues include, ease to deliver improve plant growth, activate resistance mechanism in the host, increase biomass production and yield. A wide range of antagonistic bacteria, fungi and actinomycetes are used for the management of diseases of horticultural crops. *Antagonist* like *Bacillus subtilis, B. cereus, Bradyrhizobium* *japanicum, Enterobacter cloacae, Pseudomonas aeruginosa, P. cepacia, P. fluorescens, P. corrugata, P. putida, Rhizobium* spp., *Streptomyces* spp., *S. griseoviridis, Laetisaria arvalis, Penicillium* spp., *Pythium oligandrum, P. nunn, Rhizoctonia* spp., *Trichoderma harzianum, Chaetomium globosum, Cladorrhinum foecundissimum, Coniothyrium minitans,* non-pathogenic *Fusarium oxysporum* F047, *F. oxysporum* f. sp. *niveum* (non-pathogenic), *Gliocladium virens, Glomus* spp., *Glomus fasciculatum, G. intraradices, G. mosseae, Penicillium oxalicum, Sporidesmium sclerotivorum, Talaromyces flavus, Vertilicillium biguttatum* and *Trichoderma* spp., are used as potential biocontrol agents against the pathogens causing diseases in horticultural crops.

7.7 Managing quiescent infections : The phenomenon of latent or quiescent infection in *Colletotrichum* was defined as 'the inhibition of the pathogen through physiological conditions imposed by the host until some stage of maturation has been accomplished'. Latency is defined as a quiescent or dormant parasitic relationship, which after a time changes to an active one. Molecular and physiological basis of quiescent infection during transition from immature to ripening stage was clearly described. Latent infections have been reported in stone fruits, apples, avocados, mangoes, papayas, citrus fruits, grapes and strawberries. In strawberries, *Botrytis* rots start on infected, senescent floral parts that become latent infections which develop into active rots on ripe fruits. The yeast-like fungus *Aureobasidium pullulans* and the yeast *Candida oleophila* was effective against *B. cinerea* storage rots on strawberries when applied in the field at bloom compared to application immediately after harvesting. Also, the application of *A. pullulans* and *Epicoccum purpurascens* to sweet cherry blossoms reduced the number of latent infections by *Monilinia laxa* in green strawberry fruits. The importance of establishment of a biocontrol agent before pathogen arrival, especially, when shortlived colonizable substrates (e.g. petals, stamens) acts as infection foci.

Repeated applications of antagonists was required to control postharvest decays originating from latent infections in the orchard as demonstrated by three sprays of *Bacillus subtilis* to control avocado anthracnose. In addition, continuous application of *B. subtilis* over a four-year period resulted in a gradual buildup of the antagonist population, reduction in the level of pathogen inoculum and more effective disease control. The population dynamics of *A. pullullans, Rhodotorula glutinis* and *B. subtilis* on apple fruits for the antagonists were applied to trees in the field late in the growing season. After transfer from field to cold storage, the population of the introduced antagonists remained higher than in control treatments, except for *B. subtilis* which declined. The efficacy of preharvest application of *A. pullulans* against *Botrytis* rot of sweet cherry and table grapes and R. glutinis against Botrytis rot and blue mold of pear were also reported. Field sprays of *C. guilliermondii* A and *Acremonium cephalosporium* B two to five times at 7-10 days interval on table and wine grapes could reduce decay caused by *B. cinerea* and *A. niger*.

Managing wound infection during harvest and postharvest handling: Mechanical injuries created during harvest and postharvest handling operations may result in infection by wound pathogens. In such cases, a near harvest treatment with an effective antagonist seems an appropriate strategy to prevent the colonization of wounds by weak resident antagonists and by pathogens. The efficacy of preharvest application of *B. subtilis* against anthracnose, stem-end rot and *Colletotrichum* fruit rot of avocado. On strawberries treated with *A. pullulans* immediately before harvest, *Rhizopus* rot was reduced by 72% while only slight activity was observed against *Botrytis* rot. This difference in efficacy signifies the need of knowledge on the epidemiology of the target disease for choosing the right time of application of biocontrol agents. *B. cinerea* being present inside the fruits as latent infection could not be effectively checked, but *Rhizopus stolonifer*, being an external contaminating pathogen could easily be inhibited by *A. pullulans*. Postharvest apples treated two days prior to harvest with *C. sake* had less blue mold after four months of storage. Pre-harvest application of antagonists permits a preemptive colonization of the wounds inflicted during harvest or handling, and thus allows more time to the antagonist to colonize and saturate the fresh wound before the arrival of the pathogen (Seethapathy Parthasarathy *et al.*, 2017). Such type of competitive or pre-emptive exclusion is considered as one of the mechanisms of protection by the antagonists against the pathogen infection.

Cases of unadvisable postharvest handling: In certain cases, postharvest applications are not feasible. For example, fruits that is easily damaged or cannot be exposed to water-based treatment such as strawberries and fruits with a waxy bloom on the surface such as table grapes. Postharvest treatment with *Pichia guilliermondii* or *Hanseniaspora uvarum*markedly suppressed the postharvest decay of table grapes, but it tended to remove the surface bloom; this problem was avoided by applying *P. guilliermondii* on plants three days before harvest. To avoid handling and related deleterious effects, it was preferred to harvest strawberries and pack them directly into plastic baskets where the fruit remained until it reached the consumer. In such conditions spraying suspensions of antagonist propagules, even one day before harvest seemed to be the chief method of applying the bioagents.

Essential traits of antagonists for pre-harvest application : A successful antagonist must be able to tolerate low nutrient availability, ultraviolet radiation, desiccation, rapid climatic changes and presence of agrochemicals. Attachment to the fruit surface is probably an important trait for the antagonists since persistent attachment would contribute to a better colonization and avoids dislodging due to wind, rain or water level fluctuations. A crucial trait of a good biocontrol agent is its ability to colonize and to survive on the target host tissue. Colonization of the host surface is an important and multifaceted process needed for the competitive and pre-emptive exclusion of the pathogen. The antagonist should be well established in the carposphere before the pathogen arrives, or that when the antagonist population declines, products triggering the infection have been removed from the host surface.

7.8 Biofumigation: The development and utilization of volatile organic compounds (VOCs) from microorganisms have been of increasing interest since they are naturally produced without chemical synthesis. VOCs are mixtures of carbon-based compounds that are either vapors or highly volatiles (Li Q *et al.*, 2015). The identified VOCs generated by fungi and bacteria are mainly primary (from the synthesis of DNA, amino and fatty acids, etc.) and secondary metabolites (from intermediates of primary metabolism). Microbial VOCs may be signaling substances for regulating and controlling certain physiological actions of some plants. The growth of *Arabidopsis thaliana* was promoted after the exposure to the volatiles from bacteria, and the systemic resistance in *A. thaliana*against *Erwinia carotovora* was induced. Root length, shoot length and fresh weight were increased by treating lettuce seedlings with VOCs produced by *Fusarium oxysporum* MSA 35. Volatile compounds from plant growth promoting fungi (PGPF) significantly reduced disease severity in *Arabidopsis* plants. VOCs can attract or deter insects and other invertebrates as "semiochemicals", which are defined as chemicals involved in the interactions between organisms. Some volatiles have been considered as indicators of biocontamination in the food processing industry or the basis of taxonomic research.

Recently, fungal VOCs have been developed as green chemicals and fuel sources called "mycodiesel". Additionally, fungal VOCs have been used as biological control agents ("biofumigant") for plant disease and pest management and have been described as eco-friendly due to a reduction in the need for application of synthetic fungicides. *Trichoderma* species have been used as biological agents for more than 80 years and produce anti-microbial compounds that kill and degrade other pathogenic fungi, induce systemic resistance and promote the growth of plants. Volatile substances from *Streptomyces philanthi* inhibited the growth of *Rhizoctonia solani* on rice leaves. *Bacillus subtilis*produced antagonistic volatile compounds that caused structural deformations in six pathogenic fungi, such as *Alternaria alternata*, *Cladosporium oxysporum*, *F. oxysporum*, *Paecilomyces lilacinus*, *P. variotii*, and *Pythium afertile*. Compounds from *B. vallismortis* showed strong growth inhibition activity *in vitro* against some plant pathogens. *Serratia rubidaea* generating rhamnolipids under solid-state fermentation (SSF) was used as a biocontrol agent]. Another successful case is *Muscodor* species, whose volatiles inhibit or kill a wide range of plant pathogenic fungi, bacteria, nematodes, insects and even human pathogens. These volatiles have been developed into bio-fumigants for controlling various types of plant diseases, weeds, nematodes, insects and molds developing within buildings.

Ceratocystis fimbriata Ellis & Halsted is a soilborne ascomycete fungus. In addition to being a plant pathogen, it can produce a variety of VOCs]. These volatiles are affected by the nutrition in the culture media, incubation temperature, and water activity (aw). In our preliminary study, the produced VOCs were identified by gas chromatography-mass spectrometry (HS GC-MS). Better understanding the bioactivity of VOCs from *C.fimbriata* not only can provide a tool for controlling postharvest diseases, but also further reveal the relationships among the microorganisms in natural ecological system. The aim of this study is to detect the bioactivity of *C.fimbriata* volatiles on 16 plant pathogens *in vitro*, investigate their effect in controlling peach brown rot and citrus green mold, and probe the possible mechanisms of inhibition.

Mechanisms involved in biocontrol : Relatively few attempts have been made on studying the mechanisms involved in biocontrol of postharvest diseases of fruits. Various mechanisms viz., antibiosis, competition for limiting nutrients and space, production of lytic enzymes, parasitism and induced resistance have been reported (Seethapathy Parthasarathy *et al.*, 2017).

i. Antibiotic production: The antibiotic iturin produced by *B. subtilis* B-3 and pyrrolnitrin produced by *Pseudomonas cepacia* LT-4-12W reduced *in vitro* growth and conidial germination of the stone fruit pathogen *M. fructicola* and pome fruit pathogens *P. expansum* and *B. cinerea* respectively. Both these strains could control fruit decays caused by the respective pathogens. The fruit decays were also controlled by applications of the respective antibiotics alone. *A. pullulans* may produce the antibiotic aureobasidin. P. syringae strains ESC-10 and ESC-11 produced the antibiotic Syringomycin E on some media and that the purified compound could control green mold of lemons.

ii. Competition for nutrients and space*:* Competition for nutrients has been elucidated between *Enterobacter cloacae* and *R. stolonifer* on peach and *P. guilliermondii* and *P. italicum* on citrus. The antagonists *Cryptococcus laurentii* and *Sporobolomyces roseus* used the apple volatile butyl acetate as food source and reduced the stimulatory effects of this chemical compound on germination of *B. cinerea* conidia *in vitro* . Also, it was noticed that these two antagonists exhibited greater uptake and utilization of 14C fructose suggesting that competition for nutrients might have played a major role. The same phenomenon was also revealed in the interaction of *E. herbicola* B66 and B 90 with *B. cinerea* and *P. expansum* in diluted apple juice. The very high density (1010 cfu/ml) of *E. cloacae* needed to inhibit conidial germination of *R. stolonifer* and decay of peach suggested exclusion as a mechanism of biocontrol.

iii. Production of lytic enzymes: The lysis of *B. cinerea* hyphae by β-1,3-glucanase produced by *P. guilliermiondii* on apple. *P. anomala* strain K, effective in control of grey mold of apple, increased production of exo-β-1,3-glucanase threefold in the presence of cell wall preparation of *B. cinerea* in apple wounds, reducing lesion size by more than half compared to the antagonist alone. Higher β-1,3-glucanase and chitinase activity was also detected in apple wounds treated with strains of *A. pullulans*, effective in controlling various decays of apple, and table grapes.

iv. Induced systemic resistance: The antagonistic yeast *P. guilliermondii* US-7 was shown to induce PAL and the phytoalexin scoparone in citrus fruit peel. Also, treatment with *C. saitoana* induced chitinase and caused deposition of papillae along host cell walls in apple and citrus. *A. pullulans* L47 upon wound application on apple increased β-1,3-glucanase, chitinase and PO activities 24h after treatment which reached maximum levels at 48 and 96h after treatment. *C. saitoana* when applied on apple fruit 48h to 72h pre-inoculation with *B. cinerea*, reduced lesion diameter by more than 50% and 70% respectively. In addition, treatment with the yeast antagonist increased chitinase and β-1,3-glucanase activities in fresh apples. *C. famata* reduced green mold (*P. digitatum*) decay on oranges and increased phytoalexins scoparone and scopoletin 12-fold in fruit wounds after four days when inoculated alone. The application of *Pseudomonas fluorescens* FP7 water in oil emulsion formulation increase in the activity of defense related enzymes viz., phenyl alanine ammonia lyase, peroxidase, polyphenol oxidase, catalase and b-1,3-glucanase was observed. Postharvest application of *Pseudomonas fluorescens* Pf1 and *Bacillus subtilis* EPCO16 induced defence related compounds *viz.*, peroxidase, polyphenol oxidase, phenylalanine ammonia lyase, superoxide dismutase and catalase was observed on treated mango fruits against stem end rot disease.

Bioformulations for the management: Some of the antagonistic microorganisms have been patented and commercial products such as Aspire recently named Decco I-182 (*C. oleophila* I-182, Ecogen Corp., Langhorne PA) as wettable powder, Bio-Save 100, 110 and 1000 (*P. syringae* ESC- 10 and 11, EcoScience Corp., Orlando, FL) as frozen cell concentrated pellets and Yield *Plus* (*Cryptococcus albidus*, Anchor Biotechnologies, Cape Town, South Africa) have been registered for use against postharvest decay of citrus and pome fruits (Seethapathy Parthasarathy *et al.*, 2017). Viable cells of *C. sake* CPA-1 grown in sorbitol modified liquid molasses-based medium and preserved with isotonic solution of trehalose (0.96 M) after storage for seven months maintained their viability and efficacy against P. expansum infection of apples. Coating forming dispersions with *C. sake* improves the inhibition of *B. cinerea* on grapes. Additional benefits, such as enhancement of fruits firmness and extended shelf-life, have also been correlated with microbial antagonists.

Biomanagement of mycotoxigenic fungi : The use of microbial antagonists for the control of postharvest pathogens, including mycotoxigenic fungi is most important. Recently, experiments showed some *A. pullulans* strains are highly suppressive to rot caused by *Aspergillus carbonarius* on detached and wounded wine grapes. Interestingly, these strains are pertaining potential to degrade ochratoxin A to the less toxic compound ochratoxin *in vitro*. Furthermore, in the few grape berries that were contaminated by the fungus and were pretreated with the biocontrol agents, ochratoxin. A contamination was lower than in infected nontreated berries. In apples pretreated with the microbial antagonist, the overall accumulation of patulin in the low percentage of P. expansum infected fruits was significantly lower than in infected, non-pretreated apples. Interestingly, the major patulin degradation product formed by *Rhodotorula glutinis* LS11 characterized structurally by NMR analyses was identified as desoxypatulinic acid which was reported to be nontoxic to multiple microorganisms that are inhibited by patulin.

Challenges for transformation of bioagents into practical usage : The first and rational step to prevent pathogen contamination is a good agricultural practice in the field and at the storage. The inherent ability of microbial bio-agents to prevent postharvest diseases in perishable commodities relies on their potential to outcompete the hazardous pathogen for required nutrients in the wound site and/or to form a biofilm layer that prevents pathogens physically from receiving nutrients and make starvation. Many postharvest pathogens need an infection site to settle them and initiate an infection. Despite their ability to use as an alternative approach, the use of postharvest biocontrol agents as commercial products remains narrowed. This lack of widespread commercial application has been associated to a number of factors, including unbalanced performance, lack of eradicative potential, lack of resources, varying regulatory requirements in different nations which restricts their widespread adoptability and usage and issues associated with the environmental safety. In order to improve the potential of microbial antagonists, research has been directed to combine their use with other alternative approaches.

Postharvest application of synthetic chemicals has been the traditional scheme for the management of postharvest diseases. The increasing concern for health hazards and environmental pollution due to chemical use has necessitated the development of eco-friendly strategies for the control of postharvest diseases of perishable fruits. Management of postharvest diseases by microbial antagonists has been manifested to be most suitable strategy to alternate the chemicals which are either being banned or recommended for limited use. The recent approach of postharvest disease management in fruits and vegetables is based on the knowledge of natural process of the antagonist pathogen interaction. Although there are several success stories in commercialization of antagonist against postharvest pathogens. In the controlled environment of a postharvest system, there are specific opportunities to apply microbial antagonists as delivery system. In future, it may be possible to use only strains adapted to postharvest conditions and introduce targeted genes for antagonistic activity as needed. However, if the similar pace follows, the use of microbial antagonists for the control of postharvest diseases of fruits will be greatly expanded in the future and will definitely become a globally adopted practice.

7.9 Natural microbial antagonists : Natural occurring antagonists are those, which are present naturally on the surface of fruits and vegetables, and after isolation, antagonists are used for the control of postharvest diseases . When concentrated washings from the surface of citrus fruit were plated out on agar medium, only bacteria and yeast appeared while after dilution of these washings, several rot fungi appeared on the agar, suggesting that yeast and bacteria may be suppressing fungal growth. Thus, it indicates that when fruits and vegetables are washed, they are more susceptible to decay than those, which are not washed at all.

Artificially introduced microbial antagonists : Although the first reported use of a microbial antagonist was the control of *Botrytis* rot of strawberry) with *Trichoderma* spp., the first classical work was the control of brown rot of stone fruits by *Bacillus subtilis*. Since then, several antagonists have been identified, and used for controlling postharvest diseases of different fruits and vegetables. Artificial introduction of microbial antagonists is more effective in controlling postharvest diseases of fruits and vegetables than other means of biological control.

Mode of action of microbial antagonists : A significant amount of research on the use of the microbial antagonists has been reviewed by several workers (El-Ghaouth *et al.*, 2004). However, the mechanism(s) by which microbial antagonists exert their influence on the pathogens has not yet been fully understood. It is important to understand the mode of action of the microbial antagonists because, it will help in developing some additional means and procedures for better results from the known antagonists, and it will also help in selecting more effective and desirable antagonists or strains of antagonists.

Understanding the mode of action of antagonists is important for two reasons, it will allow the development of more reliable procedures for the effective application of.known antagonists and it should provide a rationale for selecting more effective antagonists. Chalutz and Wilson (1990) have summarized the available information on the mode of action of antagonists that have been used in the postharvest environment. Antibiotic production appears to be a major mode of action of many of the antagonists identified so far. This is inherent in the in vitro method that most investigators have used to screen isolates. This methodology favors the selection of organisms that produce antibiotic zones in petri dishes when challenged with the pathogen. Other than instances where antibiosis has been reported as the source of biocontrol activity and the actual antibiotics have been identified, little information exists on other modes of action. Undoubtedly, many of the evolving concepts for modes of action for biocontrol of plant diseases in general would also potentially apply to biocontrol in the postharvest arena. Also, as Baker has recently stated, "there remain many uncharted mechanisms that have evolved and are functioning undetected in the natural world". Elucidation of modes of action is hampered by lack of knowledge of host/parasite/antagonist interactions in wounds and on the surfaces of fruits and vegetables. As research expands on postharvest biological control, however, it is anticipated that our knowledge will increase rapidly by drawing and building upon the foundation of information developed in other areas of biocontrol and phylloplane research.

Several modes of action have been suggested to explain the biocontrol activity of microbial antagonists. Still, competition for nutrient and space between the pathogen and the antagonist is considered as the major modes of action by which microbial agents control pathogens causing postharvest decay. In addition, production of antibiotics (antibiosis), direct parasitism, and possibly induced resistance are other modes of action of the microbial antagonists by which they suppress the activity of postharvest pathogens on fruits and vegetables.

i. **Competition for space and nutrients** : Competition for nutrition and space between the microbial antagonist and the pathogen is considered as the major mode of action by which microbial antagonists suppress pathogens causing decay in harvested fruits and vegetables. To compete successfully with pathogen at the wound site, the microbial antagonist should be better adapted to various environmental and nutritional conditions than the pathogen.

ii. **Space** : Rapid colonization of fruit wound by the antagonist is critical for decay control, and manipulations leading to improved colonization enhance biocontrol. Thus, microbial antagonists should have the ability to grow more rapidly than the pathogen. Similarly, it should have the ability to survive even under conditions that are unfavorable to the pathogen). The biocontrol activity of microbial antagonists with most harvested commodities increased with the increasing concentrations of antagonists and decreasing concentrations of pathogen. For example, *Candida saitona* was effective at a concentration of 107 CFU/ml for controlling *Penicillium expansum* on apples.

iii. **Attachment** : Attachment by microbial antagonist to the pathogen hyphae appears to be an important factor necessary for competition for nutrients as shown by the interactions of *Enterobacter cloacae* and *Rhizopus stolonifer* , *Pichia guilliermondii* and *Penicillium italicum* . *In vitro* studies conducted on such interactions revealed that due to direct attachment, antagonistic yeasts and bacteria take nutrients more rapidly than target pathogens and thereby prevent spore germination and growth of the pathogens.

iv. **Nutrient competition** : Research work conducted on this mode of action of microbial antagonists supports the hypothesis that competition for nutrients plays a major role in the mode of action of *Pichia guilliermondii* against *Penicillium digitatum* in citrus, *Enterobacter cloacae* against *Rhizopus stolonifer* on peach, *Cryptococcus laurentii* against *Botrytis cinerea* on apple and *Rhodotorula glutinis* and *Cryptococcus laurentii* against *Penicillium expansum* and *Botrytis cinerea*, respectively and *Metschnikowia pulcherrima* on apples. *M. pulcherrima* out competes pathogens like *Botrytis cinerea* and *Penicillium expansum* in apple through iron depletion (Saravanakumar *et al.*, 2008). It is a potential yeast for controlling fruit rots. Biocontrol of gray mold on apple by *Metschnikowia pulcherrima* was reduced or totally suppressed by the addition of several nutrients suggesting that competition for nutrients plays a role in the biocontrol capability of *Metschnikowia pulcherrima* against *Botrytis cinerea*.

v. **Populations of the microbial antagonist** : The level of control provided by the microbial antagonists is also highly dependent on the initial concentration of the antagonists applied on the wound site and the ability of the antagonist to rapidly colonize the wound site. In general, microbial antagonists are most effective in controlling postharvest decay on fruits and vegetables when applied at a concentration of 107 –108 CFU/ml (El-Ghaouth *et al.*, 2004) and rarely, higher concentrations are required. Currently, there is only fragmented data regarding the antagonist-pathogen interaction in terms of competitions for limiting nutrients essential for pathogenesis. Once more information regarding the specificity of competition between antagonistic and pathogens in fruit wounds is available and genes responses of antagonism of biocontrol agents have been characterized, it will be possible to develop antagonistic stains with a higher rate of transport and/or metabolism of limiting nutrient essential for pathogenesis. This may allow us to circumvent some of the limitations of microbial antagonists.

vi. **Production of antibiotics** : Production of antibiotics is the second important mechanism by which microbial antagonists suppress the pathogens of harvested fruits and vegetables. For instance, bacterial antagonists like *Bacillus subtilis* and *Pseudomonas cepacia* are known to kill pathogens by producing the antibiotic iturin. The antagonism so produced by *Bacillus subtilis* was effective in controlling fungal rot in citrus and *Monilinia fructicola* in peaches and cherries. Further, *Pseudomonas cepacia* inhibited the growth of postharvest pathogens like *Botrytis cinerea* and *Penicillium expansum* in apple by producing an antibiotic, pyrrolnitrin. *Pseudomonas cepacia* was also effective in controlling green mold (*Penicillium digitatum*) in lemon by producing antibiotics. Similarly, the bacterial antagonist, *Pseudomonas syringae*, controlled green mold of citrus and gray mold of apple, by producing an antibiotic syringomycin .

vii. **Direct parasitism**: *Pichia guilliermondii* cells has the ability to attach to the hyphae of *Botrytis cinerea* and *Penicillium.* After yeast cells were dislodged from the hyphae, the hyphal surface appeared to be concave and there was partial degradation of the cell wall of *Botrytis cinerea* at the attachment sites. In contrast, co-culturing *Botrytis cinerea* with non-antagonistic yeast elicited only a loose attachment to the fungus with no pitting in the hyphae. Similarly, *Candida saitona* attached strongly to the hyphae of *Botrytis cinerea* and caused swelling. Microbial antagonists also produce lytic enzymes such as gluconase, chitinase, and proteinases that help in the cell wall degradation of the pathogenic fungi. Direct parasitism was a major factor that permitted *Pantoea agglomerans* to control *Monilinia laxa* or *Rhizopus stolonifer* decay on stone fruits. Thus, strong attachment of microbial antagonist with enhanced activity of cell wall degradation enzymes may be responsible for enhancing the efficacy of microbial agents in controlling the postharvest diseases of fruits and vegetables.

viii. **Induced resistance :** Induction of defense responses in the harvested fruits and vegetables by the microbial antagonists has been suggested and shown as another mode of action of microbial antagonists for controlling postharvest decay in them. For example, *Cryptococcus saitona* induced chitinase activity and formed structural barrier (papillae) on host cell walls in apple against *Penicillium expansum*. Similarly, *Aureobasidium pullulans* caused a transient increase in the activity of 1,3-gluconase, peroxidase, and chitinase enzymes in apple wounds which stimulated wound healing processes and induced defense mechanisms against *Penicillium expansum* (Ippolito *et al.*, 2000). Induction of disease resistance responses was also reported in pineapple, avocado, and citrus fruits.

Microbial antagonists induced disease resistance in the harvested commodities by the production of antifungal compounds, as in avocado fruit (Yakoby *et al.*, 2001) and accumulation of phytoalexins, like scoparone and scopoletin in citrus fruit. Production of such antifungal compounds by microbial antagonists in the host cells help in inducing defense mechanism and hence provide biocontrol on the harvested commodities. Although a causal connection between the accumulation of host defense responses and bioprotection by antagonistic yeasts has not yet been clearly established, the occurrence of high levels of host antifungal compounds in protected tissue suggests their implication in diseases resistance.

The pre-harvest pathogens like *Colletotrichum gloeosporioides, C. acutatum, Botryodiplodia theobromae, Alternaria citri, Phomopsis citri* etc. attack the citrus fruit from fruit set till harvest and cause considerable damage to its production and quality. The incipient infection of pre-harvest pathogens subsequently also manifest in the form of postharvest diseases besides the attack of other postharvest wound pathogens *viz. Penicillium digitatum, P. italicum, Geotrichum candidum* etc. during postharvest handling, transport, storage and marketing (Naqvi, 2007). The viable technologies have been developed to reduce these losses through pre-harvest management of diseases, improved handling, transport, storage, packing and marketing. The postharvest treatments with fungicides, biocontrol agents, and wax on mechanical citrus packing line and certain physical therapies have been suggested to further minimize the postharvest decay and to extend the shelf life of fruits for successful trade of citrus in domestic and distant export market.

Fungal diseases result in significant losses of fruits and vegetables during handling, transportation and storage. At present, post-production fungal spoilage is predominantly controlled by using synthetic fungicides. Under the global climate change scenario and with the need for sustainable agriculture, biological control methods of fungal diseases, using antagonistic microorganisms, are emerging as ecofriendly alternatives to the use of fungicides. The potential of microbial antagonists, isolated from a diversity of natural habitats, for postharvest disease suppression has been investigated. Postharvest biocontrol systems involve tripartite interaction between microbial antagonists, the pathogen and the host, affected by environmental conditions (Dukare *et al.*, 2018). Several modes for fungistatic activities of microbial antagonists have been suggested, including competition for nutrients and space, mycoparasitism, secretion of antifungal antibiotics and volatile metabolites and induction of host resistance. Postharvest application of microbial antagonists is more successful for efficient disease control in comparison to pre-harvest application. Attempts have also been made to improve the overall efficacy of antagonists by combining them with different physical and chemical substances and methods. Globally, many microbe-based biocontrol products have been developed and registered for commercial use. Dukare *et al.*, (2018) reviewed on the use of microbial antagonists as postharvest biocontrol agents and summarises information on their isolation, mechanisms of action, application methods, efficacy enhancement, product formulation and commercialisation.

Two psycotrophic yeasts isolated from the surface of cold stored lemons have been selected as biocontrol agents of most common postharvest diseases of apples and citrus, during cold storage. they were identified as *Cystofilobasidium infirmominiatum* and *Leucosporidium scottii,* both yeasts resulted resistant to three of the fungicides most frequently used in postharvest applications, so a combination of the antagonists with fungicide could be used as part of an integrated management practice (Vero, *et al.*, 2011). However, the growth of both of them resulted affected by the presence of chitosan, so a combination of treatments would be impracticable. Both antagonists were characterized but mechanisms involved in bioocntrol activity could not be fully understood. Nitrogen competition seemed to play a role.

Fungal decay during postharvest storage of fruit leads to significant losses worldwide. As pathogens become resistant to synthetic fungicides microbial antagonists are promising alternatives, but only one postharvest biocontrol agent has been registered for use in Canada (Nelson, *et al.*, 2011). Plant root-associated rhizobacteria adapted to cold climates may be a potential source of novel biocontrol agents. Four rhizobacteria that were isolated from legumes grown in Saskatchewan soils and suppressed the growth of three major postharvest pathogens of pome fruit, *Penicillium expansum, Botrytis cinerea* and *Mucor piriformis* during cold storage were selected for further characterization. Three *Pseudomonas* isolates and one Serratia isolate grew well at temperatures from 1 to 28°C and at pH 5-7 and were able to utilize the major sugars, organic acids and amino acids found in apples. Cell-free extracts from one pseudomonad isolate inhibited growth of *Penicillium expansum.* This isolate consistently increased medium pH following growth on several media, a factor which may contribute to the suppression of fungal pathogens. These rhizobacteria demonstrate potential for control of fungal pathogens responsible for postharvest decay of pome fruit.

Fruit surface microflora has been the most productive source of antagonists against fungi causing postharvest decays of fruit. The establishment of populations of these organisms and their interaction with decay causing pathogens in fruit wounds has been the basis for the development of current postharvest biological control systems. However, many pathogens, including *Monilinia fructicola*, the causal agent of brown rot of stone fruits, can also infect fruit through undamaged tissue in the orchard by producing latent infections that are activated as fruit ripens in storage, resulting in decay. We characterized culturable bacteria and yeast resident microflora of nectarine fruit from early development until harvest over a two year period. Time of isolation was a significant factor in the frequency of occurrence of different bacteria during fruit development, indicating a succession of the genera.

However, for yeasts, only the last sampling time was distinct from the earlier samplings, indicating the presence of more specialized yeasts on mature fruit. Conventional screening of these microorganisms, for biocontrol of brown rot originating from wound infections on nectarines and plums, resulted in the isolation of several effective antagonists, which constituted only about 2% of all microorganisms tested. Janisiewicz, *et al.*, (2011) developed a new technique, involving in vitro and in situ tests, for detecting bacteria and yeasts able to control latent infections of *M. fructicola* on stone fruits. This new approach can also be used in other fruit systems, including latent infections caused by *Colletotrichum* spp., and may result in the development of the next generation of biocontrol agents.

A new yeast antagonist, *Metschnikowia andauensis* PBC-2, isolated from apple fruit, was evaluated for its biological control activity against the most important postharvest pathogens of pome and citrus fruit (Manso and Nunes,

2011). This yeast at 10^7 cfu ml^{-1} was very effective in reducing blue mold caused by *Penicillium expansum* on different cultivars of apples and in 'Rocha' pear, and against *P. digitatum* and *P. italicum* on citrus fruit. High control of *B. cinerea, P. expansum* and *R. stolonifer* was obtained with the application of *M. andauensis* PBC-2 on 'Golden Delicious' apple. Over more than four years of experiments in semi-commercial trials *M. andauensis* PBC-2 provided excellent control against *P. expansum* under cold storage. The mechanisms of action, as antibiosis, production of chitinolytic enzymes and iron competition were studied. Results showed that none of these mechanisms were responsible for the antagonistic ability of *M. andauensis*. Research is now in progress to examine different media, in order to optimize large scale production.

In Europe, the pome fruit growing industry is currently confronted with residue restrictions imposed by marketing chains as retail companies and import countries (Vorstermans. and Creemers 2011). Next to a high quality product also a limited residue loading on the fruit surface is requested. Particularly the number of active ingredients on the fruit surface is restricted, and also some standard applied plant protection products (PPPs) are prohibited. The chemical fungicides applied shortly before harvest towards fruit rot decay during storage are frequently detected after analytical analysis. Therefore, these chemicals are, at the moment, in the focus of interest of the industry. To this end, research concerning alternative approaches has exploded in the past years. The European government is also stimulating this kind of research indirectly as each member state is demanded to adopt a national action plan to reduce risks and impacts of pesticide use on human health and the environment (Directive 2009/128/EC). In collaboration with the plant protection industry research was done to investigate non-chemical alternatives for postharvest disease control. Several yeast strains were selected and formulated to verify their potential as a biological control agent (BCA).

In order to enhance the efficacy, the supplementary effect of natural products and physical treatments is also implemented in this study. As for the application technique in general, a postharvest treatment by dipping, which is a more standardized application where all parameters can be quantified and controlled, is selected. In addition to the fruit surface, also wounds, caused by picking and transport, will be covered resulting in a higher biological efficacy towards key wound pathogens as *Botrytis cinerea* and *Penicillium* spp. Depending on the type of BCA some promising results were obtained. Nevertheless, more research under practical conditions is requested to implement these non-chemical products into an integrated system.

Four isolates of *Bacillus* spp. were tested in a 2-year field study for biological control of pre- and postharvest diseases of apple (Poleatewich *et al*., 2012). For the preharvest test, bacteria were applied to 'Golden Delicious' and 'Rome Beauty' trees in May or May and June. Foliar apple scab severity was assessed weekly. After harvest, fruit were wounded and then either left untreated or given a postharvest application of the bacteria. Wounded fruit were then inoculated with the bitter rot pathogen and lesion size was measured over 8 days. *Bacillus megaterium* isolate A3-6, B. mycoides isolate A1-1, and B. cereus isolate FLS-5 applied in May or May and June significantly reduced fruit and foliar apple scab severity in both years. A postharvest application of the bacteria (alone or in combination with a preharvest application) resulted in the greatest suppression of bitter rot on both cultivars ($P < 0.04$). The May + June + postharvest application of isolate A3-6 resulted in the greatest suppression of bitter rot, with an average of 45 and 95% reduction in lesion size compared with nontreated controls on 'Golden Delicious' and 'Rome Beauty' fruit, respectively. Results from this study indicated that preharvest applications of the bacteria were able to reduce foliar and fruit scab and an additional postharvest application was effective in reducing bitter rot severity.

Natural plant products for biocontrol : Environmentally friendly plant extract agents have shown great potential as alternatives to synthetic fungicides (Harbant Singh *et al*., 2012). Recently, the antimicrobial activity of biodegradable and safe higher plant products has attracted the attention of microbiologists. However, the actual use of these products to control postharvest pathogens of fruits, particularly citrus pathogens, is still limited.

Fruits and vegetables have a number of constitutive and inducible compounds that are antimicrobial and that have not been fully explored as biological control agents . Such compounds could be used in or on the plant where they are produced or extracted and applied to other harvested crops. Essential oils from leaves of *Melanleuca leucadendron, Ocimum canum* and *Citrus medica* were able to protect several stored commodities from biodegradation by *Aspergillus flavus* and *A. versicolor*. These oils were active between 500-2000 ppm. A number of the fruit volatiles produced by peaches as they ripen are highly fungicidal. Benzaldehyde totally inhibited spore germination of *Botrytis cinerea* at 25µllL and germination of *M. fructicola* at 125µllL. Three compounds (benzaldehyde, methyl salicylate, and ethyl benzoate) completely inhibited growth of *Monilinia fructicola* and *B. cinerea* at 370µl/L. *Benzaldehyde* has been used in the laboratory to fumigate peaches and protect them against *Rhizopus* rot . Plant volatiles have been implicated in plant resistance mechanisms to postharvest diseases. Resistance of fruit to rot in high-CO2 storage was due to the production of high levels of acetaldehyde and ethyl acetate by the fruit in response to these conditions. Acetaldehyde vapor controlled *Botrytis cinerea* and *Rhizopus stolonifer* rots of strawberries and raspberries. Acetaldehyde has been used also as a fumigant to control the green peach aphid on head lettuce. With the broad range of natural fungicidal plant volatiles, numerous opportunities exist to explore their usefulness in controlling postharvest diseases.

Two alcoholic extracts from chilly and ginger (ranging between 500 and 3000 ppm) were tested for antifungal activity in vitro on *Penicillium digitatum, Aspergillus niger* and *Fusarium* sp isolated from naturally infected citrus fruit (Harbant Singh *et al.*, 2012). The water extracts served as control and it was observed that the alcoholic extracts concentrations were more effective than the water extract control in showing antifungal activity against the test pathogens. Results showed the alcoholic extract concentrations were more effective than the water extract control in showing antifungal activity against test pathogens. All 3000ppm concentration from *Capsicum frutescence* L. and *Zingier officinale* L. showed a 100% and 85% inhibition zone for all the three fungi respectively.

7.10 Chemical fungicides: Postharvest dips/drenches with chemicals are used to protect fruits and vegetables against postharvest diseases. Postharvest diseases of fruits and vegetables have been controlled by synthetic postharvestfungicides. Fungicides were more effective when pathogen populations were low, environmental conditions for infection development were least favorable and the fruit had some intrinsic resistance to infection. The benzimidazole derivatives were the first systemic fungicides that became available for the control of postharvest pathogens. The fungicide, 2(4-thiazol-4-yl) benzimidazole (thiabendazole, TBZ), belonging to the chemical class of benzimidazoles, remains stable in an aqueous suspension throughout the storage life of treated fruit. TBZ is one of the few fungicides approved for postharvest use on apples, pears, and citrus as a drench, dip, or line spray. Following harvest, prestorage chemical control is commonly applied to reduce the incidence of blue mold in apples that are stored for long periods of time in CA storages in Canada, Australia, and the USA (Eckert and Ogawa, 1988). In commercial operations, some of the cultivars of apples destined for long-term storage are also treated with an antioxidant and scald inhibitor, diphenylamine (DPA). In pome fruits, postharvest fungicide treatments are applied as a postharvest and prestorage drench, with or without the DPA. Following storage, fungicides are applied as a spray just before packing and sometimes the fungicide is combined with wax.

The effectiveness of benzimidazole fungicides, including TBZ, promoted increased fruit production because long-term storage became less risky. In a 2-year study, TBZ applied to fruit without artificial wounding or inoculation effectively reduced decay when applied within 3 weeks or 6 weeks. The TBZ controlled blue mold in apples for up to 6 months in cold storage The TBZ fungicide was highly effective in controlling blue mold caused by TBZ-sensitive isolates of *P. expansum* on 'Empire' apples for up to 105 days in CA storage, however, it was not effective against TBZ-resistant *P. expansum*. Three fungicides, imazalil, TBZ, benomyl, controlled postharvest blue mold in apples. The success of a chemical treatment for disease control depends on the initial spore load, the depth of the infection within the host tissues, the growth rate of the infection, the temperature and humidity and the depth to which the chemical can penetrate the host tissues. Moreover, the applied chemical must not be phototoxic (i.e. it must not cause injury to the host tissues) and must fall within the ambit of the local food additive laws (Nabi, *et al*., 2017). For controlling postharvest diseases of fruits, fungicides can be applied in different ways as under:

Pre-harvest chemical treatments: In most cases, control of postharvest diseases should start before harvest in the field of orchard itself. The possibility of controlling well established pathogens by postharvest disinfection is very less since most fungicides are unable to penetrate deeply into the tissues, and effective concentrations of the fungicide would not reach deep seated infections. Hence, the effective way to reduce infections initiated in the field, including, quiescent infections is the application of broad spectrum protective fungicides to the fruits on the plant itself. The developing fruits are sprayed to prevent spore germination and subsequent formation of appressoria and infective hyphae, which are quiescent stages of the fungus.

Postharvest chemical treatments: An efficient disinfection process should reach the pathogenic microorganisms accumulated in all those sites. Different chemicals/fungicides, are being used to manage postharvest decay in developed fruits and vegetables.

In fruits: During the 1960s, pre- and postharvest application of captan and dithiocarbamates was the principal means for controlling postharvest decay of apples and pears applied as a dip or a fumigant, was shown to be effective against *Penicillium expansum* and *Gloeosporium*, but the treatment was not utilized extensively in practice (Joseph W. Eckert and Joseph M. Ogawa 1988). Since 1970, the benzimidazole fungicides, benomyl, thiabendazole thiophanate methyl, and carbendazim have been extensively evaluated as pre- and postharvest treatments for the control of decay of apples and pears during storage. The highly successful use of these compounds as orchard sprays in the late summer to control lenticel rots in Europe has been reviewed. Postharvest treatments of unsprayed apples with 300-500 mg/l thiabendazole or benomyl have provided excellent control of lenticel rots during storage. Treatment of harvested apples and pears with suspensions containing 250-500 mg/l benomyl, 500-2000 mg/l thiabendazole or 1000 mg/l carbendazim controlled decay by *Penicillium expansum* and *Botrytis cinerea, Cladosporium herbarum* and *Aspergillus niger*.

Benomyl or thiabendazole suspensions (500 mg/l) at ambient temperatures controlled *Penicillium* infection of puncture injuries on apples, but the fungicide suspensions had to be heated to 29-45°C to control bruiseinoculations. However, the treatment at ambient temperatures gave good control of bruise-inoculations in pears, suggesting that the skin of pear is more easily penetrated by benzimidazole fungicides than that of the apple. Thiabendazole (TBZ) applied to harvested fruit in bins before degreening also was very effective against green mold (Smilanick *et al*., 2006). Fungistatic residues of both benomyl and thiabendazole were found in the flesh of pears following treatment of the fruit with 500-1200 mg/l of these fungicides . Benomyl appears to penetrate the skin of the pear and apple more readily than thiabendazole, although the latter is as effective as benomyl in controlling *P.expansum* and Botrytis cinerea on pears. Preharvest sprays of thiabendazole and benomyl gave good control of *Penicillium* and *Botrytis* on pears in storage, but the treatment appears superfluous because most pears are drenched with a solution of ethoxyquin (anti-scald agent) after harvest and a benzimidazole fungicide could be applied at this time. The benzimidazole fungicides do not prevent the development of storage decays caused by *Alternaria, Mucor* spp. or *Phytophthora* spp.. A mixture of benomyl (500 mg/l) and prochloraz (1000 mg/l) gave good control of *Penicillium* and *Alternaria* on pears stored at -0.5°C for 6-7 months .

Chlorothalonil, iprodione, and triadimenol were less effective than prochloraz in combination with benomyl. Iprodione alone applied pre- and post-inoculation suppressed the development of *Alternaria* lesions. Imazalil controlled *A lternaria* on wound inoculated apples and naturally inoculated pears during 0°C storage for 6 months. Since imazalil controls *P. expansum,* this fungicidealone should be as effective as the prochloraz + benomyl combination for control of *Penicillium* and *Alternaria*. Imazalil and iprodione controlled *Gloeosporium, Penicillium* and *Botrytis* on apples as well as benomyl, including benzimidazole-resistant strains. Etaconazole, benomyl, and benomyl + captan gave good control of *Penicillium* during storage of apples for 11 weeks at OSC. Imazalil has potential value for controlling postharvest diseases caused by Alternaria alternata in pears, apples and persimmons. Imazalil was active in vitro on germination and hyphal growth (ED50) at 1.4 and 0.5 μg/ml, respectively. Protectant action of the fungicide lasted only 1 day after treatment (Prusky *et al.,* 2008).. At 1000 μg/ml the fungicide inhibited development of Alternaria rot in inoculated apples at 0 °C and in naturally infected pears and persimmons at–1 °C.

Although captan did not eradicate wound inoculations of *Penicillium* the addition of this fungicide to the mixture was recommended to inactivate spores of benzimidazole resistant strains that might accumulate in the anti-scald solutions. *Mucor piriformis* can be a difficult problem in apple storage because this pathogen can cause decay at low temperatures and can spread from fruit to fruit. Most fungicide treatments have not eradicated inoculations of *M. piriformis* . Sanitation is the most practical method for controlling this disease. Chlorine or SOPP added to the water in the dump tank can prevent fruit inoculation. The spread of this disease in storage is decreased by the use of copper impregnated fruit wraps. *Phytophthora* rot of apples induced by *Phytophthora cactorum* and *P. syringae* has become a significant problem in recent years. The build-up of the disease is favored by drenching fruit before storage with a soil contaminated solution of an anti scald agent and fungicides that are not active against *Phytophthor*a. Treatment of orchard soil with furalaxyl in August completely suppressed *P. syringae* in the top 10 cm of soil until mid October.

Pre-inoculation treatment of apples with captan, dichlorophen, copper sulfate, fosetyl AI, and metalaxyl provided good protection against infection by *Phytophthora* spp.. Dipping fruit in solutions of metalaxyl and fosetyl-AI eradicated infections on apples inoculated with zoospores and incubated for 5 days at 10-15°C before application of the fungicide treatment. The strong suppressive action of the benzimidazoles has been attributed to their penetration through the wax and cuticle at the host surface to reach and inhibit the pathogen beneath. This behavior of the benzimidazoles contributes to the persistence of the fungicide on the fruit throughout the storage period, exerting continuous pressure for the selection of benzimidazole resistant strains of *P. expansum* in the population.

Years of continuous use of thiabendazole and benomyl in orchards have resulted in the emergence of resistant strains of *P. expansum* in most apple production areas with a simultaneous deterioration in the efficacy of the postharvest benzimidazole treatments, except in the states of Oregon and Washington where benzimidazole fungicides are not applied in the field. Several investigators have demonstrated that benzimidazole-resistant strains of *P. expansum* can be controlled on pome fruits by postharvest application of imazalil, etaconazole, propiconazole, prochloraz, iprodione, and vinclozolin. However, strains of *P. expansum* that are resistant to both iprodione and vinclozolin have been isolated, and they are not well controlled by postharvest treatment of the fruit with these fungicides. Although imazalil has provided good control of benzimidazole resistant blue mold on pome fruits, a pathogenic imazalil resistant *Penicillium crustosum* has been isolated from decaying pears. Since imazalil, etaconazole, and prochloraz are not currently registered for postharvest treatment of pome fruits, treatment of apples before storage with a mixture of captan and benomyl is advised. It must be noted that the development of etaconazole as an agricultural fungicide has been discontinued, and prochloraz is alledged to taint the odorlflavor of treated apples.

The captan-benomyl mixture gave excellent control of benzimidazole-sensitive *P. expansum* and about 80% decay reduction in apples that were inoculated with a benzimidazole-resistant strain. The mixture was effective when the treated apples were stored at O°C, but not when they were held at 20oe. Benzimidazole-resistant strains of *P. expansum* characteristically have a lower parasitic fitness than do wild (benzimidazolesensitive) strains and their frequency in the population declined when pears were treated by a fungicide program which combined two fungicides (e.g. thiabendazole + captan) and when this combination alternated with a "new" fungicide (imazalil, etaconazole, or iprodione). Molecular analysis of the β-tubulin gene in moderately resistant isolates did not reveal any amino acid substitution (Malandrakis *et al.*, 2013). This is the first report on the existence and distribution of highly mycotoxigenic field isolates of *P. expansum* resistant to the benzimidazoles indicating a high potential risk of increased mycotoxin contamination of pome fruit and by-products.

Most benzimidazole-resistant strains of *P. expansum* were usually sensitive to diphenylamine, an anti-scald agent which does not influence fruit decay by benzimidazole sensitive strains of the pathogen. Combinations of diphenylamine and benomyl, thiabendazole, and thiophanate methyl provided better control of benomyl-resistant *P.expansum* in inoculated apples stored at 2-4°C than did either diphenylamine or the fungicides used alone. Preharvest sprays of dicloran applied one week before harvest have provided good control of *Rhizopus* of peaches inoculated after harvest. The preharvest treatments may be advantageous for peaches destined for local markets, but a postharvest treatment is more effective when peaches are brushed or waxed before shipment as fresh fruit. A postharvest treatment with dicloran was superior to a preharvest spray treatment for control of *Rhizopus* and *Monilinia* on freestone peaches during controlled ripening at 200e. The preharvest spray of dicloran should be necessary only for those special situations where a postharvest treatment is not feasible. A summary of several years' trials comparing different methods of applying dicloran to peaches has been published. Postharvest treatments of dicloran are usually combined with another fungicide such as benomyl, triforine, iprodione, or captan which have high activity against *M. fructicola* .

Several investigators have reported that the activity of dicloran against *M. fructicola* can be greatly improved by heating the fungicide suspension to 46-51°C . Treatment of peaches, nectarines, and sweet cherries after harvest with dips, sprays, or wax formulations containing 100-300 mg/l benamyl or 1000--2000 mg/l thiabendazole have provided excellent control of *M. fructicola* but have not been effective against *Mucors* or *Alternaria* Carbendazim and thiophanate-methyl are somewhat Jess effective against brown rot than is benomyl, on a weight basis. The superiority of benomyl over the other benzimidazole fungicides can be attributed in part to its ability to penetrate through the epidermis and into the mesocarp of the fruit to suppress infections of brown rot. The effectiveness of the benzimidazole fungicides against brown rot has been improved by heating the fungicide suspensions to 46-52°C or by incorporating the fungicide in a wax formulation. The concentration of the benzimidazole fungicide in the heated suspension may be reduced to 25% of the recommended dosage without sacrificing the effectiveness of the treatment. Both brown rot and *Rhizopus* rot can be controlled by a single treatment combining dicloran with a benzimidazole fungicide in the same treating fonnulation. Several investigators have demonstrated that postharvest treatment of stone fruits with triforine controlled brown rot as well as benomyl. Triforine is not active against *Rhizopus*; therefore it must be combined with dicloran for broad spectrum control of postharvest diseases of stone fruits. lprodione inhibits *Rhizopus, Alternaria, Botrytis* and *Monilinia* and may be applied as a single fungicide treatment to control the major postharvest diseases of stone fruits. Imazalil and prochloraz are also effective as postharvest treatment against brown rot but apparently are not active against *Rhizopus* .

In recent years, the control of brown rot on stone fruits has been seriously impaired by the appearance of be no my I-resistant strains of *M. fructicola* which appear to be selected by the intensive use of benzimidazole fungicides in the orchard. The postharvest use of benomyl, carbendazim, and thiophanate methyl should be avoided on lots of fruit known to be contaminated with benzimidazole-resistant strains of *M. fructicola*. lprodione or triforine, which do not show cross-resistance to benzimidazole fungicides, should be substituted for benomyl in situations where benzimidazole-resistant strains of *M. fructicola* have been identified. However, resistance of *Monilinia* to vinclozolin has been reported, and laboratory resistance of some isolates of this fungus to triforine have been described. Dicloran resistant isolates of *R.*

stolonifer and *B. cinerea* have been investigated in the laboratory, but isolates of this type have not diminished the effectiveness of the postharvest treatment because this fungicide is rarely applied in the orchard.

Results on the Kent mangoes showed that the addition of fludioxonil significantly reduced stem-end rot in either hot or cold treatments. Fludioxonil, can therefore be a very valuable fungicide for control of stem end and soft brown rots on mangoes exported from South Africa (Swart, *et al.*, 2009). The phenomenon regarding improved control of stem-end rot with reduced concentrations of fludioxonil in prochloraz mixtures indicate that their might be some synergistic mode of action but cannot be proved with this data and should be explored further. Poor control of anthracnose was obtained on Kent mangoes with ambient solutions probably due to poor control in the orchard and high disease pressure. In Keitt mangoes results showed that anthracnose was better controlled with prochloraz and fludioxonil plus prochloraz mixtures. In Kent mangoes anthracnose was better controlled when fruit were treated in heated fungicide solutions. The efficacy of fludioxonil regarding control of stem-end rot, and prochloraz regarding control of anthracnose as well as improved control of postharvest diseases of mangoes in heated fungicide solutions was demonstrated in both trials. The value of using two different, chemically non-related, fungicides in a mixture, each with good activity against a specific disease and some against another, can be an excellent strategy for total disease control especially if concentrations can be lowered, without losing efficacy, in heated solutions.

In vegetables: Postharvest decay problems are usually related to the severity of the disease in the field and to inoculum levels . The microorganisms most often responsible for postharvest disease are *Botrytis cinerea, Alternaria alternata, Erwinia carotovora*, and *Phytophthora parasitica.* Diseases caused by *Geotrichum candidum, Rhizopus stolonifer, Mucor* spp., *Fusarium* spp., *Phytophthora infestans*, and *Rhizoctonia solani* may be of local or occasional importance (Joseph W. Eckert and Joseph M. Ogawa, 1988). Recent investigations have implicated *Geotrichum penicil/atum* (206), *Mucor mucedo, M. piriformis*, and *Pseudomonas marginalis* as agents responsible for postharvest disease of tomatoes. The importance of these pathogens may vary from season to season, but a forecasting method based on field conditions has not yet been developed. Control of *Botrytis* rot on staked tomatoes in California begins with a fungicide spray at flowering to protect the flowers, fruitlets, and stems from *Botrytis* infection. Fungicide sprays are applied thereafter every two weeks during wet periods until harvest. With tomatoes grown in the semi-arid valleys, fungicide sprays for *Botrytis* control are delayed until late in the harvest season when night temperatures drop to 16°C and dew forms on the fruit. Chlorothalonil and anilazine have been extensively used to protect plants and the fruits against *Botrytis* infection. Chlorothalonil is also effective against *A. alternata*. Benomyl and thiophanate methyl have provided control of *Botrytis*, but these fungicides should be used judiciously because of the existence of resistant isolates of the pathogen in the field. Iprodione and vinclozolin have been used experimentally as alternatives to the benzimidazole for *Botrytis* control.

The major fungal pathogens of cuccrbits and cantaloups are *Fusarium* spp., G. *candidum, Rhizopus oryzae, Cladosporium* spp., and *Alternaria* spp. *Alternaria* and *Cladosporium* usually cause only unsightly blemishes on the rind and the stem-scar of the fruit. Hot water treatments (SrC, 30 sec.) reduced stem-scar and surface molds caused by *Alternaria, Fusarium*, and *Rhizopus,* but the principal benefit of the hot water treatment was to increase the effectiveness of protective fungicides such an maneb, ziram, sodium dimethyldithiocarbamate, SOPP, and captan which are only weakly effective at room temperature. Some "systemic" fungicides were effective when applied at ambient temperature: thiabendazole, benomyl, imazalil, guazatine, and prochloraz. Most fungicides were rather specific for certain fungal species in the disease complex. Benomyl was very active against *Fusarium*; imazalil against *Fusarium* and *Alternaria;* guazatine against *Geotrichum* and *Alternaria.* Fenaminosulf, but not dicloran, was effective against *Rhizopus oryzae.* Prochloraz, but not etaconozole, provided acceptable control of all diseases. A mixture of benomyl (250 mg/l) and guazatine (500 mg/I) controlled all diseases (*Fusarium, Geotrichum, Alternaria, Cladosporium*, and *Rhizopus*) on naturally inoculated cantaloups in Australia.

Benomyl and carbendazim dip treatments reduced several postharvest diseases on cucumber, eggplant, pepper, and squash, but not those caused by *Alternaria.* Preharvest sprays of benomyl, thiabendazole, or dicarboximide fungicides have reduced *Botrytis* decay during storage. Pentachloronitrobenzene dust has been a commercial treatment to control *Botrytis* on stored cabbage in the Netherlands and Germany. Thiabendazole and dicloran dusts applied to cabbage before loading controlled *Botrytis* during seven month storage at ooe, but the incidence of Alternaria rot was not affected. Thiabendazole and dicloran may be used for storage of crops in alternate years to minimize the development of benzimidazole-resistant strains of *Botrytis*.. A postharvest drench of benomyl, thiabendazole, or a dicarboximide fungicide over cabbage heads in bins reduced *Botrytis* rot considerably. Benomyl alone increased the level of *Alternaria* . The benomyl treatment reduced the rate of senescence (292) but increased the incidence of black speck on stored cabbage . A mixture of thiabendazole or benomyl and iprodione (dust or water spray) controlled both *Botrytis* and *Alternaria*, but not *Phytophthora* or bacterial pathogens. A smoke of vinclozolin provided good control of *Botrytis* and *Alternaria* on stored white cabbage.

Botrytis and *Sclerotinia* rots in carrots during cold storage were substantially reduced by spraying the tops of the plants with benomyl, thiophanate methyl, or vinclozolin on the day before harvest. A postharvest dip of sodium o-phenylphenate (SOPP) controlled *Botrytis* and other storage rots in the United States, but failed to control *M. acerina* in the United Kingdom. A postharvest dip of benomyl (250-500 mg/l) resulted in 96% marketable carrots after 5 months storage (average of 3 storage seasons). Substantially the same results were obtained in Canada, where it was demonstrated that benomyl and thiabendazole sprays before storage gave better control of storage diseases than SOPP. Imazalil and prochloraz have provided control of dry rot and gangrene in potato equivalent to that of the benzimidazole fungicides. At 15°C (curing temperature), all fungicides should be applied within 3 days after wound inoculation . The benzimidazole fungicides were somewhat more effective than imazalil and prochloraz in decreasing the incidence of skin spot. Imazalil provided control of all major potato storage diseases at least equal to the benzimidazole fungicides.

7.11 Bactericides: Progress in controlling bacterial diseases of fruits and vegetables has been slow compared to the rapid advances in the chemical treatment of postharvest fungal diseases. Current strategy against bacterial diseases involves the destruction of inoculum and making the host and environment unfavorable for disease development (Joseph W. Eckert and Joseph M. Ogawa, 1988). The opportunities for eradicating a bacterial pathogen after inoculation of a susceptible host are quite limited because the bacteria multiply rapidly and few bactericides can penetrate in an active form to the infection site without damaging the healthy tissue of the host. Chlorine (hypochlorites) has been used widely in packinghouses to wash, hydrocool, and transport fresh fruits and vegetables. It is assumed that a low concentration of chlorine (e.g., 50-100 ppm free available chlorine) will either kill bacteria suspended in the water and on the surface of the produce or prevent infection if contaminated chlorinated water is drawn into openings on the surface of the produce and have little effect upon the development of bacteria or fungi present in lenticels, stomata or wounded tissue of the host.

Several factors that can influence the effectiveness of active chlorine as a bactericide have been evaluated in other reviews , but it is advisable to evaluate the effectiveness of each chlorinated water supply. Cells of *Erwinia carotovora* suspended in water with 1000 ppm active chlorine caused bacterial soft rot in

18% of the tomatoes that were pressure infiltrated with the chlorinated bacterial suspension. Active chlorine in water is not usually effective in controlling *Penicillium* decay in wound inoculated apples and oranges, and the assumption is that the active chlorine oxidizes the wounded tissue before it reaches the spores. The effectiveness of chlorine solutions depended to a great extent upon the presence of different surfactants in the chlorine solution. Sodium o-phenylphenate (SOPP) has antibacterial properties, although it is used mostly against fungal pathogens on fruit crops. SOPP is registered as a treatment to kill *Xanthomonas campestris* pv. *citri* on the surface of harvested citrus fruits in Florida. Several investigations during the 1950s showed the effectiveness of antibacterial antibiotics (streptorriycin and several tetracyclines) for the control of bacterial soft rot of potatoes and leafy vegetables.

More recently, the effectiveness of streptomycin in controlling bacterial soft rot of potatoes and bell peppers has been confirmed. Although antibiotics important in human medicine have not been used for control of postharvest diseases in the United States, streptomycin, sometimes in mixture with oxytetracycline, has been used in Japan for the control of bacterial soft rot of vegetables. The American Phytopathological Society has urged caution in the use of medically important antibiotics for the control of bacterial plant diseases because of concerns that extrachromosomal genetic elements for antibiotic resistance could be selected in plant pathogenic bacteria by the agricultural exposure to these agents, and these hereditary factors might be transfered by conjugation to human pathogenic bacteria. Thc antibiotic virginiamycin, which has very limited use in human and veterinary medicine, provided good control of bacterial soft rot of potatoes dipped in solutions of this antibiotic.

The successful application of several organic fungicides to control fungal diseases of potatoes during storage, has encouraged further testing of bactericides for control of bacterial soft rot. Two antibacterials, dichlorephen and iodophor, have been approved under the British Pesticide Safety Precaution Scheme. However, these agents have not been effective for control of bacterial soft rot of potatoes . Hydroxyquinoline decreased bacterial soft rot significantly in treated potatoes. The derivative 5-nitro 8-hydroxyquinoline was substantially more effective than the parent compound. Bronopol (2-bromo-2-nitropropane-I ,3-diol) is used as a potato seed treatment to control Erwinia spp. . Gaseous acetaldehyde applied to tubers for 4 hr gave a significant reduction in bacterial soft rot (304). A novel synthetic bactericide, 7 -chloro-I-ethyl-6-fluoro-I, 4-dihydro4-oxo-3-quinoline carboxylic acid (eGA 78039), developed mainly for control of fire blight on pears, inhibited the development of *Erwinia carotovora* on carrot slices and controlled bacterial soft rot on potatoes.

First generation postharvest fungicides prevent infection of tropical fruits by the wound-pathogens, *Penicillium*, *Rhizopus*, *Aspergillus*, and *Ceratocystis* (Eckert,1990). Fungicides developed since 1965, mainly benzimidazoles (thiabendazole, benomyl, and carbendazim) and sterol inhibitors (imazalil, prochloraz, and propiconazole) have given excellent control of anthracnose and stem-end rot on citrus, banana, mango, papaya, pineapple, and avocado. The intensive use of fungicides on citrus fruits has resulted in a serious problem of resistance to all organic fungicides used commercially. Imazalil-resistant biotypes are widely known in California but have not been detected in Florida. Benzimidazole-resistant biotypes of *Botryodiplodia* and *Colletotrichum* have been recognized in harvested mangos and bananas treated pre-harvest with this group of fungicides.

Methods of application :There are several methods by which the fungicides can be applied for controlling postharvest diseases. Most common methods include drenches (High volume sprayers - Low volume sprayers), dips flooders, foamers, brushes, fumigators, dusters, electrostatic sprays, fogging paper wraps and box liners.

i. **Dipping :** In this method, fruits are immersed in water containing an appropriate concentration of a chemical, which is toxic to the disease causing fungi. The produce may be passed below a shower of the diluted chemical, which is called as "ascade application". For improving the effectiveness of dips, additives may be included in the formulation. These include wetting agents (e.g. teepol or Triton-x-100), which reduce the surface tension and allow a better coating of the chemical on the produce, and acids such as citric acid, which lowers the pH of the fungicide and can, make it more effective.

ii. **Electrostatic sprays:** Breaking up the pesticide solution into fine droplets and then giving them an effective electric charge for field sprays. The main advantage of this system is the increased uniformity of application of spraying materials. Similarly, there is no loss of biological activity of the materials with such sprays. The principle on which they work is that all particles have the same electrical charge and thus they repel each other. They are attracted towards crop, and then form a thin even layer/ coat. This method is being used in several fruits in many advanced countries.

ii. **Fumigation** : Fumigation is also considered as an effective method of chemical application in some fruits. It has several promising applications. It can be carried out immediately after harvest to prevent infection of injuries on the fruit to be transported to long distances, degreened before processing or to be held for several days before processing.

iv. **Chemical pads** : Paper pads impregnated with fungicidal chemicals first time were developed to prevent banana infections during transportation. These are also known as crown pads, and are used to prevent the fungal infections on the cut crowns of fruits. The pads are made from several layers of soft paper previously soaked in a fungicide (often thiobendazole) and then dried.

7.12 Alternative eco-friendly approaches: Postharvest disease management is mainly led by the application of chemicals. However, awareness about pathogen resistance, environmental, and health issues has led to the research of eco-friendly and effective alternatives for disease management. The pathogen establishment on fruits can be affected by the application of GRAS compounds like chitosan, essential oils, salts, among others; besides, their efficacy can be enhanced by their combination with other technologies like ultrasound. Thus, the applications of these alternatives are suitable approaches for postharvest management of fruits. the worldwide trend for both consumers and researchers is the reduction of the use of chemical fungicides and the research for biological, organic, and environmentally friendly alternatives. All this is under certification systems to guarantee the implementation, improvement, integration, and harmonization of all mechanisms to ensure a production of healthy and good quality fruit, with high traceability (the worldwide trend for both consumers and researchers is the reduction of the use of chemical fungicides and the research for biological, organic, and environmentally friendly alternatives. All this is under certification systems to guarantee the implementation, improvement, integration, and harmonization of all mechanisms to ensure a production of healthy and good quality fruit, with high traceability (Ramsés González-Estrada *et al.*, 2019).

7.12.1 Nanoparticles: Increasing concerns continue to be expressed about health hazards and environmental pollution resulting from the use of conventional fungicides for postharvest disease control. Nanoparticles represent an alternative solution for postharvest disease management. Sardella *et al.,* (2017) assessed the physiological effects and the antifungal efficiency of ZnO nanoparticles (ZnO NPs) against a number of fungal contaminants. The efficacy of ZnO NPs was qualitatively and quantitatively assessed against: *Penicillium expansum, Alternaria alternata, Botrytis cinerea* and *Rhizopus stolonifer*. Mycelium growth diameters were measured onto Potato Dextrose Agar (PDA) plates loaded with different ZnO NPs concentrations (from 0mM to 15mM). Hereafter, the rate of the fungal diameter increase was quantified by linear regression modelling. Microscopic analysis was performed by scanning electron microscopy (SEM) images of agar plugs excised from plates with 0mM and 12mM ZnO. All the fungi were inhibited by ZnO NPs at concentrations higher than 6mM. SEM images showed clear morphological aberrations in the fungal structures of all the isolates grown in presence of ZnO. Additionally, knowing that the chelating agent EDTA sequesters metal ions, it was added to fungal inoculated PDA plates with ZnO to study the NPs' mode of action. Cultures where ZnO was mixed with EDTA showed a decrease in the antifungal effect of the nanoparticles. In conclusion, ZnO NPs are therefore a good candidate as an effective postharvest disease control antifungal agent.

Air filters support fungal growth, leading to generation of conidia and volatile organic compounds, causing allergies, infections and food spoilage. Filters that inhibit fungi are therefore necessary. Zinc oxide (ZnO) nanoparticles have anti-fungal properties and therefore are good candidates for inhibiting growth (Decelis *et al.*,2017). Two concentrations (0.012 M and 0.12 M) were used to coat two types of filters (melt-blown and needle-punched) for three different periods (0.5, 5 and 50 min). *Rhizopus stolonifer* and *Penicillium expansum* isolated from spoiled pears were used as test organisms. Conidial suspensions of 105 to 103 spores ml^{-1} were prepared in Sabouraud dextrose agar at 50°C, and a modified slide-culture technique was used to test the anti-fungal properties of the filters. *P. expansum* was the more sensitive organism, with inhibition at 0.012 M at only 0.5 min coating time on the needle-punched filter. The longer the coating time, the more effective inhibition was for both organisms. Furthermore, it was also determined that the coating process had only a slight effect on the Young's Moduli of the needle punched filters, while the Young's Moduli of the melt-blown filters is more susceptible to the coating method. This work contributes to the assessment of the efficacy of filter coating with ZnO nanopaticles aimed at inhibiting fungal growth.

Air is a vehicle for bio aerosols, which include among other entities, fungal spores. These fungal entities are highly resistant to adverse environmental conditions. Fungal spores are undesired air particles in particular environments as these can cause problems such as allergies, infections and food spoilage. Air filters are therefore essential in various settings, including healthcare facilities food manufacturing facilities and other buildings, as it must be ensured that the desired air quality is achieved in these settings. There are various types of air filters, depending on the air quality that is required. These range from high efficiency particulate air (HEPA) filters, found in specific hospital settings where patients at high risk of infection are treated , to simple filters that trap larger particles in domestic and vehicle air conditioning units. Other examples include the use of polyurethane foams, nanofibre mats and other fibre filters. Complex buildings, such as hotels and office blocks make use of heating ventilating and air conditioning (HVAC) systems, which also incorporate air filters. Food manufacturing and processing units are another example of specialized facilities, where air quality in terms of bio aerosols is essential. Air filters are usually disposable, and are designed to filter a fixed volume of air. Therefore, such air filtration equipment needs to be changed regularly to ensure effectiveness.

The material used for filters is inert; however, trapped material may include organic particles and microorganisms. Fungal spores and conidia trapped in a filter are not killed and may encounter the ideal environmental conditions, which include optimal moisture, temperature and nutrients, to germinate and form a mycelium on the filter itself. This can lead to the generation of conidia that may be dispersed in the environment Thus a filter may be responsible for seeding the environment with microorganisms it should be filtering out. Apart from allergies, infections and food spoilage, such growth may also release volatile organic compounds such as acetone, carbonyl sulfide, ketones, aldehydes and alcohols. These compounds are known to be irritants to humans and chemical contaminants in food products. It is therefore important to inhibit the growth of fungi on filters. A number of methods can be used to inactivate spores and conidia trapped in a filter, in order to prevent mycelium from developing. These range from UV irradiation (with or without a photocatalyst), fumigation, and silver, copper and zinc oxide nanoparticle coating among others. Nanocoating seems to be the method with the least disadvantages, as for example UV radiation may have penetration problems, and fumigation may lead to toxic vapour accumulation.

Zinc oxide nanoparticles are very important due to their utilization in gas sensors, biosensors, cosmetics, drug delivery systems, *etc*. ZnO nanoparticles were shown to possess antimicrobial properties. Mechanisms of action have been proposed for bacterial cells, but these can also be considered for fungal cells. These properties are thought to be both chemical and physical. Chemical antimicrobial properties are thought to be due to the ability of these particles to generate reactive oxygen species (ROS), which would then result in the formation of hydrogen peroxide. This would then interact with the cell membrane of the microorganism, leading to oxidation of the lipids, rendering the membrane dysfunctional. Physical antimicrobial properties are thought to be brought about by physical blockage of transport channels in the membrane and direct membrane damage due to the shape and size of these particles.

Efficacy of green nanoparticles (ajwain and neem leaf extract) of copper silver, nickel and magnesium was studied *in vitro* at different concentrations against *Colletotrichum musae* (Divya Jagana *et al*., 2017) Silver-Neem nano particles (Ag-Neem Nps) were tested *in vivo* for their bioefficacy against anthracnose as postharvest spraying at 0.01, 0.02, 0.05, 0.1 and 0.2 per cent concentrations. *In vitro* studies revealed that Ajwain-Mg NPs, Ajwain-Ni NPs at 0.2 per cent and Neem-Ag NPs at 0.1 and 0.2 per cent have shown complete inhibition of spore germination of C. musae. Spraying of Ag-Neem np at 0.2 per cent concentration recorded lowest disease severity (6.67 PDI) which was on par with 0.1 and 0.05 per cent concentrations (8.89 PDI). Mode of action of nanaparticles was explained by different researchers. The nanoparticles efficiently penetrate into microbial cells, which implies lower concentrations of nano-silver would be sufficient for microbial control.

Nano-silver was proposed for modifying postharvest life of cut flowers. Silver is considered an inhibiting agent against microorganisms. It was used with relative safely inspection of different plant pathogens, compared to fungicides. Silver Nano-particle one capable of inhibiting eighteen various plant pathogenic fungi include *Alternaria alternata, Botrytis cinerea* and *Fusarium* sp. in tomato, potato, strawberry, pepper and eggplant. Silver Nanoparticle in combination with benzyl adenine or salicylic acid may enhance efficacy of them to extend the vase life of gerbera cut flower. Silver Nano particles (AgNPs) were evaluated for its possible effects on the incidence and severity diseases gray mold for tomato fruits caused by the *Botrytis cinerea* (Salem *et al.,* 2019). Treatments were [(healthy tomato fruits (control), inoculated tomato fruits by *B. cinerea*, healthy tomato fruit with silver nanoparticles and combination inoculated tomato fruits with *B. cinerea* and silver nanoparticles). The obtained results were: (a) there aren't incidences and severity diseases of gray mold in case of tomato fruits with silver nan particles during forty days, (b) the lowest incidence (0.3) and severity (1.5%) diseases were after thirty days of cold storage in case of treatment tomato fruits were treated with Nano–Silver + Inoculated. While, the highest incidence (0.6) and severity (7.5%) diseases percentages were after ten days of cold storage in case of treatment tomato fruits with inoculated by *Botrytis cinerea*, comparison with incidence (0.3) and severity (1%) diseases percentages were after twenty days of cold storage in case of un-treatment tomato fruits (control), (c) the highest (TSS) percentage of tomato fruits which treated by Nanoparticles followed by untreated samples (control). While, the lowest was sample which treated by inoculated + silver Nanoparticles, followed by inoculated samples by *B. cinerea* after forty days of cold storage and (d) the highest pH value of tomato fruits which treated by Nano particles, followed by untreated samples (control), while, the lowest sample which treated by inoculated + silver nano particles, followed by inoculated samples by *B. cinerea* after forty days of cold storage.

Silver nanoparticles (AgNPs) were studied for its effects on the incidence of seed rot disease of Tomato, Faba bean and Barley, caused by the seed and soil fungus *Fusarium oxysporium* (Rabab and Raida, 2014). The lowest of silver nanoparticles lead to improve germination of seed and vigour index, also lead to reduce the disease incidence. The highest of Silver nanoparticles 0.5 mM showed a slight effect on germination of seed and vigor index in tomato seedlings. Silver nanoparticles with a concentration of 0.12 mM were less toxic to the Tomato, Faba bean and Barley seedlings. As well as, it was reduced of the disease incidence.

Three different fungi were isolated from samples of citrus. *Alternaria alternata* was the most abundant fungal pathogen isolated from citrus leaf and fruit spots followed by *Alternaria citri* and *Penicillium digitatum*. Silver nanoparticles were tested as a potent fungicide that may replace the dangerous cytotoxic fungicides (Gehan A.M.Abdelmalek and Taher A Salah Eldin, 2016). Silver nanoparticles were synthesized with 10 ± 5 nm average diameter size. In vitro assay was carried out on Potato Dextrose Agar (PDA) media treated with 50, 100 and 150 ppm of silver nanoparticles in addition to two control traditional fungicides products, namely Iprodione and Difenoconazole at 150 ppm concentrations. Results revealed that Silver nanoparticle, 150 ppm, showed potent antifungal activity against the isolated fungi that will open the gate of new generation of less harmful fungicides replacing the cytotoxic dangerous currently used fungicides.

7.12.2. Chitosan: Chitosan, which is a natural compound present in the cell wall of many fungi, is another elicitor of host defence responses which is currently being investigated. Chitosan can stimulate a number of processes including production of chitinase, accumulation of phytoalcxins and increase in lignification. A wide range of other compounds (eg salicylic acid, methyl jasmonate and phosphonates) and treatments (e.g. heat) can induce host defences in harvested commodities.

Many postharvest diseases are caused by wound pathogens, and the complete control of these agents is done through the use of an agent which grows rapidly and is environmental friendly. One of the commonly used agents in regulating postharvest diseases is a polysaccharide called chitosan. It is a common polysaccharide mostly found in crustaceans shells and fungi cell wall. It is mainly produced by the deacetylation of chitin. A great number of researches are being done on chitosan to know its mode of action. Chitosan has been proven to have various effects on plants, bacteria, viruses, and fungi (Garba Betchem, *et al.*, 2019).

Chitosan is a natural biopolymer from crab shells that is known for its biocompatibility, biodegradability, and bioactivity. In human medicine, chitosan is used as a stabilizer for active ingredients in tablets, and is popular in slimming diets. Due to its low toxicity, it was the first basic substance approved by the European Union for plant protection (Reg. EU 2014/563), for both organic agriculture and integrated pest management (Gianfranco Romanazzi *et al.*, 2018). When applied to plants, chitosan shows triple activity, *viz.*, elicitation of host defenses; antimicrobial activity; and film formation on the treated surface. The eliciting activity of chitosan has been studied since the 1990's, which started with monitoring of enzyme activities linked to defense mechanisms (e.g., chitinase, β-1,3 glucanase, phenylalanine ammonia-lyase) in different fruit (e.g., strawberry, other berries, citrus fruit, table grapes). This continued with investigations with qRT-PCR (Quantitative Real-Time Polymerase Chain Reaction), and more recently, with RNA-Seq.

The antimicrobial activity of chitosan against a wide range of plant pathogens has been confirmed through many *in-vitro* and *in-vivo* studies. Once applied to a plant surface (e.g., dipping, spraying), chitosan forms an edible coating, the properties of which (e.g., thickness, viscosity, gas and water permeability) depend on the acid in which it is dissolved. Based on data in literature, we propose that overall, the eliciting represents 30 to 40% of the chitosan activity, its antimicrobial activity 35 to 45%, and its film forming activity 20 to 30%, in terms of its effectiveness in the control of postharvest decay of fresh fruit. As well as being used alone, chitosan can be applied together with many other alternatives to synthetic fungicides, to boost its eliciting, antimicrobial and film-forming properties, with additive, and at times synergistic, interactions. Several commercial chitosan formulations are available as biopesticides, with their effectiveness due to the integrated combination of these three mechanisms of action of chitosan.

It has been reported that both soil and foliar plant pathogens fungal, bacterial and viral may be controlled by chitosan application. Microscopical observations indicate that chitosan has a direct effect on the morphology of the chitosan treated microorganism reflecting its fungistatic or fungicidal potential. In addition to its direct microbial activity, other studies strongly suggest that chitosan induces a series of defence reactions correlated with enzymatic activities. Chitosan has been shown to increase the production of glucanohydrolases, phenolic compounds and synthesis of specific phytoalexins with antifungal activity, and also reduces macerating enzymes such as polygalacturonases, pectin metil esterase etc. In addition, chitosan induces structural barriers for example inducing the synthesis of lignin like material. For some horticultural and ornamental commodities, chitosan increased harvested yield (Bautista-Ban˜osa, *et al.*, 2006). Due to its ability to form a semipermeable coating, chitosan extends the shelf life of treated fruit and vegetables by minimizing the rate of respiration and reducing water loss. As a nontoxic biodegradable material, as well as an elicitor, chitosan has the potential to become a new class of plant protectant, assisting towards the goal of sustainable agriculture.

The antifungal potential of chitosan was evaluated for the control of postharvest anthracnose in 'Breda' avocado fruit (Marques *et al.*, 2016). For *in vitro* assessment, chitosan was incorporated into a melting PDA culture medium to obtain desired concentrations (0; 0.25; 0.50; 1.0; 1.5 and 2.0%), with six replications. For *in vivo* testing, fruits were infected with 25 μL *C. gloeosporioides* conidial suspension, and stored for 10 h at 22°C. The infected fruits were immersed for 1 min in a chitosan solution at above concentrations. After, the fruits were stored for seven days at 22±2°C and 80±3% relative humidity (RH). The experiment was performed in a totally randomized design, factorial scheme, four replications with five fruits each. To analyze the variance of the data, an F test was conducted, and the average values of each variable were compared with the Tukey test ($P \leq 0.05$). The concentration from 1.0% chitosan inhibited fungal growth and conidia germination *in vitro* of *C. gloeosporioides.* Chitosan application to fruit at 1.0% provided lower disease severity and incidence, reduced activity of phenylalanine ammonia lyase, peroxidase, and polyphenol oxidase enzymes. In addition, chitosan decreased respiratory rate, ethylene production, and fresh mass loss; however, increased pulp firmness. It was concluded that 1.0% of chitosan is a potential natural covering for avocados 'Breda' during postharvest storage due to its fungicidal action against anthracnose and fruit quality maintenance.

Chitosan was tested on mycelia growth, germination of spores, enzyme activity and gene expression for different postharvest fungi and fruit tissues under *in vitro* and *in vivo* conditions . It has been found that chitosan controls postharvest infections by *Colletotrichum gloeosporioides*, *Alternaria alternata*, *Rhizopus stolonifer* and *Fusarium oxisporum*. Furthermore, the semipermeable characteristic of chitosan originates favourable physic-chemical changes on fruits metabolism, extending their storage life and the effects also include changes in enzymatic activity and genetic expression. The effects of Chitosan, on grey mold and blue mold in tomato fruit stored at 25 and 2 °C, respectively, were investigated. Chitosan provided an effective control of both diseases of tomato fruit stored at 25 and 2 °C (Jial Liu, *et al.*, 2007). Chitosan strongly inhibited spore germination , germ tube elongation, and mycelial growth of *B. cinerea* and *P. expansum in vitro*, and damaged the plasma membranes of spores of both pathogens. Chitosan treatment induced a significant increase in the activities of polyphenoloxidase (PPO), peroxidase (POD), and enhanced the content of phenolic compounds in tomato fruit. These findings suggested that the effects of chitosan on gray mould and blue mould in tomato fruit may be associated with the direct fungitoxic property against the pathogens and the elicitation of biochemical defense responses in fruit.

7.12.3. Essential oils: Essential oils are the volatile constituents of chemical compounds with multiple mode of action that enhances their activity due to the synergistic effects of the constituents. The majority of them serve as chemical messengers for 'postharvest disease management'. The volatility of EOs makes them suitable as fumigants, in protected environments, for protection of postharvest diseases of horticultural crops, but they also have demonstrated utility for management of 'pests and diseases' of field crops. Many EOs and their major active constituents, monoterpenes and sesquiterpenes, have shown contact toxicity to postharvest insect pests but their utility is broadened due to their sublethal behavioral effects as deterrents and repellents. These bioactivities result from synergistic effect among particular terpenes, and with the neurotransmitter-mediated toxic action of the terpenes. Many EOs and their active constituents have shown potential bioefficacy against the insect pests when tested in laboratory, but only few of them - those used extensively in 'flavor and fragrance industries' have been used for developing pesticidal formulations (Shukla 2018). These include the essential oils from the family *Asteraceae*, *Myrtaceae*, *Lamiaceae*, *Lauraceae*, *Poaceae*, etc.,. and have begun to establish its market presence in some countries, as essential oil-based pesticides.

Essential oils, (Generally Regarded as Safe, GRAS) may provide ''greener'' alternative fungicides for horticultural industries. Fungi may also be less likely to develop resistance to essential oil treatments because of their very complexity (made up of many constituents) which makes it very difficult for organisms to evolve coping mechanisms. This compares favourably to the synthetic fungicides currently in use which often have only a few constituents. In-vitro studies have demonstrated the antifungal activity of essential oils against a range of postharvest pathogens including *B. cinerea*, *M. fructicola*, *G. candidum*, and *Fusarium spp*. A potential advantage of using an essential oil treatment is their volatile nature which implies that no or little residue is left on the produce after treatment.

The antifungal effects of *Salvia officinalis* and *Picea abies* essential oils were studied against strawberry *Colletotrichum* spp. isolate (Morkeliūnė *et al*., 2019)) There were evaluated several EO concentrations from 1000 till 1800 µl/l. Single-spore isolate fragment placed in the centre of PDA with different concentrations. Plates were incubated at 25 °C in dark and evaluated after 2, 4, 7 days. *S. officinalis* and *P. abies* EO showed inhibitory effect on *Colletotrichum* spp. mycelial development. The *S. officinalis* inhibitory effect was more than 50% in all concentrations. The present study revealed that highest 1800 µl/l *S. officinalis* EO concentration *Colletotrichum* spp. colony diameter was significantly lower (1.84 cm) compared with control (5.75 cm).

However, *P. abies* inhibition more than 50 % were only in concentrations from 1600 µl/l. According to our results, treatment with EO can reduce the growth of *Colletotrichum* spp. and EO could be an effective potential bio-fungicide to control strawberry anthracnose. EO as bio-fungicides characteristics is their natural origin and low risk for resistance development.

The efficacy of volatiles (single compounds) and essential oils (EOs, group of compounds) produced by plants was studied to control food-borne bacteria (*Salmonella* spp., *Escherichia coli*, and *Listeria*) and fruit pathogens (mainly *Botrytis* and *Colletotrichum*) on extending the shelf life of small fruits, such as strawberries (Fragaria × ananassa Duch), raspberries, blueberries) and grapes. Insight into their *in vitro*, *in vivo* (in storage) and in situ (in the field) applications and their mechanisms of action wee also provided (Toktam Taghavi *et al.*, 2018). The efficacy of multi-compound essential oils is even more variable due to genetic variation, environmental conditions, and synergistic effects of multiple compounds. Major compounds play an essential role in the antimicrobial activity of essential oils, however, minor compounds in the mixture may have synergistic effects. Future studies are needed to explain the role of single natural compounds or their combination in plant-fungal interactions. (E)-hex-2-enal and hexanal were the most effective volatiles and thyme the most effective essential oil in controlling fungal diseases of small fruits both *in vitro* and in storage. Thyme oil has the potential as a postharvest treatment to extend shelf life of fresh fruits and vegetable and to replace commercial fungicides or controlled atmosphere storage. However, active packaging and improved formulation techniques are needed to prolong activity, reduce volatility and improve coverage. Several researchers have mentioned the toxicity of volatiles and essential oils after extended exposure. Research is needed to study the effect of volatiles and essential oils on vegetative and reproductive phases of fungal development, fruit quality (firmness, sensory evaluations), cell wall structure and integrity and their role in plant self-defense and postharvest storage.

The glandular trichomes of plants are the important site for biosynthesis of essential oils and they act as defence system against herbivores and pathogens. Since essential oils are volatile aromatic compounds, the easily vaporised property has been exploited in postharvest application as fumigants. In addition, essential oils have also been incorporated into chitosan and alginate-based materials as edible coatings.Besides applying directly to intact fruits and vegetables, essential oils have also been applied to fresh-cut produces. Lemongrass essential oil (0.3% w/v) incorporated in alginate-based edible coating showed positive result on fresh-cut pineapple where respiration rate, weight loss and yeast and mould counts were reduced while firmness, colour and sensory characteristics were maintained In another fresh-cut study, 'Bravo de Esmolfe' apple could be stored for seven days at 4°C with good attractive appearance, sensory and nutritional quality when the essential oils of eugenol and citral were incorporated into coatings which had sodium alginate or pectin as base.

The antifungal activity of plant essential oils was evaluated as postharvest treatment on stone fruit against brown rot and grey mold rot of stone fruit caused by Monilinia laxa and Botrytis cinerea, respectively. The essential oils from basil, fennel, lavender, marjoram, oregano, peppermint , rosemary , sage, savory, thyme and wild mint were tested at two different concentrations on apricots (cv. Kyoto and cv. Tonda di Costigliole), nectarines (cv. Big Top and cv. Nectaross) and plums (cv. Italia and cv. TC Sun). The volatile composition of the essential oils tested was determined by gas chromatography–mass spectrometry analysis. The treatments containing essential oils from oregano, savory, and thyme at 1% (vol/vol) controlled both *B. cinerea* and *M. laxa* growing on apricots cv. Tonda di Costigliole and plums cv. Italia and cv. TC Sun; however, the same treatments were phytotoxic for the carposphere of nectarines cv. Big Top and cv. Nectaross. Treatments with 10% (vol/vol) essential oils were highly phytotoxic, notwithstanding their efficacy against the pathogens tested. The essential oils containing as major components a-pinene, p-cymene, carvacrol, and thymol showed similar results on stone fruit, so their antimicrobial activity and the phytotoxicity produced could be based on the concentration of their principal compounds and their synergistic activity (Jorge Giovanny Lopez-Reyes *et al.,* 2013). The efficacy of the essential oil treatments on control of fungal pathogens in postharvest depended on the fruit cultivar, the composition and concentration of the essential oil applied, and the length of storage.

A study on evaluation of essential oils of plant species (*Litsea cubeba, Piper mullesua* and *Pogostemon cablin*) was carried out for their antifungal activity against phytopathogenic fungi viz. Alternaria alternata, *Botrytis cinerea, Fusarium oxysporum* and *Penicillium expansum* following Poisoned food technique method (Habung Yami *et al.,* 2016). There was 100% inhibition in the growth of phytopathogenic fungi at 5000 and 1000ppm concentration by essential oil of *L. cubeba* and *P. cablin*. The decrease in colony diameter or growth of fungus was corresponding to the concentration of oil. In comparison essential oil of *P. mullesua* was reported less effective against phytopathogenic fungi. Minimum inhibitory concentration for oil of *L. cubeb*a and *P. cablin* was found 500ppm. At higher concentration essential oil of both species were reported fungicidal in nature. Results depicted that *L. cubeba* and *P. cablin* essential oils may be used as botanical pesticides to control postharvest fungal pathogens after further investigation.

Eight substances that are main components of the essential oils from three Mediterranean aromatic plants (*Verbena officinalis*, *Thymus vulgaris* and *Origanum vulgare*), previously found active against some phytopathogenic Fungi and Stramenopila, have been tested *in vitro* against five etiological agents of postharvest fruit decay, *Botrytis cinerea*, *Penicillium italicum*, *P. expansum*, *Phytophthora citrophthora* and *Rhizopus stolonifer*. The tested compounds were β-fellandrene, β-pinene, camphene, carvacrol, citral, *o*-cymene, γ-terpinene and thymol. Citral exhibited a fungicidal action against *P. citrophthora*; carvacrol and thymol showed a fungistatic activity against *P. citrophthora* and *R. stolonifer*. Citral and carvacrol at 250 ppm, and thymol at 150 and 250 ppm stopped the growth of *B. cinerea*. Moreover, thymol showed fungistatic and fungicidal action against *P. italicum*. Finally, the mycelium growth of *P. expansum* was inhibited in the presence of 250 ppm of thymol and carvacrol (Ippolito Camele *et al.*, 2012) These results represent an important step toward the goal to use some essential oils or their components as natural preservatives for fruits and foodstuffs, due to their safety for consumer healthy and positive effect on shelf life extension of agricultural fresh products.

The mechanisms of antifungal action of lemongrass oil against *B. cinerea* were investigated. Light microscopy and scanning electron microscopy observations indicated that *B. cinerea* hyphae exposed to lemongrass oil undergo morphological damage such as vesiculation, cytoplasmic disruption and collapsed hyphae. Glucose-induced reduction in external pH of *B. cinerea* was inhibited by lemongrass oil in a time and concentration dependent manner. *Botrytis cinerea* spores treated with lemongrass oil showed strong propidium iodide fluorescence in the cytosol. Lemongrass oil significantly altered the plasma membrane, the release of cell constituents, and the total lipid content of *B. cinerea*. These observations indicate that the antifungal activity of lemongrass oil can be attributed to the disruption of the cell membrane integrity and membrane permeability (Nokwazi Carol Mbili, 2015). It was highlighted the potentiality of essential oils in combination with cold storage regimes as an alternative control strategy against grey mould, blue mould and bull's eye rot in the apple industry. This suggested the application of essential oils by incorporating the oil into wax or coating treatments that are being used in the packing-line and the possibility of fumigation and thermal fogging during the storage period.

The use of GRAS substances like essential oils (EOs) has increased in the last years, due to the research of alternatives to chemical treatments for disease control . Several investigations have reported the efficacy of essential oils *in vitro* and *vivo tests*. The application of EOs in fruits has advantages such as: high effectiveness against several pathogens and low toxicity (nontarget microorganism and humans). EOs can be applied on fruits directly (vapor phase) or incorporated as active microbial agents in different matrices (films, coatings, among others) at pre-harvest or postharvest stages. In a recent investigation, cinnamon essential oil was added into biodegradable polyester nets for controlling *A. alternata* (Black-Solis *et al.*, 2019). The results were promising due to their efficacy *in vitro* tests by stopping the development of fungus on mycelial growth (72% of inhibition) and the total control of germination. The presence of essential oil did not alter the biodegradability of nets as well as their efficiency to maintain fruit quality and the disease control on infected fruits by reducing the disease incidence. Incorporation of EOs into edible films for active packaging has demonstrated high efficacy against important postharvest pathogens. Recently, soy protein isolate was used as a carrier for limonene with good results against *P. italicum* (isolated from infected limes).

The liberation and efficacy (*in vitro* tests) of the active agents were evaluated simulating the storage conditions of limes (13 and 28 °C). The mycelial growth and germination process were successfully inhibited by the incorporation of limonene into the protein matrix. The results *in vitro* were confirmed in infected limes with *P. italicum* by reducing the severity and blue mold incidence with the application of soy protein isolate with limonene added . In another study, films based on chitosan, oleic acid/beeswax, and lemon essential oils were tested on tomatoes for preserving their quality. According to their results, the applications of these films improve the fruit quality by reducing water losses and the maintenance of appearance of fruit . In a recent study, anthracnose and stem-end rot in the greenskinned on avocado fruits was successfully controlled by the application of thyme oil in combination with a prochloraz solution; besides, this treatment improved fruit quality (firmness) during the storage time .

Essential oils of copaiba and eucalyptus were tested against *Alternaria alternata* and *Colletotrichum musae* under in vitro, and the results showed good efficacy at low concentrations of the treatments (0.0–1.0%) . The efficacy of black caraway (*Carum carvi*) and anise (*Pimpinella anisum*) essential oils was tested against *Penicillium digitatum in vitro* as well as *in vivo* (on oranges) evaluations. The results showed that treatments were capable to control fungi development *in vivo* and *in vitro* tests; besides, the quality of oranges was preserved by the application of the treatment of 600 μL/L (Ramsés González-Estrada *et al.*, 2019). In a study, lemongrass oil was tested against *Colletotrichum coccodes*, *Botrytis cinerea*, *Cladosporium herbarum*, *Rhizopus stolonifer*, and *Aspergillus niger in vitro*. According to their results, fungal development, sporulation process as well as spore germination of fungi was affected in different levels (depending on the concentration of treatment) when they were exposed to the treatments by stopping their development .

7.12.4. Plant extracts: Currently, the study of fruits and vegetables at postharvest stage has focused on the development of alternative control for conservation with better efficiency, sustainability, and lower cost than conventional methods. In this sense, some extracts from plants have proven to be a viable alternative for the extraction of substances with antimicrobial activity with high efficiency and low toxicity (Chen *et al.*, 2019). In a study, aqueous and ethanolic extracts from garlic (*Allium sativum*) were evaluated on citrus fruits (*Citrus sinensis*, *Jaffa*, and *Valencia*) with good results and it is suggested that the highest antifungal activity of the extract can be produced by the presence of allicin in the soluble fractions of extracts from garlic . In the same way, extracts of garlic, ginger, and celery have been shown to have different effects on the control of the incidence of *Penicillium* sp. in fruits of the species *Citrus reticulata* Blanco. In a similar study, soybean extracts were evaluated as protection method on oranges infected with *P. digitatum*, and green in the soy protein fraction . In recent years, extracts of some angiosperm species have been studied as natural fungicides, such as *Orobanche crenata* and *Sanguisorba minor*, which have shown high efficiency in the control of the diseases produced by *Monilinia laxa* in stone fruits such as apricot, cherry, and nectrine.

For the case *O. crenata* extracts, the antifungal activity is attributed for the presence of the phenolic compound verbacoside, and for *S. minor* extracts, the efficacy is related to the presence of a combination of phenolic compounds like caffeic acid, quercetin, luteolin, and kaempferol. An important approach in fruits is quality. In this sense, the application of some plant extracts can help preserve the fruit quality and improve their shelf life. In a recent investigation, the application of guava extracts (from leaves) and lemon on banana fruits (*Musa sapientum* L.) at postharvest stage considerably improved the shelf life of the fruits (up to 8 days) compared to untreated fruits (only 4 days), having a positive effect on the conservation of the physicochemical characteristics of fruit during the storage time. Utilization of plant extracts for disease management in both pre- and postharvest stages can be another alternative to the use of chemical treatments, due to their effectiveness and eco-friendly characteristics.

Plant products are an important source of agrochemicals used for the control various postharvest loses which include diseases as well as insect pests. The widely studied plants in this context are the neem tree (*Azadirachta indica*), chinaberry (*Melia azadrach*) and marigold (*Tagetes* spp). They are being used to manufacture natural or bio insecticides, which are environmental friendly and do not have any toxic effects on plants and soil (Anjum Malik *et al.*, 2016). More ever they possess fungicidal and insecticidal properties. Various extracts such as neem leaf extract, castor oil and neem oil on citrus fruits and reported that, among these extracts neem was best in retaining most of biochemical characteristics such as TSS (16.01°B), acidity (0.38%), pectin (0.98%) and ascorbic acid content (20.56 mg/100 ml juice).

Neem and chinaberry are two trees belonging to the mahogany family (Meliaceae). Neem is an evergreen tree , whereas chinaberry is a deciduous tree native to South Asia Australia. The efficacy of leaf extract of neem and Chinaberry against two tomato pathogenic fungi *Alternaria solani* and *Fusarium oxysporium*, the causal agents of early blight and wilt disease of tomato plants respectively were studied. Leaf extracts of different concentration (5%, 10%, 15% and 20%) of aqueous, ethanol and ethyl acetate of neem and chinaberry were added to growth media prior to inoculation. The results showed that, the extracts significantly ($P < 0.05$) inhibited the mycelial growth of the fungal pathogens at the different concentrations tested and the rate of inhibition differed from one extract to the other. Some plants contain phenolic substances and essential oils, which are inhibitory to micro-organisms. The presence of these compounds in these extracts has been reported to be responsible for their antifungal properties. These antifungal properties control various pests including fungi while the extract of ginger rhizomes is specially valued for their effectiveness against fungi.

An experiment was conducted with aqueous extracts of leaves of three plants of Sunderbans i.e. Sundari (*Heritiera fomes*), Keora (*Sonneratia apetala*) and Kankra (*Bruguiera gymnorrhiza*) to find out the efficacy of controlling postharvest disease and extending shelf life of mango. In this study, the efficacy of the three plant extracts at four different concentrations viz., 20%, 30%, 40% and 50% along with control were used. (Joydeb Gomasta *et al.,* 2017) The highest disease index (100%) was observed at control (0% concentration) within 12 days after treatment while at 50% concentration disease index reached to 100% after 18 days of treatment in case of Sundari and Kankra extract and after 20 days of treatment for Keora extract. An increasing trend in postharvest disease control was noticed with the increase in concentration of leaf extracts. Thus, maximum shelf life was observed at 50% concentration of all the three botanicals. In comparison to other leaf extracts, Keora at 50% level also found more effective in extending shelf life of mango up to 18 days having statistical similarity with that of Kankra extract (17.25 days). Now, it is obvious that botanical extracts efficiently control postharvest disease and extend shelf life of mango up to certain degree of concentration.

Two experiments were conducted to evaluate the antimicrobial activity of alternative products to the synthetic fungicides (Cruz *et al.,* 2020). First berries naturally infected by anthracnose were immersed into *Azadirachta indica* and citric extracts at 2 and 4% (v/v) for 3 minutes and stored for 11 days under environmental conditions. Next, other berries were immersed into essential oil emulsions of *Allium sativum*, *Copaifera langsdorfii*, *Cinnamomum zeylanicum* and *Eugenia caryophyllata* at 5% for 3 minutes but stored for 11 days. Berries immersed into distilled water were used as control-treatments. The percentage of disease incidence observed in the control-treatment was similar to the ones observed in the extract of *A. indica* at 2%. The control-treatment showed disease severity of 75.13% and the percentage of disease control was 20.85%. Fruit immersed into distilled water presented less effectiveness than the ones immersed into citric extracts, which promoted the highest effectiveness. Citric extract at 4% was the most efficient treatment because the disease incidence was 19.44%, the disease severity was 9.34% and the disease control was 90.16%. Less severity and, consequently, more disease control were achieved by immersing the berries into the emulsion of essential oil of *A. sativum*, followed by treatments with *C. langsdorfii*, *E. caryophyllata* and *C. zeylanicum.*

The effect of citrus extracts was evaluated for the control of pathogens that cause postharvest diseases in tomato fruit (Cristina *et al.*, 2018). The product to be evaluated is of botanical origin from citrus extracts. Doses were evaluated (0, 666, 1000, 2000, 4000, 8000 ppm). The treatments were located at a temperature of 25°C±2 and 45% relative humidity (rH). The design used corresponded to a completely random design. The least signifi cant difference was estimated by Tukey Multiple Range test at P=0.05. The statistical tests were performed through the SAS computer program. The results indicate that the pathogens detected and identifi ed correspond to *Alternaria tenuissima*; *Botrytis cinerea*; *Cladosporium fulvum*; *Colletotrichum coccodes*; *Fusarium oxysporum*; *Geotrichum candidum*; *Rhizopus stolonifer* and *Stemphylium macrosporoideum*. It was concluded is that the effi cient doses correspond to 666, 2000 and 8000 ppm. With the application of citrus extracts, the damage percentage of tomato fruit was reduced in relation to the control treatments. Based on the results with the application of citrus extracts, the shelf life of the tomato was lengthened.

Green and blue molds, caused by *Penicillium digitatum* and *Penicillium italicum*, respectively, are two major postharvest citrus diseases and cause significant economic losses during the commercialization phase. Currently the control of postharvest citrus diseases relies mainly on the use of synthetic fungicides, which usually result in the resistance against fungal attack environment pollution and health hazards. In recent years, much attention has been given to the preservation of citrus by naturally isolated edible plant extracts, medicinal plant extracts, *Citrus* extracts and volatiles, *et al.* (Jinyin Chen, *et al.*, 2019). Scientists worldwide devote their time and energy to discover the high effect, low toxicity, safety and inexpensive plant-derived fungicides. The review highlighted plant-derived fungicides and chemical constituents that aim to inhibit *P. digitatum* and *P. italicum in vitro* and *in vivo*. Coatings enriched with plant extracts could be good alternative methods for *Citrus* fruits preservation.

An investigation was undertaken to study effect of medicinal plant extracts/ bulb extracts viz bael (50 and 70%), tulsi (50 and 100%), Garlic (5 and 10%) on shelf life of mango. The physiological loss of weight in mango fruits showed an increasing trend in all the treatments (Ashwini and Nikhita Desai, 2018). Contrary to this acidity decreased with storage. However, there was an increase in pH with extension of storage period. The total soluble solids of fruits first increased upto a certain period and there after decreased. Bulb extracts of garlic @ 10 per cent were most effective of all the treatments with minimum spoilage percentage followed by garlic extracts @ 5 per cent concentration and tulsi extracts at 100 per cent concentration. In the present study bael extracts (50 and 100 per cent concentrations) were not effective in extending the shelf life of mango (Raspuri). The investigation concludes that the shelf life of mango fruits can be extended from 3 to 7 days by treating with 10 per cent garlic extract.

The exploitation of natural products to control decay and prolong storage life of perishables has received more and more attention. Biologically active natural products have the potential to replace synthetic fungicides. (Pramila Tripathi and Dubey, 2004) reviewed exploitation of some natural products such as flavour compounds, acetic acid, jasmonates, glucosinolates, propolis, fusapyrone and deoxyfusapyrone, chitosan, essential oils and plant extracts for the management of fungal rotting of fruit and vegetables, thereby prolonging shelf life.

Ten Crudes plant extracts of Neem, Pong-pong, Chilly, Lemon grass, Tumeric, Clove, Green chirayta, Mahogani, Curry leaf and Ginger were test against *Penicillium digitatum*, the causes of green mould disease of orange fruits, was evaluated for their inhibitory effect *in vitro* and *in vivo* (500 to 3000ppm) during storage conditions(2000 to 5000 ppm) (Ghassan *et al.*, 2013). Followed to lethality test LC50 (BST) to determine the lethal dose. The concentration at 3000 ppm from Neem, Pong-pong and Chilly showed the highest inhibition growth reached the (90%) followed by Lemon grass and Ginger by inhibition reached (70%). *In vivo* experiments, treating fruits 21 days crude extraction from Neem, Chilly and Pong-pong at conc. of 4000 and 5000ppm showed significantly reduced the disease severity of fruits (100%) compared with untreated fruits (control) at 25°C, through significantly reduced the undesirable fruits percentage, fruit weight loss percentage during storage for 21 days. Lethal concentration (LC50) values of the crude extracts at 20, 5 and 30 μg/ml from Neem, Pong-pong and 1 Chilly showed very high lethal toxicity on brine and effect Lemon grass and Ginger in killed 50% was at 495 and 473 μg/ml respectively. Plant extracts or botanicals have a bright future in modern plant protection to replace 1 conventional synthetic fungicides.

7.12.5 Acetic acid, peracetic acid and hydrogen peroxide: The use of sanitizing agents such as acetic acid ($C_2H_4O_2$), peracetic acid ($C_2H_4O_3$), and hydrogen peroxide (H_2O_2) on the processing and marketing industries of fruits and vegetables have been considered a useful tool to the control of different kinds of pathogens (Feliziani *et al*., 2016). The FDA has approved the use of these compounds because their decomposition products are water, oxygen, and acetic acid, and these are not toxic compounds and are friendly with the environment. For these reasons, they are classified as GRAS. The acetic acid, peracetic acid, and hydrogen peroxide are recognized as antimicrobial and antifungal agents, because they have high spectrum of attack on bacteria, fungi, spores, and viruses. *In vivo* studies realized on fresh and fresh-cut horticultural products confirm their antimicrobial capacity with the reduction of human pathogens such as *Staphylococcus aureus*, *Escherichia coli*, *Streptococcus mutant*, *Salmonella Thompson*, and *Listeria monocytogenes*. The application of acetic acid, peracetic acid, and hydrogen peroxide also decreases the microbial pollution of aerobic mesophiles, molds, and yeast, obtaining with these innocuous and acceptable products to the consumers. In fruits like guavas, peaches, and tomatoes, the antifungal effect of these sanitizers has been confirmed with the inhibition of phytopathogens such as *Rhizopus stolonifer*, *Monilinia fructicola*, *Alternaria alternata*, *Botrytis cinerea*, *Fusarium solani*, and *Rhizoctonia solani* .

The inactivation capacity of these compounds on pathogens is based on their high oxidizing power, producing reactive oxygen species (ROS) that generate instability in biomolecules such as DNA, lipids, and proteins, which are vital for the correct cellular functioning of pathogens . Usually, the application of these sanitizers at postharvest stage in fruits and vegetables is by spraying, dipping, and fumigation. Several authors have investigated the pathogen inhibitory effect of each application system, with the finality to know which application system has better efficiency. The inhibition of *A. alternata* and *B. cinerea* in tomato fruits treated with acetic acid by immersion (50 ml/L by 3 min) and fumigation (50 µl/L by 30 min) was evaluated . The results of this investigation showed the efficacy of both application systems (Ramsés González-Estrada *et al.*, 2019). The growth inhibition of the pathogens tested ranged from 90 to 100% by immersion and fumigation, respectively. In strawberry fruits infected with *B. cinerea* and treated with peracetic acid, where the lower incidence of gray mold disease was obtained with the fumigation system (66%) compared to immersion method (80%).

The application of acetic acid, peracetic acid, and hydrogen peroxide can also improve the quality of the products, according to different in *vivo* studies carried out. In peppers, fruits treated with the solution of hydrogen peroxide and applied by dipping (15 mM by 30 min) increased their shelf life and the fruits maintained their appearance after 2 weeks stored at 20°C compared to control fruits. In according to their results, fruits and vegetables such as tomatoes, grapes, nectarines, apples, and cherries have been maintained the quality. Thus, the application of these sanitizers in postharvest is a practical strategy to maintain the safe and quality of the postharvest products.

7.12.6. Salts: (organic and inorganic): Nowadays, there is a tendency to reduce the use of synthetic fungicides in agriculture. In this sense, organic and inorganic salts are chemical substances that are food additives generally recognized as safe (GRAS). They are widely used in the food industry due to the low toxicity and environmental impact, besides they can be combined with other control systems to control phytopathogens in postharvest stage in different fruits and vegetables (Deliopoulos *et al.,* 2010). It has been shown that salts such as calcium propionate completely inhibit the mycelial growth of *B. cinerea* at a concentration of 5% (w/v). This may be due to the fact that it changes the plasma membrane, thus inhibiting essential metabolic functions. Some authors have described that high concentrations of sodium benzoate inhibited the growth of *A. Alternata* and this is attributed to the fact that weak acids within the cell create a dissociation, causing that protons and anions accumulate and cannot cross the plasma membrane again. Bicarbonate salts, also were effective to inhibit the growth of various pathogens, the efficiency of this salt is attributed to the fact that it creates cellularionic imbalances affecting the synthesis of polyamines and DNA during cell division.

With the accessibility of these compounds and their effectiveness, it is possible that they can be adopted to reduce the use of traditional fungal agents. In a study, potassium sorbate was applied at low concentrations (1%) on infected tomato fruits, and significant reduction of disease incidence was obtained against *Rhizoctonia solani*, *Colletotrichum coccodes*, *Botrytis cinerea*, and *Alternaria solani*. In another study, gray mold caused by *Botrytis cinerea* was totally controlled by the application of potassium sorbate at 1% . The effectiveness of potassium sorbate is related to its undissociated form (sorbic acid), which has the ability to penetrate the cell membranes, causing an internal imbalance affecting enzymes related to the growth of microorganisms. The effectiveness of sodium benzoate is related to its undissociated form of benzoic acid, which can enter the cell membrane, and its neutralization within the cell leads to acidification of the intracellular space, thus affecting the growth of fungus. On the other hand, another alternative to avoid pathogen development is the use of silicates, and the most used in the fruit industry is: potassium, calcium, and sodium. It has been reported that the action of these organic salts causes, among other effects, inhibition of mycelial growth and alteration of the morphology of the hyphae, and in addition, the germination of conidia is inhibited and causes alterations in their external morphology. The application of sodium silicate induces alterations in the cell wall and in the morphology of the hyphae on pathogenic fungi. It is important to mention that the application of silicate as postharvest treatment presents results similar to those reported with chitosan and tebuconazole .

The investigations of the use of inorganic salts in postharvest stage are not very frequent, even there are not yet many studies that determine the effect that this may have if applied to different types of crops. These salts have been applied on orange, melon, avocado, and papaya, with good results due to the treatments, which form a barrier against pathogens on surface fruit (Ramsés González-Estrada *et al.*, 2019). In addition, the application of silicates can improve some quality attributes of fruits like the maintenance of weight and reducing the respiration rate due to the capacity of the silicates to deposit between the cell wall and the cell membrane, thus decreasing the permeability, besides the stomata are covered, maintaining the humidity of the fruit and reducing its respiration . With the accessibility of these compounds and their effectiveness, it is possible that they can be adopted to reduce the use and applications of traditional fungal agents.

Low-toxicity chemicals classified as food preservatives or as generally recognized as safe (GRAS) compounds have known and very low toxicological effects on mammals and minimal impact on the environment. Among them, inorganic or organic salts such as carbonates & bicarbonates, borax, sorbates, benzoates, silicates, etc., show significant advantages for potential commercial use, such as their availability, low cost, and general high solubility in water (Lluis Palou, 2018). Typically, these substances are first evaluated *in vitro* against target pathogens that cause important postharvest diseases. Selected salts and concentrations are then assayed as aqueous solutions in *in vivo* tests with target fresh fruit. Laboratory and small-scale experiments are conducted with fruit artificially inoculated with pathogens, whereas naturally infected fruit are used for large-scale, semicommercial, or commercial trials. Another approach that is increasingly gaining importance is evaluating GRAS salts as antifungal ingredients of novel synthetic edible coatings. These coatings could replace the fungicide-amended commercial waxes applied to many fruit commodities and could be used for organic or "zero-residue" fresh fruit production systems.

These salts provide partial control of *Penicillium* decays by alkalization of infection sites, but are not stand-alone treatments. They can also inactivate spore inoculum and thus, reduce the amount of inoculum and the potential of resistance development. Potassium sorbate is less effective than the carbonate salts. In general, potassium sorbate is the GRAS salt that has shown the best results for fruit decay control when incorporated into matrices of novel edible coatings or when mixed with conventional postharvest waxes (Youssef, *et al.*, 2012). For this reason, several commercial formulations containing this food additive have been registered and are currently available on the market for postharvest treatment of fresh fruit, especially citrus and stone fruits. Citrashine Plus® and Citrashine Plus Nature® from the manufacturer Decco Iberica Postcosecha S.A.U. (Paterna, Valencia, Spain) and Greengard-SK 50® from the manufacturer Fomesa Fruitech S.L. (Beniparrell, Valencia, Spain) are examples of this type of product commercialized in Spain.

Lasiodiplodia theobromae, Thielaviopsis paradoxa, Colletotrichum musae, C. gloeosporioides, Fusarium verticillioides, and *F. oxysporum* were screened in vitro for sensitivity to Na_2CO_3, $NaHCO_3$, CaC12, NaCl, and NaClO (Alvindia, *et al.*, 2004). The spore germination of all pathogens was completely inhibited by Na_2CO_3 4g/l, NaClO 5g/l, and $NaHCO_3$, CaC12, and NaCl 6g/l each. Dipping the bananas for 10–15min in these concentrations reduced the incidence of crown rot (compared with the untreated fruits) 17 days after harvest in fruits treated with NaClO by 67%, with $NaHCO_3$ by 62%, with NaCl by 38%, and with CaC12 by 33%. Na_2CO_3-treated fruits had the same incidence of crown rot as untreated fruits.

To find an alternative to the fungicide treatment with guazatine currently used to control postharvest citrus sour rot caused by *Geotrichum candidum*, Idriss Talibi *et al.*, (2015) screened 32 salt compounds presently used as food additives for their activity against this fungus. The lowest Minimum Inhibitory Concentration (MIC) values were obtained with ammonium carbonate and EDTA (0.1% w/v), followed by boric acid, sodium carbonate and sodium metabisulfite (0.25% w/v). The mycelial growth was not affected by medium pH values in the range of 4.0 to 12.0. The most inhibiting salt compounds were then tested for their ability to reduce the arthrospore germination in vitro and sour rot incidence in vivo. The arthrospore germination was completely inhibited by 75 mM EDTA, boric acid, sodium metabisulfite, sodium carbonate, sodium sulfate and sodium thiosulfate. Incidence of sour rot was reduced to 25.93% when mandarin fruits where treated with 3% sodium salicylate or boric acid, and to 38.89% when they were treated with 3% EDTA, compared with 100% rotten fruits of the untreated control. These results suggested that sodium salicylate, boric acid and EDTA may be useful compounds to control citrus sour rot, thus being a sustainable alternative to the use of guazatine mainly in organic production.

7.12.7. Jasmonic and salicylic acid: Resistance induced to disease in plants by biotic and abiotic elicitors is a veryeffective method for restricting the spread of fungal infection . In general, pathogen resistance processes in plants are based on their own defense mechanisms, such as pre-existing antimicrobial compounds and inducible defense mechanisms. Resistance to diseases induced in plants and fruits by biotic or abiotic treatments is a very attractive strategy to control diseases. The signal molecules salicylic acid (SA), jasmonic acid (JA), and methyl jasmonate (MeJA) are endogenous plant growth substances that play key roles in development and responses to environmental stresses. These signal molecules are involved in some signal transduction systems in plants and fruits, which induce particular enzymes catalyzing biosynthetic reactions to form defense compounds such as polyphenols, alkaloids, or pathogenesis-related (PR) proteins. This can result in induction of defense responses and provide protection for plants and fruits from pathogen attack. Salicylic acid activates induction of acquired systemic resistance (SAR) response in plants, proving that in the plant-microorganism interaction, the enzyme phenylalanine ammonia lyase (PAL) is induced, which is the key in the biosynthesis of phenolic compounds. Peroxide has an antibiotic activity against pathogens; it could intervene in the signaling cascade for the expression of defense genes. SA regulates activities of enzymes, peroxidase (POD), and polyphenoloxidase (PPO), that are related to induced defense of plants and fruits against biotic and abiotic stress.

Jasmonic acid (JA) and methyl jasmonate (MeJA) have been found to occur naturally in a wide range of higher plants acquired systemic resistance depends on signaling mediated by MeJA and is associated with some signal transduction systems, which induce particular enzymes that catalyze biosynthetic reactions to form defense compounds such as polyphenols, alkaloids, reactive oxygen species (ROS), or PR proteins . The exogenous application of MeJA induces and increases the activity of defense enzymes such as β-1,3-glucanase (β-Gluc), chitinase, polyphenoloxidase (PPO), and phenylalanine ammonia lyase (PAL), which are enzymes associated with resistance to diseases (Haggag *et al.*, 2010). Application of MeJA effectively suppressed gray mold rot caused by Botrytis cinerea in strawberry and decreased fruit decay on papaya fruit infected by *C. gloeosporioides* and *Alternaria alternata*. For grape fruits inoculated with *Botrytis cinerea*, the application of MJ (0.01 mM) increased the enzymatic activity of PAL and PPO and the same behavior was reported on Hass avocado fruits with an increase in the activity of the resistance enzymes, chitinase, β-1,3-glucanase and PAL . The application of MeJA (10 mM) in cranberry fruit inoculated with *Penicillium citrinum* maintained greater POD and PAL activity.

7.12.8. Coatings and edible film from natural sources: Edible coatings on fruits and films made from natural sources are a novelty method and an alternative to the use of postharvest chemical treatments, particularly in highly perishable fruits. The coatings act as a barrier during processing, handling and storage, delaying the deterioration of food, improving its quality and extending their shelf life. The functional properties of edible coatings and films depend on their application and the characteristics of the fruit in which they are going to be used (Ramsés González-Estrada *et al.*, 2019). The nature of the compound that is used to produce the coating strongly influences its efficiency, thus it must take into account both its physical and chemical properties, as well as its mechanical and permeable properties. Based on this, if the edible coatings are used properly, they may be able to delay the ripening of the fruit, slow the decomposition of the chlorophyll, reduce the weight loss, retain the ascorbic acid, improve the appearance of the fruit, and especially prolong the shelf life.

To date, the functional properties of different compounds for the production of edible coatings have been studied. Compounds from natural sources such as *Aloe vera* and waxes have shown promising properties to be applied in the preservation of tropical fruits, and therefore coatings based on these compounds can represent an innovation in the commercial application and exploitation of these resources. *Aloe vera* gel is one of the natural compounds, which has gained a great interest. Because of its nature mainly constituted by polysaccharides, it is capable to form a uniform layer on the surface of the fruit and be easily applied . *Aloe vera* coating can improve the postharvest qualitative and quantitative traits, thus it can be an alternative for chemicals preservative in the commercial storage of tropical and subtropical fruits.

A coating with a concentration of 4% potato starch and 20% *A. vera* showed a favorable effect by reducing the weight loss and the respiratory rate of the fruit and increasing the firmness and retention of the vitamin C content after 10 days of storage. In another study, the effect of the *Aloe vera* gel coating on the store ability of peach fruits was evaluated. The Carnauba wax stems from the leaves of the Brazilian palm *Copernicia cerifera*. It is produced as a protection method to prevent dehydration and damage. The Carnauba wax as an edible coating is being extensively studied because it has been demonstrated to reduce water loss, improve appearance, and prolong shelf life in a wide variety of fruits. An edible coating based on cassava starch and carnauba wax adding organic acids and calcium chloride was evaluated in mangoes cv Tommy Atkins minimally processed . According to the results, the attributes of sensory, physical, and chemical quality were maintained, and the useful life of fruit was possible to prolong up to 24 days under refrigeration conditions (5 ± 1°C and 90 ± 2% RH).

An edible coating was developed based on candelilla wax to improve avocado quality (Saucedo-Pompa *et al.* 2009). Furthermore, they studied the effect of the ellagic acid addition in the shelf life of the fruit. The results showing the application of edible films based on candelilla wax improved the quality of the avocado fruits and extended its shelf life compared to the control fruits. Also, the addition of ellagic acid to the edible film showed an important effect, since it reduced the damage caused by the fungus *C. gloeosporioides* (the main phytopathogenic fungus for avocados) and significantly improved the quality and shelf life of avocado. Another coating based on mesquite gum-candelilla wax was evaluated in Persian limes . The results showed that coatings decreased the weight loss of the fruit. In addition, by adding mineral oil (33%) to the emulsion, they observed that water vapor permeability was significantly improved, as well as its appearance.

7.12.9. Ultrasound : Ultrasonic is an economically and environmentally viable alternative for the processing of fruit and vegetable postharvest . Low intensity ultrasound has been used for quality control of fresh fruit and vegetables in pre- and postharvest processes. Harvesting time and storage period can be indirectly assessed by ultrasound measurements that are linked physicochemical measurements such as firmness, mealiness, dry weight percentage, oil contents, total soluble solids and acidity. On the other hand, decontamination of fresh product by ultrasound is relatively recent. The inactivation of microorganism caused by cavitation phenomenon has promoted high intensity ultrasound as method to decontaminate fruits and vegetables. The efficiency of the ultrasound process is affected by several factors such as power level, treatment time, and temperature (Pinheiro *et al.*, 2015). Additionally, ultrasound can be applied directly to the medium (water) or in combination with some compounds (organic salts, organic acids, chitosan, among others) to achieve better results. Concerning individual ultrasound application, ultrasound at low frequencies (20 and 40 kHz) has demonstrated to decrease the microbial load of mesophilic aerobes in lettuce (0.9 log CFU/g) and strawberry (1.49 log

CFU/g). At the present time, ultrasound is being implemented in combination with various aqueous sanitizers in order to improve microbial safety and maintain food quality on organic fresh produce. In peach fruit, the effect of ultrasound (40 kHz, 10 min) and salicylic acid (0.05 mM) either separately, or combined on blue mold caused by *Penicillium expansum* was investigated. The results showed that the application of salicylic acid alone could reduce blue mold, while the use of ultrasound had no effect. Results also revealed that salicylic acid combined with ultrasound treatment was more effective in inhibiting fungal decay during storage than the salicylic acid treatment alone. The combined treatment increased the activities of defense enzymes such as chitinase, β-1,3-glucanase, phenylalanine ammonia lyase, polyphenol oxidase, and peroxidase, which were associated with higher disease resistance induced by the combined treatment. Furthermore, the combined treatment did not impair the quality parameters of peach fruit after 6 days of storage at 20°C. The incorporation of ultrasound alone or in combination with other agents in decontamination process could be a useful preservation technique for postharvest fruits and vegetables. Combination of ultrasound and sanitizers could increase pathogen reduction without affecting the product quality, while concentration of sanitizers could be reduced as well as treatment time required, saving time and money and avoiding significant risks to consumers.

7.12.10. Fogging: In order to prolong the shelf life of fruits and vegetables in postharvest periods, various technologies have been developed that maintain their integrity as well as their nutritional properties. One of the technologies little explored at present is the use of ultrasonic nebulization (Fogging) as a method of distribution of compounds that serve to prevent or control pathogenic diseases in the postharvest period. Fogging has been used successfully for the spraying of disinfectants such as chlorine dioxide, sodium hypochlorite, hydrogen peroxide, acetic acid, and ethanol, achieving the control of epiphytic microorganisms on the surface of the strawberry (fungi and bacteria), thus reducing the decay index by up to 83.2%, demonstrating that nebulization is an effective method for the reduction of diseases in the postharvest stage. Regarding fruit quality, in a study with strawberry, peracetic acid was applied as a disinfectant by ultrasonic nebulization, and the results showed that the anthocyanin and phenolic compound contents were preserved even when the fruits were exposed to low concentrations of peracetic acid . In the postharvest and fruit storage period, it is necessary to minimize chemical products as environmental precautions and avoid adapting pathogens to various fungicides, causing high losses of between 30 and 50% in vegetables and fruits (Vardar *et al.*, 2010). The use of ultrasonic nebulization offers advantages such as the reduction of the amount of disinfectant and a better distribution of the treatment on the fruits, and in a study of figs, the inhibition of gray mold disease *(B. cinerea)* 80% of control was achieved with only the application of chlorine dioxide (1000 μL/L). Ultrasonic nebulization as a conservation method in the postharvest period gives high benefits in different ways, by reducing the quantity of substances applied, the exposure time as well as a better distribution of the treatments, its application in fruits and vegetables has not been explored, thus the development of this technology can offer an alternative to the use of chemical fungicides for the control of diseases.

7.12.11 Plant latex: Natural plant products contain antimicrobial activities with low toxicity and broad spectrum activity. Plant secondary metabolites have a potential to control postharvest diseases in fruits and vegetables.About 10% of flowering plants produce latex and are found in over 40 plant families including Euphorbiaceae, Moraceae, Cannabinaceae, Apocynaceae, Astereacea, Papavereceae, Sapotaceae and Asclepidiaceae. Latex is an emulsion or suspension of many solid particles and has a clear or variously colored often milky appearance which exudes upon plant tissue damage. Latex flows inside laticifers of roots, stems, leaves and fruits. Sub cellular organelles in plant laticifer cells synthesize great varieties of defense proteins and secondary metabolites which are acting as a barrier to microorganisms. Latexes contains various biologically active compounds and antimicrobial activities of plant latex has been well documented.

Bioactive compounds from plant latex are potential source of antifungic against postharvest pathogens. Latex from a total of seven plant species was investigated for its phytochemical and antifungal properties (Sibi *et al.,* 2013). Six fungi namely *Aspergillus fumigatus*, *A. niger*, *A. terreus*, *F. solani*, *P. digitatum* and *R. arrhizus* were isolated from infected fruits and vegetables and tested against various solvent extracts of latex. Analysis of latex extracts with phytochemical tests showed the presence of alkaloids, flavonoids, glycosides, phenols, saponins, steroids, tannins and terpenoids. Antifungal assay revealed the potential inhibitory activity of petroleum ether extracts against the postharvest fungal isolates. Various degree of sensitivity was observed irrespective of plant species studied with *A. terreus* and *P. digitatum* as the most susceptible ones. *F. solani* and *A. fumigatus*were moderately sensitive to the latex extracts tested. Among the plants, latex of *Thevetia peruviana* (75.2%) and *Artocarpus heterophyllus* (64.8%) were having potential antifungal activity against the isolates followed by *Manilkara zapota* (51.1%). In conclusion, use of plant latex makes interest to control postharvest fungal diseases and is fitting well with the concept of safety for human health and environment.

Petroleum ether extracts of latexes have exhibited stronger and consistent antifungal activity against the most of the isolates tested.. Aqueous extracts were completely failed to inhibit the growth of test fungi throughout the study. *A. heterophyllus* has exhibited significant antifungal activity (≥3.52 mm) with complete inhibition of *A. terreus* and *P. digitatum*. Methanolic extract showed moderate activity against all the three species of *Aspergillus* (7.4 to 10.3 mm) and acetone extract recorded inhibitory activity against *A. terreus* (8.8 mm) (Sibi, *et al.*, 2013). Various degree of inhibition (7.5 to 45.3 mm) was observed with petroleum ether extract of *C. papaya* latex with no activity against *A. fumigatus*. Methanolic extract also exhibited inhibitory activity against three of the total fungi tested. *F. religiosa* was effective against all the isolates (16.89 to 45.7 mm) and the other solvent exhibited antifungal activity apart from petroleum ether was methanolic extract against *A. niger* (20.2 mm). *A. terreus* and *F. solani* were the most sensitive organisms which were completely inhibited by petroleum ether extract of *M. zapota* throughout the incubation period. *P. digitatum* was the other fungi controlled by its latex (47 mm).

Antimicrobial activities of plant latexes have been well documented (Guerrero and Guzman, 2004) has characterized chitin binding lectins from *Artocarpus* and found its antifungal activity against *Fusarium moniliforme*. Potential anti mycotic activity against postharvest fungal pathogens by *A. heterophyllus* latex was seen throughout this investigation. Potential antifungal activity of Papaya seeds and latex were studied earlier. A significant inhibitory action against *F. solani* (41.7 mm) and *P. digitatum* (45.3 mm) were exhibited by *C. papaya* latex from our findings. Antifungal activity of *F. religiosa* leaves were reported by previous studies and a significant inhibitory action of its latex was documented in the present findings. *M. zapota* is documented for its antimicrobial activities and the antifungal activities of its latex were demonstrated in this investigation. Ethanolic extracts of *Jatropha curcas* was reported against phytofungal pathogens and a moderate inhibition (11.5 to 34 mm) was observed from *Jatropha* latex in this study. The present findings revealed the antifungal activity of *N. oleander* latex against plant pathogens. Antimicrobial activities of *Thevetia peruviana* have also been reported. In this study, *T. peruviana* latex was observed with potential antifungal activity against all the isolates tested.

7.12.12 Seaweed extracts: Seaweed extracts are natural compounds, sources rich of nutrients and bioactive compounds. An expressive number of brown seaweeds is already used in agriculture and sold as plant fertilizers. Many works have already been developed with seaweeds for disease control, but few have focused on the postharvest phase of fruits. Studies aimed at verifying the effect of these new management possibilities on the physicochemical and organoleptic characteristics of fruits are even scarcer. A study was carried out by Bahammou *et al*., (2017) to evaluate the effect of brown alga crude extracts of *Cystoseira tamariscifolia* and *Bifurcaria bifurcata* against the mycelium (growth of) *Botrytis cinerea* inducing grey mold during tomato postharvest. The results showed that the hexanic extract (91,000ppm) and the aqueous one (49,600ppm) have a very important inhibitory potency against the tested fungi with a very small amount of 100 μL. The important activity was obtained with the hexanic extract of *C. tamariscifolia* against B. cinerea with an inhibition zone up to 2 ± 0.18cm. The treatment of tomato fruits inoculated with the fungal species, with the aqueous extract of the brown algae ensures their protection against the phytopathogenicity of the fungus. An important protection was obtained also with the aqueous extract of *B. bifurcata*. Metal contents analysis show that these two seaweeds are not harmful. The practical use of seaweeds during postharvest opens a new way in the biological control. Their extracts could be an ecofriendly alternative for the protection of crops against pathogens.

The three products (hexanic extract, aqueous extracts, and powder) from the two brown algae *C. tamariscifolia* and *B. bifurcata* showed their antifungal activity against *Botrytis cinerea*. Among these products the ecological aqueous extract should be considered as a potential source of bioactive compounds and could serve as a biological control in Morocco coastal regions known for their production and tomato storage (e.g. Agadir and Dakhla). This may encourage the use of natural products as substitution of harmful chemical fungicides used along the agricultural food chain. Therefore, the investigation of heavy metal concentrations in the algae species may provide useful information on the transfer of potentially toxic elements from abiotic compartments (water and sediments) to higher consumers, including human beings. Additional studies need to be performed to define and characterize, at the chemical and biochemical levels, the preferential effect of algal extracts on microorganisms that have adverse effects. A determination of the molecules really liable for this antifungal activity is also necessary to promote the product.

The three products (hexanic extract, aqueous extracts, and powder) from the two brown algae *C. tamariscifolia* and *B. bifurcata* showed their antifungal activity against *B. cinerea*. Among these products the ecological aqueous extract should be considered as a potential source of bioactive compounds and could serve as a biological control in Morocco coastal regions known for their production and tomato storage (e.g. Agadir and Dakhla). This may encourage the use of natural products as substitution of harmful chemical fungicides used along the agricultural food chain. Therefore, the investigation of heavy metal concentrations in the algae species may provide useful information on the transfer of potentially toxic elements from abiotic compartments (water and sediments) to higher consumers, including human beings. Additional studies need to be performed to define and characterize, at the chemical and biochemical levels, the preferential effect of algal extracts on microorganisms that have adverse effects. A determination of the molecules really liable for this antifungal activity is also necessary to promote the product.

Greenhouse-grown carrot plants were sprayed with an extract (0.2%) of the seaweed Ascophyllum nodosum (SW) and then inoculated 6 h later with the fungal pathogens *Alternaria radicina* and *Botrytis cinerea.* Additional applications of SW were made 10 and 20 d after inoculation. Treated plants showed significantly reduced disease severity at 10 and 25 d after inoculation compared to control plants sprayed with water (Jayaraman *et al.*, 2008). SW was more effective than salicylic acid (SA) (100 μM) in reducing infection. Activity of certain defence-related enzymes, including peroxidase (PO), polyphenoloxidase, phenylalanine ammonia lyase, chitinase and β-1,3-glucanase, were significantly increased in plants treated with SW and SA compared to the control 12 h after treatment. The treated plants also had higher transcript levels of pathogenesis-related protein I (PR-1), chitinase, lipid transfer protein (Ltp), phenylalanine ammonia lyase (Pal), chalcone synthase, non-expressing pathogenesis-related protein (NPR-1) and pathogenesis-related protein 5 (PR-5) genes compared to control plants. These results showed that SW enhances disease resistance in carrot, likely through induction of defence genes or proteins.

The effect of *Ascophyllum nodosum* extract (AN), applied in different doses was determined on the physicochemical characteristics of postharvest 'Tommy Atkins' mangoes (Melo *et al.*, 2018). The selected mango fruits were treated with different extract concentrations (0%, 0.1%, 0.3%, 0.5% and 1.0%) and stored for 12 days at 25±2 °C and 75±5% RH. Every 3 days, fruit mass loss (PM), color angle (Hue) (AH), firmness, pH, titratable acidity (AT), content of total soluble solids in pulp (°Brix) and content of carbohydrates in shell were assessed. The experiments were carried out in CRD in a 5 x 5 factorial arrangement (concentrations x days of storage) with four replications. AN was effective in preventing PM, with a reduction of up to 40.2%, as observed at concentration 0.1% compared to control. There were no differences among concentrations on AH, however, all of them differed significantly from control with higher AH, especially on the 12nd day. Fruit firmness was lower in control, the pH increased with the storage time and AT decreased. For all these parameters, mangoes treated with the extract differed from control. The pulp sample °Brix increased with the storage time as well as the content of reducing sugars, with consequent reduction of non-reducing sugars in fruit shell. These tests indicated the possibility of using AN as an alternative management in preserving mangoes in the postharvest stage.

The application of *A. nodosum* extract in strawberries, both by fertigation and foliar spraying, decreased the content of soluble solids in relation to control. This decrease suggested the deceleration of fruit metabolism during storage, with lower sucrose hydrolysis and consequent use of reducing sugars in respiratory processes.

The antifungal activity of the wild edible seaweed, *Osmundea pinnatifida* (Rhodophyta) was explored collected from the Portuguese coast, which is used as a food seasoning in Scotland, Ireland and Portugal (Paulo Silva *et al.*, 2018)). They performed a sequential extraction of the seaweed components with methanol, dichloromethane and *n*-hexane. These extracts showed an antifungal activity against *Alternaria infectoria* and *Aspergillus fumigatus.* The *n*-hexane fraction of the seaweed inhibited the sporulation of *Alternaria infectoria* at 30 $\mu g\ mL^{-1}$ and induced a statistically significant ($P < 0.001$) decrease in β-glucan content. Furthermore, liquid cultures of *Aspergillus fumigatus* supplemented with 10 $\mu g\ mL^{-1}$ of the *n*-hexane fraction showed abnormal conidiophores, completely devoid of phialides and conidia associated with a decrease of 18.3% in the chitin content ($P < 0.01$). The *n*-hexane fraction analysis by GC-MS revealed that it includes palmitic acid (29.6%), phytol isomer 1 (12.8%), oleic acid (9.6%), stearic acid (6.2%) and D-(−)-tagatofuranose (4.1%), among other compounds present at lower concentrations. The present study reveals *Osmundea pinnatifida* as a promising source of biologically active compounds inhibiting fungal growth and conidiation, the main dispersal mechanism of filamentous fungi as *Aspergillus fumigatus* and *Alternaria alternata*, revealing its utility both as an environmental fungicide against fungal diseases and as a food preservative against fungal postharvest food contamination.

7.12.13. Photosensitization: In general, photosensitization involves a photoactive organic compound (for instance acridine orange, methylene blue or hematoporphyrin) that usually accumulates in target microorganism and subsequent irradiation with visible light (Luksiene *et al.,* 2005).. This combination of two absolutely nontoxic elements, dye and light, in oxygenated environment induces damage and total destruction of microorganisms or the target cell it accumulates in. In 1924 this phenomenon was applied to cure skin cancer. It is important to note that really major advances have been made in photosensitized antimicrobial chemotherapy, in particular disinfection of the blood and blood products, treating locally infected wounds or oral candidiasis. Moreover, numerous investigators demonstrated possible practical usefulness of photosensitization in the broad field of different sciences: virology, microbiology, immunology or dermatology.

Photosensitization is an innovative method for inhibition of pathogens in berry and is based on the interaction of light (λ = 400 nm, 20 mW cm-2) and photoactive compound (chlorophyllin derivative). Neringa Rasiukevičiūtė *et al.,* (2015) evaluated the effect of the chlorophyllin-based photosensitization on the reduction of grey mould (*B. cinerea*) in postharvest strawberry and raspberry fruits. The berries were soaked in a treatment solution for 1 hour and then exposed to prototype (light) for 30 minutes. The obtained data indicated that chlorophyllin derivative inhibited grey mould infection on fruits. The infection of *B. cinerea* in strawberries after a 6-day storage decreased by 8% and in raspberries after 1-day storage it decreased by 3%. The chlorophyllin had no impact on the visual quality and nutritional properties of strawberries and raspberries. Photosensitization is an innovative method for eliminating fruit pathogens based on simultaneous use of light and a photosensitizer. Chlorophyllin-photoactive compound is known as a photosensitizer, it is a watersoluble food additive (E140), and chlorophyllin derivative is safe as well.The chlorophyllin derivative-based photosensitization is an innovative postharvest treatment for the inhibition of *Botrytis cinerea* on strawberries and raspberries without any negative influence on the colour, antioxidant activity and phenolic compounds. The chlorophyllin derivative-based photosensitization reduced the contamination of strawberries with *B. cinerea* by 8% (after a 6-day storage) and that of raspberries by 3% (after 1-day storage). Photosensitization has potential as a nonchemical plant protection tool, which could be used for the development of new plant protection technologies in agriculture.

Photosensitization is based on the interaction of two nontoxic, nonmutagenic and non carcinogenic agents photosensitizer, accumulated in the microorganism, and visible light. This interaction in the presence of oxygen induces radical based citotoxic events (Luksiene *et al*., 2005). The study has been carried out to define a new tool to improve microbial food safety by pho-tosensitization for inactivation of several fungi, which are harmful for food industry and sometimes resistant to other treatments. The obtained data indicate that several microfun-gi such as *Alternaria alternata*, *Fusarium avenaceum*, *Acremonium strictum* and *Rhizopus oryzae* might be effectively inactivated by this new technology. Clear correlation was observed between the efficiency of inhibition of germination and the amount of photosensitizer, accumulated by the fungus.

7.12.14. Biomolecules: In a study by Wang *et al.*, (2018), the activity of the cecropin A-melittin hybrid peptide BP21 (Ac-FKLFKKILKVL-NH_2) in controlling of citrus postharvest green and blue molds and sour rot and its involved mechanism was studied. The minimum inhibitory concentrations of BP21 against *Penicillium digitatum*, *Penicillium italicum*, and *Geotrichum candidum* were 8, 8, and 4 μmol L^{-1}, respectively. BP21 could inhibit the growth of mycelia, the scanning electron microscopy results clearly showed that the mycelia treated with BP21 shrank, formed a rough surface, became distorted and collapsed. Fluorescent staining with SYTOX Green (SG) indicated that BP21 could disintegrate membranes. Membrane permeability parameters, including extracellular conductivity, the leakage of potassium ions, and the release of cellular constituents, visibly increased as the BP21 concentration increased. Gross and irreversible damage to the cytoplasm and membranes was observed. There was a positive correlation between hemolytic activity and the concentration of BP21. These results suggest peptide BP21 could be used to control citrus postharvest diseases.

The effect of β-aminobutyric acid (BABA) on *Rhizopus* rot produced by *Rhizopus stolonifer* in harvested peaches and the possible action modes were investigated (Wang *et al.*, 2018). Treatment with 50 mmol L^{-1} of BABA resulted in significantly lower lesion diameter and disease incidence compared with the control. The activities of defense-related enzymes chitinase and β-1,3-glucanase were notably enhanced by this treatment. Meanwhile, BABA treatment also increased lignin accumulation and maintained higher energy status in peaches by enhancing activities of enzymes in the phenylpropanoid and energy metabolism pathways. Semiquantitative reverse transcription PCR results indicated that the transcription of four defense-related genes was substantially and rapidly enhanced only in that BABA-treated fruit upon inoculation with the pathogen. The results demonstrated that BABA was effective on controlling *Rhizopus* rot by inducing disease resistance, which includes the increase in gene transcription and activity of defense-related enzymes, the enhancement of cell wall strength, and the maintenance of high energy status in *Prunus persica* fruit. Moreover, the disease resistance induced by BABA was demonstrated through priming model rather than direct induction.

Diphenylamine (DPA) has been used for more than 50 years to control the physiological disorder of apples known as storage scald or superficial scald. It also controls external CO_2 injury (David A. Rosenberger, 2011). DPA has traditionally been applied by drenching fruit immediately after harvest with a recycling solution that contains DPA. Fungicides have been applied with DPA in drench solutions to control blue mold caused by *Penicillium expansum* and gray mold caused by *Botrytis cinerea*. Postharvest drenches are still effective for controlling scald, CO2 injury, and postharvest decays, but recycling drenches are increasingly viewed as posing food safety concerns because human pathogens introduced into the solution could be recycled over thousands of bushels of apples. Biocides such as chlorine cannot be used in combination with DPA because DPA is an antioxidant whereas most biocides are strong oxidizers. Combining the two in a single tank or even in successive treatments would inactivate either DPA or the sanitizer, or perhaps both. DPA applied via the NRD method was just as effective for controlling scald as DPA applied in a recycling drench even though the NRD treatments resulted in incomplete coverage of fruit with treatment solutions. Vapor action of DPA is apparently sufficient to suppress scald on the portions of fruit not contacted directly by the DPA solution. Results from the 2010 trial suggest that the NRD treatment method should provide acceptable control of scald in any situation where recycling drenches of DPA would be expected to provide protection, including situations where only selected lots of fruit in large storage rooms are treated with DPA. Fruit in these trials did not develop any external CO_2 injury, so we were unable to determine if DPA applied via NRD will provide consistent control of CO_2 injury. However, we suspect that CO_2 injury will be controlled in any situation where DPA treatments provide effective scald control. None of the fungicides applied via NRD provided acceptable decay control for fruit in the middle and bottoms of bins.

References

Adel A. Kader and Rosa S. Rolle, 2004. The role of postharvest management in assuring the quality and safety of horticultural produce. FAO Agricultural Services Bulletin 152, Food And Agriculture Organization Of The United Nations, Rome.

Agurash Seid, Debrie Tasew and Binyam Tsedaley, 2017. Effect of hot water treatment on development of anthracnose (*Colletotrichum gloeosporioides*) and quality of mango fruit at Jimma southwest Ethiopia, *Archives Phytopathol. & Pl. Protn.*, **50**: 303-316]

Alvindia, Dionisio & Kobayashi, Takao & Natsuaki, Keiko & Tanda, Seinosuke, 2004. Inhibitory influence of inorganic salts on banana postharvest pathogens and preliminary application to control crown rot. *J. Gen. Pl. Pathol.*, **70:** 61-65.

Anjum Malik, A., Ahmed N. Babita, Chauhan, H. and Gupta, P.,2016. Plant Extracts in Postharvest Disease Management of Fruits and Vegetables-A Review. *J. Food Pro. Technol.*, **7**:592.

Ashwini M and Nikhita Desai , 2018. Biopreservative effect of plant extracts on the shelf life of mango cv Raspuri . *J. Pharmacognosy & Phytochem.*, **7**: 2245-2248

Bahammou , N., Cherifi , O., Bouamama , H., Cherifi , K., Moubchir, T..and Bertrand, M., 2013. Postharvest control of gray mold of tomato using seaweed extracts. *JMES*, **8 :** 831-836

Bautista-Ban˜osa, S., Herna´ndez-Lauzardoa , A.N., Vela´zquez-del Vallea , M.G., Herna´ndez-Lo´peza , M., Ait Barkab , E., Bosquez-Molinac , E., Wilsond, C.L., 2006. Chitosan as a potential natural compound to control pre and postharvest diseases of horticultural commodities. *Crop Protn.*, **25** : 108–118

Black-Solis, J., Ventura-Aguilar, R.I., Correa-Pacheco, Z., Corona-Rangel, M.L. and Bautista-Baños, S., 2019. Preharvest use of biodegradable polyester nets added with cinnamon essential oil and the effect on the storage life of tomatoes and the development of *Alternaria alternata*. *Sci Hortic* (Amsterdam). **245**: 65-73

Carl E. Sams, 1994. Management of Postharvest Disease Resistance in Horticultural Crops: Introduction to the Colloquium. *Hort. Sci.,* **29:** 746-747

Chalutz, E. and Wilson, C.L.,1990. Postharvest biocontrol of green and blue mould and sour rot of citrus fruit by *Debaryomyces hansenii*. *Pl. Dis*., **74**:134-137.

Chen, J., Shen, Y., Chen, C., Wan, C., 2019. Inhibition of key citrus postharvest fungal strains by plant extracts. *Plants (Basel),*. **8**(2). pii: E26. doi: 10.3390/plants8020026.

Chumyam, A., Kunthawun, D., Bussaban, B., Uthaibutra , J. and Saengnil, K.,2015. Effects of ClO_2 fumigation on postharvest fungi and disease development of longan fruit . *Acta Hortic*. **1088** : 339-344

Civello, P.M., Villarreal, N., Lobato, M.G. and Martinez, G.A. 2014. Physiological effects of postharvest UV treatments: recent progress. *Stewart Postharvest Rev*.,**10**: 1-6.

Clara Montesinos-Herreroa, Joseph L. Smilanick b,, J. Steven Tebbets b, Spencer Walse b, Lluís Paloua, 2011. Control of citrus postharvest decay by ammonia gas fumigation and its influence on the efficacy of the fungicide imazalil. *Postharvest Biol. and Technol*., **59** 85–93

Cristina, R.A., Pérez, J.J.R., Guillermo, H.M.L., Bernardo, M.A. and Omar, R.P.E., 2018. Control of phytopathogenic microorganisms of postharvest in tomato (*Lycopersicon esculentum* Mill.) with the use of citrus extract. *J. Plant Sci. Phytopathol*., **2**: 037-043

Cruz, M.E.S., Schwan-Estrada, K..R.F., Clemente, E.., Itako, A.T.; Stangarlin, J.R.. and Cruz, M.J.S., .2020. Plant extracts for controlling the postharvest anthracnose of banana fruit. *Rev. bras. plantas med.,*.**15**:.727-733

David A. Rosenberger , 2011. Controlling Postharvest Diseases and Disorders of Apples with Non-Recycling Drenches. New York Fruit Quarterly . Volume 19 . Number 2 . Summer 2011 23

Decelis, S., Sardella, D., Triganza ,T., Brincat, J.P., Gatt, R and, Valdramidis, V.P., 2017 Assessing the anti-fungal efficiency of filters coated with zinc oxide nanoparticles. *R. Soc. Open Sci*., **4**: 161032.

Deepshikha, Bimla Kumari , Elangbam Premabati Devi , Geeta Sharma , Shilpi Rawat and J.P. Jaiswal , 2017. Irradiation as an Alternative Method for Post–harvest Disease Management: An Overview . *Intl. J. Agri. Envt. & Biotech.,* **10** : 625-633,

Deliopoulos, T., Kettlewell, P.S.and Hare, M.C., 2010. Fungal disease suppression by inorganic salts: A review. *Crop Protn.,* **29**:1059-1075

Divya Jagana, Yashoda R., Hegde and Rajasekhar Lella, 2017. Green Nanoparticles - A Novel Approach for the Management of Banana Anthracnose Caused by Colletotrichum musae. *Intl.J.Curr.microbiol. & Appl.Sci.,* **10**: 1749-1756

Domingo Martínez-Romero, Gloria Bailén, María Serrano, Fabián Guillén, Juan Miguel Valverde, Pedro Zapata, Salvador Castillo and Daniel Valero, 2007. Tools to Maintain Postharvest Fruit and Vegetable Quality through the Inhibition of Ethylene Action: A Review. *Critical Rev. Food Sci.& Nutrn.,* **47**: 543-560,

Dukare, A.S., Paul, S., Nambi, V.E., Gupta, R.K., Singh, R., Sharma, K., Vishwakarma, R.K.., 2018. Exploitation of microbial antagonists for the control of postharvest diseases of fruits: a review. *Crit. Rev. Food Sci. Nutr*., **16**: 1-16

Eckert, J.W.,1990. Recent developments in the chemical control of postharvest diseases. *Acta Hortic*., **269**: 477-494

El-Ghaouth, A., Wilson, C.L., Wisniewski, M.E., 2004. Biologically based alternatives to synthetic fungicides for the postharvest diseases of fruit and vegetables. In: Naqvi, S.A.M.H. (Ed.), Diseases of Fruit and Vegetables, vol. 2. Kluwer Academic Publishers, The Netherlands, pp. 511–535.

El-Ramady, H. R., Domokos-Szabolcsy, É., Fári, M..,. Abdalla, N. A., . Taha, H. S E., Lichtfouse (ed.), 2015. *Sustainable Agriculture Reviews*, Sustainable Agriculture Reviews 15, Springer International Publishing, Switzerland

Feliziani, E., Lichter, A., Smilanick, J.L.and , Ippolito, A., 2016. Disinfecting agents for controlling fruit and vegetable diseases after harvest. *Postharvest Biol. &Technol.*, **122**:53-69

Fernandes, A., Antonio, A.L., Oliveira, M.B., Martins, A. and Ferreira, I.C. 2012. Effect of gamma and electron beam irradiation on the physico-chemical and nutritional properties of mushrooms: a review. *Food Chem.*, **135**: 641-650.

Garba Betchem, Nana Adwoa Nkuma Johnson and Yun Wang, 2019. The application of chitosan in the control of postharvest diseases: a review. *J.Pl.Dis Protn.*, 126 : 495–507

Gehan A.M.Abdelmalek and Taher A Salah Eldin, 2016. Silver Nanoparticles as a Potent Fungicide for Citrus Phytopathogenic Fungi. *J. Nanomed.Res.*, **3**: 2-8

Ghassan F. Al-Samarrai, Harbant Singh and Mohamed Syarhabil, 2013. Extracts Some Plants on Controlling Green Mold of Orange and on Postharvest Quality Parameters. *World Appl. Sci. J.,* **22** : 564-570, 2013

Gianfranco Romanazzi, Erica Feliziani and Dharini Sivakumar, 2018. Chitosan, a Biopolymer With Triple Action on Postharvest Decay of Fruit and Vegetables: Eliciting, Antimicrobial and Film-Forming Properties. *Front Microbio.,* **9**: 2745

Golden, K.D., Williams, O.J. and Dunkley, H.M., 2014. Ethylene in Postharvest Technology: A Review. *Asian J. Biol.Sci.*, **7:** 135-143

Guerrero, R.O. and A.L. Guzman, 2004. Bioactivities of latexes from selected tropical plants. *Rev. Cubana Plant Med.*, **9**: 1-1.

Habung Yami , Pramila Tripathi and Shukla, A.K., 2016. Efficacy of Some Essential Oils against Postharvest Fungal Diseases of Kiwifruits. *Intl. J. Adv. Agri. Sci. & Technol.*, **3:** 01-12

Haggag, W.M., Mahmoud, Y.S. and Farag, E.M., 2010. Signaling necessities and function of polyamines/jasmonatedependent induced resistance in sugar beet against beet mosaic virus (BtMV) infection. *New York Sci.J.,* **3**: 95-103

Harbant Singh, Ghassan Al-samarai, and Mohd.Syarhabil, 2012. Exploitation of Natural Products as an Alternative Strategy to Control Postharvest Fungal Rotting of Citrus. *Intl. J.Sci & Res.Publ.*, **2:** 2250-3153

Hertog, Maarten & Rudell, David & Pedreschi, Romina & Schaffer, Robert & Geerard, Annemie & Nicolaï, Bart & Ferguson, Ian., 2011. Where systems biology meets postharvest. *Postharvest Biol. and Technol.,*. **62**:. 223. 10.

Idriss Talibi, Latifa Askarne, Hassan Boubaker, El Hassane Boudyach and Abdellah Ait Ben Oumar , 2015. Effect of Organic and Inorganic Salts as Alternative Strategy for the Control of Postharvest Citrus Sour Rot Agent *Geotrichum Candidum. Acta Hortic.* **1065 :** 1577-1583

Ingole, Manohar & Gade, R & Charpe, Ashwini & T Dhule, D.,2018. Efficacy of Fungicides, Chemicals, Oils, Wax and Bio-Agent Against Postharvest Pathogens of Nagpur Mandarin. *Pl.Dis.Sci.,* **3**: 80-85.

Ippolito, A. and Sanzani, S.M., 2011. Control of Postharvest Decay by the integration of Pre- and Postharvest application of nonchemical compounds. *Acta Hortic.* **905:** 135-143

Ippolito, A., El-Ghaouth, A., Wilson, C.L., Wisniewski, M.A., 2000. Control of postharvest decay of apple fruit by Aureobasidium pullulans and induction of defense responses. *Postharvest Bio.& Technol.*, **19**, 265–272.

Ippolito Camele, Luciana Altieri, Laura De Martino, Vincenzo De Feo, Emilia Mancini and Gian Luigi Rana, 2012. *In Vitro* Control of Postharvest Fruit Rot Fungi by Some Plant Essential Oil Components. *Int. J. Mol. Sci.,* **13**: 2290–2300

Janisiewicz, W.J., Buyer, J.S. and Kurtzman, C.P.,2011. Identifying and characterizing microflora of stone fruits to select antagonists for control of brown rot with emphasis on latent infections. *Acta Hortic.* **905**: 241-250

Jayaraman, Jayaraj & Wan, A. & Rahman, Misbahur & Punja, Zamir.,2008 . Seaweed extract reduces foliar fungal diseases on carrot. *Crop Protn..* **27**: 1360-1366.

Jeong, M.A. and Jeong, R.D., 2018, Applications of ionizing radiation for the control of postharvest diseases in fresh produce: recent advances. *Pl.Path.*, **67**: 18-29

Jeong, R.D., Shin, E.J. and Chu, E.H., 2015. Effects of ionizing radiation on postharvest fungal pathogens. *The Pl. Pathol. J.,* **31:** 1– 5.

Jial Liu, Shipping Tian, Xianghong Meng and Yong Xu, 2007. Effects of chitosan on control of postharvest diseases and physiological responses of tomato fruit. *Postharvest Biol. Technol.*, **44:** 300-306

Jinyin Chen, Yuting Shen, Chuying Chen and Chunpeng Wan, 2019. Inhibition of Key Citrus Postharvest Fungal Strains by Plant Extracts In Vitro and In Vivo: A Review. *Plants (Basel),* **8**: 26

Jorge Giovanny Lopez-Reyes, Davide Spadaro, Ambra Prelle, Angelo Garibaldi, And Maria Lodovica Gullino, 2013. Efficacy of Plant Essential Oils on Postharvest Control of Rots Caused by Fungi on Different Stone Fruits *in vivo. J.Food Protn.,* **76** : 631–639

Joseph, W., Eckert and Joseph, M. Ogawa,, 1988. The Chemical Control of Postharvest Diseases: Deciduous Fruits, Berries, Vegetables and Root/Tuber Crops . *Ann. Rev. Phytopathol.,* **26:** 433-69

Joydeb Gomasta, Md. Resaul Islam, Sabiha Sultana, Ashraful Alam, 2017. Postharvest application of botanical extracts from the sunderbans for controlling disease and extending shelf life of mango (*Mangifera indica* L.). *Intl. J. Multidisci. Res.& Dev.*, **4**; 220-223

Jun Zhao, Zhong Mei, Xu Zhang, Chao Xue, Chenzhi Zhang, Tenqfei Ma and Shusheng Zhang, 2017. Suppression of Fusarium wilt of cucumber by ammonia gas fumigation via reduction of *Fusarium* population in the field. *Sci. Rep.,* volume 7, Article number: 43103

Kellerman Mareli, Liebenberg Elbie, Njombolwana, Ncumisa Erasmus, Arno Fourie and Paul, 2018. Postharvest dip, drench and wax coating application of pyrimethanil on citrus fruit: Residue loading and green mould control. *Crop Protn.*, **103 :** 115-129.

KhamisYoussef, AngelaLigorio, Simona Marianna Sanzani, Franco Nigro and AntonioIppolito. 2012. Control of storage diseases of citrus by pre- and postharvest application of salts. *Postharvest Biol & Technol.*, **72**: 57-63

Li Xueping , Zhu, Xiaoyang, Wang Hailan, Lin Xuefen, Lin Hongwei and Chen Weixin, 2018. Postharvest application of wax controls pineapple fruit ripening and improves fruit quality. *Postharvest Biol. Technol.,* **136:** 99-110.

Li, Q., Wu, L., Hao, J., Luo, L., Cao, Y. and Li, J.,2015. Biofumigation on Postharvest Diseases of Fruits Using a New Volatile-Producing Fungus of *Ceratocystis fimbriata. PLoS ONE,* **10** (7): e0132009.

Lise Korsten, 2006. Advances in control of postharvest diseases in tropical fresh produce.. *Int. J. Postharvest Technology and Innovation,* **1**, No. 1,

Lluis Palou, 2018. Postharvest Treatments with GRAS Salts to Control Fresh Fruit Decay. *Hort.*, **4**: 46

Luksiene, Zivile & Pečiulytė, Dalia & Jurkoniene, Sigita & Puras, Romualdas. 2005. Inactivation of Possible Fungal Food Contaminants by Photosensitization. *Food Technol. & Biotechnol.,.* **43**: 335–341

Macarisin, D., Bauchan, G., Droby, S. and Wisniewski, M. (2011). Novel Role for Reactive Oxygen Species (Ros) in Host-Antagonistic Yeast-Pathogen Interactions in Postharvest Biocontrol Systems . *Acta Hortic.* **905**: 113-119

Madani B., Tengku Muda Mohamed M., Biggs A.R., Kadir J., Awang Y., Tayebimeigooni A., RoodBar Shojaei T., 2014. Effect of pre-harvest calcium chloride applications on fruit calcium level and postharvest anthracnose disease of papaya, *Crop Protn.*, **55** : 66-60.

Malandrakis, A.A., Markoglou,A.N., Konstantinou, S., Doukas,E.G., Kalampokis, J.F. and Karaoglandis,G.S., 2013. Molecular characterization, fitness and mycotoxin production of benzimidazole-resistant isolates of *Penicillium expansum. Int. J. Food Microbial.*, **162**: 237-44.

Manso, T. and Nunes, C., 2011. Metschnikowia andauensis: A novel biocontrol agent of fruit postharvest diseases. *Acta Hortic.* **905** : 261-268

Marques, K.M., Galati, V.C., Fernandes, J.D.R., Guimarães, J.E.R., Silva, J.P., Mattiuz, B.H. and Mattiuz, C.F.M. 2016. Use of chitosan for the control of postharvest anthracnose and quality in avocados. *Acta Hortic.* **1120** : 225-232

Melo, Thiago Anchieta de, Serra, Ilka Márcia Ribeiro de Souza, Sousa, Adryelle Anchieta, Sousa, Thamires Yslanny Oliveira, & Pascholati, Sérgio Florentino, 2018. Effect of Ascophyllum nodosum seaweed extract on postharvest 'Tommy Atkins' mangoes. *Revista Brasileira de Fruticultura*, **40** : e-621

Miceli, A., Ippolito, A., Linsalata, V. and Nigro, F.,1999: Effect of pre-harvest calcium treatment on decay and biochemical changes in table grape during storage. *Phytopathol. Mediterr.*, **38:** 48-53.

Mike Boyette, Dave Ritchie, S. J. Carballo and D. C. Sanders, 1993**.** *Sponsored by the Energy Division, North Carolina Department of Commerce, with State Energy Conservation Program funds, in cooperation with North Carolina State University.*

Morkeliūnė, Armina & Rasiukevičiūtė, Neringa & Burokiene, Daiva & Valiuskaite, Alma. (2019). Application of *Salvia officinalis* and *Picea abies* essential oils for controlling *Colletotrichum* spp. 10.15544/RD.2019.002.

Mostafavi, H.A., Mirmajlessi, S.M., Mirjalili, S.M., Fathollahi, H. and Askari, H., 2012. Gamma radiation effects on physico□chemical parameters of apple fruit during commercial post□ harvest preservation. *Radiation Phys.& Chem.*, **81:** 666– 71.

Mudasir Magray, M , Khurshid Hussain, Malik Aijaz, Singh, P.K. and Ambreen Nabi, 2017. Postharvest management of vegetable crops: Introduction. *J. Pharmacognosy & Phytochem.*, **6:** 182-185

Nabi, S.U., Raja, W.H., Kumawat, K.L., Mir, J.I., Sharma, O.C., Singh, D.B. and Sheikh, M.A., 2017. Postharvest Diseases of Temperate Fruits and their Management Strategies-A Review, *Int. J. Pure App. Biosci.* **5**: 885-898

Naqvi, A. M. H., 2007. Diagnosis and Management of Pre and Postharvest Diseases of Citrus fruit. 10.1007/1-4020-2606-4_8.

Nelson, L.M., Nagel, C., Bose-Roberts, N., Mantyka, D., Hirkala, D. and Sholberg, P.L. (2011). Characterization of Cold-Adapted Rhizobacteria for Control Of Postharvest Fungal Decay of Pome Fruit. *Acta Hortic.* **905**: 181-188

Neringa Rasiukevičiūtė , Alma Valiuškaitė , Nobertas Uselis , Loreta Buskienė , Jonas Viškelis and Živilė Lukšienė., 2015. New non-chemical postharvest technologies reducing berry contamination. *Zemdirbyste-Agri.*, **102**: 411–416

Nokwazi Carol Mbili, 2015. Management of postharvest diseases of apples using essential OILS. Thesis presented in partial fulfillment of the requirements for the degree of Doctor of Philosophy in the Faculty of AgriSciences at Stellenbosch University

Paulo Silva, Chantal Fernandes, Lillian Barros, Isabel c.F.R.Ferrreira, Leonel Pereira and Teresa Goncalves, 2018. The antifungal activity of extracts of *Osmundea pinnatifida*, an edible seaweed, indicates its usage as a safe environmental fungicide or as a food additive preventing postharvest fungal food contamination. *Food Funct.*, ,**9** : 6187-6195

Pinheiro, J., Alegria, C., Abreu, M.,Gonçalves, E.M. and Silva, C.L.M., 2015. Influence of postharvest ultrasounds treatments on tomato (*Solanum lycopersicum*, cv. Zinac) quality and microbial load during storage. *Ultrasonics Sonochem.,*. **27**:552-559

Poleatewich, A. M., Ngugi, H. K., and Backman, P. A. 2012. Assessment of application timing of Bacillus spp. suppress pre- and postharvest diseases of apple. *Plant Dis.* **96**:211-220.

Pramila Tripathi and Dubey, N., 2004. Exploitation of natural products as an alternative strategy to control postharvest fungal rotting of fruit and vegetables. *Postharvest Biol.& Technol.*,**32:** 235-245

Prusky, Dov and Ben-Arie, Ruth, 2008. Control by imazalil of fruit storage rots caused by *Alternaria alternata. Annals Appl. Biol.*, **98:** 87 - 92.

Rabab, M., Elamawi. and Raida, E Al-Harbi, 2014. Effect of biosynthesized silver nanoparticles on Fusarium oxysporum fungus the cause of seed rot disease of faba bean, tomato and barley. *J. Pl. Prot. & Path.*, **5** : 225-237.

Rae-Dong Jeong, Eun-Jung Shin, Eun-Hee Chu and Hae-Jun Park, 2015. Effects of ionizing radiation on postharvest fungal pathogens. *Plant Pathol .J.* , **31**: 176–180.

Ramsés González-Estrada, Francisco Blancas-Benítez, Rita M. Velázquez-Estrada, Beatriz Montaño-Leyva, Anelsy Ramos-Guerrero, Lizet Aguirre-Güitrón, Cristina Moreno-Hernández, Leonardo Coronado-Partida, Juan A. Herrera-González, Carlos A. Rodríguez-Guzmán, José A. Del Ángel-Cruz, Edson Rayón-Díaz, Hector J. Cortés-Rivera, Miguel A. Santoyo-González and Porfirio Gutierrez-Martinez, 2019. Alternative Eco-Friendly Methods in the Control of Postharvest Decay of Tropical and Subtropical

Rivka Barkai-Gorivka Barkai-Golanlan, Postharvest Diseases of Fruits and Vegetables Development and Control. Elsevier B.V. 442 pp

Romanazzi, G., Gabler, F.M., Lichter, A. and Smilanick, J.L. (2011). Natural and Safe Alternatives to Conventional Methods to Control Gray Mold of Table Grapes in Storage. *Acta Hortic.* **905:** 161-168.

Salem, E.H., Nawito, M.A.S.and El-Raouf Ahmed, A.E.A., 2019. Effect of Silver Nano-Particles on Gray Mold of Tomato Fruits. *J. Nanotechnol. Res.*, **1** : 108-118.

Sardella, D., Gatt, R. and Valdramidis, V.P., 2017. Physiological effects and mode of action of ZnO nanoparticles against postharvest fungal contaminants. *Food Res.Intl.*, **101**: 274-279

Saucedo-Pompa, S., Rojas-Molina, R., Aguilera-Carbó, A.F., Saenz-Galindo, A., de La Garza, H.and Jasso-Cantú, D., 2009.. Edible film based on candelilla wax to improve the shelf life and quality of avocado. *Food Res. Intl.*, **42**:511-515.

Seethapathy, Parthasarathy & Jayaraman, Rajalakshmi & Palani, Narayanan & Kathaperumal, Arunkumar & Kuppusami, Prabakar, 2017. Bio-Control Potential of Microbial Antagonists against Postharvest Diseases of Fruit Crops: A Review. *J. Bot. Sci.*, **6:** 17-23.

Shama, G, Alderson, P, 2005. UV hormesis in fruits: a concept ripe for commercialization. *Trends in Food Sci. & Tech.,* **16**, 128– 36.

Sharma , R.R, Dinesh Singh B and Rajbir Singh, 2009. Biological control of postharvest diseases of fruits and vegetables by microbial antagonists: A review. *Biol. Control* , **50 :** 205–221

Shukla, A.C.,2018. Essential oils as green pesticides for postharvest disease management. *Acta Hortic..*, **1210 :** 199-206

Sibi,G., Rashmi Wadhavan, Sneha Singh, Abhilasha Shukla, K. Dhananjaya, K.R. Ravikumar and Mallesha, H., 2013. Plant Latex: A Promising Antifungal Agent for Postharvest Disease Control. *Pakistan J. Biol. Sci.,* **16:** 1737-1743.

Smilanick, J. L., Mansour, M. F., and Sorenson, D., 2006. Pre- and postharvest treatments to control green mold of citrus fruit during ethylene degreening. *Plant Dis*., **90**:89-96

Smilanick, J.L.,2011. Integrated approaches to postharvest disease management in California citrus packinghouses. *Acta Hortic*. **905 :** 145-148

Spadaro, D., Zhang, D., Garibaldi, A. and Gullino, M.L. (2011). The Role of Competition for Iron and Cell Wall Degrading Enzymes in Mechanism of Action of Postharvest Biocontrol Agents. *Acta Hortic.* **905 :** 87-102

Stefan Stošić , Saša Stojanović , Anja Milosavljević , Erika Pfaf Dolovac , Svetlana Živković , 2014. Effect of calcium salts on postharvest fungal pathogens *in vitro* . *Pl. Protn*., **65**: 40-46

Susan Lurie and Romania Pedreschi, 2014. Fundamental aspects of postharvest heat treatments. *Hort. Res*., Vol. 1, Article number: 14030

Swart, S.H., Serfontein, J.J., Swart, G. and Labuschagne, C. 2009. Chemical Control of Postharvest Diseases of Mango: The Effect of Fludioxonil and Prochloraz on Soft Brown Rot, Stem-End Rot and Anthracnose. *Acta Hortic*. **820** : 503-510

Temur, C. and Tiryaki, O., 2013. Irradiation alone or combined with other alternative treatments to control postharvest diseases. *African J. Agri. Res.,* **8 :** 421– 34.

Toktam Taghavi , Chyer Kim and Alireza Rahemi, 2018. Role of Natural Volatiles and Essential Oils in Extending Shelf Life and Controlling Postharvest Microorganisms of Small Fruits. *Microorganisms* , **6 :** 2-15

Trevor Suslow , 2015. Chlorination in The Production and Postharvest Handling of Fresh Fruits and Vegetables. Postharvest Pathology, Transportation & Distribution University of California - Davis

Vardar, C., Ilhan, K. and Karabulut, O.A., The application of various disinfectants by fogging for decreasing postharvest diseases of strawberry. *Postharvest Biol.& Technol*., **66** ; 30-34

Vero, S., Garmendia, G., Garat, M.F., de Aurrecoechea, I. and Wisniewski, M. (2011). *Cystofilobasidium infirmominiatum* as a biocontrol agent of postharvest diseases on apples and citrus. *Acta Hortic*. **905** : 169-180

Vorstermans, B. and Creemers, P. ,2011. Current Status of Development Of Biological Control Products For Postharvest Use In Europe. *Acta Hortic*. **905:** 313-318

Wang, J., Cao, S., Wang, L., Wang, X., Jin, P. and Zheng, Y.,2018. Effect of β-Aminobutyric Acid on Disease Resistance against *Rhizopus* Rot in Harvested Peaches. *Front. Microbiol.* **9**:1505

Wang, W., Liu, S., Deng, L., Ming, J., Yao, S. and Zeng, K.,2018. Control of Citrus Postharvest Green Molds, Blue Molds, and Sour Rot by the Cecropin A-Melittin Hybrid Peptide BP21. *Front. Microbiol*., **9**: 2455.

Waskar, D.P. and. Gaikwad, R.S ,, 2005. Postharvest hot water treatment for disease control in kesar mango fruits. *Indian J.Agril. Res*., **39**:186 - 191

Wilson, C.L., Wisniewski, M., Droby, S. and Chalutz, E.,2011. Historical Perspective on Biological Control of Postharvest Diseases - Past, Present, And Future. *Acta Hortic*. **905:** 23-28

Wonsu Cheon, Young Soo Kim, Kotnala Balaraju, Bong-Su Kim, Byeong-Ho Lee and Yongho Jeon, 2016. Postharvest Disease Control of *Colletotrichum gloeosporioides* and *Penicillium expansum* on Stored Apples by Gamma Irradiation Combined with Fumigation. *Plant Pathol. J.,* **32**: 460–468

Yakoby, N., Zhou, R., Koblier, I., Dinoor, A. and Prusky, D., 2001. Development of Colletotrichum gloeosporioides restriction enzyme-mediated integration mutants as biocontrol agents against anthracnose in avocado fruit. *Phytopathology*, **91**: 143–148

Youssef, K., Ligorio, A., Nigro, F. and Ippolito, A., 2012. Activity of salts incorporated in wax in controlling postharvest diseases of citrus fruit. *Postharvest Biol. Technol.*, **65:** 39–43

8

Integrated Management of Postharvest Diseases

A wide variety of fungal and bacterial pathogens cause postharvest disease in fruits and vegetables. Some of these infect produce before harvest and then remain quiescent until conditions are more favourable for disease development after harvest. Other pathogens infect produce during and after harvest through surface injuries. In the development of strategies for postharvest disease control, it is imperative to take a step back and consider the production and postharvest handling systems in their entirety. Traditionally fungicides have played a central role in postharvest disease control. However, trends towards reduced chemical usage in horticulture are forcing the development of new strategies. Controlling or reducing disease relies on integrated crop and postharvest management, with attention to fungicide application, crop hygiene and nutrition, and the management of ripening, to optimise advantages conferred by the plant's natural resistance factors that prevent and delay disease development. Agricultural production and food distribution systems have evolved over many years to a complex system which allows for nearly year-round supply of most fresh commodities through longterm storage and long distance shipments to consumers. The expectation of year round supplies has resulted in increased challenges for postharvest handling systems. The availability of fungicides for disease management likely played a major role in the evolution of our current system of food distribution (Mitcham, 1999). With the recent loss of registration of many postharvestfungicides and more expected under the Food Quality Protection Act (FQPA), as well as consumer pressure to reduce the amount of chemicals used on food, the challenge to maintain our current system of food distribution increases.

Many of the fungi which cause postharvest disease belong to the phylum Ascomycota and the associated Fungi Anamorphici (Fungi Imperfecti) pathogen inhibition is greater when the antagonist is applied prior to infection taking place (Coates and Johnson, 1997) The genera Phytophthora and Pythium are important postharvest pathogens, causing a number of diseases such as brown rot in citrus (*Phytophthora citrophthora* and *P. parasitica*) and cottony leak of cucurbits (*Pythium* spp.). *Rhizopus* and *Mucor* are important genera of postharvest pathogens in the *Phylum Zygornycota*. *Rhizopus stolonifer* is a common wound pathogen of a very wide range of fruits and vegetables, causing a rapidly spreading watery soft rot. Genera within the *phylum Basidiomycota* are generally not important causal agents of postharvest disease, although fungi such as *Sclerotium rolfsii* and *Rhizoctonia solanr*, which have basidiomycete sexual stages, can causesignificant postharvest losses of vegetable crops such as tomato and potato. The major causal agents of bacterial soft rots are various species of *Erwinia, Pseudomonas, Bacillus*, *Lactobacillus* and *Xanthomonas*. Bacterial soft rots are very important postharvest diseases of many vegetables, although they are generally of less importance in most fruits. This is because most fruits have a low pH which is inhibitory to the majority of bacterial plant pathogens.

8.1 Categories: Postharvest diseases of fruits, in general, fall into threse general categorie.

i. **Blossom infections :** A number of fungi cause fruit infection by first colonizing or infecting the blossoms. The incidence of infection is much more prevalent in a wet spring. There are often no visible symptoms of disease until after harvest. In some cases infected fruit may colour and fall prematurely in the orchard. The spores of these fungi are prevalent in the ground litter and in the tree. Therefore, mowing and clearing of leaves, fallen fruits and prunings can help reduce potential sources of infection. These fungal pathogens can be reduced with spring sprays. Although there are no chemicals registered specifically for these fungi, carefully selected scab fungicides at the pink to petal-fall stages will provide some control. See table below for more information.

ii. **Fall lenticel infections:** Several fungi can enter the fruit through weak lenticels and wounds,and may not show symptoms until after a period of storage. Cleaning up the cankers in the trees is the most effective control for Bull's Eye and Phacidiopycnis rot. Since snow mold rot has been associated with soil and grasses or weeds being transferred from the orchard floor to the bin, careful handling of bins in the orchard should help reduce this problem. There are no fungicides registered specifically for these diseases. Some pin point scab sprays will reduce the incidence of these diseases in storage.

The key diseases are anthracnose in mango and banana, caused by *Colletotrichum gloeosporioides* and *C. musae*, respectively, and stem-end rot in mango, cause by fungi in the family Botryosphaeriaceae (*Neofusicoccum* spp., *Lasiodiplodia theobromae* etc.).A significant component was the identification and evaluation of activators of plant defences under field conditions. The activators were known resistance inducing agents, including acibenzolar-S-methyl (Bion), and elicitors derived from fungal pathogens (in banana). Another component was to characterise some of the key biochemical defences contributing to the resistance, and to identify treatments, varietal properties or other agronomic practices that may influence their relative effectiveness (Johnson, 2007). The most serious diseases of avocado fruit are all caused by fungi. In Australia, anthracnose is predominantly caused by the fungus *Colletotrichum gloeosporioides*, although *C. acutatum* is a minor causal agent of the disease. Although symptoms of anthracnose do not normally appear in "Hass" fruit until ripening, the initial stages of infection by the pathogen actually occur in the field on developing fruit. Once the fungus penetrates the outer skin layer of fruit, it remains there in a dormant or 'quiescent' state until changes occur in the fruit skin which allows infection to proceed. In particular, changes in the concentration of antifungal compounds called 'dienes' are thought to be important in the regulation of anthracnose quiescence in avocado.

One theory states that the fungi occurs "endophytes" in avocado stem tissue, then graduallycolonising inflorescences, fruit pedicels and fruit stemend tissue are inoculated. Evidence for this mode of infection has been found in the case of mango stem-end rot. Another theory is that spores of the stem-end rot fungi infect avocado at flowering, eventually leading to colonisation of the fruit stem-end tissue. A third possible mode of infection for the stem-end rot fungi is at harvest time via the freshly cut surface of the fruit. *Colletotrichum gloeosporioides* has been shown to cause pepper spot. *Cercospora* spot caused by *Pseudocercospora purpurea* and sooty blotch caused by an *Akaropeltopsis* sp, are important pre-harvest diseases. The major postharvest diseases are stem-end rot, caused by any of 10 organisms, with *Thyronectria pseudotricha* the most important; anthracnose caused by *Colletotrichum gloeosporioides* and the *Dothiorella*/ *Colletotrichum* rot complex. In all the above instances, infection of fruit takes place in the orchard prior to harvest. Sooty blotch is a field disease that is generally adequately controlled by the copper fungicide spray program used for anthracnose control. Although symptoms of thedisease are only superficial, they can reduce the market value of fruits. Recently, the causal agent of this disease in Australia was shown to be *Stomiopeltis* sp. (Pegg, 2001).

During storage and transportation of melon fruits, postharvest diseases occur primarily caused by *A. alternata*, *Fusarium semitectum*, *Rhizopus stolonifer* and *Trichothecium roseum*. Currently, application of synthetic fungicides is the primary means of controlling these postharvest diseases. However, use of these fungicides has been progressively restricted, due to increasing concerns on the protection of the environment and human health, together with increased pathogen resistance to fungicides.

Biological control of postharvest disease (BCPD) has emerged as an effective non-chemical alternative. *Bacillus species*, including *B. subtilis*, produce spores that are resistant to various physical and chemical treatments, such as desiccation, heat, UV irradiation, and organic solvents, and serve as excellent biological control agents against a wide range of plant pathogens by their production of antibiotics (iturin, surfactin, and fengycin), cell wall-degrading enzymes (chitinase and b-1,3 glucanase), and antifungal volatiles. At the same time, *B. subtilis* can occupy the same niche as many pathogens and play an antagonistic role. Thus, *B. subtilis* has been recommended as a generally safe ('GRAS') microorganism by the United States Food and Drug Administration, which promotes its use in the food industry (Wang *et al.*, 2010).The development of new fungicides based on natural compounds should include detailed investigations on their degradation in foods or in biological systems, although their natural origin makes them relatively biodegradable and with negligible residues.

Over the last two decades the increasing interest towards control measures alternative to synthetic fungicides against postharvest pathogens has produced numerous studies with encouraging results. On the other hand, almost all the measures described above, taken individually, are unable to obtain an economically acceptable control level. So, some studies are in progress to improve their effectiveness. In the case of biocontrol agents, the formulation is still a key issue, together with the compatibility with the normal practice of fruit processing and storage. Moreover, research has highlighted some critical points that like difficulty in the registration of biofungicides, potential toxicity to humans and the environment of so called natural molecules and their possible phytotoxicity when used at high concentrations, need for exhaustive study on the environmental fate and dispersal of the antagonist

Integration of disease management strategies in fruits and vegetables emerged an answer to shortcomings of reliance on chemical control. It deals with integration of all available disease control methods to manage the diseases effectively and economically to satisfy human needs at the same time guarding the quality of harvested produce and environment. Several studies have highlighted the advantages of Postharvest application of biological control agents over field or soil application. The major advantages are the convenience of bringing the antagonist in contact with the commodity as compared to its addition to the soil and the possibility of acting under controlled conditions created and maintained during storage. Likewise, it is possible to introduce

Bacillus subtilis into the wax applied to peaches in the packinghouse to protect them from the brown rot caused by *Monilinia fructicola* (Sadegh Mohajer *et.al.,* 2015). The antagonistic yeast *Pichia guilliermondii* can be introduced into the wax mixture applied to citrus fruit in packinghouse. Compatibility between a microbial antagonist and a synthetic fungicide offers the option of using the antagonist in combination with reduced levels of the fungicide. Applying the yeast antagonist, *Pichia guilliermondii* to citrus fruit in combination with substantially reduced concentrations of thiabendazole reduced *Penicillium digitatum* decay to a level similar to that achieved by the currently recommended concentration of thiabendazole applied alone.

Thus, an integrated pest management system adaptation provides effective pest control and maintains very low levels of chemical residues. The intensive studies on biocontrol of postharvest diseases have led to the registration of two biological products for commercial postharvest applications on citrus fruits i.e. Aspire, which is *Candida oleophila,* and Bio-Save 1000 from *Pseudomonas syringae*. The biological agent must have low sensitivity to any of the supplemented chemical fungicides. Combining Aspire with each of the chemicals used improved the results; sometimes combinations with a low rate of fungicide were sufficient to achieve effects similar to those obtained by the chemicals at standard rates. Safe compounds or natural products of plant origin have been suggested as alternatives to synthetic and conventional fungicides could also be used in combination with biocontrol agents, complementing their activity. Pathogens treated with such antifungal substances might be weakened and become more vulnerable to the antagonist activity.

The integration of Postharvest biocontrol into modern production, storage and handling systems must begin before harvest. Several preharvest factors that affect fruit quality may have profound effects on the efficacy of Postharvest biological control agents. Preharvest calcium sprays during the growing season of apples and pears can increase fruit firmness, decrease the incidence of certain disorders and enhance resistance to Postharvest infection. Calcium amendments and postharvest application of some antagonistic yeasts can be additive in reducing fruit decay and significantly increase disease control compared with either treatment alone. The advantages in increased firmness, enhanced resistance to postharvest decay and enhanced biocontrol efficacy under some circumstances reflect the multiple benefits of integrating postharvest biological control with cultural and production practices. postharvest factors may have a major impact on the effectiveness of biological control. Fruit maturity at harvest and the application of antagonists is affecting Postharvest biological control. Delayed picked and over matured fruits are more susceptible to decay than are fruits picked at optimal storage maturity. Fruit maturity in apples and pears markedly affected biocontrol efficacy of antagonistic yeast *Cryptococcus*, while excellent control was achieved on freshly harvested fruit and treatments of ripened fruit gave much lower levels of control (Mangosteen White Sapote, 2011).

Temperature management is a critical factor in the maintenance of fruit quality and pathogen development, may enhance biological control of storage decay. As the storage temperature of apples and pears decreased, there was a reduction in the concentration of pyrrolnitrin (a metabolite of *Pseudomonas cepacia* and other *Pseudomonas* spp.) needed to protect the fruit from gray mould and blue mould caused. An integrated strategy to control Postharvest decay in pome and stone fruits has been advanced in recent years comprises several pre and postharvest components *viz*., alteration of fruit nutrient status, calcium applications as sprays during the growing season, lower fruit nitrogen content have been associated with reduced disease severity and influence the susceptibility of pome fruits to decay. Controlled atmosphere storage with reduced O_2 and elevated CO_2 can reduce the severity of postharvest fungal decay in apples by inhibiting fruit senescence and maintaining host resistance to infection.

Integration of early harvest, low fruit nitrogen, high fruit calcium, yeast or yeast + one-tenth of the dose of thiabendazole along with a controlled atmosphere (2% O_2, 0.6% CO_2) were found to reduce blue mould severity in picked pears. All the components of the integrated approach were found to be compatible with thiabendazole fungicides used for the management of postharvest decay and blue mould in harvested pears. Several yeast species such as *Cryptococcus laurentii*, are capable of colonizing wounds of pear fruit under conditions of low temperature, ambient or reduced O_2 and CO_2 therefore, can be integrate into postharvest strategies. The two postharvest treatments, biological control and modified atmosphere packaging were combined with preharvest iprodione spray; the incidence of the brown rot was reduced from 41.5 per cent in the control to only 0.4 per cent (Mangosteen White Sapote, 2011).

A combined strategic approach was elaborated by various worker to control wound infecting pathogens by a series of treatments including disinfection of the fruit surface and environment, eradication or suppression of fungal spore germination at wound sites by a combination of fungicides and reduction of wound susceptibility to infection by the addition of biocontrol antagonists which act as protecting agents. It is being apparent that a multifaceted integrated approach has been advantageous over the use of solitary measures. Thus, the application of antagonists, induced resistance, physical treatments, chemicals, natural and safe fungicides etc. can be utilized in a unified way to provide greater consistency and efficacy to management of Postharvest diseases. Postharvest diseases caused by the necrotrophic ascomycetes *P. digitatum* and *P. italicum* affect all citrus species and cultivars and are strictly originated from infections in peel wounds. Successful infection and disease development depend on intrinsic characteristics of the pathogen and the fruit host and on their interactions, modulated by the environmental conditions. Recent research on molecular and biochemical aspects of host–pathogen interaction has considerably increased the knowledge on the mechanisms of natural and induced fruit resistance to disease. On the one hand, the mitochondrial genome and, later, the complete genome of *P. digitatum* have been entirely sequenced for the first time for a phytopathogenic *Penicillium* species. On the other hand, the construction of cDNA libraries and subsequent transcriptomic analyses of fruit wounded, *P. digitatum*-infected or with resistance elicited by postharvest physical treatments have revealed genes that are involved in fruit responses to fungal attack (Mangosteen White Sapote, 2011). Mechanisms of resistance that have been studied include the biosynthesis of lignin and derivatives, the accumulation of antifungal flavonoids, and the production of PRPs like chitinases or β-1,3-glucanases. The ability of the pathogen in compatible interactions to suppress the burst of ROS and the accumulation of H_2O_2 that occurs in the peel tissue as a previous step to the activation of resistance mechanisms, as well as the role of antioxidant enzymes such as SOD, CAT, APX or GR have also been documented.

Traditional postharvest chemical fungicides such as IMZ, TBZ, SOPP and new reduced risk active ingredients such as PYR, FLU or AZX are the principal means used worldwide for effective control of citrus green and blue molds. Nevertheless, concerns about environmental contamination and human health risks, restricted markets, as well as the proliferation of resistant fungal strains increasingly lead to important usage restrictions. Appropriate integrated disease management approaches for effective control of green and blue molds should consider the influence on disease incidence of commercial fruit handling operations and procedures. Harvest should not be conducted under humid conditions and conscientious and well-trained pickers are required to minimize the production of rind wounds. The effect of ethylene degreening is controversial and greatly dependent on treatment characteristics and fruit condition. Preharvest field treatments could be of use to reduce inoculum levels or induce resistance in the fruit. Interesting research is being conducted on early detection of infections by *Penicillium* spp. through either machine vision systems or olfactometric analysis. Adequate procedures and technologies for fruit and packing house sanitation are important to reduce inoculum levels and potential reinfections.

Substantial progress has been accomplished in evaluating new physical, low-toxicity chemical, and biological alternative control methods as part of IDM programs. Curing, hot water, HWRB, irradiation and illumination (UV-C, blue light), CO_2 and O_2 based antifungal atmospheres, food additives and GRAS compounds, natural compounds (i.e. essential oils and other volatiles, plant extracts, antifungal peptides, chitosan), resistance inducers (e.g. PSi), antifungal edible coatings, or phosphite salts are nonpolluting alternative control treatments that have shown variable value. However, the lack of either curative or preventive activity, low persistence, inconsistency, high variability and dependence on fruit condition, excessive specificity, risk of adverse effects on fruit quality, or technological problems for cost-effective application are general limitations associated with the nature of these alternatives that hinder their general commercial implementation as stand alone treatments (Mangosteen White Sapote, 2011). Research work to find new and more effective alternatives and especially to integrate different treatments in a multifaceted approach is in progress in many public and private laboratories. As advances in the knowledge on the molecular and biochemical mechanisms underlying host pathogen interactions and how they are influenced by potential postharvest treatments are achieved, more tools will be available for the postharvest citrus industry to design tailored cost effective integrated management approaches against molds due to *Penicillium*.

Factors deciding the integration of management strategies : Effective and consistent control of storage diseases is dependent upon integration of the practices *viz.*, select disease resistant cultivars where possible, maintain correct crop nutrition by use of leaf and soil analysis, irrigate based on crop requirements and avoid overhead irrigation, apply pre-harvest treatments to control insects and diseases, harvest the crop at the correct maturity for storage, apply postharvest treatments to disinfest and control diseases and disorders on produce, maintain good sanitation in packing areas and keep dump water free of contamination and store produce under conditions least conducive to growth of pathogens (Bijendra Kumar Singh *et.al.*, 2017). Integration of cultural methods and biological treatments with yeast bio-controls has been studied on pears These results demonstrated that unrelated cultural and biological methods that influenced pear decay susceptibility can be combined into an integrated program to substantially reduce decay. As a general rule, alternatives to chemical control are often less effective than many fungicides. Therefore, it will generally be necessary to combine several alternative methods to develop an integrated strategy to successfully reduce postharvest decay.

Postharvest disease, disorders or losses in quality have economic impacts vastly greater than the actual losses caused by frequency and intensity of their occurrence. For example there are direct financial losses incurred by the grower from batches of fruit expressing the disease and disorder. Direct losses can also cause financial losses for postharvest operators and marketers. Another point to remember is that the loss of value of a downgraded product is likely to be substantially greater for highly differentiated branded products which sell at a premium in the market. All the hard work that has gone into promoting and raising the profile of a branded product can be quickly eroded if there are postharvest quality problems with some lines of that product.

8.2 Management options: Horticultural produce is alive and has to stay alive long after harvest. Like other living material it uses up oxygen and gives out carbon dioxide. It also means that it has to receive intensive care. For a plant harvesting is a kind of amputation. In the field it is connected to roots that give it water and leaves which provide it with the food energy it needs to live. Once harvested and separated from its sources of water and nourishment it must inevitably die. Horticultural managers must possess many skills to succeed in this. They need a keen appreciation of horticultural diversity. For example, apples and bananas each have their own requirements (Wang, *et al.* 1996).

Harvest handling: The care taken during harvesting is repaid later, because fewer bruises and other injuries mean less disease and enhanced value. Good managers train their pickers so that they select the product at the correct stage of maturity with adequate care. It is worthwhile reducing the amount of hard physical work required in picking fruit as far as possible.

Pre-cooling: The harvested produce has to be transported to the packing shed without delay. In the field the heat of the sun and the respiration of the produce combine to heat up the produce, especially in the centre of field bins. This accumulation of "field heat" reduces the postharvest life of the product and has to be removed quickly. Strawberries for example, respire nearly eight times faster at a field temperature of 25°C as they do in a storage temperature of 0°C. Pre-cooling requires a greater refrigeration capacity than does cool storage and is often best done as a separate step (Bijendra Kumar Singh *et.al.*, 2017). Hydro-cooling with cold water drenches, forced air cooling through stacks that ensure proper air distribution and packing with ice are the systems most commonly used, with the choice depending on the individual requirements of the commodity.

Natural fungicides: Many compounds produced naturally by microorganisms and plants have fungicidal properties. Chitosan, for example, is not only an elicitor of host defence responses but also has direct fungicidal action against a range of postharvest pathogens. Antibiotics produced by various species of *Trichoderrna* have potent antifungal activity against *Botrytis* *cinerea, Sclerotinia sclerotiorum, Corticium rolfsii* and other important plant pathogens. There are many other natural compounds which have been isolated and shown to possess considerable antifungal activity (Dubey and Jalal, 2013). Postharvest fungicides can be applied as dips, sprays, fumigants, treated wraps and box liners or in waxes and coatings. Dips and sprays are very commonly used and depending on the compound, can take the form of aqueous solutions, suspensions or emulsions. Fungicides commonly applied as dips or sprays include the benzimidazoles (e.g. benomyl and thiabendazole) and the triazoles (e.g. prochloraz and imazalil). The benzimidazole groups of fungicides are very useful for the control of many important postharvest pathogens such as *Penicillium* and *Colletotrichum*, Fumigants, such as sulphur dioxide for the control of grey mould (*Botrytis cinerea*) of grape and various postharvest diseases of litchi, are sometimes used for disease control. Other fumigants used in certain situations include carbon dioxide, ozone and ammonia. Fruit wraps or box liners impregnated with the fungicide biphenyl are used in some countries for the control of *Penicillium* in citrus. Fungicides to control postharvest diseases of citrus and some other fruit are often applied to the fruit in wax on the packing Iine.

Refrigeration: Refrigeration is the most important tool for extending the life of fruit. In a typical cool store. To maintain a storage temperature of 0°C the temperature of the coils will have to be appreciably below 0°C. Moisture is therefore removed from the air and this accumulates as ice on the coils. The lower the average temperature of the cooling coils, the more moisture will be removed. The drier and cooler air then circulates around the room where it warms and picks up moisture. The more moisture that freezes on the refrigerator coils, the greater the frequency of defrosts cycles and these make good temperature management control more difficult to attain (Bijendra Kumar Singh *et.al.*, 2017). An even distribution of air produces a room with a consistent temperature, but if the flow of cooled air is "short circuited" back to the coiling coils, the areas starved of circulation will become warmer.

Quality control: Most consumers have been disappointed with the quality of fresh produce they have purchased at one time or another. Fruit are to be stored or transported over long distances may have to be picked in an immature state so that the fruit are firm and store or travel well. In recent years, much work has been done to improve the quality of fruit. New varieties have been introduced which gives consumers a wider choice and some, such as new varieties of nectarine have improved flavor. However, despite the best efforts to handle fresh produce in the optimum way there will always be compromises to be made which affect the final quality of the product. Much of the fresh produce we eat must travel many kilometers to reach the central or wholesale market or distribution centre and from there it travels to local or distant retail outlets. Good handling will ensure that the final consumers are satisfied and so will return again to buy that product (Vander Steen, *et al.* 2001).

A wide variety of fungal and bacterial pathogens cause postharvest disease in fruits. Some of these infect produce before harvest and then remain quiescent until conditions are more favourable for disease development after harvest (Bijendra Kumar Singh *et.al.*, 2017). Other pathogens infect produce during and after harvest through surface injuries. In the development of strategies for postharvest disease control, it is imperative to take a step back and consider the production and postharvest handling systems in their entirety. Many preharvest factors directly and indirectly influence the development of postharvest disease, even in the case of infections initiated after harvest. Traditionally fungicides have played a central role in postharvest disease control. However, trends towards reduced chemical usage in horticulture are forcing the development of new strategies. All fruits are storage as the average temperature for the good marketing whereas fruit crops are affected by the many pathogens on postharvest. Thus, proper growth of postharvest technology of fruits is vital for development of India's economy.

Gray mold, incited by *Botrytis cinerea*, causes major postharvest losses in a wide range of crops. Some infections that occur in the field remain quiescent during the growing season and develop after harvest (Gianfranco Romanazzia *et.al.*, 2016). The pathogen is also capable of infecting plant tissues through surface injuries inflicted during harvesting and subsequent handling; these develop during storage, even at 0°C, and spread among products by aerial mycelial growth and conidia. The postharvest decay by this pathogen is controlled by a combination of preharvest and postharvest practices. To minimize postharvest gray mold, control programs rely mainly on applications of fungicides. However, mounting concerns of consumers and regulatory authorities about risks associated with chemical residues in food have led to imposition of strict regulations, the banning of use of certain chemical groups, and preferences by wholesaler, retailers and consumers to avoid chemically treated produce. These developments have driven the search for alternative management strategies that are effective and not reliant on conventional fungicide applications. In this review, conventional and alternative control strategies are discussed including their advantages and disadvantages. They include the use of conventional fungicides, biocontrol agents, physical treatments, natural antimicrobials, and disinfecting agents. Based on examples to control gray mold on specific crops, it is concluded that an integrated management program where adoption of a holistic approach is the key for meeting the challenge of minimizing postharvest losses caused by *B. cinerea*. To optimize the efficacy of treatments, it is essential to understand their mechanism of action as much as possible.

Although applications of conventional fungicides constitute the most common practice for controlling gray mold in the field/orchard or in the packinghouse, their use after harvest on fruits is not allowed in many countries. Their continued use as preharvest treatments has come under increased scrutiny and their future as a control strategy is somewhat questionable. This is because of problems associated with failure to effectively control pre and postharvest gray mold due to development of fungicide resistance, consumers desire to reduce human and environmental exposure to chemicals and increased restrictions imposed by marketing chains and governmental regulatory agencies on the use and food residues of agro chemicals in fresh agricultural commodities. These have been the driving forces for the development of postharvest disease control measures that do not rely on conventional fungicides. Currently, the use of alternative methods as stand-alone treatments for the control of postharvest gray mold, however, does not provide the efficacy and consistency required for commercial situations.

B. cinerea uses several modes of infection to attack fruit and vegetables before and after harvest. To increase control of these infections, it is important to influence the process of infection at different levels: the pathogen, the micro environment, and the host. For example, application of a biocontrol agent or any other alternative method at a time that prevents establishment of the pathogen in the host tissue, given that the attachment of pathogen propagules to the host surfaces and the early stages of germination are critical to successful infection. The micro environment (e.g., surface wounds) can also be altered to directly or indirectly affect the pathogen. The pH and nutritional composition of the infection site can be manipulated by the addition of salts, organic acids, or surfactants/adjuvants. In certain crops, surface injuries can be cured to resist infection by various thermal treatments, and subsequently the chances for infection are lowered. Susceptibility of the commodity (host) may also be reduced by changing its physiology using various treatments to either retard senescence or induce natural resistance (Gianfranco Romanazzia *et.al.*, 2016). It is anticipated that the continuing withdrawal of key synthetic postharvest fungicides from the market, due to exclusion by regulatory agencies or the high-cost of registration, will lead to an absence of effective conventional chemical tools for reducing postharvest losses due to gray mold.

Hence, the use of alternative management methods is expected to gain popularity in the coming years and become more widely accepted as a component of an integrated strategy to manage postharvest diseases. Along with this approach, effective alternative control strategies would rely on elements such as classical microbial antagonists, natural plant resistance, natural antimicrobials which are the product of a biological process and combinations among the above cited methods such as thermal curing treatments, plant growth regulators, ethylene inhibitors, MA, CA and heat treatments. It is essential to reduce the inoculum load and conditions conducive to establishment of infections through well-established cultural and management practices.

The integrative strategies for control of postharvest diseases in pome fruits include effectively inhibiting pathogens growth, enhancing resistance of hosts and improving environmental conditions resulting favourable to the host and unfavourable to the pathogen growth. The strategies that can directly act upon the microbial pathogens may be integrated as, physical + chemical, physical+ biocontrol, biocontrol + chemical and resistance+biocontrol + physical + chemical methods. The possibility of integrating the different effective strategies to achieve higher level of control of postharvest pathogens and to minimize or replace the use of synthetic fungicides has to be explored in certain host–pathogen systems (Nabi, *et.al*., 2017). It is essential to ensure that close coordination, constant surveillance, and efficient technical support for rapid detection and precise identification of microbial pathogens, and feedback on the effectiveness of corrective measures taken to restrict the incidence and subsequent spread of diseases, are available. Avoidance of wounds to the harvested produce is the basic precaution, to be strictly enforced in all cases. Though these methods are nonspecific, they will effectively reduce the chances of infection by microbial pathogens. The strategies that can directly act on the microbial pathogens may be integrated broadly into four combinations viz., physical + chemical, physical + biocontrol, biocontrol + chemical and Resistance + biocontrol + physical + chemical methods.

Various strategies for the management of postharvest diseases have been found to be effective to varying levels under wide range of conditions which interact with each other. The possibility of integrating the different effective strategies to achieve higher level of control of postharvest pathogens and to minimize or replace the use of synthetic fungicides has to be explored in certain host–pathogen systems. The usefulness of integrating different strategies to provide better control of diseases and to obtain safe, disease- and residue-free food products is the need of hour.

8.3 Postharvest IDM Tools: Genetic (resistance), Biological (yeasts, antagonistic fungi/bacteria), Cultural (sanitation, decrease mechanical damage), Physical (sorting, heat treatment, temperature and RH management and Chemical) (chlorinated wash water, fungicides/pesticides)

General do's and don'ts for postharvest IDM

Consider the entire system (production, harvest, postharvest and marketing) while developing disease /pest management strategies.

Pre-harvest: Begin with cultivars offering some natural resistance to the diseases expect to have to deal with in a particular region; Plant only good quality, clean seed or stock; Use appropriate cultural practices during production to assist the produce to avoid and/or resist pest/disease attack (proper planting density. fertilization, irrigation. pH modification, weeding, pruning, thinning, ventilation/air movement through the canopy); Monitor fields/orchards to determine actual pest/disease levels before implementing control; Use a combination of appropriate pest/disease control methods (biological, chemicals, protectants, sanitation practices); Keep fields and orchards free of debris and discarded produce; Eradicate diseased produce.

Harvest: Avoid damage during harvest by handling produce gently ; Harvest at the proper maturity for produce to have the maximum resistance against diseases; Use sharp, clean tools for harvest and trimming processes.

Curing: Cure root, tuber and bulb crops to heal harvest wounds and increase resistance to pests.

Packing house: Sort to remove any damaged, decayed, over mature or under-ripe produce; Wash or clean produce to remove soil and debris and to reduce the amount of innoculum on surfaces; Trim senescent leaves from vegetables and remove dried flower parts from fruits; Use appropriate postharvest treatments to manage pest problems (chemicals, heat, hot water, pesticides).

Packing : Avoid over use of liners that constrict air flow in the package and contribute to condensation (free moisture) and poor cooling efficiency. Use ventilated plastic bags as liners for produce highly susceptible to water loss.

Storage : Avoid ethylene damage to sensitive commodities by using ethylene scrubbers and avoiding mixed lots of produce in storage; Keep produce at its lowest safe temperature for maximum pest management; Avoid chilling injury by keeping sensitive commodities at appropriate moderate temperatures; Keep leafy vegetables, carrots, and cool season vegetables at very high relative humidity (98-100%) to reduce incidence of decay; Store certain fruits at slightly lower RH than commonly recommended in order to reduce decay; Store onions and garlic at low humidity to reduce decay (60-70% RH).

In papaya, Bazie *et al*., (2014) determined the integrated effect of plant extracts and hot water treatment on postharvest anthracnose and banana fruit quality. Artificially inoculated banana fruit were dipped in 10% and 20% (w/v) extracts of four plant species, *viz. Acacia albida, Argemone mexicana, Dovalis abyssinica* and *Prosopis juliflora* at room temperature (20°C), 45°C and 50°C. Fruit treated separately with 20% extracts of *A. albida* and *P. juliflora* at 50°C had reduced anthracnose incidence and severity. These treatment combinations also had the highest percentage of marketable ripe banana fruits. Applying aqueous plant extracts at 50°C enhanced the antifungal activity of extracts without affecting the physico-chemical properties of banana.

Strawberry is delicate and succulent and after harvest it can commonly undergo fungal spoilage. The main strawberry pathogen is *Botrytis cinerea*, followed by *Rhizopus stolonifer*, *Mucor* spp., *Colletotrichum* spp., *Penicillium* spp., which are the major pathogens responsible for postharvest decay of strawberry fruit. The traditional strategy of control of postharvest strawberry decay relies on the application of fungicides during the crop growing cycle. Conventional fungicides are applied around flowering, and treatments can be repeated up to harvest. Nowadays, there are many alternatives to conventional fungicides that are characterized by low impact on the environment and on human health. These include biological control agents, natural compounds, decontaminating agents, physical methods, and their combinations. According to an integrated pest-management approach, management of postharvest diseases starts before cultivation in the field, with the suitable choice of the strawberry variety, place of cultivation, and cultivation technique (Feliziani and Romanazzi, 2016). This continues during the crop development, with weed control, soil sanitation, and management of water and humidity. Careful picking at harvest and then rapid and constant cold storage are the key factors to assure good quality of harvested strawberry fruit. A modified atmosphere that is enriched in ozone, oxygen, or carbon dioxide is frequently used during strawberry storage, to slow down senescence and reduce decay.

In banana, De Costa and Erabadupitiya (2005) developed an integrated method to manage postharvest diseases by combining antagonistic bacteria occurring naturally in the fructosphere with hot water treatment. Twenty-five temperature and exposure time combinations were tested for their ability to control anthracnose and crown rot and their influence on postharvest quality parameters such as peel colour, hardness and Brix value of fruit. Optimum temperature and exposure time for postharvest hot water treatment of banana were determined to be 50 °C and 3 min, respectively. Temperatures higher than 50 °C reduced the colour of fruit peel and gave a pale appearance to fruit even after ripening. Exposure times longer than 5 min substantially reduced the Brix value. A member of the *Burkholderia cepacia* complex, isolated from the fructosphere of banana was effective as an antagonist of postharvest pathogens even after 5 years of storage in sterile distilled water at ambient temperature. The isolate of *B. cepacia* used was different from the already identified species of the *B. cepacia* complex. The most effective concentration of *B. cepacia* was determined to be 10^{10} CFU/ml for in vivo control of anthracnose and crown rot. The most effective control was achieved by repeated dipping of bananas in bacterial solution or by adding a wetter (Tween 20) to the bacterial solution.

Combining hot water treatment with the bacterial antagonist also gave more effective control of anthracnose, crown rot and blossom end rot than using the two treatments individually. Higher efficacy of this integrated method was confirmed further by its greater control of artificially infected anthracnose. This study determined the most effective cell concentration of the antagonistic bacterium to achieve maximum disease control as 10^{10} cfu/ml. Although 10^{11} cfu/ml gave an equal level of control as 10^{10} cfu/ml, the lower concentration was selected because there was no need to use a higher bacterial concentration when a similar result could be achieved with a lower concentration. It also provided valuable options for non-chemical control of postharvest diseases of banana using an antagonistic bacterium and hot water treatment either individually or together.

In citrus against green mold or blue mold, integrated management practices include fungicide bin drenching of harvested fruit on transit trailers, the use of fungicides in aqueous solutions rather than in waxes to improve their performance, the application of sodium bicarbonate or potassium sorbate in heated tanks or through high-pressure washer nozzles, ozonation of storage rooms to oxidize ethylene and retard sporulation of *Penicillium digitatum* and *P. italicum*, and the introduction of new fungicides (pyrimethanil, azoxystrobin, fludioxonil) (Smilanick, 2011). Recently evaluated experimental practices include pre-harvest applications of fungicides and growth regulators, mycofumigation with natural volatiles from the fungus *Muscodor albus*, UV-assisted sorting to remove fruit with mechanical injuries before storage (since these later decay at a high frequency), thermal and chemical disinfection of packinghouses and storage rooms, and the use of phosphites before and after harvest.

The integrated management of postharvest anthracnose in mango under tropical and sub-tropical production conditions requires a sound knowledge of the biology of the pathosystem, the technologies available for field and postharvest control in any given area or region, their economical feasibility and ecological acceptability (Chrys N.Akem, 2006).

A successful integrated management program of mango anthracnose must take into account the factors like the system is rain moisture driven. Infection criteria have been identified and should be used to time fungicide applications rather than continue to rely on calendar based spray schedules; the severity of postharvest anthracnose on mango is the result of cumulative quiescent infections that develop after harvest, as fruit ripens. These are subject to eradication by postharvest treatments; flowering can be managed so that susceptible tissue is produced during the drier months of the year.

This will decrease the probability of fruit infections; several pre harvest and postharvest treatments are commercially available and should always be used in combination wherever and whenever possible and the efficacy of the various treatments is dependent on infection severity and this could vary from season to season and from one locality to another (Arauz, 2000).

Pre and postharvest management strategies were integrated to reduce stem end rot (SER) incidence and extend saleable life of mango fruits. Preharvest management focused on the development and improvement of fungicide spray program, while postharvest management aimed to develop alternative interventions aside from hot water treatment (HWT). Field evaluation of systemic fungicides, namely azoxystrobin, tebuconazole, carbendazim difenoconazole and azoxystrobin+difenoconazole, reduced blossom blight severity and improved fruit setting and retention, resulting in higher fruit yield but failed to sufficiently suppress SER incidence. Based on these findings an improved fungicide spray program was developed taking into account the infection process of SER pathogens and fungicide resistance. Timely application of protectant (mancozeb) and systemic fungicides (azoxystrobin, carbendazim and difenoconazole) during the most critical stages of mango flower and fruit development ensured higher harvestable fruit yield and minimally lowered SER incidence. Control of SER was also achieved by employing postharvest treatment such as HWT (52-55°C for 10 min), which significantly prolonged the saleable life of mango fruits (Akem *et.al.*, 2013). However, extended hot water treatment (EHWT; 46°C pulp temperature for 15 min), rapid heat treatment (RHT; 59°C for 30-60 sec), fungicide dip and promising biological control agents failed to satisfactorily reduce SER and prolong saleable life. In contrast, the integration of the improved spray program as preharvest management practice, and postharvest treatments such as HWT and fungicide dips (azoxystrobin, 150-175 ppm; carbendazim, 312.5 ppm; and tebuconazole, 125-156 ppm), significantly reduced disease and extended marketable life for utmost 8 days.

An integrated approach to postharvest disease management tries to reduce fruit susceptibility to infection, reduce pathogen inoculum potential, and apply control materials effectively. Orchard sanitation, removing decaying organic matter from the orchard floor and from within trees is critical to reducing pathogen inoculum. Reduction of fruit nitrogen content through minimal and timely fertilizer application, increasing fruit calcium content through foliar sprays, applying fruit resistance stimulants, harvesting without injuring the fruit and harvesting early in the maturity range all can enhance fruit resistance to infection (David Sugar, 2016).The efficacy of fungicide or biocontrol treatments diminishes rapidly as the time between fruit inoculation and protective treatment increases. Fungicide application in the orchard before harvest or promptly after harvest can significantly reduce incidence of postharvest diseases.

The integrated management of storage rot of ginger was carried out by using fungicides, biocontrol and hot water treatment (Sushma Sharma *et.al.*, 2017). In present investigation, the samples were collected different region of Himachal Pradesh, A total of six fungi were isolated from stored diseased rhizomes and identified using ITS gene sequencing as *Aspergillus niger, Fusarium oxysporum, Penicillium digitatum, Pythium ultimum, Rhizopus oryzae* and *Verticillium chlamydosporium*. The relative frequency distributions of fungi causing storage rot of ginger have been studied. Among different fungi causing storage rot of ginger in storage pits and heaps, the frequency occurrence of *Fusarium oxysporum* was found maximum (51%) followed by *Pythium ultimum*. In integrated management studies, the minimum incidence of storage rot was recorded in treatments when used in combination, Carbendazim + HWT (Hot water treatment) + *Trichoderma harzianum* (7.20%) and HWT + Mancozeb + *T. harzianum* (12.23%) after 30 days of storage as compared to control (39.30%). Similarly, highest rhizome recovery (86.53%) was obtained in treatment Carbendazim + HWT + *T. harzianum* in comparison to control. Since, there are no single effective management practices against the storage rot disease so far, a well coordinated effort is required to develop an integrated disease management program that will help to minimize the damage caused by the disease.

The frequency occurrence of *Fusarium* was most dominant in ginger rots. Among various diseases storage rot is a serious menace of ginger due to its worldwide occurrence. Storage rot of ginger caused by *Fusarium oxysporum* f. sp. *zingiberi* in storage is one of the major problems for successful ginger cultivation. Seven fungi were isolated from diseased rhizomes and frequency occurrence of *F. oxysporum* was maximum (50%) among fungi causing storage rot of ginger in storage pits and heaps. Disease management practice alone is not sufficient until integrated with other components, as disease is a complex phenomenon governed by many factors. The integration of effective fungicides with plant oils, botanicals and antagonists were important for the successful management of plant diseases. Taking into consideration we use integrated disease management practice and results indicate that minimum incidence of disease was recorded with combination of carbendazim + HWT + *T. harzianum* in a compatible manner.

8.4 Integrated approach in a nut shell

I. Pre harvest care

a. Phytosanitation: Most of the primary inoculums of much postharvest disease are carried from the orchard, therefore preharvest cultural practices will influence postharvest diseases problem, and phytosanitation would provide a simple and effective measure to keep the incidence of diseases low, fallen fruits, infected leaves and dead twigs can harbor a large quantities of inoculums. The reproductive bodies of the pathogen on the plant debris can hibernate and can perpetuate the infection cycle (e.g. Acervulus, Pycnidia, Chlamydospores, Sclerotia etc.). e.g. Anthracnose of Guava (*Gloeosporium psidii*), Anthracnose of mango, Crown rot of banana, *Colletotrichum* and *Botryosphaeria* rot of apple (Yadav *et.al.,* 2013).

b. Pre-harvest chemical treatments: The effective method to reduce infections initiated in the field, including quiescent infections, is the application of broad spectrum protective fungicides to the developing fruit on the plant, in order to prevent the infection e.g. Copper oxychloride for citrus brown rot; Benomyl for oranges stem end rot, Carbendazim for Anthracnose and stem-end rot of mango, banana and other tropical fruit crops.

c. Resistant cultivars : The first pre harvest factor which may affect postharvest quality is the cultivar. In fact, one of the aims of plant breeding and genetic engineering is to incorporate resistance genes in new varieties of crop plants. Differences in cultivar characteristics can markedly affect the keeping quality of the fresh produce e.g. melons with a thick skin and raspberries with a firm texture are better than others to withstand the rigors of harvesting and handling and have longer storage lives (Yadav et.al., 2013).

II. At harvest care

a. Maturity at harvesting : In general the maturity at the time of harvesting is a factor which can influence the postharvest losses. Both over mature as well as pre-mature harvesting can make the fruits more prone to post- harvest infection. Harvest the fruits at proper stage by considering thesize, shape, colour, flesh firmness, sugar, starch and oil content.

b. Harvesting technique : Choosing the most appropriate time and technique of harvesting is most important. Fruits and vegetables require a careful harvesting technique because of their delicate nature; simple manual harvesting methods however, are usually quite effective and satisfactory. For example the bamboo poleharvester with a net attached to its top, used in India is excellent for harvesting mango fruits. Whatever harvesting method is adopted, the basic care that needs to be exercised is avoidance of causing punctures or injuries to the fruits during the process, since the tissue thus injured often becomes the focal point of infection at later stages. Harvesting by hand is the predominant method for fruits and vegetables.

III. Postharvest care

a. Handling and packaging : Careful handling of fruit is the most important step that has to be taken at all postharvest stages. Sorting and grading of the harvested fruits is good farm house practice. Eliminating the fruits showing injury or early symptoms of a disease, revealed at this stage by careful inspection can substantially reduce the intensity of disease at later stages. Packing practice has to be evolved to suit the individual consignment, aiming at the maximum perfection as well as cost effectiveness. Before packing remove superficial tissues from banana bunches for reducing the incidence of crown-rot disease. Discourage the common Indian practice of using the leaves from respective fruit plants for packaging. Instead of paddy straw, shredded paper or saw dust has been suggested. In modern practice corrugated boxes/ fiber board packs/molded plastic trays/ wooden, crates are findings their places. Use impregnated paper for wrapping individual fruits such as Mango, Papaya Banana, Guava with paper (Yadav *et.al.,* 2013).

b. Care during transport : Horticultural produce may be transported by road, rail, sea or air, sometimes for many thousands of miles. A well planned schedule should be worked out for transportation of fruits according to the mode of journey, the nature of fruits and distance as well as duration of the journey. The fruits of guava and papaya are comparatively more prone to injury and hence need special care during transportation and if possible temperature and other ambient conditions may be controlled. High value perishables can be sent by air, either in wide bodied freights or in the cargo holds of passenger aircraft (Yadav *et.al.,* 2013).

c. Storage : Low pressure storage, cold storage and modified atmosphere storage are aimed out preventing the perpetuation of pathogen and spread of the disease.

i. **Storage disinfection of warehouses:** Debris from warehouses/cold storages should be removed, the walls and floors should be washed with bleaching powder, copper sulfate, etc to eradicate the pathogens which are surviving in the store structures.

ii. **Low temperature storage:** Low temperature storage is less suitable for tropical fruits like mango, banana, sapota and guava which are sensitive to chilling. Papaya fruit inoculated with *Colletotrichum gloeosporiodes* and when stored at low temperature (10°C for 21 days) develops less anthracnose. Suitable temperature for Mango storage is more than 13°C, temperature below 13°C causes chilling injury, for Guava more than 15°C, for Banana more than 10°C and for Papaya more than 6°C is required.

iii. **Cold storage:** Storage at low temperature is the main method for reducing deterioration of harvested fruits and vegetables. Low temperature affects both the host and the pathogen simultaneously. They prevent moisture loss from the host tissues and consequent shriveling; they retard metabolic activity and delay physiological changes that lead to ripening and senescence.

iv. **Modified atmosphere storage:** This method have found effective in extending the shelf life of harvested fruits. In storage maintaining a low temperature, an adequate humidity, low oxygen, low ethylene concentration are favorable. If atmosphere of Mango storage is modified as 2% CO_2, 6% O_2 and 13°C it can be stored for longer.

d. Postharvest treatment with chemicals: Since open wounds, created during harvesting, handling and packaging are the major sites of invasion by postharvest wound pathogens, the protection of wounds by chemicals will considerably decrease decay in storage. Many chemical compounds have been used as part ofpostharvest treatment of tropical fruits for the retardation of microbial infection. e.g. Inorganic copper or sulphur compound, Phenolic compounds, Dithiocarbamates, Antibiotics, Systemic fungicides (benomy/ TBZ), Iprodine, Imazalil, Prochloraz, Fosetyl, etc. Contents of major and minor nutrients also influence on storage rots, for example higher N in apple increase the decay due to Gloeosporium and Boron deficiency in Aonla causes blackening of fruits (Yadav *et.al.,* 2013)..

e. Physical approaches

i. **Fruit irradiation:** Radiation treatments are meant for killing or weakening the quiescent pathogen thereby improving the shelf life of the fruits. Among the ionizing radiation gamma irradiation has been most successfully used for inhibiting the growth of pathogens inside fruit tissue. Irradiation ranging between 0.3 KGY and 1.2 KGY reduces the incidence of postharvest storage disease of mango (*Gloesporium & Botryodiplodia)* but a dose above 0.6 KGY results in lenticels spotting and surface discoloration of the fruits. e.g. peaches, strawberries guava fruits (infected by *C. gloeosporides* to radiation at 100 Krad (1 KGY). *Alternaria, Botrytis* and *Stemphylium* can be controlled in vegetables in greenhouses with special UV-absorbing polyvinyl film.

ii. **Washing fruits:** After harvest before packing/marketing individual fruits or fruit bunch should be washed in abundant flow of clean water to remove plant debris/trash/latex and pathogen which are responsible for causing diseases in transit (Yadav *et al.*, 2013).

iii. **Hot water treatment:** Hot water treatment is promising and has been used with success in eradicating or suppressing the development of fungi/bacteria on the fruit surface as well as those situated just below the surface as a result of pre harvest infection.

iv. **Heated forced air treat/Aerated steam/hot air:** Recently heated forced air treatment has gained some importance in the postharvest treatment of some fruits. e.g. Mango at 48°C for 15 minutes considerably reduce the anthraonose /stem end rot (*B. theobromae*) and also eradicate the infection of fruit fly.

v. **Drying fruits:** Many fruits can be stored for longer time after drying and can be kept free of disease because moisture is kept below a certain level during storage e.g Grapes, plums, dates and figs, slices of fleshy fruitsapples, peaches and apricots, etc.

f. Biological approach: Employing microbial antagonism offers one of the most effective means of controlling the postharvest diseases without any adverse effect on the environment and the consumer. Control of pathogen through employing antagonist is popularly known as bio-control. Spraying with suspensions of *T. harzinum, T. viride, Gliocladium roseum* and *Paceilomyces variotii* resulted in a partial control of *Botrytis* in strawberry fruits and *Alternaria* rot in lemons.

g. Fruit Coating with Vegetable / Edible Oil: Use of vegetable oils in plant disease control is a relatively recent development in the field of plant pathology. e.g. Castor, Linseed, Mustard, Sunflower, Safflower, Groundnut, mineraloils, Palmarosa, red thyme / liquid paraffin, etc. coated on harvested fruits to prevent entry of pathogen as well as decreasing respiration due to their antifungal activity.

h. Homoeopathic drugs/growth regulator: In recent years some homoeopathic drugs have been used for the control of post- harvest spoilage of fruits. For example, Lyeopodium 140 for *Pestalotia mangiferae*, Kaliodatum 200 for Stem-end rot and *Rhizopus* rot of mango fruits and Methyljasmonate-200 for *Fusarium* fruit rot of banana.

i. Natural plant products: The study of green plants of their antimicrobial activity has received little attention though such activity was known since ancient times (Yadav *et.al.,* 2013). Neem leaves were kept in woolly cloths and grain store houses for preventing deterioration by molds & pests and still today it is common practice. Baskets of fruits and vegetables are lined with neem leaves for their protection against microbial attacks. It is believed that toxic substance emitted by the leaves keep the air remarkably free of pathogenic microorganisms.

Disease management programs have to be developed specifically for each crop and largely depend on epidemiological and etiological information. Because this data is lacking for many tropical commodities, a great deal of research still needs to be conducted. Lacking this information, the availability and utilization of simple cultural methods previously discussed (adequate sanitization, refrigeration, controlled atmosphere facilities, etc.) could greatly reduce postharvest losses due to postharvest disease. The use of fungicides as an effective tool again requires a basic understanding of the pathogens involved in the disease and a variety of factors that affect pathogenesis. Chemical fungicides, however, have come under scrutiny and their use as a control strategy in the long term is somewhat questionable. This is because of problems associated with *(1)* failure to effectively control decay in several harvested fruit commodities due to the development of fungicidal resistance, *(2)* consumers' desire to reduce human and environmental exposure to chemicals, and *(3)* increased restrictions imposed by governmental regulatory agencies on the use of agro-chemicals and/or allowable residues. These issues have been the driving force for the development of postharvest disease control measures that do not rely on chemical fungicides. Indeed, reduced chemical postharvest disease control measures have now become an economic imperative, not just an option. Currently, the use of alternative non-chemical control methods as stand-alone treatments, however, does not provide the efficacy and consistency required for commercial postharvest situations. At this time, a more realistic scenario to combat decay of harvested commodities is the use of an integrated control approach combining biological, GRAS chemicals, and physical.

8.5 Holestic approach: Integration of disease management strategies in fruits and vegetables emerged an answer to shortcomings of reliance on chemical control. It deals with integration of all available disease management methods to manage the diseases effectively and economically to satisfy human needs at the same time guarding the quality of harvested produce and environment. Several studies have highlighted the advantages of postharvest application of biological control agents over field or soil application. The major advantages are the convenience of bringing the antagonist in contact with the commodity as compared to its addition to the soil and the possibility of acting under controlled conditions, created and maintained during storage. Likewise, it is possible to introduce *Bacillus subtilis* into the wax applied to peaches in the packing house to protect them from the brown rot caused by *Monilinia fructicola.*

The antagonistic yeast *Pichia guilliermondii* can be introduced into the wax mixture applied to citrus fruit in packing house. Compatibility between a microbial antagonist and a synthetic fungicide offers the option of using the antagonist in combination with reduced levels of the fungicide.Applying the yeast antagonist, *Pichia guilliermondii* to citrus fruit in combination with substantially reduced concentrations of thiabendazole reduced *Penicillium digitatum* decay to a level similar to that achieved by the currently recommended concentration of thiabendazole applied alone. Thus, an integrated pest management system adaptation provides effective pest control and maintains very low levels of chemical residues.

The intensive studies on biocontrol of postharvest diseases have led to the registration of two biological products for commercial postharvest applications on citrus fruits i.e. Aspire, which is *Candida oleophila*, and Bio-Save 1000 from *Pseudomonas syringae*. The biological agent must have low sensitivity to any of the supplemented chemical fungicides. Combining Aspire with each of the chemicals used improved the results; sometimes combinations with a low rate of fungicide were sufficient to achieve effects similar to those obtained by the chemicals at standard rates. Safe compounds or natural products of plant origin have been suggested as alternatives to synthetic and conventional fungicides could also be used in combination with biocontrol agents, complementing their activity. Pathogens treated with such antifungal substances might be weakened and become more vulnerable to the antagonist activity. The integration of postharvest biocontrol into modern production, storage and handling systems must begin before harvest. Several preharvest factors that affect fruit quality may have profound effects on the efficacy of postharvest biological control agents.

Preharvest calcium sprays during the growing season of apples and pears can increase fruit firmness, decrease the incidence of certain disorders and enhance resistance to postharvest infection. Calcium amendments and postharvest application of some antagonistic yeasts can be additive in reducing fruit decay and significantly increase disease control compared with either treatment alone. The advantages in increased firmness, enhanced resistance to Postharvest decay and enhanced biocontrol efficacy under some circumstances reflect the multiple benefits of integrating postharvest biological control with cultural and production practices.

Postharvest factors may have a major impact on the effectiveness of biological control. Fruit maturity at harvest and the application of antagonists is affecting postharvest biological control. Delayed picked and over matured fruits are more susceptible to decay than are fruits picked at optimal storage maturity. Roberts (1990, 1994) found that fruit maturity in apples and pears markedly affected biocontrol efficacy of antagonistic yeast *Cryptococcus,* while excellent control was achieved on freshly harvested fruit and treatments of ripened fruit gave much lower levels of control. Temperature management is a critical factor in the maintenance of fruit quality and pathogen development, may enhance biological control of storage decay. Janisiewicz (1991) demonstrated that as the storage temperature of apples and pears decreased, there was a reduction in the concentration of pyrrolnitrin (a metabolite of *Pseudomonas cepacia* and other *Pseudomonas* spp.) needed to protect the fruit from gray mould (*Botrytis cinerea*) and blue mould caused by *Penicillium expansum.*

An integrated strategy to control postharvest decay in pome and stone fruits has been advanced in recent years comprises several pre and postharvest components *viz*., alteration of fruit nutrient status, calcium applications as sprays during the growing season, lower fruit nitrogen content have been associated with reduced disease severity and influence the susceptibility of pome fruits to decay. Controlled atmosphere storage with reduced O_2 and elevated CO_2 can reduce the severity of Postharvest fungal decay in apples by inhibiting fruit senescence and maintaining host resistance to infection. Integration of early harvest, low fruit nitrogen, high fruit calcium, yeast or yeast + one-tenth of the dose of thiabendazole along with a controlled atmosphere (2% O_2, 0.6% CO_2) were found to reduce blue mould severity in picked pears. All the components of the integrated approach were found to be compatible with thiabendazole fungicides used for the management of Postharvest decay and blue mould (*P. expansum*) in harvested pears.

Several yeast species such as *Cryptococcus laurentii*, are capable of colonizing wounds of pear fruit under conditions of low temperature, ambient or reduced O_2 and CO_2 therefore, can be integrate into Postharvest strategies. The two postharvest treatments, biological control and modified atmosphere packaging were combined with preharvest iprodione spray; the incidence of the brown rot was reduced from 41.5 per cent in the control to only 0.4 per cent. A combined strategic approach was elaborated by various worker to control wound infecting pathogens by a series of treatments including disinfection of the fruit surface and environment, eradication or suppression of fungal spore germination at wound sites by a combination of fungicides and reduction of wound susceptibility to infection by the addition of biocontrol antagonists which act as protecting agents. It is being apparent that a multifaceted integrated approach has been advantageous over the use of solitary measures. Thus, the application of antagonists, induced resistance, physical treatments, chemicals, natural and safe fungicides etc. can be utilized in a unified way to provide greater consistency and efficacy to management of postharvest diseases.

The effect of fungicides, chemical, oils, wax and bio-agent against postharvest pathogens of Nagpur mandarin was evaluated for their inhibitory effect in vitro (Ingole *et al*., 2018) Cent per cent mycelial growth inhibition of *Colletotrichum gloeosporioides*, *Penicillium digitatum* and *Trichoderma viride* were recorded in Benomyl @ 0.05 % and Benomyl @ 0.05 % + Potassium sorbate @ 2% and Hydrogen peroxide @ 10 per cent. Hydrogen peroxide @ 10 per cent recorded maximum mycelial growth inhibition (87.90%) of *Geotrichum candidum*. Complete growth inhibition (100%) of *T.viride* was achieved in Carbendazim @ 0.1%. Carbendazim @ 0.1%, Benomyl @ 0.05%, and Hydrogen peroxide @ 10% checked the cent per cent mycelial growth of *Aspergillus Niger*. Carbendazim @ 0.1% inhibited the mycelial growth of the *C. gloeosporioides* (64.07%). *Saccharomyces cerevisiae* recorded maximum growth restriction of *Geotrichum candidum* (14.94%). Benomyl @ 0.05 per cent inhibited 64.81 per cent mycelial growth of P. digitatum. *S. cerevisiae* 10^9 cfu/ml restricted mycelial growth (25.43%) of *A. Niger*. Carbendazim @ 0.1 per cent was found superior in inhibiting the mycelial growth of *T.viride* (32.96%)

The effectiveness of sodium bicarbonate (SB), sodium carbonate (SC), sodium silicate (SS), potassium bicarbonate (PB), potassium carbonate (PC), potassium sorbate (PS), calcium chloride (CC), and calcium chelate (CCh) against naturally occurring postharvest decay on 'Comune' clementine and 'Valencia late' orange fruit was investigated (Khamis Youssef *et.al*., 2012).. Aqueous salt solutions (2%, w/v, 20 hl ha^{-1}) were applied according to three strategies: (i) by spraying before harvest, (ii) by dipping after harvest, and (iii) by the combination of pre- and postharvest applications. Decay was assessed after two months at 4 ± 1 °C (oranges) or 6 ± 1 °C (clementines) and 95–98% RH, followed by 7 days of shelf life at 20 ± 2 °C. For both species, preharvest sprays and the combination of pre- and postharvest applications were more effective in suppressing decay than postharvest dipping. With regard to preharvest application, several salts completely inhibited the incidence of decay as compared to the water control, namely, SC and PC on both species, and SS on 'Valencia late' oranges. In combined applications, all salts were effective in reducing the decay as compared to the water control with an efficacy varying between 66–100 and 78–100% for oranges and clementines, respectively. When salts were applied after harvest, the activity was in general less pronounced, SC and PC being the most effective on both species. In *in vitro* tests, the minimum inhibitory concentration (MIC) for both *Penicillium digitatum* and *P. italicum*, was achieved at 0.25% SB, SC, PB, PC, PS, and SS. The filamentous fungal population on fruit treated once in the field and with the double treatment was reduced as compared to the water control, whereas no statistical differences were observed for postharvest application. Based on these results, field application of salts can be considered a useful strategy to be included in an integrated approach for controlling postharvest diseases of citrus fruit.

Practices in the citrus industry evolve rapidly to minimize postharvest decay losses, mostly a result of green mold or blue mold, caused by *Penicillium digitatum* or *P. italicum*, respectively. Current problems include the continued development of fungicide resistance among these pathogens, and compliance with emerging secondary residue standards and other privatized regulation of production and postharvest practices, the rising popularity of new cultivars (such as mandarin oranges, where natural decay resistance is low) and the decline of others (such as Valencia oranges, where natural decay resistance is high), and marketing issues, including exceptionally long storage for distant markets or a requirement for food safety interventions, such as chlorination at high rates. (Smilanick., 2011). Recently (since 2000) introduced practices in California include fungicide bin drenching of harvested fruit on transit trailers, the use of fungicides in aqueous solutions rather than in waxes to improve their performance, the application of sodium bicarbonate or potassium sorbate in heated tanks or through high pressure washer nozzles, ozonation of storage rooms to oxidize ethylene and retard sporulation of *Penicillium digitatum* and *P. italicum*, and the introduction of new fungicides (pyrimethanil, azoxystrobin, fludioxonil). Recently evaluated experimental practices include pre-harvest applications of fungicides and growth regulators, mycofumigation with natural volatiles from the fungus *Muscodor albus*, UV-assisted sorting to remove fruit with mechanical injuries before storage (since these later decay at a high frequency), thermal and chemical disinfection of packing houses and storage rooms, and the use of phosphites before and after harvest.

Practices in the citrus industry evolve rapidly to minimize postharvest decay losses, mostly a result of green mold or blue mold, caused by *Penicillium digitatum* or *P. italicum*, respectively. Current problems include the continued development of fungicide resistance among these pathogens, and compliance with emerging secondary residue standards and other privatized regulation of production and postharvest practices, the rising popularity of new cultivars (such as mandarin oranges, where natural decay resistance is low) and the decline of others (such as Valencia oranges, where natural decay resistance is high), and marketing issues, including exceptionally long storage for distant markets or a requirement for food safety interventions, such as chlorination at high rates.(Smilanick.,2011). Recently (since 2000) introduced practices in California include fungicide bin drenching of harvested fruit on transit trailers, the use of fungicides in aqueous solutions rather than in waxes to improve their performance, the application of sodium bicarbonate or potassium sorbate in heated tanks or through high pressure washer nozzles, ozonation of storage rooms to oxidize ethylene and retard sporulation of *Penicillium digitatum* and *P. italicum*, and the introduction of new fungicides (pyrimethanil, azoxystrobin, fludioxonil). Recently evaluated experimental practices include pre-harvest applications of fungicides and growth regulators, mycofumigation with natural volatiles from the fungus *Muscodor albus*, UV-assisted sorting to remove fruit with mechanical injuries before storage (since these later decay at a high frequency), thermal and chemical disinfection of packing houses and storage rooms, and the use of phosphites before and after harvest.

The role of competition for iron and production of cell wall degrading enzymes by the antagonists *Metschnikowia pulcherrima* MACH1 and *Aureobasidium pullulans* PL5 against *Botrytis cinerea, Monilinia laxa* and *Penicillium expansum* were studied on postharvest fruits by Spadaro,et.al., (2011). In presence of lower concentration of $Fe3+$, MACH1 showed higher biocontrol activity. In absence of $Fe3+$, MACH1 exhibited the highest antimicrobial activity, but sufficient $Fe3+$ enabled the disappearance of the activity, suggesting that competition for iron played a key role in the biocontrol activity of MACH1 against the pathogens. In Lilly-Barnett minimal salt medium with the fungal cell walls of the pathogens as sole carbon source, PL5 produced exo-chitinase, endo-chitinase, and ®-1,3-glucanase. The extracted crude enzymes produced by the antagonists showed a high activity in inhibiting the growth of the pathogens in vitro. It was found that competition for iron was the main mode of action of MACH1 and that production of chitinase, glucanase were involved in the biocontrol activities of PL5.

To achieve the full potential of postharvest biocontrol microorganisms as a viable commercial technology, fundamental knowledge on their mechanisms of action is crucial. After more than 20 years of postharvest biocontrol research there is still limited understanding of interactions taking place between the host, the pathogen, and the antagonist mainly due to difficulties associated with the study of this complex system. Recent research by Macarisin *et.al.,* (2011) showed that various antagonistic yeasts used to control postharvest diseases (*Metschnikowia fructicola, Candida oleophila, Candida saitoana,* and *Pichia guilliermondii*) secreted high, transient levels of superoxide anion as they colonized wounded host tissue and intact surfaces of apple, peach and citrus fruits. The application of *Metschnikowia fructicola* and *Candida oleophila* into citrus and apple fruit wounds induced a significant ($P \leq 0.05$) increase in hydrogen peroxide accumulation in host tissue. Importantly, living cell of *M. fructicola* were detected on the third day after inoculation in fruit wounds exhibiting a high level of host-generated hydrogen peroxide indicating the ability of this yeast to tolerate elevated levels of host ROS. ROS-tolerance may be an essential characteristic of effective yeast antagonists. These data, together with our earlier finding on the importance of hydrogen peroxide in the mechanism of fruit defense against postharvest pathogens, indicate that the ability of yeast antagonists to self-generate and possibly stimulate an oxidative response in host tissue could be one of the major factors underlying the performance of a biocontrol agent.

Traditionally, chemical fungicides have been used to preserve the quality of fruit and vegetables over extended periods of storage or transportation. However, increased global concern about pesticide residues and the reduced efficacy of chemicals due to pathogen resistant strains, have forced producers to evaluate safer alternatives for controlling postharvest diseases in the context of sustainable agriculture. Several means, such as natural compounds of animal and plant origin, organic and inorganic salts, antagonistic microorganisms, and physical means, represent the approaches recently evaluated to ensure fruit quality and safety. Unfortunately, under commercial conditions these means as standalone treatments rarely yield satisfactory control levels. Lack of a high activity, low persistence, a narrow spectrum of activity, and a failure to control previously established infections can be considered the main limiting factors. Thus, it is generally accepted that a combination of various methods is necessary to improve the efficacy of these alternatives. In this regard, the use of alternative control practices involves the use of both pre- and postharvest strategies that have the potential to reduce latent, quiescent, and incipient infections; pathogen inoculum in the environment; and increase innate resistance in fruit. It has been shown that integrated approaches involving applications of microbial antagonists or other non-chemical means both before and after harvest can provide rates of disease control comparable to or better than synthetic fungicides, due to additive or synergic effects (Ippolito *et al.*, 2011).

Postharvest diseases constitute a serious problem for avocado commercialization. Colhetta *et.al.,* (2018) evaluated the effect of conventional and alternative products in controlling diseases affecting 'Hass' avocados in the field and in the postharvest by carrying out physicochemical characterization of fruits subjected to postharvest treatments. In the field, besides the management adopted by the farmer, seven products were sprayed three times during fruiting for evaluation. Postharvest products were diluted in water or in oxidized polyethylene wax and shellac. Water treatments with potassium phosphite, Soil-Set®, chlorine dioxide, thyme essential oil, sodium bicarbonate, lemon grass essential oil and thiabendazole reduced the incidence of diseased fruits, and anthracnose, the main disease, was controlled with sodium bicarbonate, lemon grass essential oil and thiabendazole. Greater soluble solids content was found for control (water), chlorine dioxide, acibenzolar-S-methyl and thiabendazole. For the products that reduced anthracnose, there was no correlation between the disease and the physicochemical parameters, evidencing that the disease control is not associated with delayed ripening. For wax treatments, diseases were not controlled, and the fruits presented lower titratable acidity with thyme essential oil, sodium bicarbonate, control (wax), acibenzolar-S-methyl and lemon grass essential oil. Control and thyme essential oil were highlighted for maintaining the green coloration of the fruit skin for the shortest period. Under field conditions, azoxystrobin, thiabendazole, difenoconazole+azoxystrobin and acibenzolar-S-methyl+azoxystrobin reduced the occurrence of diseased fruits, while anthracnose control was only obtained with azoxystrobin.

Two biocontrol agents, *Muscodor albus* and *Hanseniaspora uvarum*, have shown equal or better effectiveness than conventional means to control gray mold of table grapes in laboratory scale experiments (Romanazzi, *et.al.*, 2011). Currently, the bottleneck regarding the commercial use of biocontrol agents is that the registration process is comparable to that of fungicides, with similar costs but often with a narrower market. Natural antimicrobials, such as salts, chitosan, and plant extracts, demonstrated good results and often were applied in semi-practical or practical conditions. Several GRAS-classified sanitizers were tested to extend postharvest storage of table grapes, including acetic acid, electrolyzed oxidizing water, ozone, and ethanol, although the GRAS status of some of these compounds is dose-dependent or questionable. Physical means in reference to variations in temperature, radiation, pressure or changing atmospheric composition are all postharvest practices which require significant adaptation by the industry which is accustomed to minimal intervention during harvest. Overall, the use of ozone and calcium chloride are two successful examples moved to practical application. Improved understanding of the various methodologies and their mode of action will lead to further optimization of the treatments and to generation of novel combinations to control postharvest decay of table grapes.

In summary, effective and consistent control of storage diseases depends on integration of the following practices:

- Select disease-resistant cultivars wherever possible.
- Maintain correct crop nutrition by use of leaf and soil analysis.
- Irrigate based on crop requirements and avoid overhead irrigation.
- Apply preharvest treatments to control insects and diseases.

- Harvest crop at the correct maturity for storage.
- Apply postharvest treatments to disinfest and control diseases and disorders on produce.
- Maintain good sanitation in packing areas and keep dump-water free of contamination.
- Store produce under conditions least conducive to growth of pathogens. Integration of cultural methods and biological treatments with yeast bio-controls has been studied (Bijendra Kumar Singh *et.al.*, 2017). These results demonstrated that unrelated cultural and biological methods that influenced fruit decay susceptibility can be combined into an integrated program to substantially reduce decay. As a general rule, alternatives to chemical control are often less effective than many fungicides. Therefore, it will generally be necessary to combine several alternative methods to develop an integrated strategy to successfully reduce postharvest decay.

8.6 Integrated management in stored grains: Effective management of stored-grain pests and diseases with minimal pesticide use requires an integrated approach combining sanitation, monitoring, and other preventive practices. Cereal panicles (the grain bearing heads) are infected and damaged by field fungal pathogens during crop growth and development, and harvested grain is constantly at risk of infection from postharvest pathogens during the entire grain processing and storage period (Terry Mabbett, 1996). Loss in grain yield from fungal infection and associated physical deterioration is high but the ability of some fungi to synthesise chemicals called mycotoxins and left as 'calling card' contaminants in grain is a much more sinister and deadly dimension. Mycotoxins are fungal metabolites, the 'signature' chemicals for specific plant pathogens. They are toxic to livestock and humans at trace concentrations, measured in parts per million (ppm) and parts per billion (ppb), and cause a wide range of diseases generically known as 'mycotoxicosis'. Mycotoxins are supremely stable even withstanding digestion and passing along the food chain to appear in damaging concentrations in milk and other farm animal products.

Fungi like *Fusarium* species are versatile pathogens which infect the panicles and are subsequently carried into store on harvested grain along with their mycotoxins. Some continue to grow and produce mycotoxins on stored grain alongside the classic postharvest, storage moulds like *Aspergillus* and *Penicillium* each with their own signature mycotoxins. Store hygiene including the programmed disinfection and disinfestation of all transporting, loading and storage equipment is essential to prevent any infection carryover from one grain store load and into the next.

i. **Grain cleaning :** Grain cleaning begins with threshing during combine harvesting and is completed postharvest by grain cleaning using aspiration equipment and vibrating screens or by gravity separation. Physically intact and healthy cereal grains are separated and isolated from chaff, straw, weed seeds and damaged or diseased cereal grains. Physical contamination will invariably reduce the market value of grain, but the really serious damage is caused by the transfer of mycotoxin producing fungi from the field and into the grain store.Ergot disease is caused by *Claviceps purpurea* with a signature mycotoxin that has hallucinatory effects. Ergotism is one of the oldest known mycotoxicoses. One of the currently most widespread and damaging mycotoxin-producing fungal pathogens is *Fusarium graminearum* responsible for 'head blight' or 'scab' disease in wheat, barley, oats and rice or ear-rot disease in maize. In addition to reducing grain yield and quality the pathogen produces high levels of mycotoxins, namely the trichothecenes e.g. T-2 and HT-2 toxins [Type A] and deoxynivalenol [Type B] and the oestrogenic mycotoxin zearalenone (Terry Mabbett, 1996).

ii. **Equilibrium moisture content of stored grain :** Damage to grain in store happens is due to interacting biotic and abiotic factors. The biotic factors are fungal pathogens (and bacteria, mites and insect pests) and the grain itself which are all affected by various abiotic factors that control metabolic activity and growth. Temperature and moisture are the two key abiotic factors governing grain integrity and longevity. Maintaining these at optimum levels inside and outside of the grain is the secret to successful and cost effective storage. Two salient points underpin the safe storage of cereal grain include, during storage, moisture inside the grain kernels establishes equilibrium with the air outside to generate a relative humidity level that may favour and encourage the growth and metabolic activity of deteriorative organisms, including those fungal moulds which synthesise mycotoxins. Secondly, different deteriorative organisms which affect stored cereal grains need different levels of relative humidity for normal growth and development. Bacteria, storage moulds (including mycotoxin producers) and storage mites require relative humidity level above 90 per cent, 70 per cent, 60 per cent, respectively, while insects depending on species need relative humidity levels within the range of 30 to 50 per cent.

However, both grain moisture level and the relative humidity of the air surrounding grain are dependent on temperature. For example, if the temperature of an air sample having a relative humidity of 50 per cent is increased by 5°C from 25°C to 30°C then it's original relative humidity of 50 per cent will fall to 38 per cent. Grain storage specialists used this information to construct a relationship between grain equilibrium moisture, relative humidity and temperature for selected cereals as shown in Table 12. Grain moisture contents in equilibrium with 70 per cent relative humidity are shown in bold type. These are the levels beyond which most stored grain begins to show microbial spoilage and damage.

iii. **Grain drying:** Grain moisture contents in equilibrium with 70 per cent RH are shown in Table 12 (in bold). These are the levels (at 25°C) beyond which most stored grain begins to show microbial spoilage and damage (Terry Mabbett, 1996). In practical terms this means that grain intended for storage at 25°C must be dried to the moisture level indicated in bold type and maintained at this level. Acceptable equilibrium grain moisture and relative humidity levels are reassessed accordingly with higher and lower grain storage temperatures.

Table 12: Equilibrium moisture levels (g/kg) for selected cereal grains at 25°C*

Cereal grain at 25°C	Relative Humidity (%)					
	40	**50**	**60**	**70**	**80**	**90**
Barley	97*	108*	121*	135*	158*	195
Maize, shelled, yellow dent	98	112	129	140	156	196
Rice, whole grain	109	122	133	141	152	191
Sorghum	98	110	120	138	158	188
Wheat, durum	94	105	115	131	154	199
Wheat, soft red winter	97	109	119	136	157	197

**Equilibrium moisture level of grain expressed as g/kg; divide by 10 to express as per cent moisture on a weight to weight basis; for example 135 g/kg is equivalent to 13.5% w/w (weight/weight)*

Cereal crops can be harvested by combine and be successfully threshed at quite high grain moisture levels (over 50% w/w) but harvesting at this level is not advised. Yield will be lost because maximum grain dry weight has not been attained and moreover farmers will not want the expenditure of artificial drying of grain down to moisture levels of around 13-14% w/w. One exception to this rule is if grain is treated with an appropriate acid as a mould inhibitor but this will essentially restrict the grain to use as animal feed. Other problems associated with harvesting too early, when the grain is wet, relate to milling. For instance, immature kernels in early, high-moisture rice harvests do not mill well, while maize harvested at high moisture content is soft and easily punctured or split.

Most farmers are forced to carry out at least some artificial drying of grain to reduce the moisture content in the crop and control relative humidity levels, and therefore grain spoilage, is essentially zero (Terry Mabbett, 1996). The two main types of commercial drying utilise air that is at near-ambient temperature or air which has been heated. Near-ambient air drying methods restrict the rate of mould growth so that drying is completed before any substantial mould growth takes place. Heated-air drying methods essentially avoid the mould problem altogether because the high temperatures used to rapidly dry the grain also kill the pathogens.

iv. **Mycotoxin testing:** Mycotoxins can mean 'life or death' for cereal grain as a marketable and consumable commodity. Mycotoxins are difficult to remove or neutralise, so prevention is always better than cure. A dull appearance and musty smell of grain invariably means mould growth, but not necessarily the presence of mycotoxins because not all fungal pathogens and storage moulds are mycotoxin producers. Even so, presence or absence of mycotoxins is insufficient for national and international grain trading which demands to know the specific mycotoxin(s) present and their exact concentrations. Concentrations of mycotoxins that make grain unfit and dangerous for human and livestock consumption are measured in ppm and ppb and therefore require measurement down to same exacting trace levels. Mycotoxin testing is essential to determine if there is a problem requiring action or whether a particular treatment has worked (Terry Mabbett, 1996). It must be highly sensitive and provide reliable and reproducible results whether carried out onsite or in the laboratory. Mycotoxins are not randomly or uniformly distributed throughout grain bulks and loads but tend to occur in 'hot spots' corresponding to the growth of fungal colonies. For this reason sampling from a moving stream of commodity reduces the selection biases associated with sampling a static load. Laboratory testing equipment is highly specific for mycotoxin type and ultra sensitive, detecting mycotoxins in parts per billion at least. Results are delivered with speed, accuracy and repeatability. These attributes are now available in practical 'hands-on' kits that can be used onsite by farmers and grain store managers to get fast and accurate results all with an absolute minimum of training.

References

Arauz, L.F., 2000. Mango anthracnose: Economic impact and current options for integrated management. *Plant Dis.*, **84:** 600-611.

Bazie, S., Ayalew, A. and Woldetsadik, K.,2014. Integrated management of postharvest banana anthracnose (*Colletotrichum musae*) through plant extracts and hot water treatment. *Crop Protn.*, **66:** 14–18

Bijendra Kumar Singh, Kulveer Singh Yadav and Akhilendra Verm, 2017. Impact of Postharvest Diseases and Their Management in Fruit Crops: An Overview. *J.Bio.Innov.,* **6 :** 749-760

Chrys N. Akem, 2006. Mango Anthracnose Disease: Present Status and Future Research Priorities. *Pl.Pathol. J.,* **5**: 266-273

David Sugar, 2016. Integrated Management of Postharvest Diseases of Pome Fruit. *Oregon State University. Southern Oregon Research and Extension Center, Medford*

De Costa, D.M. and Erabadupitiya, R.U.T., 2005. An integrated method to control postharvest diseases of banana using a member of the *Burkholderia cepacia* complex. *Postharvest Bio & Technol.*, **36**: 31-39

Dubey, S.R. and Jalal, A.S. 2013. Species and Variety Detection of Fruits and Vegetables from Images. *Intl. J. Appl. Pattern Recognition,* **1 :** 108-126.

Feliziani, E. and Romanazzi, G., .2016. Postharvest decay of strawberry fruit: Etiology, epidemiology, and disease management. *J. Berry Res.,***6** : 47–63

Gianfranco Romanazzia,, Joseph L. Smilanickb, Erica Feliziania, Samir Droby, 2016. Integrated management of postharvest gray mold on fruit crops.. *Postharvest Biol. & Technol.*, **113**: 69-76.

Johnson, G., 2007. Adoption of ACIAR project outputs: studies of projects completed in 2006/7. Management of postharvest diseases of tropical and subtropical fruit using their natural resistance mechanisms in Sri Lanka, the Philippines and Australia (HORT/ 1997/094), *J. Pl.Pathol.*, **92** : 645-652,

Mangosteen and White Sapote, 2011. Postharvest Biology and Technology of Tropical and Subtropical Fruits. Woodhead Publishing Series in Food Science, *Technology and Nutrition*, Pages 312-333, 334e-335e

Nabi, S.U., Raja, W.H., Kumawat, K.L., Mir, J.I., Sharma, O.C., Singh, D.B. and Sheikh, M.A.,2017. Postharvest Diseases of Temperate Fruits and their Management Strategies-A Review, *Int. J. Pure App. Biosci*. **5**: 885-898.

Pegg, K., 2001. Management Strategies in Disease Control. Queensland Hortficullure Institute. *Future J.*, 32: 147-152

Philips, D. J.; Mycotoxins as a postharvest problem. *In: Moline H.E.(ed). Postharvest Pathology of Fruits and Vegetables: Postharvest Losses in Perishable Crops*, pp. 50-54. Agricultural Experimental Station,

Sadegh Mohajer, Rosna Mat Taha, Jamilah Syafawati Yaacob and Aloka Kumari, 2015. Postharvest Management of the Pests and Some Important Diseases of the Fruits. 1-12 PP

Smilanick, J.L.,2011. Integrated Approaches to Postharvest Disease Management in California Citrus Packing houses. *Acta Hortic*. **905 :** 145-148

Sushma Sharma, Netra Prakash Dohroo, Shanmugam Veerubommu, Sumitra Phurailatpam, Neelam Thakur and Ajar Nath Yadav, 2017. Integrated Disease Management of Storage Rot of Ginger (*Zingiber officinale*) caused by *Fusarium* sp. in Himachal Pradesh, India . *Int.J.Curr.Microbiol.App.Sci.,* **6**: 3580-3592

Terry Mabbett, 1996. Integrated management of stored grain. *In:* Cereal grain quality. Henry, R.J. and Kettlewell, P.S. (Edrs.), Chapman & Hall. 488 pages,

Van der Steen, C., Jacxsens, L., Devlieghere, F. and Debevere, J. 2001. Combining high oxygen atmospheres with low oxygen modified atmosphere packaging to improve the keeping quality of strawberries and raspberries. *Postharvest Biology and Technology*, **26**:49-58.

Wang, H., Cao, G. and Prior, R.L. 1996. Totalantioxidant capacity of fruits. *J. Agril.& Food Chem,* **44**:701-705

Wang, Y., Xu, Z., Zhu, P., Liu, Y., Zhang, Z., Mastuda, Y., Toyoda, H. and Xu, L., Postharvest biological control of melon pathogens using *bacillus subtilis* exwb1, The Macmilan Press., Basingstoke, UK.

Yadav, S.M., Patil, R.K. Balai, L.P. and Ram Niwas, 2013. Postharvest Diseases of Horticultural Crops and Their Management. *Popular Kheti,* Volume -1, Issue-3 (July-September).

9

Postharvest Nematode Diseases and Their Management

Postharvest food loss has been considered an unavoidable affliction since food storage first began. These losses are much more painful and costlier in terms of money and man-hours. Such losses due to plant parasitic nematodes, in particular, is severe in tropical regions especially in developing countries like India, where proper storage, transmit and marketing facilities are inadequate. In tropical agriculture, increase in production rate is generally low and, by itself, unable to keep pace with the requirements of the increasing population. Hence, such food shortage demands the need to reduce postharvest loss due to nematodes. Postharvest food losses also constitute a vast complex of physical and biological changes, which makes it very difficult to represent in a single figure. As suggested by different sources, an overall loss of at least 10 per cent occurs in the developing countries. However, the magnitude may range from negligible to cent percent in particular nematode species and crop.

In addition to causing direct losses individually, phytonematodes also favour the establishment of secondary, passive micro- organisms like fungi, bacteria, viruses etc. Nematodes cause mechanical wounds, which serve as avenue for the entry of secondary microorganisms (Ravichandra, 2008). This complex association results in severe damage to planting material.

Postharvest diseases due to nematodes can be managed through restricting the movement of infected planting material to uninfected areas; use of uninfected, healthy planting materials; use of treated planting materials; improvements in the cultural practices at the growing site; timely and careful harvesting; use of nematicides both at pre- and postharvest stages; employment of some physical means like heat, radiation etc; improved storage techniques and integration of these methods. By maintaining hygienic conditions during harvest, transportation , storage and packing most of the postharvest damage due to nematodes can be minimized. Of all the methods, however, regulatory measures like quarantine is more practical and effective in restricting the movement of nematode infected plant materials and contaminated soil into a state or country. Contaminated plant material should be treated (hot water or nematicides) to kill the plant parasitic nematodes before planting to prevent the further postharvest damage. These various measures are primarily aimed at either reduction or wiping out of stages of nematode from the site or protection of host from infection or creation of unfavorable host and/or environmental conditions for the nematodes.

Future should be focused on minimum usage of chemicals to avoid the residual and other hazards, by employing safer and eco-friendly approaches like bio-agents for the successful management of postharvest diseases. Emphasis should also be given on the use of genetically modified planting material, which are free from parasitic nematodes.

9.1 General characteristics of phytonematodes: Nematodes are second only to insects in the number of species in the animal kingdom. However, only about 3 percent of all nematode species have been studied and identified. One cubic foot of soil may contain millions of individual nematodes belonging to several different taxonomic groups. Plant parasitic nematodes are nearly microscopic, worm-shaped animals virtually invisible to the naked eye when in the soil. Phytonematodes can cause significant plant damage ranging from negligible injury to total destruction of plant material. The severity of plant injury resulting from nematode activity depends on several factors such as the combination of plant and nematode species and prevailing environmental factors including rainfall, soil types, land contour, and culture practices.

Although a few nematode species feed on above ground plant parts, such as leaves, stems, flowers and seeds, the majority of these parasites feed on underground parts of plants, including roots, bulbs, and tubers. Because of this below ground, "hidden" feeding activity, nematode damage to plants cannot always be diagnosed readily. Nematodes are microscopic roundworms that live in many habitats (Ravichandra, 2014). At least 2,500 species of plant-parasitic nematodes have been described, characterized by the presence of a stylet, which is used for penetration of host plant tissue. Most attack roots and underground parts of plants, but some are able to feed on leaves and flowers.

Phytonematodes are of great economic importance. However, because most of them live in the soil, they represent one of the most difficult pest problems to identify, demonstrate and manage. Often, plant damage caused by nematodes is overlooked because the resulting nonspecific symptoms, such as slow growth, stunting, and yellowing, can also be attributed to nutritional and water-associated disorders (Ravichandra, 2014). Their effects are commonly underestimated by growers, horticulturists and nematode management consultants, but it has been estimated that around 10 percent of world crop production is lost as a result of plant nematode damage. Most nematodes feed on bacteria, fungi, or other microscopic creatures. As such, they are a major component of soil and sediment ecosystems. Of late, nematodes are being considered as a very successful class of animals. Four out of five multi-cellular animals on earth are nematodes. They are present almost everywhere: in cultivated fields, in sand dunes, in the sediments beneath the ocean floor, in groundwater, in plants, animals and even in humans.

Although over 4,100 species of plant-parasitic nematodes have been identified (Decraemer and Hunt 2006), new species are continually being described while others, earlier considered as minor or non-damaging, are becoming pests as cropping patterns change . However, the phytonematodes of economic importance can be grouped into relatively restricted specialized groups that either cause direct damage to their host or act as virus vectors. Most nematodes affect crops through feeding on or in plant roots, whilst a minority are aerial feeders. It is a known fact that in addition to direct feeding and migration damage, nematode feeding facilitates subsequent infestation by secondary pathogens, including soil fungi and bacteria that complicates the condition of the host.

On a global scale the distribution of nematode species varies greatly. Some are cosmopolitan, such as certain *Meloidogyne* spp., while others are particularly restricted geographically as *Nacobbus* spp., or are highly host specific, such as *Heterodera carotae* which attacks only carrots. Some crops may have very few nematode pests while others have a particularly wide range of genera and species associated with them leading to difficulties for nematode management strategies. Distribution maps and host range data are available and updated regularly as a useful source for determining nematode damage potential. One difficulty with assessing nematode impact is that damage resulting from nematode infection is often less obvious than that caused by many other pests or diseases. Losses that result from nematode attack may not necessarily be as a consequence of direct cell death, necrosis or 'diseased' tissue but may derive from other more insidious aspects, such as interference with the root system, reducing their efficiency in terms of access and uptake of nutrients and water; to the unaware, nematode-affected plants present typical drought and nutrient stress symptoms, which are easily and often misdiagnosed.

The nematodes are a very successful group of animals that can destroy up to 10% of all arable crops, a figure that is comparable with the destructive effects of insects. Despite the tremendous economic importance of nematodes, the great majority of them are free living and they occur in every conceivable environment. It has been claimed that "if all matter was destroyed except nematodes, one would still be able to see the world outlined by nematodes". Nematodes are 'hidden and unseen enemies' of crop plants as they are microscopic and present in soil, particularly, in rhizosphere (Ravichandra, 2014). They are microscopic animals, which are vermiform, bilaterally symmetrical, unsegmented, pseudocoelomatic and triploblastic. Their body is not segmented as in earthworms. There are no respiratory or circulatory organs present in nematodes. All phytonematodes possess a structure 'stylet' or spear, which is the unique feature. Stylet is the major feeding organ present in the mouth of phytonematodes that penetrates into the host tissue. Non-plant parasitic and other forms do not possess the stylet.

Body Size: There is a great variability in the size of nematodes. Free living nematodes in soil and freshwater vary in length from about 150 microns to 10 mm. Marine species vary from 83 microns to about 50 mm in length. The animal parasites range from about 1 mm to 7 meters in length, the largest being nematodes parasitic in whales. However, most plant parasites are microscopic and range from 0.25 mm to approximately 12 mm in length with a body width from 0.01-0.5 mm. Longest phytonematode is *Paralongidorus* and the smallest is *Paratylenchus.*

Body Shape: Nematodes generally, are elongate, cylindrical, unsegmented worms. They are bilaterally symmetrical. Though most of them are vermiform, in some genera the body of the adult female may be greatly modified (*Meloidogyne, Heterodera, Rotylenchulus*, *Tylenchulus etc*.,). In these genera, the body may be melon shaped, spindle shaped (fusiform), pear shaped, kidney shaped, lemon shaped or variations of saccate. The entire nematode body may be divided into three regions, the outer body tube or body wall, the inner body tube and the body cavity of pseudocoelom.

The outer body tube constitutes cuticle (exoskeleton), hypodermis and muscle layer. The body wall is composed of a thick cuticle, which may be considered to be an exoskeleton, a syncitial hypodermis and the mucus layer composed of a single layer of longitudinal muscle only. Pseudocoelom is the false body cavity (the area between the body wall and the digestive tract), which is not lined with epithelial tissue. It is filled with pseudocoelomic fluid, which bathes the muscles, the digestive system, the reproductive system and some or all of the excretory system. The fluid in the pseudocoelm serves as a transport system for oxygen, food materials and products of metabolism. The gut of nematodes consists of a terminal mouth, buccal cavity with six lips, pharynx, intestine, rectum and sub-terminal anus. In the male, there is cloaca, with the rectum and vas deferens opening into the same cavity (Ravichandra, 2008). The inner body tube consists of digestive system, esophagus, intestine, rectum and cloaca. Nervous system and reproductive systems are well developed, where as excretory system is not very well developed in phytonematodes. Nervous system is well equipped with a range of sophisticated sense organs. Reproduction in nematodes is well developed and of diverse types, which plays a major role in the pathogenicity of a phytonematode.

Reproduction in nematodes: The reproductive systems are major organs of the nematodes and can occupy a large portion of the body cavity in males and females. There are many morphological and physiological differences between the species and so they are separated here.

Males: Males are either monorchic (most Secernentea) or diorchic (most Adenophera), with regards to the number of testis present. The testes are tubular structures lined with epithelium and glandular tissue, sperms are produced at the end and mature as they migrate towards the shared opening of the cloaca. Many males have paired chitinous protrusion from the cloaca known as spicules, which are maincopulatory organs and are used for attaching to a female during copulation. The spicules are easily seen under a microscope due to their chitinous structure and their position and shape may be used as an identifying feature. Sperms are ejaculated from the cloaca around the spicule, rather than through it. The sperm produced by nematodes is amoeboid and is very motile, employing the same locomotion mechanism as seen in amoeboid species.

One of the most distinctive features of some male nematodes is the presence of a copulatory bursa, seen in nematodes of the order Strongylida. Nematodes with a copulatory bursa are known as bursae, whilst those without are non-bursae. This bursa is at the posterior end of the nematode and is formed from alae with laeral rays that are used for grasping onto the body of the female during copulation.

Females: Female nematodes usually have a single genital pore through which sperm may enter the uterus and oviduct, this pore is also referred to as a vulva and may be covered by a vulval flap. The uterus may take many forms from being short and straight, long with a single bend or a coiled form. Eggs produced in the ovaries populate the oviducts and uterus and may be released as embryonated or non embroynated eggs once fertilisation has occured. Nematodes may have either one (monodelphic) or two ovaries (didelphic) with various typles of orientation in case of two ovaries. A small muscular organ exists at the vulval opening of some species known as the 'ovijector', which aids in the expulsion of eggs from the vulva. Unlike in males, the end of the female is usually blunt ended with the anus being positioned proximally on the body wall.

Male and female nematodes are usually similar in appearance except for the reproductive systems. It mainly includes one or two tubular gonads which may vary in length and may be straight, reflexed or coiled back and forth. In males and females the genital tubes are lined with a single layer of epithelium which covers the germ cells and forms the ducts. In majority of nematodes the germ cells are proliferated only in the distal end of the gonad i.e., telogonic. In few nematodes the germ cells are formed along the entire length of the gonad, i.e., hologonic. Most phytonematodes are characterized by an increase in the size of the entire reproductive system. Th uteri and the growth zone of the ovary are lengthened that leads to more egg production. However, pronounced dimorphism occurs in some species. Females become swollen and males remain slender and cylindrical.

Major types of reproduction

1. **Bisexual (Amphimixis):** Male and female reproductive structures are present on two different individuals. Ex: Burrowing nematode, Root-knot nematodes etc.
2. **Parthenogenesis:** Fertilization of eggs occurs without the help of males in this type. Mating and genetic recombination do not occur. Ex: Root-knot nematodes, Citrus nematode, Cyst nematodes etc.,
3. **Hermophroditism:** Both male and female reproductive structures are present on the same individual. Ex: Root-knot nematodes, Stem and bulb nematode etc.

Intersexes are also common in some nematodes like root-knot nematodes. It is a variation in sex characteristics including chromosomes, gonads, and / or genitals that do not allow an individual to be distinctly identified as male or female. These intersexes have normal male head and tail structures and exhibit some mating behavior, but possess hermaphrodite-like gonads which produce no sperm and usually contain a few oocytes. All nematodes pass through an embryonic stage, four larval / juvenile stages and an adult stage (Ravichandra, 2014). Juveniles hatch from eggs as vermiform, second-stage juveniles, the first moult having occurred within the egg. Newly-hatched juveniles may reinvade the host plants of their parent or migrate through the soil to find a new host root. In most nematode genera, second stage juveniles and adult females only are plant parasitic. The duration of life cycle depends on important factors like nematode species, soil temperature, soil moisture, soil structure, soil texture, type of host etc.

Most phytonematodes attack roots / underground parts (bulbs, suckers, rhizomes, corms *etc*.,) of host plants while some attack aerial plant parts too.

General above ground symptoms of nematode infestation on horticultural crops include stunted growth, wilting and yellowing in patches. However, characteristic symptoms are expressed in below ground, which may comprise presence of knots/galls, cysts, lesions and cavities/tunnels on the roots. Excessive branching of roots, deformed/abnormal roots/rootlets, rotting and discoloration of roots, decreased/complete destruction of feeder roots are some of the symptoms of nematode damage, which contribute to overall reduction in the growth and development of plants. This affects the quality and quantity of the produce that ultimately affects the grower. Each phytonematode possesses a common and a scientific name. Such names for the major nematodes of postharvest significance are listed in table 13.

Table 13 : Scientific and common names of major phytonematodes of postharvest significance

Sl. No.	Scientific name	Common name
1.	*Anguina* spp.	Seed and leaf gall nematode
2.	*A. agrostis*	Bentgrass nematode
3.	*Aphelenchoides* spp.	Bud and leaf nematode, Foliar nematode
4.	*A. besseyi*	Rice white tip nematode, Strawberry bud nematode, Summer crimp, Summer dwarf nematode
5.	*A. fragariae*	Spring crimp nematode, Spring dwarf nematode, Strawberry bud nematode
6.	*A. ritzemabosi*	Chrysanthemum foliar nematode
7.	*Bursaphelenchus (Rhadinaphelenchus) cocophilus*	Coconut palm nematode, Red-ring nematode
8.	*Bursaphelenchus xylophilus*	Pinewood nematode
9.	*Cacopaurus pestis*	Walnut nematode
10.	*C. cacti*	Cactus cyst nematode
11.	*Ditylenchus angustus*	Rice Ufra nematode
12.	*Ditylenchus destructor*	Potato rot nematode
13.	*D. dipsaci*	Stem and bulb nematode, Alfalfa stem nematode
14.	*Globodera pallida*	Pale / White potato cyst nematode
15.	*G. rostochiensis*	Golden nematode, Golden potato cyst nematode
16.	*Helicotylenchus multicinctus*	Banana spiral nematode, Spiral nematode
17.	*Heterodera avenae*	Cereal cyst nematode
18.	*H. carotae*	Carrot cyst nematode
19.	*H.glycines*	Soybean cyst nematode
20.	*H. goetingiana*	Pea cyst nematode
21.	*H. schachtii*	Sugarbeet cyst nematode
22.	*Hirschmaniella oryzae*	Rice root nematode
23.	*Meloidogyne* spp.	Root-knot nematode

24.	*M.arenaria*	Peanut root-knot nematode
25.	*Meloidogyne graminicola*	Rice root-knot nematode
26.	*M. hapla*	Northern root-knot nematode
27.	*M. incognita*	Southern root-knot nematode
28.	*M. javanica*	Javanese root-knot nematode
29.	*Pratylenchus* spp.	Lesion nematode
30.	*P. vulnus*	Boxwood lesion nematode, Walnut lesion nematode
31.	*Radopholus similis*	Banana burrowing nematode, Burrowing nematode
32.	*Tylenchulus semipenetrans*	Citrus nematode

9.2 Major nematodes of postharvest significance: Important crops suffer from postharvest diseases due to potential plant parasitic nematodes include staple food crops like rice (*Aphelenchoides besseyi, Ditylenchus angustus);* wheat (*Anguina tritici*); fruit crops like banana (*Radopholus similis*); vegetables like potato (*Globodera rostochiensis*, *Ditylenchus destructor*, *Meloidogyne* spp. and *Pratylenchus* spp.); oil palms including coconut *(Rhadinaphelenchus cocophilus*) and sugar beet (*Meloidogyne* spp., *D. dipsaci* and *Heterodera schactii*); several bulb crops like onion (*D. dipsaci*); soybean (*H.glycines*); pea (*H.goettingiana*) and various flower crops(*D. dipsaci*). Most of these are seed borne in nature and several of them are dispersed through seed/propagating material (Table 14) (Ravichandra, 2014).

All these plant parasitic nematodes affect the plant products-no matter, if they carry nematodes from field just before harvest or become infected during harvest or afterwards. Such infections essentially develop after harvest and cause loss in quantity and/or deterioration in quality and value of the plant product. The special feature of postharvest nematode diseases is that most of the plant products affected are mostly storage organs and therefore dormant and their tissues are physiologically different from those of actively growing plants. Postharvest disease inciting nematodes employ mechanical energy and bio- chemical components (enzymes and toxins) to cause diseases. These molecules are required to penetrate, reproduce and to degrade host tissues and consume the substances/nutrients degraded from the host.

Seed borne nematodes may be either internal or seed infestation, as external. Seed infestations by several saprophytic nematodes occur in soil beds or mixed with seed. They may not be able to exhibit specific symptoms on seed, but endoparasitic nematodes affect seed in drastic manners, *viz.*, floral malformations and galls in wheat by *Anguina tritici*, discolouration of testa in groundnut by *Ditylenchus destructo*r, symptomless in paddy kernels infested by *Aphelecnhoides besseyi. Anguina, Aphelenchoides, Ditylenchus, Heterodera , Meloidogyne* in some crops, *Radopholus* and *Rhadinaphelenchus* are major seed transmitted phytonematodes, although large amount of work has been carried out with *Meloidogyne,* and *Heterodera. Heterodera* and *Rhadinaphelenchus* normally occur as endoparasites in stems, foliage or roots. *Rhadinaphelenchus* also affects growing tips, inflorescence and husks or dropped nuts.

Fortunately, the active dispersal of nematodes is rather limited but this is more than compensated by the ease of passive dispersal with movement of infested soil, plant material and water. This Chapter pertains only to phytonematodes which are seed borne or borne on propagating material. Man has aided in the indiscriminate spread of nematode while carrying the plant materials and other items infested with nematodes for research, trade, charity or other purposes. With the recent trade liberalization, the bulk movement of commodities has become common and consequently the risk of entry of nematodes has increased many folds. India has already suffered enormously from the entry of serious exotic nematode pests such as the burrowing nematode, *Radopholus similis* playing havoc with banana and other plantation crops and the potato cyst nematode, *Globodera rostochiensis* and *G.pallida.*

Table 14: Major seed borne nematodes

Nematode species	Crop
Globodera rostochiensis, Ditylenchus destructor *Meloidogyne spp., Pratylenchus* spp.	Potato
Radopholus similis , Helicotylenchus multicinctus	Banana
Anguina tritici, Heterodera avenae	Wheat
Ditylenchus angustus, Aphelenchoides besseyi, Meloidogyne graminicola, Hirschmaniella oryzae	Rice
Radinaphelenchus cocophilus	Palms
Heterodera schachtii	Sugar beet
Heterodera geotingiana	Vegetables
Heterodera glycines	Soybean
Aphelenchoides fragariae	Narcissus
Meloidogyne arenaria, Pratylenchus penetrans	Peanut
Pratylenchus vulnus	Temperate fruits
Ditylenchus dipsaci	Onion, Garlic, Ornamentals
Tylenchulus semipenetrans, Meloidogyne incognita	Citrus

9.2.1. Cyst nematodes of potato (*Globodera rostochiensis* & *G.pallida*): *Globodera* is a genus with its cyst globose in shape. Golden cyst nematode / golden eelworm / yellow potato cyst nematode, *Globodera rostochiensis* and pale cyst nematode, *G. pallida* are major pests on potato. *G. rostochiensis* is a pest of plants in the family Solanaceae, primarily infesting potato and tomato as well as a variety of other root crops.

Altogether 90 species of the genus *Solanum* are known to be hosts for this group of nematodes. Both species are confined to the cooler places. Apart from potato, these are also pests on tomato, eggplant, members of the Solanaceae plant family (such as nightshade) and other root crops. Both species feed on and produce cysts on potato roots, thus causing substantial damage to potato crops. The cysts are typically yellow-brown/golden yellow in color. Both the golden and pale cyst nematodes are regulated federally. In India, golden cyst nematode is under Quarantine Act. Major species include *G. rostochiensis, G. pallida* and *G. tabacum.*

Classification

Kingdom: Animalia
Phylum: Nematoda
Class: Secernentea
Order: Tylenchida
Family: Heteroderidae
Genus: *Globodera*

Symptoms: The golden nematode forms cysts on the roots of susceptible hosts. The cysts, which are composed of dead nematodes, are formed to protect the female's eggs and are typically yellow-brown in color. The first symptoms of infestation are typically poor plant growth, chlorosis and wilting. Heavy infestations can lead to reduced root systems, water stress, and nutrient deficiencies, while indirect effects of an infestation include premature senescence and increased susceptibility to fungal infections. Symptoms of golden nematode infestation are not unique, and thus identification of the pest is usually performed through testing of soil samples. An infestation may take several years to develop, and can often go unnoticed for between five and seven years. After detection, however, it may take up to thirty years for the pest to be effectively eradicated.

Management

- Quarantine measures to restrict the movement of infested tubers
- Check that machinery is thoroughly clean and free from plant debris
- Do not return soil to fields as it may cause infestation of potato cyst nematode to spread.
- Clean soil from potato tubers and have the soil tested to be sure of non-transference of potato cyst nematode.
- Grow susceptible and resistant varieties of potato alternately, thus reducing the possibility of selecting a highly virulent or new pathotypes

- To reduce nematode population densities, non-host crops such as barley are grown between host crops
- Solarization is a good method of killing nematodes in very hot climates
- 1,3-Dichloropropene,dazomet,.Organophosphatesandoximecarbamates are very effective nematicides.
- Integration of *Pseudomonas fluorescens* / *Purpureocillium lilacinum* or *Pochonia chlamydosporia*+ neem cake + mustard intercrop + carbofuran 3 G is effective.
- Since potato, tomato and brinjal crops are the only important host crops, these should not be grown more than once in 3-4 years in the infested field. Nursery of these crops should not be raised in such field
- Potato grown in the infested fields should not be used for seed purposes not the produce allowed to be carried out of the notified infested area unless it is certified and fully ensured for the absence of adhering soil and nematode cysts on potato tubers by the competent authorities.
- Avoid the movement of irrigation and rain water from infested fields to other fields to check the spread of cysts.
- The nematode populations usually decrease by 80 per cent when grown with resistant varieties. Resistant varieties like 'Maris Piper' and 'Ulster Glade' have been released in Europe. The source of resistance was from the crosses between the *S. tuberosum* and *S. tuberosum sub species andifena..* A large number of germplasm collection accessions and cultures were screened for cyst nematodes and the resistance was found in S. *varnei, S. acaule, S. microdontum, S multidissectum, S. oplocense, S. sparsiplum, S, phureja, A. fendleri, S. chacoense* and *S. sucrense.* Complete resistance has been observed in CN-312 which is a cross between *S. tuberosum sp. Andigena* and *S. multidissectum.*
- Since cysts can persist in soil for 10-15 years in the absence of host, it is difficult to eradicate this nematode completely. Non host crops like wheat, maize, beans etc., can be used to minimize the population in soil. Crop rotation with cabbage, peas, cauliflower, carrot etc. during autumn season brings down the population of cyst. Potato French beans-peas=peas sequence in a 4 year rotation is effective. Vegetable growing in Nilgiri hills has been reported to bring down soil population *viz.*, garlic, radish, cabbage, peas, French beans, and carrot

Spread of cyst nematodes has been prevented by following these measures at Nilgiri Hills, Tamil Nadu, India. Potato grown in infested soil should not be used for seed. ; Potato, tomato and brinjal should not be grown for 3-4 years in infested fields.; Volunteer potato plants should be removed from the infested fields ; Movement of water and soil from infested fields should be avoided ;Crop rotation may be followed even in healthy fields so that potato is grown only once in 3-4 years.

- Fensulfothion at 30kg ai/ha, aldicarb @ 2 kg a.i./ha as spot treatment, carbofuran at 2 kg ai/ha, have been effective to manage the cyst nematodes and to achieve increased yields. Tuber disinfection by fumigation with methyl bromide fr 5-6 hrs. formaldehyde vapor for 12 hrs and soaking seed tubers in 10% formalin for 1 hr, 9% calcium hypochlorite for 30 minutes, 0.4%-1.6% solutions of aldicarb/aldicarb sulfone or carbofuran for 3-60 minutes is effective.

9.2.2. Potato rot nematode (*Ditylenchus destructor*) : *Ditylenchus destructor* is a major phytonematode, commonly known as the potato rot nematode. Other common names include the iris nematode, the potato tuber eelworm and the potato tuber nematode. It is an endoparasitic, migratory nematode. In addition to potatoes, potato rot nematodes have a wide range of over 100 species of hosts from numerous families. Some of these include alfalfa, beets, carrots, garlic, hops, mint, parsnips, peanuts, rhubarb, tomatoes, and flowering plants such as irises and tulip.-The nematodes only attack the below ground, non-aerial tissues of plants such as the roots, bulbs, rhizomes, and tubers.

Kingdom: Animalia
Phylum: Nematoda
Class: Tylenchoidea
Order: Tylenchida
Family: Anguinidae
Genus: *Ditylenchus*

Symptoms: The main symptoms of *Ditylenchus destructor*, common to potatoes and its other hosts, are the rotting and discoloration of subterranean plant tissue. In potatoes, early infection can be detected by small white spots underneath the potato's skin. As the disease progresses, these spots become larger and darker with a spongy or hollow appearance. Tubers develop sunken areas and the skin becomes dry, cracked, and detached from the underlying flesh. There is further discoloration at this stage that is often due to secondary invasions of fungi, bacteria, and free-living nematodes. Above ground symptoms are usually not seen, although heavily infected plants are often weaker, smaller and can have curled or discolored leaves (De Waele *et.al.*, 1989). Symptoms of *Ditylenchus destructor* on the bulbs of flowering plants such as irises and tulips are similar to those of potatoes, except infection usually occurs at the bulb's base and moves upwards. The fleshy scales develop discolored yellow to black lesions, the roots become blackened, and leaves can develop yellow tips. Potato rot nematodes in groundnuts, such as peanuts, develop blackened hulls, shrunken kernels, and embryos with a brown discoloration.

Management

- Planting materials and locations free of these nematodes is crucial, so the soil, seeds, and farm machinery must all be carefully controlled. This is done by disinfecting machinery and removing all potentially infected plant debris from farm equipment when transferring from field to field.
- Seeds certified to be free of *Ditylenchus destructor* should be planted.
- Seed potatoes carry the infection and spread the disease. It is essential, therefore, to use clean seed tubers.
- Crop rotation as a form of control is difficult due to the nematodes' wide host range; therefore, non-host crops must be selected carefully to use for rotation each season
- Weeds must also be eradicated, as they often act as alternative hosts for potato rot nematodes.
- Chemical control of *Ditylenchus destructor* can be achieved with soil-applied nematicides such as carbofuran, ethylene dibromide, Vapam HL, and Telone. Fumigation with these nematicides is often paired with mechanical measures to attain optimal control.

9.2.3. Root-knot nematodes (*Meloidogyne* spp.): Most common sedentary endoparasite on crops, can damage and cause severe economic loss. It has worldwide distribution though more common in temperate, subtropical and tropical areas. *Meloidogyne incognita* is probably the most economically important phytonematode species among the tropical and subtropical regions. This nematode is extremely polyphagous, attacking both monocotyledons and dicotyledons. It is estimated that more than 3,000 plant species can be affected (Diez and Dusenbery. 1989). Female embedded in root tissue, melon-shaped, induces characteristic knots/galls on the roots of affected plants. This genus includes more than 60 species, with some species having several races. However, four species *M. incognita, M. javanica, M. arenaria* and *M. hapla* are major pests worldwide. These nematodes are more destructive when they associate with other soil borne fungal and bacterial pathogens to form complex diseases.

Major hosts of root-knot nematodes include most horticultural crops *viz.* vegetables, fruits, ornamentals, polyhouse grown crops, plantation, medicinal aromatic and spice crops. They modify normal cells into giant cells before feeding. Female produces large number of eggs embedded in a gelatinous egg mass attached to the knot on the root, which is produced by six rectal glands and secreted before and during egg laying. Normally, 500-1000 eggs are laid per female. Knot / gall formation is a visible sign of root-knot nematode infection resulting from the hypertrophy and hyperplasia indicative of feeding site formation within the roots. The matrix initially forms a canal through the outer layers of root tissue and later surrounds the eggs, providing a barrier to water loss by maintaining a high moisture level around the eggs (Ravichandra, 2014). As the gelatinous matrix ages, it becomes tanned, turning from a sticky, colourless jelly to an orange-brown substance which appears layered. This egg-mass serves as a dormant structure for this group of nematodes.

Classification

Kingdom: Animalia
Phylum: Nematoda
Class: Secernentea
Order: Tylenchida
Family: Heteroderidae
Genus: *Meloidogyne*

Symptoms: Root knot nematodes cause abnormal, destructive swellings, called galls / knots on the roots of affected plants. Infestations of these nematodes are fairly easy to recognize. The nematodes feed and develop within the galls, which can grow as large as 1 inch in diameter on some plants but usually are much smaller. The formation of these galls damages the water- and nutrient-conducting abilities of the roots. Galls can crack or split open, especially on the roots of vegetable plants, allowing the entry of soil-borne, disease-causing microorganisms. Root knot nematode galls are true swellings and can't be rubbed off the roots as can the beneficial, nitrogen-fixing nodules on the roots of legumes. Root knot nematodes can feed on the roots of grasses and certain legumes without causing galling. Plants affected by *M. incognita* presents above ground symptoms of water and nutrient stress, yellowing, wilting and stunting. Below ground galling on roots, bulbs, tubers is the typical symptom. Plant death may occur in high infestation level.

Management

- Though root-knot nematodes are polyphagous in nature, there are a few crops like cereals and millets which are not favorable hosts of *M.incognita*. The nematode can be managed by growing maize, wheat, potato, wheat as rotational sequences under Simla conditions. Wherever possible, non-host crops may be taken up to reduce infestation levels. Use of resistant varieties is another best option.

- Soil solarization is a method of heating moist soil by covering it with transparent plastic sheets to trap solar radiation during hottest period or summer season of the year. This is technology of thermal disinfestations. Linear Low Density Polyethylene (LLDPE) clear films were efficient to manage root knot nematode incidence.
- Formulations of fungal (*Trichoderma harzianum, T. viride, Purpureocillium lilacinum*) and bacterial antagonists (*Pseudomonas fluorescens, Bacillus subtilis*) @ 0.5% can be drenched through manually or drip irrigation at regular interval of 30 days after sowing or transplanting.
- Application of neem cake at 0.5 kg/m^2 area followed by solarization of the raised nursery beds with polyethylene mulch and optimum 1-2 t/ ha neem cake was efficient to manage root-knot nematodes under field condition. In addition to neem cake, other available oilcakes such as castor , Mustard cake , Pongamia cake and Karanja cake have also been proved their efficacy against root knot nematodes.
- Use of trap (*Crotalaria)* and antagonistic (*Tagetes* spp.) crops.
- Application of organic manure such as poultry, farm yard manure, and vermicompost are efficient to increase plant growth parameters as well as to reduce the root knot nematode inoculum in vegetable crops. The combined application of FYM @ 10 t/ha + poultry manure @ 2.5 t/ ha + bio-fertilizers (Rhizobium and Phosphate Solubilizing Bacteria) effectively reduced root knot incidence.
- Application of Carbofuran @ @ 0.3 g a.i. /m^2 area of nursery bed. ; 3 kg a.i./ha or Aldicarb @ 2 kg a.i. / ha have proved effective. The efficacy increases when applied in two split doses, once at planting and later at earthling time.
- Bare root treatment of seedlings with carbosulfan 25 EC @ 2 ml/litre during transplanting crops.
- Seed dressing of directly-seeded crop like okra and cucurbits with carbosulfan 25 DS @ 3% a.i. (w/w) effectively manage root knot nematode incidence in vegetable crops.
- Seed treatment with carbosulfan (25 EC) at 0.1% for overnight or root dipping 0.05% for 6 hours in cucurbitaceous crops.

- Application of Carbofuran 3G @ 1 kg a.i/ha is recommended to nematode infested vegetable crops under field condition.

9.2.4. Burrowing nematode (*Radopholus similis*) & Spiral nematode (*Helicotylenchus multicinctus*): Burrowing nematode constitutes an important group of endoparasitic nematodes of plant roots and tubers, which is migratory. The major species is *R. similis* with two host races that differ in parasitism of citrus. It completes its entire life cycle within the root tissue. Infected plants experience malnutrition. It induces lesions and burrows in the roots, which later turn into tunnels and cavities causing destructive damage. Infection by burrowing nematode causes toppling-over disease in banana, yellows disease in pepper and spreading decline in citrus. Because of the damage it causes worldwide, burrowing nematode is considered as one of the most regulated nematode pests. Burrowing nematode although, is native to Australasia, is found worldwide in tropical and subtropical regions of Africa, Asia, Australia, North and South America and many island regions. The widespread range of this nematode is due to its dissemination with propagative plant material, especially infected banana corms/rhizomes/suckers. This devastating nematode has been observed infecting more than 300 plant species including banana, coconut, areca nut, tea, ginger, pepper, betel vine, turmeric, avocado, coffee, ornamentals, sugarcane, citrus etc.. *R. similis* infects both banana and citurs while *R. citrophilus* infects only citrus. However, these two species are categorised as the "banana race" and "citrus race," respectively, within *Radopholus similis*.

Spiral nematode attains usually spiral shape and are distributed worldwide and endoparasitic nematodes on roots of several crops viz., banana, maize, ornamentals, plantation crops, fruit trees, turf etc., They are also ectoparasites of several plant roots, which insert their stylets into root epidermis to feed. Some species live half-buried in the root tissue, while others penetrate the root and live inside. Out of 200, the most damaging species is *H. multicinctus* apart from *H. mucronatus, H. dihystera, H. pseudorobustus* and *H. vulgaris.*

Classification

Kingdom: Animalia
Phylum: Nematoda
Class: Secernentea
Order: Tylenchida
Family: Pratylenchidae
Genus: *Radopholus*

Kingdom: Animalia
Phylum: Nematoda
Class: Secernentea
Order: Tylenchida
Family: Hoplolaimidae
Genus: *Helicotylenchus*

Symptoms: Symptoms of burrowing nematode are most readily observable as dark and necrotic lesions on the root system , similar to those caused by pathogenic fung, and other endoparasitic nematodes that may infect banana roots. Root systems can become stunted, unthrifty and necrotic Citrus roots that are growing within the nematode conducive sandy soil have few to no small feeder roots and are severely damaged by the nematode feeding and tunneling. Above ground symptoms of citrus and other plants infected by burrowing nematode include yellowing, stunting, dieback, reduced fruit size, and thinning of the canopy. In banana, 'toppling-over' disease due to this nematode depicts the bending of bunch bearing plants towards ground. Spreading decline of citrus can begin in small areas of citrus groves where a few trees demonstrate reduced thriftiness, canopy thinning near the crown and a reduction in fruit size. Banana toppling occurs when heavy infestations cause trees to fall over during periods of heavy rain or wind due to their severely reduced and damaged root system, which is unable to anchor the plant into the ground (Thorne,1961.). Pepper fields suffering from burrowing nematode infestation will display yellowing symptoms coupled with root necrosis and canopy dieback.

In spiral nematode infestation, the symptoms are more subtle than those of certain other nematodes such as root-knot or sting nematodes. Heavy infection by *Helicotylenchus* spp. can cause a reduction in the root system, leading to unhealthy plants. On turfgrasses, *Helicotylenchus paxilli* has been shown to reduce root length and cause thinning of turf. Generally the turf decline will occur in patches and is often accompanied by proliferation of weeds in the affected areas . *H. multicinctus* causes small necrotic lesions on roots, and with heavy infestation these lesions coalesce, leading to root necrosis and die-back of roots. Plant stunting and reduction in yield can result.

Management

- Use of healthy tubers for planting
- Application of FYM/oil cakes to the pits before planting suckers
- Paring & pralinage operation before planting

Paring: Removal / chopping of infested discoloured portion on the sucker

Pralinage: Smear clay slurry /cow dung onto the pared sucker, apply 10 to 20 g of carbofuran 3G depending upon the size of the sucker, before planting.

- *Purpureocillium lilacinum / Pesudomonas fluorescens /Trichoderma viride* @ 20 g/pit
- Soil application with carbofuran 3G @ 20 g / pit / cartap hydrochloride @ 20 g or neem cake @ 250 g
- Growing marigold around the basin and incorporating before flowering could rectify the damage caused by *R. similis*

9.2.5. Seed gall nematode (*Anguina tritici*): *Anguina tritici* (ear-cockle nematode, seed-gall nematode, seed and leaf gall nematode, wheat gall nematode, wheat seed gall nematode, wheat seed-gall nematode, wheat seed and leaf gall nematode) is a potential plant pathogenic nematode. *Anguina tritici* was the first plant parasitic nematode to be described in the literature in 1743. It causes a disease in wheat and rye called "ear-cockle" or seed gall.

Classification

Kingdom: Animalia
Phylum: Nematoda
Class: Secernentea
Order: Tylenchida
Family: Anguinidae
Genus: *Anguina*

Symptoms: It is an ectoparasite that becomes endoparasitic invading inflorescence and developing seeds. It causes a disease called "ear-cockle", "gout" or seed gall on wheat and rye. It is not a host of oat, maize and sorghum. On wheat it causes stunted plants, distorted leaves, seeds are transformed into galls which contain a dried mass of nematodes. If compared to normal wheat seeds, galls are smaller in size, lighter, and their color ranges from light brown to black (normal wheat seeds are tan in color).

Management

- Crop rotation is most effective to reduce the nematode infection. In general, fields that previously had crops other than wheat for one or two years reveal less infection than fields that had been planted with wheat continuously. Less infestation has been reported in fields with wheat followed by rice.

- Kanred, Hsuchow Resistant No.2 winther varieties Concho, Fanone, bison Dica, Marquis and Magurele, Larma Rojo, Sonara 64 and Kalyan Sona, Nainari 60, Pictic 62, E 6049 are either tolerant or resistant to *A.tritici.*
- Presoaking wheat galls in cold water and then treating them for 10-20 minutes in water at 54-56°C can destroy the nematode.
- Seeds containing galls are dipped in 30 % brine solution in which galls float on the surface and can be removed. The difference in specific gravity of healthy grains and nematode galls makes it possible to separate galls from healthy kernels. Healthy seeds are rinsed with water and dried to prevent premature germination. This is the most effective method of control (Ravichandra, 2008).
- Screening with sieves of definite pore size allows the galls to pass through, leaving healthy gains on the sieves. Small grains also sometimes pass through and some compound galls (several small galls fused to form a compound all) being large, remained on the sieve along with healthy wheat grains.
- Use of chemicals may not be very economical and not much information is available on chemical control of *A. tritici*. Although several chemicals have been evaluated, some of the effective ones are ethyl parathion as spray, lindane, dimethoate, fumigation of the seeds with methyl bromide, chloropicrin, carbon disulphide and phostoxin etc. Methyl bromide applied at 40-80 g/m3 seed for 24 hrs is effective against *A. tritici.*

9.2.6. Cereal cyst nematode *(Heterodera avenae)* : *Heterodera* is a sedentary endoparasite and a potential pathogen on several crop plants. Different species of this genus attack several crops, often causing great economic damage. The genus is unique among nematode genera because of the ability of the female to transform into a tough, brown, cyst. Cysts are either partially enclosed in root tissue or in the soil. It is called a cyst nematode because the greatly swollen, egg-filled adult female is referred to as the "cyst stage". Female is swollen or obese, lemon shaped with a distinct neck. Females produce several hundred eggs and after death, the female cuticle forms a protective cyst. Eggs are retained within the cyst. Males are vermiform which are present in soil. The genus has worldwide distribution with a wide host range including horticultural crops. *H. avenae* is a pest of graminaceous crops worldwide.

Classification

Kingdom: Animalia
Phylum: Nematoda
Class: Secernentea
Order: Tylenchida
Family: Heteroderidae
Genus: *Heterodera*

Symptoms: The symptoms of infection can be readily recognised. Above ground, patches of unthrifty yellowed and stunted plants can be observed. Planting a susceptible crop in successive years will result in these patches becoming larger with time. Closer examination of the roots will reveal symptoms that are typical of *H. avenae.* Below ground, wheat and barley roots will be 'knotted' and oat roots appear 'ropey' and swollen. Development of root systems is retarded and shallow. In spring, characteristic 'white cysts' (about the size of a pin head) can be seen with the naked eye if roots are carefully dug and washed free of soil. These are the swollen bodies of the female, each containing several hundred eggs.

Management

- Once cereal cyst nematodes have been introduced into a field, eradication is nearly impossible. Therefore, efforts to minimize the transmission of soil from infested to noninfested fields or regions are of critical importance. The use of a break crop (e.g. canola, lupins, chickpeas) ensures a large proportion of the *H. avenae* population is removed. In serious outbreaks of *H. avenae*, it may be important to avoid cereals for two years to ensure an adequate reduction in the population.
- Crop rotations that include broadleaf crops, corn, fallow, and resistant wheat, barley, or oat varieties can greatly reduce the nematode density
- **Biofumigaiton:** When green tissue from a biofumigation crop of mustard, rapeseed, or sudangrass is thoroughly macerated and incorporated into soil, the toxic products generated during degradation of that tissue are sometimes toxic to soil nematodes.

9.2.7. Nematodes infesting rice

i. Rice Ufra (*Ditylenchus angustus*) : The nematode is also referred to as the 'ufra' nematode deriving its name from the disease it causes in rice (ufra). Ufra disease is known locally in Bangladesh as "dak pora" (because the field damage caused by the nematode resembles that caused by a lightning strike); in Myanmar this nematode is known as "akhet-pet" and in Vietnam as "tiem dot san".

Classification

Kingdom: Animalia
Phylum: Nematoda
Class: Tylenchoidea
Order: Tylenchida
Family: Anguinidae
Genus: *Ditylenchus*

Symptoms : During vegetative growth from seedling to flag leaf, the principal symptom of infection is leaf chlorosis. In light infections, the chlorosis will be discrete white spots, less than 1 mm in diameter. As nematode numbers increase, these spots become more numerous, intensify and coalesce towards the base of the penultimate leaf and lower part of the emerging leaf. *D. angustus* does feed on the inner surface of the leaf sheaths, but these rarely show obvious symptoms. In time the chlorotic areas will show some localized browning. Depending on the severity of infection, chlorotic leaf areas, tillers or whole plants will wither and die, attaining a light-brown appearance. In severe infections, the entire crop takes on this appearance. If infected tillers survive to flowering then the panicles remain partially or completely enclosed within the leaf sheath. Partially emerged panicles bear few, if any, true seed and the empty spikelets are distorted and discoloured.

Management

- **Sanitation**: The destruction or removal of infested stubble and straw, crop rotation, control of weeds and ratoon crops, the control of water flow and the use of escape cropping methods.
- An economical control of ufra disease could be obtained by applying carbofuran at 0.75–1.0 kg ai/ha upon transplantation or during land preparation
- Root-dip treatment of ufra-infected seedlings with 10% Miral (isazofos) and Tecto 60FL (thiabendazole) suspensions for 18 hours
- The wild rice species O. subulata and the cultivar RDA16-06 exhibited complete resistance to D. angustus. The cultivars RDA B3, RDA 14, RDA 4, RDA 2, RDA B4, RDA B8, RDA 3, Bazail 65 and RDA B5 also demonstrated ufra resistance
- The cultivars Padmapani and Digha grown in parts of India and Bangladesh, however, escape ufra disease in the field due to their short growth cycle (early crop maturity) (Bora and Medhi, 1992)

ii. Rice root-knot nematode (*Meloidogyne graminicola*): *Meloidogyne graminicola* is known to infect and cause serious damage to cereals, especially rice, in many countries. *M. graminicola* is the major species attacking both upland (rainfed) and lowland (irrigated) rice. It is well adapted to flooded conditions and yield losses of up to 87% have been reported. It is an economically important pest in most rice growing tracts (Ravichandra, 2008). In India, it is reported to cause 17-30% yield loss due to poorly filled kernels (Jain *et al.*, 2007).

Classification

Kingdom: Animalia
Phylum: Nematoda
Class: Secernentea
Order: Tylenchida
Family: Heteroderidae
Genus: *Meloidogyne*

Symptoms: The rice crop in nursery-beds and main field appears uneven with patches of yellowish, stunted plants. The leaf size is reduced, tillering is poor and earhead emergence is delayed. In case of heavy infestation, no earheads may be produced. The earheads have poorly filled or no grains. The crop in infested patches dries early during moisture stress. The symptoms produced due to the infestation of *M. graminicola* are manifested in the form of characteristic terminal hook or typical ring like spindle or bead/nodule shaped galls on the roots. The thicker roots of transplanted crop may appear irregularly thickened with no apparent cortical swellings but with stele swollen at various places

Management

- Tkm 6, Patnai 6, KMP- 179 and N 136 cultivars are resistant to this nematode
- Integration of *P. fluorescens* (2.5 kg/ha), neem cake (1 ton/ha) and carbofuran (1 kg a.i/ha)
- Crop rotation with non-host crops viz., sweet potato, cowpea, sesamum, castor, sunflower, soybean, turnip and cauliflower inhibit nematode development (Rao *et al.*, 1986)
- Bio-agents (*Pseudomonas fluorescens*, *Glomus fasciculatum* , *Paecilomyces lilacinus, Pochonia chlamydosporia* and *Trichoderma viride*) and were advised to apply each to the nursery at 20 g.sq. m
- Soil solarization in rice nursery

- Carbofuran 3G at 3Kg a.i. /ha incorporated at sowing in deepwater rice.
- Carbofuran, phorate, isazophos, cartap and carbosulfan or quinalphos when given as soil application at 1 kg a.i. /ha significantly reduces the root galling byM. g

iii. White tip nematode (*Aphelenchoides besseyii*): It is sometimes referred to as rice white tip, spring dwarf, strawberry bud, flying strawberry nematode or strawberry crimp disease nematode. It is a foliar nematode. The nematodes survival stage is to remain anhydrobiotic in seed until planting. As surrounding plants grow the nematodes become active and feed on meristematic tissue (Hoshino, *et.al.,* 2000). *A. besseyi* reproduces amphimicticly although parthenogenesis can take place. As the plant begins to reach reproductive maturity the number of nematodes increases dramatically. The nematodes migrate to feed on reproductive structures eventually settling in the developing rice seed. As the kernel dries the nematode slowly desiccates and can remain viable in the kernel for up to three years.

Classification

Kingdom: Animalia
Phylum: Nematoda
Class: Secernentea
Order: Tylenchida
Family: Aphelenchoididae
Genus: *Aphelenchoides*

Symptoms : Feeding by nematodes at leaf tips in rice results in whitening of the top 3-5 cm of the leaf, leading to necrosis known as "White Tip". Tips of developing leaves become twisted and crinkled. Diseased plants are stunted, lack vigor and produce small panicles. Affected panicles show high sterility, distorted glumes and small and distorted kernels. In extreme cases the affected grains become chaffy. When nematode carrying seeds are sown, the immature forms become active, move up the plant along a thin film of moisture and feed on the foliage ectoparasitically.

Management of rice nematodes

- Fallowing, use of resistant varieties viz., Supriya, CR-1014
- Crop rotation with non-hosts viz., sesamum and *Basella rubra*;
- Foliar spray with chlorpyriphos 0.05%
- Soil application with carbofuran or phorate at 1 kg a.i/ha

- Application of 225 l/ha of DD prior to sowing or DBCP at 18 l/ha at planting or to growing crops was reported to be effective as soil treatment, foliar sprays of oxamyl or chlorpyriphos at 50 ppm also are effective against this nematode.
- **Integrated approach:** The above methods individually provide some level of control but none of them is sufficient. A combination of cultural practices, soil solarization and a small quantity of nematicides has been designed and demonstrated to be effective in giving effective, safe and economical method of nematode management. This package also provides good control of other soil-borne pathogens and weeds and can fit in well into an IPM schedule. The package comprises of;

i. Nursery bed

Location of nursery-bed at a site where no rice has been grown during previous two years, or

- Soil solarization of the area marked for nursery- beds using colorless, thin (25-60um) polythene sheet mulch for three weeks in May, and
- Application of carbofuran or phorate @ 0.2 g a.i. m^2 mixed with upper 5 cm layer of soil at the time of sowing.

ii. Main field

- Ploughing twice at 15 day interval with a soil turning harrow, in May-June
- Good puddling before transplanting
- Use of healthy nematode-free seedlings from treated nursery-beds.
- In case of heavy infestation replacing rice with a non-host viz.,
- Increasing soil fertility can compensate for some damage by *M. graminicola*. Resistant cultivars hold out the most promise for effective and economic control, and some resistance to the different species has been found. Chemical control on the field scale is generally uneconomic, particularly with low-yielding upland rice, but could be an economic proposition for nursery soils.
- Certain crops are resistant or poor hosts of *M. graminicola* and could be used in rotation to reduce nematode populations e.g. castor, cowpea, sweet potatoes, soybean, sunflower, sesame, onion, turnip, *Phaseolus vulgaris*, jute and okra (Rao *et.al.*, 1986)

- Seed priming (5 g or ml kg^{-1} seed) and soil application (2 kg or l ha^{-1}) of phorate 10G /carbofuran 3G
- One or two consecutive crops of cowpea or seasons of fallow before a rice crop lowered nematode populations and improved rice yield by 30-80%
- Carbofuran @ 0.3 g a.i/m^2 in the nursery bed and field application of fungal bio-agent *Trichoderma viride* @2.5 kg /ha incubated with 25 kg FYM at 45 days after transplanting is quite effective

9.2.8. Root-lesion nematode (*Pratylenchus* spp.): Lesion nematode an important group of migratory endoparasite on roots of several crops. It causes serious damage to many economic plants worldwide including coffee. Root lesion nematodes infect a great variety of hosts. *Pratylenchus penetrans* alone has over 350 host plants. Major hosts include coffee, potato, yam, ornamentals, mint, plantation crops, corn, banana, turf, peanut and wheat. They nematodes induce lesions on root surface. Major species are *P. penetrans, P. brachyurus, P. coffeae, P. pratensis, P. zeae, P. teres, P. goodeyi, P. thornei* and *P. vulnus.*

Classification

Kingdom: Animalia
Phylum: Nematoda
Class: Secernentea
Order: Tylenchida
Family: Pratylenchidae
Genus: *Pratylenchus*

Symptoms: Symptoms of disease are similar in most plants and generally include necrotic lesions of the roots. The lesions can also be entrances for pathogenic bacteria and fungi, which produce secondary infections. Above ground the plant becomes stunted, chlorotic and wilted, and it often dies. A crop field may be patchy as plants wither and die. Root lesion nematodes can cause significant yield losses. It occurs in temperate regions worldwide, regions between the subtropics and the polar circles. It is an animal that inhabits the roots of a wide variety of plants and results in necrotic lesions on the roots. Symptoms of *P. penetrans* make it hard to distinguish from other plant pathogens; only an assay of soil can conclusively diagnose a nematode problem in the field.

As their common name implies, lesion nematodes produce characteristic necrotic lesions (darkened areas of dead tissue) on the surface and throughout the cortex of infected roots. The lesions turn from reddish-brown to black and are initially spotty along the root surface. As the nematodes continue to migrate and feed within the roots, the lesions can coalesce to become large necrotic areas of tissue that may eventually girdle the root. Tissue distal to the lesion is frequently sloughed off. Severe damage from high populations of lesion nematodes can result in a stunted and necrotic plant root system. The extent of lesion formation can be accelerated during concomitant root invasion by other soil borne plant pathogens, and sometimes these interactions can develop into synergistic disease complexes. Lesion nematode damage to roots not only causes losses in yield , but also can cause losses in marketable quality of infected plant products.

Management

- Crop rotation with black gram or green gram
- Soil application with carbofuran 3G or phorate @ 1 kg a.i/ha
- Use of resistant varieties viz., CR 115-5029, TNAU 6464 and CRHP-1
- Aldicarb or carbofuran 3G at 4 kg/ha could reduce the nematode population

9.2.9. Citrus nematode *(Tylenchulus semipenetrans*) : It is a major pest on all members of *Citrus* group fruits. Their feeding strategy is semi-endoparasitic and has a very narrow host range among commonly grown crops. Apart from *Citurs* group, they can also attack olive, grape, persimmon and lilac. It is the causal agent of slow decline in citrus. Their feeding strategy is semi-endoparasitic and has a very narrow host range among commonly grown crops. These nematodes are considered as major plant-parasitic nematode because they can cause 10-30% losses reported on citrus trees. Mature females are non-vermiform, anterior part is embedded in root tissues while the slender posterior part protrudes from roots and is swollen. Males and juveniles are vermiform and slender.

Classification

Kingdom : Animalia
Phylum: Nematoda
Class: Secernentea
Order: Tylenchida
Family: Tylenchulidae
Genus: *Tylenchulus*

Symptoms: High population densities of the citrus nematode can result in severe damage on the citrus tree. Some above ground symptoms can be observed such as suppression of citrus tree growth, lack of vigor or decline symptoms, yellowing of foliage and small size of fruit. The young adult females penetrate into the cortex cells, become sedentary and form multiple 'nurse' cells. The nematode feeding from these nurse cells reduces the amount of water and nutrients available to the growing plant. For below ground symptoms, the infected roots are thicker, darker, decayed and appear dirty. This is caused by soil particles sticking the gelatinous matrices which have been excreted by the females. The infected root systems due to the nematode damage lose the ability to absorb enough water and nutrients for normal growth (Duncan *et.al.*, 1990). Yellowing of foliage, leaf curling and dieback are consequences of insufficient root development and decayed young roots

Management

- Use of healthy seedlings/cuttings, never to establish nurseries near old citrus orchards, sterilization of nursery soil before planting, avoid spread through tools, machinery and irrigation water, soil application of oil cakes-castor, neem, mahua, groundnut, mustard etc., are some of the important cultural practices
- Bare root dips in hot water at 45°C for 25 min. or at 46.7°C for 10 minutes.
- Use of resistant root stocks like trifoliate orange (*Pocirus trifoliata*) , and its hybrids (citrumello), root stocks evolved at India Horticultural Research Institute, Bangalore, Karnataka, India, by crossing Rangpur lime (*Citrus limonia*) with *P.trifoliata* viz., CRH-3, CRH-5 and CRH-41 are resistant ones.

The hybrid rootstock called Swingle citrumelo (*Citrus paradisi* x *P. trifoliata*) is highly resistant to the citrus nematode.

- Carbofuran 3G/aldicarb at 6kg ai/ha are effective against the nematode
- *P. fluorescens* @ 20 g/tree is effective
- Avoidance of infested planting materials and contaminated farm implements.
- Provision of adequate light and water drainage

9.2.10. Mushroom nematodes *(Ditylenchus myceliophagus, Aphelenchoides composticola)*

i. *Ditylenchus myceliophagus:* These are pests on commercial mushroom cultivation. They feed on a wide range of fungal hyphae. They swarm and aggregate on mushroom beds.

Classification

Kingdom : Animalia
Phylum: Nematoda
Class: Secernentea
Order: Tylenchida
Family: Anguinidae
Genus: *Ditylenchus*

ii. *Aphelenchoides composticola*

Kingdom : Animalia
Phylum: Nematoda
Class: Secernentea
Order: Tylenchida
Family: Aphelenchoididae
Genus: *Aphelenchoides*

Symptoms: These nematodes withdraw contents of fungal cells, mycelium disappears nnd mushroom beds become soggy and foul smelling. They can remain viable for 3 years and become cryptobiotic if drying is slow.

Management

- A chemical nematicide for use in mushroom growing requires to be; harmless to crop and man; quick, safe and easy to apply, persistent and protect ant for the crop duration and safe to the environment. Of the many pesticides tested all over the world, thionazin at a concentration of 80 ppm in compost has been reported to be very effective against *D. myceliophagus* and *A. composticola* without any toxic and residual effect. The most commonly used nematicides of the present day viz. aldicarb, carbofuran, phorate and quinalphos have not given better results and also there was not much improvement in the yield and also high amount of residues were present in the fruiting bodies.

- Mushrooms are sensitive to many chemicals. So pesticides to exterminate the nematodes should be recommended after a great caution.

- Bacteria, nematode trapping fungi and predaceous nematodes have been found effective in controlling mushroom nematodes. *Bacillus thuringiensis* has been reported as useful in controlling the nematode population. In France, "Antipolis"-a strain of trapping fungus, *Arthrobotrys robusta* has been reported to traps the mycophages and saprophages and improves the cropping.

- Resistance in mushroom to nematodes is characterized by the lower rate of nematode multiplication and mycelial destruction and is generally attributed to inability of the nematodes to feed on mycelium with certain hyphal size but it is now suggested that resistance depends on unidentified biochemical attributes of the spawn. Reported the susceptibility of Strain 1005 and 252 allow feeding only upto 4th stage and 3rd stage respectively and, therefore, suffer less damage. *Agaricus edulis* and *A. sylvicola* have been reported resistant to the nematode has been proved to be very efficient for complete nematode destruction. In the Mushroom Research Laboratory,, Solan steam heating of badly infested mushroom trays for 10-12 hours at 60*C resulted in very good crop.
- All the mycophage and saprophage nematodes should be regarded as undesirable in the mushroom beds. Once established, they cannot be eradicated from the beds without destroying the crops. With experience, it has been learnt that nematode operation in the mushroom growing can be comprehensively checked if proper sanitation conditions are maintained from compost to "cookout" stage. Complete nematode destruction and bumper mushroom crop are inevitable if the following measures are adopted in the mushroom growing;
- Always keep the mushroom houses and its surroundings clean and free from flies and other dirt by spraying 5% formalin solution or any other recommended disinfectant
- Never prepare compost on a naked uncemented platform so that the compost does not come in contact with soil
- The handling tools, wooden trays, etc. should be treated with 5% formalin or dipped in boiling water for 5 minutes before use
- During compost-making, the air as well as the bed temperature in all the corners of pasteurization room, especially near the ventilator should be taken to two hours.
- The casing material should be properly sterilized before use
- The water used for irrigation should be free from nematodes
- After the crop is over, the mushroom house along with trays should be "cooked out" with steam either at 70° C for 5-6 hours or at 80° C for ½ to 1 hour.

9.2.11. Red ring nematode of coconut *(Rhadinaphelenchus/Bursaphelenchus cocophilus)*: Red ring nematode causes red ring disease of palms. It parasitizes the palm weevil *Rhynchophorus palmarum* L., which is attracted to fresh trunk wounds and acts as a vector for *B. cocophilus* to uninfected trees. It has not been reported to be present in some regions including India. *R. cocophilus* is found in tropical environments, closely associated with palms. It is found mainly in the stems of palm trees, but also in leaf petioles and occasionally in roots. The palm weevil, *Rhynchophorus palmarum*, is the vector of the nematode and life stages of *R. cocophilus* can be found in the gut, body cavity and the region of the ovipositor of the weevil.

Classification

Kingdom : Animalia
Phylum: Nematoda
Class: Secernentea
Order: Aphelenchida
Family: Aphelenchoididae
Genus: *Rhadinaphelenchus*

Symptoms: Young coconut palms easily succumb to *R. cocophilus* attack. There is no record of any tree, once affected, having recovered. The disease occurs more commonly in trees 2.5-10 years old, with greatest incidence in those 4-7 years old. Occasionally, a palm as young as 1.5 or as old as 20 years or more may be attacked. The major internal symptom of red ring infection is the 'red ring' for which the disease is named. A cross cut through the trunk of an infected palm one to seven feet above the soil line usually exhibits a circular, colored band around 3 to 5 cm wide. External symptoms include shortening and deformation of already established leaves that might turn yellow-bronze, later deep reddish-brown. The discolouration usually begins at the tip of each leaf and starts in the older leaves before moving to the younger ones. As the leaves change color and dry up, they wilt and die.

The disease is not recognizable externally in its very early stages. The roots, stems and leaf petioles are already infested and there is full development of internal symptoms before the first external symptoms become visible. In the dwarf cultivars, the red colour gives way to shades of brown. Thus, instead of a red ring internally, there is a brown band. The discrete spots are also brown and the yellow discoloration of the leaves is not often apparent. Generally, the leaves become dried and brown, beginning at the tips of the leaflets and progressing downwards. The yellow dwarf cultivars respond in the same way as the green and the crosses between talls and dwarfs, or between Panama tall and any dwarf. They show a browning instead of a characteristic reddening of the leaves and stem tissue.

Management

- Phytosanitation is still the best method of red ring disease management. This strategy is directed at reducing the vector population as well as the number of sources for nematode inoculum.
- As soon as palms with red ring disease or *B. cocophilus*-induced little leaf symptoms have been detected they should be destroyed. In coconut palm, the disease can be confirmed by examining stem tissue extracted with a coring device for evidence of discoloured tissue and red ring nematodes. In African oil palm, stem coring for evidence of necrosis and red ring nematodes is not reliable.
- Trees should be sprayed with an insecticide (ie. methomyl) and killed with 100-150 mls (48.3% a.i.) of the herbicide monosodium acid methanearsonate (MSMA) or other herbicide that is injected or placed into the trunk. Occasionally trees injected with MSMA will harbor weevil larvae. Therefore, the tree should be cut and sectioned to make sure that weevils are not present.
- Palms that are heavily infested with weevils should be cut, sectioned, and treated with an insecticide such as methomyl, trichlorfon, monocrotophos, carbofuran, cabaryl, or lindane.
- Injections of systemic nematicides, such as fenamiphos, oxamyl, and carbofuran into little leaf symptomatic palms can help with palm recovery. However, because of the damage to the very young leaves in little leaf palms the recovery can take between 6-8 months.
- Mass-trapping with traps baited with sugarcane and synthetic aggregation pheromone (Rhyncholure; racemic 6-methyl-2-hepten-4-ol; ChemTica International) reduce the numbers of *R. palmarum* and change their distribution patterns (from highly aggregated to random) while reducing red ring disease incidence. At mass trapping onset, most *R. palmarum* were captured in "border" traps of the test site suggesting removal of potential immigrants into the study area. A combination of perimeter and "internal" traps is most effective for mass trapping (Anonymous, 2019).

9.2.12 Cyst nematodes (*Heterodera* spp.): These sedentary endoparasites are potential pathogens on several crop plants. Different species of this genus attack several crops, often causing great economic damage. The genus is unique among nematode genera because of the ability of the female to transform into a tough, brown, cyst. Cysts are either partially enclosed in root tissue or in the soil. It is called a cyst nematode because the greatly swollen, egg-filled adult female is referred to as the "cyst stage". Female is swollen or obese, lemon shaped with a distinct neck. Females produce several hundred eggs and after death, the female cuticle forms a protective cyst. Eggs are retained within the cyst. Males are vermiform which are present in soil. The genus has world-wide distribution with a wide host range including horticultural crops. Major species include *H. avenae, H. glycines, H. cajani, H. schachtii, H. trifolii, H. gottingiana, H. carotae, H. cruciferae, H. ciceri, H. oryzae, H. oryzicola, H. sacchari, H. polygoni etc.,* However, *H. glycines* and *H. schachtii* are of postharvest importance.

i. Sugar beet cyst nematode, *Heterodera schochtii:* It infects more than 200 different plants including economically important crops such as sugar beets, cabbage, broccoli, and radish. *H. schachtii* is found worldwide.

Classification

Kingdom : Animalia
Phylum: Nematoda
Class: Secernentea
Order: Tylenchida
Family: Heteroderidae
Genus: *Heterodera*

Symptoms : *H. schochtii* modifies the host root cells and forms a hypermetabolic feeding site which nematodes withdraw nutrients from. It is assumed that this parasite manipulate the production and signaling of the host plant hormone cytokinin to activate cell division. However, the mechanism of how the cytokinin modifies host plant's cells is still unknown. It is thought that syncytium is the result of repeated mitosis since there are multiple enlarged nuclei inside the cells. These feeding sites serve as the nematode's sole source of nutrients throughout its life cycle for several weeks.

Affected plants are marked by stunted growth, wilting, yellowing, decreased yields, and death. Infested roots and tubers will have cysts attached to them. Infested plants will have poor root growth devoid of feeder roots.

Management

- Ethylene dibromide and metham-sodium are effective at controlling the nematode, but economic and environmental concerns generally prevent the use of fumigants as a form of control.
- *H. schachtii* is susceptible to infection by *Verticillium suchlasporium* and other fungi. When sugar beets are grown on a virgin field and a given amount of *H. schachtii* is introduced, beets are most vulnerable during

the first several seasons. With the continued presence of the nematode in a field, infectious fungi levels rise, nematode concentrations decrease, and crop production increase

- By growing non-susceptible plants (vegetables, pulses, cereals) for 2 or 3 years between sugar beet planting, nematode levels can be dramatically reduced

ii. Soybean cyst nematode, *Heterodera glycines:* It is a devastating pest of the soybean worldwide. The nematode infects the roots of soybean, and the female nematode eventually becomes a cyst.

Classification

Kingdom : Animalia
Phylum: Nematoda
Class: Secernentea
Order: Tylenchida
Family: Heteroderidae
Genus: *Heterodera*

Symptoms : Infection causes various symptoms that may include chlorosis of the leaves and stems, root necrosis, loss in seed yield and suppression of root and shoot growth. It is also a significant problem in the soybean growing areas of South America and Asia.. There are usually three generations in the year. In the autumn or in unfavorable conditions, the cysts containing dormant larvae may remain intact in the soil for several years

Symptoms could be confused with nutrient deficiency, particularly iron deficiency, stress from drought, herbicide injury or another disease. The first signs of infection are groups of plants with yellowing leaves that have stunted growth. The pathogen may also be difficult to detect on the roots, since stunted roots are also a common symptom of stress or plant disease. Observation of adult females and cysts on the roots is the only accurate way to detect and diagnose nematode infection in the field.

Management

- The most important form of management is prevention of introduction of the nematode to the environment. One should avoid planting infected plants,
- *H. glycines* has a relatively limited host range and can thus be managed by appropriate crop rotation with non-hosts. A two year rotation with a non-host is usually sufficient to reduce populations to a non-damaging level.

- Soybean seed-coating strategy through combination of three rhizobacterial strains, including *Bacillus simple*, *B. megaterium* and *Sinarhizobium fredii*
- Resistant varieties of soyabean. The two common sources of resistance are PI88788 and Peking, both of which are effective in North Dakota.
- Chemicals may be used wherever it is economically viable. The best management strategy involves rotation with non-hosts and use of resistant varieties of soyabean
- Diazinon at 3.4 g a.i./liter was sprayed at foliage until runoff
- Mother strawberry dipping in solution of nematicides (fosthiazate or ethoprophos at 2.5 g a.i./liter in 20□ or 46□) for 10 min. and planted in a greenhouse for dipping treatment
- Hot water treatment: 20–30 minutes at 44.4□C or 10–15 minutes at 46.1□C or 8–10 minutes at 47.7□.
- One of the best and most practiced forms of management to reduce the local and geographical spread of the disease is sanitation. Removing the infected leaves of the plant can reduce spread in the individual plant, but because the nematode is found in most foliar tissue the nematodes may already be present in other tissues before the leaf symptoms appear.

9.2.13 Stem & bulb nematode (*Ditylenchus dipsaci*): It is slender, vermiform migratory endoparasite of plant stems and leaves. *D. dipsaci* is one of the most devastating phytonematodes in the world. Its' races are very diverse and found in most temperate areas of the world. In this nematode, fourth stage juveniles are infective but not second stage, as in most other phytonematodes. It produces a dormant structure known as 'nematode-wool', which is a bundle of juveniles to survivie under adverse climatic conditions. This pathogen infects 400-500 plant species worldwide including onion, narcissus, potato garlic and other bulb crops, alfalfa, strawberry, carrot, tobacco, tulips and feva beans. *D. dipsaci* has an extensive host range. Major damage occurs in garlic onion, carrot, fava bean, alfalfa, oats, and strawberry. Ornamental plants can also be infected including hyacinth and tulip. It is estimated that this pathogen infects 400–500 plant species worldwide.

Classification

Kingdom : Animalia
Phylum: Nematoda
Class: Tylenchoidea
Order: Tylenchida
Family: Anguinidae
Genus: *Ditylenchus*

Symptoms: Symptoms of infection include stunted growth, discoloration of bulbs, and swollen stems. *D. dipsaci* is a migratory endoparasite that has a five-stage lifecycle and the ability to enter into a dormancy stage. *D. dipsaci* enters through stomata or plant wounds and creates galls or malformations in plant growth. This allows for the entrance of secondary pathogens such as fungi and bacteria.

In onions, garlic, and leeks, infected plants show characteristic symptoms including stunted growth, yellow spots, leaf curl, and foliage lesions. Stems often have swollen regions called "spikkles." As adult nematodes migrate into the scales of the bulb, and the scales become soft, grey, and loosely packed. Highly infected bulbs can also split apart or show malformed bloating. The leaves of the plant become flaccid and may collapse. This can lead to defoliated plants. Garlic shows similar symptoms of leaf yellowing and stunted bulbs. When harvested, the infected garlic may be missing portions of the root system.

In fava beans (*Viva faba*) symptoms of infection include reddish-brown stem lesions that can turn black. Young bean pods are dark-brown. Infected seeds are smaller and distorted compared to healthy beans. Speckles and spots are also commonly seen on infected fava beans. Up to 30 biological races occur within *D. dipsaci* that are mostly distinguished by their host preferences. Very little morphological differences are seen between the races which makes diagnosis difficult (Vanstone and Russell, 2011) Seed material samples from infected plants can be dissected and viewed under a microscope to confirm the correct race.

Management

Certified nematode-free seeds and planting material are most essential to prevent crop damage by *D. dipsaci*

- Infection can be prevented by ensuring that only clean seeds and bulbs are planted. Bulbs and seeds can be disinfected by hot-water treatments. Soaking them in 110 to 115 °F water with formalin, a formaldehyde solution, for two to three hours can successfully kill nematodes

- Proper sanitation in fields and of tools is essential in preventing and controlling the spread of *D. dipsaci*, because they can survive and reproduce in infected plants and residues. The fourth-stage juvenile is the most resilient and can survive repeated desiccation or drying and recover upon rehydration. All infected tissues should be removed from growing sites and destroyed to control populations and all farm tools and equipment should be cleaned of potentially contaminated soil before moving them to a new location.
- Races of *D. dipsaci* are highly host-specific, so employing a three-year crop rotation can deprive the nematodes of a suitable host and starve the population. Because some weeds serve as hosts for nematodes, controlling weeds in fields decreases the number of susceptible hosts and the ability of the nematodes to survive and spread.
- The time a susceptible host crop is planted also infects the severity of nematode damage. Cooler temperatures and lower humidity can suppress the reproduction and infestation rates of *D. dipsaci*. Growers should avoid planting susceptible bulbs, seeds, or seedling during seasons of peak nematode infection. Soil fumigation in fields during fall can control nematodes on a susceptible crop in the spring. A nematicide fumigant that is specific to the genus *Ditylenchus* should be used. Selectively fumigate only the regions of the fields that are infected to ensure that the high cost of fumigating does not mitigate the economic gain from saving crops from nematode damage. Fumigants are usually applied before planting and subsequently after emergence
- Postharvest treatment of planting bulbs for 3 hours in a 1% solution of formalin at 440 C combined with preplant soil fumigation with 1,3-dicbloropropene and (or) at-planting applications of phorate
- Hot-water treatment of narcissus bulbs infected with stem nematodes consist either of storage for 1-2 weeks at 25-30°C, followed by soaking in water for 24 h and hot-water treatment at 45°C for 4 h, or storage for 1-2 weeks at 25-30°C, followed by hot-water treatment at 47°C for 4 h. Routine hot-water treatment for the control of other pests and diseases should be carried out at 43.5°C for 3 h to control any slight, possibly unnoticed nematode infection.

9.3. Symptoms of major postharvest nematode diseases

Carrot

Beetroot

Yam

Turmeric

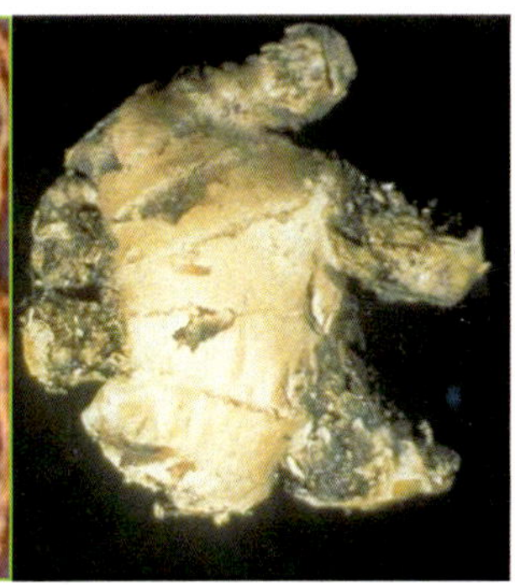

Ginger

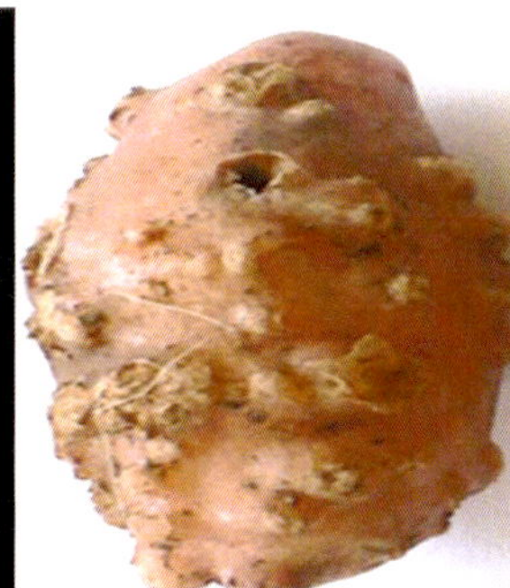

Potato

Symptoms due to Root-knot nematode (*Meloidogyne* spp,)

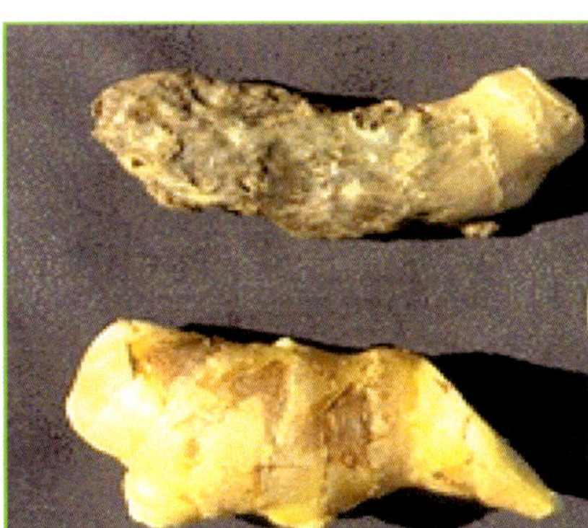

Ginger

Turmeric

Symptoms due to Burrowing nematode (*Radopholus similis*)

Seed gall nematode in Wheat (*Anguina trtici*)

Rice 'Ufra' (*Ditylenchus angustus*)

Rice white tip nematode (*Aphelenchoides besseyii*)

Lesion nematode on Yam (*Pratylenchus* coffeae)

Sugar beet cyst nematode (*Heterodera schachtii*)

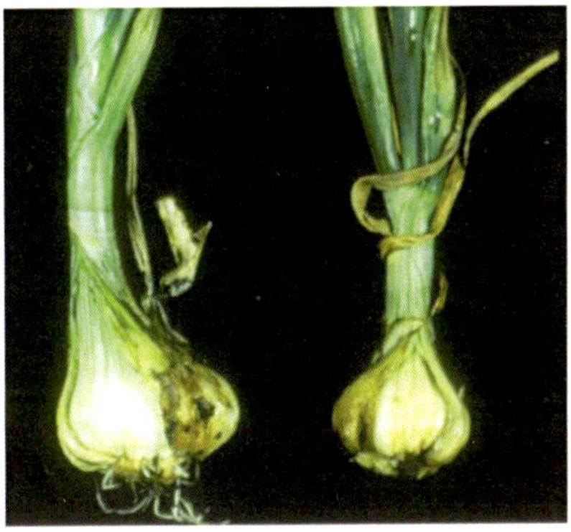

Onion Stem & bulb nematode (*Ditylenchus dipsaci*) Potato rot nematode (*D.destructor*)

References

Anonymous, 2019. Current year. Invasive Species Compendium. CABI, Wallingford, UK: CAB International.

Bora, L. C., and Medhi. B. N., 1992. Resistance of deepwater rice DWR varieties to ufra disease in Assam. *Intl. Rice Res. Newl.*, **17** :12-12

Carlos H. Crisosto, 1999. Effect of Preharvest Factors and Bruising in Stone Fruit Decay. Perishables Handling Quarterly Issue Publishers. Issue No. 99

De Waele, D,M., Jones, B,L, Bolton, C. and Van den Berg, E., 1989. Ditylenchus destructor in hulls and seeds of peanut. *J. Nematol.*, **21**:10-15.

Decraemer, Wilfrida and Hunt, D.J., 2006. Structure and classification. *Plant Nematology*, 3-32. 10.1079/9781845930561.0003.

Diez and Dusenbery. 1989. Repellent of root-knot nematodes from the exudate of host roots. *J. Chem. Ecol.* **15**:2445-245

Duncan, L.W., and Cohn. E., 1990. Nematode parasites of citrus. Pp. 321-346 . *In. Plant-parasitic nematodes in subtropical and tropical agriculture.* M.Luc, R.A. Sikora, and J.Bridge, (Eds.,), Wallingford, UK: CAB International.

Hoshino, S. and Togashi, K., 2000. Effect of Water-Soaking and Air-Drying on Survival of Aphelenchoides besseyi in *Oryza sativa* Seeds. *J Nematol*, **32** :. 303-8.

Lennox,C.L. and Spotts,R.A., 2004. Timing of Preharvest Infection of Pear Fruit by Botrytis cinerea and the Relationship to Postharvest Decay. *Pl. Dis*., **88**:468-473. .

Naqvi S.A.M.H.,2004. Diagnosis and Management of Pre and Postharvest Diseases of Citrus fruit. In: Naqvi S.A.M.H. (eds) Diseases of Fruits and Vegetables Volume I. Springer, Dordrecht

Rao, Y.S., Prasad, J.S. and Panwar, M.S., 1986. Nematode problems in rice: crop losses, symptomatology and management. *In:. Plant Parasitic Nematodes of India: Problems and Progress. Swarup G, Dasgupta DR, (Eds* .), New Delhi, India: Indian Agricultural Research Institute, 279-299

Ravichandra, N.G., 2008. Plant Nematology, I.K. International Publishing House Pvt. Ltd., New Delhi, India, 693 pp.

Ravichandra, N.G., 2014.Horticultural Nematology, Springer Publications, Netherlands, 412 pp.

Sdiri, S., Navarro, P., Monterde, A., Benabda, J. and Salvador, A., 2012. New degreening treatments to improve the quality of citrus fruit combining different periods with and without ethylene exposure. *Postharvest Biol. Technol/.*, **63**:25–32

Stephen F. Marsh, 1984. Preharvest Infection of Corn Silks and Kernels by *Aspergillus flavus*. *Phytopathology,* **74:**1284-1289.

Syama Chatterton, Andrew C.Wylie and Zamir K.Punja, 2012. Fruit infection and postharvest decay of greenhouse tomatoes caused by *Penicillium* species in British Columbia. *Canadian J. Pl. Pathol.*, **34**: 524-535

Thorne, G., 1961. Principles of Nematology. McGraw-Hill Book Company, Inc. New York, NY.

Vanstone, V. and Russell, J., 2011. Pathogen of the Month-March 2011, Date accessed October 24, 2012, www.appsnet.org/publications/potm/March2011%20POTM.pdf American Society for Microbiology. DOI**:** 10.1128/AEM.02972-14

Wei Zhao, Jinhe Bai, Greg McCollum, Elizabeth Baldwin, 2015. High Incidence of Preharvest Colonization of Huanglongbing-Symptomatic *Citrus sinensis* Fruit by *Lasiodiplodia theobromae* (*Diplodia natalensis*) and Exacerbation of Postharvest Fruit Decay by That Fungus.

William R. Jarvis, 1994. Latent Infections in the Pre- and Postharvest Environment. *Hort.Sci.*, **29**: 7449-7751

10

Biochemical and Molecular Aspects of Postharvest Diseases

Postharvest losses of fruits and vegetables are even higher and are estimated to be 40–50%. Postharvest fruit rotting are a major cause of those losses and are chiefly caused by fungal pathogens after fruit ripening. In a manner similar to foliar diseases that occur in the field, several factors as fungal pathogenicity, host response and environment determine the outcome of host resistance or susceptibility. However, in postharvest diseases fruit ripening is another major component that will determine fruit resistance. The molecular trigger induces the germ tube to differentiate an appressorium. It has been shown that excess nutrients over ride it and protein synthesis is not required once germination has occurred. Protein synthesis during conidium germination and appressorium formation on cellulose membranes were observed (Pathak, 1997). A specific polypeptide (95 kDa) was synthesized only when appressoria matured (a period lasting from 12-24 h after germination). This molecule is involved in the synthesis of the melanin precursor, Scytalone .

Appressoria grown in the presence of cyclohexamide do not mature nor they subsequently differentiate a penetration peg. Appressorium maturation involves the formation of melanin. It has been shown that melanization of appressoria is necessary for formation of a successful penetration peg. The melanin appears to give the appressorium mechanical strength, but pigmentation is only one factor required for penetration competent appressoria. At restrictive temperatures, pigmentation of conidia of *C. lagenarium* mutants could be restored with dihydroxyphenylalanine without restoration of appressorial competence. Non-pathogenic mutants have been used to determine factors important for pathogenicity. Cutinase production was essential for penetration of papaya fruits. The cutinase produced during infection of papaya fruit by *C. gloeosporioides* has been purified and characterized as a glycoprotein of 24 kDa. Polyclonal antibodies raised against this enzyme do not cross react with cutinase from Fusarium so/ani, and lesions do not develop when papaya fruits are inoculated with spore suspensions containing the antibody preparation.

Interestingly, papayas pretreated with the purified cutinase of *C. gloeosporioides* and subsequently inoculated with *Mycosphaerella* sp. get infected and show lesion formation in the absence of wounding (*Mycosphaerella* is normally a wound pathogen of papaya fruits, causing a stem end rot, and cannot invade the intact cuticle). Role of enzymes and toxins in various rots has been demonstrated by several workers. A number of organophosphorus pesticides, as well as the anionic detergent, sodium dodecyl sulphate are known to inhibit cutinase activity in vitro in addition to preventing infection of papaya fruits by *C. gloeosporioides*. It is likely that some pre-harvest fungicides and wetting agents may operate in the field partially through this mechanism.

10.1 Postharvest disease development: Fruits infected by postharvest fungal pathogens develop, in general, disease symptoms after harvest and during storage. Postharvest fungal pathogens germinate and enter the fruit by breaching the host cuticle. This is achieved by: degrading host cuticle, entering through natural openings of the host and wounds, or by living endophytically in the stem end. When those particularly insidious pathogens encounter unripe fruit these fungi often remain quiescent and confined to the initial site of introduction. They are unnoticed to visual examination, for as long as months of storage, until the harvested fruit ripen (Noam Alkan and Ana M. Fortes, 2015). Several species of fungal pathogens, such as *Colletotrichum, Alternaria, Botrytis, Monilinia, Lasiodiplodia, Phomopsis*, and *Botryosphaeria* have been reported to live quiescently in their hosts until the fruits ripen. As fruit ripen, postharvest fungal pathogens switch to aggressive growth. At this aggressive stage, the fungi are necrotrophs, which kill the host cell and obtain nutrients from the host, leading to decomposed fruit tissue and decay (Prusky *et al.*, 2013) . However, before this devastating stage those fungi adopt different types of life styles. Some fungi as *Lasiodiplodia, Phomopsis, Colletotrichum, Alternaria* and others, cause stem-end-rot and colonize the stem-end by adopting endophytic-like lifestyle before fruit ripening. Other fungi, e.g., *Colletotrichum* are defined as hemibiotroph, those fungi live quiescently as biotrophs in unripe fruit cells without killing them. In a parallel manner, fully necrotrophic fungi as *Botrytis* can infect and live in a restricted 1–3 cells of unripe fruit without damaging the surrounding tissue.

10.1.1 *Botrytis* and *Colletotrichum* Models: Better understood *Colletotrichum* (anthracnose) representing hemibiotrophic fungi and on *Botrytis* (gray mold) as necrotrophic fungi were discussed as these fungi are two of the most common postharvest fruit disease agents that are known to attack many economically important fruits and present problems world-wide (Noam Alkan and Ana M.Fortes, 2015). *Colletotrichum gloeosporioides* causes the anthracnose disease to at least 470 host genera and *Botrytis cinerea* causes the gray mold disease on over 200 species of fruit. On unripe fruit, *Colletotrichum* conidia germinate and develop appressoria which penetrate the fruit cuticle via an infection peg. *C. gloeosporioides* enters the quiescent stage whereupon two distinct structures develop: dendritic-like protrusions which form within the fruit cuticle and swollen hyphae which colonize the first epidermal cell layer but advance no further (Alkan *et al.*, 2015). Simultaneous transcriptome analysis of *Colletotrichum gloeosporioides* and tomato fruit pathosystem reveals novel fungal pathogenicity and fruit defense strategies.

A mechanistic model for *Botrytis cinerea* on grapevine was developed, which accounts for conidia production on various inoculum sources and for multiple infection pathways, considers two infection periods. During the first period ("inflorescences clearly visible" to "berries groat-sized"), the model calculates the infection severity on inflorescences and young clusters caused by conidia (*SEV1*). During the second period ("majority of berries touching" to "berries ripe for harvest"), the model calculates the infection severity of ripening berries by conidia (*SEV2*) and the severity of berry-to-berry infection caused by mycelium (*SEV3*). The model was validated in 21 epidemics (vineyard × year combinations) between 2009 and 2014 in Italy and France. A discriminant function analysis (DFA) was used to: *i)* evaluate the ability of the model to predict mild, intermediate, and severe epidemics; and *ii)* assess how *SEV1*, *SEV2*, and *SEV3* contribute to epidemics. The model correctly classified the severity of 17 of 21 epidemics. Results from DFA were also used to calculate the daily probabilities that an ongoing epidemic would be mild, intermediate, or severe. *SEV1* was the most influential variable in discriminating between mild and intermediate epidemics, whereas *SEV2* and *SEV3* were relevant for discriminating between intermediate and severe epidemics (González-Domínguez *et al.*, 2015). The model represents an improvement of previous *B. cinerea* models in viticulture and could be useful for making decisions about Botrytis bunch rot control.

When *C. gloeosporioides* germinates on the cuticle of ripe fruit it germinates as on green fruit and goes through a short biotrophic stage. Only this time it is much more rapid and the quiescent structures immediately switch to necrotrophic growth. This indicates that hemibiotrophic growth in *C. gloeosporioides* is developmentally cued when encounter with fruit cuticle. On the other hand, *Botrytis* spore germlings tend to penetrate through small wounds or cracks in the epidermis or cuticle of unripe fruit and remain confined within the lumen of the wounds. When the hemibiotrophic *C. gloeosporioides* germinates on small wounds of unripe fruits, its colonization skips the biotrophic-like stage and it adopts the necrotrophic strategy, similarly to *B. cinerea*. Growth of either pathogen on wounds in unripe fruit is limited for long periods, and upon ripening, both pathogens become necrotrophic, degrade host tissues and produce symptoms of disease (Prusky *et al.,* 2013).

Species of *Colletotrichum* use diverse strategies for invading host tissue, ranging from intracellular hemibiotrophy to subcuticular intramural necrotrophy. In addition, these pathogens develop a series of specialized infection structures, including germ tubes, appressoria, intracellular hyphae, and secondary necrotrophic hyphae. *Colletotrichum* species provide excellent models for studying the molecular basis of infection structure differentiation and fungal-plant interactions. Various stages of the infection processes of *Colletotrichum* species, including spore adhesion and germination, germ tube and appressorium differentiation and functions, and biotrophic and necrotrophic development were discussed (Perfect *et al.,* 1999). The contribution of molecular, biochemical, and immunological approaches to the identification of genes and proteins relevant to each stage of fungal development were considered. As well as reviewing results from several groups, the hemibiotrophic pathogen, *C. lindemuthianum* was also discussed.

10.1.2 Strawberry-*Botrytis cinerea* pathosystem : Grey mould in strawberries can result from *B. cinerea* infections of open flowers (primary infections) or by penetration of fruit receptacle tissues (secondary infections) (Stefan Petrasch *et al.,* 2019). In primary infections, *B. cinerea* infects flower organs during or right after flowering, allowing hyphae to grow into the receptacle . The sources of primary inoculum range from overwintering sclerotia to conidia or mycelium from infected neighbouring plants. Infected senescent petals, stamens and calyxes can facilitate primary infections in fruit. Histological studies have shown that even though styles are frequently infected, fungal growth appears to be strongly inhibited and never reaches the receptacle. In contrast, fungal growth in colonized stamens can reach the receptacle in some cultivars. Following infection of the unripe receptacle by *B. cinerea*, fungal growth is usually arrested and a symptomless quiescent phase occurs. The mechanisms that lead to quiescent infections are not yet fully understood. Proanthocyanins (PAs) appear to induce *B. cinerea* quiescence in unripe fruit by restricting the activity of fungal enzymes like polygalacturonases (PGs) that are necessary for aggressive infection of hosts (50% inhibition in unripe fruit compared to 8% inhibition in ripe fruit). Even though PA content in fruit remains constant during ripening, increasing polymerization of PAs leads to lower inhibitory activity in ripe fruit. Similarly, anthocyanins might delay *B. cinerea* infections or cause quiescence. For instance, strawberries illuminated with white fluorescent light showed increased anthocyanin content and delayed development of grey mould .

Reduced fruit decay has also been observed in raspberries with high pigmentation and in transgenic tomatoes that accumulate anthocyanins. Other small phenolics, especially catechins, may have a role in quiescence. High levels of catechins inhibit fungal growth, and a decrease in catechins is correlated with a reduction of other antifungal compounds such as lipoxygenases. Interestingly, young and ripe fruit have low catechin concentration, suggesting that initial infections of young receptacles are possible because they do not yet accumulate enough catechins to stop colonization. *B. cinerea* quiescence is complex and involves additional factors besides the accumulation of phenolic compounds. It has been proposed that quiescence in unripe fruit is initiated by lack of nutrients such as sugars (e.g. mono- and disaccharides) from the host, presence of preformed antifungal compounds, unsuitable environment for fungal virulence factors (Prusky and Lichter, 2007). In unripe strawberries, factors from all three categories are present, including lack of available sugars, preformed antifungal compounds, and high activity of PG-inhibiting proteins (PGIPs). Induction of the necrotrophic phase in ripe strawberries could be triggered by changes in biochemical composition of the host tissues associated with the ripening process, such as increased sugar content, volatile production and alteration of plant defences.

These modifications promote not only fungal growth but also host susceptibility, e.g. via the release of oxalic acid and efflux of toxins (Stefan Petrasch *et al.*, 2019). During secondary infections, the fungus initiates the necrotrophic phase without quiescence. The sources of conidia for secondary infections can also be diverse, from senescent leaves to infected fruit. Conidia from *B. cinerea*-infected flower parts are major sources of secondary inoculum. It has been estimated that more than 64% of the strawberry infections result from organic fragments that are in contact with the fruit, such as petals and stamens. Contrary to other fruit (e.g. raspberries), senescent flower parts often adhere to strawberries long enough to retain water films for at least 8 h, which is the time needed for *B. cinerea* conidia germination. Secondary infections can also result from nesting, which corresponds to direct penetration of mycelia growing on neighbouring plant organs such as infected leaves and fruit. Generally, secondary infections proceed rapidly and *B. cinerea* can complete its germination and infection as fast as 16 h post-inoculation (hpi) with a rapid increase in fungal biomass at 48 hpi . Early responses of strawberries to infection include higher expression of the defence genes FaPGIP and FaChi 2-1 (Class II Chitinase), whereas lower expression of the reference gene DNA Binding Protein – FaDBP indicates extensive cell death induced by *B. cinerea* at late stages of infection.

10.1.3 Unripe fruit tolerance and changes occurring during ripening: During fruit ripening, significant physiological shifts occur: cell wall remodeling soluble sugar accumulation, decrease in the amount of phytoanticipins and phytoalexins; decline of inducible host defense responses; cuticle biosynthesis and changes in the ambient host pH. Most of those changes are thought to be governed by complex hormonal signals including ethylene, ABA, jasmonic acid (JA), and salicylic acid (SA), which occur during natural fruit ripening. Interestingly, similar phytohormones are regulated in the host in response to pathogens. In response to the changes in the host, pathogens alter the enzymes and compounds they produce which allow them to infect and break down or macerate the fruit tissue (Noam Alkan and Ana M.Fortes, 2015). Signals for release from quiescence probably occur during fruit ripening and may include disassembled cell wall substrates, alterations in cuticle and other signals. When the fungi are re-activated they induce rotting disease that impairs crop quantity, quality and appearance. Changes in the fruit occur during ripening and senescence that regulate susceptibility. Morphological and biochemical changes taking place during the transition from unripe and tolerant to ripe and susceptible fruit. These changes could stimulate the release from quiescent/ latent or confined colonization to the aggressive colonization of the necrotrophic fungi in ripe fruits.

Concentrations of different gases and volatiles present or produced inside a fruit are determined by the permeability of the fruit tissue to these compounds. Primarily, surface morphology and anatomical features of a given fruit determine the degree of permeance across the fruit. Species and varietal variability in surface characteristics and anatomical features therefore influence not only the diffusibility of gases and volatiles across the fruits but also the activity and response of various metabolic and physiological reactions/processes regulated by these compounds. Besides the well-known role of ethylene, gases and volatiles; O_2, CO_2, ethanol, acetaldehyde, water vapours, methyl salicylate, methyl jasmonate and nitric oxide (NO) have the potential to regulate the process of ripening individually and also in various interactive ways. Differences in the prevailing internal atmosphere of the fruits may therefore be considered as one of the causes behind the existing varietal variability of fruits in terms of rate of ripening, qualitative changes, firmness, shelf-life, ideal storage requirement, extent of tolerance towards reduced O_2 and/or elevated CO_2, transpirational loss and susceptibility to various physiological disorders. In this way, internal atmosphere of a fruit (in terms of different gases and volatiles) plays a critical regulatory role in the process of fruit ripening. So, better and holistic understanding of this internal atmosphere along with its exact regulatory role on various aspects of fruit ripening will facilitate the development of more meaningful, refined and effective approaches in postharvest management of fruits (Vijay Paul and Rakesh Pandey, 2014). Its applicability, specially for the climacteric fruits, at various stages of the supply chain from growers to consumers would assist in reducing postharvest losses not only in quantity but also in quality.

10.2 Host factors modulating postharvest fungal development : Recently, with the expansion of omics technique several observations elaborated on the involvement of differential response of ripe and unripe fruit to fungal pathogens(Noam Alkan and Ana M.Fortes, 2015). Storage of freshly harvested fruit is a key factor in modulating their supply for several months after harvest; however, their quality can be reduced by pathogen attack. Fruit pathogens may infect their host through damaged surfaces, such as mechanical injuries occurring during growing, harvesting, and packing, leading to increased colonization as the fruit ripens. Of particular concern are fungal pathogens that not only macerate the host tissue but also secrete significant amounts of mycotoxins. Many studies have described the importance of physiological factors, including stage of fruit development, biochemical factors (ripening, C and N content), and environmental factors (humidity, temperature, water deficit) on the occurrence of mycotoxins. However, those factors usually show a correlative effect on fungal growth and mycotoxin accumulation. Recent reports have suggested that host factors can induce fungal metabolism, leading to the synthesis and accumulation of mycotoxins.

Alternaria alternata is the causal organism of core rot decay symptoms in susceptible cv. Red Delicious but not in resistant cv. Golden Delicious. The two cultivars did not differ in natural colonization of the style and ovary during the first week after full bloom; colonization of the ovary in the susceptible cultivar subsequently decreased with increasing distance from the calycine tube (Niem *et al.*, 2007). By 30 days after full bloom, *Alternaria* recovery from ovary 1, adjacent to the end of the calycine tube, was 100 and 40% in the susceptible and resistant cultivars, respectively. In the susceptible cultivar, *Alternaria* recovery decreased from 75 to 20% in ovaries 2, 3, and 4, while there was only minor incidence in the resistant cultivar. Inoculation of the mesoderms of the two cultivars induced similar decay symptoms, but inoculated locules of Red Delicious were more susceptible than those of Golden Delicious. Increased inoculum concentration or isolate virulence enhanced the difference in locule susceptibility between the cultivars. Inoculation on isolated seed locules or on media amended with susceptible locule tissue as a carbon source induced greater transcript levels of several genes than the inoculation on resistant tissue. Endo- and exoglucanase activity levels were higher at pH 4.8 than at 4.2, conditions typical of the mesoderm adjacent to the seed locules of the susceptible and resistant cultivars, respectively. Results suggested that susceptibility of Red Delicious apples to core rot decay is dependent on the sensitivity to locule colonization and on mesoderm pH, a factor that enhances fungal virulence.

Penicillium expansum is one of the most important postharvest rots of apple fruit and is also of great concern to fruit processing industries. Elucidating the pathogenicity mechanism of this pathogen is of utmost importance for the development of effective and safe management strategies (Levin *et al.*, 2017). Although, many studies on modification of the host environment by the pathogen were done, its interactions with fruit during the early stages of infection and the virulence factors that mediate pathogenicity have not been fully defined. Effectors carrying LysM domain have been identified in numerous pathogenic fungi and their role in the first stages of infection has been established. In this study, they identified 18 LysM genes in the *P. expansum* genome. Amino acid sequence analysis indicated that *P. expansum* LysM proteins belong to a clade of fungal-specific LysM. Eleven of the discovered LysM genes were found to have secretory pathway signal peptide, among them, 4 (*PeLysM1 PeLysM2*, *PeLysM3* and *PeLysM4*) were found to be highly expressed during the infection and development of decay of apple fruit. Effect of targeted deletion of the four putative PeLysM effectors on the growth and pathogenicity was studied. Possible interactions of PeLysM with host proteins was investigated using the yeast-two-hybrid system.

Pathogens can penetrate fruit and foliar plants directly through the cuticle following the development of appressoria, or through wounds during fruit growth and postharvest life. In both cases, disease symptoms may not occur until long after infection. Maturation (when plant or fruit has attained maximum growth and size) and ripening (the process by which mature fruit become ready for consumption) are critical factors in fruit susceptibility to postharvest pathogens (Prusky and Wilson, 2018). During these periods, fruit undergo physiological and metabolic changes that affect nutritional composition and cause a decline in resistance mechanisms. Recent research suggests that host nutrients during ripening may modulate signaling processes in the pathogen that lead to metabolic responses, triggering the accumulation of pH-modulating molecules. The pH shift in the surrounding host tissue induces optimal gene-expression conditions for the fungus to use specific pathogenicity factors at each particular pH. Leaf pathogens usually penetrate via appressoria and also respond to carbon-regulation signaling. The importance of carbon regulation in fruit and leaf tissue and its contribution to pathogen colonization via facilitation of the transition from quiescent to necrotrophic lifestyle was highlighted..

Fruits are fresh products that undergo significant biochemical and chemical changes during postharvest storage. Fruit nutritional components that undergo changes during fruit maturation and ripening include sugars (levels may increase from 5 to 20%) and organic acids, coupled with an increase in fruit pH (Dilip Kumar *et al.*, 2017). This raises the question of whether such changes in the host can modulate the expression of fungal mycotoxin biosynthetic gene clusters (BGCs). With respect to host-produced metabolites, Barkai-Golan, Paster, Jackson, and Al-Taher claimed that the high content of carbohydrates in figs, dates, citrus fruit, and raisins probably enhances aflatoxin production by *Aspergillus flavus* and *Aspergillus parasiticus*; however, no modulatory mechanism for cluster activation was indicated.

On the other hand, recent studies of patulin synthesis suggest that several host factors are involved in the inhibition of mycotoxin accumulation in apples. Patulin accumulation during colonization of apples by *Penicillium expansum* was inhibited by the addition of extracts of the mushroom *Lentinula edodes* and of the antioxidants quercetin and umbelliferone suggesting that metabolites, produced by the host or other organisms, can alter patulin synthesis *in vivo*. Two possible mechanisms that might modulate fungal mycotoxin cluster activation and the accumulation of mycotoxins are pathogen-modified nutritional factors present in the host that become precursors for activation of the mechanism of mycotoxin production; and natural chemical factors present in the host that directly modulate the transcription factors that induce mycotoxin production (Dilip Kumar *et al.*, 2017).

10.3 Pathogen-modified factors present in the host: Accumulation of patulin by P. expansum was found to be strongly dependent on the secretion of D-gluconic acid (GLA) resulting from the oxidation of host sugars by the pathogen. While the acidification of host tissue contributes to the activation of pathogenicity factors and colonization, it was also found that GLA accumulation may contribute both to pH reduction of the host tissue and as a precursor for patulin (Barad, *et al.*, 2014).

Most postharvest pathogens exhibit necrotrophic or saprotrophic lifestyles, resulting in decomposition of the host tissues and loss of marketable commodities. Necrotrophic fungi can readily infect ripe fruit leading to the rapid establishment of disease symptoms. However, these pathogens generally fail to infect unripe fruit or remain quiescent until host conditions stimulate a successful infection. Previous research on infections of fruit has mainly been focused on the host's genetic and physicochemical factors that inhibit or promote disease. If fruit pathogens can modify their own infection strategies in response to the ripening stage of the host was studied. To test this hypothesis,

Petrasch *et al.,* (2019) profiled global gene expression of three fungal pathogens that display necrotrophic behavior, *Botrytis cinerea*, *Fusarium acuminatum*, and *Rhizopus stolonifer,* during interactions with unripe and ripe tomato fruit. They assembled and functionally annotated the transcriptomes of *F. acuminatum* and *R. stolonifer* as no genomic resources were available. Differential gene expression analysis was conducted to compare each pathogen during inoculations versus *in vitro* conditions. Through characterizing patterns of overrepresented pathogenicity and virulence functions (e.g., phytotoxin production, cell wall degradation, and proteolysis) among the differentially expressed genes, they were able to determine shared strategies among the three fungi during infections of compatible (ripe) and incompatible (unripe) fruit tissues. Though each pathogen's strategy differed in the details, interactions with unripe fruit were commonly characterized by an emphasis on the degradation of cell wall components, particularly pectin, while colonization of ripe fruit featured more heavily redox processes, proteolysis, metabolism of simple sugars, and chitin biosynthesis. The three fungi were unable to infect fruit from the *non-ripening* (*nor*) tomato mutant, confirming that to cause disease, these pathogens require the host tissues to undergo specific ripening processes. By enabling a better understanding of fungal necrotrophic infection strategies, they generated accurate models of fruit diseases and the development of early detection tools and effective management strategies.

Soft rot is one of the destructive diseases of vegetables and occurs worldwide wherever fleshy storage tissues of vegetables and ornamentals are found. It causes a greater total loss of produce than any other bacterial disease (Bhat *et al*., 2010). The disease can be found on crops in the field, in transit, in storage and during marketing; resulting in great economic losses. It is primarily caused by *Erwinia carotovora* sub-sp. *carotovora* and sometimes by *Erwinia carotovora* sub-sp. *atroseptica*. The soft rot disease has a very wide host range infecting vegetable species belonging to all families. Name of the disease aroused from the characteristic soft decay of fleshy tissue which terminates into watery or slimy mass. The bacteria enters the host tissue through injuries. The decay is aggravated when high humidity is coupled with high temperature which results in fast rate of multiplication of the pathogen. For this reason much of the loss due to this disease occurs during middle of the summer. Increased amounts of pecteolytic enzymes released by the pathogen results in maceration of tissue and are of great significance to the pathogenesis of the disease.

10.4 Natural host factors affecting the pathogen: Efforts to identify host factors that are potentially involved in patulin regulation have centered on the role of sucrose in impacting patulin levels. Kumar *et al.* (2016) demonstrated that sucrose, the main sugar usually present in fruit, modulates patulin accumulation in a dose-responsive pattern by directly regulating the expression of the global regulator of secondary metabolism, *laeA*. An increase in sucrose culture amendment from 15 to 175 mM decreased both patulin accumulation and *laeA* expression by 175- and 5-fold, respectively. These results may tie into the ammonia levels, since the highest patulin accumulation was observed in the presence of low sucrose level, conditions that induce ammonia accumulation by the pathogen. This suggests that limiting sugar levels, probably resulting from intact cells in unripe fruit, may be a mechanism for activating mycotoxin synthesis that differs from that in cell walls from mature ripe cells, which activates different processes of fungal metabolism. Interestingly, negative regulation of *creA* compared to *laeA* was observed at different sucrose levels, also indicating the importance of sugar regulation in these transcription factors. These data support the view that host nutritional factors, as a result of fruit maturity, may differentially contribute to regulation of the patulin BGC

Using freshly harvested fruit sampled at increasing stages of maturity, the ability of *P. expansum* to colonize fruit and accumulate patulin was analyzed in apple fruit at increasing maturity. The wild-type strain of *P. expansum* (WT) showed increasing patulin accumulation, from 0.2 to 1.5 mg g^{-1} fresh weight as the fruit matured, and the total soluble solids of the fruits (TSS) increased from 12.5% and 13.5% to 13.96% in the first, second, and third harvests, respectively. Thus the WT Pe-21 strain showed an increasing trend for patulin accumulation in apples in progressive harvests with progressive aggressiveness of the pathogen. The differential levels of sugar, found both *in vitro* and *in vivo*, suggest that internal metabolic factors may further contribute to the regulation of the metabolic cluster and patulin accumulation.

Changes in fruit maturity and ripening are accompanied by significant nutritional factors that determine the quality of the fruit. While many reports indicate that mycotoxin accumulation is dependent on a variety of environmental factors (Drusch and Ragab, 2003), there are few examples showing how changes in natural host metabolites during ripening affect fungal metabolites and the global regulation of mycotoxin synthesis. This is important given the differential levels of mycotoxin accumulation during maturation and ripening. Interestingly, until now, there have been few reports indicating the importance of changes in the level of organic acids and other natural fruit compounds, such as phenols. In *Aspergillus*, the phenolic antioxidants gallic acid, vanillic acid, protocatechuic acid, 4-hydroxybenzoic acid, catechin, caffeic acid, and chlorogenic acid, which are present in fruit, were studied for their effects on OTA production. Whereas *Aspergillus ochraceus* was not inhibited by any of these compound, the effects of each compound on OTA production were variable, suggesting that species-specific OTA production and response to phenolic compounds may be influenced by different and as yet unreported mechanisms affecting OTA accumulation (Palumbo, *et al.*, 2007). A similar approach should be taken for organic acids, such as malic and citric acids, which are present in ripening fruit. A recent report indicates that organic acids may have antifungal activity against mycotoxigenic pathogens and an inhibitory effect on aflatoxin B1 accumulation (Hassan *et al.*, 2015)]. Future research should focus on understanding how these natural host factors impact BGC expression and mycotoxin production during fruit colonization.

10.4.1. Phytohormones: Phytohormones are well-known to affect fruit ripening and the defense responses to pathogens. Important signaling roles have been ascribed to classical defense hormones SA, JA, abscisic acid (ABA), and ethylene (ET) in molding plant–pathogen interactions. Gibberellic acid (GA), auxin (IAA), brassinosteroids (BRs), and cytokinines (CK) have recently been emerged as important modulators of plant defenses against microorganisms based mostly on vegetative tissues data and on the lifestyle of the infecting pathogen (Noam Alkan and Ana M.Fortes, 2015).

Until recently, most studies on the role of hormones in plant-pathogen interactions focused on salicylic acid (SA), jasmonic acid (JA), and ethylene (ET). It is now clear that pathogen-induced modulation of signaling via other hormones contributes to virulence. A picture is emerging of complex crosstalk and induced hormonal changes that modulate disease and resistance, with outcomes dependent on pathogen lifestyles and the genetic constitution of the host. Recent progress has revealed intriguing similarities between hormone signaling mechanisms, with gene induction responses often achieved by derepression. Robert-Seilaniantz *et al.,* (2010) reported on recent advances, updating current knowledge on classical defense hormones SA, JA, and ET, and the roles of auxin, abscisic acid (ABA), cytokinins (CKs), and brassinosteroids in molding plant-pathogen interactions. an emerging theme that positive and negative regulators of these disparate hormone signaling pathways are crucial regulatory targets of hormonal crosstalk in disease and defense were highlighted.

Phytohormone crosstalk is crucial for plant defenses against pathogens and insects in which salicylic acid (SA), jasmonic acid (JA), and ethylene (ET) play key roles. These low molecular mass signals critically trigger and modulate plant resistance against biotrophic as well as necrotrophic pathogens through a complex signaling network that even involves participation of other hormones (Yang *et al.*, 2015). Crosstalk among SA, JA and ET is mediated by different molecular players, considered as integral part of these crosscommunicating signal transduction pathways. Recent progress has revealed that the positive versus negative interactions among those pathways ultimately enable a plant to fine-tune its defense against specific aggressors. On the other hand, pathogens have evolved strategies to manipulate the signaling network to their favour in order to intensify virulence on host plant. Here they reviewed recent advances and current knowledge on the role of classical primary defense hormones SA, JA and ET as well as their synergistic and antagonistic interaction in plant disease and immune responses. Crosstalk with other hormones such as abscisic acid, auxin, brassinosteroids, cytokinins and melatonin is also discussed mainly in plant disease resistance.

Understanding the molecular mechanisms underlying coffee-pathogen interactions are of key importance to aid disease resistance breeding efforts. In this work the expression of genes involved in salicylic acid (SA), jasmonic acid (JA) and ethylene (ET) pathways were studied in hypocotyls of two coffee varieties challenged with the hemibiotrophic fungus *Colletotrichum kahawae*, the causal agent of Coffee Berry Disease. Based on a cytological analysis, key time-points of the infection process were selected and qPCR was used to evaluate the expression of phytohormones biosynthesis, reception and responsive-related genes (Diniz *et al.*, 2017). The resistance to *C. kahawae* was characterized by restricted fungal growth associated with early accumulation of phenolic compounds in the cell walls and cytoplasmic contents, and deployment of hypersensitive reaction. Similar responses were detected in the susceptible variety, but in a significantly lower percentage of infection sites and with no apparent effect on disease development. Gene expression analysis suggests a more relevant involvement of JA and ET phytohormones than SA in this pathosystem. An earlier and stronger activation of the JA pathway observed in the resistant variety, when compared with the susceptible one, seems to be responsible for the successful activation of defense responses and inhibition of fungal growth. For the ET pathway, the down or non-regulation of ET receptors in the resistant variety, together with a moderate expression of the responsive-related gene *ERF1*, indicates that this phytohormone may be related with other functions besides the resistance response. However, in the susceptible variety, the stronger activation of *ERF1* gene at the beginning of the necrotrophic phase, suggests the involvement of ET in tissue senescence. This is the first attempt to unveil the role of phytohormones in coffee-*C. kahawae* interactions, thus contributing to deepen our understanding on the complex mechanisms of plant signaling and defense.

Although fruit decay in sweet cherries has been studied in some detail, information is still scarce about the possible role of phytohormones during postharvest. Tijero *et al.*, (2019)examined whether or not changes in endogenous contents of phytohormones occur during fruit decay of sweet cherries stored at room temperature. They evaluated the endogenous variations in the contents of phytohormones, including abscisic acid, gibberellins, cytokinins, auxin, jasmonic acid, salicylic acid and melatonin, by UHPLC–ESI–MS/MS during fruit decay at room temperature and to what extent these changes in phytohormone contents were associated with alterations in water contents, soluble sugars and acidity. Endogenous contents of abscisic acid, cytokinins and gibberellins decreased in parallel with fruit decay, thus suggesting a protective role against over ripening for these compounds. Among cytokinins and gibberellins, free cytokinin bases (zeatin and 2-isopentenyl adenine, rather than their ribosides), and gibberellin 3, changed in parallel with fruit decay. It was concluded that abscisic acid, free cytokinin bases and gibberellin 3 may prevent fruit decay during storage of sweet cherries at room temperature.

The effects of the jasmonic acid derivative n-propyl dihydrojasmonate (PDJ) on ethylene signal transduction and endogenous jasmonic acid (JA) in apples infected with *Botrytis cinerea* were investigated (Sirinan Suktawee *et al.*, 2019). Apples were dipped into 400 μM PDJ solution and then inoculated with B. cinerea. The fruit were stored at 25°C and 95% relative humidity for 16 days after PDJ treatment. The inoculation without PDJ application (PDJ– Ino+) showed larger B. cinerea lesion diameters compared to the PDJ application with inoculation (PDJ+ Ino+) and the untreated controls. In contrast, the PDJ+ Ino+ group showed a higher ethylene production rate, higher 1-aminocyclo-propane-1-carboxylic acid (ACC) concentration, and greater expressions of ethylenerelated genes [MdACS1 (ACC synthase), MdACO1 (ACC oxidase), MdETR1 (Ethylene receptor 1), MdERS1 (Ethylene response sensor 1), and MdCTR1 (Constitutive triple response 1)], and endogenous JA, MdAOS1 (Allene oxide synthase 1) gene. However, the abscisic acid concentrations were decreased in the PDJ+ Ino+ group. The results suggested that PDJ application induces ethylene production through ethylene-related genes and endogenous JA, resulting in *B. cinerea* infection inhibition.

10.4.2. Jasmonate-salicylate crosstalk: Salicylic acid and JA signaling pathways are generally considered as antagonistic dependent on NPR1 and hormone concentration. This interplay between SA and JA was suggested to optimize host-response to pathogen's lifestyle. In vegetative tissue it is commonly postulated that an effective responses to biotrophic pathogens are typically mediated by SA and programmed cell death, and responses to necrotrophic pathogens, which benefit from host cell death, involve JA and ethylene signaling (Noam Alkan and Ana M.Fortes, 2015). During normal fruit ripening many phytohormones as ethylene, IAA, ABA, GA, JA, and SA are regulated (Seymour, *et al*., 2013) , which complicate the effective fruit response to pathogens. Indeed, in infections of grapes and tomato fruit with *Botrytis* and tomato fruit infections with *Colletotrichum* several stress hormone responsive pathways including ethylene, ABA, JA, and SA were regulated. In fruit, high levels of ET and ABA, which stimulate senescence/ ripening processes, may facilitate colonization by necrotrophs. The balance between SA and JA responses seems to be crucial for fruit resistance .

During the biotrophic-like quiescence stage of *C. gloeosporioides* the resistant unripe fruit mainly responded through activation of JA, ethylene, and ABA (Alkan *et al.,* 2015). However, the ripe tomato is more susceptible to disease and the fungi adopt a necrotrophic mode of host colonization. In the fruit response to necrotrophic infections, SA biosynthetic, signaling and response pathways were activated. SA signaling was shown to have an important role in necrotrophic colonization stage of *Colletotrichum* on tomato fruit by inducing cell death and likely as a means to suppress JA mediated defense response. The activation of this pathway was dependent on NADPH oxidase activity that was induced by ammonia secreted by the fungus. Indeed, JA application leads to increase tolerance and SA application leads to PCD and increased susceptibility to *C. gloeosporioides* at it necrotrophic stage. Similarly, ripe *NahG* tomato fruit mutant, lacking SA responses, showed increased tolerance to *C. gloeosporioides.* In a reciprocal manner, the *Spr1* mutant, deficient in JA signaling, showed increased susceptibility.

The responses of tomato fruit to *Botrytis* involved several stress hormone responsive pathways including ethylene, ABA, JA, and SA in a complex manner that differ during infections of ripe and unripe tomato fruit. In general, both grape and tomato defense response to *Botrytis* were mainly mediated by JA and ET (Agudelo-Romero *et al*., 2015). Similarly, *Arabidopsis* leaves response to *Botrytis* was mediated mainly by JA and ET. However, in unripe tomato fruit the *NahG* gene showed susceptibility to *Botrytis* . This result suggests that SA has an important role in unripe fruit resistance to *Botrytis*. In grapes of susceptible cultivar infected with *B. cinerea*, the pathogen causes shutdown of defenses, which are regulated by SA and are expressed during the normal ripening. It remains to be established whether SA-mediated defense system is inhibited or not in resistant cultivars (Noam Alkan and Ana M.Fortes, 2015). In climacteric fruit ET and ABA stimulate ripening and may affect the host defense response. Thus, resistance of *sitiens* mutants at ripe fruit stage to *Botrytis* and ABA elevation in response to *Botrytis* was correlated to susceptibility. The ethylene and ABA elevation seems to play a dual role, on the one hand it correlated with JA and ethylene resistance response, and on the other hand it induced ripening and increase susceptibility.

Two cultivars Buram-3-ho (susceptible) and CR-Hagwang (moderate resistant) of kimchi cabbage seedlings showed differential defense responses to anthracnose *(Colletotrichum higginsianum*), black spot (*Alternaria brassicicola*) and black rot (*Xanthomonas campestris* pv. *campestris*, Xcc) diseases (Lee *et al*., 2014). Defense-related hormones salicylic acid (SA), jasmonic acid (JA) and ethylene led to different transcriptional regulation of pathogenesis-related (PR) gene expression in both cultivars. Exogenous application of SA suppressed basal defenses to *C. higginsianum* in the 1st leaves of the susceptible cultivar and cultivar resistance of the 2nd leaves of the resistant cultivar. SA also enhanced susceptibility of the susceptible cultivar to *A. brassicicola*. By contrast, SA elevated disease resistance to Xcc in the resistant cultivar, but not in the susceptible cultivar. Methyl jasmonate (MJ) treatment did not affect the disease resistance to *C. higginsianum* and Xcc in either cultivar, but it compromised the disease resistance to A. brassicicola in the resistant cultivar. Treatment with 1-aminocyclopropane-1-carboxylic acid (ACC) ethylene precursor did not change resistance of the either cultivar to *C. higginsianum* and Xcc. Effect of ACC pretreatment on the resistance to *A. brassicicola* was not distinguished between susceptible and resistant cultivars, because cultivar resistance of the resistant cultivar was lost by prolonged moist dark conditions. Taken together, exogenously applied SA, JA and ethylene altered defense signaling crosstalk to three diseases of anthracnose, black spot and black rot in a cultivar-dependent manner.

The yeast *Kloeckera apiculata* strain 34–9 is an antagonist that shows biological control activity against the postharvest fungal pathogens of citrus. An antifungal compound, 2-phenylethanol (PEA), has been identified from the extract of *K. apiculata* (Liu *et al*., 2016). To better understand the molecular processes underlying the response of citrus fruit tissue to *K. apiculata*, the extract and PEA, microarray analyses were performed on navel oranges using an Affymetrix Citrus GeneChip. As many as 801, 339 and 608 differentially expressed genes (DEGs) were identified after the application of *K. apiculata,* the extract and PEA, respectively. In general, *K. apiculata* induced the expression of defence-related genes. In addition to chitinase and β-1,3-glucanase, genes involved in ethylene (ET), jasmonic acid (JA), calcium signalling, MAPK signalling and phenylalanine metabolism were induced. In contrast, monodehydroascorbate reductase, superoxide dismutase (SOD), catalase (CAT), peroxidase (POD) and carotenoid biosynthesis genes were down-regulated. The expression profiles for the extract- and PEA-treated samples were similar to that found for yeast (sharing 57.4 % DEGs), with a significant increase in the transcript levels of defence-related genes.

This study provided a global picture of the gene expression changes in navel oranges after the application of the antagonist yeast *K. apiculata*, its extract and PEA. The interpretation of the DEGs revealed new insight into the molecular processes that regulate the defence responses in orange tissue.

Plants use different immune pathways to combat pathogens. The activation of the jasmonic acid (JA)-signaling pathway is required for resistance against necrotrophic pathogens; however, to combat biotrophic pathogens, the plants activate mainly the salicylic acid (SA)-signaling pathway (Taha Abd *et al.*, 2012). SA can antagonize JA signaling and vice versa. NPR1 (noninducible pathogenesis-related 1) is considered a master regulator of SA signaling. NPR1 interacts with TGA transcription factors, ultimately leading to the activation of SA-dependent responses. SA has been shown to promote disease development caused by the necrotrophic pathogen *Botrytis cinerea* through NPR1, by suppressing the expression of two JA-dependent defense genes, proteinase inhibitors I and II. It was shown that the transcription factor TGA1.a contributes to disease development caused by *B. cinerea* in tomato by suppressing the expression of proteinase inhibitors I and II. Evidence was provided to the fact that SA-signaling pathway contributes to disease development caused by another necrotrophic pathogen, *Alternaria solani*, in tomato. Disease development promoted by SA through NPR1 requires the TGA1.a transcription factor. These data highlighted how necrotrophs manipulate the SAsignaling pathway to promote their disease in tomato.

To respond to pathogen attack, selection and associated evolution has led to the creation of plant immune system that are a highly effective and inducible defense system. Central to this system are the plant defense hormones jasmonic acid (JA) and salicylic acid (SA) and crosstalk between the two, which may play an important role in defense responses to specific pathogens or even genotypes. The *Arabidopsis–B. cinerea* pathosystem was used to test how the host's defense system functions against genetic variation in a pathogen (Wei Zhang *et al.* 2017). Defense-related phenotypes and transcriptomic responses were measured in Arabidopsis wild-type Col-0 and JA- and SA-signaling mutants, coi1-1 and npr1-1, individually challenged with 96 diverse *B. cinerea* isolates. Those data showed genetic variation in the pathogen influences on all components within the plant defense system at the transcriptional level. Four gene co-expression networks and two vectors of defense variation triggered by genetic variation in *B. cinerea* were identified. This showed that the JA and SA signaling pathways functioned to constrain/canalize the range of virulence in the pathogen population, but the underlying transcriptomic response was highly plastic. These data showed that plants utilize major defense hormone pathways to buffer disease resistance, but not the metabolic or transcriptional responses to genetic variation within a pathogen.

The effects of the jasmonic acid derivative *n*-propyl dihydrojasmonate (PDJ) on ethylene signal transduction and endogenous jasmonic acid (JA) in apples infected with *Botrytis cinerea* (gray mold) were investigated (Sirinan Suktawee *et al*., 2019). Apples were dipped into 400 μM PDJ solution and then inoculated with *B. cinerea*. The fruits were stored at 25°C and 95% relative humidity for 16 days after PDJ treatment. The inoculation without PDJ application (PDJ^- Ino^+) showed larger *B. cinerea* lesion diameters compared to the PDJ application with inoculation (PDJ^+ Ino^+) and the untreated controls. In contrast, the PDJ^+ Ino^+ group showed a higher ethylene production rate, higher 1-aminocyclo-propane-1-carboxylic acid (ACC) concentration, and greater expressions of ethylene-related genes [*MdACS1* (*ACC synthase*), *MdACO1* (*ACC oxidase*), *MdETR1* (*Ethylene receptor 1*), *MdERS1* (*Ethylene response sensor 1*), and *MdCTR1* (*Constitutive triple response 1*)], and endogenous JA, *MdAOS1* (*Allene oxide synthase 1*) gene. However, the abscisic acid concentrations were decreased in the PDJ^+ Ino^+ group. The results suggested that PDJ application induces ethylene production through ethylene-related genes and endogenous JA, resulting in *B. cinerea* infection inhibition.

10.4.3 Not only SA and JA crosstalk

Major responses to fungal pathogens are mediated by ethylene, ABA, JA, and SA phytohormones. Additional phytohormones and growth regulators such as gibberellin, cytokinins, steroid, polyamines, and BRs were reported to affect fruit response to fungal pathogens. For example gibberellin-treated persimmon fruit had increased resistance of to *Alternaria alternata* by delaying fruit maturation and reducing cuticle cracks (Biton *et al.,* 2014). Gibberellin seems to act also in an alternative way, the DELLA transcription factors enable plants to respond to gibberellin; this mechanism seems to maintain transient growth arrest and lead to plant response to biotic and abiotic stress. DELLA transcription factors were shown to promote susceptibility to biotrophs and resistance to necrotrophs in leaves. Polyamines may have a role in response to postharvest pathogens. In fact, genes involved in polyamine biosynthesis are up-regulated in *Botrytis*-grapes pathosystem . Over-expression of *ARGAH2* in *Arabidopsis* leads to enhanced resistance to *B. cinerea*, thus suggesting a role for the polyamine arginase in plant resistance(Noam Alkan and Ana M.Fortes, 2015). Furthermore, polyamines seem to be co-regulated with ethylene biosynthesis and seem to co-work with ABA.

Steroids such as β-sitosterol and stigmasterol have been previously shown to be involved in plant–pathogen interactions. Indeed, several genes involved in steroid biosynthesis were also upregulated in grape and tomato response to *Botrytis* . Furthermore, BRs play a role in grape ripening and are known to change in plants response pathogen infection. BAK1, known for BRs signaling, contributes to pathogen associated molecular pattern-triggered immunity (PAMP) to necrotrophic pathogens. Loss of BAK1 results in increased susceptibility to the necrotrophic fungi. Several genes coding for BAK1 were down-regulated in grapes infected with *Botrytis* at the onset of ripening (Noam Alkan and Ana M.Fortes, 2015). Application of BRs reduced *Penicillium expansum* decay in jujube fruit and delayed fruit senescence.

10.4.4. Ethylene dual role in ripening and defense response: Ethylene is a key postharvest hormone with a dual role. In climacteric fruit ripening (Seymour *et al*, 2013), the presence of ethylene will lead to susceptibility, however, ethylene could also acts as a defense hormone together with JA . In tomato, a climacteric fruit, ethylene is known to commence during the breaker stage of fruit ripening. Postharvest fungal pathogen as *B. cinerea* and *C. gloeosporioides* infections induce the ethylene biosynthesis pathway, transcription factors as non-ripening (*NOR*), ripening-inhibitor *(RIN)* and never-ripe *(NR)* and the ethylene-regulated defense genes in both ripe and unripe tomato fruit. Therefore, they enhance the ripening process and hasten the release from quiescence. The increased susceptibility of ripe fruit to *B. cinerea* depends on the ripening regulator *NOR*, but not on *RIN*, and only partially on the fruit's perception of ethylene. The *rin* mutant fruits and those treated with 1-MCP did not ripe but, nevertheless, were susceptible to *B. cinerea* . The differential effect of *NOR* and *RIN* indicate the existence of a specific susceptibility factor. Although *nor* and *rin* act together in a cascade for ripening *nor* appears to have a more global effect indicating that it operates upstream of rin. Inspection of the differential expression of genes in these mutants may point to a source for susceptibility.

Applications of 1-MCP, an inhibitor of ethylene perception, are widely used due to the beneficial effects in delaying fruit senescence and prolonging storage. Indeed, in apples, peaches, and plums 1-MCP treatment reduced fungal pathogens rotting. However, in tomato, avocado, custard apple, mango, and papaya, citrus, apples 1-MCP promoted susceptibility to pathogens. It seems that 1-MCP could affect citrus, strawberry and tomato fruits susceptibility in a concentration dependent manner (Noam Alkan and Ana M.Fortes, 2015). At low concentrations it enhances tolerance, but reduces it at high concentrations. Thus, small amounts of endogenous ethylene may be necessary to maintain basic levels of resistance to pathogens due to ethylene involvement in regulation of plant defense genes. These results support the conclusion that ethylene has dual opposing roles: as a ripening hormone it promotes susceptibility and as a participant in hormone-activated defense responses ethylene provide resistance. Thus, the timing of ethylene release, perception and the levels of ethylene are likely to be crucial for the outcomes of resistance or susceptibility.

Ripening marks several changes in physiological and biochemical attributes of fruit. Several hundred genes are recruited to express differentially during this process and their cumulative effect brings about ripening and softening in the fruit. Using suppression subtractive hybridization (SSH), 37 EST-unigenes from banana which are expressed differentially during ripening were identified (Kesari Ravi *et al*., 2007) About 50% of these belong to processes such as stress, defense and detoxification. Besides these, we have also identified genes which are known to involve regulation of gene expression and other processes, although their expression has not been reported during fruit ripening. Expression of some of these genes was studied during ethylene-induced ripening with or without the ethylene action inhibitor, 1-methylcyclopropene (1-MCP), in fruit pulp and other tissues. Expression studies clearly suggest that most of these genes are ethylene-regulated and related to banana fruit ripening. It is concluded that ethylene-induced banana ripening evokes a stress-like response and several genes belonging to stress/defense are expressed in addition to genes related to ethylene biosynthesis, cell wall hydrolysis, secondary plant product biosynthesis, fatty acid biosynthesis, metabolite transport and transcription/translation machinery.

Kiwifruit (*Actinidia deliciosa*) is a climacteric fruit sensitive to low concentrations of ethylene. To investigate the transcriptional mechanisms underlying kiwifruit ethylene response, transcription factors encoding four EIN3-Like (*EIL*s) and 14 Ethylene Response Factors (*ERF*s) were cloned from kiwifruit (Xue-ren Yin, *et al.,* 2010). Expression of these transcription factors was examined during fruit development. The expression of transcripts of most *AdERF*s was higher during early fruit development, with the exception of *AdERF3*, which increased with maturity. Several *AdERF*s were apparently down-regulated by ethylene, as they were affected by the ethylene inhibitor 1-methylcyclopropene and by antisense suppression of *ACO* (for 1-aminocyclopropane-1-carboxylic acid oxidase) in the fruit. In contrast, *AdEIL*s were constitutively expressed during fruit development and ripening. The transcription factors *AdEIL2* and *AdEIL3* activated transcription of the ripening-related genes *AdACO1* and *AdXET5* (xyloglucan endotransglycosylase gene) and, when overexpressed in Arabidopsis (*Arabidopsis thaliana*), stimulated ethylene production. The potential repressor *AdERF9* suppressed this promoter activity. These results supported a role for kiwifruit *EIL*s and *ERF*s in transcriptional regulation of ripening-related genes and in the regulation of kiwifruit fruit-ripening processes.

10.4.5 Reactive oxygen species (ROS) role: Reactive oxygen species play a dual role in fruit–pathogen interaction, which largely depends on their different levels in cells. Fruit recognition of a pathogen immediately triggers an oxidative burst that is considered an integral part of the fruit defense response. ROS are also necessary for the virulence of pathogenic fungi. However, the accumulation of ROS in cells causes molecular damage and finally leads to cell death. On the basis of data regarding ROS production and the scavenging systems determining ROS homeostasis, Ying Wang *et al.*, (2019) focussed on the role of ROS in fruit defense reactions against pathogens and in fungi pathogenicity during fruit–pathogen interaction.

Fruits can withstand pathogens by means of physical barriers (presence of thick cuticular or trichome layers) and chemical ones, or through induced defenses that are activated once the host detects the presence of the pathogen, triggering the oxidative burst during the early hours of interaction. This burst entails the generation of reactive oxygen species (ROS), such as superoxide (O^{2-}), hydroxyl radical (OH^-) or hydrogen peroxide (H_2O_2), and the activation of genes involved in several metabolic pathways. The study of such mechanisms may allow detecting disease-resistant genetic materials, thus reducing the use of toxic products (Sandoval-Chávez *et al.*, 2015).

a. In ripening and defense response : Accumulation of reactive oxygen species (ROS) are the result of the balance between ROS production and antioxidant activity. The mitochondria chloroplast and peroxisome are all potential sources for ROS production. ROS and particularly hydrogen peroxide (H_2O_2) and singlet oxygen ($^1O_2$) contribute to fruit ripening and senescence (Tian *et al.*, 2013). In grape and tomato, ROS, lipid peroxidation, and protein oxidation were increased at breaker stage. In many fruits, storage is associated with an increase in ROS, which results either from increased ROS production or from a decrease in antioxidative makeup. Antioxidants inhibit fruit ripening and senescence, while, high O_2 or application of H_2O_2 leads to increase ROS and senescence. Oxidative damage of several mitochondrial proteins, which involved in fruit senescence would result in impairment of mitochondrial function and lead to fruit senescence.

Many postharvest fungi can modulate host ROS production. Fungi secrete elicitors, toxins and antioxidants in order to modify the plant ROS production. For example, oxalate secreted by *Sclerotinia* generates reducing conditions and inhibited plant oxidative burst at pH 3–4. In contrast, oxalic acid also induced plant NADPH oxidase and ROS production that correlated with a pH increase to 5–6, that later induced host PCD. Another example is ammonia that is secreted by *Colletotrichum*, which activated the tomato NADPH oxidase this resulted in oxidative burst leading to induction of SA mediated defense response, host cell death and enhanced necrotrophic colonization. How these small molecules manipulate the NADPH oxidase is unknown (Noam Alkan and Ana M.Fortes, 2015). Plant ROS can be toxic, e.g., in photo-oxidative stress, but are also known to play an important role in eliciting a wide range of defense mechanisms. One of the earliest plant cellular responses following successful pathogen recognition is the production of ROS, also called oxidative burst. ROS also play a central role in redox-dependent defense signaling and in creating toxic environments that induce cell death. In this regard, at the cellular juncture of pathogenesis, NPR1, the master regulator of SA-mediated defense, is actually reduced by thioredoxin, which removes the effect of nitrosylation. Once reduced, the NPR1 oligomer is disrupted and its monomers enter the nucleus, and activate SA-mediated defense and PCD. It should be noted that in a manner that is not fully understood such signaling can take place even under conditions of cellular oxidative stress.

Reactive oxygen species levels can regulate the host cells fate. High ROS levels in plant cells results in a spreading cell death, which provides nutrients to necrotrophic pathogens. Intermediate concentration of ROS usually results in restricted PCD, and low concentration of ROS can act as a signaling molecule, including the activation of antioxidant enzymes usually leading to host protection against necrotrophs. Mitochondria ROS was shown to be involved in plant PCD in both biotic and abiotic stress responses. A recent report revealed another ROS, singlet oxygen, which plays a major role in plant response to both biotic and abiotic stress (Mor *et al.*, 2014). Signaling of ROS is important in both the ripening process and defense response. However, the exact roles of ROS can appear counterintuitive and there is a need to better understand their compartmentalization.

Various antagonistic yeasts used to control postharvest diseases (*Metschnikowia fructicola, Candida oleophila, Candida saitoana*, and *Pichia guilliermondii*) secreted high, transient levels of superoxide anion as they colonized wounded host tissue and intact surfaces of apple, peach and citrus fruits (Macarisin *et al.*, 2011). The application of *Metschnikowia fructicola* and *Candida oleophila* into citrus and apple fruit wounds induced a significant ($P \leq 0.05$) increase in hydrogen peroxide accumulation in host tissue. Importantly, living cell of *M. fructicola* were detected on the third day after inoculation in fruit wounds exhibiting a high level of host-generated hydrogen peroxide indicating the ability of this yeast to tolerate elevated levels of host ROS. ROS-tolerance may be an essential characteristic of effective yeast antagonists. These data indicated that the ability of yeast antagonists to self-generate and possibly stimulate an oxidative response in host tissue could be one of the major factors underlying the performance of a biocontrol agent.

b. Roles of reactive oxygen species in pathogenesis : *Botrytis cinerea* is a significant necrotrophic plant pathogen causing devastating diseases on more than 500 plant species, especially on fresh fruits and vegetables, resulting in the economic losses ranging from \$10 billion to \$100 billion worldwide. This fungal pathogen invades nearly all parts of plants including stems, leaves, flowers, fruits, and seeds at both pre-harvest and postharvest stages. Due to its wide host range and the huge economic losses that it causes, extensive investigations have been carried out to effectively control this plant pathogen. It is beneficial for exploring the pathogenic mechanisms of *B. cinerea* to provide fundamental basis for control strategies. In recent years, tremendous progress has been made in understanding these pathogenic genes and regulatory pathways, as well as the control strategies of *B. cinerea* (Li Hua, *et al.*, 2018). It is important to deeply understand the molecular basis of pathogenesis of *B. cinerea* and develop new strategies to prevent grey mold rot caused by *B. cinerea* in fresh fruits. In recent years, great efforts have been put on exploring the molecular mechanisms of *B. cinerea,* since its genome information (strain B05.10) is available (Amselem *et al.*, 2011), and the functions of various genes especially those pathogenesis-related proteins are unravelled.

Reactive oxygen species (ROS) are a collective of highly reactive molecules including superoxide anion (O^{2-}), hydroxyl radical (OH), and certain non-radical oxidizing agents such as hydrogen peroxide (H_2O_2) and ozone (O_3) that can be converted into radicals. ROS have an ambivalent role since they damage DNA, causing lipid peroxidation and protein oxidation (Noam Alkan and Ana M. Fortes, 2015), but also function as diffusible second messengers. In the early stage of infection, plant hosts usually trigger oxidative burst which generate large amounts of ROS transiently to counteract the invasive pathogen (Tian *et al.*, 2013). However, as a necrotrophic fungus, *B. cinerea* can exploit the oxidative burst and even contribute to it by producing its own ROS.

ROS can be generated in *B. cinerea* either as unavoidable byproducts of metabolic processes or as the major products of NADPH oxidase (NOX) (Li *et al.*, 2016). NOX is a multi-subunit complex which reduces oxygen to superoxide with the electron supplied by NADPH. The function of the subunits of NOX in *B. cinerea* has been extensively investigated. Both the catalytic subunits BcNoxA and BcNoxB are responsible for pathogenicity and the formation of sclerotia, which allow the fungi to survive under adverse environmental conditions and are fundamental for sexual reproduction . Interestingly, BcNoxA and BcNoxB were shown to play different roles in the pathogenicity of *B. cinerea*. BcNoxA is essential for colonizing the host tissue, whereas BcNoxB contributes to the primary infection. The regulatory subunit BcNoxR has a phenotype consistent with that of *ΔbcnoxA/B* double mutant. Elimination of BcNoxR showed reduced growth rate, sporulation, and impaired virulence on French bean/tomato leaves and various fruits (Li *et al.*, 2016). Based on the comparative proteomic approach to unravel the potential downstream targets of BcNoxR, they identified a total of 49 unique proteins whose abundance changed in the deletion mutant of *bcnoxR* (Δ*bcnoxR*) and found that BcNoxR could affect the expression of proteins with various functions, such as stress response, carbohydrate metabolism, translation, and intracellular signalling. Further analysis showed that 6-phosphogluconate dehydrogenase (BcPGD), whose abundance decreased in the deletion mutant of *bcnoxR*, was responsible for growth, sporulation, and virulence of *B. cinerea* (Li *et al.*, 2016). Moreover, it was observed that small GTPase BcRho3 was contributed to the regulation of mycelial growth, conidiation production, and virulence of *B. cinerea*, and deletion mutant of *bcrho3* (Δ*bcrho3*) showed reduced virulence to apple, tomato fruits, and tomato leaves, and proved that the reduction in virulence of Δ*bcrho3* mutant might be due to the impaired penetration ability.

Modulation of postharvest host pathogen interactions by pH and reactive oxygen species (ROS) was studied (Li Hua *et al.*, 2018). Modulation of host pH in response to a host signal is bidirectional and includes either alkalinisation by ammonification of the host tissue, or acidification by secretion of organic acids. These changes sensitise the host and activate the transcription and secretion of fungal hydrolases that promote maceration of the host tissue. This sensitisation is further enhanced at various stages by the accumulation of host or fungal ROS that can further weaken host tissue and amplify fungal development.

As signalling molecules, it is necessary for ROS to move from the place of origin to the site of action. The transformation molecules of ROS are obscure since ROS are composed of various highly reactive molecules including radicals (O^{2-} and .OH) and molecules (H_2O_2 and O_3), which are difficult to investigate. Among the ROS molecules, H_2O_2 is stable and suitable for investigation (Noam Alkan and Ana M.Fortes, 2015). H_2O_2 can function as both intracellular and intercellular signal molecules. However, H_2O_2 is unable to cross the membrane lipid bilayer freely by simple diffusion, it needs to the aid of membrane lipid compositions or channel proteins to cross over plasma membranes. Aquaporins, which are known as efficient water channels, have been demonstrated to mediate the transportation of H_2O_2 across membranes. These studies indicated that aquaporin8 (BcAQP8) in *B. cinerea* could play a crucial role in the transmembrane transportation of ROS, and the distribution of mitochondria, which is the main source of ROS. Besides to regulate ROS transportation, BcAQP8 also affects the expression of BcNoxR, the regulatory subunit of NOX. Deletion of *bcaqp8* results in decrease in growth, sporulation, and pathogenicity of *B. cinerea.*

The effect of exogenous melatonin treatment of citrus fruit to postharvest green mold caused by *Penicillium digitatum* (*Pd*) was studied (Yunlong Lin *et al.*,2019). Melatonin had no obvious effect on *Pd* germination and growth, but melatonin treatment of citrus fruit significantly promoted disease phenotype of green mold. In addition, the content and associated enzyme activity of hydrogen peroxide (H_2O_2) in citrus fruit was induced by *Pd* infection and H_2O_2 treatment of citrus fruit decreased disease phenotype of green mold. H_2O_2 content and associated enzyme activity in citrus fruit was declined after melatonin treatment. Transcriptome analysis identified a lot of differentially expressed genes from citrus upon melatonin treatment, which are mainly involved in cell wall, reactive oxygen species metabolism and plant defense. Results indicated that melatonin decreases resistance to green mold on citrus fruit by scavenging defense-related reactive oxygen species, which provides significant insights into melatonin function in fruit resistance and also molecular mechanisms of fruit resistance to postharvest diseases.

Effects of β-aminobutyric acid (BABA) on diseases-tolerance of harvested banana (Musa AAA Group cv. Brazil) and the possible mechanism were investigated (TAN Wei-ping *et al.*, 2014). Diphenylene iodonium (DPI), a NADPH oxidase specific inhibitor, was used to inhibit ROS generation in BABA treated fruits. Reactive oxygen species was involved in the BABA-induced disease tolerance. BABA treatment started the protection mechanism of reactive oxygen species in banana fruits, including the production of reactive oxygen, the increasing of free radical scavenging enzyme activities, and the response of defense-related gene expression.

10.4.6. Cuticle and fatty acid biosynthesis : Necrotrophic plant pathogens such as *B. cinerea* and *C. gloeosporioides*, at the pathogenic stage, produce cutinases and pectinolytic enzymes to penetrate plant cuticle and epidermal cell wall. Cuticle and cuticular wax are known to regulate fungal cutinase gene expression, leading to the release of cutin monomers from the plant cuticle . Cutin monomers induce typical PAMP-triggered immunity (PTI) responses including medium alkalization, ethylene production, ROS, and upregulation of defense-related genes. Furthermore, treatments with cutin monomers or their production *in vivo* enhances resistance to both biotrophic and necrotrophic fungi. *Arabidopsis* plants expressing a fungal cutinase or mutants with a defective cuticle, such as *long-chain acyl-CoA synthetase2* which is involved in cutin biosynthesis, are surprisingly resistant to *B. cinerea* . This may be due to faster perception and response to fungal elicitors, easier diffusion of defense signals and antifungal compounds to the infection site and faster oxidative burst (Noam Alkan and Ana M.Fortes, 2015). *Bdg* and *lacs2.3* mutants impaired in cutin synthesis are known to display a high level of resistance to *B. cinerea* produced ROS even in the absence of wounding of the leaves. Moreover, *aba2* and *aba3* mutants together with *bdg* and *lacs2.3* mutants presented increased permeability of the cuticle and enhanced ROS production. In fact, ABA was reported to play an essential role in cuticular permeability, which may influence tomato fruit resistance to *B. cinerea* and may lead to the termination of quiescence.

Cuticle related mutants that alter cuticle development and composition were shown to modify plant defenses response and resistance. Therefore, the processes of degradation of cuticle and cell wall may have an effect on quiescence. Thus, cell wall and cuticle can constitute valuable targets for improvement of early sensing of pathogen and activation of immune responses accompanied with fruit quality traits. Due to this reason the regulatory mechanisms involved in cuticle deposition have been investigated. The first identified cuticle – associated transcription factor was SHINE1/WAXINDUCER1 (SHN1/WIN; Aharoni *et al.*, 2004; Broun *et al.,* 2004). Recently, it was shown that silencing a tomato ortholog (*SlSHN3*) in the fruit resulted in a dramatic reduction in cuticle formation (Shi *et al.*, 2013). It was suggested in this study that *SlSHN3* regulates not only the genes involved in cutin metabolism but also controls the expression of regulatory genes associated with epidermal cell patterning including tomato genes similar to *GLABRA2* and *MIXTA* . Recent data showed that *SlMIXTA-like* is a major transcriptional regulator of cutin biosynthesis, likely acting downstream of SlSHINE3. *SlMIXTA-like* not only promoted conical-type epidermal cell development in tomato fruit but also positively regulated cuticular lipids in particular cutin monomer biosynthesis as well as cuticle assembly. Tomato fruit silenced in either *SlSHN3* or *SlMIXTA-like* had a significant increase susceptibility to *Colletotrichum* . In another study, leaves of *SlSHN3* over-expressing plants were shown to be more resistant to *B. cinerea* than wild-type leaves, highlighting the importance of cuticle in plant–pathogen interactions. The thickness and composition of the cuticle has been shown to influence infection of grape berries by *B. cinerea.*

Hexacosanoic acid, an important component of wax, is present in higher amounts in Touriga Nacional berries than in Trincadeira berries and this may be involved in tolerance of Touriga Nacional cultivar to *B. cinerea*. Recently, it was reported that Trincadeira berries infected with *B. cinerea* accumulate saturated long-chain fatty acids which are major constituents of grape waxes accompanied with up-regulation of several acyl-CoA synthetases and wax synthases (Noam Alkan and Ana M.Fortes, 2015). In this study, significant changes were observed in the contents of glycerol and fatty acids such as oleic acid. In addition, a gene encoding a stearyl acyl carrier protein desaturase, which catalyzes the desaturation of stearic acid to oleic acid, was up-regulated in infected berries. It has been shown that a reduction in oleic acid levels results in constitutive activation of the SA-dependent pathway and repression of the JA-dependent pathway. The results in grapes supported these data, since increased oleic acid levels were correlated with activation of JA biosynthesis and signaling in infected berries . Thus, changes in lipid composition likely represent fruit response to infection.

Interestingly, transcriptome analysis of *C. gloeosporioides* and tomato fruit pathosystem revealed that during appressoria formation and before fungal penetration the tomato fruit host recognizes the fungus and activates fatty acid biosynthesis, elongation, and synthesis of cutin and waxes. Specifically, genes involved in synthesis of very long chain fatty acids that are components of cutin and waxes were up-regulated namely 3-ketoacyl CoA synthase and CYP86A cytochrome p450. Interestingly, tomato fruit mutated in CYP86 were much more susceptive to *Colletotrichum* infection. These genes were also up-regulated in grape berries infected with *B. cinerea* along with a decrease in expression of genes involved in glycerolipid catabolism). Both pathosystems also reported the induction of β-oxidation fatty acid degradation pathway (Agudelo-Romero *et al.*, 2015). This would provide both reducing power and carbon components for metabolism of very long chain fatty acids or alternatively would provide more sugars that might be metabolized by the fungus or serve as precursors of plant secondary metabolites involved in defense. These observations further emphasize the critical role of fatty acids, wax and cutin during infection. This topic deserves further attention, since pathogen and host lipids and lipid metabolites have a critical role in the dynamics of pathogenesis and in plants defense.

To elucidate the role of fruit cuticle in fungal infection, changes in cuticle composition and morphology of 'Satsuma' mandarin during ambient (at 25 °C) storage and their role in *Penicillium digitatum* infection were investigated (Shengua Ding *et al.,* 2020) Results showed that the epicuticular wax yield increased from 1.11 μg cm^{-2} to 4.21 μg cm^{-2} during storage for 20 days and then decreased to 1.35 μg cm^{-2} as storage time prolonged to 40 days. Intracuticular wax content of fruits stored for 20 days showed a peak value that was 1.7-fold higher than that of fruits stored for 40 days. The contents of cutin monomers of fruits showed a decreased trend during storage, while their proportions in the cutin stayed stable. Acids were identified as the most abundant components in epicuticular wax independently of the storage time, followed by alkanes and terpenoids. Terpenoids were found as the predominant components in intracuticular wax during the whole storage, followed by alkanes and acids. The flattened platelets crystals of fruits at harvest changed into small granule-like wax ones after 10 days of storage then gradually distributed across the surface of the fruits as stored for 40 days. Results of *in vitro* tests showed that mycelial growth of *Penicillium digitatum* could be promoted by epicuticular wax and conidial germination could be inhibited by cutin at different storage stages. These results shed new light on the chemical basis for cuticle involvement in fungal infection.

A barrier for *B. cinerea* infection is the fruit cuticle. During fruit expansion and ripening the cuticle gets thinner, which makes strawberries more susceptible to initial penetration by germinating conidia. *B. cinerea* can penetrate the plant cuticle by secretion of cutinases (Stefan Petrasch *et al.*, 2019). Additionally, cuticle properties can result in higher incidence of cracks and other damages through which *B. cinerea* can enter the fruit without the need of cutinases. Studies on strawberry cuticles are scarce and only exist for leaf tissues. In tomato fruit, thicker and stiffer cuticles lead to higher resistance to initial *B. cinerea* infections. Moreover, it is known that the chemical composition of the cuticle changes during tomato ripening, and this is likely to be the case in strawberry.

10.4.7.Cell wall remodeling and soluble sugar accumulation: Cell walls are structurally complex network of polysaccharides, including cellulose, hemicelluloses, and pectin. They serve as a physical barrier that limits pathogen access, but are also involved in pathogen recognition and in the deployment of plant responses to pathogens. In order to break down the cell wall barrier pathogens use mechanical force as appressoria or release cell wall degrading enzymes (CWDEs), which serve as a pathogenicity factors. Also postharvest pathogens as *Botrytis* and *Colletotrichum* produces CWDEs during pathogenic colonization of tomato fruit (Noam Alkan and Ana M.Fortes, 2015). The degradation of the plant cell wall matrix by pathogens may affect the proteins embedded in the cell wall and are likely to activate PAMP-triggered immunity, which often leads to callose deposition at sites of penetration, accumulate phenolic compounds and various toxins in the cell wall and synthesize lignin-like polymers to reinforce the wall.

During natural fleshy fruit ripening the fruit soften as a result of fruit activation of CWDEs as polygalacturonase (PG), pectin methylesterase, pectate lyase, β-galactosidase, cellulase, and expansin. Phytohormones as ethylene, ABA, SA and JA, are known to influence the expression of CWDEs which contribute to fruit softening. Because, fruit CWDEs are normally activated during ripening, it was commonly assumed that fruit softening contributes to the transition to susceptibility to pathogens. In tomato, the suppression of softening-associated CWDEs, *SlPG* and *SlExp1*, reduced susceptibility to *B. cinerea* infection during ripening, indicating that PG and Exp support both softening and susceptibility to *B. cinerea* . Interestingly, *B. cinerea* infections induce *SlPG* and *SlExp1* expression suggesting that *Botrytis* induces similar softening to the softening that occurs during fruit ripening. Endo-β-1,4-glucanase (EGase) is another CWDE that have a role in fruit softening. Tomato fruit EGase antisense had enhanced callose deposition and was more resistant against *Botrytis* infection. Plants PG inhibiting proteins (PGIPs) reduce the pathogen pectin degradation. PGIPs inhibit most of the *Botrytis* PGs during pear pathogenesis. An over-expression of PGIPs enhances ripe tomato fruit tolerance to *Botrytis* (Noam Alkan and Ana M.Fortes, 2015). To conclude, fruit cell wall is a complex and dynamic barrier which changes during ripening and its interaction with fungal pathogens plays a major role in the defense response against pathogens.

10.4.8.Cell wall modifications: The plant cell wall is typically divided into three regions based on composition and design: the middle lamella, primary cell wall, and secondary cell wall. No clear boundary exists between the three regions and, in fact, the cell wall is somewhat of a continuum. The middle lamella, consisting primarily of pectic polysaccharides, cements cell walls together. Pathogen-produced pectolytic enzymes break down these pectic polymers, leading to loss of structural integrity and cell separation, a process known as maceration (Wood, 1967). Due to the accessibility of pectic polymers in the primary cell wall to enzymatic attack and the resulting release of pectic inducers, pectic enzymes are the first polysaccharides to be produced in tissue infected by fungi (Collmer and Keen, 1986). Postharvest diseases of fungal origin produce one or more of the following pectic enzymes: pectin esterase, endo- or exopolygalacturonase, endo- or exopectate lyase, and endopectin lyase (Dennis, 1987). The endopectin lyases are produced almost exclusively by fungi, while the endopectate lyases are produced primarily by bacteria (Chesson, 1980). Almost every pathogen-enzyme system interacts in a specific manner, depending on the plant tissue encountered. Plant pathogenic and nonpathogenic organisms are capable of producing a variety of cellwall-degrading enzymes. The mere fact that a fungus can produce cell-wall-degrading enzymes does not mean it will be a successful pathogen. There are many factors involved in host–pathogen interaction, ranging from illicitors to phytoalexins.

The plant cell wall, a dynamic and complex structure surrounding every plant cell, has emerged recently as an essential component of plant monitoring systems, thus expanding its function as a passive defensive barrier. Plants have a dedicated mechanism for maintaining cell wall integrity (CWI) which comprises a diverse set of plasma membrane-resident sensors and pattern recognition receptors (PRRs). The PRRs perceive plant-derived ligands, such as peptides or wall glycans, known as damage associated molecular patterns (DAMPs). These DAMPs function as 'danger' alert signals activating DAMP triggered immunity (DTI), which shares signalling components and responses with the immune pathways triggered by non-self microbe associated molecular patterns that mediate disease resistance. Alteration of CWI by impairment of the expression or activity of proteins involved in cell wall biosynthesis and/or remodelling, as occurs in some plant cell wall mutants, or by wall damage due to colonization by pathogens/pests, activates specific defensive and growth responses. Our current understanding of how these alterations of CWI are perceived by the wall monitoring systems is scarce and few plant sensors/PRRs and DAMPs have been characterized. The identification of these CWI sensors and PRR–DAMP pairs will help us to understand the immune functions of the wall monitoring system, and might allow the breeding of crop varieties and the design of agricultural strategies that would enhance crop disease resistance.

Postharvest (detached) and in planta (attached) fruits of pepper plants, *Capsicum annuum* cv. Jejujaerae (susceptible) and *Capsicum baccatum* cv. PBC80 (resistant), inoculated with the anthracnose pathogen Colletotrichum *gloeosporioides* were examined using light, confocal laser scanning, and electron microscopy to compare the cytological differences between the compatible and incompatible interactions (Kim *et al.*, 2004). In nonwound inoculation of postharvest pepper fruit, resistant pepper tissues showed a significant increase in the thickness of the cuticle layer compared with that of the susceptible and noninoculated fruit. Cytological features of programmed cell death (PCD) were observed in the resistant pepper fruit with postharvest inoculation, and these were characterized by positive responses to terminal deoxynucleotidyl transferase-mediated dUTP nick end labeling. The oligonucleosomal fragments of DNA were confirmed electrophoretically as DNA laddering. The PCD-positive responses occurred around the inoculation sites early in in planta wound inoculation in the resistant pepper. Nuclear modifications and structural changes of hypersensitivity were also observed in the resistant fruit, including separation of the plasma membrane from the cell wall, dilation of the endoplasmic reticulum, accumulation of electron-dense inclusions in vacuoles, and cytoplasmic vacuolization accompanying fragmentation of the cytoplasm. These structural changes may also implicate PCD-like host responses. In addition, in planta wound inoculation resulted in cell enlargement and cell division during the later stages of infection to form a periderm-like boundary layer around the inoculation site.

The fungal cell wall is made predominantly of carbohydrates and glycoproteins, many of which serve as binding receptors for host defence molecules or activate host immune responses through interactions with membrane-bound receptors.

Plant defensins are a large family of cationic antifungal peptides that protect plants against fungal disease. Binding of the plant defensin NaD1 to the fungal cell wall has been described but the specific component of the cell wall with which this interaction occurred was unknown. The effect of binding was also unclear, that is whether the plant defensin used fungal cell wall components as a recognition motif for the plant to identify potential pathogens or if the cell wall acted to protect the fungus against the defensin. Here they described describe the interaction between the fungal cell wall polysaccharides chitin and β-glucan with NaD1 and other plant defensins. It was discovered that the β-glucan layer protects the fungus against plant defensins and the loss of activity experienced by many cationic antifungal peptides at elevated salt concentrations is due to sequestration by fungal cell wall polysaccharides. This has limited the development of cationic antifungal peptides for the treatment of systemic fungal diseases in humans as the level of salt in serum is enough to inactivate most cationic peptides.

Ronald P. de Vries and Jaap Visser (2001) have presented an overview of the large number of enzymes and genes from Aspergilli involved in the degradation of plant cell wall polysaccharides. Although the enzymes reported cover most of the functions necessary for the complete degradation of plant cell wall polysaccharides, some functions remain to be isolated (e.g., enzymes removing fucose residues from xyloglucan). Also, the substrate specificity and mode of action of many of the enzymes have not been studied in detail. A good example of this is β-galactosidase. So far, only a single β-galactosidase has been found in any*Aspergillus* species, although β-linked galactose residues are found in xylan, xyloglucan, and pectin side chains. Whether this single enzyme is able to hydrolyze all these linkages or whether other β-galactosidases are produced by aspergilli is unclear at this point. The availability of genome sequences for several *Aspergillus*species will accelerate the discovery in the near future of new genes encoding polysaccharide-degrading enzymes.

Only recently, plant proteins have been identified that are able to inhibit the action of some fungal plant cell wal polysaccharide-degrading enzymes. The polygalacturonase inhibitor protein (PGIP) has been studied in the most detail and has been proposed to form a part of the plant "immune system". Evidence for the production of PGIPs in plants as a response to wounding but not to fungal infection was obtained. Fungal pathogens produce a family of polygalacturonases, suggesting the neccessity of several PGIPs with different specificities. Recent studies confirmed this by demonstrating that inhibition depends on the type of PGIP and the type of endopolygalacturonase. A glycoprotein was identified in kiwi fruit that inhibits the action of a different pectin-degrading enzyme, pectin methylesterase. In addition, several plant proteins have been isolated that inhibit the activity of endoxylanases. One of these studies also produced evidence for the production of an arabinofuranosidase inhibitor by wheat. It is likely that future studies will identify other plant proteins capable of inhibiting the action of fungal plant cell wall polysaccharide-degrading enzymes.

A number of studies report the production of *Aspergillus* plant cell wall polysaccharide-degrading enzymes in other organisms, in particular *S. cerevisiae*. Most plant cell wall-degrading enzymes as produced by *Aspergillus* are highly glycosylated proteins. When these enzymes are produced by *S. cerevisiae*, they are usually overglycosylated and one should keep in mind that the properties might differ from those of the native enzyme.

Degradation of plant cell wall polysaccharides is of major importance in the food and feed, beverage, textile, and paper and pulp industries, as well as in several other industrial production processes. Enzymatic degradation of these polymers has received attention for many years and is becoming a more and more attractive alternative to chemical and mechanical processes. Over the past 15 years, much progress has been made in elucidating the structural characteristics of these polysaccharides and in characterizing the enzymes involved in their degradation and the genes of biotechnologically relevant microorganisms encoding these enzymes. The members of the fungal genus Aspergillus are commonly used for the production of polysaccharide-degrading enzymes. This genus produces a wide spectrum of cell wall-degrading enzymes, allowing not only complete degradation of the polysaccharides but also tailored modifications by using specific enzymes purified from these fungi. This review summarizes our current knowledge of the cell wall polysaccharide-degrading enzymes from aspergilli and the genes by which they are encoded.

Ripening is associated with the disassembly of the fruit cell walls, which leads to tissue softening. Cell wall degradation benefits pathogens like *B. cinerea* as it reduces mechanical barriers to infection and spread, increases the possibilities of bruising (e.g. leading to more wounds for pathogen entry) and provides the fungus with access to simple sugars as a carbon source (Stefan Petrasch *et al.*, 2019). In strawberry, cell wall solubilization occurs early in fruit development when the walls start to swell. Cell wall solubilization correlates with an increase in fruit sugar content, resulting from polysaccharide breakdown. A decrease of acid-soluble pectins and the alcohol-insoluble fraction of cell walls occur during ripening, whereas the water-soluble content increases (i.e. enriched in non-covalent bound pectins). The degree of pectin solubilization and depolymerization is highlyrelated to strawberry fruit firmness. Silencing of an endogenous pectin lyase (PL) gene in strawberry resulted in fruit with higher external and internal firmness, mostly due to low pectin solubilization, stiffer cell walls, and increased cell to cell adhesion. Besides PL, other enzymes that may have affected strawberry firmness include PGs, β-galactosidases, endoglucanases, α-arabinofuranosidases and β-xylosidases. In addition to the fruit endogenous cell wall disassembly, *B. cinerea* secretes an extensive array of CWDEs that target most polysaccharides in the fruit cell walls, particularly pectins. These CWDEs include fungal PGs, such as Bcpg2, a gene that is mainly active in the early penetration stage. The expression of *B. cinerea* PGs is dependent on the host species, the plant tissue, temperature and the stage of infection.

Fruit ripening is characterized by processes that modify texture and flavor but also by a dramatic increase in susceptibility to necrotrophic pathogens, such as *Botrytis cinerea*. Disassembly of the major structural polysaccharides of the cell wall (CW) is a significant process associated with ripening and contributes to fruit softening (Cantu *et al*., 2008). In tomato, polygalacturonase (PG) and expansin (Exp) are among the CW proteins that cooperatively participate in ripening-associated CW disassembly. To determine whether endogenous CW disassembly influences the ripening-regulated increase in necrotropic pathogen susceptibility, *B. cinerea* susceptibility was assessed in transgenic fruit with suppressed polygalacturonase (*LePG*) and expansin (*LeExp1*) expression. Suppression of either *LePG* or *LeExp1* alone did not reduce susceptibility but simultaneous suppression of both dramatically reduced the susceptibility of ripening fruit to *B. cinerea*, as measured by fungal biomass accumulation and by macerating lesion development. These results demonstrate that altering endogenous plant CW disassembly during ripening influences the course of infection by *B. cinerea*, perhaps by changing the structure or the accessibility of CW substrates to pathogen CW-degrading enzymes. Recognition of the role of ripening-associated CW metabolism in postharvest pathogen susceptibility may be useful in the design and development of strategies to limit pathogen losses during fruit storage, handling, and distribution.

Infection by necrotrophs is a complex process that starts with the breakdown of the cell wall (CW) matrix initiated by CW-degrading enzymes and results in an extensive tissue maceration. Plants exploit induced defense mechanisms based on biochemical modification of the CW components to protect themselves from enzymatic degradation. The pectin matrix is the main CW target of *Botrytis cinerea*, and pectin methylesterification status is strongly altered in response to infection (Vincenzo Lionetti, *et al.,* 2017). The methylesterification of pectin is controlled mainly by pectin methylesterases (PMEs), whose activity is posttranscriptionally regulated by endogenous protein inhibitors (PMEIs). Here, AtPMEI10, AtPMEI11, and AtPMEI12 are identified as functional PMEIs induced in Arabidopsis (*Arabidopsis thaliana*) during *B. cinerea* infection. AtPMEI expression is strictly regulated by jasmonic acid and ethylene signaling, while only AtPMEI11 expression is controlled by PME-related damage-associated molecular patterns, such as oligogalacturonides and methanol. The decrease of pectin methylesterification during infection is higher and the immunity to *B. cinerea* is compromised in *pmei10*, *pmei11*, and *pmei12* mutants with respect to the control plants. A higher stimulation of the fungal oxalic acid biosynthetic pathway also can contribute to the higher susceptibility of *pmei* mutants. The lack of *PMEI* expression does not affect hemicellulose strengthening, callose deposition, and the synthesis of structural defense proteins, proposed as CW-remodeling mechanisms exploited by *Arabidopsis* to resist CW degradation upon *B. cinerea* infection. They showed that PME activity and pectin methylesterification are dynamically modulated by PMEIs during *B. cinerea* infection. These findings point to AtPMEI10, AtPMEI11, and AtPMEI12 as mediators of CW integrity maintenance in plant immunity.

Plants infected by *Fusarium oxysporum* show severe symptoms of dehydration due to the growth of the fungus in vascular tissues. As the disease develops, vascular browning and leaf yellowing can be observed. In the case of more virulent strains, plants die (Wojtasik, *et al*., 2016). The pathogen's attack starts with secretion of enzymes degrading the host cell wall. The main aim of the study was to evaluate the role of the cell wall polymers in the flax plant response to the infection in order to better understand the process of resistance and develop new ways to protect plants against infection. For this purpose, the expression of genes involved in cell wall polymer metabolism and corresponding polymer levels were investigated in flax seedlings after incubation with *Fusarium oxysporum*.

10.4.9 pH change during fruit ripening and fungal colonization

a. *Colletotrichum* and *Alternaria* : The pH change plays an important role in three different aspects of fruit-fungal interaction *viz*., the pH change during fruit ripening, fungal-dependent pH modulation of the local infection court, and the local host pH modulation during the activation of defense responses. The combined pH changes were suggested to trigger defense related processes as ROS and activate cell wall hydrolases in the fruit. The ratio between sugar and pH are determinants of the fruit taste. During fruit ripening total soluble sugars (TSSs) increase and organic acid usually decreases leading to increase in pH. For instance, the pH of avocado fruit increases from 5.2 to 6.0 during ripening. However, in tomato fruit the apoplastic pH decreases during ripening from 6.7 to 4.4. Also fungal pathogens alter their local pH by secretion of ammonia or organic acids to optimize the environment to each fungus enzymatic arsenal. Interestingly, fruit pH greatly affects fungal pathogenicity. Acidified environment induce ammonia secretion in alkalizing fungi as *Colletotrichum* and *Alternaria*, while alkaline environment activate organic acid secretion in acidifying fungi as *Penicillium* and *Botrytis*. The pH dependent fungal pathogenicity factors are controlled in filamentous fungi by Pal signaling pathway which activates PACC transcription factor. PACC activates transcription of those pathogenicity factors at alkaline pH and the AREB transcription factor, which represses acidic expressed pathogenicity factors at alkaline environment (Noam Alkan and Ana M.Fortes, 2015). In this way, fungi adjust their ambient pH in order to optimize the activity of their enzymatic arsenal.

Changes in apoplastic pH, could lead to oxidative burst. For example, exposing bean cells to alkaline condition resulted in oxidative burst. Medium alkalization activate NADPH oxidase, probably as a result of induced K^+/ H^+ exchange, followed by Ca^{2+} influx/Cl_2 efflux. Transient extracellular alkalization is an essential factor in induction of defense response and PCD. In this connection, changes in *Arabidopsis thaliana* roots external pH rapidly alter plant gene expression and modulate host responses, similarly to elicitors. Similarly, tomato fruit apoplastic alkalization by *Colletotrichum* or application of ammonia lead to activation of fruit NADPH oxidase, oxidative burst and SA mediated defense response that ended with extended cell death. *P. expansum* secrets gluconic acid and acidify its ambient pH in apples; this acidification was correlated with oxidative burst. Taken together both pathogen and fruit modulate their ambient pH in order to optimize respectively their attack and responses. Host pH can be raised or lowered in response to host signals, including alkalization by ammonification of the host tissue as observed in *Colletotrichum* and *Alternaria*, or acidification by secretion of organic acids as observed in *Penicillium*, *Botrytis* and *Sclerotinia*. These changes sensitize the host and activate transcription and secretion of fungal hydrolases that promote maceration of the host tissue (Prusky *et al*., 2009).

Growth of *Colletotrichum gloeosporioides* in pectolytic enzyme-inducing medium (PEIM) increased the pH of the medium from 3.8 to 6.5. Pectate lyase (PL) secretion was detected when the pH reached 5.8, and the level of secretion increased up to pH 6.5. PL gene (*pel*) transcript production began at pH 5.0 and increased up to pH 5.7. PL secretion was never detected when the pH of the inducing medium was lower than 5.8 or when *C. gloeosporioides* hyphae were transferred from PL-secreting conditions at pH 6.5 to pH 3.8. This behavior differed from that of polygalacturonase (PG), where *pg* transcripts and protein secretion were detected at pH 5.0 and continued up to 5.7 (Nit

Yakoby *et al.*, 2000). Under in vivo conditions, the pH of unripe pericarp of freshly harvested avocado (*Persea americana* cv. Fuerte) fruits, resistant to *C. gloeosporioides* attack, was 5.2, whereas in ripe fruits, when decay symptoms were expressed, the pericarp pH had increased to 6.3. Two avocado cultivars, Ardit and Ettinger, which are resistant to *C. gloeosporioides* attack, had pericarp pHs of less than 5.5, which did not increase during ripening. The present results suggest that host pH regulates the secretion of PL and may affect *C. gloeosporioides* pathogenicity. The mechanism found in avocado may have equivalents in other postharvest pathosystems and suggests new approaches for breeding against and controlling postharvest diseases.

b. *Phomopsis mangiferae* : Colonization of mango and other deciduous and tropical fruits by *Phomopsis mangiferae* was accompanied by local acidification of the host tissue (Maayan Davidzon *et al.*, 2010). The fungus acidified the host tissue in mango and grape from pH 5.1 and 4.1, respectively, to 3.8 and 2.5. Analysis of the acidification process in colonized fruit showed that gluconic acid was the main organic acid accumulated at the infection site and in the lesion tissue. In liquid culture conditions the relative induction of transcripts of pmgox1, encoding for glucose oxidase (GOX) was 8–12 times greater at pH 7.0 and 8.0 than at pH 4.0. In infected fruit the detection of high levels of transcripts of pmgox1 and the accumulation of gluconic acid and H2O2 in the decayed tissue, suggested that glucose oxidase contributed to the acidification of the tissue. At the same time, transcripts encoding the endopolygalacturonase gene, pmpg1 accumulated greatly under acidic culture conditions, signifying the importance of the acidification in enhancing the pathogenicity of *P. mangiferae.* Results indicated that ambient pH is a regulatory cue for processes linked to pathogenicity of postharvest pathogens, and that specific genes are expressed as a result of the environmental pH created by the pathogen.

c. *Penicillium expansum* : It is a destructive phytopathogen that causes decay in deciduous fruits during postharvest handling and storage. During colonization the fungus secretes D-gluconic acid (GLA), which modulates environmental pH and regulates mycotoxin accumulation in colonized tissue. Till now no transcriptomic analysis has addressed the specific contribution of the pathogen's pH regulation to the *P. expansum* colonization process. For this purpose total RNA from the leading edge of *P. expansum*-colonized apple tissue of cv. 'Golden Delicious' and from fungal cultures grown under pH 4 or 7 were sequenced and their gene expression patterns were compared (Barad *et al.*, 2016). They presented a large-scale analysis of the transcriptome data of *P. expansum* and apple response to fungal colonization. The fungal analysis revealed nine different clusters of gene expression patterns that were divided among three major groups in which the colonized tissue showed, respectively, differing transcript expression patterns between mycelial growth at pH 4 and pH 7; similar transcript expression patterns of mycelial growth at pH 4; and similar transcript expression patterns of mycelial growth at pH 7. Each group was functionally characterized in order to decipher genes that are important for pH regulation and also for colonization of apple fruits by *Penicillium*. Furthermore, comparison of gene expression of healthy apple tissue with that of colonized tissue showed that differentially expressed genes revealed up-regulation of the jasmonic acid and mevalonate pathways, and also down-regulation of the glycogen and starch biosynthesis pathways.

Barad *et al.*, (2016) identified important genes and functionalities of *P. expansum* that were controlled by the environmental pH. Differential expression patterns of genes belonging to the same gene family suggest that genes were selectively activated according to their optimal environmental conditions (pH, *in vitro* or *in vivo*) to enable the fungus to cope with varying conditions and to make optimal use of available enzymes. Comparison between the activation of the colonized host's gene responses by alkalizing *Colletotrichum gloeosporioides* and acidifying *P. expansum* pathogens indicated similar gene response patterns, but stronger responses to *P. expansum*, suggesting the importance of acidification by *P. expansum* as a factor in its increased aggressiveness.

10.4.10 Preformed and inducible antifungal resistance: Plants possess a range of tools for combating fungal infections. Defence mechanisms include physical barriers and chemical defence includes many naturally occurring compounds and systems to prevent microbial attacks by using several types of pre-formed compounds and induced response plant defence compounds. Plants have defensive response by means of some formed or pre-formed substances belong to the commonly called plant secondary compounds (secondary metabolism). Secondary metabolism is known as all those compounds that they are not involved in primary metabolic processes such as respiration and photosynthesis. However, many compounds do have roles intermediate between primary and secondary metabolism. For instance, some plant secondary compounds like plant growth substances are associated with plant growth and development (Juan A. Martínez, 2012). In fact, secondary metabolites are compounds with a restricted occurrence in taxonomic groups that are not necessary for a cell (organism) to live with its environment, ensuring the survival of the organism in its ecosystem. There has been much speculation over many years about why organisms take the trouble to produce secondary metabolites or natural products which often have considerable structural complexity. All such structures serve the producing organisms by improving their survival fitness (Williams et al., 1989). In this case, it seems likely that the fungi and bacteria which produce strobilurins, oudemansins, myxothiazols and/or melithiazols, which are substances with antifungal properties synthesized by fungi or bacteria give themselves an advantage by doing so, presumably by suppressing other organisms which compete for nutrients in the same environment. Secondary metabolites are now thought to mediate chemical defence mechanisms by providing chemical barriers against animal and microbial predators. Plants produce numerous chemicals for defence and communication, and can elicit their own form of offensive chemical warfare by targeting the proliferation of pathogens. These chemicals may have general or specific activity against key target sites in bacteria, fungi and viruses.

These substances are classified into different ways: based on chemical characteristics, plant origin or biosynthetic origin. From a chemical point of view, the compounds can be divided into a number of groups based on typical characteristics such as alkaloids and phenolic compounds, principally, Based on biosynthetic origin, there are terpenoids, phenylpropanoids, polyketides and other compounds belong to other minor groups must be considered. Therefore, the total number of these compounds is very elevated. A great number of those are used to prevent animal consumption. However, many of these defensive substances are toxic for the plants themselves and they are inactivated as glycosides or by polymeration or they are confined in intercellular spaces.

Preformed resistance involves the presence in healthy tissues of biologically active, lowmolecular-weight compounds whose activity affords protection against infection (BarkaiGolan, 2001). These compounds belong to the secondary metabolism which includes all the compounds belong to the three categories described above. It could also be included other defensive physical structures (physical resistance) like structural barriers (cuticle, epicuticular wax, cell wall and processes like lignification and suberization of cells at the wound areas).

Plants contain preformed secondary metabolites of a defensive nature such as phenolics, sulfur compounds, saponins, cyanogenic glycosides, and glucosinolates. Phenolic compounds play an important role in non-host resistance to fungi. They can either be performed occurring constitutively in healthy plants (phytoanticipins) or synthesized from precursors in response to pathogen attack, being more restricted to the damaged tissue (phytoalexins). Some antibiotic phenolics are stored in plant cells as inactive bound forms but are readily converted into biologically active antibiotics by plant glycosidases in response to pathogen attack. Since these compounds do not involve *de novo* transcription of gene products they are also considered phytoanticipins. Concentrations of preformed phytoanticipins and inducible phytoalexins were found to decline during fruit ripening and this occurred more rapidly in susceptible cultivars than in resistant cultivars (Noam Alkan and Ana M.Fortes, 2015).

The grape berry cultivar Trincadeira is susceptible to *B. cinerea* and downy mildew. It presents lower phenolics content than the tolerant cultivars Touriga Nacional and Aragonês. The green and *veraison* stages of Aragonês and Touriga Nacional showed higher levels of quercetin glucoside, catechin and hydroxycinnamic acid derivatives such as caftaric acid and coutaric acid than the ripe grape berry. A decrease in caffeic acid was also detected in ripe berries of all the three varieties (Agudelo-Romero *et al.*, 2015). This decline may be related to increased susceptibility of ripe fruits to pathogenic fungi. In fact, caffeic acid presents antimicrobial activity. Further, constitutive secondary metabolites of the bitter orange *Citrus aurantium* are the flavanone-naringin and the polymethoxyflavone-tangeretin, which showed antifungal activity against *Penicillium digitatum*.

Other widely reported preformed antifungal compounds are the family of mono-, di-, and triene compounds in avocado; the resorcinol derivates in mango; the tannins in banana peel and tomatine in tomato fruits. These compounds were shown to decline dramatically during fruit ripening, thus enabling development of penetrating mycelia. Mass Spectrometry Imaging (MSI) was used to screen for potential antifungal metabolites involved in the interaction between *Penicillium digitatum* and *Penicillium citrinum*. MSI revealed a chemical warfare between the fungi: two tetrapeptides, deoxycitrinadin A, citrinadin A, chrysogenamide A and tryptoquialanines are produced in the fungi confrontation zone. Antimicrobial assays confirmed the antifungal activity of the investigated metabolites. Also, tryptoquialanines inhibited sporulation of *P. citrinum*. The fungal metabolites reported here were never described as antimicrobials until this date, demonstrating that co-cultures involving phytopathogens that compete for the same host is a positive strategy to discover new antifungal agents (Costa *et. al.,* 2019).

10.4.11 Inducible phenylpropanoid metabolism : Phytoalexins are generally induced after infection. They accumulate rapidly in response to infection and reach high antimicrobial levels in resistant plants, while there is either lesser or slower accumulation in susceptible plants. When the accumulation of phytoalexins is either increased or decreased by manipulation of the experimental conditions such as postharvest stress treatments, the plant and fruit become either more resistant or more susceptible reported that silicon is involved in the increased resistance of cucumber to powdery mildew by enhancing the antifungal activity due to the presence of metabolites such as flavonol aglycone rhamnetin (3,5,3′,4′-tetrahydroxy-7-*O*-methoxyflavone).

Pathogens often remain quiescent in unripe fruits. During ripening the concentrations of pathogen-induced and pre-formed antifungal phenolics decrease to subtoxic levels; this chemical decline occurs more rapidly in susceptible cultivars than in resistant ones. The principal phenolics in the peach fruit include chlorogenic acid, catechin, and epicatechin. The decline in chlorogenic acid and other endogenous phenolics during fruit ripening correspond to the transition to susceptibility. In immature strawberry fruits with a high content of proanthocyanidins *B. cinerea* remains quiescent. When the inhibitory activity of proanthocyanidins in fruits decreases due to maturation, the quiescent fungus can switch to the necrotrophic stage and progress further into the fruit tissue. In addition, inducible antifungal compounds, such as capsicannol in pepper, scoparone in citrus, resveratrol in grapes, and others, have been reported to be activated in unripe fruits but not always in ripe fruits(Noam Alkan and Ana M.Fortes, 2015). These compounds were hypothesized to be quiescence modulating factors.

The resistance of *Vitis* sp. to *B. cinerea* infection has been shown to correlate with *trans*-resveratrol content. Touriga Nacional is not infected by *B. cinerea* under normal field conditions; this cultivar presents higher content in *trans*-resveratrol than the susceptible cultivar Trincadeira (Agudelo-Romero *et al.*, 2015). Recently, in-field infections of Trincadeira cultivar with *B. cinerea* led to profound alterations in secondary metabolism linked to stress response together with a significant increase in *trans*-resveratrol. Indeed, several genes involved in phytoalexin biosynthesis and coding for stilbene synthase and resveratrol synthase were up-regulated at pre-*veraison* in infected grapes (Agudelo-Romero *et al.*, 2015). Therefore, resveratrol was considered a potential positive metabolic marker of *B. cinerea* infection at this stage. Other identified positive markers of infection were gallic acid and 3,4-dihydroxybenzoic acid which present antifungal properties. In fact, plant benzoic acids and their derivatives are common and widespread mediators of plant responses to biotic and abiotic stress. In another work with grape, flavonoid compounds were only found in botrytized berries of botrytized bunches at harvest stage.

Trincadeira, a *Botrytis*-susceptible variety, is able to initiate to some extent a basal defense reprogramming of the transcriptome and metabolome that is unable to slow down disease progression. This can be due to the fact that the pathogen can shut down host defenses. For example, sesquiterpenoid biosynthesis, as measured by genes involved in their synthesis; namely *beta*-amyrin synthase and (-)-germacrene D synthase, was down-regulated at the *veraison* stage. Oleanolic acid, a triterpenoid, decreased in infected grapes at pre-*veraison* and *veraison* stages. Unconjugated triterpenoids, such as oleanolic acid, are often found in the epicuticular waxes of plants serving as a first defense barrier against pathogens. This result suggested that infection renders the fruit to be more susceptible by down-regulating defense compounds.

Recently, it was showed through a combined analysis of the transcriptomes of *C. gloeosporioides* and tomato fruit pathosystem that during the quiescent stage, defense pathways were up-regulated including the phenypropanoid pathway for phytoalexin and lignin precursors such as cinnamic, cumarayl, coniferyl, caffeoyl, shikimic, quinic, and sinapyl derivatives. The authors suggested that phytoalexin biosynthesis and lignification comprise a major ongoing fruit defense pathway employed by the fruit in response to the persistent presence of quiescent fungi. Such host responses may effectively restrain the pathogen. Indeed, the number of infection sites emerging from quiescence appears to be below the potential primary infection sites, indicating successful containment of the infection. In this study, genes involved in the synthesis of sesquiterpenoids (e.g., rishitin) were also down-regulated together with the up-regulation of key steroid glycoalkaloid (e.g., tomatine) transcripts. Previously, α-tomatine was shown to inhibit *Colletotrichum* fungal growth and germination. Hence, transcript expression suggests the occurrence of shifts from rishitin to α -tomatine biosynthetic pathway as an effective response to this fungus (Noam Alkan and Ana M.Fortes, 2015). All of those antifungal compounds decline in ripe fruits, which may permit emergence of fungi from quiescence.

Ripening of fleshy fruit is a differentiation process involving biochemical and biophysical changes that lead to the accumulation of sugars and subsequent changes in tissue texture (Singh *et al.,* 2010). Also affected are phenolic compounds, which confer color, flavor/aroma, and resistance to pathogen invasion and adverse environmental conditions. These phenolic compounds, which are the products of branches of the phenylpropanoid pathway, appear to be closely linked to fruit ripening processes. Three key enzymes of the phenylpropanoid pathway, namely phenylalanine ammonia lyase, O-methyltransferase, and cinnamyl alcohol dehydrogenase (CAD) have been reported to modulate various end products including lignin and protect plants against adverse conditions. In addition, peroxidase, the enzyme following CAD in the phenylpropanoid pathway, has also been associated with injury, wound repair, and disease resistance. However, the role of these enzymes in fruit ripening is a matter of only recent investigation and information is lacking on the relationships between phenylpropanoid metabolism and fruit ripening processes. Understanding the role of these enzymes in fruit ripening and their manipulation may possibly be valuable for delineating the regulatory network that controls the expression of ripening genes in fruit. They further elucidated the functional characterization of these key phenylpropanoid biosynthetic enzymes/genes during fruit ripening processes

Olive anthracnose caused by the hemibiotrophic fungal pathogen *Colletotrichum acutatum* is a serious threat to the olive sector. Olive oil and fruit production is severely constrained by *Colletotrichum* spp. infection, being *C. acutatum* the most distributed pathogen in Portuguese olive orchards. To understand the impact of *C. acutatum* on phenylpropanoids biosynthesis, the enzyme activity, phenolic compounds, *ortho*-diphenols, and flavonoids content were determined and correlated with the expression of gene encoding key enzymes within phenylpropanoids metabolism in susceptible and tolerant olive fruits, during maturation and when infected with *C. acutatum* (Irene Gouvinhas *et al.*, 2019). Differences between cultivars was observed, the tolerant olive cv. Picual presented a higher basal value and a stable phenolic content throughout the infection process, supporting its high *C. acutatum* tolerance, whereas in the susceptible olive cv. Galega these secondary metabolites were significantly increased only after the elicitation with *C. acutatum*.

Muskmelon fruit were dipped in 0.1 g/L acibenzolar-S-methyl (ASM) solution for 10 min and then stored at room temperature (22 ± 2 °C, relative humidity 55–60%). The results indicated that ASM significantly ($P \leq 0.05$) increased the activities of phenylalanine ammonia lyase, tyrosine ammonia lyase, cinnamate-4-hydroxylase, 4-coumarate/coenzyme A ligase, peroxidase, and laccase. ASM also enhanced cinnamic acid dehydrogenase activity. Furthermore, ASM significantly ($P \leq 0.05$) enhanced the contents of caffeic acid and ferulic acid, the precursors of lignin biosynthesis, and decreased the content of p-coumaric acid during the late assay time. In addition, significant accumulation of total phenolics, flavonoids and lignin were observed in ASM-treated fruit (Liu *et al.,* 2014). These results suggested that ASM treatment activates the phenylpropanoid pathway to enhance the activity of related enzymes as well as the phenylpropanoid content in the fruit to strengthen the cell wall and prevent pathogenic invasion.

The pectin-degrading enzyme polygalacturonase (PG) normally accumulates to high levels in ripe tomato fruit. Processing and fresh market tomato genotypes were transformed with an antisense PG construct and lines were developed with fruit which had PG levels reduced by more than 99%. To determine the effect of reduced PG activity on both processing and fresh market tomato products, transgenic tomato plants were field-tested. Analysis of field-grown material demonstrated a significant increase in the serum viscosity of processed juice and paste, and a significant decrease in softening during storage of fresh market fruit relative to non-transgenic controls. Fruit with reduced PG were also shown, in laboratory tests, to have increased resistance to *Geotrichum candidum* and *Rhizopus stolonifer*, fungi which normally infect ripening fruit (Matthew Kramer *et al.,* 1992). Thus, manipulation of PG levels results in multiple phenotypic changes which have important bearings on fruit quality through their effects on the pectic substances.

10.4.12 Roles of extracellular proteins in pathogenesis: Plant cell wall is among the first lines of defence that an invasive pathogen encounters. They are heterogeneous structures mainly composed of polysaccharides and proteins (Tian *et al.*, 2016). As a necrotrophic fungus, the appressorium of *B. cinerea* was not strong enough to breach the plant cell wall. Therefore, it is necessary for *B. cinerea* to secrete a series of cell wall–degrading enzymes (CWDEs) to degrade the structural polysaccharides of the host cell wall. In *B. cinerea*, 1155 genes are predicted to encode enzymes to degrade, modify, or create glycosidic bonds. Among them, 275 have signal peptide sequence indicating their function in extracellular matrix. A variety of proteins encoded by the predicted genes were detected through comparative proteomics. However, only few of them have been confirmed to have function in the pathogenicity of *B. cinerea*. Two endopolygalacturonase (BcPG1 and BcPG2) are involved in the virulence of *B. cinerea*. The *Bcpg1* gene is not required for primary infection but necessary for further colonization on apple fruits, tomato fruits, and leaves, whereas *Bcpg2* affects both primary infection and lesion expansion on tomato and broad bean. The endo-β-1,4-xylanases (BcXYN11A), which degrade plant cell wall content xylan, were proven to have a pronounced effect on virulence. Lots of the CWDEs are demonstrated to be not essential for the virulence of *B. cinerea.* Pectin methyl esterase induces the demethylesterification of cell wall components polygalacturonans. Cutinase has the potential to hydrolyse cutin, thus facilitating pathogen penetration through the cuticle. However, deletion mutants of genes of pectin methylesterase (BcPME1 and BcPME2) or cutinase (BcCUTA) exhibit no effects on the virulence of *B. cinerea* . Considering the high redundancy of CWDEs, these enzymes might have an overlapped function with others and contributed to the overall pathogenicity of this fungus.

The important roles of extracellular proteins necessitated a precise regulation mechanism. Extracellular proteins usually initiate with the process of endoplasmic reticulum transported to the golgi compartment for further modifications and then transported to the membrane by secretory vesicles.

Rab family proteins, which belong to the Ras superfamily of small GTPase, have been extensively reported to play important roles in the secretory pathway (Tian *et al.*, 2016). The first identified Rab protein is SEC4 in yeast. SEC4 has been suggested to participate in the growth and protein secretion of *Candida albicans*. The homologue of SEC4-like Rab in *Aspergillus niger* and *Colletotrichum lindemuthianum* has also been identified, namely, srgA and CLPT1. Disruption of srgA in *A. niger* resulted in reduced protein secretion and abnormal apical branching . In contrast, deletion of CLPT1 in *C. lindemuthianum* led to a lethal phenotype . A mutant with a dominant-negative allele of CLPT1 was further used to investigate the function of CLPT1, and ascertained that CLPT1 was essential for secretory vesicles transportation, infectious structures formation, and pathogenicity. In *B. cinerea*, BcSAS1 was determined to be a Rab/GTPase family gene. Deletion of *bcsas1* (Δ*bcsas1*) resulted in reduced virulence in apple, tomato fruits, and tomato leaves. The Δ*bcsas1* mutant exhibits an accumulation of trafficking vesicles at the hyphal tips. A comparative approach was applied to investigate the secretome of Δ*bcsas1*, and the secretion of polysaccharide hydrolases and proteases were significantly depressed.

Gray mold, caused by the fungus *Botrytis cinerea*, is the most significant postharvest disease of kiwifruit., iTRAQ with LC-ESI-MS/MS was used to identify the kiwifruit proteins associated with the response to *B. cinerea* (Jia Liu, *et al.,* 2018). A total of 2,487 proteins in kiwifruit were identified. Among them, 292 represented differentially accumulated proteins (DAPs), with 196 DAPs having increased, and 96 DAPs having decreased in accumulation in *B. cinerea*-inoculated vs. water-inoculated, control kiwifruits. DAPs were associated with penetration site reorganization, cell wall degradation, MAPK cascades, ROS signaling, and PR proteins. In order to examine the corresponding transcriptional levels of the DAPs, RT-qPCR was conducted on a subset of 9 DAPs. In addition, virus-induced gene silencing was used to examine the role of *myosin 10* in kiwifruit, a gene modulating host penetration resistance to fungal infection, in response to *B. cinerea* infection. The present study provides new insight on the understanding of the interaction between kiwifruit and *B. cinerea.*

Comparision of the cellular and extracellular proteomes of *P. expansum* in the absence and presence of borate, which affects the virulence of the fungal pathogen was carried out (Guozheng Qin *et al.,* 2017). The differentially expressed proteins were identified using ESI-Q-TOF-MS/MS. Several proteins related to stress response (glutathione *S*-transferase, catalase, and heat shock protein 60) and basic metabolism (glyceraldehyde-3-phosphate dehydrogenase, dihydroxy-acid dehydratase, and arginase) were identified in the cellular proteome. Catalase and glutathione *S*-transferase, the two antioxidant enzymes, exhibited reduced levels of expression upon exposure to borate. Because catalase and glutathione *S*-transferase are related to oxidative stress response, we further investigated the reactive oxygen species (ROS) levels and oxidative protein carbonylation (damaged proteins) in *P. expansum*. Higher amounts of ROS and carbonylated proteins were observed after borate treatment, indicating that catalase and glutathione *S*-transferase are important in scavenging ROS and protecting cellular proteins from oxidative damage. Additionally to find secretory proteins that contribute to the virulence, we studied the extracellular proteome of *P. expansum* under stress condition with reduced virulence. The expression of three protein spots were repressed in the presence of borate and identified as the same hydrolytic enzyme, polygalacturonase.

The *Botrytis cinere*a VELVET complex regulates light dependent development and virulence. The goal of this study was to identify common virulence defects of several VELVET mutants and to reveal their molecular basis (Nathalie Muller *et al.*, 2018). Growth, differentiation, physiology, gene expression and infection of fungal strains were analyzed, and quantitative comparisons of *in planta* transcriptomes and secretomes were performed. VELVET mutants showed reduced release of citric acid, the major acid secreted by the wild type, whereas no significant role for oxalic acid was observed. Furthermore, a common set of infection related and secreted proteins was strongly underexpressed in the mutants. Quantitative secretome analysis with ^{15}N metabolic labeling revealed a correlation of changes in protein and mRNA levels between wild type and mutants, indicating that transcript levels determine the abundance of secreted proteins. Infection sites kept at low pH partially restored lesion expansion and expression of virulence genes by the mutants. Drastic downregulation of proteases in the mutants was correlated with incomplete degradation of cellular host proteins at the infection site, but no evidence was obtained that aspartyl proteases are required for lesion formation. The *B. cinerea* VELVET complex controls pathogenic differentiation by regulating organic acid secretion, host tissue acidification, gene expression and protein secretion.

Secreted proteins are vital for the pathogenicity of many fungi through manipulating their hosts for efficient colonization. *Fusarium proliferatum* is a phytopathogenic fungus infecting banana fruit. To access the proteins involved in pathogen–host interaction, Taotao Li *et al.*, (2019) used label-free quantitative proteomics technology to comparatively analyze the secretomes of F. proliferatum cultured with and without banana peel in Czapek's broth medium. By analyzing the secretomes of *F. proliferatum*, we have identified 105 proteins with 40 exclusively secreted and 65 increased in abundance in response to a banana peel. These proteins were involved in the promotion of invasion of banana fruit, and they were mainly categorized into virulence factors, cell wall degradation, metabolic process, response to stress, regulation, and another unknown biological process. The expressions of corresponding genes confirmed the existence of these secreted proteins in the banana peel. Furthermore, expression pattern suggested variable roles for these genes at different infection stages. This study expanded the current database of *F. proliferatum* secreted proteins which might be involved in the infection strategy of this fungus. Additionally, this study warranted the further attention of some secreted proteins that might initiate infection of *F. proliferatum* on banana fruit.

10.4.13 Significance of salicylic acid in postharvest physiology of crops: Salicylic acid (SA), an endogenous plant growth regulator, has been found to generate a wide range of metabolic and physiological responses in plants thereby affecting their growth and development. SA as a natural and safe phenolic compound exhibits a high potential in controlling postharvest losses of horticultural crops. Various intrinsic biosynthetic pathways and effects of exogenous salicylic acid on postharvest decay and disease resistance, oxidative stress, fruit ripening, ethylene biosynthesis and action, fruit firmness, respiration, antioxidant systems and nutritional quality were discussed by Mohammadreza Asghari and Morteza Sloeimani Aghdam (2010).

SA or acetyl salicylic acid (ASA), a synthetic analogue of SA, when applied exogenously induced the expression of *PR* (pathogenesis related) genes and also conferred resistance against various pathogens. Exogenous application of SA at nontoxic concentrations to susceptible fruits and vegetables could enhance resistance to pathogens and control postharvest decay. MeSA triggers disease resistance and mediates the expression of defense related genes in neighboring plants and in healthy tissue of infected plants. SA also effectively enhances the biocontrol efficacy of antagonist yeasts. 0.5 mmol L^{-1} SA significantly reduced the incidence of *P. expansum* and *A. alternata* in sweet cherry without any surface injury. Adding SA significantly improved the activity of *R. glutinis* against both pathogens. Salicylates are a major component in the signal transduction pathways of plants playing an important role in disease resistance. Once plant defense responses are activated at the site of infection (LAR), a systemic defense response is often triggered in distal plant parts to protect these undamaged tissues against subsequent invasion by the pathogen. This long-lasting and broad-spectrum induced disease resistance is referred to as SAR and is characterized by the coordinate activation of a specific set of *PR*-genes, many of which encode for proteins with antimicrobial activity (Van Loon *et al.*, 2006).

SA can induce the accumulation of H_2O_2 levels in plant tissues which acts as a signal activating the SAR (Tian *et al.*, 2007). Mutant and transgenic plants that are impaired in SA signaling are incapable of developing SAR and don't show *PR gene activation* upon pathogen infection, which indicates that SA is a necessary intermediate in the SAR signaling pathway. The SA induced defense responses are probably involved in the expression of a range of defense genes, especially those encoding *PR*-proteins such as chitinase, β-1, 3-glucanase and peroxidase. It has been demonstrated that, SA inhibits the H_2O_2-scavenging activity of cytosolic ascorbate peroxidase (APX), and H_2O_2 levels concomitantly rise upon SA treatment of tobacco leaves. Thus, SA may also facilitate H_2O_2 accumulation during the oxidative burst (OB) induced by infection with the virulent pathogens. The increased ROS associated with the OB may contribute to resistance via several mechanisms, including directly killing the invading pathogen and/or activating cell wall crosslinking and lignification, thereby strengthening the cell wall and helping confine the pathogen to the infection site (Tian *et al.*, 2016).

SA prevents postharvest oxidative stress and alleviates chilling injury during cold storage. Plant defense system against oxidative stress consists of two lines; The first line of defense is termed ROS avoidance genes includes alternative oxidase (AOX) and the second is termed as ROS scavenging genes includes SOD, CAT, the ascorbate/glutathione cycle, the glutathione peroxidase system and thioredoxin system (Mohammadreza Asghari and Morteza Sloeimani Aghdam,2010). . SA has been shown to induce expression of AOX and increase the antioxidant capacity of the cells. Consequent application of SA at three stages of vegetative growth, fruit development and postharvest stage was the most effective strategy in improving fruit TAC. Postharvest treatment of sweet cherry fruits with SA significantly inhibited CAT activity, but stimulated the activity of SOD and POD. After inoculation with *P. expansum*, CAT activity decreased and SOD activity increased in SA-treated fruits. SA treatment also changed the expression of POD isozymes, indicating that SA directly or indirectly activates antioxidant enzymes.

SA, as a natural and safe phenolic compound, exhibits a high potential in controlling postharvest losses of horticultural crops. Decrease in ethylene production and action, induction of disease resistance, prevention of oxidative stresses, induction of crop tolerance to chilling injury, decrease in respiration rate, decrease in ripening and senescence rate, prevention of cell wall degrading enzymes and maintaining crop firmness are of main results obtained following SA treatment. The application of SA and especially pre harvest, for inducing the defense resistance systems against postharvest diseases may be a useful and promising measure for controlling postharvest decay on a commercial scale. Since it effectively enhances the effects of other postharvest treatments, such as heat treatments and biocontrol agents, use of SA in combination with other postharvest treatments may give better results in controlling postharvest losses. SA can be used as an appropriate alternative to chemicals in postharvest technology of horticultural crops to assure food safety. Since SA, like any other postharvest treatment, may have different effects on different crops at different circumstances, it is necessary to determine the best and safe concentration for each crop cultivar.

The biocontrol agents and chemical inducers of resistance tested had direct effects against *Penicillium expansum*, significantly reducing the percentages of infection and/or lesion sizes when applied directly against this pathogen. The antifungal activities of acibenzolar-S-methyl, β-aminobutyric acid and methyl jasmonate were also confirmed *in vitro* against *P. expansum* and *Trichoderma atroviride* P1. Treatments with BCAs and chemical inducers of resistance did not control infection on the untreated wounds that were made 1 cm from the original treated wounds in apple. However, reverse-transcription PCR showed that treatments with *Pseudomonas syringae* pv. *syringae* 1,3S and acibenzolar-S-methyl induced significant increases in the levels of gene transcripts of the pathogenesis-related proteins PR-1a, PR-2, PR-5 and PR-8 in apple peel, as compared to controls. Moreover, immunoblotting showed an accumulation of PR-2 and PR-5 proteins in treated samples (Mara Quaglia, *et al.,* 2011). These data indicate that these treatments trigger resistance responses in the apple fruit, although these responses do not limit the infection.

Pear fruit treated by different elicitors, such as salicylic acid (SA), oxalic acid, calcium chloride, and antagonistic yeast *Cryptococcus laurentii*, were investigated to determine the induction of defense responses. The possible mechanism by which elicitors induced the resistance of pear fruit against postharvest disease was also evaluated. The results indicated that all the elicitors could significantly enhance defense-related enzyme activities, such as β-1,3-glucanase, phenylalanine ammonia lyase, peroxidase, and polyphenol oxidase activity, and reduce the disease incidence caused by *Alternaria alternata* in pear fruit (Tian *et al.*, 2006). Among these different elicitors, SA treatment showed the best result in inducing the defense responses and reducing the decay in pear fruit.

The mechanism by which various elicitors induce the resistance against postharvest diseases of fruit is not completely understood. Generally, antifungal glucanohydrolases and peroxidases are considered potentially important in host resistance mechanism as POD can catalyze the last step of lignin biosynthesis and may have an important role in the defense mechanism of plants. In addition, PPO, β-1,3-glucanase, and PAL also showed an important role in induced resistance of plants. The phenolic compounds can be oxidized by PPO and produce antimicrobial phenolic substances, such as quinines, which are more toxic to pathogens than the former. β-1,3-Glucanase has been proven to hydrolyze the β-1,3-linked glucans, major components of the cell wall of fungi, and synergistically acts with chitinase to inhibit fungal growth. PAL is the first enzyme of the phenylpropanoid pathway and is involved in the biosyntheses of phenolics, phytoalexins, and lignins. Therefore, the increased PAL activity will contribute to the reduced percentage of incidence. A better understanding of the mechanisms of elicitors on induced resistance of harvested fruit is an important prerequisite for their commercial application. Although the induction of pathogenesis-related proteins, such as β-1,3-glucanase, POD, PPO, and PAL, is correlated with induced resistance of fruit, that is, as a multicomponent phenomenon, other mechanisms, alone or collectively, may also be involved.

10.4.14 Hijacking of ripening regulation by *Botrytis cinerea* : *B. cinerea* releases enzymes and metabolites that act as virulence factors but may also induce plant responses that are beneficial for fungal infection (van Kan *et al*.,, 2017). A relevant example of the manipulation of physiological processes in the host by *B. cinerea* is the interference with specific developmental processes. In tomato plants, *B. cinerea* infections modified host gene expression to increase susceptibility, such as the induction of senescence in leaves. Moreover, infected unripe tomato fruit show premature expression of genes involved in ethylene synthesis during tomato ripening. These findings suggest that *B. cinerea* could initiate ethylene production and thereby stimulate early ripening. As strawberries are non-climacteric fruit, ethylene production of *B. cinerea* may not have substantial effects on strawberry ripening; however, the fungus was also shown to induce genes involved in the biosynthesis of other plant hormones such as ABA. Moreover, *B. cinerea* can synthesize and secrete ABA that functions as a virulence factor. Besides hormones, increased oxidative reactions caused by the pathogen may influence ripening progression (Stefan Petrasch *et al*., 2019).

Fruit ripening is a developmental process that is associated with increased susceptibility to the necrotrophic pathogen *Botrytis cinerea*. Histochemical observations demonstrate that unripe tomato fruit activate pathogen defense responses, but these responses are attenuated in ripe fruit infected by *B. cinerea* (Dario Cantu *et al.,* 2009) Tomato fruit ripening is regulated independently and cooperatively by ethylene and transcription factors, including NON-RIPENING (NOR) and RIPENING-INHIBITOR (RIN). Mutations in NOR or RIN or interference with ethylene perception prevent fruit from ripening and, thereby, would be expected to influence susceptibility. We show, however, that the susceptibility of ripe fruit is dependent on NOR but not on RIN and only partially on ethylene perception, leading to the conclusion that not all of the pathways and events that constitute ripening render fruit susceptible. Additionally, on unripe fruit, *B. cinerea* induces the expression of genes also expressed as uninfected fruit ripen. Among the ripening-associated genes induced by *B. cinerea* are LePG (for polygalacturonase) and LeExp1 (for expansin), which encode cell wall-modifying proteins and have been shown to facilitate susceptibility. LePG and LeExp1 are induced only in susceptible rin fruit and not in resistant nor fruit. Thus, to infect fruit, *B. cinerea* relies on some of the processes and events that occur during ripening, and the fungus induces these pathways in unripe fruit, suggesting that the pathogen itself can initiate the induction of susceptibility by exploiting endogenous developmental programs. These results demonstrated the developmental plasticity of plant responses to the fungus and indicate how known regulators of fruit ripening participate in regulating ripening-associated pathogen susceptibility.

Noble rot results from exceptional infections of ripe grape berries by *Botrytis cinerea*. Unlike bunch rot, noble rot promotes favorable changes in grape berries and the accumulation of secondary metabolites that enhance wine grape composition. Noble rot-infected berries of cv Sémillon, a white-skinned variety, were collected over 3 years from a commercial vineyard at the same time that fruit were harvested for botrytized wine production (Barbara Blanco-Ulate *et al.*, 2015). Using an integrated transcriptomics and metabolomics approach, they demonstrated that noble rot alters the metabolism of cv Sémillon berries by inducing biotic and abiotic stress responses as well as ripening processes. During noble rot, *B. cinerea* induced the expression of key regulators of ripening-associated pathways, some of which were distinctive to the normal ripening of red-skinned cultivars. Enhancement of phenylpropanoid metabolism, characterized by a restricted flux in white-skinned berries, was a common outcome of noble rot and red-skinned berry ripening. Transcript and metabolite analyses together with enzymatic assays determined that the biosynthesis of anthocyanins is a consistent hallmark of noble rot in cv Sémillon berries. The biosynthesis of terpenes and fatty acid aroma precursors also increased during noble rot. The impact of noble rot in botrytized wines was studied. Altogether, the results of this work demonstrated that noble rot causes a major reprogramming of berry development and metabolism. This desirable interaction between a fruit and a fungus stimulated pathways otherwise inactive in white-skinned berries, leading to a greater accumulation of compounds involved in the unique flavor and aroma of botrytized wines.

MicroRNAs are endogenous small non-coding RNAs that negatively regulate mRNAs, mainly at the post-transcriptional level, and play an important role in resistance response of plants. To date, there are few reports on resistance response of strawberry miRNAs to pathogens (Liang *et al.*, 2018). In this study, using high-throughput sequencing, 134 conserved and 35 novel miRNAs were identified in six libraries within the treatment of *Botrytis cinerea.* A total 497 potential target genes were predicted using *Fragaria vesca* genome. Most of the differential expressed miRNAs in strawberry fruits were up-regulated in early libraries and down-regulated in late libraries. *PIRL*, the target gene of miR5290a, showed the opposite expressed trend compared with miR5290 from T1 to T3 libraries, and functional analysis of the *PIRL* gene shows that it has obvious resistance to *B. cinerea* in the strawberry fruits with overexpressed *PIRL* gene. It was speculated that miR5290a negatively regulates its target gene *PIRL* to increase resistance to pathogen infection, and further analysis of *PIRL* function is meaningful for studying the plant-pathogen relationship and improving strawberry fruit quality and yield.

Fruit–pathogen interactions are a valuable biological system to study the role of plant development in the transition from resistance to susceptibility. In general, unripe fruit are resistant to pathogen infection but become increasingly more susceptible as they ripen. During ripening, fruit undergo significant physiological and biochemical changes that are coordinated by complex regulatory and hormonal signaling networks. The interplay between multiple plant stress hormones in the interaction between plant vegetative tissues and microbial pathogens has been documented extensively, but the relevance of these hormones during infections of fruit is unclear. Barbara Blanco-Ulate *et al.*, (2013) analyzed a transcriptome study of tomato fruit infected with *Botrytis cinerea* in order to profile the expression of genes for the biosynthesis, modification and signal transduction of ethylene (ET), salicylic acid (SA), jasmonic acid (JA), and abscisic acid (ABA), hormones that may be not only involved in ripening, but also in fruit interactions with pathogens. The changes in relative expression of key genes during infection and assays of susceptibility of fruit with impaired synthesis or perception of these hormones were used to formulate hypotheses regarding the involvement of these regulators in the outcome of the tomato fruit–*B. cinerea* interaction.

It was shown that some *B. cinerea* small RNAs (Bc-sRNAs) can silence *Arabidopsis* and tomato genes involved in immunity (Ame weiberg *et al.*, 2013). These Bc-sRNAs hijack the host RNA interference (RNAi) machinery by binding to *Arabidopsis* Argonaute 1 (AGO1) and selectively silencing host immunity genes. The *Arabidopsis* ago1 mutant exhibits reduced susceptibility to *B. cinerea*, and the *B. cinerea* dcl1 dcl2 double mutant that can no longer produce these Bc-sRNAs displays reduced pathogenicity on *Arabidopsis* and tomato. Thus, this fungal pathogen transfers "virulent" sRNA effectors into host plant cells to suppress host immunity and achieve infection, which demonstrates a naturally occurring cross-kingdom RNAi as an advanced virulence mechanism.

10.4.15 Mechanisms of defence and avoidance against *Botrytis cinerea* : Defence mechanisms can be divided into preformed and induced defences. In strawberries, preformed defence compounds are especially abundant in the unripe stage, as reviewed in the section on quiescence of *B. cinerea.* Even though plants accumulate defence compounds, B. cinerea has mechanisms to cope with these metabolites by efflux and detoxification of inhibitory substances. ATP-binding cassette (ABC) transporters are used by *B. cinerea* to facilitate the efflux of antifungal compounds, such as stilbenes. *B. cinerea* is capable of detoxifying inhibitory substances, like epicatechin by secretion of laccases. Active *B. cinerea* infections can result in a reduction of specific secondary metabolites. It has been reported that levels of flavan-3-ol, benzoic acid and phenylpropanoids drop in *B. cinerea*-infected strawberries. Strawberries respond to *B. cinerea* infection by triggering defences. In some cases, preformed and induced defences can overlap such as in the case of PGIPs. An endogenous PGIP appears to be constitutively expressed in fruit from various strawberry cultivars . However, this PGIP and six additional ones show higher expression levels upon infection with *B. cinerea.* Overexpression of FaPGIP1a and FaPGIP2a in cisgenic plants conferred enhanced resistance to grey mould. Other enzymes induced by *B. cinerea* infections are chitinases (Stefan Petrasch *et al.*, 2019). Expression of the chitinases FaChi2-1 and FaChi2-2 peaked 16 hpi in *B. cinerea*-infected strawberries. Furthermore, heterologous expression of *Phaseolus vulgaris* chitinase cH5B in strawberry resulted in higher resistance to infection Another study demonstrated that application of heat-inactivated cells of the yeast *Aureobasidium pullulans* promoted tolerance to *B. cinerea* in strawberries. This primed resistance is probably due to the fruit's perception of chitin from the yeast leading to induction of chitinases or other plant immune responses.

Fruit defence responses may be primed using mechanical stimulation as it was reported for strawberry leaves. Induced defences include accumulation of secondary metabolites and ROS. For instance, strawberries accumulate proanthocyanins around infection zones possibly to restrict fungal growth . The surroundings of infection sites generally display higher ROS production. ROS can serve as an effective defence against pathogens but also can lead to cell death, which is considered beneficial for necrotrophic fungi (Prusky and Lichter, 2007). *B. cinerea* itself produces ROS to induce host cell death, deplete plant antioxidants and increase lipid peroxidation. It is therefore interesting that, in unripe tomato fruit ROS production leads to resistance against *B. cinerea*, whereas in ripe fruit it seems to promote susceptibility. Future research will likely shed more light on the role of ROS in induced defences of strawberry fruit. Basal immunity is activated upon fungal infection. Degradation of fruit cell wall pectins can produce demethylated oligogalacturonides that trigger basal immune responses.

Expression of the *F. x ananassa pectin methylesterase 1 FaPE1* in *Fragaria vesca* resulted in reduced methyl-esterification of oligogalacturonides in fruit. This reduced esterification activated basal defences via the salicylic acid (SA) signalling pathway that led to a higher resistance to *B. cinerea* (Osorio *et al.*, 2011). Involvement of SA signalling in responses against *B. cinerea* was previously suggested when strawberry plants and fruit treated with SA showed decreased postharvest decay. *B. cinerea* can suppress the expression of plant defence responses by hijacking the host sRNA regulatory pathways. In strawberry fruits, *B. cinerea* infections can alter the expression of microRNAs involved in the regulation of defence genes, including the plant intracellular Ras group-related LRR protein 9-like gene. Interestingly, *B. cinerea* can also take up plant sRNAs during its interaction with the host. For instance, transgenic plants expressing sRNA that targets *B. cinerea* DCL1 and DCL2 show significantly reduced fungal growth in strawberries (Stefan Petrasch *et al.*, 2019). The suppression of fungal growth via host sRNA is not well understood, and it is yet to be demonstrated that this mechanism of defence naturally occurs in plants.

Cation salts and fungal growth: Growth of *Colletotrichum gloeosporioides* in the presence of cation salts NaCl and KCl inhibited fungal growth and anthracnose symptom of colonization (Dubey *et al.*, 2016). Previous reports indicate that adaptation of *Aspergillus nidulans* to salt- and osmotic-stress conditions revealed the role of zinc-finger transcription factors SltA and CrzA in cation homeostasis. Homologs of *A. nidulans* SltA and CrzA were identified in *C. gloeosporioides*. The *C. gloeosporioides* CrzA homolog is a 682-amino acid protein, which contains a C_2H_2 zinc finger DNA-binding domain that is highly conserved among CrzA proteins from yeast and filamentous fungi. The *C. gloeosporioides* SltA homolog encodes a 775-amino acid protein with strong similarity to *A. nidulans* SltA and *Trichoderma reesei* ACE1, and highest conservation in the three zinc-finger regions with almost no changes compared to ACE1 sequences. Knockout of *C. gloeosporioides crzA* (Δ*crzA*) resulted in a phenotype with inhibited growth, sporulation, germination and appressorium formation, indicating the importance of this calciu006D-activated transcription factor in regulating these morphogenetic processes. In contrast, knockout of *C. gloeosporioides sltA* (Δ*sltA*) mainly inhibited appressorium formation. Both mutants had reduced pathogenicity on mango and avocado fruit. Inhibition of the different morphogenetic stages in the Δ*crzA* mutant was accompanied by drastic inhibition of chitin synthase A and B and glucan synthase, which was partially restored with Ca^{2+} supplementation. Inhibition of appressorium formation in Δ*sltA* mutants was accompanied by downregulation of the MAP kinase *pmk1* and carnitine acetyl transferase (*cat1*), genes involved in appressorium formation and colonization, which was restored by Ca^{2+} supplementation. Furthermore, exposure of *C. gloeosporioides* Δ*crzA* or Δ*sltA* mutants to cations such as Na^+, K^+ and Li^+ at concentrations that the wild type *C. gloeosporioides* is not affected had further adverse morphogenetic effects on *C. gloeosporioides* which were partially or fully restored by Ca^{2+}. Overall results suggest that both genes modulating alkali cation homeostasis have significant morphogenetic effects that reduce *C. gloeosporioides* colonization.

10.4.16 Molecular aspects in pathogen-fruit interactions: Virulence and Resistance: Fruits and vegetables can withstand pathogens by means of physical barriers (presence of thick cuticular or trichome layers) and chemical ones, or through induced defenses that are activated once the host detects the presence of the pathogen, triggering the oxidative burst during the early hours of interaction. This burst entails the generation of reactive oxygen species (ROS), such as superoxide (O_2^-), hydroxyl radical (OH) or hydrogen peroxide (H_2O_2), and the activation of genes involved in several metabolic pathways. The study of such mechanisms may allow detecting disease-resistant genetic materials, thus reducing the use of toxic products.

Various stages of postharvest disease development include contamination, the inoculum reaches the host, usually by the spreading of spores; penetration, the pathogen enters through the tissue of the host by means of penetration structures, specialized enzymes or wounds; infection, the process by which the pathogen comes in contact with the cells of the host which it later feeds on; incubation, the time between infection and the appearance of disease symptoms; diffusion or invasion, the pathogen extends beyond the colonized tissues; reproduction and dissemination, the pathogen colonizes new tissues and forms resistance structures, and survival, the resistance structures formed remain in the environment until finding conditions conducive to infect new hosts (Sandoval-Chávez *et al.*, 2015). Not all pathogens cause disease. This inability is because, in many cases, fruits have different defense mechanisms, viz., constitutive or noninduced, which involve intrinsic factors that can foster a hostile environment for pathogens and activated or induced, including pathogen recognition by the fruit and activation of different biochemical pathways to counteract the infection.

Postharvest pathogens can start its attack process immediately after spores land on wounded tissue, whereas other pathogens can forcibly breach the unripe fruit cuticle and then remain quiescent for months until fruit ripens and then cause major losses. Postharvest fungal pathogens activate their development by secreting organic acids or ammonia that acidify or alkalinize the host ambient surroundings. These fungal pH modulations of host environment regulate an arsenal of enzymes to increase fungal pathogenicity (Eduardo A.Espeso and Dov Prusky, 2013). This arsenal includes genes and processes that compromise host defenses, contribute to intracellular signaling, produce cell wall-degrading enzymes, regulate specific transporters, induce redox protectant systems, and generate factors needed by the pathogen to effectively cope with the hostile environment found within the host. Further, evidence is accumulating that the secreted molecules (organic acids and ammonia) are multifunctional and together with effect of the ambient pH, they activate virulence factors and simultaneously hijack the plant defense response and induce program cell death to further enhance their necrotrophic attack.

With the increased focus on healthy eating and consuming raw vegetables, this study assessed the extent of contamination of fresh vegetables by *Pseudomonas aeruginosa* in Jamaica and examined the antibiotic susceptibility profiles and the presence of various virulence associated determinants of *P. aeruginosa*. Analyses indicated that vegetables from retail markets and supermarkets were widely contaminated by *P. aeruginosa*; produce from markets were more frequently contaminated, but the difference was not significant. Lettuce and carrots were the most frequently contaminated vegetables, while tomatoes were the least. Pigment production (Pyoverdine, pyocyanin, pyomelanin and pyorubin), fluorescein and alginate were common in these isolates. Imipenem, gentamicin and ciprofloxacin were the most inhibitory antimicrobial agents. However, isolates were resistant or showed reduced susceptibility to ampicillin, chloramphenicol, sulphamethoxazole/trimethoprim and aztreonam, and up to 35% of the isolates were resistant to four antimicrobial agents. As many as 30% of the isolates were positive for the *fpv1* gene, and 13% had multiple genes (Kashina Allydice-Francis and Paul D. Brown, 2012). Sixty-four percent of the isolates harboured an exoenzyme gene (*exoS*, *exoT*, *exoU* or *exoY*), and multiple exo genes were common. It was concluded that *P. aeruginosa* is a major contaminant of fresh vegetables, which might be a source of infection for susceptible persons within the community.

Each year, fungal storage rots cause serious losses in the apple industry. The problem is especially prominent in production systems with little or no use of pre- or postharvest fungicides. Consequently, a high level of tolerance to storage rots is very desirable. A set of 81 apple cultivars were studied in two different production areas (Njøs in Norway and Balsgård in Sweden), to investigate genetically determined susceptibility to blue mold, *Penicillium expansum* (Tahir *et al.*,2015). Lesion diameter, measured on symptoms developing after artificial inoculation with this pathogen and cold storage for 6–12 weeks, and decay index, calculated as lesion diameter divided by number of weeks in storage, varied significantly among cultivars. Associations between disease evaluation data and different fruit characteristics were also investigated. Decay index was negatively correlated with harvest date, estimated as number of days since full bloom, and fruit firmness at harvest. A positive correlation was instead found between decay index and amount of fruit softening during storage. These results provided data about genetically determined level of resistance to blue mold for apple cultivars adapted to a cool climate, and will be valuable for further research on the genetic control of resistance, as well as for choice of breeding material.

Postharvest decay of fruits, vegetables, and grains by fungal pathogens causes significant economic losses. Infected produce presents a potential health risk since some decay fungi produce mycotoxins that are hazardous to human health. Infections are the result of the interplay between host resistance and pathogen virulence. Both of these processes, however, are significantly impacted by environmental factors, such as temperature, UV, oxidative stress, and water activity. In a review, the impact of various physical postharvest treatments (e.g., heat and UV) on the viability and virulence of postharvest pathogens was reviewed and discussed (Jia Liu, *et al.,* 2018). Oxidative injury, protein impairment, and cell wall degradation have all been proposed as the mechanisms by which these abiotic stresses reduce fungal viability and pathogenicity. The response of decay fungi to pH and the ability of pathogens to modulate the pH of the host environment also affect pathogenicity. The effects of the manipulation of the postharvest environment by ethylene, natural edible coatings, and controlled atmosphere storage on fungal viability were also discussed and avenues of future research were proposed.

To study how crop domestication impacts plant resistance to pathogens, and correspondingly how this interacts with the pathogen's genetics, Nicole E. Soltis *et al*., (2018) infected a collection of wild and domesticated tomato accessions with a genetically diverse population of the generalist pathogen *Botrytis cinerea*. They quantified variation in lesion size of *B. cinerea* genotypes (isolates) on domesticated *Solanum lycopersicum* and 6 wild *S. pimpinellifolium* genotypes. This showed that lesion size was significantly controlled by plant domestication, plant genetic variation, and the pathogen's genotype. Overall, resistance was slightly elevated in the wild germplasm in comparison to domesticated tomato accessions. Genome-wide association

(GWA) mapping in *B. cinerea* identified a highly polygenic collection of genes. This suggested that breeding against this pathogen would need to utilize a diversity of isolates to capture all possible mechanisms. Critically, a discrete subset of *B. cinerea* genes was identified where the allelic variation was linked to altered virulence against the wild versus domesticated tomato accessions. This indicates that this generalist pathogen already has the necessary allelic variation in place to handle the introgression of wild resistance mechanisms into the domesticated crop.

Pathogenicity denotes the ability of a pathogen to cause disease on a particular host (a qualitative trait), whereas virulence is the degree of damage caused to the host (a quantitative trait). Virulence factors may be defined as those pathogen components that are non-essential for in vitro growth in a culture medium but contribute to disease. When one of these factors is required for pathogenicity it can be considered a pathogenicity factor. As virulence factors are important for infection, preventing pathogens from producing them constitutes an interesting alternative strategy for disease control. This strategy has been termed "antivirulence therapy" (Cegelski *et al.*, 2008). As this therapy aims to disarm the pathogen rather than kill it or halt its growth, it is presumed that antivirulence compounds will generate much less selection pressure than traditional antibiotics in the pathogen to regain resistance, a problem faced with commonly used fungicides. However, before any antivirulence therapy can be developed there is a need to identify the virulence factors.

In a broad sense, these factors can be as diverse as plant cell wall degrading enzymes (CWDEs), effectors (molecules that modify the physiology of the host in order to allow pathogen infection), or mechanisms that permit the rapid adaptation of the pathogen to the host environment. The regulatory systems that govern all these processes can also be considered as virulence factors. (Tian *et al.*, 2016). One of the earliest responses detected in many 17 incompatible interactions (i.e., where the plant is resistant to the pathogen) is the 18 activation of an oxidative burst characterized by an increase in the levels of reactive 19 oxygen species (ROS). The oxidative burst precedes the synthesis 20 of antimicrobial compounds like phytoalexins. This burst leads up to alterations in the 21 synthesis of cell-wall structural proteins and the transcriptional activation of specific 22 genes leading to the synthesis of pathogenesis-related (PR) proteins such as chitinases, β23 1,3-glucanases and peroxidases (POD). Although these defense factors in host tissues have been extensively studied, the mechanisms that regulate these responses have not been elucidated.

The storage facilities applying fungicides and storing diverse fruit species selected for *P. expansum* isolates with a larger lesion diameter were bserved on apples (Bochra A. Bahri *et al.*, 2019). In addition, application of fungicides and diversification in stored fruit species significantly select for *P. expansum* isolates with higher levels of mycelial growth and spore production, respectively. Moreover, the diversity of host species of stored fruit accounted for 38% of the variability observed between storage facilities for the measured fitness parameters in *P. expansum* isolates and had a stronger effect on *P. expansum* population structure than fungicide treatment. Essential oils from *Mentha pulegium* and *Syzygium aromaticum* significantly decreased mycelial growth and spore production of *P. expansum* isolates *in vitro. Mentha pulegium* essential oil also significantly decreased the size of lesions associated with the blue mold rot of apples. Reducing the diversity of stored host species and applying *M. pulegium* essential oil may be useful in counter-selecting for aggressive *P. expansum* isolates and reducing losses due to blue mold rot during fruit storage.

Pathogenic fungi including *Penicillium digitatum* and *Penicillium italicum* are the main destructive pathogens in the citrus industry, causing great losses during postharvest process. Only one mycovirus from *P. digitatum* has been reported and the prevalence of such mycoviruses against citrus postharvest pathogenic fungi and their genotyping were still under investigation (Niu *et al.*, 2018). It was shown that 39 of 152 *Penicillium* isolates from main citrus-growing areas in China were infected with various mycoviruses belonging to polymycoviruses, Narna-like viruses, and families *Totiviridae*, *Partitivirdae* and *Chrysoviridae*. The next generation sequencing (NGS) towards virus genome library and the following molecular analysis revealed two novel mycoviruses *Penicillium digitatum* polymycovirus 1 (PdPmV1) and *Penicillium digitatum* Narna-like virus 1 (PdNLV1), coexisting in *P. digitatum* strain HS-RH2. The fungicide-resistant *P. digitatum* strains HS-F6 and HS-E9 coinfected by PdPmV1 and PdNLV1 exhibited obvious reduction in triazole drug prochloraz resistance by mycelial growth analysis on both PDA plates and citrus fruit epidermis with given prochloraz concentration. This report at the first time characterized two novel mycoviruses from *P. digitatum* and revealed the mycovirus-induced reduction of fungicide resistance.

In order to gain insight into the genetic bases of virulence mechanisms of *Penicillium digitatum* and its high degree of host-specificity, the genomes of two *P. digitatum* strains that differ in their antifungal resistance traits have been sequenced and compared with those of 28 other Pezizomycotina (Marcet-Houben *et al.,* 2012). The two sequenced genomes are highly similar, but important differences between them include the presence of a unique gene cluster in the resistant strain, and mutations previously shown to confer fungicide resistance. The two strains, which were isolated in Spain, and another isolated in China have identical mitochondrial genome sequences suggesting a recent worldwide expansion of the species. Comparison with the closely-related but non-phytopathogenic *P. chrysogenum* reveals a much smaller gene content in *P. digitatum*, consistent with a more specialized lifestyle. We show that large regions of the *P. chrysogenum* genome, including entire supercontigs, are absent from *P. digitatum*, and that this is the result of large gene family expansions rather than acquisition through horizontal gene transfer. Our analysis of the *P. digitatum* genome is indicative of heterothallic sexual reproduction and reveals the molecular basis for the inability of this species to assimilate nitrate or produce the metabolites patulin and penicillin. Finally, we identify the predicted secretome, which provides a first approximation to the protein repertoire used during invasive growth.

The complete genome of *P. digitatum*, the first of a phytopathogenic *Penicillium* species, is a valuable tool for understanding the virulence mechanisms and host-specificity of this economically important pest.

10.4.17 New technologies for studying host-pathogen interactions in postharvest fruit systems: During the last ten years we have witnessed an "omics" revolution that has involved all fields of biological sciences. This change is reflected in the way researchers now go about the study of biological processes, from a single gene/enzyme level to a holistic approach through which the researchers try to examine problems from a global point of view. Sequencing technology and its applications have expanded from the sequencing of individual genes to the assembly of complete genomes, from the analysis of the expression of a few genes to global transcriptomic analyses, and so on. However, the implementation of these new technologies in the field of postharvest pathology is lagging behind their uses in other closely related field.

The genomes of several postharvest pathogens are now available, including *Botrytis cinerea* and several different species of *Alternaria* and *Colletotrichum*. The genomes of the following specific postharvest fruit pathogens from the genus *Penicillium* have also been recently sequenced: *P. digitatum* (Tian *et al*., 2016); *P. griseofulvum* and *P. expansum* and *P. italicum*. There are also ongoing genome sequencing projects for other relevant postharvest fungi, such as *Monilinia fructicola* and *M. laxa*. The genetic information gathered in these projects provides a foundation for an in-depth analysis of the virulence factors of these important postharvest pathogens. Data are available on the whole set of putative genes present in these pathogens and on how many of these genes are secreted, which ones are candidate effectors and the arsenal of carbohydrate-active enzymes (CAZymes). There is also data on the array of proteases that each fungus can produce. In order to decipher which genes are relevant to pathogenesis and virulence, however, more studies are needed. Genes involved in pathogenicity and virulence are actively expressed during the colonization of the host. A transcriptomic analysis of the pathogen-fruit interaction will thus reveal these genes, as well as other genes that are necessary for the normal growth of the fungus and are not necessarily related to pathogenicity/virulence. One advantage of directly analyzing the host/pathogen interaction is that information on host defense responses deployed in response to the fungal attack can also be obtained.

i.Unravelling fruit defense mechanisms : Several species of postharvest pathogens, such as *Colletotrichum, Alternaria, Botrytis* and *Monilinia*, have been reported to remain quiescent in their hosts until the fruit ripens. In a recent study focused on tomato fruit defense responses against the fungus *C. gloeosporioides*, researchers showed that the expression of several fruit genes was induced even before appressoria penetration. Such genes included PAMP receptors and genes related with fatty acid biosynthesis, elongation, and the synthesis of cutin and waxes (Alkan *et al*., 2015). Moreover, the host defense reaction was further intensified during the quiescent stage by the activation of lignification, glycoalkaloid and phenylalanine pathways.

To unravel fruit responses against pathogen attack, it should be considered that fruits induce cellular defenses that prevent further colonization of the fruit tissue once the structural barriers have been breached by the pathogen. The structural or physical barriers "that pathogens have to break" are known as the passive (or constitutive) defense mechanisms of fruits to avoid pathogen ingress and/or progress. It has been reported that *Trincadeira* berries infected with *B. cinerea* accumulate saturated long-chain fatty acids, which are major constituents of grape waxes, and that this accumulation is accompanied by the up-regulation of several acyl-CoA synthetases and wax synthases (Agudelo Romero *et al*., 2015). These studies show that the formation of barriers that are commonly recognized as constitutive can be induced within the framework of a host's active response to fungal attack. These findings warrant further attention and future research given the critical role host lipids and lipid metabolites play in the dynamics of fruit defense.

Other fruit defense mechanisms, described as active (or inducible) defense mechanisms, require host mechanisms to function. Knowledge of plant-fungal interaction indicates that a crucial role in host defense is played by a rapid and massive generation of ROS within the host cells, a process known as an oxidative burst . The oxidative burst is one of the earliest responses to pathogen attack and involves the production of ROS including the superoxide anion, the hydroxyl radical and H_2O_2. In spite of the important role of ROS in plant defense, there are few reports on the role that they might play in fruits (Tian *et al.*, 2016). The production of H_2O_2 in plants is also involved in lignification and H_2O_2 is a signaling molecule mediating the induction of hypersensitive cell death and the expression of a wide array of defense-related genes in surrounding cells. The lignification process may contribute to fruit resistance in different ways, viz., making the fruit more resistant to degradation by enzymes secreted by an invading pathogen, constituting a barrier preventing free nutrient movement and therefore helping to starve the pathogen and slowing down the penetration process

The role of phenylalanine ammonia lyase (PAL), a key enzyme in the phenylpropanoid pathway, in regulating lignin accumulation (Tian *et al.*,2016) and synthesizing phenols has been described. A recent study on the orange-*P. digitatum* pathosystem detected the maximum expression of several phenylpropanoid-related genes at 48 h after inoculation. PAL1, caffeic acid O-methyl-transferase (COMT1) and POX1 gene expression were induced to a greater extent by *P. digitatum* than by *P. expansum,* although the maximum expression was observed at 24 h, which may be due to the inoculum concentration used. A further active defense mechanism is related to pathogenesis-related proteins (PR), which have diverse functions and most of which are induced by stress (biotic or abiotic) . In terms of the fruit defense response, diverse works have described both the enzyme activity and gene expression of PR-proteins. For example, an increase in chitinase and glucanase enzymatic activities was detected in tomato fruit as a response to infection caused by *A. alternata* and in grapes infected by *B. cinerea.*

In plums, researchers observed an increase in the transcript level of PR10 and in H_2O_2 production after *M. fructicola* infection. Additionally, the gene expression of PR proteins [chitinase (CHI), PR10 and β-glucanase] was detected earlier in a susceptible variety after 24 h of *C. acutatum* inoculation (Miles *et al.,* 2011). Furthermore, two PR genes (PR5 and PR10) were found to be over-expressed in *Fragaria chiloensis* and *Fragaria x ananassa* after *B. cinerea* infection and showed different expression profiles depending on the tissue analyzed .

Plant defenses are regulated by complex signaling pathways involving salicylic acid, jasmonic acid and ethylene. Specifically, ethylene has been implicated in biotic stress, both as a virulence factor of fungal and bacterial pathogens and as a signaling compound in disease resistance (Tian *et al.,* 2016). However, the role of this hormone in the postharvest disease resistance of fruit is still unknown. In both ripe and unripe tomato fruit, *B. cinerea* and *C. gloeosporioides* infections induce the ethylene biosynthesis pathway; transcription factors like non-ripening (NOR), ripening-inhibitor (RIN) and never-ripe (NR); and the ethylene-regulated defense genes (Alkan et al., 2015). In citrus fruit, ethylene production has been found to increase with *P. digitatum* infection, and this hormone is synthesized by both the fruit and the pathogen . Moreover, the fact that many genes induced in citrus fruit upon *P. digitatum* attack are also up-regulated by ethylene highlights the role this hormone plays in the defense response of citrus to this pathogen.

During recent years, a growing number of studies have sought to identify cultivars with a high level of resistance to pathogens. This increase can be attributed to the fact that more and more breeding programs are seeking out new tools for the early selection of seedlings with enhanced resistance to pathogens. Several studies have provided data on genetically determined levels of resistance to *B. cinerea* in strawberry (Bestfleisch *et al.*, 2015) and to *P. expansum* in apple cultivars with a high level of resistance to fungal pathogens. In peach breeding programs. QTLs in a few genomic regions underlying brown rot response traits and providing a starting point for the development of marker-assisted selection for increased resistance to this pathogen. Preliminary results from an apple breeding program have already made it possible to identify QTLs for blue mold resistance in Malus sieversii (Norelli *et al.,* 2014).

Among the major postharvest phytopathogenic fungi, *B. cinerea* is the best characterized from a molecular point of view, a fact that reflects the importance of this fungus as a general plant pathogen. In recent years, however, extensive research has also been conducted on other important postharvest pathogens, such as *P. expansum, P. digitatum, C. gloeosporioides* and *A. alternata.* Interestingly, most of these fungi are necrotrophs that kill the cell to obtain nutrients. For a long time, these fungi were thought to rely on CWDEs and other extracellular hydrolases, such as proteases, as the major virulence factors. However, the characterization of gene knockout mutants indicates that in most instances all these enzymes are just necessary for full virulence, but they are not essential for pathogenesis. A similar situation is being uncovered with pathogen enzymes involved in ROS production/detoxification. These results are probably not so surprising, however, because many of these enzymes belong to families with redundant enzymes, and the loss of one member of the family can be compensated for by another enzyme that carries out a similar biological function. Although the biological functions of the majority of the virulence genes identified and characterized by molecular biology techniques are still not clear, there is increasing optimism supported by recent experimental evidence (An *et al.*, 2016).

ii. Effectors : One of the research topics that is probably gaining the most attention in many pathogenic fungi, and which we think is likely to gain traction in the near future in postharvest pathogens, is the study of effectors: pathogen molecules that modulate host defense responses favoring the development of the pathogen. Although most effectors characterized so far are small cysteine rich secreted proteins, there are new players entering the scene. Small RNAs are emerging as very important regulators of gene expression, and not only in the organism in which they are synthesized, since they can be translocated to the host cells where they manipulate defense signaling pathways favoring pathogen infection. The study of small RNAs will probably open new perspectives in the study of fruit-pathogen interactions.

In order to be able to infect, pathogens have to adapt to the environmental conditions found within the host. As we have seen, these factors, such as ambient pH, light or the presence of exogenously added substances, also play a fundamental role in the outcome of fruit-pathogen interactions. The manipulation of environmental conditions thus constitutes another tool that can be exploited for the control of postharvest diseases. Furthermore, other environmental factors deserve further research, like the manipulation of nutrient availability or increasing the concentration of compounds that either promote fruit resistance or that are toxic to the pathogen. Besides inducing resistance, there are clearly two major strategies for increasing the resistance of fruits to pathogens. One involves the genetic manipulation of the fruit crop following similar strategies already used to increase resistance in whole plants. The other involves breeding programs that aim to introgress resistance QTLs from wild varieties into commercial germplasms. Such programs are already underway in important crops like apple, peach and strawberry. Hopefully these long-term initiatives will provide new varieties with improved resistance against major postharvest pathogens.

During ripening, fruit undergo major changes such as activation of ethylene synthesis and other phytohormones, pH change, cuticular changes, cell-wall loosening and increase of soluble sugars, decline of antifungal compounds, which release the fungus from its quiescent state and promote a necrotrophic and pathogenic life style. Knowledge on the molecular and metabolic events responsible for the onset of necrotrophic stage, occurring both in the host and in fungi, is important key in order to develop strategies to enhance fruit defense and decrease of fungal virulence that ultimately will result in increased quality of fruits (Noam Alkan and Ana M.Fortes, 2015).. This knowledge can be considered in breeding programs, pre and postharvest treatments or alternatively provide a framework for biotechnological approaches.

iii.Ultra cellular changes : Most of the pathogens have the potential for attacking only a few different products from related species of plants. For example, *Penicillium digitatum* causes green mold on citrus fruits but does not cause disease in apples and pears. *P. expansum* on the other hand, attacks apples and pears, but not citrus fruits (Eckert *et al.*, 1983). Successful pathogens must have the capability to overcome the host defenses, must be able to initiate growth under conditions of water potential, pH and nutrients present in the host tissues, and must elaborate enzymes that macerate host tissue and cause a release of nutrients required to sustain the indeterminate parasitic development of the pathogen in the host tissues

Chitosan, a given name to a deacetylated form of chitin, is a natural biodegradable compound derived from crustaceous shells such as crabs and shrimps, whose main attributes corresponds to its polycationic nature. Chitosan has been proven to control numerous pre and postharvest diseases on various horticultural commodities. It has been reported that both soil and foliar plant pathogens fungal, bacterial and viral may be controlled by chitosan application. Microscopical observations indicate that chitosan has a direct effect on the morphology of the chitosan-treated microorganism reflecting its fungistatic or fungicidal potential. In addition to its direct microbial activity, other studies strongly suggest that chitosan induces a series of defence reactions correlated with enzymatic activities. Chitosan has been shown to increase the production of glucanohydrolases, phenolic compounds and synthesis of specific phytoalexins with antifungal activity, and also reduces macerating enzymes such as polygalacturonases, pectin metil esterase *etc.* In addition, chitosan induces structural barriers for example inducing the synthesis of lignin-like material. For some horticultural and ornamental commodities, chitosan increased harvested yield.

Due to its ability to form a semipermeable coating, chitosan extends the shelf life of treated fruit and vegetables by minimizing the rate of respiration and reducing water loss. As a nontoxic biodegradable material, as well as an elicitor, chitosan has the potential to become a new class of plant protectant, assisting towards the goal of sustainable agriculture (Chitosan, a given name to a deacetylated form of chitin, is a natural biodegradable compound derived from crustaceous shells such as crabs and shrimps, whose main attributes corresponds to its polycationic nature). Chitosan has been proven to control numerous pre and postharvest diseases on various horticultural commodities. It has been reported that both soil and foliar plant pathogens fungal, bacterial and viral may be controlled by chitosan application. Microscopical observations indicate that chitosan has a direct effect on the morphology of the chitosan-treated microorganism reflecting its fungistatic or fungicidal potential. In addition to its direct microbial activity, other studies strongly suggest that chitosan induces a series of defence reactions correlated with enzymatic activities. Chitosan has been shown to increase the production of glucanohydrolases, phenolic compounds and synthesis of specific phytoalexins with antifungal activity, and also reduces macerating enzymes such as polygalacturonases, pectin metil esterase etc. In addition, chitosan induces structural barriers for example inducing the synthesis of lignin-like material. For some horticultural and ornamental commodities, chitosan increased harvested yield. Due to its ability to form a semipermeable coating, chitosan extends the shelf life of treated fruit and vegetables by minimizing the rate of respiration and reducing water loss (Bautista-Banos, *et al.,* 2006). As a nontoxic biodegradable material, as well as an elicitor, chitosan has the potential to become a new class of plant protectant, assisting towards the goal of sustainable agriculture.

Effect of chitosan treatment on the infection process of bell pepper fruit by *Botrytis cinerea* was investigated at the ultrastructural level. In inoculated control tissues, fungal colonization proceeded rapidly, causing extensive degradation of host walls and middle lamellae. Such alterations were also observed over walls distant from invading hyphae. In chitosan-treated tissue, however, fungal cells were mainly restricted to wound cavities and ruptured epidermal cells (Ghaouthi, *et al.,* 1994). Host walls, even when appressed against invading hyphae, appeared well preserved and showed no sign of alteration. While in inoculated control tissues, invading hyphae appeared normal, the fungal cells detected in chitosan-treated tissue displayed various degrees of cellular disorganization from wall loosening to protoplasm degradation. Structural host defence responses such as formation of wall appositions and plugging of intercellular spaces with fibrillar material were frequently observed in chitosan-treated tissues. Such structure reacted intensely with an *Aplysia gonad* lectingold complex, indicating the presence of pectic residues.

iv.The role of biofilm formation and quorum sensing : To successfully colonize intact and injured fruit surfaces, the antagonist should have the ability to use specific features facilitating its adherence, colonization and multiplication. In most cases this feature is associated with the formation of a biofilm, where micro colonies are enclosed in a hydrated matrix of microbe produced proteins, nucleic acids, and polysaccharides (Davide Spadaro and Samir Droby, 2016). The importance of biofilm formation and quorum sensing (QS) in biocontrol systems was reported. Yeast cell attachment is often mediated by specific cell wall adhesive glycophosphatidyl inositol (GPI)-anchored proteins. Environmental sensing and signal transduction pathways regulating morphogenetic transformations have been studied in depth in *Candida albicans*. Two families of adhesin genes (HWP1and ALS) showed to play a critical role in host cell recognition, adhesion, invasion, and biofilm formation. Different yeast species carry different families of adhesins that reflect their species lifestyle. In *Saccharomyces cerevisiae*, five FLO (flocculation) genes are responsible for adhesion. Different aromatic alcohols exert different effects on morphogenesis in *S. cerevisiae* and *C. albicans* . Two QS regulatory molecules, tyrosol and farnesol, coordinating phenotype switching (yeast-to-hypha and vice versa), have been identified in *C. albicans*. Recently, the aromatic alcohol phenylethanol was identified as a QS molecule stimulating pseudohyphal growth in *S. cerevisiae* and *Debaryomyces hansenii* .

However, little is known about the role of biofilms in the biocontrol activity of yeast antagonists used to manage postharvest diseases and the mechanisms involved in their formation. Experiments carried out on *S. cerevisiae,* capable of forming a biofilm in liquid culture, demonstrated its effectiveness against some postharvest pathogens. The biocontrol activity of a biofilm-forming *S. cerevisiae* strain is tightly related to the morphological phase of cell harvesting after growing in liquid culture. Only yeast cells collected from the biofilm phase are effective in limiting pathogen growth, apparently being able to colonize more efficiently the inner surface of artificial wounds (Ianiri *et al.*, 2013). In this relation, the ability to form biofilms and filamentous growth are often correlated

The ability to form biofilms has been also proposed as an effective mechanism of action in some biocontrol yeasts. Interestingly, a strain of *Pichia fermentans*, which controls brown rot on apple fruit, showed to become a destructive pathogen when applied to peach fruit. On apple surfaces and within apple wounds, the antagonist retained its yeast-like shape whereas colonization of peach fruit tissue was always characterized by a transition from budding growth to pseudohyphal growth, suggesting that pseudohyphal growth could play a major role in governing the potential pathogenicity of *P. fermentans* on peaches

Phenylethanol, as a QS molecule, can induce *Kloeckera apiculata* to adhere and form biofilm on citrus fruit and embed in an extracellular matrix, by creating a mechanical barrier interposed between the wound surface and the pathogen (Davide Spadaro and Samir Droby, 2016). suggested that biofilm formation is one of the main features of yeast antagonists against pear postharvest decay, involved in their efficacy and adaptation to low temperatures.

Understanding the mechanisms of biofilm formation as well as the environmental cues regulating morphogenetic transformations in postharvest BCAs will lead to the selection of more effective antagonists and new methods of optimizing their activity.

v. Production of diffusible and volatile antimicrobial compounds: Yeasts can produce antifungal compounds, such as killer toxins, peptides and antibiotic metabolites (Davide Spadaro and Samir Droby, 2016). Certain yeast strains with a killer phenotype produce extracellular protein toxins designated as killer toxins or killer proteins, which are lethal to sensitive microbial cells belonging to either the same or a different species. Producers of these toxins are able to kill each other, but are immune to killer toxins of their own class. The most studied examples are the toxins K1, K2 and K28 of *S. cerevisiae* . The killer toxins confer an ecological advantage to yeast cells over their competitors. Most killer toxins are stable and active at pH values ranging from 3 to 5.5, typical of wounded or injured fruits, and they are protease-sensitive and heat labile (the killer toxin K1 is unstable at temperatures above 25 °C). *Pichia membranifaciens* can produce two killer toxins (PMKT and PMKT2) that are active against spoilage yeast and fungi. Though there is diversity in the mode of action of killer toxins, several killer toxins (K1, PMKT) seem to be membrane pore forming-related toxins.

Among the antibiotic metabolites, the most thoroughly studied example is farnesol from *C. albicans*, which can inhibit *in vitro* various bacteria and fungi. Another antifungal volatile substance, 2-phenylethanol, was isolated from *K. apiculata* and demonstrated to have inhibitory activity against green and blue mould on citrus caused by *P. digitatum and P. italicum* (Liu, *et al.*, 2014).

A. pullulans may produce aureobasidin A, a cyclic depsipeptide, with antifungal and antibiotic properties, particularly against *Botrytis* spp., *Monilinia* spp., and *Penicillium* spp. both *in vitro* and *in vivo* . Aureobasidin A is able to block the activity of inositol phosphorylceramide synthase, an essential enzyme for fungal sphingolipid biosynthesis. The main concern, related to the use of antifungal and antibiotic compounds in food products, is the development of human pathogens resistant to these compounds and the possible development of resistance in fruit pathogens. Particular care should be taken in using BCAs producing antimicrobials on fruit, though antibiotic producers may be able to control also wound infections established before antagonist application.

Many fungal species, such as *Trichoderma harzianum, Fusarium oxysporum,* and *A. pullulans,* are known to produce low concentrations of volatile antifungal substances (Davide Spadaro and Samir Droby, 2016). Fungi capable of producing antifungal volatile organic compounds (VOCs) have the potential of being used as biofumigants and to be potential new BCAs for controlling postharvest diseases. The production of VOCs, including 2-phenethyl alcohol, could play an essential role in the antagonistic activity of *A. pullulans* against fruit postharvest pathogens both *in vitro* and *in vivo* *P. anomala*, when applied in fruit wounds, initially may consume a large part of the available oxygen, but later, during the stationary phase, it could use alcoholic fermentation to produce antifungal VOCs, such as ethanol or ethyl acetate, which could be involved in the antimicrobial activity against *B. cinerea*. The main issue, when elucidating the role of diffusible or volatile antimicrobial compounds, is the assessment of the real impact of VOCs on biocontrol efficacy. Based on the results obtained *in vitro*, it is difficult to relate with the antagonistic activity on fruit surfaces under real situations. When tests are conducted *in vivo*, however, it is imperative to know the relevance of VOC concentration applied to that produced under commercial conditions. *Muscodor albus* and *Muscodor crispans* are endophytic fungi that produce mixtures of VOCs with antimicrobial activity. The potency and types of VOCs, that include 3-methyl-1-butanol, 2-nonanone, and phenylethyl alcohol, vary among isolates. An isolate of *M. albus,* obtained from a cinnamon tree, was able to kill a wide spectrum of plant pathogens and other microorganisms both *in vitro* and *in vivo*.

Another antifungal volatile-producing species, *Oxyporus latemarginatus,* was able to inhibit the mycelial growth of *A. alternata, B. cinerea* and *Colletotrichum gloeosporioides*, by mycofumigation (Lee *et al*., 2009). Mycofumigation with solid cultures of this strain effectively reduced also the development of postharvest apple decay caused by *B. cinerea*, due to the production of 5-pentyl-2-furaldehyde. *Candida intermedia* and *Sporodiobolus pararoseus* were able to suppress conidial germination and mycelial growth of *B. cinerea* and control grey mould of strawberry due to the release of a variety of VOCs, including 2-nonanone and 2-ethyl-1-hexanol. VOC-producing microorganisms open new possibilities to control microbial decays in the agro-food chain, as biofumigation does not require physical contact with the product or commodity to be treated. Volatile compounds are ideal antimicrobials because their spectrum of activity extends from proximal interaction through water diffusion to greater distances via air diffusion. In this regard, however, the safety of using such approach needs to be thoroughly evaluated.

vi.Parasitism and release of hydrolases: Parasitism occurs when the antagonist feeds on the pathogen, resulting in a direct destruction or lysis of fungal propagules and structures. a strong adhesion was observed *in vitro* of *P. guilliermondii* antagonist cells to *B. cinerea* mycelium, perhaps due to a lectin like interaction (Davide Spadaro and Samir Droby, 2016). Such adhesion was blocked by exposure to compounds able to alter the protein integrity and the respiration process. When *Candida saitoana* cells were cultivated together with *B. cinerea* mycelium, the fungal hyphae showed cytological damages, such as formation of papillae and other protuberances in the cell wall, as well as cytoplasm degeneration.

In the fungal cell walls, chitin as a structural backbone is arranged in regularly ordered layers, and β-1,3-glucan as a filling material is arranged in an amorphous manner. Glucan is the major structural polysaccharide of the fungal cell wall, constituting approximately 50–60% by dry weight of the wall. β-1,3-glucan is considered the main structural constituent to which other cell wall components are covalently attached, providing the cell wall with mechanical strength and integrity. Chitin is a linear, insoluble homopolymer composed of beta-1,4-linked subunits of the acetylated amino sugar N-acetylglucosamine. The cell walls of filamentous fungi consist of up to 20% or more chitin (Seidl, 2008). In addition, proteins represent approximately 20–30% of the cell wall. Most of the cell wall proteins are glycoproteins extensively modified with O-linked and N-linked oligosaccharides. Cell wall proteins play an important role in maintaining cell shape, mediating adhesion for cell migration and fusion, protecting the cell against foreign substances, mediating the absorption of molecules and synthesizing and remodelling cell wall components. Breakdown of fungal cell wall requires the participation of different enzymes, especially β-1,3-glucanase and chitinase, but also proteases. Glucanases, chitinases and proteases can be directly or indirectly involved in the mechanism of several yeast antagonists of postharvest pathogens

vii. The potential of omics to study antagonist-pathogen-host interactions: The availability of more efficient DNA-based and proteomics technologies, along with bioinformatics, has provided new opportunities and tools to gain deeper and more accurate insights about the interactions already indicated (Davide Spadaro and Samir Droby, 2016). Though omics technologies have been widely used to elucidate the mechanisms of action of BCAs against soilborne and foliar pathogens, seldom they have been used to clarify the modes of action of postharvest BCAs. Developments in deep sequencing, transcriptomics, proteomics, metagenomics, comparative and functional genomics could be utilized to determine changes in the physiological status of BCAs, and the effect of environmental stress on its intracellular machinery.

Global changes in gene expression both in host tissue and antagonist cells have been reported (Hershkovitz *et al.,* 2013). New omic tools can be used to evaluate the global effect of the application of BCAs on the transcriptome and/or proteome of fruit. To obtain an overview on transcript modification during the interaction of cherry tomato fruit with *C. laurentii*, a microarray analysis, using Affymetrix Tomato Genechip arrays, representing approximately 10,000 genes, was performed. The results showed that 194 and 312 genes were up- or downregulated, respectively, more than ten time fold in BCA-treated tomato fruit as compared with control fruit. Up-regulated genes included genes involved in metabolism, signal transduction, and stress response. Conversely, genes related to energy metabolism and photosynthesis were generally down-regulated. BCA treatment induces fruit resistance response and it suppresses energy metabolism and photosynthesis.

In grapefruit surface wounds treated with *M. fructicola* cells, there was significant expression of PRPs genes and MAPK cascade genes involved in defence signalling, and down-regulation in antioxidant genes, like POD, superoxide dismutase (SOD) and CAT. The genes up-regulated by *M. fructicola* in grapefruit were consistent with an induced resistance response and it was suggested that the induced response played a role in the efficacy of *M. fructicola* against postharvest pathogens like *P. digitatum* indicated that *P. membranifaciens* induced antioxidant and PR proteins in peach fruit, and it was suggested that these proteins played an essential role in the control of *P. expansum* in this biocontrol system. In an investigation to study the responses of cherry tomato to the yeast antagonist *C. laurentii*, Jiang *et al.*, (2009) showed that genes involved in metabolism, signal transduction, and stress response were up-regulated while genes related to energy metabolism and photosynthesis were suppressed. A transcriptomic analysis was carried out using RNA-Seq, to examine changes in gene expression in *M. fructicola* when it was exposed to citrus tissues and the postharvest pathogen *P. digitatum*. Results indicated that more than 250 genes exhibited expression responses specifically associated with the yeast-citrus vs. the yeast–pathogen interaction. Genes related to transmembrane, multidrug transport, and amino acid metabolism were induced in the yeast–pathogen interaction, while expression of genes involved in oxidative stress, iron homeostasis, zinc homeostasis, and lipid metabolism were induced in the yeast–fruit interaction. Collectively, these reports indicate that different gene/protein profiles are involved in different antagonistic yeast–host–pathogen interactions, demonstrating the dynamics of different biocontrol system and how “omic” technologies can provide insights into the modes of action of antagonistic yeast. The above reported studies were the first to report molecular changes at the biocontrol system. Determination of changes in the level of expression of “biocontrol genes” during mass production, formulation and storage, or in response to exposure and contact with host plant tissue after application can be now easily studied. It is expected however, that many more results will be reported in the near future about interactions between antagonistic yeast, host tissue, the pathogen, and also the epiphytic microflora.

To date, there are hundreds of reports about using of yeast antagonists to biologically control postharvest diseases. Very few of these antagonists, however, have reached the commercial development stage and launched as commercial products. In most cases, there are inherit problems in the biocontrol systems related to poor performance and inconsistency under commercial conditions. Among the reasons for these shortcoming is the lack of understanding of mechanisms of actions of these BCAs. It is apparent that the performance of yeast BCA is the result of complex interactions taking place between all the components of the biocontrol system (plant host, the antagonist, the pathogen, and resident microflora). Although these interactions have been the subject of research for over thirty years, our understanding is still incomplete. This because of the difficulties associated with the study of complex interactions and the lack of appropriate research tools and technologies.

In recent years there has been a phenomenal advancement in the use of molecular techniques contributing to the development of innovative tools for improving knowledge on the antagonistic mechanisms of BCAs. In particular, the omics techniques, including genomics, transcriptomics, proteomics, metagenomics, and metabolomics are providing a powerful tool to dissect the complex interactions between the antagonist, the pathogen, the fruit host, the natural microflora, and the environmental conditions. Induced resistance has been suggested to be one of the mechanisms of action of postharvest BCAs. However, information about elicitors/effectors of the antagonist involved and our ability to genetically and physiologically manipulate them is still lacking. Fundamental knowledge on the physiology, genetic traits and molecular basis of colonization, survival and differentiation of BCAs on plant tissue is needed. From a commercial point of view, complex modes of action make antagonistic performance and efficacy more dependent on production, formulation, packing, application, and storage. A deep understanding of the mode of action is essential to develop appropriate formulation and methods of application, and to obtain registration.

References

Agudelo-Romero, P., Erban, A., Rego, C., Carbonell-Bejerano, P., Nascimento, T., Sousa, L., Martínez-Zapater, J.M., Kopka, J., Fortes, A.M., 2015. Transcriptome and metabolome reprogramming in *Vitis vinifera* cv. Trincadeira berries upon infection with *Botrytis cinerea*. *J. Exp. Bot.*, **66:** 1769-1785.

Aharoni, A., Dixit, S., Jetter, R., Thoenes, E., van Arkel, G., and Pereira, A. (2004). The SHINE clade of AP2 domain transcription factors activates wax biosynthesis, alters cuticle properties, and confers drought tolerance when overexpressed in *Arabidopsis*. *Plant Cell,* **16**: 2463–2480.

Alkan, N., Friedlander, G., Ment, D., Prusky, D. and Fluhr, R., 2015. Simultaneous transcriptome analysis of *Colletotrichum gloeosporioides* and tomato fruit pathosystem reveals novel fungal pathogenicity and fruit defense strategies. *New Phytol,*. **205 :** 801–815

An, B., Li, B.Q., Li, H., Zhang, Z.Q., Qin, G.Z., Tian, S.P., 2016. Aquaporin regulates cellular development and reactive oxygen species production, a critical component of virulence in *Botrytis cinerea. New Phytol.*, **209**: 1668-1680.

Arne Weiberg, Ming Wang, Feng-Mao Lin, Hongwei Zhao, Zhihong Zhang, Isgouhi Kaloshian, Hsien-Da Huang and Hailing Jin, 2013.. Fungal small RNAs suppress plant immunity by hijacking host RNA interference pathways. *Science* (New York, N.Y.)

Barad, S., Horowitz, S.B., Kobiler, I., Sherman, A..and Prusky, D., 2014. Accumulation of the mycotoxin patulin in the presence of gluconic acid contributes to pathogenicity of *Penicillium expansum. Mol. Plant-Microbe Interact.,* **27:** 66–77.

Barad, S., Sela, N., Kumar, D.,2016.*l.* Fungal and host transcriptome analysis of pH-regulated genes during colonization of apple fruits by *Penicillium expansum. BMC Genomics* , **17:** 330.

Barbara Blanco-Ulate, Estefania Vincenti, Ann L. T. Powell and Dario Cantu, 2013. Tomato transcriptome and mutant analyses suggest a role for plant stress hormones in the interaction between fruit and *Botrytis cinerea* . *Front. Plant Sci.*, **4**: 142.

Barbara Blanco-Ulate, Katherine C.H. Amrine, Thomas S. Collins, Rosa M. Rivero, Ariel R. Vicente, Abraham Morales-Cruz, Carolyn L. Doyle, Zirou Ye, Greg Allen, Hildegarde Heymann and Susan E. Ebeler, Dario Cantu, 2015. Developmental and Metabolic Plasticity of White-Skinned Grape Berries in Response to *Botrytis cinerea* during Noble Rot. *Pl. Physiol.*, **169**: 2422-2445

Bautista-Banos, S., Hernandez-Lauzardo, A.N., Velazquez-del Valle, M.G., Hernandez-Lopez, M., Ait Barka, E. and Bosquez-Molina, E.Wilson., C.L ., 2006. Chitosan as a potential natural compound to control pre and postharvest diseases of horticultural commodities. *Crop Protn.,* **25** : 108–118

Bestfleisch, M., Luderer□Pflimpfl, M., Höfer, M., Schulte, E., Wünsche, J., Hanke, M.V., 20 Flachowsky, H., 2015. Evaluation of strawberry (*Fragaria* L.) genetic resources for resistance to *Botrytis cinerea. Plant Pathol.*, **64**: 396-405

Bhat, K.A., Masood, S.D., Bhat, N.A., Ashraf Bhat, M., Razvi, S.M., Mir, M.R., Sabina Akhtar, Wani, N. and Habib, M., 2010. Current Status of Postharvest Soft Rot in Vegetables: A Review. *Asian J. Pl. Sci.,,* **9***:* 200-208.

Biton, E., Kobiler, I., Feygenberg, O., Yaari, M., Friedman, H. and Prusky, D., 2014. Control of Alternaria black spot in persimmon fruit by a mixture of gibberellin and benzyl adenine, and its mode of action. *Postharvest Biol. Technol.,* **94:** 82–88.

Bochra A. Bahri, Ghaya Mechichi, Wafa Rouissi, Imtinen Ben Haj Jilani, Zeineb Ghrabi-Gammar, 2019. Effects of cold-storage facility characteristics on the virulence and sporulation of *Penicillium expansum* and the efficacy of essential oils against blue mold rot of apples. *Folia Hort.*, **31** : 301-317

Broun, P., Poindexter, P., Osborne, E., Jiang, C. Z., and Riechmann, J. L.., 2004. WIN1, a transcriptional activator of epidermal wax accumulation in *Arabidopsis. Proc. Natl. Acad. Sci. U.S.A.* **101**: 4706–4711

Cantu, D., Vicente, A. R., Greve, L. C., Dewey, F. M., Bennett, A. B., . Labavitch, J. M. and Powell, A. L. T., 2008. The intersection between cell wall disassembly, ripening, and fruit susceptibility to *Botrytis cinerea* . *Proc. National Acad.Sci.,* **105** : 859-864

Cegelski, L., Marshall, G.R., Eldridge, G.R., Hultgren, S.J., 2008. The biology and future 17 prospects of antivirulence therapies. *Nat. Rev. Microbiol.*, **6:** 17-27.

Costa, J.H., Wassano, C.I., Angolini, C.F.F. ,2019.. Antifungal potential of secondary metabolites involved in the interaction between citrus pathogens. *Sci Rep.,* **9:** 18647

Dario Cantu, Barbara Blanco-Ulate, Liya Yang, John M. Labavitch, Alan B. Bennett, and Ann L.T. Powell, 2009. Ripening-Regulated Susceptibility of Tomato Fruit to Botrytis cinerea Requires NOR But Not RIN or Ethylene. *Pl.Physiol.*, **150:** 1434–1449

Davide Spadaro and Samir Droby, 2016. Development of biocontrol products for postharvest diseases of fruit: The importance of elucidating the mechanisms of action of yeast antagonists. *Tr.Food Sci.&Technol.*, **47**: 39-49

Dempsey, D.A. , Shah, J. and Klessig, D.F., 1999. Salicylic acid and disease resistance in plants

Dilip Kumar, Shiri Barad, Edward Sionov, Nancy P.Keller and Dov.B.Prusky, 2017. Does the Host Contribute to Modulation of Mycotoxin Production by Fruit Pathogens? *Toxins*, **9:** 280.

Diniz, I., Figueiredo, A., Loureiro, A., Batista, D., Azinheira, H., Várzea, V., 2017. A first insight into the involvement of phytohormones pathways in coffee resistance and susceptibility to *Colletotrichum kahawae*. *PLoS ONE*, **12**(5): e0178159.

Drusch, S. and Ragab, W., 2003. Mycotoxins in fruits, fruit juices, and dried fruits. *J. Food Prot.* , **66:** 1514–1527

Dubey, A.K., Barad, S., Luria, N., Kumar, D., Espeso, E.A. and Prusky, D.B., 2016. Cation-Stress-Responsive Transcription Factors SltA and CrzA Regulate Morphogenetic Processes and Pathogenicity of *Colletotrichum gloeosporioides*. *PLoS ONE,* **11**(12): e0168561.

Eckert, J.W.and Ratnayake, M.,1983. Host-Pathogen Interactions in Postharvest Diseases. In: Lieberman M. (eds) Postharvest Physiology and Crop Preservation. Nato Advanced Study Institutes Series (Series A: Life Sciences), vol 46. Springer, Boston, MA

Eduardo A.Espeso and Dov Prusky, 2013. Virulence refulaiton of phytopahtogenic fungi by pH. *Antioxid. Redox Signal.*, **19**:1012–1025.

Ghaouthi, A.El., Arul, J., Wilson,C and Benhamou, N., 1994. Ultrastructural and cytochemical aspects of the effect of chitosan on decay of bell pepper fruit. *Physiol. & Mol.Pl.Pathol.*, **44**: 417-432

González-Domínguez, E., Caffi, T., Ciliberti, N. and Rossi, V.,2015. A Mechanistic Model of *Botrytis cinerea* on Grapevines That Includes Weather, Vine Growth Stage, and the Main Infection Pathways. *PLoS ONE*, **10** (10): e0140444.

Guozheng Qin, Shiping Tian, Zhulong Chan and Boqiang Li, 2007. Crucial Role of Antioxidant Proteins and Hydrolytic Enzymes in Pathogenicity of *Penicillium expansum*-Analysis Based on Proteomics Approach. *Mol. & Cell. Proteomics*, **6** : 425-438

Hassan, A.M., Abdel-Aziem, S.H., El-Nekeety, A.A.. and Abdel-Wahhab, M.A., 2015. Panax ginseng extract modulates oxidative stress, DNA fragmentation and up-regulate gene expression in rats sub chronically treated with aflatoxin B1 and fumonisin B1. *Cytotechnology,* **67**: 861–871

Hershkovitz, V., Sela, N., Taha-Salaime, L., Liu, J., Rafael, G., Kessler, C., 2013. De-novo assemble and characterization of the transcriptome of *Metschnikowia fructicola* reveals differences in gene expression following interaction with *Penicillium digitatum* and grapefruit peel. *BMC Genomics*, **14** : 168

Ianiri, G. , Idnurm, A., Wright, S.A.I. , Duran-Patron, R., Mannina, L. and Ferracane, R., 2013. Searching for genes responsible for patulin degradation in a biological control yeast provides insights into the basis for resistance to this mycotoxins. *Appl. Env. Microbiol.*, **79** : 3101-3115

Irene Gouvinhas, Paula martinsLopes, Teresa Carvalho, Ana Barros and Sonia Gomes, 2019. Impact of Colletotrichum acutatum Pathogen on Olive Phenylpropanoid Metabolism, *Agriculture,* **9**: 173

Jia Liu, Yuan Sui, Huizhen Chen, Yiging Liu and Yongsheng Liu, 2018. Proteomic Analysis of Kiwifruit in Response to the Postharvest Pathogen, *Botrytis cinerea. Front Pl.Sci.*, **9**: 158

Jia Liu, Yuan Sui, Michael Wisniewski, Zhigang Xie, Yiqing Liu, Yuming You, Xiaojing Zhang, Zhiqiang Sun, Wenhua Li, Yan Li & Qi Wang, 2018 The impact of the postharvest environment on the viability and virulence of decay fungi, *Crit Rev. in Food Sci. & Nutr.,* **58**: 1681-1687

Jiang, F., Zheng, X. and Chen, J., 2009. Microarray analysis of gene expression profile induced by the biocontrol yeast *Cryptococcus laurentii* in cherry tomato fruit. *Gene*, **430** :12-16

Juan A. Martínez, 2012. Natural Fungicides Obtained from Plants, Fungicides for Plant and Animal Diseases, Dr. Dharumadurai Dhanasekaran (Ed.), ISBN: 978-953-307-804-5

Kashina Allydice-Francis and Paul D. Brown, 2012. Diversity of Antimicrobial Resistance and Virulence Determinants in *Pseudomonas aeruginosa* Associated with Fresh Vegetables. *Intl.J.Microbiol.*, Volume 2012 |Article ID 426241 | 7 pages

Kesari Ravi & Trivedi, Prabodh & Nath, Pravendra, 2007. Ethylene-induced ripening in banana evokes expression of defense and stress related genes in fruit tissue. *Postharvest Biol. &Technol.*, **46:** 136-143.

Kim, K.-H., Yoon, J.-B., Park, H.-G., Park, E. W., and Kim, Y. H. 2004. Structural modifications and programmed cell death of chili pepper fruit related to resistance responses to Colletotrichum gloeosporioides infection. *Phytopathology,* **94**:1295-1304

Kumar, D., Barad, S., Chen, Y., Luo, X., Tannous, J., Dubey, A., Matana, N.G., Tian, S., Li, B. and Keller, N.,2016. LaeA regulation of secondary metabolism modulates virulence in *Penicillium expansum* and is mediated by sucrose. *Mol. Plant Pathol.,* **18:** *1150-1163*

Lee, Y.H., Kim, S.H., Yun, B.W.and Hong, J.K.., 2014. Altered cultivar resistance of kimchi cabbage seedlings mediated by salicylic Acid, jasmonic Acid and ethylene. *The Pl. Pathol.J.,.* **30**:323-329

Lee, S.O. , Kim, H.Y. , Choi, G.J., Lee, H.B., Jang, K.S. Choi, , Y.H., 2009. Mycofumigation with *Oxyporus latemarginatus* EF069 for control of postharvest apple decay and Rhizoctonia root rot on moth orchid. *J. Appl.Microbiol.,* **106** :1213-1219

Levin, E., Ballester, A.R., Raphael, G., Feigenberg, O., Liu, Y.and Norelli,J., 2017. Identification and characterization of LysM effectors in *Penicillium expansum. PLoS ONE,* **12**(10): e0186023.

Li Hua, Chen Yong, Zhang Zhanquan, Li Boqiang, Qin Guozheng, Tian Shiping, Pathogenic mechanisms and control strategies of *Botrytis cinerea* causing postharvest decay in fruits and vegetables, *Food Quality and Safety*, Volume 2, Issue 3, September 2018, Pages 111–119.

Liang, Y., Guan, Y.and Wang, S., 2018.. Identification and characterization of known and novel microRNAs in strawberry fruits induced by *Botrytis cinerea. Sci Rep* .,**8:** 10921 .

Liu, P., Chen, K., Li, G*., 2016.*. Comparative transcriptional profiling of orange fruit in response to the biocontrol yeast *Kloeckera apiculata* and its active compounds. *BMC Genomics,* **17:**17.

Liu, Yaoyao & Ge, Yonghong & Bi, Yang & Li, Canying & Deng, Huiwen & Hu, Lingang & Dong, Boyu, 2014. Effect of postharvest acibenzolar-S-methyl dipping on phenylpropanoid pathway metabolism in muskmelon (Cucumis melo L.) fruits. *Scientia Horticulturae*, **168**: 113–119.

Liu, P., Cheng, Y. , Yang, M. , Liu, Y., Chen, K., Long, C., 2014. .Mechanisms of action for 2-phenylethanol isolated from *Kloeckera apiculata* in control of *Penicillium* molds of citrus fruits BMC Microbiol., 2014 Sep 19;14:242. doi: 10.1186/s12866-014-0242-2.

Maayan Davidzon, Noam Alkan, Ilana Kobiler, Dov Prusky, 2010. Acidification by gluconic acid of mango fruit tissue during colonization via stem end infection by Phomopsis mangiferae. *Postharvest Biol. &Technol.*, **55**: 71–77

Macarisin, D., Bauchan, G., Droby, S. and Wisniewski, M., 2011. Novel Role For Reactive Oxygen Species (Ros) In Host-Antagonistic Yeast-Pathogen Interactions In Postharvest Biocontrol Systems . *Acta Hortic.* **905**: 113-119

Mara Quaglia, Luisa Ederli, Stefania Pasualini and Antonio Zazzerini, 2011. Biological control agents and chemical inducers of resistance for postharvest control of *Penicillium expansum* Link. on apple fruit. *Postharvest Biol. & Technol.*, **59** : 307-315

Marcet-Houben, M., Ballester, A., de la Fuente, B., 2012.. Genome sequence of the necrotrophic fungus *Penicillium digitatum*, the main postharvest pathogen of citrus. *BMC Genomics* , **13:** 646 .

Matthew Kramer, Rick Sanders, Hassan Bolkan, Curtis Waters, Raymond E.Sheeny abd William R. Hiatt, 1992. Postharvest evaluation of transgenic tomatoes with reduced levels of polygalacturonase: processing, firmness and disease resistance. *Postharvest Biol. & Technol.*, **1:** 241-155

Melatonin decreases resistance to postharvest green mold on citrus fruit by scavenging defense-related reactive oxygen species. *Postharvest Biol. & Technol.*, **153**: 21-30

Miles, T.D., Day, B., Schilder, A.C., 2011. Identification of differentially expressed genes in a resistant versus a susceptible blueberry cultivar after infection by *Colletotrichum acutatum. Mol. Plant Pathol.*,**12**, 463-477.

Mohammadreza Asghari and Morteza Sloeimani Aghdam, 2010. Impact of salicylic acid on postharvest physiology of horticultural crops. *Tre. Food & Sci. Tecnol.*, **21** : 502-509

Mor, A., Koh, E., Weiner, L., Rosenwasser, S., Sibony-Benyamini, H., and Fluhr, R.,(2014. Singlet oxygen signatures are detected independent of light or chloroplasts in response to multiple stresses. *Plant Physiol..*, **165:** 249–261.

Nathalie Muller, Michaela Leroch, Julia Schumacher, David Zimmer, Anne K onnel, Klaus Klug, Thomas Leisen, David Scheuring, Frederik Sommer, Timo M uhlhaus, Michael Schroda and Matthias Hahn, 2018. Investigations on VELVET regulatory mutants confirm the role of host tissue acidification and secretion of proteins in the pathogenesis of *Botrytis cinerea. New Phytologist,* **219**: 1062–1074

Nicole E. Soltis, Susanna Atwell, Gongjun Shi,, Rachel Fordyce, Raoni Gwinner, Dihan Gao, Aysha Shafi, Daniel J. Kliebenstein, 2018. Crop domestication and pathogen virulence: Interactions of tomato and Botrytis genetic diversity. Publication from University of 21 California, Davis, One Shields Ave, Davis, CA, 95616, USA.

Niem, J., Miyara, I., Ettedgui, Y., Reuveni, M., Flaishman, M., and Prusky, D. 2007. Core rot development in red delicious apples is affected by susceptibility of the seed locule to *Alternaria alternata* colonization. *Phytopathology*, **97**:1415-1421.

Nit Yakoby, Ilana Kobiler,Amos Dinoor and Dov Prusky, 2000. pH Regulation of Pectate Lyase Secretion Modulates the Attack of *Colletotrichum gloeosporioides* on Avocado Fruits. *App.Env. Microbiol.*, **66**: 1026-1030

Niu, Y., Yuan, Y., Mao, J. , 2018. Characterization of two novel mycoviruses from *Penicillium digitatum* and the related fungicide resistance analysis. *Sci Rep.,* **8:** 5513 .

Noam Alkan and Ana M.Fortes, 2015. Insights into molecular and metabolic events associated with fruit response to postharvest fungal pathogens. *Front.Plant Sci.*, **20**: 1-14

Norelli, J.L., Wisniewski, M. and Droby, S., 2014. Identification of a QTL for postharvest disease resistance to *Penicillium expansum* in *Malus sieversii. Acta Hort.*, **1053:** 199-203.

Olson, P. and Varner, J., 1993. Hydrogen peroxide and lignification. *Plant J.*, **4 :** 887-892

Osorio, S., Alba, R., Damasceno, C.M.B., Lopez-Casado, G., Lohse, M., Zanor, M.I., Tohge, T., Usadel, B., Rose, J.K., Fei, Z. and Giovannoni, J.J., 2011. Systems biology of tomato fruit development: Combined transcript, protein, and metabolite analysis of tomato transcription factor nor, rin and ethylene receptor Nr mutants reveals novel regulatory interactions. *Plant Physiol.*, **157:** 405–425

Palumbo, J.D., O'Keeffe, T.L. and Mahoney, N.E., 2007.Inhibition of ochratoxin A production and growth of *Aspergillus* species by phenolic antioxidant compounds. *Mycopathologia,* **164:** 241–248.

Pathak, V.N., 1997. Postharvest Fruit Pathology - Present Status and Future Possibilities. *Indian Phytopath*., **50**: 161-185

Perfect, S.E., Hughes,H.B., O'Connell,R.J. and Green,J.R., 1999. Colletotrichum: A model genus for studies on pathology and fungal-plant interactions. *Fungal Genet.Biol.,* **27**: 186-198

Petrasch, S., Silva, C,J. and Mesquida-Pesci, S.D., 2019. Infection Strategies Deployed by *Botrytis cinerea, Fusarium acuminatum*, and *Rhizopus stolonifer* as a Function of Tomato Fruit Ripening Stage. *Front Plant Sci*., **10**:223.

Prusky, D. and Lichter, A.,2007. Activation of quiescent infections by postharvest pathogens during transition from the biotrophic to the necrotrophic stage. *FEMS Microbiol. Lett.,* **268**:1–8.

Prusky, D. B. and Wilson, R. A., 2018. Does increased nutritional carbon availability in fruit and foliar hosts contribute to modulation of pathogen colonization? *Postharvest Biol. & Technol.,* **145***:* 27-32.

Prusky, D., Alkan, N., Mengiste, T., and Fluhr, R.,2013. Quiescent and necrotrophic lifestyle choice during postharvest disease development. *Annu. Rev. Phytopathol.,* **51 :** 55–76

Prusky, Dov & Alkan, Noam & Miyara, Itay & Barad, Shiri & Davidzon, Maayan & Kobiler, Ilana & Braun Miara, Sigal & Lichter, Amnon & Sherman, Amir & Fluhr, Robert,2009. Mechanisms Modulating Postharvest Pathogen Colonization of Decaying Fruits. 10.1007/978-1-4020-8930-5_4.

Robert-Seilaniantz, Alexandre & Grant, Murray & Jones, Jonathan,2010. Hormone Crosstalk in Plant Disease and Defense: More Than Just Jasmonate-Salicylate Antagonism. *Ann. Rev. Phytopathol*., **49**. 317-343.

Ronald P. de Vries and Jaap Visser, 2001. *Aspergillus* Enzymes Involved in Degradation of Plant Cell Wall Polysaccharides. *Microbiol. & Mol. Biol. Rev*.,**65** : 497-522

Sandoval-Chávez, R. A., Martínez-Peniche, R. Á., HernándezIturriaga, M., Teixidó-Espasa, N., Usall-Rodié, J., Viñas-Almenar, I., & Torres-Sanchis, R., (2015. Mechanisms of resistance in postharvest fruit-pathogen interaction, Mexico. *Revista Chapingo Serie Horticultura,* **21**: 185-198

Sandoval-Chávez, R. A., Martínez-Peniche, R. Á., HernándezIturriaga, M., Teixidó-Espasa, N., Usall-Rodié, J., Viñas-Almenar, I., & Torres-Sanchis, R., 2015. Mechanisms of resistance in postharvest fruit-pathogen interaction, Mexico. *Revista Chapingo Serie Horticultura,* **21:**185-198

Seidl, V. , 2008. Chitinases of filamentous fungi: a large group of diverse proteins with multiple physiological functions. *Fungal Biol. Rev*., **22** : 36-42

Seymour, G. B., Østergaard, L., Chapman, N. H., Knapp, S. and Martin, C.,2013. Fruit development and ripening. *Annu. Rev. Plant Biol.,* **64**, 219–241.

Shengua Ding, Jing Zhang, Lvzhu Yang and Xinyu Wang, 2020. Changes in Cuticle Components and Morphology of 'Satsuma' Mandarin (*Citrus unshiu*) during Ambient Storage and Their Potential Role on *Penicillium digitatum* Infection. *Molecules* , **25***:* 412

Shi, J. X., Adato, A., Alkan, N., He, Y. H., Lashbrooke, J., Matas, A. J., 2013. The tomato SlSHINE3 transcription factor regulates fruit cuticle formation and epidermal patterning. *New Phyto.,.* **197:** 468–480.

Singh, Rupinder & Rastogi, Smita & Dwivedi, Upendra, 2010. Phenylpropanoid Metabolism in Ripening Fruits. *Compr. Rev.in Food Sci.& Food Safety,* **9:** 398 - 416

Sirinan Suktawee, Masahiro Shishido, Shanshan Wang, Takanori Saito, Katsuya Okawa, Hitoshi Ohara, Hataitip Nimitkeatkai, Hiromi Ikeura and Satoru Kondo, 2019. n-Propyl dihydrojasmonates influence ethylene signal transduction in infected apple fruit by *Botrytis cinerea . The Hort. J.,* **88**: 41–49.

Sirinan Suktawee, Masahiro Shishido, Shanshan Wang, Takanori Saito, Katsuya Okawa, Hitoshi Ohara, Hataitip Nimitkeatkai, Hiromi Ikeura, Satoru Kondo, 2019. *n*-Propyl dihydrojasmonates influence ethylene signal transduction in infected apple fruit by *Botrytis cinerea. Hort. J.,* **88:** 41-49

Stefan Petrasch, Steven J. Knapp, Jan A. L. Van Kan and Barbara Blanco-Ulate, 2019. Grey mould of strawberry, a devastating disease caused by the ubiquitous necrotrophic fungal pathogen *Botrytis cinerea . Mol.Pl.Pathol.*, **20**: 877-892

Taha Abd El Rahman, Mohamed El Oirdi, Rocio Gonzalez-Lamothe, and Kamal Bouarab, 2012. Necrotrophic Pathogens Use the Salicylic Acid Signaling Pathway to Promote Disease Development in Tomato. *Mol. Pl. Micr. Int.,* **25**: 1584–1593

Tahir, I.I., Nybom, H., Ahmadi-Afzadi, M., Røen, K.., Sehic, J. and Røen, D., 2015. Susceptibility to blue mold caused by *Penicillium expansum* in apple cultivars adapted to a cool climate. *Eur.J.Hortic.Sci.*, **80**: 117-127

Tan Wei-ping, Pang Xue-qun, Zhang Zhao-qi, Huang Xue-Mei, 2014. Accumulation of Reactive Oxygen Species Related to Disease Resistance Induced by BABA in Postharvest Banana (Musa AAA. cv. Brazil) Fruit[J].*Scientia Agricultura Sinica,* **47:** 3290-3299

Taotao Li, Yu Wu, Yong Wang, Haiyan Gao, Vijai Kumar Gupta, Xuewu Duan, Hongxia Qu and Yueming Jiang, 2019. Secretome Profiling Reveals Virulence-Associated Proteins of *Fusarium proliferatum* during Interaction with Banana Fruit. *Biomolecules,* **9** : 246

Tian, S., Torres, R., Ballester, A-R., Li, L B., Vilanova, González-Candelas, L., 2016. Molecular aspects in pathogen-fruit interactions: Virulence and resistance. *Postharvest Biol. & Technol.,* **122**: 11-21

Tian, S., Qin, G., and Li, B., 2013. Reactive oxygen species involved in regulating fruit senescence and fungal pathogenicity. *Plant Mol. Biol.,* **82:** 593–602.

Tian, S., Wan, Y. and Qin, G. ,2006.. Induction of defense responses against *Alternaria* rot by different elicitors in harvested pear fruit. *Appl Microbiol Biotechnol* ., **70:** 729 .

Tian, S. , Qin, G. , Li, B., Wang, Q. and Meng, X., 2007. Effects of salicylic acid on disease resistance and postharvest decay control of fruits. *Stewart Postharvest Rev.***,** **6** : 1-7

Tijero, V., Teribia, N. & Munné-Bosch, S., 2019. Hormonal Profiling Reveals a Hormonal Cross-Talk During Fruit Decay in Sweet Cherries. *J. Pl Gr. Regul.*, **38**: 431–437 .

Torres, M.A., Jones, J.D.G., Dangl, J.L., 2006. Reactive oxygen species signaling in 10 response to pathogens. *Plant Physiol.*, **141**: 373-378.

Van Kan, J.A.L., Stassen, J.H.M., Mosbach, A., Van Der Lee, T.A.J., Faino, L., Farmer, A.D., Papasotiriou, D.G., Zhou, S., Seidl, M.F. and Cottam, E.,2017. A gapless genome sequence of the fungus *Botrytis cinerea. Mol. Plant Pathol.,* **18**: 75–89.

Van Loon, L.C., Rep, M.and Pieterse, C.M.J., 2006. Significance of inducible defense-related proteins in infected plants. *Ann. Rev. Phytopathol.,* **44 :** 135-162

Vijay Paul and Rakesh Pandey, 2014. Role of internal atmosphere on fruit ripening and storability—a review. *J.Food Sci.Technol.*, **51**: 1223-1250

Vilanova, L., Torres, R., Viñas, I., González-Candelas, L., Usall, J., Fiori, S., Solsona, C., Teixidó, N., 2013. Wound response in orange as a resistance mechanism against *Penicillium digitatum* (pathogen) and *P. expansum* (non-host pathogen). *Postharvest Biol. Technol.*, **78:** 113-122.

Vincenzo Lionetti, Eleonora Fabri, Monica De Caroli, Aleksander, R., Hansen, William,G.T., Willats, Gabriella Piro and Daniela Bellincampi, 2017. Three Pectin Methylesterase Inhibitors Protect Cell Wall Integrity for Arabidopsis Immunity to *Botrytis . Pl.Pathol.*, **173**: 11844-1863

Wei Zhang, Jason A. Corwin, Daniel Copeland, Julie Feusier, Robert Eshbaugh, Fang Chen, Susana Atwell, and Daniel J. Kliebenstein, 2017. Plastic Transcriptomes Stabilize Immunity to Pathogen Diversity: The Jasmonic Acid and Salicylic Acid Networks Within the Arabidopsis/Botrytis Pathosystem . Plant Cell Advance Publication, American Society of Plant Biologists

Wojtasik, W., Kulma, A., Dymińska, L.. 2016. Evaluation of the significance of cell wall polymers in flax infected with a pathogenic strain of *Fusarium oxysporum*. *BMC Plant Biol* **16:** 75

Xue-ren Yin, Andrew C. Allan, Kun-song Chen, Ian B. Ferguson, 2010. Kiwifruit *EIL* and *ERF* Genes Involved in Regulating Fruit Ripening. Research Article, Systems Biology, Molecular Biology, And Gene Regulation, American Society of Plant Biologists.

Yang, Y.X., Ahammed, G.J., Wu, C., Fan, S.Y. and Zhou, Y.H., 2015. Crosstalk among Jasmonate, Salicylate and Ethylene Signaling Pathways in Plant Disease and Immune Responses. *Curr.Protein Pept.Sci.*, **16**: 450-451

Ying Wang, Dongchao Ji, Tong Chen Boqiang Li Zhanquan Zhang, Guozheng Qin and Shiping Tian, 2019. Production, Signaling, and Scavenging Mechanisms of Reactive Oxygen Species in Fruit–Pathogen Interactions. *Int. J. Mol. Sci.*, **20**: 2-12

Yunlong Lin, Liqin Fan, Xuehan Xia, Zhongkang Wang, Youping Yin, Yulin Cheng, Zhengguo Li, 2019. Melatonin decreases resistance to postharvest green mold on citrus fruit by scavenging defense-related reactive oxygen species. *Postharvest Biol & Technol.,* **153:** 21-30